Lecture Notes in Networks and Systems

Volume 188

D1744437

Series Editor

Janusz Kacprzyk, Systems Research Institute, Polish Academy of Sciences, Warsaw, Poland

Advisory Editors

Fernando Gomide, Department of Computer Engineering and Automation—DCA, School of Electrical and Computer Engineering—FEEC, University of Campinas—UNICAMP, São Paulo, Brazil

Okyay Kaynak, Department of Electrical and Electronic Engineering, Bogazici University, Istanbul, Turkey

Derong Liu, Department of Electrical and Computer Engineering, University of Illinois at Chicago, Chicago, USA; Institute of Automation, Chinese Academy of Sciences, Beijing, China

Witold Pedrycz, Department of Electrical and Computer Engineering, University of Alberta, Alberta, Canada; Systems Research Institute, Polish Academy of Sciences, Warsaw, Poland

Marios M. Polycarpou, Department of Electrical and Computer Engineering, KIOS Research Center for Intelligent Systems and Networks, University of Cyprus, Nicosia, Cyprus

Imre J. Rudas, Óbuda University, Budapest, Hungary

Jun Wang, Department of Computer Science, City University of Hong Kong, Kowloon, Hong Kong

The series "Lecture Notes in Networks and Systems" publishes the latest developments in Networks and Systems—quickly, informally and with high quality. Original research reported in proceedings and post-proceedings represents the core of LNNS.

Volumes published in LNNS embrace all aspects and subfields of, as well as new challenges in, Networks and Systems.

The series contains proceedings and edited volumes in systems and networks, spanning the areas of Cyber-Physical Systems, Autonomous Systems, Sensor Networks, Control Systems, Energy Systems, Automotive Systems, Biological Systems, Vehicular Networking and Connected Vehicles, Aerospace Systems, Automation, Manufacturing, Smart Grids, Nonlinear Systems, Power Systems, Robotics, Social Systems, Economic Systems and other. Of particular value to both the contributors and the readership are the short publication timeframe and the world-wide distribution and exposure which enable both a wide and rapid dissemination of research output.

The series covers the theory, applications, and perspectives on the state of the art and future developments relevant to systems and networks, decision making, control, complex processes and related areas, as embedded in the fields of interdisciplinary and applied sciences, engineering, computer science, physics, economics, social, and life sciences, as well as the paradigms and methodologies behind them.

Indexed by SCOPUS, INSPEC, WTI Frankfurt eG, zbMATH, SCImago.

All books published in the series are submitted for consideration in Web of Science.

More information about this series at http://www.springer.com/series/15179

Mykola Nechyporuk · Vladimir Pavlikov ·
Dmitriy Kritskiy
Editors

Integrated Computer Technologies in Mechanical Engineering - 2020

Synergetic Engineering

 Springer

Editors
Mykola Nechyporuk
National Aerospace University
"Kharkiv Aviation Institute"
Kharkov, Ukraine

Vladimir Pavlikov 🅞
National Aerospace University
"Kharkiv Aviation Institute"
Kharkov, Ukraine

Dmitriy Kritskiy
National Aerospace University
"Kharkiv Aviation Institute"
Kharkov, Ukraine

ISSN 2367-3370   ISSN 2367-3389   (electronic)
Lecture Notes in Networks and Systems
ISBN 978-3-030-66716-0   ISBN 978-3-030-66717-7   (eBook)
https://doi.org/10.1007/978-3-030-66717-7

This Springer imprint is published by the registered company Springer Nature Switzerland AG
The registered company address is: Gewerbestrasse 11, 6330 Cham, Switzerland

About ICTM

The International Scientific and Technical Conference "Integrated Computer Technologies in Mechanical Engineering"—Synergetic Engineering (ICTM) was formed to bring together outstanding researchers and practitioners in the field of information technology in the design and manufacture of engines, creation of rocket space systems, aerospace engineering from all over the world to share their experience and expertise. It was established by National Aerospace University "Kharkiv Aviation Institute."

The Conference ICTM'2020 was held in Kharkiv, Ukraine, during October 29–30, 2020. During this conference, technical exchanges between the research community were carried out in the forms of keynote speeches, panel discussions, as well as special session. In addition, participants were treated to a series of receptions, which forge collaborations among fellow researchers. ICTM'2020 received 156 papers submissions from different countries. This book contains papers devoted to relevant topics including:

- Information technology in the design and manufacture of engines;
- Information technology in the creation of rocket space systems;
- Aerospace engineering;
- Transport systems and logistics;
- Big data and data science;
- Nano-modeling;
- Artificial intelligence and smart systems;
- Networks and communication;
- Cyber-physical system and IoE;
- Software engineering and IT-infrastructure.

All of these offer us plenty of valuable information and would be of great benefit to experience exchange among scientists in modeling and simulation. The organizers of ICTM'2020 made great efforts to ensure the success of this conference. We hereby would like to thank all the members of ICTM'2020 Advisory Committee for their guidance and advice, the members of program committee and organizing committee, and the referees for their effort in reviewing and soliciting the papers, and all authors for their contribution to the formation of a common intellectual environment for solving relevant scientific problems. Also, we grateful to Springer-Janusz Kacprzyk and Thomas Ditzinger as the editor responsible for the series "Lecture Notes in Networks and Systems" for their great support in publishing these selected papers.

Contents

Big Data and Data Science

Information Technology in Creation of Rocket Space Systems

Project Management and Business Informatics

Artificial Intelligence and Smart Systems

Synthesis Method of Robust Neural Network Models of Systems and Processes

Nina Bakumenko[1(✉)] , Viktoriia Strilets[1] , Ievgen Meniailov[2] ,
Serhii Chernysh[2] , Mykhaylo Ugryumov[1] ,
and Tamara Goncharova[3]

[1] V. N. Karazin, Kharkiv National University,
4 Svobody Sq., Kharkiv 61022, Ukraine
n.bakumenko@karazin.ua, striletsvictoria@gmail.com,
ugryumov.mykhaylo52@gmail.com
[2] National Aerospace University "Kharkiv Aviation Institute",
17 Chkalova Street, Kharkiv 61070, Ukraine
j.menyailov@khai.edu, mr.serhii.chernysh@gmail.com
[3] National University of Civil Defence of Ukraine, 94 Chernyshevska Street,
Kharkiv 61023, Ukraine
super-gusenichka@ukr.net

Abstract. The work deals with the study of some problems of reconstruction multidimensional statistical dependences on the basis of empirical data by means of artificial neural networks. To provide stability (robustness) of systems and processes statistical model parameters on the basis of trained artificial neural networks (ANN) at the a priori input data uncertainty as well as practically sufficient accuracy of data approximation, it is appropriate to use stable (robust) methods of deep ANN training methods. The work uses the cricking model for statistical data since we couldn't get precise parameter values; therefore, to achieve the required accuracy, some probability was introduced. The synthesis method of scalar convolution of selection functions for mathematical model identification, based on the law of requisite variety (Ashby law), Kolmogorov power overage concentration and the maximum likelihood principle, where Student and Romanovski statistics, are used as the proximity measure of true multidimensional samples. It makes it possible to structure preference systems of a person who makes decisions for multi-criterial problems to identify mathematical models in determinate and stochastic formulations (MV-, MH-problem). Neural network identification was made by the stochastic approximation method on the basis of ravine conjugate gradient method. The method of effective robust estimation of statistical model systems parameters was worked out by employing a regularizing sequential (adaption) algorithm for synthesis of solutions with deferred correction. Samples of Rosenbrock function data and corresponding parameters of aerodynamic characteristics of the jet engine multistage axial compressor were taken as examples.

Keywords: Artificial neural network · Deep learning · Robust evaluation · Regularizing method

1 Introduction

Modern models based on the basis of artificial neural network learning (ANN) are regarded as most needed in various spheres of practical activity. A wide range of their use creates various practical problems with different input data settings and types concerning optimum system designing, systems and process state diagnosis such as analysis of information of state variables, cluster analysis, object classification (image recognition, syntactic analysis of texts, state monitoring etc.).

However, we have mathematical problems which hamper a wide application of these models for solving practical problems. The first problem is that neural model identification problems are treated as ill-conditioned. However, these problems can be considered as conditionally correct of structural parameter optimization, when model structure is known, but its parameters are not known. In this case, to find learning parameters (ANN), the machine learning methods relying on the use of solutions synthesis regulating algorithms, including deep learning methods, are widely used.

The second problem is that the input data are stochastic values and can have gross errors. This problem is also solvable if we use the solutions synthesis based on invariance concept (the concept is widely used in the theory of automatic control). In this case, effective stable parameter assessments can be obtained. Thus, the robust neural network models can be formed.

The backpropagation method, in which the selection function minimization is made by gradient descent is known to be the most frequently used learning method ANN [1]. The primary advantage of the gradient descent method is the simplicity of its realization and the fact that this method is guaranteed converged to the global or local minimum for convex and non-convex functions respectively. However, there are many disadvantages of this method; therefore this method is seldom in practice:

- gradient descent can be very slow at big data samples, since each iteration demands calculating the gradient for all training set vectors;
- doesn't permit to renovate the model on-the-run and to add new training samples in the process also because the selection function updating is made for the whole initial data set at once;
- for non-convex functions, the problem of getting into local minimums arises because the method guarantees the exact solutions only for the error convex selection functions;
- choice of the optimal learning rate may be a difficult problem. Low learning rate can lead to a very slow convergence. On the contrary, high learning rate can impede convergence, and, hence, the error function will vary around it without reaching it;
- steady updating of all parameters with the same learning rate results in learning quality deterioration in the case when a data set is not balanced, i.e. the selected data contain classes represented by fewer objects.

At present, various variations of the classic gradient descent method are used. They have less time complexity for a synthesis of real practical problems to solve. It is due to the realized mechanisms of elimination of defects of the classic gradient method. Let's consider some of them.

The method of stochastic gradient descent [2, 3] assumes updating of neural network weight coefficients using just one example from training set at each step. Thus, we avoid excess calculations, since, unlike the classic gradient descent, the error function is calculated not through the whole sample, but only for one example. Hence, ANN is learnt much faster, and new examples can be added to the network input directly in the process of training.

A common disadvantage of the gradient descent method is the problem of finding a global minimum as an optimal point in the case of non-convex selection function; this peculiarity is taken into consideration in the group of impulse methods due to the accumulation of previous gradient values [4].

Stochastic gradient descent doesn't often work either in the case of the ravine error function. In this case, output values of the majority of neurons become close to asymptotic values of the activation function long before the end of training: the weight ratio practically doesn't change any longer. As a result, training becomes unacceptably slow. The simplest improvement of the gradient descent method is the introduction of momentum when the gradient influence on the weight change is gained with time [5].

Despite essential reductions in time complexity, the impulse methods don't include the integrated optimization mechanisms for unbalanced datasets, i.e. the data selections with rarely found signs. The problem of neural network training for solving classification problems for non-balanced selections was particularly considered in the works [6, 7].

Adaptive gradient (Adagrad [8]) is the method in which the updating rate of the weight coefficients of the neural network is adapted dynamically, i.e. significant renovations are made for the sign values, which are represented in minority, and weaker renovations – for frequent values. Adadelta [9, 10] is an Adagrad extension in which the problem of rapidly decrease of learning rate is solved. The sum of squared gradients in this method is replaced by exponentially damping average of all previous squared gradients, i.e. the later partial derivatives are mainly taken into account.

In the adaptive assessment method (Adam), the rule of weight updating is defined on the basis of evaluation of two different moments. The method is described in the paper [11–13].

There is also a group of second-order methods, founded on the second partial derivative calculations of the selection function error. Such methods possess a more accurate and fast convergence, but they are more complex to realize and demand big memory consumption.

The Broyden-Fletcher-Goldfarb-Shanno method (BFGS, [14]) is a quasi-Newton method, in which the weight coefficient updating occurs thanks to the hessian evaluation of the error function, but it is still a first-order method since the direct calculation and matrix inversion of the second partial derivatives of the inverse matrix are not derived.

The Broyden-Fletcher-Goldfarb-Shanno method with restricted memory (L-BFGS [15]) is a variation of BFGS, developed specially for solving optimal problems on a large datasets. For L-BFGS, the estimation of the inverse hessian is made only on the basis of the latest m iteration data. In this method, movement in the quasi-Newton direction is effected without using matrices, by forming a ring buffer.

At works [16, 17] used the cricking models. It was caused by the lack of stable parameter values, using statistical data; therefore, to achieve a required accuracy, some probability is introduced.

This work is devoted to the study of incorrect problems of multidimensional statistical dependence reconstruction under a prior uncertain data, neural network model construction using examples of multilayer feedforward and radial basis function artificial neural network training by advanced deep learning methods on the invariance concept basis. We present the examples of practical building of robust neural models to solve analysis problems of the systems and processes behavior on the basis of the interactive computer system for decision-making support «ROD&IDS®», which realizes the presented methodology.

2 General Problem Statement

The vector function is given by a training sample $\left(\vec{Y}^{(0)}, \vec{F}\right)_p$, $p = 1..P$, where $\vec{Y}^{(0)}, \vec{F}$ are the input vectors, dimension H_0, output dimension H_{K+1}, respectively. We must approximate the given set. The problem can be solved with a resultant mathematical mechanism, which may give any value of the vector function $\vec{Y}_p^{(K+1)}\left(\vec{Y}_p^{(0)}\right)$, represented by this training set at a fixed input vector within the range, limited by the input data.

3 Method of Multilayer Feedforward Artificial Neural Network Learning

A multilayer feedforward artificial neural network (MFFN), used for data approximation, is a parallel distributed processor, which is capable of saving acquired knowledge and processing information between local processor elements (neuroelements or neurons), bound by special links (synaptic links).

MFFN includes three types of neuron layers:

– the layer of input data which are known from the problem;
– the layers of intermediate data which take the corresponding data values from previous neurons, they form data and pass them to the follow-up layers;
– the layer of output data which had to be derived in the learning process.
– the scored information about the data is distributed through the network as weight parameters of these combinations, and MFFN potential development is carried out by MFFN learning.

The input data for the data approximation with MFFN are: input parameters and prototype (analog) control variables $\left\{\vec{Y}_{ph}^{(0)}\right\}$; output data – selection function values $\{f_{pi}\}$.

Initial data are usually reduced to a dimensionless form. In our case, a direct transformation was used:

$$f^0 = \frac{2l_f(f - <f>)}{(f_{max} - f_{min})},$$

where $\langle f \rangle = (f_{max} - f_{min})/2, f^0 \in [-1, 1]$, inverse:

$$\langle f \rangle = \left[(f_{max} - f_{min})f^0/l_w + (f_{max} + f_{min})/2 \right],$$

where $l_f = th(\beta)$ for MFFN, $l_f = 1$ for another network type.

The MFFN backpropagation algorithm is focused on finding the error value between the actual network output data and desirable ones. An error value can be reduced by modifying network characteristics. The process is repeated till the network becomes able to make a desirable type of "input – output" transformation. As a result of MFFN learning, we evaluate MFFN layer connection weights, and its output data are evaluated by input parameters and control variables of the designed object.

The simplest MFFN was used with one hidden layer ($K = 1$). Here $\left\{ \vec{Y}_{ph}^{(0)} \right\}$ is a multitude of input data, $\left\{ \vec{Y}_i^{(k)} \right\}$ is a multitude of k layer output data; k is the layer number, $k = 1...(K + 1)$, K is the number of hidden layers; $p = 1...P$, P is the number of analogs; $\left\{ w_{ij}^{(k)} \right\}$ is a multitude of k layer weights; i is the element of k layer; j is the element of $(k - 1)$ layer. The analytic presentation of the unknown functions for MFNN has the following structure:

$$Y_i^{(2)} = f\left(s_i^{(2)}\right), s_i^{(2)} = w_{i0}^{(2)} + \sum_{j=1}^{H_1} w_{ij}^{(2)} Y_j^{(1)}, i = 1..H_2, , j = 1..H_1;$$

$$Y_j^{(1)} = f\left(s_j^{(1)}\right), s_j^{(1)} = w_{j0}^{(1)} + \sum_{h=1}^{H_0} w_{jh}^{(1)} Y_h^{(0)}, h = 1..H_0,$$

where $f(s) = th(\beta_s) = \frac{e^{\beta_s} - e^{-\beta_s}}{e^{\beta_s} + e^{-\beta_s}}$ is the selected activation function, $f' = \beta[1 - f^2(S)]$ is the derivative of the activation function.

To provide parameter stability (robustness) and informative capability of statistical systems and processes models on the basis of learning ANN at the a priori input data uncertainty and also practically sufficient data approximation, it is reasonable to use advanced deep learning methods – stable (robust) statistical assessment of their parameters with adaptive learning rate as the ANN learning method.

The function (MV- problem) was used as a scalar convolution of selection functions, considering $f_i \equiv Y_i^{(2)}, x_h \equiv Y_h^{(0)}$ [18]:

$$E = \frac{1}{2PI} \sum_{p=1}^{P} \gamma^{P-p} \sum_{i=1}^{I} \left\{ f_{fit} \left[4 \left(\frac{\Delta_{f_i,p}}{f_i^*} \right)^2 \left(1 + \sigma_{f_i,p}^0 \right)^{-2} \right] + \beta_{i+1} \cdot f_{fit} \left[\left(\sigma_{f_i,p}^0 \right)^2 - 1 \right] \right\} \quad (1)$$

here $I = H_{K+1}$, where γ is the significance level ($\gamma = [0.95, 0.99]$), $t = 1..T$ is the epoch number learning,

$$f_{fit}(d_i) = 1 - exp\left[-\frac{L_{fit}}{4} d_i\right], L_{fit} \geq 4(d_i > 0),$$

$$\Delta_{f_i} = Y_i^{(K+1)}\left(\vec{Y}^{(0)}\right) - f_i\left(\vec{Y}^{(0)}\right), \left(\sigma_{f_i}^0\right)^2 = \frac{\left(\sigma_{f_i}\right)^2}{\left(\sigma_{f_i}^*\right)^2},$$

$f_i\left(\vec{Y}^{(0)}\right), \sigma_{f_i}^*$ is the mathematical expectation value and standard deviation of variables value $f_i \in F^0$ (index * is the required values);

$$\sigma_{f_i}^* = \beta^4 \left[1 - f^2\left(s_i^{(2)}\right)\right]^2 \cdot \sum_{j=1}^{H_1}\left\{\left(w_{ij}^{(2)}\right)^2\left[1 - f^2\left(s_j^{(1)}\right)\right]^2 \cdot \sum_{h=1}^{H_0}\left[\left(w_{jh}^{(1)}\right)^2\left(\sigma_{Y_h^{(0)}}^*\right)^2\right]\right\},$$

$$\left(\sigma_{f_i}^*\right)^2 = \left(\frac{2 \cdot l_f}{f_{i,max} - f_{i,min}}\right)^2 \cdot \left[\frac{\Delta_{f_i}^0}{300} \cdot f_{i,max}\right]^2 n_\alpha, \cdot_{f_i}^0 = \frac{\Delta_{f_i}}{f_{i,max}} \cdot 100\%.$$

The approximating functions $\vec{Y}_i^{(K+1)}\left(\vec{Y}^{(0)}\right)$ will be solved for the deep learning method, using the regular successive (adaptive) algorithm of solutions synthesis with suspended correction by the stochastic approximation method on the basis of ravine conjugate gradient method [19] with adaptive learning rate [20]. Bond weight correction will be made by the following formula (the recurrent learning algorithm, corresponding to the stochastic approximation method which provides convergence $w_{ij}^{(k)}(t) \underset{t \to \infty}{\to} \hat{w}_{ij}^{(k)}$ with probability $P = 1$):

$$w_{ij}^{(k)}(t+1) = w_{ij}^{(k)}(t) + \mu(t)\left\{\eta_{ij}^{(k)}(t)r_{ij}^{(k)}(t) - \upsilon(t)\alpha_{ij}^{(k)}(t)\left[w_{ij}^{(k)}(t) - \hat{w}_{ij}^{(k)}\right]\right\} + \hat{w}_{ij}^{(k)}(t+1),$$

$$(2)$$

where $\mu(t), \eta_{ij}^{(k)}, \vartheta(t), \alpha_{ij}^{(k)}$ are the learning and moment coefficients, respectively: $\mu(t) = \frac{\mu(0)}{1+t}$ is the learning coefficient $t = 1..P \cdot T$, T is a learning epoch number;

$$v(t) = \sqrt{\frac{2}{\ln(2+t)}},$$

$$r_{jk}^{(k)}(t) = S_{ij}^{(k)}(t) + \frac{\left(S_i^{(k)}(t)\right)^T\left(S_i^{(k)}(t) - S_i^{(k)}(t-1)\right)}{\left|\left(r_i^{(k)}(t-1)\right)^T S_i^{(k)}(t-1)\right|} r_{ij}^{(k)}(t-1),$$

are the conjugate search direction vector projections, defined in accordance with the ravine conjugate gradient method $r_{jk}^{(k)}(0) = 0$; the projection of the scalar convolution of selection functions gradient (according to the method of exponential smoothing) is

$$S_{ij}^{(k)}(t) = (1 - \vartheta)S_{ij}^{(k)}(t-1) + \vartheta\left(-\frac{\partial E}{\partial w_{ij}^{(k)}}\right)_p, \tag{3}$$

the quantities $\hat{w}_{ij}^{(k)}$ were determined according to the method of simulating the movement of bee colonies by the formula:

$$\hat{W} = \arg \inf_{\substack{W \in D_w \\ \tau \in [1,t]}} E(W, \tau), W = \left\{w_{ij}^{(k)}\right\}.$$

The learning and moment coefficients were determined by:

$$\eta_{ij}^{(k)}(t) = \rho_{ij}^{(k)}(t)\eta_{ij}^{(k)}(t-1), \eta_{ij}^{(k)}(0) = \eta_{max},$$

where $\rho_{ij}^{(k)}(t) = -\frac{1}{\frac{\partial S_{ij}^{(k)}}{\partial w_{ij}^{(k)}}\Big|_t}\left(1 - exp\left(h \cdot \frac{\partial S_{ij}^{(k)}}{\partial w_{ij}^{(k)}}\Big|_t\right)\right)$; $\alpha_{ij}^{(k)}(t) = -\frac{S_{ij}^{(k)}(t)}{S_{ij}^{(k)}(t)-S_{ij}^{(k)}(t-1)}$.

Learning used the regularizing algorithm, realizing the iteration process interruptions in the case of calculation error accumulation (the upper T index, shown below, stands for the transposing operation of a vector into line):

$$\text{if } \frac{\left(S_i^{(k)}(t)\right)^T\left(S_i^{(k)}(t) - S_i^{(k)}(t-1)\right)}{\left|\left(r_i^{(k)}(t-1)\right)^T S_i^{(k)}(t-1)\right|} \geq r_{max} \text{ then } r_{ij}^{(k)}(t-1) = 0;$$

$$\text{if } \left(S_i^{(k)}(t)\right)^T S_i^{(k)}(t-1) \geq r_{min}\left(S_i^{(k)}(t)\right)^T S_i^{(k)}(t) \text{ then } r_{ij}^{(k)}(t-1) = 0$$

(assumed: $r_{max} = 5, r_{min} = 0.2$).

When passing to a new epoch, the order of presenting new learning pairs $\left(\vec{Y}^{(0)}, \vec{F}\right)_p$, $p = 1..P$, was renewed in the recurrent algorithm (the random number generator was used at the interval $[1..P]$.

4 Learning Method of Radical Basis Function Networks

A hybrid algorithm will be used for learning radial basis function network (RBFN) in the case when the quantity of learning pairs considerably exceeds the number of neurons in the hidden layer. In the hybrid algorithm, the learning process is subdivided into two stages:

- selection of linear network parameters (output layer weights) based on the pseudoinverse method using Singular Value Decomposition (SVD);
- activation function nonlinear parameters (centers $\vec{c}_j$ and width $\vec{\sigma}_j$ of these functions).

The simplest RBFN structure with an open ($K = 1$) layer is similar to the presented MFNN. We'll introduce the following symbols:

- $\vec{Y}^{(k)} = \left[Y_1^{(k)}, \ldots, Y_{H_k}^{(k)}\right]^T, k = 0, 1, 2$ is the k-layer input data vector;
- $\vec{c}_j = \left[c_{j1}, c_{j2}, \ldots, c_{jH_0}\right]^T, j = 1..H_1$ is the vector of activation function centers for hidden layer neurons;
- $\vec{\sigma}_j = \left[\sigma_{j1}, \sigma_{j2}, \ldots, \sigma_{jH_0}\right]^T, j = 1..H_1$ is the vector which sets the window size of the j neuron hidden layer activation function;
- $\varphi_j\left(\vec{Y}_p^{(0)}, \vec{c}_j, \vec{\sigma}_j\right) = exp\left(-\frac{1}{2}\sum_{h=1}^{H_0} Z_{pjh}^2\right) \equiv \varphi_{pj}$ is the hidden layer neuron radial basis activation function $Z_{pjh} = \frac{Y_{ph}^{(0)} - c_{jh}}{\sigma_{jh}}$;
- w_{ij} is the bond weight between an output layer i neuron and a hidden layer j neuron (here, in compliance with symbols, imply that $w_{jh}^{(1)} = e_{jh} = 1$, $w_{ij}^{(2)} \equiv w_{ij}$).

Analytical presentation of the required functions for the radial basis network (RBSN) has the following structure:

$$Y_i^{(2)} = s_i^{(2)}, s_i^{(2)} = w_{i0}^{(2)} + \sum_{j=1}^{H_1} w_{ij}^{(2)} Y_j^{(1)}, i = 1..H_2, \ j = 1..H_1;$$

$$Y_j^{(1)} = \varphi\left(s_j^{(1)}\right), s_j^{(1)} = \frac{1}{2}\sum_{h=1}^{H_0} [Z_{jh}]^2, h = 1..H_0,$$

where $\varphi(s) = exp(-s)$ is the selected activation function, $\varphi'(s) = -\varphi(s)$ is the activation function derivative,

$$\sigma_{f_i}^2 = \sum_{j=1}^{H_1}\left\{\left[w_{ij}^{(2)}\right]^2 \cdot \varphi^2\left(s_j^{(1)}\right) \cdot \sum_{h=1}^{H_0}\left[Z_{jh}^2 \cdot \left(\frac{\sigma_{Y_h^{(0)}}^*}{\sigma_{jh}}\right)^2\right]\right\},$$

To define neuron weights of the network output layer $\vec{w}_i = [w_{i1}, w_{i2}, \ldots, w_{iH_1}]^T$, $i = 1..H_2, j = 1..P$, the SVD decomposition, proposed in the work of Golub and Kohan, was used.

Further on (the second stage), at the fixed output weigh data values, the input data are put through the network to the output layer; it makes it possible to evaluate the scalar convolution of selection functions values for vector succession $\left\{ \vec{Y}_p^{(0)} \right\}$. After this, we return to the hidden layer (backpropagation). The scalar convolution of selection functions gradient vector relative to particular centers $\vec{c}_j$ and width $\vec{\sigma}_j$ is determined by $\triangle_{f_i} = Y_i^{(K+1)} \left(\vec{Y}^{(0)} - f_i(\vec{Y}^{(0)}) \right)$.

To determine coordinate centers matrix values $C = \left\{ c_{jh} \right\}$ of the activation functions for the hidden layer neurons, the c-means algorithm was used at $K = H_1$ [21]; it makes it possible to reduce radial basis function network learning time.

The initial approximation was selected as: $\sigma_{jh} = \frac{\rho}{\sqrt{2H_1}}$, where ρ is the maximal distance between centers $\vec{c}_j$, $j = 1..H_1$, $h = 1..H_0$; $\rho = \max_j \max_k \left(\sqrt{\sum_{h=1}^{H_0} \left(c_{jh} - c_{kh} \right)^2} \right)$, $k = 1..H_1$ (if all initial data are reduced to dimensionless form of the equation in advance, then $\rho = 2\sqrt{H_0}$, $\sigma_{jh} = \sqrt{\frac{2H_0}{H_1}}$).

To specify the values $\sigma = \left\{ \sigma_{jh} \right\}$, we used the algorithm of "coverage domain" formation by the radial basis functions allowing for K "neighbors": $\sigma_{jh}^2 = \Sigma_j = \frac{1}{K} \sum_{k=1}^{K} \sum_{h=1}^{H_0} \left(c_{jh} - c_{kh} \right)^2$, $k = 1...K$, which made it possible to reduce the learning time for radial basis function network. $K \in [3, 5]$ was assumed.

Refining the covariance matrix $\sigma = \left\{ \sigma_{jh} \right\}$ and center coordinates $C = \left\{ c_{jh} \right\}$ completes a successive learning stage. Correction of the covariance matrix elements was carried out by the above described deep learning method, applying regularizing successive (adaptive) algorithm of solutions synthesis with suspended correction, i.e. the stochastic approximation method on the basis of the ravine method of conjugate gradients with adaptive learning rate according to formulae (2), (3) with changing $w_{ij}^{(k)}$ for σ_{jh} and c_{jh}.

At the end, the reconstruction of the unknown function analytical representation $Y_{pi}^{(2)} Y_{pi}^{(2)}$, $i = 1..H_2$; in dimensionless form at the beginning and then – by using a reverse transformation in the form of physical value dependences.

Multiple repetition of both stages leads to the complete and fast network learning, especially when the initial parameter values of radial basis functions are close to the optimal values.

In practice, the evolved stages influence parameter adaptation to varying degrees.

As a rule, the SVD algorithm is faster: it finds the function local minimum within one step. To level this disproportion, one linear parameter amendment was usually followed by several cycles of non-linear parameter adaptation.

To check the ANN data prediction significance (quality), we used the following values, which were averaged as a result of 10 independent launches:

- mean-square error energy: $E_{av} = \frac{1}{2P} \sum_{p=1}^{P} \sum_{i=1}^{H_{K+1}} \Delta_{f_i,p}^2$;

- mean relative errors $\delta_i^0 = \frac{1}{P} \sum_{p=1}^{P} \left| \frac{\Delta_{f_i,p}}{f_{i,p}} \right|$;

- residual variances $S_{res,i}^2 = \frac{\sum_{p=1}^{P} \Delta_{f_i,p}^2}{P-1}$.

At a formal mathematical model comparison (FMM), we'll estimate the signal change variance which characterizes robustness of either model:

$$D_{Y_i,dB} = 10 \, log_{10} \left(\frac{D_{Y_i}^{(\lambda)}}{D_{Y_i}^{(1)}} \right), \text{ decibel, } \lambda = 1, 2.$$

Here, as the residual variances estimates were further used values of residual variances for each of the compared FMM.

5   Results of the Modeling

The methodology and supporting it interactive computer system for decision-making support "ROD&IDS®" was worked out. It is intended for wide users [22] and may be used to solve the following problems in robust optimal designing and intellectual diagnosing (ROD&IDS) systems:

- to form robust models (meta-models) of the viewed systems and processes (regression analysis);
- to estimate the information capability of the variables of obtained models (Sensitivity analysis), to perform the dimension reduction of the state's space;
- to estimate selection function confidence intervals at the parameter preset confidence intervals of the parameters and variables (analysis of variation – Monte-Carlo analysis);
- to define rational values of averages and corresponding confidence intervals of unknown parameters (project parameters, control variables) at preset average values and corresponding selection function confidence intervals (normal solutions, Pareto set).
- such problems appear at:
- robust multi-criterion optimal designing of items under conditions of uncertainty on the basis of discrete data about analogs;
- quality control of production: to reduce scrap level of the output produce (Design for Six Sigma);
- storage control in logistical problems: selection of an optimal output product program: use of resources;
- synthesis of the rational composition of medicines;
- prescription of the rational diet, medicament treatment etc.;
- forecasting risks in banking, client insurance.

MFFN and RBSN of diverse structure were used to solve the problems of designing robust models (meta-models) of the examined systems and processes (regression analysis) [19]. The same networks were built and learnt with the mathematical package MATLAB 7.0.1.

The analysis of the results of data prediction significant testing (quality) by the above viewed MFFN and RBFN learning methods revealed the following peculiarities:

- the proposed learning method of the MFFN using stochastic approximation with one open layer provides data prediction accuracy comparable with the MFFN learning with one hidden layer by the Levenberg- Markquardt method on the basis of Bayes regularization;
- RBSN learning with non-linear parameter adaptation of the radial-basis functions results in considerable informational complexity of the learning method, but, still, it doesn't provide substantial energy reduction of mean-square error E_{av} (by an order of magnitude for MFFN). Therefore, to solve practical problems, it's appropriate to do RBFN learning at $T = 1\ldots5$.

At the same time, the method of parameter assessment of structural-parametric models of systems and processes in the form of trained ANN by using the regularizing successive (adaptive) algorithm of the solutions synthesis with suspended correction (by the method of stochastic approximation on the basis of the ravine conjugate gradients method with adaptive learning rate) provides effective stable (robust) assessment of unknown values at the parametric input data ambiguity and sufficient practical data approximation accuracy in the problems of systems improvement.

The sample of the test function values of type $y = \frac{1}{4}\left[(x-2)^2 + (z-2)^2\right]$ at $\Delta_y^0 = 1\%$ was taken as the first example of realizing the presented methodology for effective stable (robust) estimation of systems and process model parameters in the form of learning ANN. To solve data approximation problems, MFFN and RBSN with identical structure were used. The results of the quality estimation of the test function value sampling robust approximation by neural models are presented in the table. It is evident that quality of robust approximation with RBSN $(T = 3)$ is higher than with MFFN.

The robust approximation results of value sampling for the Rosenbrock function at $\Delta_y^0 = 0.001\%$ with RBSN $(T = 3)$ are shown in Fig. 1 as the second example (red dot – minimum function value) (Table 1).

As the third example of realizing the presented methodology for effective stable (robust) estimation of systems and process model parameters in the form of learning ANN, the data sampling of the corresponding parameters of the jet engines multistage axial flow compressor (MSAFC) was taken. The input data for ANN included the inlet mass flow rate values (G_{air}) and the number of rotor turnovers per minute (n); the output data included the compression ratio (π_c^*) and coefficient of efficiency (η_c^*) at $\Delta_\pi^0 = 0.7\%$, $\Delta_\eta^0 = 0.1\%$. The results of robust approximation pressure ratio characteristics with RBSN $(T = 3)$ are presented in Fig. 2 (red dot – corresponds to the design point of MSAFC work). The mean approximation error was $\delta_\pi^0 = 0.625\%$.

Table 1. Results of quality estimation of data approximation with ANN.

ANN type	Mean-square error energy	Mean-square error energy, %	Change in signal variance, decibels
MFFN [2–20]	0,00222	28,9	0
RBFN [2–20]	1,63E-06	1,79	−47.0

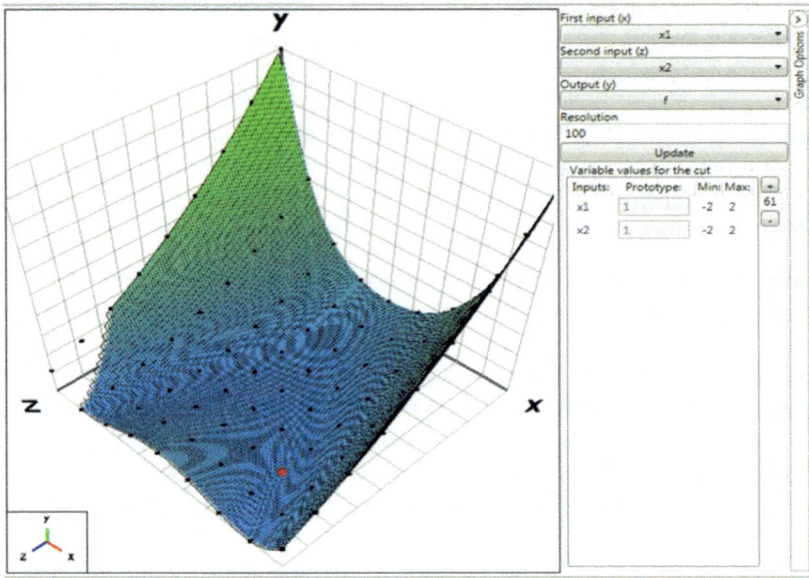

Fig. 1. Robust approximation results for Rosenbrock function data via a RBFN.

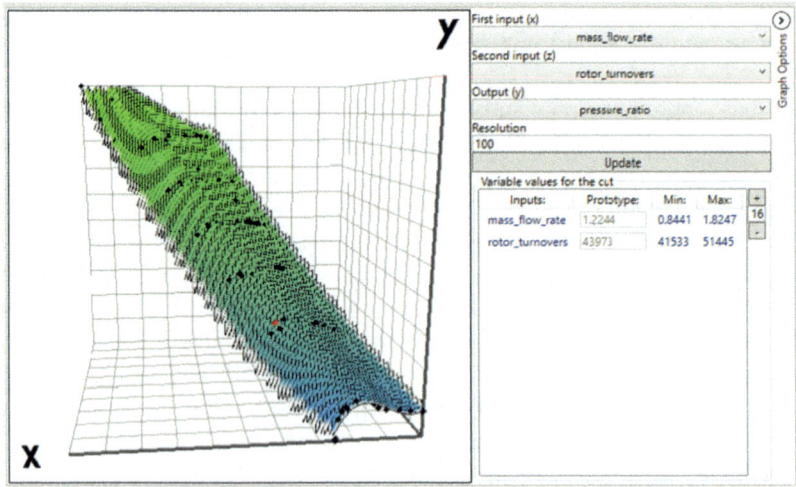

Fig. 2. Results of multistage axial flow compressor pressure ratio via a RBFN.

6   Conclusions

The paper analyzes mathematical and software solutions in the field of robust optimal designing and intellectual diagnostics of the systems (ROD&IDS).

The methods for approximating vector functions of vector variables were based on the use of the learning artificial neural network (ANN), multilayer feedforward (MFFN) and radial-basis function networks (RBFN). ANN learning was carried out by the advanced deep learning method by applying the regularization successive (adaptive) algorithm of defusing synthesis with suspended correction – by the method of stochastic approximation based on the ravine conjugated gradients method with adaptive learning rate. Application of the proposed methods avoids the appearance of false ravines or valleys on response surfaces in case of gross errors in the input data.

The methodology and interactive computer system for decision-making support "ROD&IDS®" to realize it, which is intended for a wide range of users and may be used for solving practical problems in ROD&IDS were worked out.

References

1. Kathuria, A.: Intro to optimization in deep learning: Gradient Descent. Paperspace. Ser.: Optimization. https://blog.paperspace.com/intro-to-optimization-in-deep-learning-gradient-descent/ (2018). Accessed 15 Oct 2020
2. Ruder, S.: An overview of gradient descent optimization algorithms. arXiv Preprint, arXiv: 1609.04747 (2016)
3. Robbins, H., Monro, S.: A stochastic approximation method. Ann. Math. Stat. **22**(3), 400–407 (1951)
4. Nesterov, J.E.: Method of convex function minimization with convergence rate $O(1/k2)$. Report of USSR Academy of Sciences 269(3), 543–547 (1983). [in Russian]
5. Goh, G.: Why momentum really works. Distill (2017). http://doi.org/10.23915/distill.00006
6. Kukar, M., Kononenko, I.: Cost-sensitive learning with neural networks. Machine learning and data mining. In: Prade, H. (ed.) 13th European Conference on Artificial Intelligence, pp. 445–449. Wiley–Blackwell (1998)
7. Demchenko, M.V.: Building the neural network classifier for reveling the major artery atheroscleerosis risk. In: Optimization and Modelling in Automatized Systems, pp. 29–36. Voronezh State Technical University Publ., Voronezh (2017). [in Russian]
8. Duchi, J., Hazan, E., Singer, Y.: Adaptive subgradient methods for online learning and stochastic optimization. J. Mach. Learn. Res. **12**, 2121–2159 (2011)
9. Zeiler, M.D.: Adadelta: An adaptive learning rate method. arXiv Preprint, arXiv:1212.5701 (2012)
10. Wilson, A.C., Roelofs, R., Stern, M., et al.: The marginal value of adaptive gradient methods in machine learning. In: Guyon, I. (eds.) Advances in Neural Information Processing Systems, vol. 30, pp. 4148–4158. NIPS Foundation (2017)
11. Kingma, D.P., Ba, J.: Adam: a method for stochastic optimization. arXiv Preprint, arXiv: 1412.6980 (2014)
12. Hu, J., Shen, L., Albanie, S., et al.: Squeeze-and-excitation networks. IEEE Trans. Pattern Anal. Mach. Intell. **42**(8), 2011–2023 (2020). https://doi.org/10.1109/TPAMI.2019.2913372

13. Le Roux, N., Fitzgibbon, A.: A fast natural newton method. In: Proceedings of the 27th International Conference on Machine Learning (ICML-10), pp. 623–630. Omnipress, Madison (2010)
14. Fletcher, R.: Practical Methods of Optimization, 2nd edn. Wiley, Chichester (2000)
15. Schraudolph, N.N., Yu, J., Günter, S.: A stochastic quasi-Newton method for online convex optimization. Proc. Mach. Learn. Res. **2**, 436–443 (2007)
16. Koch, P.N., Wujek, B.A., Golovidov, O., Simpson, T.W.: Facilitating probabilistic multidisciplinary design optimization using kriging approximation models. In: 9th AIAA/ISSMO Symposium on Multidisciplinary Analysis and Optimization, paper No. AIAA-2002-5415. AIAA, Atlanta (2002)
17. Gano, S.E., Renaud, J.E., Martin, J.D., Simpson, T.W.: Update strategies for kriging models used in variable fidelity optimization. Struct. Multi. Optim. **32**(4), 287–298 (2006). https://doi.org/10.1007/s00158-006-0025-y
18. Meniailov, I., Khustochka, O., Ugryumova, K., et al.: Mathematical models and methods of effective estimation in multi-objective optimization problems under uncertainties. In: Schumacher, A., et al. (eds.) Advances in Structural and Multidisciplinary Optimization. WCSMO 2017, pp. 411–427. Springer, Cham (2018). https://doi.org/10.1007/978-3-319-67988-4_32
19. Strilets, V.E., Tronchuk, A.A., Ugryumova, K.M. et al.: Systematic perfection of complex technical systen elements on the basis of inverse problems. National Aerospace University "Kharkiv Aviation Institute", Kharkiv (2013). [in Russian]
20. Strilets, V., Bakumenko, N., Chernysh, S., et al.: Application of artificial neural networks in the problems of the patient's condition diagnosis in medical monitoring systems. In: Nechyporuk, M., et al. (eds.) Integrated Computer Technologies in Mechanical Engineering. AISC, vol. 1113, pp. 173–185. Springer, Cham (2020). https://doi.org/10.1007/978-3-030-37618-5_16
21. Bakumenko, N., Strilets, V., Ugryumov, M.: Application of the C-means fuzzy clustering method for the patient's state recognition problems in the medical monitoring systems. CEUR Workshop Proceedings, vol. 2362, pp. 218–227 (2019)
22. Meniailov, I., Ugryumov, M., Chumachenko, D., et al.: Non-linear estimation methods in multi-objective problems of robust optimal design and diagnostics of systems under uncertainties. In: Nechyporuk, M., et al. (eds.) Integrated Computer Technologies in Mechanical Engineering. AISC, vol. 1113, pp. 198–207. Springer, Cham (2020). https://doi.org/10.1007/978-3-030-37618-5_18

A Fast Method for Visual Quality Prediction and Providing in Image Lossy Compression by SPIHT

Fangfang Li[1] , Sergey Krivenko[2(✉)] , and Vladimir Lukin[2]

[1] Nanchang Hangkong University, 696 Fenghe South Avenue,
Nanchang 330063, China
liff_niat@yahoo.com
[2] National Aerospace University "Kharkiv Aviation Institute",
17 Chkalova Street, Kharkiv 61070, Ukraine
krivenkos@ieee.org, lukin@ai.kharkov.com

Abstract. Wavelet-based image compression techniques, in particular, SPIHT have an obvious advantage of providing a desired compression ratio. However, for a given compression ratio, quality can vary in wide limits and it can be unsatisfactory for a given image, especially if it is highly textural. Then, one needs to choose and set another compression ratio providing an appropriate quality of the compressed image. Often, this adaptation should be done quickly and with providing high accuracy of reaching a desired quality. In this paper, we propose a fast and efficient approach to predict quality of images compressed by SPIHT using two favorable facts. Firstly, a fast procedure of quality prediction has been proposed for the coder AGU based on discrete cosine transform. Secondly, performance characteristics of SPIHT and AGU are quite similar. This allows recalculating predictions obtained for AGU to predictions for SPIHT and decision undertaking concerning parameters of image encoding by SPIHT.

Keywords: Image lossy compression · SPIHT · Full-reference quality metrics · AGU · Parameter adaptation

1 Introduction

There are numerous applications where images are used nowadays [1, 2]. The number of acquired images increases, resolution improves and, thus, average image size becomes larger. These images have to be transferred via communication lines and/or stored [2, 3]. Although performance characteristics of communication and memory means improve each year, it is often needed to carry out image compression [3–5].

Image compression techniques are divided into lossless and lossy ones. The former ones can be used in some applications but compression ratio (CR) for them is usually not large enough [3]. Because of this, lossy compression techniques are widely used. Most modern lossy compression techniques are based on orthogonal transforms [4–6]. The most popular are wavelet and discrete cosine transforms put into basis of the standards JPEG2000 and JPEG, respectively. There are also many other compression

© The Author(s), under exclusive license to Springer Nature Switzerland AG 2021
M. Nechyporuk et al. (Eds.): ICTM 2020, LNNS 188, pp. 17–29, 2021.
https://doi.org/10.1007/978-3-030-66717-7_2

techniques that employ the aforementioned orthogonal transforms as well [7]. An obvious advantage of many compression techniques based on wavelet transforms as JPEG2000 and SPIHT (Set Partitioning in Hierarchical Trees) [8–10] is that they are able to easily provide a desired CR. However, this advantage turns into a drawback if it is necessary to provide a desired quality of a compressed image. It has been shown in [7]. that, for a given CR or bits per pixel (bpp), quality of images compressed by JPEG2000 or SPIHT can vary in very wide limits depending on image complexity (in Sect. 2, some examples will be given).

Then, two questions arise. The first one is can one predict quality of an image compressed by JPEG2000 or SPIHT with a given bpp in advance. Certainly, it is possible to compress a given image with a given bpp, then decompress it and calculate a metric characterizing image quality. Then, a decision can be undertaken what to do – to remain the compressed image for further use or to look other bpp to ensure a more acceptable solution. This takes time that depends on many factors – how large a given image is, how compression and decompression are realized, what a desired CR is, etc. It is often desired to get the solution quickly enough. Recently, a two step-procedure of image compression by SPIHT has been proposed [11]. It employs compression/ decompression with the recommended bpp and metric determination at the first step and corrected bpp determination with final compression at the second step [11]. Due to the second step, the accuracy of providing a desired quality has improved sufficiently. However, it is still not high enough and worth further improving.

Meanwhile, there are two favorable obstacles that allow solving the aforementioned task in indirect but quite fast and accurate way. One obstacle is that performance characteristics of SPIHT are quite similar to those ones [12] of the DCT based coder AGU [13]. Another obstacle is that several simple and fast procedures for predicting performance of AGU have been proposed recently [14, 15]. Thus, our idea consists in exploiting experience in prediction of AGU coder characteristics for prediction of SPIHT performance parameters. The goal is to show that this way, under certain conditions, produces accuracy which is better or, at least, comparable to accuracy of the two-step procedure [11], but is considerably faster. The novelty of our paper is twofold. Firstly, we propose a very fast and quite accurate prediction procedure. Secondly, we analyze not only standard quality metrics but also visual quality ones [16–18] which has become a modern trend in image compression [18–20]. Note that analysis is carried out for grayscale images represented as 8-bit 2D data arrays. However, we hope that it can be extended for more complex practical cases.

The paper structure is as follows: In Sect. 2, the main characteristics of SPIHT are introduced. Section 3 involves the comparison of the dependences for AGU and SPIHT encoders. Section 4 explains the basis for predicting and providing the image compression quality of SPIHT. Section 5 proposes the prediction procedures. Section 6 contains analysis of the obtained results in detail. Then, the conclusions are given.

2 SPIHT Coder and Visual Quality Metrics

The SPIHT coder is a typical representative of lossy compression techniques based on discrete wavelet transform, which was developed from the early EZW (embedded zero-tree wavelet) image compression technology [10]. SPIHT uses a tree splitting method (spatial direction tree) to split the set of all points into some subsets by diversity sorting method, and performs a significance test on these subsets. The test uses a multi-level threshold that decreases by 2 times to make a judgment. The splitting algorithm continues until the significance test is performed on the significance subset of a single node to determine each significance wavelet coefficient [9]. Thus, it is easy for SPIHT to provide a desired CR. In addition to CR characterizing the encoder and a compressed image, an end user also needs to consider whether the compressed image quality meets the quality requirements, i.e. is a desired quality provided. The evaluation or prediction of compressed image quality is essential for numerous applications and many quality metrics can be used for this purpose [15, 16, 19, 21, 22]. Below, we consider one conventional quality metric (peak signal-to-noise-ratio – PSNR) and one visual quality metric (PSNR that takes into account human vision system (HVS) and masking effect – PSNR-HVS-M).

PSNR is the computationally simplest and, thus, the most widely used metric. But as image quality estimator, it does not correlate well with visual quality [22]. In the case where human perception is the ultimate task, it is necessary to use metrics that incorporate properties of HVS. PSNR-HVS-M takes into account less sensitivity to distortions in high spatial frequencies and masking effect of texture [18]. Both metrics are expressed in dB where larger values correspond to better quality. The metrics have approximately the same values for distortions due to lossy compression if they are similar to additive white Gaussian noise and masking is absent.

Table 1. Metric values (dB) for different test images and CR.

Test image	CR = 2		CR = 4		CR = 8		CR = 16	
	PSNR	PSNR-HVS-M	PSNR	PSNR-HVS-M	PSNR	PSNR-HVS-M	PSNR	PSNR-HVS-M
Lenna	55.80	67.26	45.14	52.59	40.46	44.90	37.24	39.53
Barbara	53.37	64.71	43.59	50.05	37.41	40.53	32.08	33.64
Baboon	46.06	57.23	34.96	41.05	30.16	32.62	25.96	26.70
Goldhill	52.39	64.63	42.03	48.22	36.61	40.16	33.13	33.90
Mrt_prepared	102.32	116.25	52.41	62.84	44.12	48.54	38.21	40.24

Table 1 presents PSNR and PSNR-HVS-M values for SPIHT for five common test images and four values of CR. Data in Table 1 clearly show one problem in SPIHT compression – for a given CR, quality of compressed images sufficiently depends upon image content (complexity). For example, for CR = 8, PSNR varies from 40...44 dB (invisible distortions) for images of quite simple structure (Lenna, Mrt_prepared) till 30 dB (clearly visible distortions) for the test image Baboon that is highly textural. Similarly, for CR = 16, PSNR values vary from ≈ 38 dB (almost invisible distortions) for Mrt_perpared till ≈ 26 dB (annoying distortions) for Baboon. For comparison, two images – Baboon and Mrt_prepared - are shown in Fig. 1, a and Fig. 1, b, respectively. In fact, limits of these variations can be even wider for larger image set. The visual quality data evaluated by PSNR-HVS-M also embodies such characteristics.

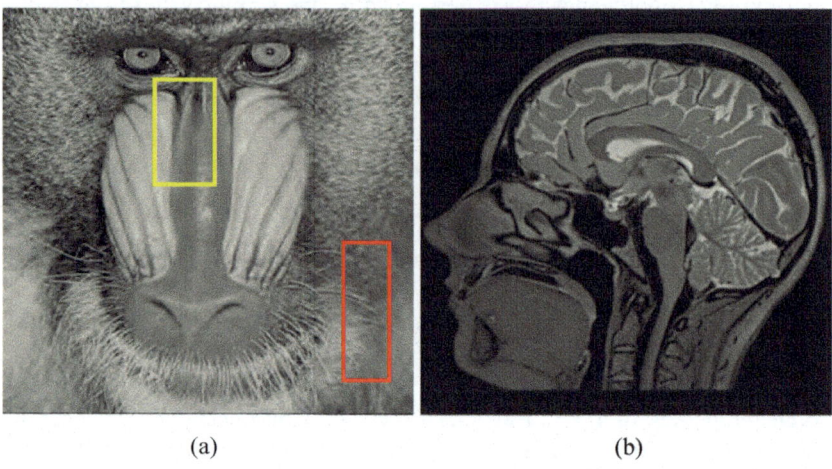

(a)   (b)

Fig. 1. Test images compressed with CR = 16: (a) high complexity image Baboon, PSNR = 25.96 dB, PSNR-HVS-M = 26.70 dB, visible distortions; (b) low complexity image Mrt_prepared, PSNR = 38.21 dB, PSNR-HVS-M = 40.24 dB, invisible distortions.

Figure 1 shows the difference in visual quality of images with different complexity when CR is equal to 16. For high-complexity image Baboon, more distortions are introduced. The PSNR and PSNR-HVS-M values are 25.96 dB and 26.97 dB respectively, which are below the visual quality threshold (∼ 35 dB), and the distortion can be clearly observed – they appear themselves as smearing in areas marked by red and yellow rectangles. For low-complexity image Mrt_prepared, less distortions are introduced, and the values of PSNR and PSNR-HVS-M are about 38 dB and 40 dB, respectively, distortions are not easy to detect by visual inspection.

It is clear that for a given bpp or CR (CR and bpp are strictly dependent where, for 8-bit grayscale images, CR ≈ 8/bpp [11]), the quality can be very different. Through the data analysis of Table 1, it can be concluded that image visual quality depends on both bpp and image complexity. This means that often in practice of lossy compression the CR or bpp should be set individually for each image to be compressed with a desired quality. Therefore, a challenging task for the SPIHT is to predict the parameters and provide the desired visual quality. The recently proposed two-step compression method [11] partially solves this problem. The two-step method is derived from a large number of image test results, which show that the visual quality of all compressed images has a monotonous dependence on control parameter (bpp), and the change trend of image quality is such that it can be approximately linearly in a limited size interval. Therefore, it is assumed that parameters of average rate-distortion curve obtained in advance can be used in predicting and providing the image quality.

The SPIHT two-step method of providing a desired quality according to a given metric [11] can be described as follows. In the first step, the procedure sets the initial value of control parameter (bpp) using the average rate-distortion curve obtained off-line (in advance) and, then, conducts the first compression/decompression of a considered image. Then, the used visual quality metric is calculated, this value is in some neighborhood of the desired one but it can be quite different (variance of the provided values after the first step can be about 40 dB^2, thus, the accuracy is very low). Having the quality estimate obtained from the first step and the slope of the average rate-distortion curve, the control parameter value corresponding to the desired visual quality is calculated using linear approximation, so as to perform the second step compression. After the parameter adjustment, the visual quality value obtained by the second compression is usually closer to the desired value. Due to the second step, variance decreases by several times but can be still too large, especially for PSNR (the error can be up to 7...8 dB). The reason is that we use the slope corresponding to the bpp point in the average distortion curve. Compared to the previously proposed iterative method [7], the two-step method performs faster, but it is still desired to decrease computation time and improve accuracy.

3   Comparison of the Dependences for SPIHT and AGU

An idea put forward in this paper consists in the following. Firstly, a very fast procedure of metric prediction in lossy compression of images by methods based on DCT, in particular AGU [13], has been proposed recently [14, 15, 23] (see details in the next Section). Simplicity of this procedure is explained by the fact that statistics of DCT coefficients in a limited number of 8 × 8 pixel blocks is obtained and very quickly analyzed. Secondly, there is a quite strict connection between performance characteristics of AGU and SPIHT that has been earlier exploited in lossy compression of noisy images [12], We propose to use this property but it should be first thoroughly analyzed for the case of compressing images that are practically noise-free.

Before coming to such a study, recall peculiarities of the AGU coder. AGU is a lossy image compression technique that uses 32 × 32 pixel fixed-size blocks in which 2D DCT is performed. Quantized DCT coefficients are then encoded and embedded deblocking is applied after decompression. The control parameter used is the quantization step (QS). Let us analyze rate-distortion curves obtained for two test images. They are presented in Fig. 2 as dependences of PSNR and PSNR-HVS-M on CR. It is possible to see that there is the strict connection between the dependences for the considered coders. The results show that the DCT-based AGU achieves slightly better metric values than SPIHT for the same CR. However, the correction algorithm (the algorithm of recalculating parameters of AGU to the corresponding SPIHT parameters) should be based on a larger volume of experimental data.

To get a better understanding, the average dependence curves of visual quality metrics on CR under for two encoders are given in Fig. 3. If the visual quality values of

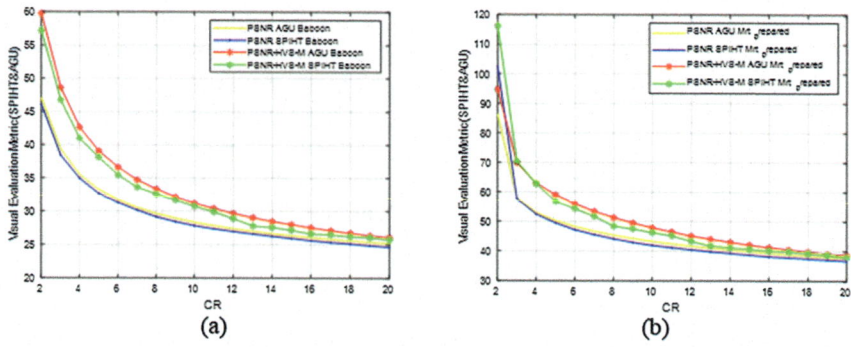

Fig. 2. Test images dependence curve (a) high complexity image Baboon; (b) low complexity image MRT_prepared.

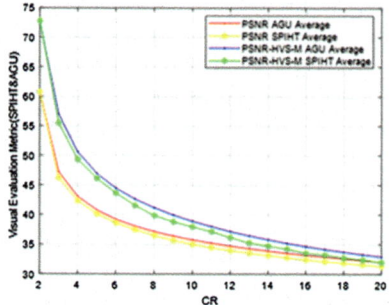

Fig. 3. Average dependences for AGU and SPIHT.

multiple images are averaged, the compressed visual quality dependence curves of SPIHT and AGU codes are still similar, and the deviation is basically fixed for each CR value. After calculation, the average deviation of PSNR (SPIHT relative to AGU) is 0.6232 dB, and the average deviation of PSNR-HVS-M (SPIHT relative to AGU) is 0.9337 dB. Then, the visual quality metrics value for SPIHT can be calculated as

$$PSNR_{SPIHT} = PSNR_{AGU} - 0.6232, dB, \tag{1}$$

$$PSNR-HVS-M_{SPIHT} = PSNR - HVS-M_{AGU} - 0.9337, dB. \tag{2}$$

4 Previous Prediction Approaches for AGU

The analysis of the experimental results of SPIHT and AGU compression shows that there is a fixed relationship between the dependence of the visual quality values of the two on CR. Since the image quality prediction after AGU compression has been successfully achieved through control parameters and image complexity in [23], it becomes possible to predict the image quality of SPIHT by this method.

Let us briefly review the method of visual quality prediction for AGU. As known, there is direct dependence between distortions due to DCT coefficient quantization and losses in compressed data. We will use the parameter P_{0q} which denotes the mean probability that quantized DCT coefficients in 8×8 blocks are equal to zero [14, 15].

$$P_{0q} = \left(\sum_{n=1}^{N_{bl}} N_n \right) / 64 N_{bl}, \tag{3}$$

where N_n denotes the number of DCT coefficients to be zeroed after quantization (this happens if DCT coefficient magnitude is less than QS/2) for an n-th block, N_{bl} denotes the number of the considered 8×8 pixel blocks randomly chosen in a considered image. The larger P_{0q} (more coefficients are assigned zero values), the greater CR. The function that approximates this relationship can be expressed as [15]:

$$CR = 0.9462 exp \left(2.895 P_{0q} \right) + 1.045 \times 10^{-13} exp (35.52 P_{0q}). \tag{4}$$

This means that P_{0q} can be predicted for a given CR or bpp. In [23], a method for predicting the PSNR for AGU is proposed. This method calculates the DCT coefficients in 8×8 pixel image blocks and quantizes them as (5).

$$D_q(n, k, l) = [D(n, k, l) / QS], k = 0, \ldots, 7; l = 0, \ldots, 7, \tag{5}$$

where $D(n, k, l), k = 0, \ldots, 7, l = 0, \ldots, 7$ denotes a set of DCT coefficients for an n-th block, and $D(n, 0, 0)$ is the direct current (DC) coefficient that relates to mean in the n-th block, $[\cdot]$ denotes the rounding-off to the nearest integer.

As one knows, standard PSNR is estimated as (6).

$$PSNR = 10 \, log_{10}(MAX^2 / MSE), MSE = \frac{1}{I_{Im} J_{Im}} \sum_{i=1}^{I_{Im}} \sum_{j=1}^{J_{Im}} (I_{ij}^c - I_{ij}^{or})^2, \tag{6}$$

where MAX defines the image dynamic range; $I_{Im} J_{Im}$ defines the image size; I_{ij}^c and I_{ij}^{or} denote the ij-th pixel value of compressed and original images, respectively. It has been shown in [15, 23] that MSE for AGU can be quickly and quite accurately predicted for a given QS in the following manner. Let us calculate differences as

$$\Delta D_q(n, k, l) = QS \times D_q(n, k, l) - D(n, k, l), k = 0, \ldots, 7; l = 0, \ldots, 7, \tag{7}$$

and then the MSE for an n-th block can be predicted by the formula (8).

$$MSE_n = \frac{1}{64} \sum_{k-0}^{7} \sum_{l=0}^{7} (\Delta D_q(n, k, l))^2. \tag{8}$$

The MSE estimate for the entire image can be calculated as follows:

$$MSE = \frac{1}{N}\sum_{n=1}^{N} MSE_n = \frac{1}{64N}\sum_{n=1}^{N}\sum_{k=0}^{7}\sum_{l=0}^{7}\left(\Delta D_q(n,k,l)\right)^2. \qquad (9)$$

and, after that, the metric PSNR can be estimated according to (6). Standard deviation of prediction errors is less than 2 dB [23], i.e., better than for two-step procedure for SPIHT. The coefficients $D(n,k,l)$, $k = 0,\ldots,7$, $l = 0,\ldots,7$ are collected for all considered blocks and, thus, one can easily predict visual quality metrics for any QS.

Concerning the visual quality metric PSNR-HVS-M, it is defined as (10).

$$PSNR-HVS-M = 10log_{10}\left(255^2/MSE_{HVS-M}\right). \qquad (10)$$

where MSE_{HVS-M} is MSE that takes into account peculiarities of HVS. The paper [14] shows how MSE_{HVS-M} can be quickly and quite accurately predicted for AGU.

5 Proposed Procedure for Prediction in SPIHT

Performance characteristics of AGU (CR, MSE, MSE_{HVS-M}) can be quickly and accurately predicted. At the same time, the dependences of average PSNR and PSNR-HVS-M on CR for AGU and SPIHT are also known (expressions (1) and (2)). These two aspects provide the realization ideas for the visual quality prediction for SPIHT.

There is one key problem to be solved. The control parameter of SPIHT is bpp, but the PCC of AGU is QS. To predict the compressed image quality of SPIHT for a given bpp, one has to determine the corresponding QS for AGU. For this purpose, let us first convert bpp to CR as CR = 8/bpp. The next stage is the conversion from CR to QS; this stage is relatively complicated, because it depends on an image at hand. Recall that the parameter P_{0q} mentioned in Sect. 4 reflects the percentage of DCT coefficient values returned to zero after quantization. Known CR allows determining P_{0q} using (4). Of course, it is not easy to find the inverse function, but P_{0q} can be derived using piecewise linear interpolation. Then, one needs an algorithm for obtaining the QS corresponding to the determined P_{0q} and, respectively, CR for the considered image. One very simple version of this algorithm is shown in Fig. 4.

First, set an initial value of QS, e.g., equal to 0, and, then, gradually increase it until the percentage of DCT coefficients with absolute values smaller than QS/2 is smaller than P_{0q}. When this happens, remember QS. The fourth stage in the prediction process is to use the method in Sect. 4 to calculate the PSNR and PSNR-HVS-M values corresponding to the QS value obtained in the third stage. After this, complete the entire prediction process of the visual quality of SPIHT compression coder via AGU using expressions (1) and (2).

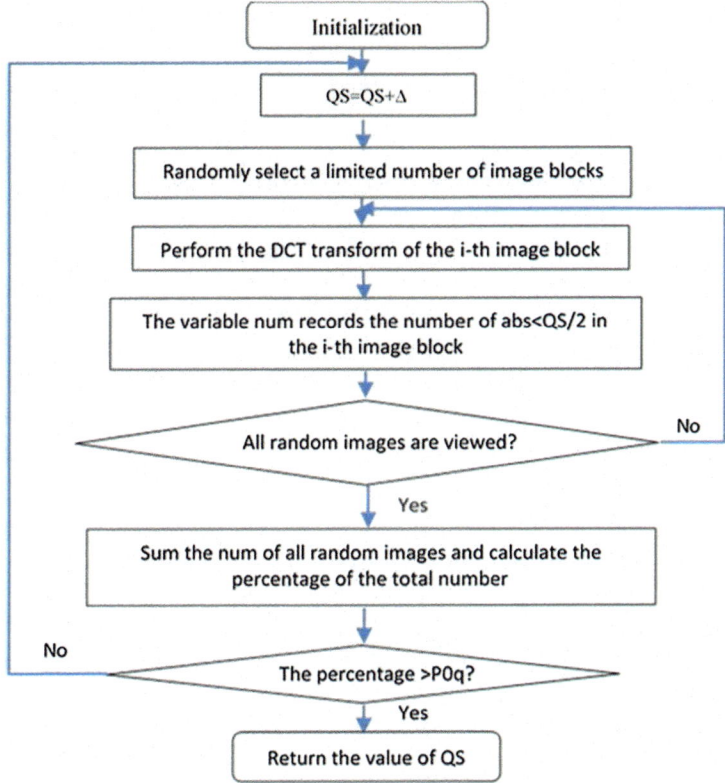

Fig. 4. Flow chart for calculating QS.

6   Analysis of Accuracy

In the prediction process, some calculations are used to replace the actual compression process to achieve the purpose of prediction, and of course deviations are inevitable. In this Section, the prediction accuracy analysis will be carried out.

We have considered two values of bpp in our experiments, namely 0.5 and 1. To analyze a prediction accuracy, two visual quality metrics, PSNR and PSNR-HVS-M, have been employed. In the experiment, we have selected 300, 500 and 1000 as the number of random image blocks. The comparison has shown that the data accuracy for 500 and 1000 blocks has not improved compared to the case of 300 blocks, but the time consumption has increased because of the increase in the amount of calculations. Therefore, in this paper, 300 is selected as the number of random image blocks for the experiment. In the experiment, we selected nine commonly used images (including highly complex texture images, moderately complex texture images and simple texture images). The PSNR prediction experimental data are presented in Tables 2, corresponding to the visual quality prediction data with bpp of 0.5 and 1, respectively.

Table 2. Statistics of PSNR data.

Test image	bpp = 0.5				bpp = 1			
	$PSNR_{pre}$	$PSNR_{real}$	Error	Time (s)	$PSNR_{pre}$	$PSNR_{real}$	Error	Time (s)
Goldhill	33.6655	33.1294	−0.5361	0.0977	36.2791	36.6047	0.3256	0.0828
Baboon	26.1117	25.6281	−0.4836	0.1964	30.7143	29.8422	−0.8721	0.1227
Lenna	37.0936	37.2353	0.1417	0.0859	40.778	40.4585	−0.3195	0.0629
Barbara	35.2759	32.0839	−3.192	0.1097	39.0924	37.4122	−1.6802	0.0689
Aerial	28.6229	28.7333	0.1104	0.1832	32.7749	33.1914	0.4165	0.1079
Airfield	29.0245	27.4969	−1.5276	0.1558	31.9097	30.3014	−1.6083	0.1056
Frisco	42.8915	42.6874	−0.2041	0.0598	48.3571	47.7694	−0.5877	0.0638
Diego	26.5196	26.6317	0.1121	0.1912	29.4761	29.4905	0.0144	0.1250
Mrt_prepared	39.4486	38.2106	−1.238	0.0869	47.7793	44.1214	−3.6579	0.0578
SSD			1.0274				1.2120	

In this Table, $PSNR_{pre}$ represents the PSNR value of the image processed by the SPIHT encoder calculated by the program, and $PSNR_{real}$ is the PSNR value of the image actually compressed/decompressed by SPIHT. Error refers to the deviation between the predicted value and the actual value, and time refers to the time spent in the prediction process. In order to statistically analyze the accuracy of the prediction, the SSD (sample standard deviation) calculation was performed for each group of errors as

$$S = \sqrt{\frac{1}{N-1} \sum_{i=1}^{N} (X_i - \bar{X})^2}. \tag{11}$$

From analysis of data in Tables 2, it can be seen that the prediction accuracy of this method is good, and the maximum deviation is about 3.66 dB, and the overall distribution of errors is reasonable. The sample standard deviations based on nine images are 1.0274 dB and 1.212 dB, respectively. Thus, accuracy is better than for the two-step method.

The data on PSNR-HVS-M are given in Tables 3. Here PSNR-HVS-M_{pre} is the PSNR-HVS-M value of the image processed by the SPIHT encoder calculated by the program, and PSNR-HVS-M_{real} is the PSNR-HVS-M value of the image actually compressed/decompressed by SPIHT.

The deviation of PSNR-HVS-M when bpp is 0.5 is slightly larger than that one for bpp is 1. The maximum deviation is 3.65 dB, and the sample standard deviations are 1.0531 dB and 2.0946 dB, respectively. At the same time, because the calculation of PSNR-HVS-M involves DCT transformation and DCT inverse transformation, the time consumed increases, about 2 times the PSNR prediction. From the statistical data of PSNR and PNSR-HVS-M, this prediction method can predict SPIHT compressed image quality very well on the basis of AGU, and the accuracy has been improved. Table 4 presents data for analysis of time efficiency. In order to more objectively

Table 3. Statistics of PSNR-HVS-M data.

Test image	bpp = 0.5				bpp = 1			
	PSNR-HVS-M$_{pre}$	PSNR-HVS-M$_{real}$	Error	Time (s)	PSNR-HVS-M$_{pre}$	PSNR-HVS-M$_{real}$	Error	Time (s)
Goldhill	36.2594	33.9034	−2.3560	0.2408	40.8164	40.1590	−0.6574	0.2105
Baboon	27.8017	26.6977	−1.1040	0.3336	35.0034	32.6220	−2.3814	0.2583
Lenna	39.0606	39.5323	0.4717	0.2125	44.1417	44.9048	0.7632	0.1926
Barbara	36.4551	33.6409	−2.8142	0.2305	42.6237	40.5339	−2.0898	0.2024
Aerial	30.2541	29.3158	−0.9383	0.3006	37.7715	36.6634	−1.1081	0.2786
Airfield	31.3978	29.1119	−2.2859	0.2706	36.7416	33.5935	−3.1481	0.2538
Frisco	42.9987	42.5198	−0.4789	0.2005	46.7146	48.9363	2.2217	0.1926
Diego	28.2534	27.4636	−0.7898	0.3229	33.7246	33.3702	−0.3544	0.2790
Mrt_prepared	42.7600	40.2362	−2.5238	0.1905	44.8965	48.5444	3.6479	0.1584
SSD			1.0531				2.0946	

Table 4. Average time consumption(s) comparison (PSNR and PSNR-HVS-M).

bpp value	Time$_{pre}$	Time$_{Two-step}$	Time$_{SPIHT}$
0.5	0.2558	0.8937	0.3914
1	0.2251	0.9220	0.3005

understand the time efficiency of the SPIHT visual quality prediction method, the average prediction time of 9 test images is used as a measurement factor in the Table 4.

In Table 4, two sets of data were measured for bpp of 0.5 and 1, and the prediction and calculation of PSNR and PSNR-HVS-M two visual quality metric systems were performed. Time$_{pre}$ refers to the time required to obtain the visual quality value through the prediction method; Time$_{two-step}$ represents the time required in the first step of compression/decompression through the two-step method and obtains the visual quality value; Time$_{SPIHT}$ is the time required to SPIHT compressed images.

As can be seen from the data in Table 4, the newly proposed prediction method has greatly improved the time efficiency because it avoids the compression/decompression step. The average time required is 1/3 of the two-step method, and is less than the time required for the actual compression process of SPIHT.

7 Conclusions

The problem of prediction and providing a desired quality of images compressed by SPIHT is considered. It is demonstrated that for a given bpp, the image quality can vary in wide limits. The two-step procedure produces better accuracy of quality providing but still it can be unsatisfactory. Using simplicity of performance of SPIHT and AGU and exploiting earlier proposed approaches to prediction of performance characteristics for AGU, the algorithm of PSNR and PSNR-HVS-M prediction is proposed. It is faster and more accurate than the two-step procedure. We hope that in future work, procedures and methods can be further optimized so that the desired visual quality of compressed images can be better provided by this method.

References

1. Xia, J., Du, P., He, X., Chanussot, J.: Hyperspectral remote sensing image classification based on rotation forest. IEEE Geosci. Remote Sens. Lett. **11**(1), 239–243 (2013). https://doi.org/10.1109/LGRS.2013.2254108
2. Pillai, D.K.: New computational models for image remote sensing and big data. In: Swarnalatha, P. (ed.) Big Data Analytics for Satellite Image Processing and Remote Sensing, pp. 1–21. IGI Global, Hershey (2018). https://doi.org/10.4018/978-1-5225-3643-7.ch001
3. Salomon, D.: Data Compression, 4th edn. Springer, London (2007)
4. Taubman, D., Marcellin, M.: JPEG2000: Image compression fundamentals. Springer Science, New York (2002)
5. Magli, E.: The JPEG family of coding standards. In: Barni, M. (ed.) Document and Image Compression, pp. 87–112. CRC Press, Boca Raton (2006)
6. Birnbaum, T., Ahar, A., Blinder, D., et al.: Wave atoms for lossy compression of digital holograms. In: 2019 Data Compression Conference (DCC), pp. 398–407. IEEE, Snowbird (2019). https://doi.org/10.1109/DCC.2019.00048
7. Zemliachenko, A., Lukin, V., Ponomarenko, N., et al.: Still image/video frame lossy compression providing a desired visual quality. Multidimension. Syst. Signal Process. **27**(3), 697–718 (2016). https://doi.org/10.1007/s11045-015-0333-8
8. Skodras, A., Christopoulos, C., Ebrahimi, T.: The JPEG 2000 still image compression standard. IEEE Signal Process. Mag. **18**(5), 36–58 (2001). https://doi.org/10.1109/79.952804
9. Doss, S., Pal, S., Akila, D., et al.: Satellite image remote sensing for identifying aircraft using SPIHT and NSCT. J. Crit. Rev. **7**(5), 631–634 (2020)
10. Said, A., Pearlman, W.A.: A new fast and efficient image codec basedon set partitioning in hierarchical trees. IEEE Trans. Circuits Syst. Video Technol. **6**(3), 243–250 (1996). https://doi.org/10.1109/76.499834
11. Li, F., Krivenko, S., Lukin, V.: Two-step providing of desired quality in lossy image compression by SPIHT. Radioelectronic Comput. Syst. **94**(2), 22–32 (2020). https://doi.org/10.32620/reks.2020.2.02
12. Lukin, V., Zemliachenko, A., Abramov, S., et al.: Automatic lossy compression of noisy images by spiht or jpeg2000 in optimal operation point neighborhood. In: 2016 6th European Workshop on Visual Information Processing (EUVIP), pp. 1–6. IEEE, Marseille (2016). https://doi.org/10.1109/EUVIP.2016.7764581
13. Ponomarenko, N., Lukin, V., Egiazarian, K., Astola, J.: DCT based high quality image compression. In: Kalviainen, H., et al. (eds.) Image Analysis. SCIA 2005. LNCS, vol. 3540, pp. 1177–1185. Springer, Heidelberg (2005). https://doi.org/10.1007/11499145_119
14. Krivenko, S., Li, F., Lukin, V., et al.: Prediction of visual quality metrics in lossy image compression. In: 2020 IEEE 40th International Conference on Electronics and Nanotechnology (ELNANO), pp. 478–483. IEEE, Kyiv (2020). https://doi.org/10.1109/ELNANO50318.2020.9088819

15. Zemliachenko, A.N., Kozhemiakin, R.A., Abramov, S.K., et al.: Prediction of compression ratio for DCT-based coders with application to remote sensing images. IEEE J. Sel. Top. Appl. Earth Obs. Remote Sens. **11**(1), 257–270 (2017). https://doi.org/10.1109/JSTARS.2017.2781906
16. Larson, E.C., Chandler, D.M.: Most apparent distortion: Full-reference image quality assessment and the role of strategy. J. Electron. Imaging **19**(1), 011006 (2010). https://doi.org/10.1117/1.3267105
17. Moorthy, A.K., Bovik, A.C.: Visual quality assessment algorithms: what does the future hold? Multimed. Tools Appl. **51**(2), 675–696 (2011). https://doi.org/10.1007/s11042-010-0640-x
18. Ponomarenko, N., Silvestri, F., Egiazarian, K., et al.: On between-coefficient contrast masking of DCT basis functions. In: Proceedings of the Third International Workshop on Video Processing and Quality Metrics (2007)
19. Jin, L., Egiazarian, K., Kuo, C.C.J.: Perceptual image quality assessment using block-based multi-metric fusion (BMMF). In: 2012 IEEE International Conference on Acoustics, Speech and Signal Processing (ICASSP), pp. 1145–1148. IEEE, Kyoto (2012). https://doi.org/10.1109/ICASSP.2012.6288089
20. Ahar, A., Barri, A., Schelkens, P.: From sparse coding significance to perceptual quality: a new approach for image quality assessment. IEEE Trans. Image Process. **27**(2), 879–893 (2017). https://doi.org/10.1109/TIP.2017.2771412
21. Bosse, S., Becker, S., Müller, K.R., et al.: Estimation of distortion sensitivity for visual quality prediction using a convolutional neural network. Digit. Signal Proc. **91**, 54–65 (2019). https://doi.org/10.1016/j.dsp.2018.12.005
22. Blau, Y., Michaeli, T.: Rethinking lossy compression: the rate-distortion-perception tradeoff. Proc. Mach. Learn. Res. **97**, 675–685 (2019)
23. Kozhemiakin, R.A., Abramov, S.K., Lukin, V.V., et al.: Output MSE and PSNR prediction in DCT-based lossy compression of remote sensing images. In: Image and Signal Processing for Remote Sensing XXIII, vol. 10427, pp. 1042721. SPIE (2017). https://doi.org/10.1117/12.2278002

A Fast and Efficient Method for Time Delay Estimation for the Wideband Signals in Non-gaussian Environment

Viacheslav Oliinyk[1]([✉]) [iD], Vladimir Lukin[1] [iD], and Igor Djurovic[2] [iD]

[1] National Aerospace University "Kharkiv Aviation Institute",
17 Chkalova Street 2, Kharkiv 61070, Ukraine
[2] University of Montenegro, Cetinjska Street, Podgorica 81000, Montenegro
igordg@ucg.ac.me

Abstract. Time delay estimation (TDE) for signals arriving at interferometric or array antennas is an operation widely used for different applications. There are efficient solutions for fixed (not moving) sources of irradiated or reflected signals under the condition of high input signal-to-noise ratio (SNR), the appropriate interval of signal reception and Gaussian noise. However, efficiency and accuracy of estimation considerably reduce if noise is not Gaussian and SNR is low. Available solutions require pre-processing of received signals that often takes a lot of time and efforts. Here, we propose another solution for the interferometric case that can be treated as looking for the robust similarity of received signal/noise mixtures for a set of possible delays. It is shown that such processing is fast and efficient providing considerable improvement of TDE accuracy in the sense of sufficient reduction of probability of abnormal estimates for the non-Gaussian environment.

Keywords: Time delay · Robust estimation · Similarity measure · Wideband signal

1 Introduction

Time delay estimation is a classical task that arises in various applications in teleconferencing, hydroacoustics (sonars), telecommunications, nondestructive control [1–3]. High accuracy of TDE (or direction of arrival (DOA) estimation) is observed if: 1) input SNR is high enough; 2) antenna size is large enough and elementary interval of signal reception (data accumulation) are large enough; 3) a source of the signal under interest practically does not change its location during this elementary interval; 4) optimal or quasi-optimal processing is applied [3, 4]. Then, all elementary estimates are normal (close to a true value of time delay or DOA and abnormal estimates are absent) and variance of estimates is low (appropriate). Otherwise, i.e. if one or several aforementioned conditions are not satisfied, the accuracy of estimation can radically decrease [5–7]. This accuracy reduction appears itself in an increase of normal estimate variance and the non-zero probability of abnormal estimates [5, 7]. In turn, this leads to problems in source tracking [7–10], providing quality of service in teleconferences

[11], etc. Thus, it is desired to improve the accuracy of elementary time delay estimates – in the first order, to reduce the probability of abnormal estimate and, if possible, to decrease the variance of normal estimates.

There are several known approaches [12–15]. They deal with special pre-processing of received signal spectra [14] or cross-spectrum [12, 13] using fractional lower-order statistics, phase transform, robust forms of discrete Fourier transform or preliminary removal of impulse noise [15] by center-weighted median filter [16, 17]. In any case, Fourier transform in different forms is used to get some estimate of cross-spectrum and its analogue and inverse Fourier transform is applied to get cross-correlation function or its analogue with further finding the location of its global maximum.

Fourier transform-based realizations provide a rather fast calculation of cross-correlation function or its analogues, but other operations, e.g., the use of preliminary filtering of received signals [15] or robust forms of discrete Fourier transform take signal processing time comparable or even larger than FFT. This decreases the computational efficiency of processing and leads to the necessity to make it faster. Besides, the standard version of processing signals received by two mutually spaced receivers which is based on three FFT loses its efficiency if noise is not Gaussian and its distribution has heavier tails as this often happens in practice [12–15, 18]. Thus, we also need a method of signal processing to be robust with respect to non-Gaussian noise that can stem from various sources.

Therefore, our goal is to design a fast and efficient method of TDE in the non-Gaussian environment for wideband information signals received by two displaced microphones (hydrophones, sensors) with a known distance between them. For this purpose, we propose to use the sum of absolute values of differences between mutually shifted received signals and find an argument of this function global minimum. We explain motivations for our idea and describe situations when this approach works well.

2 Standard and Modified Approaches

Below we consider a very simple case of two spatially displaced two receivers for which the signals can be presented as.

$$x_1(t) = s(t) + \xi_1(t), x_2(t) = s(t - \tau_0) + \xi_2(t). \tag{1}$$

where $s(t)$, $t = [T_b; T_e]$ is the wideband noise-like (WNL) signal (e.g., speech); $\xi_1(t)$ and $\xi_2(t)$ are mutually independent realizations of additive noise realizations for the first and second channels, respectively; τ_0 is time delay supposed to be constant for the observation interval which is defined by its beginning (T_b) and end (T_e) instances. Other assumptions are that WNL signal and additive noise both have zero means and τ_0 is sufficiently smaller than the observation interval length $(T_e - T_b)$. The optimal (for large SNR and Gaussian noise) approach to TDE is to get the cross-correlation function (CCF) and to carry out a search for an argument of its global maximum [5]. In most modern systems that are based on DSP, CCF which is formally defined as

$$Y(\tau) = \int_{-T/2}^{T/2} x_1(t)x_2(t+\tau)dt, \tag{2}$$

where $T = T_e - T_b$, is calculated as

$$S_{12}(f) = S_1(f)S_2^*(f), \; S_1(f) = FFT(x_1(t)), \; S_2^*(f) = IFFT(x_2(t)), \tag{3}$$

$$Y(\tau) = IFFT(S_{12}(f)), \tag{4}$$

where $\dot{S}_1(f)$ is the spectrum estimate of signal/noise mixture for the first receiver; $S_2^*(f)$ is the conjugate FFT spectrum for the second channel, $Y(\tau)$ is the CCF estimate, IFFT is inverse FFT. The method of calculating $Y(\tau)$ using (3) and (4) instead of discrete calculation according to (2) in discrete form has become popular since the availability of FFT allows decreasing the computation time by approximately one order if the sample size is large enough. Besides, it allows getting the CCF estimates for τ from $-T/2$ to $T/2$. Note that, in fact, a physically reasonable range of possible time delays is determined by distance between receivers L and speed of wave propagation in a given medium (air, water) C as $\tau_{max} = L/C_t$. If $\tau_{max} \ll T$, then $Y(\tau)$ can be calculated directly by a discrete version of (2) for τ from $-\tau_{max}$ to τ_{max} and analyzed only in this interval without reduction of computation efficiency.

Let us rewrite (2) as

$$E_1 + E_2 - 2Y(\tau) = \int_{-\frac{T}{2}}^{\frac{T}{2}} (x_1^2(t) - 2x_1(t)x_2(t+\tau) + x_2^2(t+\tau))dt, \tag{5}$$

where $E_1 = \int_{-T/2}^{T/2} x_1^2(t)dt$ and $E_2 = \int_{-T/2}^{T/2} x_2^2(t+\tau)dt$ are energies of received mixtures that can be considered constant for stationary WBNL signal and noises. Then, one has to find argument of global minimum of $\int_{-T/2}^{T/2} (x_1(t) - x_2(t+\tau))^2 dt$, i.e. to minimize Euclidian distance (similarity measure) between sampled signals $x_1(t)$ and $x_2(t+\tau)$. Thus, having $x_1(i), i = 1, \ldots, I$ and $x_2(i+j), i = 1, \ldots, I$, it is possible to measure a similarity $S(j), i = -j_{max}, \ldots, j_{max}$ in some way, where $j_{max}\Delta t \approx L/C$, Δt denotes data sampling rate. Euclidian norm is only one of distance (similarity) measures defined as $S_E(j) = \sum(x_1(i) - x_2(i+j))^2$ where summation is done for all available i.

3   Proposed Approach

Recall now that numerous distances (similarity measures) are used nowadays in different areas including signal and image processing applications [19, 20]. An idea put forward in this paper consists of the following. Consider another similarity metric that can be easily calculated and is more robust with respect to impulses (outliers) compared to Euclidian norm. Our task was just to verify this idea without giving final recommendations. Because of this, we have chosen L1 norm and calculated the distance as.

$$S_E(j) = \sum[x_1(i) - x_2(i+j)],\qquad(6)$$

where summation is carried out for all available i.

A simulation was carried out to test a new approach. Wide Noise-like (WNL) signal was taken to get closer to real conditions. A test signal was generated randomly with the help of the Mersenne twister generator algorithm as zero-mean low pass Gaussian process. Thus, WNL signal has been modelled as an AWGN passed through a lowpass filter. The upper frequency was set to 4 kHz and the sampling frequency was equal to 20 kHz. The signal variance was fixed and set to unity. For a processing convenience, the length of a fragment was set to 1024 samples (0.05 s). An example of WNL without noise is shown in Fig. 1.

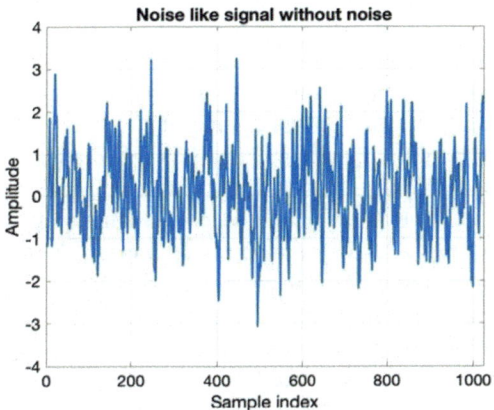

Fig. 1. Test WNL signal without noise.

To simulate processing according to (6) for WNL signal without noise, a set of shifted test signal copies with a shift from −100 samples to + 100 samples according to the original signal in Fig. 1 has been created. By calculating the sum of absolute values of differences between the original signal and its shifted copies, the following output effect could be observed (Fig. 2). We can see the symmetric plot. Furthermore, there is an obvious minimum (main lobe) where the sum of differences approaches minimum and equals to zero in the point of zero shift. Outside this main lobe, the output values are about 1180. So, the idea is to find a global minimum of the considered output. This is an opposite value to the mutual cross-correlation function (CCF), where we need to find a global maximum of the function within the interval of possible delay values.

The conventional approach to characterization of accuracy presumes that it is needed to find the main lobe width (limits) to determine what estimates are normal (a normal estimate of the time delay is such an estimate that gets into the area of main

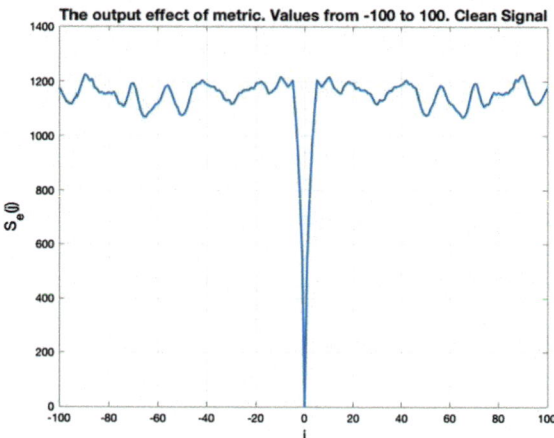

Fig. 2. The output of the proposed function (6) for noise-free signal.

lobes of CCF or its analog in noise-free case). The new approach suggests that those estimates that fall within the range from −5 to + 5 could be considered normal.

Two cases were considered to test the operability of the new approach against the standard one. Every test was simulated 1000 times to gain statistics. In the first case, the test signal was corrupted with the additive white Gaussian noise with low intensity. On every realization, the additive noise in the second channel was generated randomly. The following figure shows that the new approach works for this case when SNR is about unity. The global minimum is still in the same place of zero shift (i.e. the time delay estimate is normal), but the output value in it is about 1180. Outside the main lobe, the value fluctuates randomly around 1650.

The more interesting situation was observed when the noise model was changed to a process with the symmetric α-stable distribution. As it was described in previous researches, the noise in practical situations often cannot be considered as Gaussian. The noise distribution is heavy-tailed. Hence, we are interested in testing the new approach in conditions that are close to the real practical tasks.

The SαS noise process allows modelling a hard noise environment using two parameters α and γ to vary equivalent SNR. Theoretically, the equivalent variance can be infinite. It is inconvenient to set the desired SNR and vary it for simulations. These problems could be avoided by varying γ instead of SNR.

Figure 4 demonstrates that the output effect is sufficiently distorted even in the area of the main lobe. However, the time delay estimate is still normal. In fact, in this case, the normal estimates percentage is about 85%. The informative peak becomes less obvious if noise intensity increases (Fig. 5) but the estimate is still normal.

Finally, Fig. 6 shows a case of abnormal estimation. The global minimum is in the point −59. The percentage of abnormal estimation is about 15% (Fig. 3).

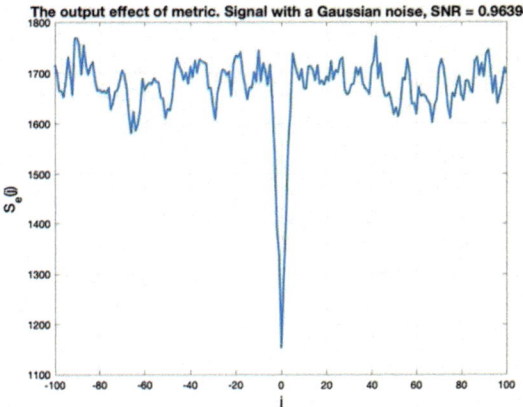

Fig. 3. The output effect of the proposed metric (6). WNL signal corrupted with a Gaussian noise.

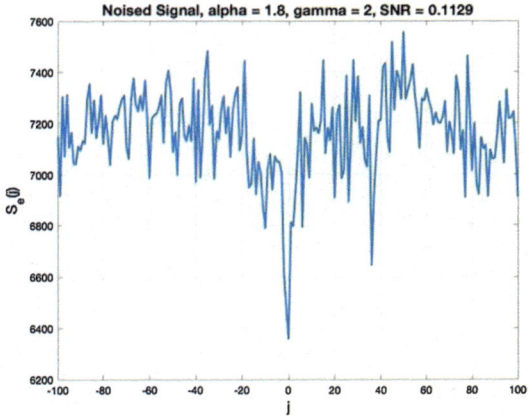

Fig. 4. The output effect of the proposed metric. WNL signal corrupted with a SαS noise with low intensity.

4   Analysis of Accuracy

Now we need to analyse the performance of the new method in different possible practical situations and to compare it to the conventional method that employs Fourier transforms (3), (4). TDE techniques are usually characterized by a mean square error of normal estimates and probability of abnormal errors. For the considered case, it is more important to produce less probability of abnormal errors, so we will pay the main attention to this characteristic in our further analysis.

In our studies, the values of $\alpha_{S\alpha S}$ have been set equal to 2, 1.8, 1.6, and 1.4 where $\alpha_{S\alpha S} = 2$ corresponds to the Gaussian PDF and it can be considered as a classical case. The dependencies of $P_{abn}(\gamma)$ are shown in Fig. 7.

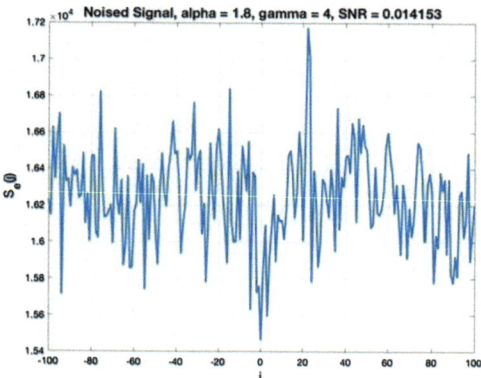

Fig. 5. The output effect of the proposed metric. WNL signal corrupted with a SαS noise with high intensity.

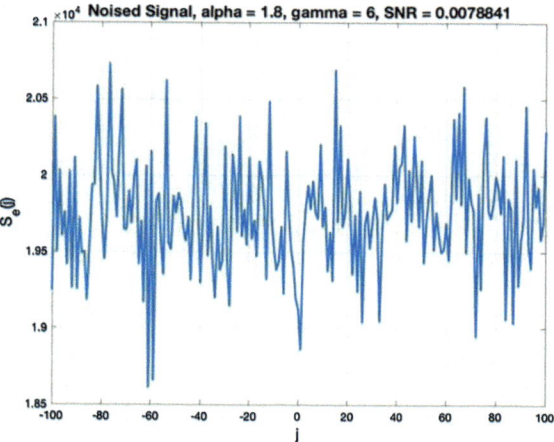

Fig. 6. The output effect of the proposed metric. WNL signal corrupted with a SαS noise with high intensity.

As one can see, abnormal estimates start to appear if γ exceeds 2. It can be seen that $P_{abn}(\gamma)$ monotonously increases if γ grows. For very large γ, the probability is expected to reach almost unity (in fact, it reaches $(2\tau_{max} - \tau_{mlw})/2\tau_{max}$ where τ_{mlw} denotes the main lobe width. As one can see from the comparison of the dependencies, even in conventional conditions the new method gives a better result than the conventional one.

The situation changes when the noise gets more intensive (Fig. 8, 9 and 10). The new method shows obviously better results. For $\alpha_{SαS} = 1.8$ (see Fig. 8), and $\alpha_{SαS} = 1.6$ (see Fig. 9), the $P_{abn}(\gamma)$ for the new method reaches the aforementioned upper limit not as fast as the conventional method. For $\alpha_{SαS} = 1.4$ (see Fig. 10), the classical method produces mostly abnormal estimates for even very small γ. Meanwhile, the proposed technique performs sufficiently better.

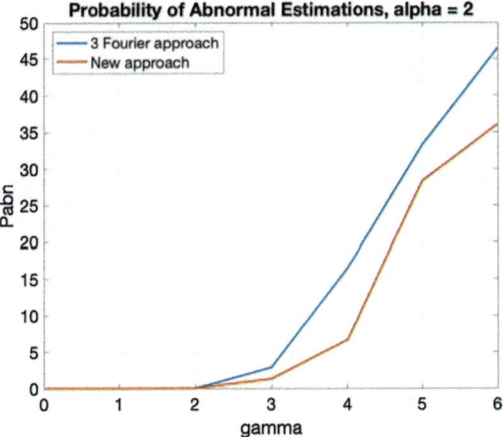

Fig. 7. Dependencies Pabn(γ) for $\alpha_{S\alpha S} = 2$.

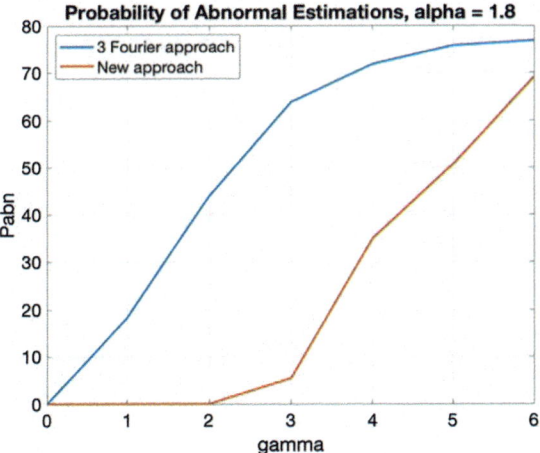

Fig. 8. Dependencies $P_{abn}(\gamma)$ for $\alpha_{S\alpha S} = 1.8$.

Thus, it can be stated that the proposed method has a better performance compared to the classical approach, at least, in terms of the probability of abnormal estimates. Meanwhile, we are also interested in MSE or RMSE of normal estimates.

The dependences of RMSE on γ for Gaussian additive noise is presented in Fig. 11. Analysis of these dependencies shows the following. As it can be expected, for the larger γ, i.e. for larger variance of additive noise and smaller SNR, RMSE increases. RMSE for both techniques is practically the same.

Figure 12 shows the dependencies for non-Gaussian additive noise ($\alpha_{S\alpha S} = 1.4$). Again, RMSE increases if γ becomes larger (equivalent SNR decreases). For $\gamma > 2$, RMSE for conventional technique reaches saturation. Comparison of dependences clearly shows the benefits provided by the proposed method. Analysis of other values

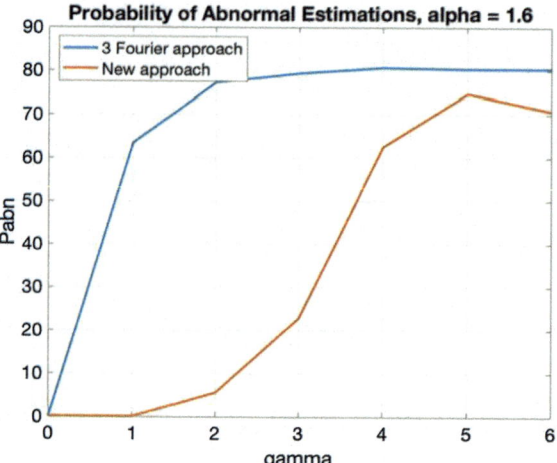

Fig. 9. Dependencies $P_{abn}(\gamma)$ for $\alpha_{S\alpha S} = 1.6$.

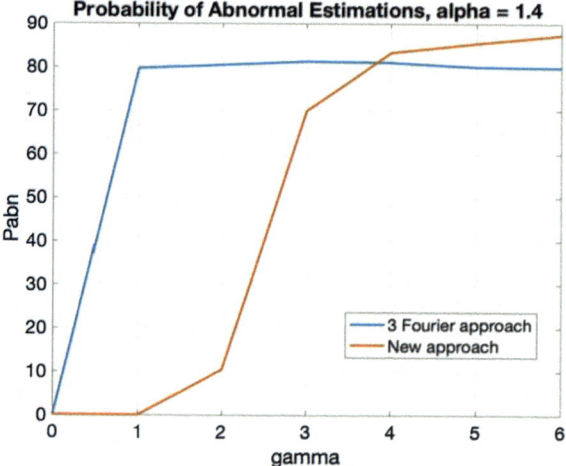

Fig. 10. Dependencies $P_{abn}(\gamma)$ for $\alpha_{S\alpha S} = 1.4$.

of $\alpha_{S\alpha S}$ (1.8, 1.6) has shown that the accuracy (in terms of RMSE of normal estimates) for the proposed method is better than for the conventional method. Thus, we can state that the proposed technique performs either not worse or better than the conventional one for a wide range of characteristics of additive noise.

One more important item is that the proposed method also works about 18 times faster than the conventional one (51 s. for 10 thousand realisations in Matlab comparing to the conventional method (941 s)). The reason is that the used mathematic operations are very simple.

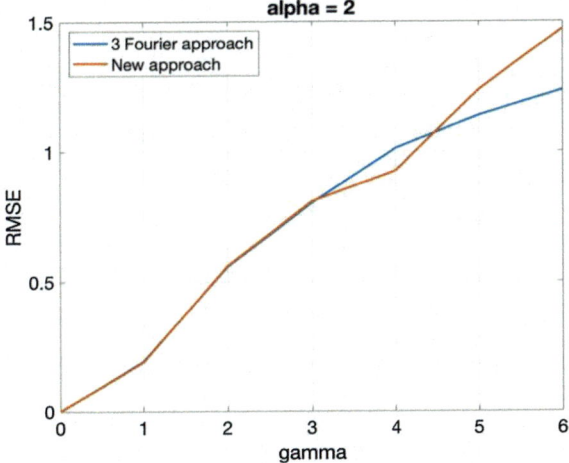

Fig. 11. Dependencies of RMSE on γ for $\alpha_{S\alpha S} = 2$ (Gaussian noise).

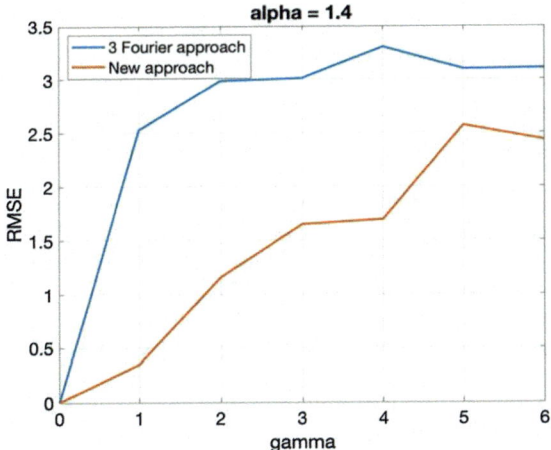

Fig. 12. Dependencies of RMSE on γ for $\alpha_{S\alpha S} = 1.4$ (non-Gaussian noise).

5   Conclusions

The new method of time delay estimation for WNL signals in non-Gaussian noise environment is proposed and described. It is based on the evaluation of received signal similarity which is more robust to non-Gaussian noise simulated as the symmetric α-stable process. Benefits of the proposed approach are three-fold. Firstly, it produces a smaller probability of abnormal estimates, especially if additive noise is essentially non-Gaussian. Secondly, RMSE of normal estimates is not greater (not worse) than for the conventional method. Thirdly, the proposed method can be realized faster than for the conventional method.

References

1. Benesty, J., Huang, Y., Chen, J.: Time delay estimation via minimum entropy. IEEE Signal Process. Lett. **14**(3), 157–160 (2007). https://doi.org/10.1109/LSP.2006.884038
2. Osman, L., Sfar, I., Gharsallah, A.: The application of high-resolution methods for DOA estimation using a linear antenna array. Int. J. Microwave Wirel. Technol. **7**(1), 87–94 (2015). https://doi.org/10.1017/S1759078714000464
3. Carter, G.C.: Coherence and time delay estimation. Proc. IEEE **75**(2), 236–255 (1987). https://doi.org/10.1109/PROC.1987.13723
4. Benesty, J., Chen, J.: Study and Design of Differential Microphone Arrays. Springer, , Heidelberg (2013)
5. Ianniello, J.: Time delay estimation via cross-correlation in the presence of large estimation errors. IEEE Trans. Acoust. Speech Signal Process. **30**(6), 998–1003 (1982). https://doi.org/10.1109/TASSP.1982.1163992
6. Champagne, B., Eizenman, M., Pasupathy, S.: Exact maximum likelihood time delay estimation for short observation intervals. IEEE Trans. Signal Process. **39**(6), 1245–1257 (1991). https://doi.org/10.1109/78.136531
7. Gustafsson, T., Rao, B.D., Trivedi, M.: Source localization in reverberant environments: modeling and statistical analysis. IEEE Trans. Speech Audio Process. **11**(6), 791–803 (2003). https://doi.org/10.1109/TSA.2003.818027
8. Fanaswala, M., Krishnamurthy, V.: Detection of anomalous trajectory patterns in target tracking via stochastic context-free grammars and reciprocal process models. IEEE J. Sel. Top. Signal Process. **7**(1), 76–90 (2013). https://doi.org/10.1109/JSTSP.2012.2233459
9. Yu, H., An, W., Zhu, R.: Extended target tracking and feature estimation for optical sensors based on the Gaussian process. Sensors **19**(7), 1704 (2019). https://doi.org/10.3390/s19071704
10. Hasan, A.H., Mnati, A.M., Obied, A.: Bayesian approach for multi-mode kalman filter for abnormal estimation. Nonlinear Dyn. Syst. Theory **18**(4), 372–391 (2018)
11. Talantzis, F., Pnevmatikakis, A., Constantinides, A.G.: Audio-Visual Person Tracking: A Practical Approach. Imperial College Press, London (2012)
12. Shao, M., Nikias, C.L.: Signal processing with fractional lower order moments: Stable processes and their applications. Proc. IEEE **81**(7), 986–1010 (1993). https://doi.org/10.1109/5.231338
13. Georgiou, P.G., Tsakalides, P., Kyriakakis, C.: Alpha-stable modeling of noise and robust time-delay estimation in the presence of impulsive noise. IEEE Trans. Multimed. **1**(3), 291–301 (1999). https://doi.org/10.1109/6046.784467
14. Oliinyk, V., Lukin, V., Djurović, I.: Time delay estimation for noise-like signals embedded in non-Gaussian noise using adaptive robust DFT. In: 2018 7th Mediterranean Conference on Embedded Computing (MECO), pp. 1–4. IEEE, Budva (2018). https://doi.org/10.1109/MECO.2018.8406054
15. Oliinyk, V., Ieremeiev, O., Djurović, I.: Center weighted median filter application to time delay estimation in non-Gaussian noise environment. In: 2019 IEEE 2nd Ukraine Conference on Electrical and Computer Engineering (UKRCON), pp. 985–989. IEEE, Lviv (2019). https://doi.org/10.1109/UKRCON.2019.8879775
16. Oliinyk, V., Lukin, V.: Time delay estimation for noise-like signals embedded in non-Gaussian noise using pre-filtering in channels. In: 2020 IEEE 15th International Conference on Advanced Trends in Radioelectronics, Telecommunications and Computer Engineering (TCSET), pp. 638–643. IEEE, Lviv-Slavske (2020). https://doi.org/10.1109/TCSET49122.2020.235510

17. Sun, T., Gabbouj, M., Neuvo, Y.: Analysis of two-dimensional center weighted median filters. Multidimension. Syst. Signal Process. **6**(2), 159–172 (1995). https://doi.org/10.1007/BF00981569
18. Nolan, J.P.: Univariate Stable Distributions. Springer, Cham (2020). https://doi.org/10.1007/978-3-030-52915-4
19. Ebdelli, M., Le Meur, O., Guillemot, C.: Analysis of patch-based similarity metrics: Application to denoising. In: 2013 IEEE International Conference on Acoustics, Speech and Signal Processing, pp. 2070–2074. IEEE, Vancouver (2013). https://doi.org/10.1109/ICASSP.2013.6638018
20. Mironică, I., Ionescu, B., Vertan, C.: The influence of the similarity measure to relevance feedback. In: 2012 Proceedings of the 20th European Signal Processing Conference (EUSIPCO), pp. 1573–1577. IEEE, Bucharest (2012)

On the Applications of the Special Class of Atomic Functions: Practical Aspects and Perspectives

Viktor Makarichev$^{(\boxtimes)}$ ⓘ, Vladimir Lukin ⓘ, and Iryna Brysina

National Aerospace University "Kharkiv Aviation Institute",
17 Chkalova Street, Kharkiv 61070, Ukraine
victor.makarichev@gmail.com, lukin@ai.kharkov.com,
iryna.brysina@gmail.com

Abstract. A special class of atomic functions, which are compactly supported solutions of linear functional differential equations with constant coefficients and linear transformations of the argument, and their basic properties are considered. Spaces of linear combinations of shifts of these functions are given. Such spaces have good approximation properties and combine smoothness of elements with an existence of compactly supported basis. For this reason, their application to solution of different problems is perspective. System of atomic wavelets, which is a basis of such spaces, is a practical tool for data analysis and processing. The aim of this paper is to analyze atomic wavelets expansion algorithm in terms of complexity. Using properties of atomic functions and atomic wavelets, it is shown that expansion algorithm has linear time and memory complexity. Moreover, the numerical scheme can be easily parallelized. This means that application of atomic wavelets is promising especially when processing digital images and video.

Keywords: Atomic function · Atomic wavelet · Data analysis · Lossy image compression · Fast computing

1 Introduction

Atomic functions theory arose in the beginning of 1970s. The emergence of this branch of mathematics was mainly due to the need for a new mathematical tool for solving different boundary value problems of mechanics and engineering. Trigonometric and algebraic polynomials, as well as splines, which were widely applied at that time, did not allow obtaining the required results. This gave a boost to the development of atomic functions theory. Its foundations and the key principles are given in [1] (see also [2]). Also, detailed analysis of the development stages of this theory and the main results obtained by 1986 can be found in [3, Chapter 2].

In general, atomic functions are solutions of some special equations. They have a number of useful properties inherent to polynomials and splines. In this sense, atomic functions have advantages of each of these classic mathematical tools and are in the middle between them. Besides, there is much in common between atomic functions and

M. Nechyporuk et al. (Eds.): ICTM 2020, LNNS 188, pp. 42–54, 2021.
https://doi.org/10.1007/978-3-030-66717-7_4

wavelets, whose rapid development started shortly after the appearance of the theory of atomic functions.

The V.A. Rvachev up-function, which is a solution of the equation

$$y'(x) = 2y(2x - 1) - 2y(2x + 1),$$

is the most famous and well-researched atomic function. This function and related objects have many applications in numerical methods [4–8], wavelet theory [9–11], modeling of physical processes and phenomena [12, 13], image processing [14, 15], generalized Taylor series [2], approximation theory [1, 2, 16], etc.

We note that there are other well-studied atomic functions. Consider the function.

$$up_s(x) = \frac{1}{2\pi} \int_{-\infty}^{\infty} e^{itx} \prod_{k=1}^{\infty} \frac{\sin^2\left(\frac{st}{(2s)^k}\right)}{s^2 \cdot \frac{t}{(2s)^k} \cdot \sin\left(\frac{t}{(2s)^k}\right)} dt. \tag{1}$$

that is a solution of the equation.

$$y'(x) = 2 \sum_{k=1}^{s} (y(2sx + 2s - 2k + 1) - y(2sx - 2k + 1)), \tag{2}$$

where s is an arbitrary natural number. This function was introduced in [17].

The system $\{up_s(x)\}$ is a set of functions including $up(x)$. It is obvious that $up_1(x) = up(x)$. Despite some similarities, atomic functions $up_s(x)$ are quite different from each other. Hence, their use provides researches and developers with wider options than the use of the atomic function $up(x)$.

The appearance of atomic functions had come at a time when computer technologies had poor capabilities. For this reason, the development of atomic functions theory and the construction of algorithms based on their application were performed according to the need for efficient use of available computing resources. Of course, technologies have been improved and computing power has increased significantly. Nevertheless, the amount of data has also increased explosively and continues to increase [18, 19]. And as a result, storing, processing and transmission of these data via networks have become more resource-intensive. Therefore, developing of new algorithms and improving of existing ones, especially using green technologies principles, are of particular interest [20]. Atomic functions $up_s(x)$ can be useful for this purpose.

It is clear that each mathematical tool, which is applied to processing especially big data, should provide fast algorithms and low memory expenses required for intermediate computations. In this paper, we consider properties of atomic functions $up_s(x)$ and discuss them in terms of the possibility of their application to fast data processing. Our aim is to show that the development of technologies using these atomic functions and related applications is promising.

This paper is organized as follows. First, we consider fundamental properties of $up_s(x)$ and related objects. Then, we present and analyze the expansion algorithm. And finally, we consider and discuss promising directions of applications and further research.

2   Atomic Functions and Their Applications

2.1   Definitions and Basic Properties

A function is called atomic if it is compactly supported solution of the linear functional differential equation with constant coefficients and linear transformations of argument:

$$a_0 y^{(n)} + a_1 y^{(n-1)} + \ldots + a_n y = c_1 y(ax + b_1) + \ldots + c_k y(ax + b_k) \tag{3}$$

where a, a_i, b_i, c_i are fixed real constants. The closure of the set $\{x : f(x) \neq 0\}$ is called a support of the function $f(x)$ and is denoted by $supp f(x)$. If $a_0 = a_1 = \ldots = a_{n-1} = 0$, we get functional equation.

$$a_n y(x) = c_1 y(ax + b_1) + \ldots + c_k y(ax + b_k),$$

which is called refinement. Such equations are widely applied in wavelet theory. Therefore, wavelet functions, which are solutions with a compact support of some of these refinement equations, can be considered as a partial case of atomic functions. In [1, Chapter 3], conditions of existence of solutions of the Eq. (3) were obtained.

In the current paper, we consider only atomic functions $up_s(x)$, $s \geq 1$. For any natural s the basic properties of these functions are:

1   $supp up_s(x) = [-1, 1]$, i.e. the function $up_s(x)$ is equal to zero outside the segment $[-1, 1]$ [2, 17].
2   $up_s(x)$ is infinitely differentiable in $(-\infty, \infty)$ [2, 17].
3   $\max\limits_{x \in [-1,1]} \left| up_s^{(n)}(x) \right| = 2^n (2s)^{\frac{n(n-1)}{2}}$. The function $up_s(x)$ is not analytic at any point of the segment $[-1, 1]$ [2, 17].
4   $up_s(x)$ is an even function [2, 17].
5   $up_s(x)$ increases on $[-1, 0]$ and decreases on $[0, 1]$ [2, 17].

The first property of the function $up_s(x)$ is very convenient from the point of view of numerical methods. It follows that this function and related objects should be computed only on the segment $[-1, 1]$. This provides fast algorithms and low memory costs, as well as small computational errors.

Combining the second and the third properties, we see that atomic functions $up_s(x)$ are in the middle between algebraic and trigonometric polynomials, which are analytic infinitely smooth functions, and splines, which have finite order of smoothness. High order of smoothness provides fast approximation and good description of smooth functions and data.

Moreover, it follows from (2) that values of each derivative $up_s^{(k)}(x)$ can be found using values of $up_s(x)$. In a sense, atomic function $up_s(x)$ is invariant with respect to differentiation, as well as integration.

In [1, Chapter 3], formulas for calculation of the function $up_1(x) = up(x)$ were obtained (see also [21]). It was also shown that values $up(k2^{-n})$ can be calculated exactly in the form of the rational number and the corresponding formulas were introduced. For the case $s \geq 2$, similar results are provided in [22].

Consequently, in spite of complexity of the expression (1), atomic function $up_s(x)$ is a convenient mathematical tool for any $s \geq 1$.

There are many ways to use atomic functions $up_s(x)$. First, these functions were used to construct the so-called generalized Taylor series, which can be applied to solving special classes of integral and differential equations [2, 17]. Another way to use $up_s(x)$ is an application of spaces constructed using them. In the current paper, we consider such functional spaces and their key features.

We also note that functions $up_s(x)$ are closely related to other classic functions such as trigonometric and algebraic polynomials. However, these atomic functions are non-quasi analytic, which makes the scope of their application wider. By a combination of properties, they are between polynomials and splines [1]. That is what makes usage of atomic functions $up_s(x)$ and related constructions promising.

2.2 Spaces of Atomic Functions

Let $UP_{s,n}$ be a space of functions $\varphi(x)$ such that $\varphi(x) = \sum_k c_k up_s\left(x - \frac{k}{(2s)^n}\right)$, $x \in [a, b]$, where $n = 0, 1, 2, \ldots$ and E is a subset of the set of real numbers. In other words, $UP_{s,n}$ is a space of linear combinations of shifts of the atomic function $up_s(x)$.

Smoothness. Since $up_s(x)$ is infinitely differentiable, it follows that each function $\varphi \in UP_{s,n}$ is infinitely smooth on $[a, b]$.

Compactness. There exists locally supported basis in $UP_{s,n}$. Consider the function.

$$Fup_{s,n}(x) = \frac{1}{2\pi} \int\limits_{-\infty}^{\infty} e^{itx} \left(\frac{\sin\frac{t}{2(2s)^n}}{\frac{t}{2(2s)^n}}\right)^n F_s\left(\frac{t}{(2s)^n}\right) dt,$$

where $F_s(t)$ is the Fourier transform of $up_s(x)$. It was shown in [1, 23] (see also [2] for the case $s = 1$) that the segment

$$\left[-\frac{n+2}{2(2s)^n}, \frac{n+2}{2(2s)^n}\right] \tag{4}$$

is a support of the function $Fup_{s,n}(x)$ and the system of functions

$$\left\{Fup_{s,n}\left(x - \frac{2k+n}{2(2s)^n}\right)\right\} \tag{5}$$

constitutes a basis of the space $UP_{s,n}$.

In addition, the function $Fup_{s,n}(x)$ is also atomic and its values can be computed using values of $up_s(x)$.

Good Approximation Properties. It was proved in [1, 2, 16, 23] that these spaces are asymptotically extremal for approximation of classes of periodic differentiable functions. This means that atomic functions $up_s(x)$ are the best approximation tools, as well as trigonometric polynomials $\{1, \sin(kx), \cos(kx)\}$. Moreover, estimates of approximation errors were obtained. This ensures getting the required results by pre-selecting the parameters such as s and n.

Fast Decomposition Procedure. From (4) it follows that the length of the function $Fup_{s,n}(x)$ support is quite small. It provides fast processing of this function, in particular computation of integrals. Besides, complexity of decomposition procedure on the basis (5) is of the order $O(N)$, where N is a number of applied elements of this basis. Actually, this is one of the most important features of the space $UP_{s,n}$. Consider it in more detail.

Let $f(x)$ should be approximated by $UP_{s,n}$: $f(x) \approx \sum_k c_k Fup_{s,n}\left(x - \frac{2k+n}{2(2s)^n}\right)$.

There are several ways to find coefficients $\{c_k\}$. Collocation method can be used, as well as projection method. The last one is often applied in different algorithms of data processing.

To find $p(x)$, which is projection of the function $f(x)$ on the linear space $UP_{s,n}$, we can use a standard algebraic approach. First, we compute integrals

$$a_{ij} = \int_a^b Fup_{s,n}\left(x - \frac{2i+n}{2(2s)^n}\right) Fup_{s,n}\left(x - \frac{2j+n}{2(2s)^n}\right) dx, \tag{6}$$

$$b_i = \int_a^b Fup_{s,n}\left(x - \frac{2i+n}{2(2s)^n}\right) f(x) dx. \tag{7}$$

We get system of linear algebraic equations.

$$Ac = b, \tag{8}$$

where A is a matrix of integrals (6), c is a column of coefficients c_j and b is the column of integrals (7). Since $Fup_{s,n}(x)$ has a local support (4), computation of each integral reduces to integration over the small interval. Furthermore, $a_{ij} = 0$ if intersection of supports of the functions $Fup_{s,n}\left(x - \frac{2i+n}{2(2s)^n}\right)$ and $Fup_{s,n}\left(x - \frac{2j+n}{2(2s)^n}\right)$ is empty. Thus, the matrix A is sparse. This provides fast solution of the system (8).

We also note that integrals a_{ij} do not depend on the function $f(x)$. So, it is reasonable to find a_{ij} once and then use them without recalculation.

Hence, decomposition procedure requires computing of the integrals over small intervals and solving the linear system with a sparse matrix.

2.3   Atomic Wavelets

Spaces of atomic functions $UP_{s,n}$, which were described in Sect. 2.2, can be used for solving different applied problems. It requires some basis of $UP_{s,n}$ for this purpose. System of functions (5), as already mentioned, is a basis of this space. This basis can be useful in the finite element method. However, in the analysis and processing of data, wavelet-like systems of functions are more useful due to the ability to distinguish orthogonal components.

In [24, 25], wavelets based on the function $up_s(x)$ were introduced. Since these wavelets are constructed using atomic functions, we call them atomic functions. Atomic wavelets provide wavelet decomposition procedure of some given function.

Now, consider the application of atomic functions spaces using atomic wavelet decomposition approach.

Let s be fixed. Also, let N be an arbitrary real number such that $N \neq 0$ (for example, N can be equal to $(2s)^n$). For any $k = 0, 1, 2, \ldots$ consider the functions

$$v_k(x) = \frac{1}{2\pi} \int_{-\infty}^{\infty} e^{itx} \frac{\sin(2^k t/N)}{2^k t/N} \prod_{j=0}^{k-1} \cos(2^j t/N) F(t/N) dt, \qquad (9)$$

and $F(t)$ is the Fourier transform of the function $up_s(x)$. Actually, the function $v_k(x)$ depends on s and N. So, this function should be denoted by $v_{s,N,k}(x)$. For the sake of brevity, we omit these parameters. Expression (9) generalizes (1).

We stress that values of the function $v_0(x)$ can be easily found using values of atomic function $up_s(x)$.

It follows that for any $k \geq 1$.

$$v_k(x) = \frac{1}{4}\left(v_{k-1}(x + 2^k/N) + 2v_{k-1}(x) + v_{k-1}(x - 2^k/N)\right). \qquad (10)$$

Furthermore,

$$v_k(x) = 0 \ \text{if} \ |x| \geq 2^{k+1}/N. \qquad (11)$$

By L_k we denote the space of functions $\varphi(x) = \sum_j c_j v_k(x - 2^{k+1}j/N)$.

From (10), it follows that these spaces are nested, i.e. L_{k+1} is a subset of L_k for any k.

Denote by W_k the orthogonal complement to the space L_k in L_{k-1}:

$$W_k = \{f \in L_{k-1} : (f, g) = 0 \ \text{for any} \ g \in L_k\}.$$

Here the inner product (f, g) is defined by formula.

$$(f, g) = \int_{-\infty}^{\infty} f(x)g(x)dx. \qquad (12)$$

This yields that $L_0 = W_1 \cup W_2 \cup \ldots \cup W_n \cup L_m$. This means that any function $p \in L_0$ can be expressed as follows:

$$p(x) = p_1(x) + p_2(x) + \ldots + p_m(x) + p_{m+1}(x), \qquad (13)$$

where $p_k \in W_k$ for any $k = 1, 2, \ldots, m$ and $p_{m+1} \in L_m$.

In (13), the natural number m defines number of levels of detail. We call this value a depth of the decomposition.

Besides, each term of the expansion (13) is orthogonal to other terms of this formula. In other words, (13) provides decomposition of $p(x)$ into orthogonal components.

Consider the function.

$$w_k(x) = \sum_{j=1}^{5} c_{k,j} v_{k-1}(x - 2^k j/N). \tag{14}$$

where $c_{k,1} = c_{k,5} = -\frac{b_{k-1}}{4}$, $c_{k,2} = c_{k,4} = \frac{a_{k-1} + 2b_{k-1}}{4}$, $c_{k,3} = -\frac{a_{k-1} + b_{k-1}}{2}$ and

$$a_{k-1} = \int_{-\infty}^{\infty} v_{k-1}^2(x)dx, \, b_{k-1} = \int_{-\infty}^{\infty} v_{k-1}(x)v_{k-1}\left(x - \frac{2^k}{N}\right)dx. \tag{15}$$

The system of functions $\left\{w_k(x - 2^{k+1}j/N)\right\}_j$ constitutes a basis of the space W_k [26]. Hence, each orthogonal component $p_k(x)$ of the expansion of (13) is a linear combination of these functions:

$$p_k(x) = \sum_j \omega_j^{[k]} w_k(x - 2^{k+1}j/N) k = 1, 2, \ldots, m, \tag{16}$$

Finally, the last term of (13) is.

$$p_{m+1}(x) = \sum_j \omega_j^{[m+1]} v_n(x - 2^{n+1}j/N). \tag{17}$$

System of functions $\left\{w_k(x - 2^{k+1}j/N), v_n(x - 2^{n+1}j/N)\right\}$ is a basis of the space L_0. We call it a system of atomic wavelets. Coefficients $\left\{\omega_j^{[k]}\right\}$ are called atomic wavelet coefficients. Each element of this system is infinitely differentiable function with a compact support. Also, these functions have good approximation properties (see Subsect. 2.2).

2.4   Atomic Wavelets Expansion Algorithm

The space L_0 and its basis of atomic wavelets were presented above. Now, we consider an algorithm for obtaining approximation of some given function $f(x)$ by these constructive tools. For this purpose, the following rather classical approach is used: we find projection $p(x)$ of the function $f(x)$ onto the space L_0. Using this way, we get $f(x) = p(x) + r(x)$, where $r(x)$ is small, since L_0 has good approximation properties. Therefore, the remainder term $r(x)$ can be dropped and analysis of the function $f(x)$ can be carried out by analyzing its projection $p(x)$.

To find $p(x)$, the method, which was described in Subsect.2.2, can be applied. However, there are several important features that should be used when developing the required algorithm.

First, since components of expansion (13) are orthogonal, they can be found separately. Hence, the procedure of determination of $p_1(x), p_2(x), \ldots, p_{m+1}(x)$ can be parallelized.

Secondly, just like approximating $f(x)$ by the space $UP_{s,n}$ (see Subsect. 2.2), we should solve systems of linear algebraic equations.

$$A^{[k]} \bullet \omega^{[k]} = b^{[k]}, k = 1, 2, \ldots, m+1, \tag{18}$$

where $A^{[k]}$ is a matrix of $a_{ij}^{[k]} = \left(w_k\left(\bullet - 2^{k+1}i/N\right), w_k\left(\bullet - 2^{k+1}j/N\right)\right)$ for any $k = 1, \ldots, m$ and $a_{ij}^{[m+1]} = \left(v_m\left(\bullet - 2^{n+1}i/N\right), v_m\left(\bullet - 2^{n+1}j/N\right)\right)$, $b^{[k]}$ is a column of inner products $b_i^{[k]} = \left(f, w_k\left(\bullet - 2^{i+1}i/N\right)\right)$ for any $k = 1, 2, 3, \ldots, m$ and $b_i^{[m+1]} = \left(f, v_m\left(\bullet - 2^{n+1}i/N\right)\right)$, $\omega^{[k]}$ is a column of atomic wavelet coefficients $\omega_j^{[k]}$ that should be found.

Despite all the apparent complexity, the process of solving these systems is quite fast and simple.

Indeed, it follows from (11), (14) and (15) that.

$$a_{ii}^{[k]} = a_{k-1} \sum_{r=1}^{5} c_{k,r}^2 + 2b_{k-1} \sum_{r=1}^{4} c_{k,r}c_{k,r+1}, \tag{19}$$

$$a_{ii-1}^{[k]} = a_{ii+1}^{[k]} = a_{k-1}\left(c_{k,1}c_{k,3} + c_{k,2}c_{k,4} + c_{k,3}c_{k,5}\right) + 2b_{k-1}\left(2c_{k,1}c_{k,2} + c_{k,2}c_{k,3}\right), \tag{20}$$

$$a_{ii+2}^{[k]} = a_{ii-2}^{[k]} = a_{k-1}c_{k,1}^2 + 2b_{k-1}c_{k,1}c_{k,2}, \tag{21}$$

$$a_{ij}^{[k]} = 0 \text{ if } |i - j| > 2. \tag{22}$$

Furthermore, consider the following notation:

$$h_{k,i} = \int_{-\infty}^{\infty} f(x)v_k\left(x - \frac{2^{k+1}i}{N}\right)dx. \tag{23}$$

It follows from (10) that.

$$h_{k+1,i} = \left(h_{k,2i-1} + 2h_{k,2i} + h_{k,2i+1}\right)/4, \tag{24}$$

for any i and $k = 0, 1, \ldots, m - 1$. Also, from (14), we get

$$b_i^{[k]} = \sum_{j=1}^{5} c_{k,j}h_{k-1,2i+j} \tag{25}$$

for each $k = 1, \ldots, m$ and.

$$b_i^{[m+1]} = h_{m,i}. \tag{26}$$

Finally, (11) provides the following

$$a_{ii}^{[m+1]} = a_n, a_{ii-1}^{[m+1]} = a_{ii+1}^{[m+1]} = b_n s \tag{27}$$

and

$$a_{ij}^{[m+1]} = 0 \text{ if } |i - j| > 1 \tag{28}$$

(here, a_n and b_n are defined by (15)).

We see that matrices $A^{[k]}$ of systems (18) are sparse. Moreover, for each matrix $A^{[1]}, \ldots, A^{[m+1]}$. only several values are required (see (19)–(22), (27), (28)). Hence, storing of these matrices requires very small memory expenses. Also, this provides fast algorithm for computing the wavelet coefficients $\omega_j^{[k]}$.

Note that computation of $b^{[k]}$. requires a huge number of integrals. Nevertheless, combining (11) and (23), we get.

$$h_{0,i} = \int_{(2i-2)/N}^{(2i+2)/N} f(x) v_0 \left(x - \frac{2i}{N} \right) dx. \tag{29}$$

In other words, values of $h_{0,i}$ can be found by integration over the segment $[(2i - 2)/N, (2i + 2)/N]$, which is rather small if N is big. Complexity of computation of these integrals depends on the applied method of numerical integration. Taking into account that intervals of integration are small, we see that calculation of $h_{0,i}$ does not require huge resources. All other integrals $h_{k,i}$ can be found using (24).

In addition, it follows from (10) and (15) that.

$$a_j = \frac{(3a_{j-1} + 4b_{j-1})}{8}, b_j = (a_{j-1} + 4b_{j-1})/16, \tag{30}$$

for any $j = 1, \ldots, m$. . Also, using (11), we get.

$$a_0 = 2 \int_0^{2/N} v_0^2(x) dx, b_0 = \int_0^{2/N} v_0(x) v_0 \left(x - \frac{2}{N} \right) dx. \tag{31}$$

Notice that the decompositioon the entire real axis R has been described above. However, in practice, functions are usually considered in the bounded interval (α, β). In this case, everything described above remains valid.

Consequently, we have described the decomposition procedure. In Fig. 1, the corresponding algorithm is given. Note that the pivotal condensation method can be used to solve systems of linear algebraic equations (18) (see Fig. 1, Line 11, of the function find_wavelet_coefficients). We also stress that the process of solving these systems can be easily parallelized [26, 27].

```
procedure preliminary_preparation(v₀(x),m)
// this procedure computes aₖ,bₖ,cₖ,ⱼ,A⁽ᵏ⁾
begin
1.   compute a₀,b₀; // use (31)
2.   for j←1 to m
3.     compute aⱼ,bⱼ; // use (30)
4.   for k←1 to m
5.     for j←1 to 5
6.        compute cₖ,ⱼ; // see (14)
7.   for k←1 to m+1
8.     compute a⁽ᵏ⁾ᵢⱼ; // use (19)-(22),(27)
end

function find_wavelet_coefficients(f(x),(α,β),v₀(x),m)
// this procedure finds atomic wavelet coefficients ωⱼ⁽ᵏ⁾
begin
1. preliminary_preparation(v₀(x),m);
2. foreach i
3.    compute h₀,ᵢ; // use (29)
4. for k←1 to m
5.   foreach i
6.      compute hₖ,ᵢ; // use (24)
7. for k←1 to m+1
8.   foreach i
9.      compute bⱼ⁽ᵏ⁾; // use (25),(26)
10. for k←1 to m+1
11.   solve A⁽ᵏ⁾·ω⁽ᵏ⁾ = b⁽ᵏ⁾; // see (18)
12. return ωⱼ⁽ᵏ⁾;
end
```

Fig. 1. Algorithm for atomic velets decomposition.

Time complexity of the decomposition procedure is of the order $O(d)$, where d is the number of atomic wavelet coefficients. Besides, the size of the used auxiliary memory is also $O(d)$. Hence, application of the presented algorithm is promising.

2.5   Discussion

The main disadvantage of the described method is the need to solve systems of linear algebraic equations (18). The reason is the non-orthogonality of the function system.

$$\{w_k(x - 2^{k+1}j/N)\}_j. \tag{32}$$

There are several ways to fix it. The first way is to orthogonalize this system. Another way is to construct and apply biorthogonal system of functions. However, these approaches violate the main advantage of (32), which is compactness or even locality of the support.

From (11) and (14), it follows that.

$$w_k\left(x - \frac{2^{k+1}j}{N}\right) = 0 \quad \text{if} \quad x < \frac{2^{k+1}j}{N} \text{ or } x > \frac{2^{k+1}(j+3)}{N}. \tag{33}$$

In other words, values function $w_k\left(x - \frac{2^{k+1}j}{N}\right)$ inside the set $\left[\frac{2^{k+1}}{N}, \frac{2^{k+1}(j+3)}{N}\right]$ are equal to zero and they can be ignored in (16). The same is true for (17). Hence, there are only a few terms in the right part of (13) for any x. This provides fast reconstruction of the function $f(x)$. Besides, it is the compactness of the basis of atomic wavelets that provides fast decomposition procedure.

In addition, using (33), we can get estimates of errors that occur when quantizing wavelet coefficients $\omega_j^{[k]}$. This provides the ability to control loss of quality in lossy processing algorithms, in particular, in image compression.

If we orthogonlize the system (32) or apply biorthogonal system of functions, we eliminate the need to solve (18). But then, to search for expansion coefficients, it will be necessary to calculate a large number of integrals over large intervals. Actually, the same number of integrals are computed in the proposed method (see Fig. 1, Lines 2–3 of the function find_wavelet_coefficients). Nevertheless, computation of each integral $h_{0,i}$ requires integration over the small interval and, hence, we get considerably smaller numerical complexity and calculation errors.

2.6 Perspectives

In [28], research of discrete atomic compression (DAC), which is a lossy image compression algorithm, is presented. This algorithm is based on the application of atomic wavelets $w_k(x)$. It was shown that DAC provides better compression than JPEG with the same quality losses measured by *PSNR*-metric.

Moreover, there exists mechanism for quality loss control [29]. Besides, privacy protection of digital images compressed by DAC can be provided [30]. If we combine this with the results obtained above, we see that application of the algorithm DAC is promising, in particular in the cases, when low expenses of memory, time and computing resources are required. In addition, development of video compression and protection technologies based on atomic wavelets is perspective.

3 Conclusions

In this paper, we showed that atomic functions $up_s(x)$ and atomic wavelets, which are constructed using these functions, have significant potential for use in data analysis and processing. A large number of convenient properties provide high speed algorithms based on these functions, as well as low memory costs. Moreover, as the research of discrete atomic compression showed, using atomic functions $up_s(x)$ and related objects, it is possible to surpass already existing widely used technologies. Finally, in current paper, we presented a clear approach to the practical use of atomic functions.

References

1. Rvachev, V.L., Rvachev, V.A.: Nonclassical methods of approximation theory in boundary value problems. Naukova dumka Publ, Kyiv (1979).in Russian)
2. Rvachev, V.A.: Compactly supported solutions of functional-differential equations and their applications. Russian Mathematical Surveys 45(1), 87–120 (1990). https://doi.org/10.1070/RM1990v045n01ABEH002324
3. Stoyan, Y.G., et al.: Theory of R-functions and current problems of applied mathematics. Naukova dumka Publ, Kyiv (1986). (in Russian)
4. Fedotova, E.A.: On the interpolation using atomic functions. Math. Methods Dyn. Syst. Anal. 1, 34–38 (1977). in Russian)
5. Dabagyan, A.A., Fedotova, E.A.: On the interpolation using atomic functions. Math. Methods Dyn. Syst. Anal. 1, 38–45 (1977). (in Russian)
6. Yarmolyuk, V.K.: Application of the generalized Taylor series to approximate computation of integrals. Math. Methods Dyn. Syst.Anal. 7, 48–50 (1983). in Russian)
7. Rvachova, T.V.: Computation of Fourier transform with the help of the atomic functions. Radioelectronic Comput. Syst. 49(1), 113–116 (2011). in Russian)
8. Rvachov, V.O., Rvachova, T.V., Tomilova, E.P.: Application of atomic generalized Taylor series to solving of integral equations of electrodynamics and antenna theory. Radioelectronic Comput. Syst. 60(1), 7–14 (2013). [in Russian]
9. Dyn, N., Ron, A.: Multiresolution analysis by infinitely differentiable compactly supported functions. Appl. Comput. Harmonic Anal. 2(1), 15–20 (1995). https://doi.org/10.1006/acha.1995.1002
10. Cooklev, T., Berbecel, G.I., Venetsanopoulos, A.N.: Wavelets and differential-dilatation equations. IEEE Trans. Signal Process. 48(8), 670–681 (2000). https://doi.org/10.1109/78.852007
11. Berkolaiko, M.Z., Novikov, I.Y.: On infinitely smooth compactly supported almost-wavelets. Math. Notes 56(3), 877–883 (1994). https://doi.org/10.1007/BF02362405
12. Gotovac, H., Cvetkovic, V., Andricevic, R.: Multi-resolution adaptive modeling of ground flow and transport problems. Adv. Water Resour. 30(5), 1105–1126 (2007). https://doi.org/10.1016/j.advwatres.2006.10.007
13. Gotovac, H., Andricevic, R., Gotovac, B.: Adaptive Fup multi-resolution approach to flow and advective transport in highly heterogeneous porous media. Adv. Water Resour. 32(6), 885–905 (2009). https://doi.org/10.1016/j.advwatres.2009.02.013
14. Landin, C.J., Reyes, M.M., Martin, A.S., et al.: Medical image processing using novel wavelet filters based on atomic functions: optimal medical image compression. In: Arabnia, H., Tran, Q.N. (eds.) Software Tools and Algorithms for Biological Systems. AEMB, vol. 696, pp. 497–504. Springer, New York (2011). https://doi.org/10.1007/978-1-4419-7046-6_50
15. Carvajal-Gamez, B.E. Gallegos-Funes, F.J., Ponomaryov, V.I., Cruz-Santiago, R.: A new steganographic method for RGB color images using estimation of variance field in the wavelet domain. In: 2013 International Kharkov Symposium on Physics and Engineering of Microwaves, Millimeter and Submillimeter Waves, pp. 23–28. IEEE, Kharkiv (2013). https://doi.org/10.1109/MSMW.2013.6622143
16. Rvachev, V.A.: On approximation by means of the function up(x). Sov. Math. Dokl. 18, 340–342 (1977). in Russian)
17. Rvachev, V.A., Starets, G.A.: Some atomic functions and their applications. Proc. Ukr. SSR Acad. Sci. 11, 22–24 (1983). in Ukrainian)

18. Desjardins, J.: How much data is generated each day?. https://www.visualcapitalist.com. Accessed 03 Jul 2020
19. Cakebread, C.: People will take 1.2 trillion digital photos this year – thanks to smartphones. https://www.businessinsider.com. Accessed 14 Oct 2019
20. Kharchenko, V., Illiashenko, O.: Concepts of green IT engineering: Taxonomy, principles and implementation. In: Kharchenko, V., et al. (eds.) Green IT Engineering: Concepts, Models, Complex Systems Architectures. SSDC, vol. 74, pp. 3–19. Springer, Cham (2017). https://doi.org/10.1007/978-3-319-44162-7_1
21. Gotovac, B., Kozulic, V.: On a selection of basis functions in numerical analyses of engineering problems. Int. J. Eng. Model. 12(1–4), 25–41 (1999)
22. Starets, G.A., Kurpa, L.I.: About moments and values of some atomic functions. Syst. Arms Milit. Equipment 23(3), 162–163 (2010). (in Russian)
23. Makarichev, V.A.: Approximation of periodic functions by mup$_s$(x). Math. Notes 93(6), 858–880 (2013). https://doi.org/10.1134/S0001434613050258
24. Makarichev, V.A.: On a nonstationary system of infinitely differentiable wavelets with a compact support. Visnyk of V.N. Karazin Kharkiv National University. Ser.: Mathematics, Applied Mathematics and mechanics 967, 63–80 (2011). (in Russian)
25. Makarichev, V.A.: Applications of the function mup$_s$(x). In: Burenkov, V.I. (ed.) Progress in Analysis, vol. 2, pp. 297–304. Peoples' Friendship University of Russia, Moscow (2012)
26. Pas, R., Stotzer, E., Terboven, C.: Using OpenMP – The next step: Affinity, accelerators, tasking, and SIMD. MIT Press, Cambridge (2017)
27. Gropp, W., Lusk, E., Skjellum, A.: Using MPI: Portable parallel programming with the message-passing interface. MIT Press, Cambridge (2014)
28. Lukin, V., Brysina, I., Makarichev, V.: Discrete atomic compression of digital images: a way to reduce memory expenses. In: Nechyporuk, M., et al. (eds.) Integrated Computer Technologies in Mechanical Engineering. AISC, vol. 1113, pp. 492–502. Springer, Cham (2020). https://doi.org/10.1007/978-3-030-37618-5_42
29. Makarichev, V.O., Lukin, V.V., Brysina, I.V.: On estimates of coefficients of generalized atomic wavelets expansions and their application to data processing. Radioelectronic Comput. Syst. 93(1), 44–57 (2020). https://doi.org/10.32620/reks.2020.1.05
30. Makarichev, V., Lukin, V., Brysina, I. Discrete atomic compression with different structures of discrete atomic transform: efficiency comparison and perspectives of application to digital images privacy protection. In: 2020 IEEE 11th International Conference on Dependable Systems, Services and Technologies (DESSERT), pp. 301–306. IEEE, Kyiv (2020). https://doi.org/10.1109/DESSERT50317.2020.9125073

Cloud IoT Platform for Creating Intelligent Industrial Automation Systems

Oleksandr Prokhorov[1]([✉]) [ID], Yurii Pronchakov[1] [ID],
and Valeriy Prokhorov[2]

[1] National Aerospace University "Kharkiv Aviation Institute",
17 Chkalova Str, Kharkiv 61070, Ukraine
o.prokhorov@khai.edu
[2] Kharkiv National University of Radio Electronics,
14 Nauky Ave, Kharkiv 61166, Ukraine

Abstract. A cloud platform is proposed within the framework of PaaS and SaaS models, which provides support for all stages of the development of intelligent decision support systems, their adaptation to solve applied problems for IoT and control applications in industrial automation systems, storage of knowledge and data bases in cloud data centers, provision access to intelligent systems as services remotely through a web interface. Use scenario and tasks to be solved are considered. The IoT platform structure of intelligent control systems is presented. Examples of creating intelligent systems in Node-RED are considered.

Keywords: Industrial 4.0 · Industrial internet of things · Artificial intelligence · Intelligent systems · Decision support systems · Cloud platform

1 Introduction

The increase in the volume of information, the complexity of management and analysis tasks, the need to take into account a large number of difficultly formalized and interrelated factors determine the changes that occur in the field of information processing and management in businesses and the need to use modern intelligent information technologies in the decision-making process.

The development of ideas of artificial intelligence and intelligent systems has lead to the possibility to implement a cognitive method that is knowledge-based, to automate decision-making and management processes in complex systems. The use of the cognitive method in the development of information systems provides a high degree of adaptability to various situations and conditions, the ability to accumulate experience, etc. The main difference between modern intelligent systems is their distribution, which is manifested in the need to ensure the processing and application of distributed knowledge.

Tasks of intellectual decision support can arise at various levels of business management (Fig. 1). In this case, the intellectualization of various information systems is carried out, which in this case are the sources of primary information.

M. Nechyporuk et al. (Eds.): ICTM 2020, LNNS 188, pp. 55–67, 2021.
https://doi.org/10.1007/978-3-030-66717-7_5

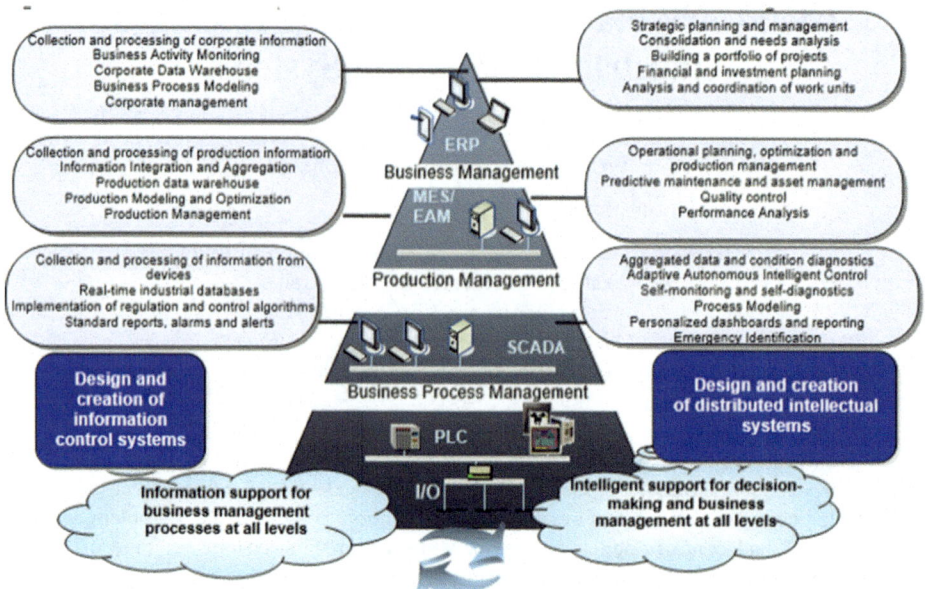

Fig. 1. Tasks of intellectual decision support and business management.

2   Industry 4.0

The Industrial Concept Industrial 4.0 is a large-scale multi-level organizational and technical system that involves the transition to fully automated digital production controlled by real-time intelligent systems, where Cyber-Physical Systems (CPS) are integrated into a single information space using cloud technologies and Industrial Internet of Things (IIoT) technologies.

The concept of Industrial 4.0 is based on the availability of complete and reliable digital data from smart devices, as well as new opportunities for their remote monitoring and processing, a picture of production and real-time control, advanced analytics, whereas advanced BigData analytics methods (using methods and tools of artificial intelligence, machine learning, data mining and predictive analytics, etc.) are the hallmark of Industrial 4.0. In fact, industrial automation connected to the Internet and enhanced by artificial intelligence is the fourth industrial revolution.

The paper [1] provides the overview of intellectual production, production with support for the Internet of things and cloud production in the context of Industrial 4.0. The term "smart" began to be applied everywhere – both by application (smart factory, smart agro, smart city, smart hotel, smart house, etc.), and by value chain (smart manufacturing, smart logistic, smart healthcare, etc.) and in case of combination of a set of new technologies. The current principal areas for implementation of Industrial 4.0: horizontal and vertical system integration; industrial internet of things; cloud technologies; big data analysis; modeling; additive manufacturing (3D printing); augmented and virtual reality; robotics; cybersecurity.

The program logic of automated control systems (ACS) is implemented as inter-acting cloud services, where a transition is made from rigidly hierarchically structured information-isolated ACSs to the direct connection of management objects to the "control cloud" without human involvement and intermediate ACS, which makes IoT Industrial Cloud Platform the main element of the new Architecture. This platform is capable to combine the entire "hardware park" of devices in a single ecosystem with effective management tools. Therein the IoT cloud platform performs all the necessary functionality (software algorithms for data processing and control) of both grass-roots management systems and enterprise-level management systems and above, which means that it simultaneously performs the functions of a universal means of integration and the execution functions of arbitrarily complex and diverse intelligent control algorithms. The paper [2] demonstrates why IIoT integration is important for pro-duction systems and how a large amount of collected data can be used. Using a specific example, the authors showed the capabilities of modern IIoT communication protocols, namely MQTT, and how quickly prototype IIoT applications can be developed with the used of IBM Node-RED tools. By using the mechanism of open application pro-gramming interfaces (API) and IoT protocols, any devices and any ACSs can be connected to the cloud platform without the need to make changes to connected devices and systems, and the processing logic of the data delivered to the "cloud" can be implemented using ready-made templates and, in the absence thereof, using built-in software development tools. In this case the BigData effect accumulated in this IoT platform and the use of artificial intelligence and machine learning technologies enable automation of the improvement processes of the algorithms executed in the cloud, that is, optimization of control algorithms as historical data is accumulated from a wide range of devices and ACSs. The world experience gained in implementing IoT shows that the transition to the IoT concept allows for quick implementation of arbitrarily complex end-to-end fully automated business processes. These processes cover many different ACSs of various businesses and organizations and involve many different devices, which, when using the traditional approach to automation, in most cases are impossible to be implemented within a reasonable time and for an economically viable budget. SCADA systems are transforming with the advent of the Internet of Things (IoT) and cloud computing technologies. In fact, taking into consideration the evolution of SCADA systems and the classification used in the world, it should be noted that this transformation in the framework of Industrial 4.0 led to fourth-generation SCADA adapted for the Internet of things. This implies an even greater degree of decentral-ization and unification, that is, the possibility to quickly move the execution points of the algorithms between SCADA servers and controllers, support for modern wireless standards, the ability of controllers without static IP addresses to connect to SCADA servers operating in the cloud themselves.

3 Artificial Intelligence Platforms

The paper [3] considers the concept of Industrial 4.0 as closely related to the achievements in the field of artificial intelligence. The authors consider the following as key elements of industrial artificial intelligence: analytics, big data, cloud technologies, knowledge and reasoning.

The paper [4] positions Intelligent Manufacturing Systems (IMS) as the next generation of manufacturing systems by creating and implementing new models, new forms and new methodologies for transforming a traditional manufacturing system into Smart Manufacturing. The authors emphasize the important role of artificial intelligence in IMS, through features such as learning and reasoning in human-machine interaction.

The use of machine learning in SCADA is considered in the paper [5]. The authors emphasize a specific scenario for forecasting emergency processes.

The paper [6] discusses the development of an additive manufacturing system that combines an expert system with IoT. The expert system is built on the basis of an artificial neural network and is used to classify the designs of input parts based on CAD data and user data. Interaction with IoT devices is implemented through Node-RED.

Thus, taking into consideration the ongoing implementation and development of artificial intelligence technologies and platforms, perhaps the concept of the fifth generation SCADA systems should be introduced. When considering the main directions of solving semi-structured or unstructured production problems using artificial intelligence two directions can be distinguished in their automation: 1) inductive training: the use of neural networks and machine learning methods (building machine learning algorithms) at the level of recognition (classification) and generalization of objects and situations, i.e.; 2) deductive learning: formalization of professional knowledge and the use of intelligent systems that have mechanisms for storing this knowledge (knowledge base), which may include mechanisms for acquiring knowledge, as well as mechanisms for reasoning or manipulating knowledge based on methods for deriving decisions.

To implement both methods, various artificial intelligence platforms have been created. AI platforms provide users with a suite of tools for building intelligent applications. These platforms provide connection of intelligent decision-making algorithms with data, which enables developers to create the necessary logical and analytical solutions. In addition, some platforms provide pre-built algorithms and workflows with functions such as modeling and visual interfaces that simplify the creation of an intelligent system, while others require more in-depth knowledge in development and programming. These algorithms may include image and speech recognition functions, natural language processing, forecasting, etc.

It should be noted that the combination of cloud technologies and artificial intelligence is implemented in two main forms: 1) cloud machine learning platforms, e.g. Google Cloud Machine Learning; 2) cloud services with built-in artificial intelligence, e.g. IBM Watson.

The latter form is found more promising since it implements various options for the application of artificial intelligence in cloud services.

Further are considered some of the artificial intelligence platforms.

Google Cloud AI is a set of cloud services that allow the user to easily create machine learning models and that work with any type of data of any size. As part of services such as Cloud Auto ML (Vision API — models for image recognition, Natural Language API – models for text recognition, Translation API – machine translation), Dialogflow Enterprise Edition – a package for creating interactive interfaces of websites, mobile applications and popular platforms messaging and IoT devices and other services. Dialogflow, operating on the basis of GCP (Google Cloud Platform) and using the capabilities of Google Cloud ML, has all the functions similar to IBM Watson, but demonstrates better performance in machine learning. Dialogflow provides users with new ways to interact, creating interesting conversational and textual interfaces, such as voice applications and chat bots, working on artificial intelligence.

Microsoft Azure AI integrates smart cloud technology and smart devices using the Azure platform and a wide range of productivity tools, including Cognitive Services. Architecture components are either Azure cloud services or microservices that provide a RESTful endpoint.

IBM Watson is an IBM supercomputer equipped with an artificial intelligence question and answer system within the framework of the IBM DeepQA project, which solves the following tasks: understanding natural language; hypothesizing based on processed data; learning while working; making recommendations accompanied with the facts which the conclusion is based on. IBM Bluemix, an open platform for developing and deploying mobile and web applications, is used to host and manage applications. Watson Services for Bluemix is a tool for quickly prototyping and building powerful cognitive applications in the cloud.

The following should be noted among the less popular, but quite interesting in their concept of artificial intelligence platforms: 1) Acumos is a new platform for open source development of artificial intelligence (Linux Foundation). This is a framework for turning artificial intelligence into microservices; 2) Rainbird is a cloud-based artificial intelligence platform that allows anyone to collect their knowledge on a favorite subject and publish a virtual online expert with human-like solutions; 3) Graph Grail Ai is the first of its kind blockchain artificial intelligence platform, built on the basis of natural language processing technologies with a marketplace of decentralized applications.

4 Cloud IoT Platform for Creating Intelligent Business Management Systems

This paper offers a platform that provides support for all stages of the development of intelligent systems, their adaptation for solving applied problems in any subject areas, storage of knowledge and data bases in cloud data centers, providing access to intelligent systems as services remotely via a web interface. The platform can be useful to specialists and experts in various subject areas, i.e. access to intelligent systems (implementation of the SaaS model), and developers of intelligent systems, i.e. access to their development tools (implementation of the PaaS model).

Intelligent cloud systems allow real-time data collection from servers or external sources, reasoning based on rules stored in the knowledge base, provision of results for visualization in natural language form for users, archiving, and delivery to other information systems. Knowledge is represented in the system on the basis of logical models for calculating first-order predicates, and logical inference is made using a modified resolution method taking into account the problem of computational solvability. Cloud services provide filling, updating and replenishing the knowledge base of the intellectual system in the process of evolutionary development thereof.

Support of various data sources and web services facilitates to integrate smart systems into heterogeneous software environments.

Intelligent systems are created to solve the following tasks: 1) analysis, assessment and recognition of situations, objects; 2) control, assessment and diagnostics of conditions, parameters; 3) assessment of the importance and priority of alternatives; 4) identification and notification of emergency situations and critical conditions and their registration; 5) forecasting the development of events, situations and actions; 6) implementation of algorithms and action scenarios; 7) formation of recommendations, tips and evaluation of solutions.

The enlarged structure of the cloud-based IoT platform for creating intelligent business management systems is provided in Fig. 2.

for creating intelligent business management systems.

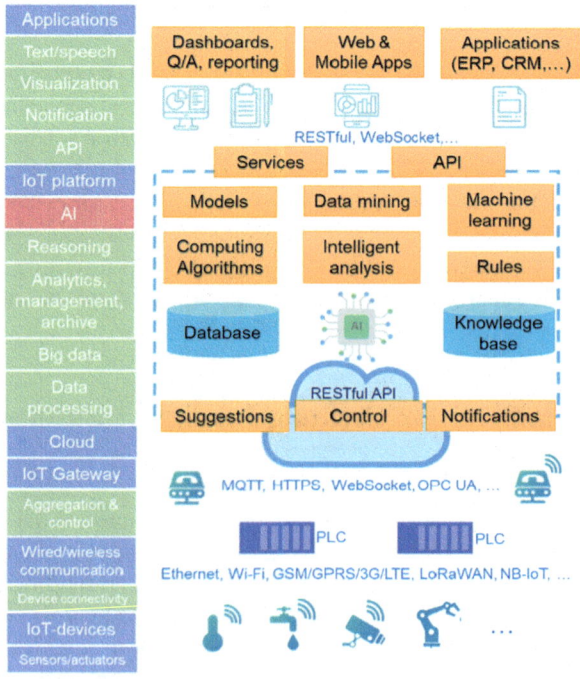

Fig. 2. The structure of the cloud-based IoT platform

Functionality of IoT platform of intelligent systems: 1) creation of applied intelligent systems for any subject areas and information systems with the goal of their intellectualization; 2) real-time data collection and analysis using cloud technologies and IoT analytics tools; 3) provision of controlled access to intelligent systems via the Internet; 4) single environment for the development, implementation and management of intelligent systems; 5) means of integration with other information systems and various DBMSs; 6) implementation and connection of various digital scenarios for processing sensory data, optimization models and intelligent control; 7) use of machine learning technologies for management (forecasting values, determining anomalies, classification and recognition, etc.); 8) implementation of deductive inference mechanism with different strategies for backtracking reduction; 9) dialogue interaction and formation of answers to the questions posed can be carried out in a natural language; 10) possibility to form a chain of events, facts, criteria and rules for explaining the proposed solutions; 11) notification through various channels, issuing recommendations and automatic control commands to controller equipment in real time.

The created intelligent systems can be conditionally assigned to one of four types:

- intelligent diagnostic systems: determine the nature of the deviation of controlled indicators from the standard and, on the basis thereof, solve the problem of recognizing states and implementing action algorithms;
- intelligent monitoring systems: focus on real-time continuous data interpretation and implementation of optimization models and intelligent control;
- intelligent forecasting systems: make conclusions about the future development of events based on the current situation, predict development with different decision schemes;
- intelligent planning systems: determine the optimal plans and action scenarios and monitor their implementation.

The platform implements an effective algorithm for reasoning with various strategies for reducing enumeration, which can be used for first-order predicate logic, as well as for all dialects of descriptive logic in ontologies. A new inference algorithm was developed on connected conjunctions and the construction of the most common replacement, which ensures the search for solutions in the minimum number of steps by the resolution method.

Platform server modules are developed and deployed on NodeJS. Node-RED is used as a programming tool for connecting IoT devices, APIs and cognitive services of the platform. Node-RED is a visual browser development environment for JavaScript runtimes for IoT (non-blocking, event-driven applications). This tool facilitates to create and connect threads using a variety of nodes from a set deployed in the runtime environment with a single click of the mouse button. Due to the large number of primitives and the ability to quickly visually configure and create new components, including directly in the JavaScript language, Node-RED can be used by both lay users and professional developers to accelerate the creation of cognitive applications in the cloud. This enables Node-RED to interact with various hardware platforms as part of the IoT approach, transferring the bulk of the computational load associated with the logical output to the cloud platform (Fig. 3). Thus, many flows or intelligent systems, where the runtime mechanism of logical inference occurs, are simultaneously launched in the cloud. In the first scenario (Fig. 4) of using a cloud platform, an intellectual system knowledge base is developed (a rule base and a question base – one for each logical task) and access to primary information sources is configured.

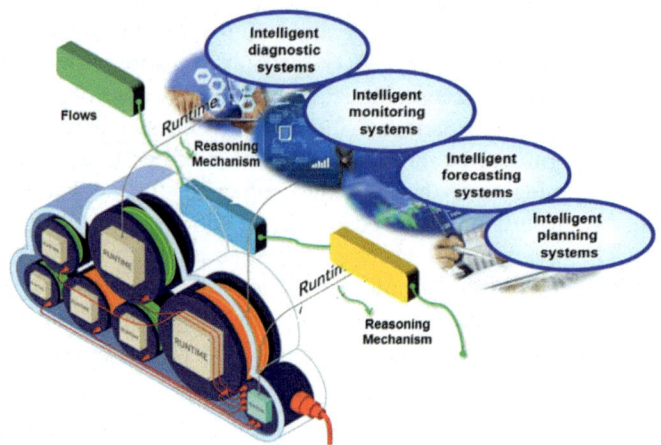

Fig. 3. Modes of platform use.

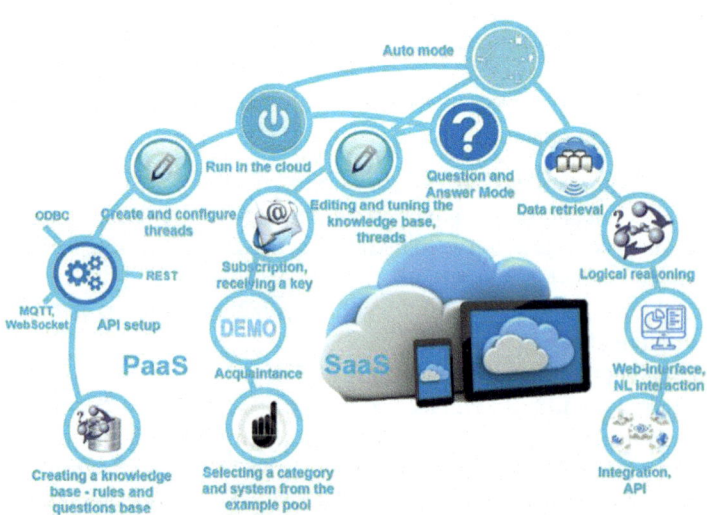

Fig. 4. Modes and stages of platform usage.

Ready-made smart systems are hosted in the cloud and can be downloaded either from the repository or using the Examples menu in Node-RED. Therefore, the user can familiarize themselves with existing ready-made intelligent systems. If any of them is of interest, the user can subscribe thereto by logging in to the system and receiving a key, then, at own discretion, make the necessary correction of the knowledge base and adjusting access to the primary sources of information, thereby obtaining a system for solving their logical and analytical tasks.

The main modes of working with an intelligent system are question-answer and automatic. In the question-answer mode, the user selects the question of interest, and then the system updates the data from the user's infrastructure or data center, performs a logical conclusion and outputs the result to the user in a natural language. In automatic mode, responses are represented in the form of JSON data that can be parsed and further used in other information processing scenarios in Node-RED flows or transmitted to other information systems through the API.

Knowledge base includes a rule base or axioms and a question base and is intended for a formalized description of logical tasks in a simple internal language for describing expert knowledge. Figure 5 shows the editing mode of the knowledge base. The knowledge base for each predicate indicates its type (the primary predicate: indicated by data from various sources (from devices using Modbus TCP, MQTT, OPCUA protocols, etc., via Websocket or HTTP, from databases, etc.), the secondary predicate: determined by the formula; and the calculated predicate: indicated by facts obtained as a result of a logical conclusion), internal representation in a special formalized language, semantic content, predicate calculation formula (for secondary predicates), fact determination formula (for calculated predicates).

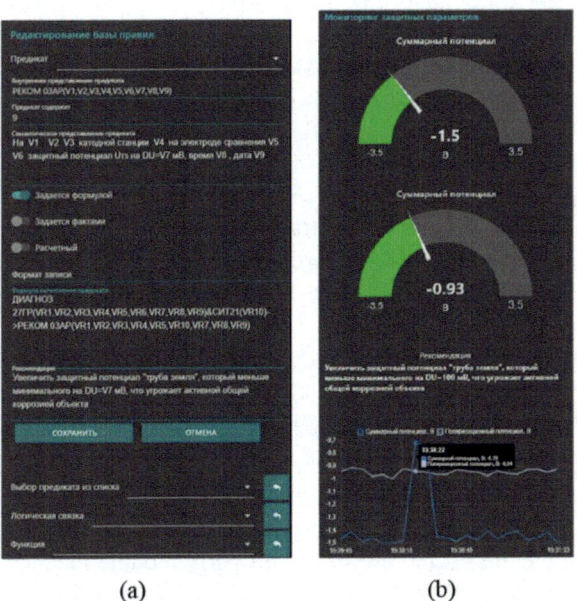

(a)   (b)

Fig. 5. System operation mode: (a) editing the rule base; (b) control panel with recommendations of the intelligent system.

The semantic content of the predicate is the text of the answer to the question. The semantic content indicates the variables where the values found as a result of the logical inference will be substituted. Besides the indicated predicates, one more type exists, i.e. order predicates ("GREATER", "LESS", "EQUAL", etc.), the truth whereof is calculated by the inference module after substituting the predicate terms for the actual values.

The main nodes added to the Node-RED panel after installing the package are as follows:

- create data source: a primary predicate attribution node capable to receive data from primary sources (Modbus TCP, MQTT, Websocket, HTTP nodes, from databases, XML, CSV and other data sources). For a predicate attribution node, the number of variables, semantic content, and other parameters are indicated depending on the type of data source;
- API call reasoning JSON: call of the logical inference service (the logical inference module is the central element of the kernel of the software complex designed to logically derive consequences from the rule system that are in the knowledge base using the modified resolution method), user access API key and name of the logical task shall be indicated for the node, whereas msg.payload at the input shall contain the identifier of the question to which the answer will be searched (it is most convenient to formulate a request to start a logical output using the template node). The node has two outputs and returns a response of a special structure in the JSON format, which can subsequently be used in the flow, for example, to form control actions on IoT devices, whereas the second output shows an error. If the explanation option is enabled, the output also provides an explanation, i.e. information about the reasons for receiving a particular answer (facts and rules) involved in the logical conclusion, which makes it easier for the expert to test the system and increases the user's confidence in the result;
- API call reasoning NL: similar to the call of the logical inference service described above, but in this case the output is natural language text for its further output to the web-based interface on the Dashboard, sending it as a notification via various channels, or implementing a voice notification mechanism using third-party services;
- rule change: using this node, by specifying the rule identifier of a particular logical task, enables on-the-fly changes adding any software component uploaded to the Node.js server as a predicate for logical rules (with AND/OR logical connectives).

A flow example in Node-RED using the cloud services of our cognitive platform is shown in Fig. 6. The flow demonstrates the formation of primary predicates from various data sources, the generation of queries and the launch of the inference service. In addition, the flow contains the use of a third-party module face-api.js to implement the task of recognizing faces on camera stream. The web-interface uses the node-red-dashboard module, as well as other packages that enable quick design of SCADA-like interfaces for the tasks of managing various smart objects, i.e. houses, hotels, businesses, etc. An intelligent system, fragments of the interface of which are shown in Fig. 7, solves the tasks of adaptive and personalized control in a smart house: 1) expanding the capabilities of video surveillance systems and face recognition by solving the problems of identifying objects, postural poses, emotions, etc. using neural networks and deep learning; 2) collection, accumulation and analysis of data to identify people's preferences and further personalization; 3) building an owner profile; 4) automatic adaptive management of Smarthouse subsystems based on people's preferences, etc.; 5) use of machine learning and artificial intelligence to promote the services of a digital assistant; 6) personalized recommendations for menus, entertainment and

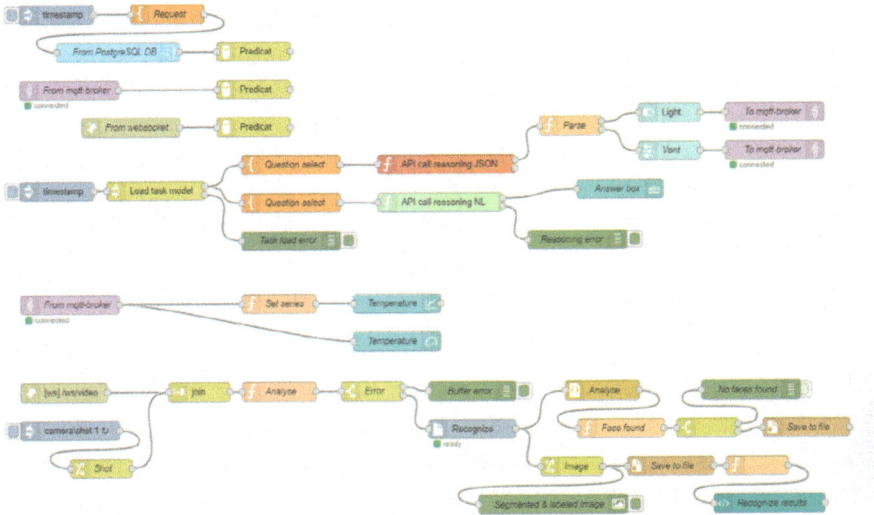

Fig. 6. Example of a flow using cognitive modules of the system.

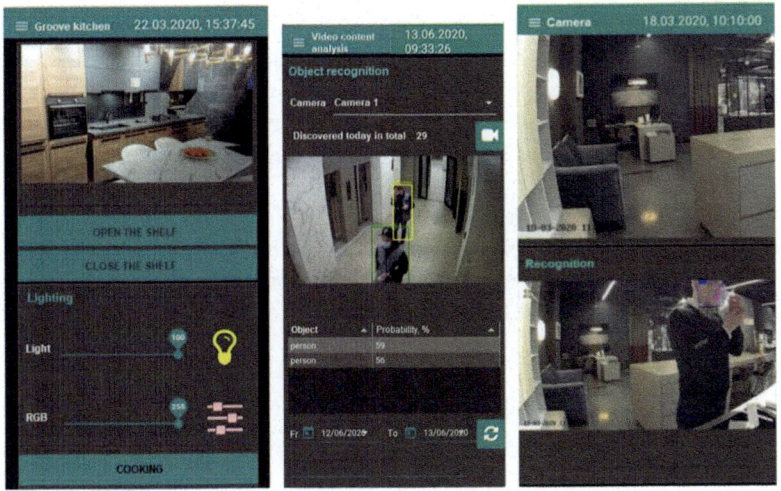

Fig. 7. Web interface fragments in the Smarthouse intelligent system.

relevant useful information based on identified preferences or owner requests; 7) notifications through various channels for the transmission of recommendations, reminders and information; 8) chatbot with question-answer mode in natural language.

Further are considered the advantages of the proposed platform. The first advantage is the intellectualization of processing and the synthesis of information from different sources, without the need to understand the features of other systems and essentially without programming. The knowledge base of the system can be easily adjusted and supplemented with new rules in the process of possible evolutionary changes without affecting the kernel software. The system provides dialogue in natural language in a question-answer mode. In automatic mode, the system generates data for further analysis, transfer to other systems or issuing control actions on equipment. The Node-RED environment enables quick integration with equipment, various DBMSs, other information systems, and cloud services through the API. Another important advantage is that processing algorithms can be embedded directly into the rules. Any software module that is already loaded on the Node-RED palette and placed in the flow can be connected to the system as a predicate. Thus, the formalization of problem solving in the form of models of knowledge (rules) leads to the synthesis of its solution carried out in the process of inference.

5   Conclusion

The proposed platform facilitates the development, testing, deployment and maintenance of applied intelligent systems without the need for investment in infrastructure and software environment; it increases the degree of intellectualization and adaptation of existing information systems to changing tasks and functioning goals; it enables accumulation and reuse of knowledge without hard-set data structures and processing algorithms; it improves the quality, reliability and reduces the time of development and decision making.

Usage of Node.js as a server for deploying the platform and the Node-RED development tool environment has the following advantages: it is easy to create processes using an intuitive drag&drop editor; the logical conclusion of decisions and the necessary artificial intelligence can be easily embedded in the solution of logical and analytical problems; the process flow can be controlled in real time; each step of the process can be debugged to quickly find and fix errors; Node-RED allows for deployment of processes with one click in a cloud environment; after the processes are created and deployed, the user can start and debug without stopping and rebooting the entire system; since all processes are collected in one place, each member of the development team can access and manage the processes in accordance with the given permissions.

References

1. Zhong, R.Y., Xu, X., Klotz, E., Newman, S.T.: Intelligent manufacturing in the context of Industry 4.0: A review. Engineering **3**(5), 616–630 (2017). https://doi.org/10.1016/J.ENG. 2017.05.015
2. Ferencz, K., Domokos, J.: Using node-RED platform in an industrial environment. In: XXXV Jubileumi Kandó Konferencia, pp. 52–63 (2019)
3. Lee, J., Davari, H., Singh, J., Pandhare, V.: Industrial Artificial Intelligence for industry 4.0-based manufacturing systems. Manuf. Lett. **18**, 20–23 (2018). https://doi.org/10.1016/j. mfglet.2018.09.002
4. Yao, X., Zhou, J., Zhang, J., Boër, C.R.: From intelligent manufacturing to smart manufacturing for Industry 4.0 driven by next generation artificial intelligence and further on. In: 2017 5th International Conference on Enterprise Systems (ES), pp. 311–318. IEEE, Beijing (2017). https://doi.org/10.1109/ES.2017.58
5. Skripcak, T., Tanuska, P.: Utilisation of on-line machine learning for SCADA system alarms forecasting. In: 2013 Science and Information Conference, pp. 477–484. IEEE, London (2013)
6. Elhoone, H., Zhang, T., Anwar, M., Desai, S.: Cyber-based design for additive manufacturing using artificial neural networks for Industry 4.0. Int. J. Production Res. **58**(9), 2841–2861 (2020). https://doi.org/10.1080/00207543.2019.1671627

Improving Vehicle Safety Through the Use of Arduino Controller-Based Automotive Voice Informants

Nataliia Kobrina$^{(\boxtimes)}$ ⓘ, Andrey Makoveckiy,
and Dmitriy Makarenko ⓘ

National Aerospace University "Kharkiv Aviation Institute",
17 Chkalova Street, Kharkiv 61070, Ukraine
n.kobrina@khai.edu

Abstract. An analysis is made of the vehicle safety from the point of view of the possibilities of increasing it. Criteria are proposed for assessing active, passive, post-accident and environmental safety. The possibility of increasing the safety of vehicle operation by enhancing the use of informative security, that is, providing the necessary information to the driver and other traffic participants, is being considered. Control schemes for general vehicle safety systems, lighting, engine and braking are offered. At the same time, the Arduino controller and the DFPlayer mini module are considered as a control unit when modeling a car voice informant.

Keywords: Safety · Vehicle · Automobile voice informant · Simulation · Program block diagram · Arduino controller

1 The Problem

One of the most important operational qualities of a car is its safety, since the life and health of people, the safety of cars, directly depend on it. Car safety is considered as a complex system with a certain amount of interconnected elements. In the simplest case, it can be distinguished from the system "man – machine – environment". For a general assessment, which allows to compare the effectiveness of the mechanisms and devices of the car that increase its safety, integral indicators are used. Most often, the concept of risk is used for these purposes. The influence of risk factors of the "man – machine – environment" system on human safety is presented in Fig. 1.

It is evident that the greatest contribution to the safety of a person in a moving car could be made by measures aimed at assisting the driver in driving a vehicle. This could be facilitated by the use of voice informants in automobiles.

2 Main Part

Analysis of Vehicle Safety in Terms of Its Improvement. Safety is an integrated quality of a car, determined by its structural properties and is usually divided into groups of passive, active, environmental and post-accident safety [1–3], Fig. 2.

M. Nechyporuk et al. (Eds.): ICTM 2020, LNNS 188, pp. 68–80, 2021.
https://doi.org/10.1007/978-3-030-66717-7_6

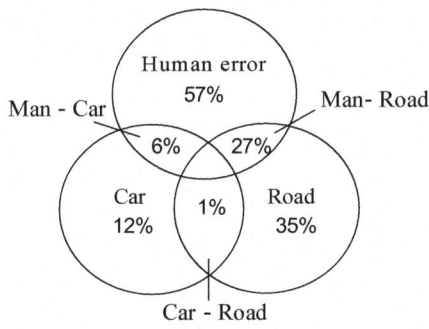

Fig. 1. The role of risk factors.

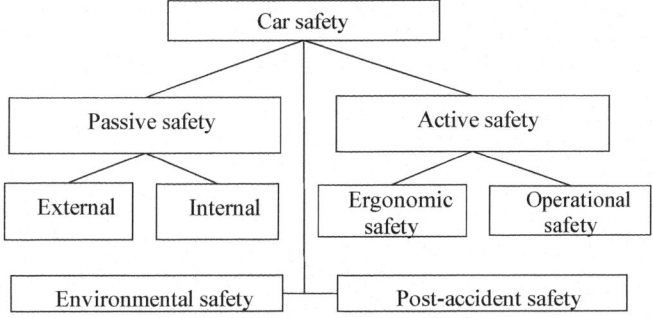

Fig. 2. Car safety structure.

Passive safety is the property of a car to reduce the severity of the consequences of a traffic accident. Passive safety is manifested in a period when the driver, despite the safety measures taken, can't change the nature of the car's movement and prevent an accident (culminating phase). There are internal passive safety of the car, which reduces the degree of injury to passengers and the driver, and external passive safety, which reduces the possibility of causing damage to other participants in the movement.

External passive safety is determined by the shape of the body, the design of the bumper, the presence of safety elements, etc. Internal passive safety is manifested in the presence of elements that reduce overload (seat belts, airbags, head restraints), elements that ensure safety (steering column, controls and all interior elements of appropriate designs), glass safety, etc.

To assess the elements of passive safety, the criterion K_p can be used, which determines the average severity of the consequences of an accident:

$$K_p = \sum_1^m \frac{P_i n_i}{m}, \tag{1}$$

where n_i is the number of injured drivers and passengers in each accident; P_i is the severity coefficient of the consequences in this accident, UAH/person; m is the total number of drivers and passengers involved in traffic accidents.

Active car safety is the property of a car to prevent a traffic accident (to reduce the likelihood of its occurrence). Active safety is manifested in the period corresponding to the initial phase of the accident, when it is still possible to change the nature of the vehicle.

Active vehicle safety is determined by operational and ergonomic safety. Operational safety is ensured by the braking, traction and speed parameters of the car, its stability, controllability, information content, reliability of structural elements, etc. Ergonomic safety is determined by the driver's workplace parameters (microclimate of the cabin, noise, vibration, electromagnetic fields in the areas where the road users are located, ergonomic qualities of the workplace).

To assess active safety, it is possible to apply the criterion K_{ac}, defined as the sum of specific indicators of the ratio of the number of accidents that occurred due to unsatisfactory operation of mechanisms during operation of the vehicle N_{ec} or due to poor performance of indicators of ergonomic safety N_{er} to vehicle mileage L:

$$K_{ac} = \frac{N_{ec} + N_{er}}{L}. \tag{2}$$

Post-accident vehicle safety is a property of the car that helps reduce the severity of the consequences of a traffic accident. It is characterized by the ability to quickly eliminate the consequences of a traffic accident and prevent the occurrence of new emergencies (for example, fire).

To assess post-accident safety, K_{pa} criterion can be applied, which determines the costs of eliminating the consequences of traffic accidents $\sum_1^n C_i$ to their total number n:

$$K_{pa} = \frac{\sum_1^n C_i}{n}. \tag{3}$$

Ecological safety of a car is a property of a car that allows to reduce the harm caused to road users and the environment during its normal operation. Environmental safety, manifested during the daily operation of the car, fundamentally differs from the above three types of safety, which are detected only in an accident.

Environmental safety is ensured by: the presence of systems and devices aimed at reducing the toxicity of exhaust and crankcase gases, fumes of fuels, oils and acids; reduction of noise, vibration and electromagnetic fields arising from the movement of cars; utilization of liquid and solid wastes from the operation of vehicles, used batteries, used tires, used oils and oil products, used industrial fluids, motor vehicles, spare parts and assemblies, which have become unusable, etc.

None of the above items are classified as particularly hazardous. However, at the current scale of the use of vehicles, the concomitant factors of its operation cause significant damage to the environment and human health.

Measures to ensure environmental safety can be assessed by the K_e criterion, which determines the costs of maintaining a given level of environmental safety of cars of a given category and model $\sum_1^m C_{ci}$ to their total number m:

$$K_e = \frac{\sum_1^m C_{ci}}{m}. \tag{4}$$

It should be noted that all types of vehicle safety are interrelated. In critical situations, the qualities of the vehicle's active safety should first come into operation and prevent accidents, if for some reason they do not work, then the qualities of passive safety are turned on and the degree of injury to the participants of the accident is activated, then the qualities of post-accident safety come into play, using which ensure the evacuation of people, fires and car explosions are prevented.

Unlike the first three safety qualities, which are included in emergency situations, environmental safety is manifested throughout the entire life of the car.

Informational Content of the Car. All types of safety should be supported by such a property of the car as its information content, that is, provide the necessary information to the driver and other road users in any conditions. Informational content of the vehicle is crucial for safe control. Information about the features of the vehicle, the nature of the behavior and intentions of its driver largely determines the safety in the actions of the participants in the movement and confidence in the implementation of their intentions. In conditions of insufficient visibility, especially at night, information content in comparison with other operational properties of the car has a major impact on traffic safety.

One of the many automobile systems providing its information content may be an automobile voice informant designed for sound notification about the operation of various sensors or, in other words, about the state of automobile systems [4–6]. The informant "interrogates" the sensors located in the most important nodes of the car, and according to the results of the survey generates speech fragments reflecting the state of the monitored nodes. Voice informants have been releasing for a long time [4]. As a permanent option, voice informants are used, for example, on Renault 25, Renault Laguna, etc. However, a relatively small number of controlled parameters, attachment to a particular car model and a rather high price limit the wide distribution of these devices.

An analysis of the dangers that may arise during the operation of a vehicle allows one to single out the information most often requested by the driver. However, this information should not be redundant, not to distract the driver from driving the car. The phrases that a voice informant should reproduce, from our point of view, can be:

- welcome, the on-board computer checks the car's systems;
- side lights are defective;
- license plate lights are defective;
- right brake light is defective;
- left brake light is defective;
- malfunction of the battery charging circuit;
- parking brake is active;

- minimum fluid level in the washer fluid reservoir;
- emergency fuel remaining in the tank, limited range;
- minimum engine oil level, check;
- drop in oil pressure, stop the car, switch off the ignition and refer to the instruction manual;
- engine overheating, stop the car, do not perform any operations on a hot engine, refer to the operation manual;
- injection system malfunction;
- malfunction of the toxicity reduction system, contact the service station;
- malfunction of the adaptive power steering has been detected, contact a service station;
- brake system malfunction, avoid sudden braking, stop the car, contact a service station;
- critical decrease in air pressure in the left front tire;
- critical decrease in air pressure in the right front tire;
- critical decrease in air pressure in the right rear tire;
- critical decrease in air pressure in the left rear tire.

For the further implementation of informing the driver about the state of car systems, taking into account the above analysis, schemes of programs for their control can be proposed, Figs. 3, 4, 5 and 6.

Simulation of Arduino Controller-Based Automobile Voice Informant. To work out and implement control programs for car systems Figs. 3, 4, 5 and 6 is possible due to the use of automobile speech informers, and, in particular, made on the basis of the Arduino controller, a trademark of hardware and software for building simple automation and robotics systems aimed at non-professional users [7–10]. In particular, this technology can be used in the process of training students in the automotive industry while working out issues of ensuring vehicle safety.

The software part consists of a free software shell (IDE) for writing programs, and compiling, and programming hardware. The hardware part is a set of mounted printed circuit boards sold by both the official manufacturer and third-party manufacturers. The completely open system architecture allows to freely copy or complement the Arduino product line.

To simulate an automobile voice informer from the Arduino product line, the DFPlayer mini module (Fig. 7) can be used. DFPlayer is a miniature MP3 module with a built-in power amplifier and the ability to connect a speaker up to 3 W. The player can be used as a separate module, or in combination with an Arduino-compatible controller. DFPlayer supports the common audio formats MP3, WAV and WMA. Supports TF cards with FAT16, FAT32 file system.

For this module to work in the voice informant mode, it is necessary to record a set of phrases on a micro SD card (it is possible to record the above phrases if the vehicle has the corresponding sensors), insert the card into the module and call the "play" function with the number of the desired record at the right time. At any time, it is possible to pause playback, and then resume from the same place.

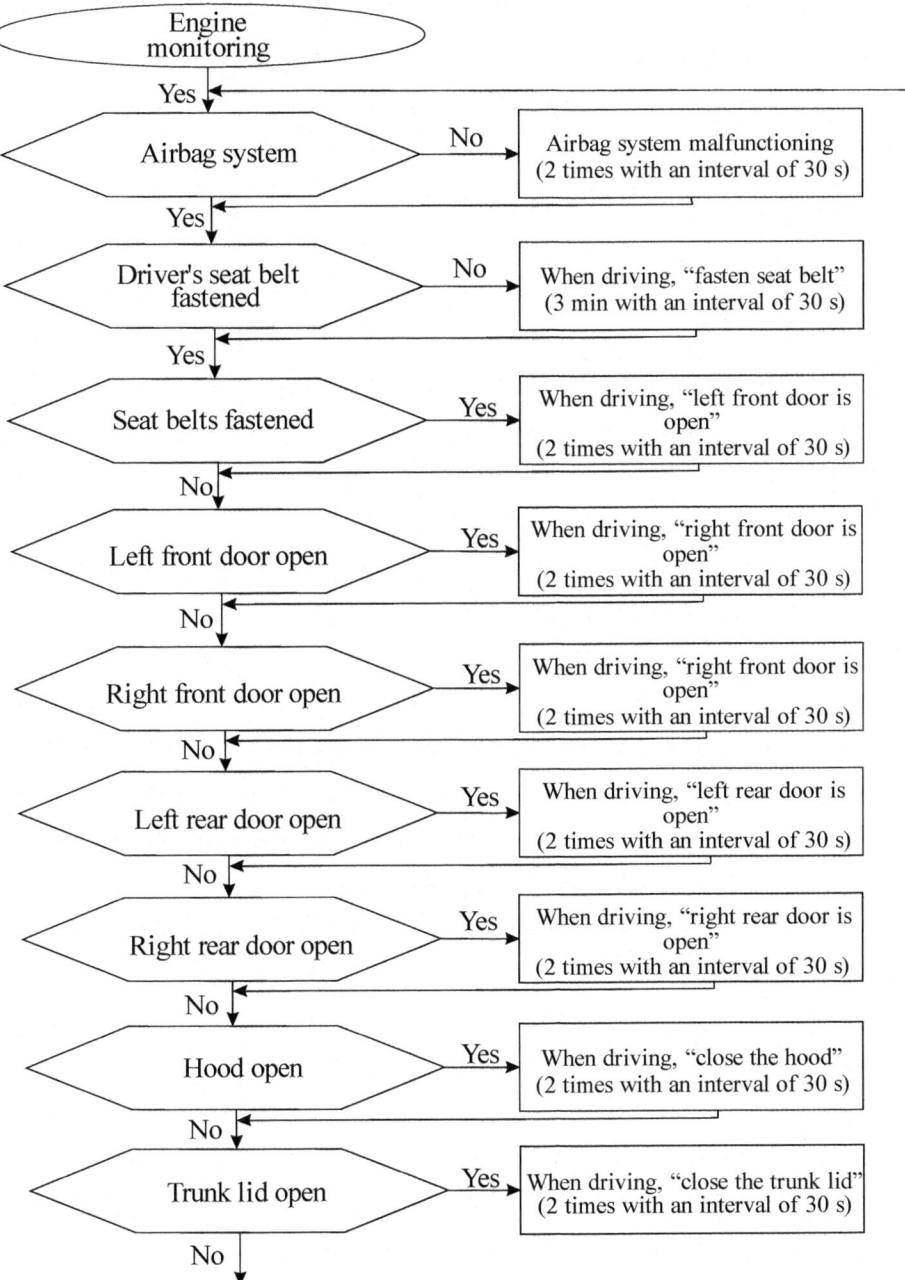

Fig. 3. Scheme of the program for monitoring the car security system.

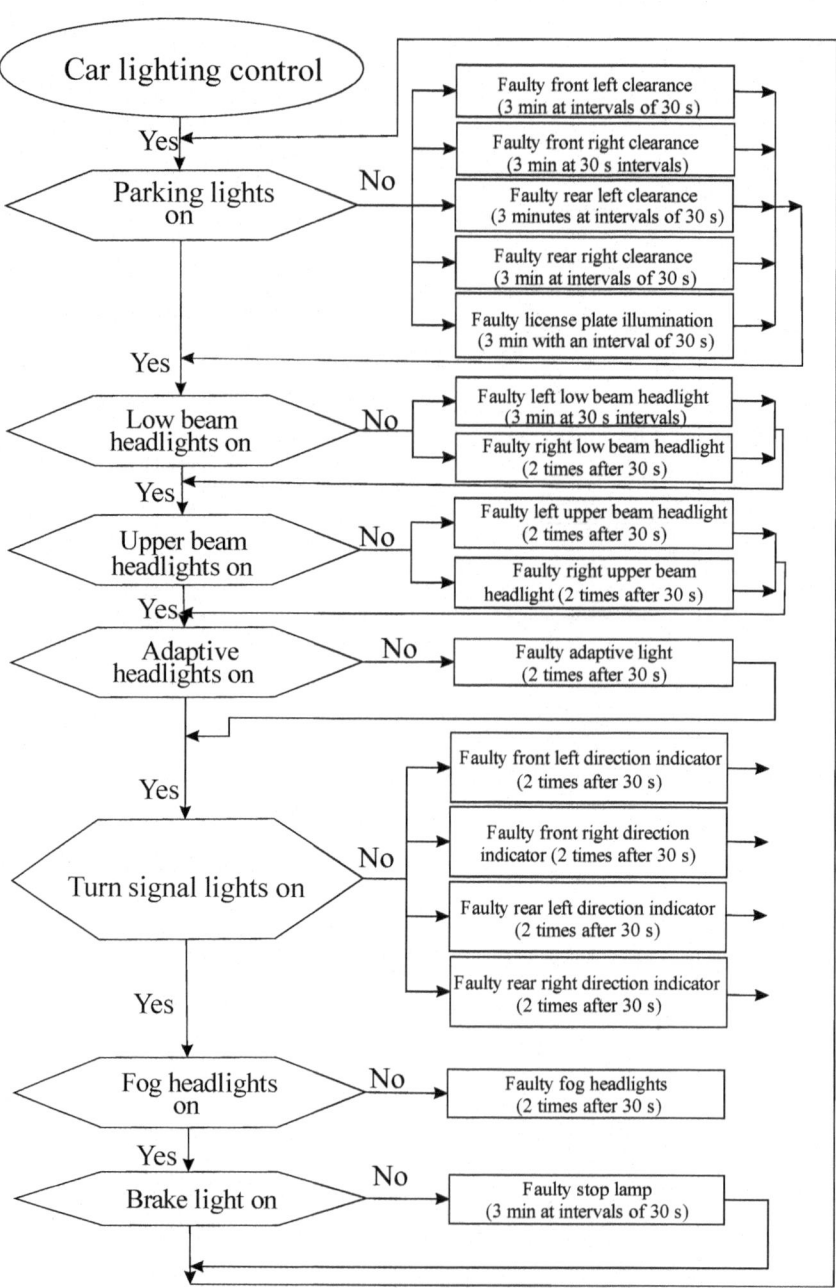

Fig. 4. Control circuit of the car lighting system.

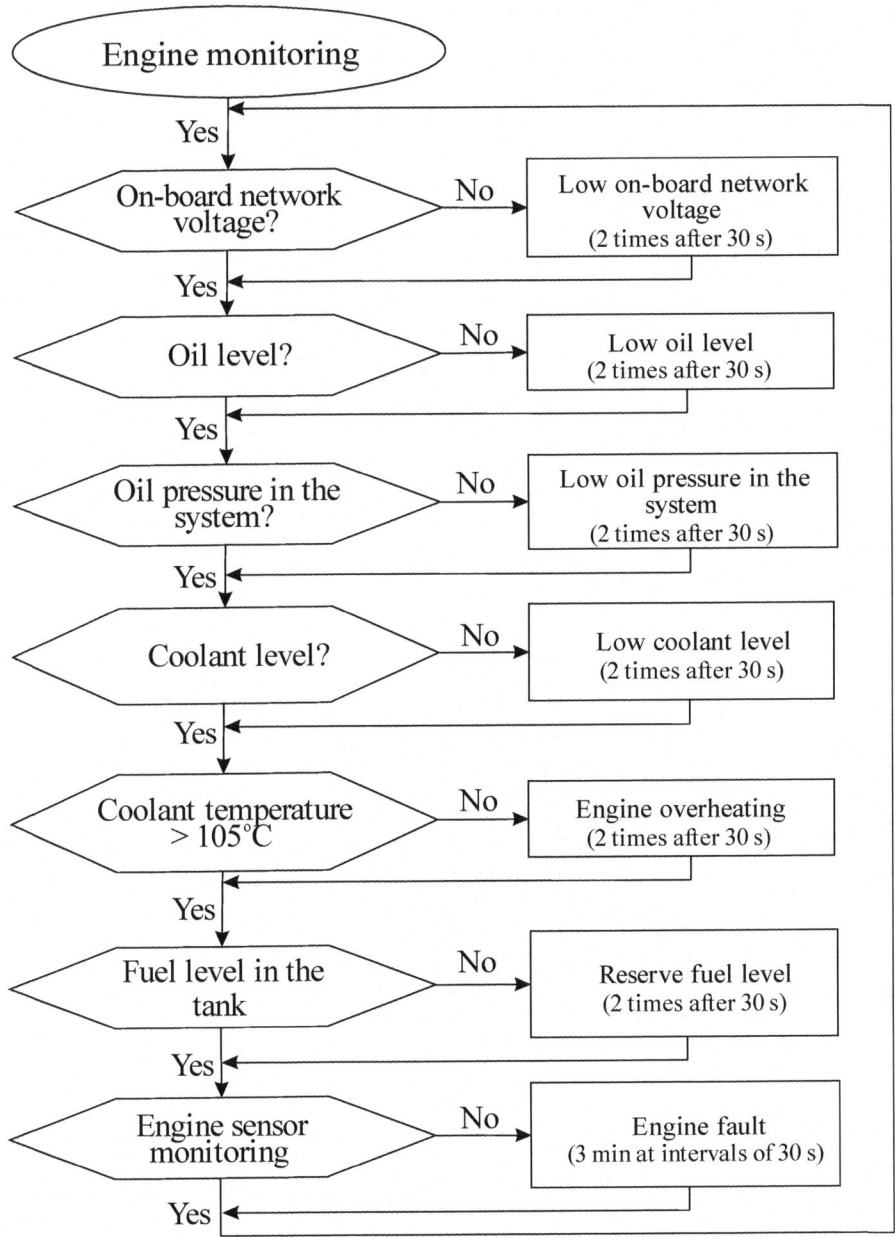

Fig. 5. Scheme of the control system for engine systems.

The module supports up to 25500 fragments of phrases, signals or melodies. All audio files can be divided into groups of 255 tracks. It is possible to choose one of 30 volume levels and one of 6 equalizer modes.

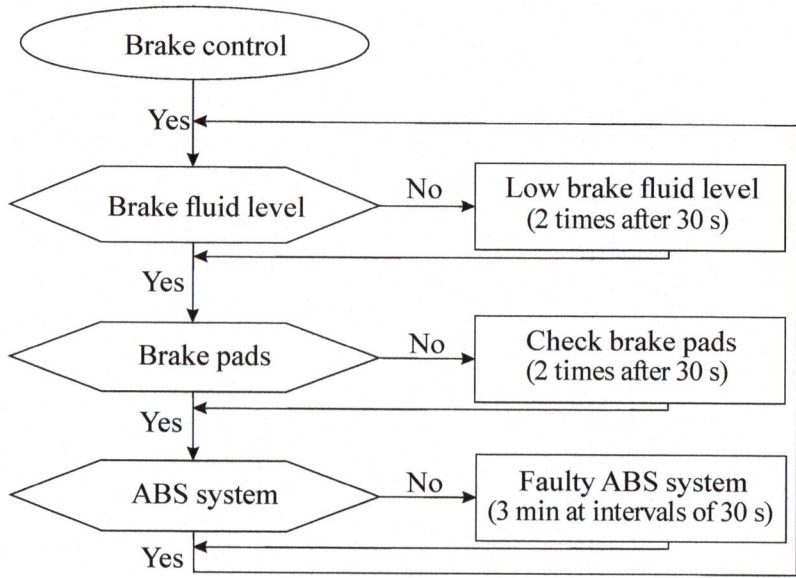

Fig. 6. Scheme of the brake control program.

Fig. 7. DFPlayer mini module.

If it is necessary to use the module as a regular player, it is not necessary to allocate additional digital inputs for the control buttons. The module has two inputs to which it is possible to connect up to 20 buttons. Pinout configuration in the DFPlayer mini module is shown in Fig. 8 and in Table 1.

DFPlayer can operate both offline and under the control of Arduino. In autonomous mode, FDPlayer is able to play files recorded on a micro SD card to the speaker connected to it. The FDPlayer connection diagram for autonomous operation is shown in Fig. 9.

A short press of the S2 button will switch to the next file, while a long press will increase the volume until it reaches its limit.

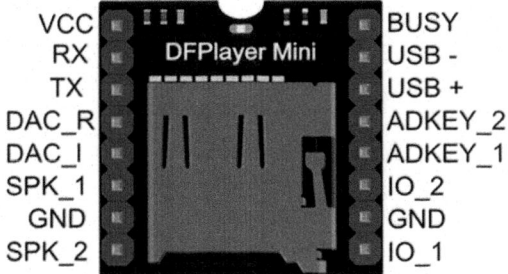

Fig. 8. Pinout configuration in the DFPlayer mini module.

Table 1. Pinout configuration in the DFPlayer mini module.

Pinout	Description
VCC	Power «+»
GND	Power «−»
RX	UART reception
TX	UART transmission
SPK1	Speaker «+»
SPK2	Speaker «−»
BUSY	Status indicator («0» – downtime, «1» – play)
DAC_R	Headphone or amplifier output («R» channel)
DAC_L	Headphone or amplifier output («L» channel)
IO1	Control input: short press – "back", long press – decrease volume
IO2	Control input: short press – "forward", long press – increase volume
ADKEY1	Port for connecting buttons and resistors, input 1
ADKEY2	Port for connecting buttons and resistors, input 2
USB+	USB port, port «+»
USB−	USB port, port «−»

A short press of the S1 button will switch to the previous file, while a long press will decrease the volume until it reaches its minimum.

It is possible to connect additional buttons, which provides a whole series of possibilities. To connect the buttons, there are two inputs ADKEY1 and ADKEY2, each of which can connect 10 buttons. The buttons are connected through resistors of different denominations, and the inputs ADKEY1 and ADKEY2 in the module itself are connected through resistors to the plus of the power source. Thus, when any of the buttons is pressed, a divider is formed, the voltage value from the output of which is a signal of which button is pressed.

When working with Arduino, the player's capabilities significantly increase, and the main ones are the ability to synthesize sound phrases from separate previously recorded words, as well as the ability to pronounce the desired phrase at the required time.

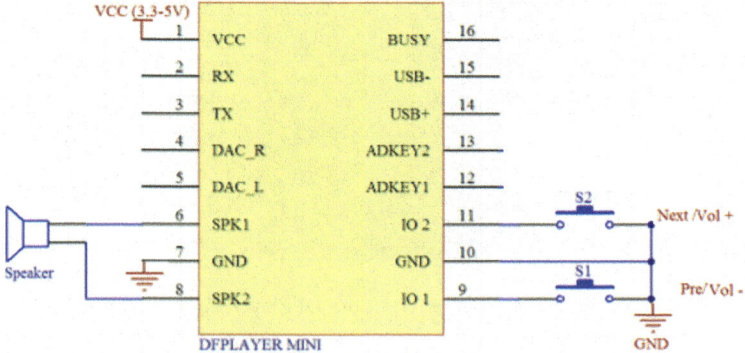

Fig. 9. Connections when using the player autonomously.

The connection diagram of the player to the Arduino Nano is shown in Fig. 10.

The simplest option for the player to work with Arduino is to emulate the contacts of the buttons in the circuit shown in Fig. 9. To do this, the player's output is connected to the Arduino digital output and uses the simplest sketch shown below, in which the button is emulated by the digital Write (7, LOW) command. When using this sketch, it is necessary to connect the ADKEY2 output of the player through a 33 kΩ resistor to the D7 output of the Arduino. As a result, record 11 will be played three times.

```
Skeych_jun17a
int F=7;
int i=0;
void setup() {
 pinMode (7,OUTPUT);
}
void loop() {
 if (i<=2){
    i=i+1;
    digitalWrite (7,LOW);
    delay (100);}
 digitalWrite (7,HIGH);
 delay (6000);
}
```

As indicated earlier, a microSD card, or a standard flash drive, or a USB hard drive can be used as a drive that can be used in conjunction with a DFPlayer player. Up to 25,500 files can be recorded on the media used. By default, the player can work with a microSD card. The player control system searches for the file to be played by its physical location, i.e. serial number of the record. Therefore, to work in conjunction with Arduino, a special order of recording files must be observed (when using the player offline, it is not necessary to adhere to this order).

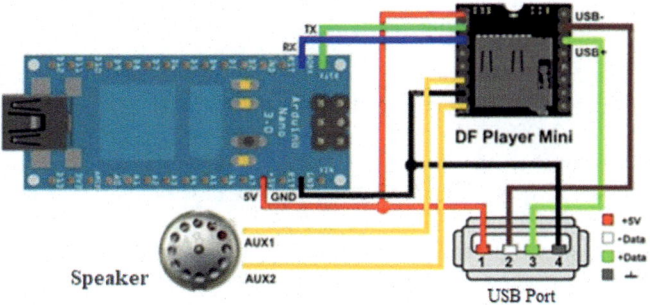

Fig. 10. Player connection to the Arduino Nano.

There are two possible recording systems. The first system provides for the creation in the root directory of the recording medium of a folder with the name mp3 into which files should be written. The file name must begin with the serial number of the entry: 0001xxxx.mp3. Thus, 9999 files can be recorded in a folder. After the serial number, any file name can be written, as it is ignored by the system.

The second system allows to increase the number of stored files. It is possible to create up to 100 folders in the root directory, each of which can store up to 255 files.

Files intended for playback can be obtained from the Internet, copied from other media or recorded independently.

When creating audio files yourself, recording is done from a microphone using standard computer tools or using one of many specialized programs, one of which is i-Sound WMA MP3 Recorder Pro.

3   Conclusion

The analysis of vehicle safety from the point of view of the possibilities of increasing it allows to propose criteria for assessing active, passive, post-accident and environmental safety.

An analysis of options for improving the safety of car operation makes it possible to identify as one of the possible ways to enhance informative safety, that is, providing the necessary information to the driver and other participants in the movement.

To develop and implement a program for monitoring vehicle systems, an automobile voice informant can be used, in particular, made on the basis of the Arduino controller.

The development of the technology for the development of Arduino controller-based voice informants can be used in the process of training students in the automotive industry while studying the issues of ensuring vehicle safety.

References

1. Gaylor, L., Junge, M., Abanteriba, S.: Effectiveness of vehicle passive safety systems in lateral fixed-object collisions. Int. J. Veh. Saf. **10**(3–4), 195–211 (2018). https://doi.org/10.1504/IJVS.2018.097705
2. Dimian, M., Chassagne, L., Andrei, P., Li, P.: Smart technologies for vehicle safety and driver assistance. J. Adv. Transp. **2019**, 2690498 (2019). https://doi.org/10.1155/2019/2690498
3. Hilmann, J.: On the development of a process chain for structural optimization in vehicle passive safety. Dissertation, Technical University Berlin (2009)
4. Kamizono, Y., Onoye, T., Kobayashi, W., et al.: Evaluation of auditory signals in an automobile for safe driving. In: 2019 International Symposium on Multimedia and Communication Technology (ISMAC), Quezon City, pp. 1–5. IEEE (2019). https://doi.org/10.1109/ISMAC.2019.8836140
5. Balogh, R., Lipková, M., Lučkanič, V., Ťapajna, P.: Natural notification system for the interior of shared car. IFAC-PapersOnLine **52**(27), 175–179 (2019). https://doi.org/10.1016/j.ifacol.2019.12.752
6. Samara-Stavr company: Voice informant Gamma GF 820. https://samara-stavr.ru/Rechevoj_informator_Gamma_GF_820.htm. Accessed 13 Aug 2020. (in Russian)
7. Kushner, D.: The making of Arduino. IEEE Spectrum (2011). https://spectrum.ieee.org/geek-life/hands-on/the-making-of-arduino. Accessed 13 Aug 2020
8. Pal, P., Gupta, R., Tiwari, S., Sharma, A.: IoT based air pollution monitoring system using Arduino. Int. Res. J. Eng. Technol. **4**(10), 1137–1140 (2017)
9. Pushpavalli, M., Sivagami, P., Sindhuja, S.: Vehicle quality check test bench using Arduino. In: 2018 IEEE International Conference on Computational Intelligence and Computing Research (ICCIC), Madurai, pp. 1–4. IEEE (2018). https://doi.org/10.1109/ICCIC.2018.8782403
10. Seelam, K., Lakshmi, C.J.: An Arduino based embedded system in passenger car for road safety. In: 2017 International Conference on Inventive Communication and Computational Technologies (ICICCT), Coimbatore, pp. 268–271. IEEE (2017). https://doi.org/10.1109/ICICCT.2017.7975201

Oil Products Moisture Measurement Using Adaptive Capacitive Instrument Measuring Transducers

Oleksandr Zabolotnyi$^{(\boxtimes)}$ ⓘ, Vitalii Zabolotnyi ⓘ, and Nikolay Koshevoy ⓘ

National Aerospace University "Kharkiv Aviation Institute", 17 Chkalova Street, Kharkiv 61070, Ukraine
`pretorian14@ukr.net`

Abstract. When measuring moisture content in oil products using capacitive moisture meters we face with method error named 'type uncertainty', caused by different values of dielectric permittivity for different oil products in dehydrated state, which depend from geographical origin, processing conditions etc. and can be hardly predicted automatically. The main task of the research is to reduce this type of method error by developing special measuring instruments. A prototype product of the instrument measuring transducer had been developed and experimentally tested together with the special method of moisture measurement which includes two additive, two multiplicative and two complementary testing influences on the substance under research. Moisture content with nominal values 0%, 10%, 20% and 30% was reproduced by two oil products with different dielectric permittivity values: transmission oil ($\varepsilon = 2.01$) and mazut ($\varepsilon = 2.67$). Experimental setup of an adaptive moisture meter was assembled using the substitution method of measurement to provide good accuracy for the conditions of capacitance measurement in substances with significant dielectric losses.

Keywords: Moisture measurement · Oil products · Instrument measuring transducer · Type uncertainty method error · Testing influence

1 Introduction

Among different methods, applied to solve the task of fuel and energy resources economy both in industry and transport, and simultaneously improve environmental indices of appropriate power equipment, is a method of oil-water emulsions burning [1]. Volume part of water in such emulsions can change from 8 to 20% to provide stable burning process.

For different boiler rooms and internal combustion engines fuel efficiency is critical and emission requirements of NO_x, CO_2 and other harmful emissions have become more and more stringent. Using emulsion (emulsified) fuel is considered as an effective way to enhance fuel efficiency and reduce harmful emissions [2]. Emulsion fuel is a blend of immiscible oil and water with surfactant agents. Due to the distinct physical properties of the oil and water such as the boiling point, emulsion fuels may show

© The Author(s), under exclusive license to Springer Nature Switzerland AG 2021
M. Nechyporuk et al. (Eds.): ICTM 2020, LNNS 188, pp. 81–91, 2021.
https://doi.org/10.1007/978-3-030-66717-7_7

particular physical phenomena when injected into the combustion chamber, such as microexplosion, which can be another dominant secondary breakup mechanism and play an important role in accelerating spray atomization. Explosive boiling occurs in a very short time and causes breakup of the parent oil droplet [3].

Emulsion fuel burning helps to save up to 10% of pure fuel. Another important benefit is that generation of evaporated water vapor reduces the local flame temperature, thereby reduces the emissions of NO_x. Addition of OH radicals due to evaporated water vapor also helps soot reduction [2].

Usually emulsion fuel is prepared by direct uninterrupted water addition into the fuel with further careful mixing to create stable substance with small diameter of water particles. Or, if the fuel already contains some amount of water, it can be enough to provide thorough mixing without water addition. Moisture content in both cases is an object of strict control [4–7].

Dielcometer (capacitance) principle of moisture measurement remains a forward-looking among all indirect methods, as its technical embodiment provides high operation speed and good accuracy, automation of measurements and data processing.

Problem Statement. Dielcometer principle of moisture measurement is based on measuring materials' dielectric properties with a help of capacitive instrument measuring transducer. A lot of researches and developers choose capacitive principle of measurement because of its relative simplicity and possibility to define moisture content as a difference of water and researched materials' dielectric permittivity.

But the problem of this method is significant variation of dielectric permittivity values in different mediums, which contain water. It makes strong influence on the result of moisture measurement when moisture meter had been calibrated for a certain type of fuel. An error, created by dielectric permittivity variation, especially when measuring small moisture values (for example in a range from 0 to 5%), can overcome the range of measurement for more than 2 times. The value of this error decreases if we increase the range of measurement, but remains significant.

When we have practical measurements, change of dielectric permittivity for different oil products used as fuels, causes method "type uncertainty" error if we have a scale of a meter, graduated only for one dielectric permittivity value. Its value can be estimated using formula:

$$\delta_m = \frac{\Delta \varepsilon_2}{\varepsilon_2} \left(1 - \frac{3\varepsilon_1 \varepsilon_2}{(\varepsilon_1 + 2\varepsilon_2)(\varepsilon_1 - \varepsilon_2)} \right) \approx \frac{\Delta \varepsilon_2}{\varepsilon_2},$$

where ε_1 is the dielectric permittivity of water; ε_2 is the dielectric permittivity of oil; $\Delta \varepsilon_2$ is the variation of oils' dielectric permittivity.

Mainly because of that further improvement of the existing methods of moisture measurement with a task to solve the problem of method error compensation is a relevant and perspective mission.

Main task of the research is to develop capacitive instrument measuring transducer, which design would provide both rapid measurements and versatility for the wide range of oil fuels without a necessity to store big number of calibrating characteristics in measuring instruments' memory. Adaptive instrument measuring transducer should provide a possibility of 'type uncertainty' method error reduction due to adaptability to different fuels with different values of dielectric permittivity.

2   Materials and Methods

One of the perspective directions had been detected to reduce 'type uncertainty' method error. The idea was to use special testing methods that allow increase the accuracy of measurements. Essentiality of these methods consists of determining the transducers' static function with a help of additional tests, functionally connected with moisture content. For the moisture control purposes testing actions should be formed as a number of water injections into the substance under consideration. Using the values of dielectric permittivity after each testing action it is possible to calculate the initial moisture content of a substance. The authors explored various testing algorithms and detected a perspective combination, consisting of independent additive and multiplicative tests [8].

Main idea of the suggested method is described below. At first we take a sample of a substance under research and get first reading from the capacitive initial transducer C_1. After that some quantity of water should be added into that sample (for example it can be $\Delta W = 10\%$ of water) to create first additive test and get a second reading C_2 from the initial transducer. At the third step value of capacitance C_1 should be increased two times to get the first multiplicative test and a third reading C_3 from the initial transducer. Step number four needs the capacitance value C_2 to be multiplied on two to get reading number four C_4. Next step foresees adding double amount of water $\Delta W' = 20\%$ into initial sample under research to form second additive test and reading number five – C_2'. Step number six means four times increasing the capacitance of the first reading C_1 to get reading number six C_3' as a second multiplicative test. To get the last reading C_4' (number seven) it is necessary to provide four times increasing the capacitance of the fifth reading C_2' [9].

To get the moisture value of the material under research all five capacitance readings should be substituted into the formula:

$$W = \left| \frac{(C_3'-C_1)\cdot\Delta W'}{(C_4'-C_2')-(C_3'-C_1)} - \frac{(C_3-C_1)\cdot\Delta W}{(C_4-C_2)-(C_3-C_1)} +0.033 \right| \cdot 833$$
$$+ \left[\left| \left(\frac{\Delta W'(C_3'-C_1)}{(k'-1)(C_2'-C_1)} - \frac{\Delta W(C_3-C_1)}{(k-1)(C_2-C_1)} \right) \right| - 0,033 \right] \cdot 800. \tag{1}$$

2.1   First Prototype Product of the Capacitive Instrument Measuring Transducer

Instrument measuring transducer consists of a system of flat electrodes 1 (Fig. 1). System 1 is fixed on the internal surface of two dielectric rings 3 with a help of dielectric joints 2. Each flat electrode is fastened on two corresponding dielectric joints.

Each joint is fixed on the external surface of axes 4 with small radius and on the internal surface of a peer of two dielectric rings 3. Another system of dielectric joints 5

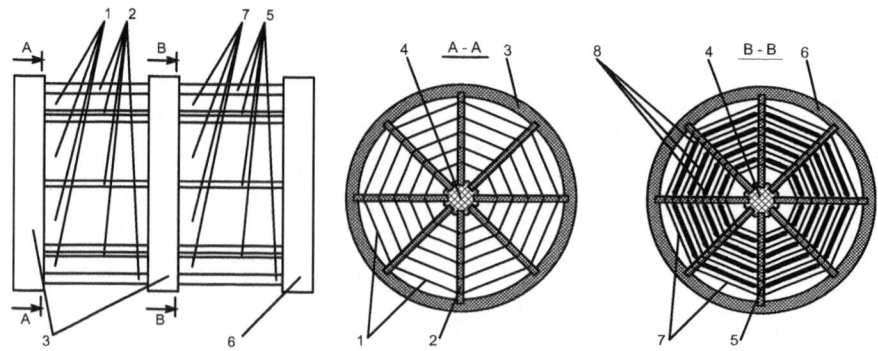

Fig. 1. Design of the first capacitive instrument measuring transducer.

is fixed on the external surface of axes 4 with small radius and on the internal surface of a peer of one dielectric ring 3 and dielectric ring 6. Second system of flat electrodes 7, identical to system 1, is assembled on the second system of dielectric joints 5. Between each peer of systems' 7 electrodes we have a capsule 8 of appropriate size filled with water.

Dielectric permittivity of a substance under research depends from presence of moisture, so, electric capacitances of all capacitive sensors would change in correspondence with water content. Presence of capsules 8 in the gap between the electrodes of system 7 would increase its capacitance in comparison with electrodes of system 1. By measuring electric capacitances of both systems it is possible to define moisture content of a substance under research.

Main advantage of such an instrument transducer is a possibility to create one additive testing influence on a substance under research without direct water addition. If more testing influences is necessary (in accordance with (1) at least one more additive test is necessary), we can increase number of sections.

But among disadvantages we have a necessity to use hermetic capsules with water, what makes the construction of an instrument measuring transducer more complicated from technological point of view.

That's why further attempts of improvement and simplification had been done.

2.2   Second Prototype Product of the Capacitive Instrument Measuring Transducer

Instrument measuring transducer consists of the system of electrodes 1 that have V-type shape (Fig. 2). System of electrodes 1 is fixed on the internal surface of two dielectric rings 2 and 3. Transducer is also provided with two internal dielectric rings 4 and 5 to ease flat metallic plates introduction into the gap between electrodes. Testing influences are provided by introduction of flat metallic plates 6 and 7 into the gap between electrodes of system 1 instead of capsules with water (Fig. 2).

If we need to provide two additive testing influences on a substance under research, par example it can be 10% and 20% of water content, instrument measuring transducer

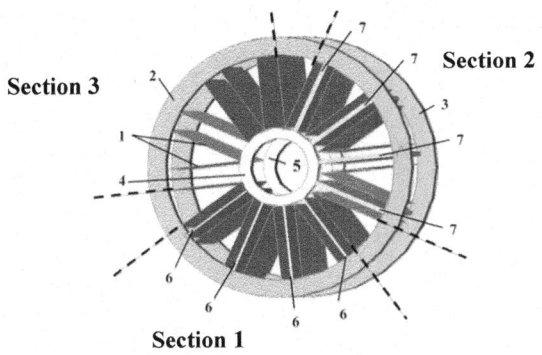

Fig. 2. 3D image of the capacitive instrument measuring transducer.

can be divided into three measuring sections, as it's done on Fig. 2. Into the gap between section 1 electrodes appropriate quantity of flat plates 6 of a certain thickness should be introduced. If we use universal Wiener equation [10] to calculate values of dielectric permittivity for water in oil emulsion, it would be possible to define that to simulate 10% of water addition it is necessary to increase section 1 electric capacitance on 27.64%. Knowing that, it is possible to calculate the thickness of metallic plates 6. In [11] it is told that capacitance C of a flat capacitor with a flat metallic plate placed accurately at the center of the gap between electrodes can be calculated using formula:

$$C = \frac{\varepsilon \varepsilon_0 S}{2d},$$

where d is a distance between the edge of the electrode and the edge of a plate inside the gap.

Using this formula, it is possible to calculate thickness of the metallic plates 6. After the performance of testing influence electric capacitance C should increase on 27.64% or in 1.2764 times. Gap between electrodes (designated like Z) should decrease on the same value. First testing influence can be fulfilled by placing four metallic plates 6 with thickness d_1 inside the gap between electrodes of section 1.

Section 2 implements second testing influence on a substance under research in a similar way, using metallic plates 7 with thickness d_2. As a result, electric capacitance C of four capacitors in section 2 should increase on 59.74% or in 1.5974 times.

Linear dimensions d_1 and d_2 can be calculated using formulas:

$$d_1 = Z - \frac{Z}{1.2764} = 0.2165 \cdot Z, \, d_2 = Z - \frac{Z}{1.5974} = 0.374 \cdot Z \tag{2}$$

After measuring signals from instrument measuring transducers' output we can get three values of dielectric permittivity: ε_1 is the dielectric permittivity of initial sample (taken from section 3); ε_2 is the dielectric permittivity of initial sample after first testing influence (taken from section 1); ε_3 is the dielectric permittivity of initial sample after second testing influence (taken from section 2).

Having these three values of dielectric permittivity, we can calculate moisture content of initial substance using formula (1). Electric capacitance C_1 can be formed by four capacitors of section 3, connected in parallel, electric capacitance C_2 – by four capacitors of section 1, and electric capacitance C_2' – by four capacitors of section 2. Rest of the capacitances: C_3, C_4 and C_3', C_4' can be calculated after the values C_1, C_2 and C_2'.

3 Theory/Calculation

To start calculations it's necessary to define linear dimensions of instrument measuring transducer on Fig. 2. Length of the electrode (distance between two dielectric rings 2 and 3) is equal to $l = 50$ mm; gap between two electrodes $Z = 20$ mm; internal diameter of dielectric rings 2 and 3 $D = 150$ mm; width of one electrode (distance between the rings 2 and 4) $L = 0,5\sqrt{D^2 - Z^2} - 1,866Z = 37$ mm.

In correspondence with formula (2) thickness of metallic plates, which simulate testing influences, would have values: $d_1 = 4.33$ mm, $d_2 = 7.48$ mm.

Then it is necessary to calculate electric capacitances of Sects. 1, 2 and 3 in empty state. Section 1 consists of four connected in parallel capacitors with a gap $Z_1 = Z - d_1$. Section 2 has four connected in parallel capacitors with a gap $Z_2 = Z - d_2$. And four capacitors of section 3 have a gap Z.

Electric capacitance of the instrument measuring transducer can be calculated using the method of calculating spatial characteristics of electric field, created by the electrodes of system 1. For section 1 it will be:

$$C_1 = \varepsilon_0\varepsilon[g_{01} + 2(g_{21} + g_{41} + g_{61} + g_{81} + 2g_{101} + 2g_{121})] \cdot 4. \tag{3}$$

In formula (3) we have (see Table 1): g_{01} is the spatial characteristic of basic electric field (inside the gap), m; g_{21} is the spatial characteristic of edge effects field in a form of half cylinder for the side l, m; g_{41} is the spatial characteristic of edge effects field in a form of half cylinder for the side L, m; g_{61} is the same in a form of half tube for the side l, m; g_{81} is the same in a form of half tube for the side L, m; g_{101} is the same in a form of spheric quadrants, m; g_{121} is the same in a form of quadrants of the spheric shells, m.

Now we have to check the workability of formula (1) for such conditions: dielectric permittivity of a substance under research $\varepsilon = 3.5$, moisture content of a substance $W = 0\%$. After that capacitances C_1, C_2, C_3, C_4, C_2', C_3', C_4' should be calculated.

Section 3 of instrument measuring transducer is free from testing influences, so:

$$C_{1(0\%)} = C_3\varepsilon_{(0\%)} = 4.87 \cdot 3.5 = 17.05 \ pF.$$

Section 1 is under the first testing influence (10% of water addition into the substance under research):

$$C_{2(10\%)} = C_1\varepsilon_{(0\%)} = 6.28 \cdot 3.5 = 21.98 \ pF,$$

Table 1. Calculation of spatial characteristics, used in formula (3).

Spatial characteristic	Formula	Result, m
Section 1		
g_{01}	$L \cdot l/(Z - d_1)$	0.1181
g_{21}	$0.26 \cdot l$	0.0130
g_{41}	$0.26 \cdot L$	0.0096
g_{61}	$\frac{l}{\pi} \ln\left(\frac{2m}{Z - d_1} + 1\right)$	0.0019 ($\rightarrow 0$)
g_{81}	$\frac{L}{\pi} \ln\left(\frac{2m}{Z - d_1} + 1\right)$	0.0014 ($\rightarrow 0$)
g_{101}	$0.077 \cdot (Z - d_1)$	0.0012 ($\rightarrow 0$)
g_{121}	$\frac{m}{4}$ (m is electrodes' thickness)	0.00025 ($\rightarrow 0$)
$C_1 = \varepsilon_0\varepsilon[g_{01} + 2(g_{21} + g_{41})] \cdot 4 = 6.28$ pF		
Section 2		
g_{01}	$L \cdot l/(Z - d_2)$	0.1478
g_{21}	$0.26 \cdot l$	0.0130
g_{41}	$0.26 \cdot L$	0.0096
$C_2 = \varepsilon_0\varepsilon[g_{01} + 2(g_{21} + g_{41})] \cdot 4 = 7.99$ pF		
Section 3		
g_{01}	$L \cdot l/Z$	0.0925
g_{21}	$0.26 \cdot l$	0.0130
g_{41}	$0.26 \cdot L$	0.0096
$C_3 = \varepsilon_0\varepsilon[g_{01} + 2(g_{21} + g_{41})] \cdot 4 = 4.87$ pF		

$$C_3 = k \cdot C_{1(0\%)} = 2 \cdot 17.05 = 34.10 \text{ pF},$$

$$C_4 = k \cdot C_{2(10\%)} = 2 \cdot 21.98 = 43.96 \text{ pF}.$$

In these formulas $k = 2$ is a multiplication coefficient of the first multiplicative test. Section 2 performs second testing influence (20% of water addition):

$$C_2' = C_2\varepsilon_{(0\%)} = 7.99 \cdot 3.5 = 27.97 \text{ pF},$$

$$C_3' = k' \cdot C_{1(0\%)} = 4 \cdot 17.05 = 68.20 \text{ pF},$$

$$C_4' = k' \cdot C_2' = 4 \cdot 27.97 = 111.86 \text{ pF}.$$

Here $k' = 4$ is a multiplication coefficient of the second multiplicative test.

After substituting these values of capacitances into formula (1) we get moisture content value $W = 0.77\%$ (in ideal case it should be $W = 0\%$), what can be an acceptable method error of formula (1).

Now workability of formula (1) will be checked for $W = 20\%$ of moisture content and the same value of dielectric permittivity $\varepsilon = 3.5$. Capacitances C_1, C_2, C_3, C_4, C_2', C_3', C_4' should be calculated again.

So, section 3 now is filled with a moist substance ($W = 20\%$):

$$C_{1(20\%)} = C_3\varepsilon_{(20\%)} = 4.87 \cdot 5.741 = 27.96 \text{ pF}.$$

In the first section complementary 10% of water should be added to the substance:

$$C_{2(20\%)} = C_1\varepsilon_{(20\%)} = 6.28 \cdot 5.741 = 36.05 \text{ pF},$$

$$C_3 = k \cdot C_{1(20\%)} = 2 \cdot 27.96 = 55.92 \text{ pF},$$

$$C_4 = k \cdot C_{2(20\%)} = 2 \cdot 36.05 = 72.11 \text{ pF}.$$

Third section would contain 40% of moisture in total:

$$C_2' = C_2\varepsilon_{(20\%)} = 7.99 \cdot 5.741 = 45.87 \text{ pF},$$

$$C_3' = k' \cdot C_{1(20\%)} = 4 \cdot 27.96 = 111.84 \text{ pF},$$

$$C_4' = k' \cdot C_2' = 4 \cdot 45.87 = 183.48 \text{ pF}.$$

If we substitute new values of C_1, C_2, C_3, C_4, C_2', C_3', C_4' capacitances into formula (1), we would get the value of moisture content W = 20.157% (ideal value is W = 20%). It makes possible to say that formula (1) is still workable.

4   Experiments

A prototype product of instrument measuring transducer, created in accordance with an image on Fig. 2 and mentioned above linear dimensions, was developed (Fig. 3). It performs simulation of two additive testing influences on the oil product under research.

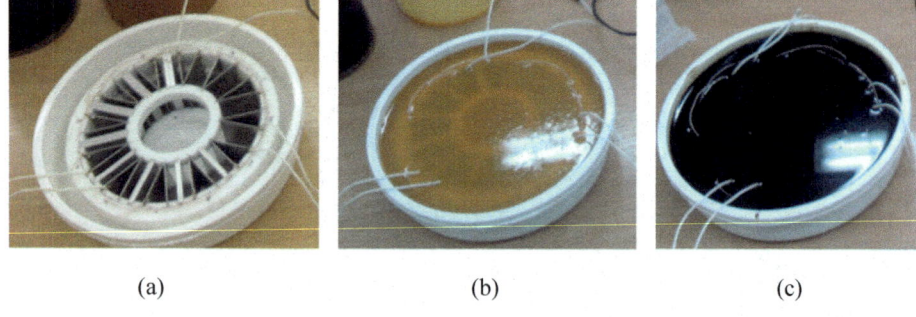

(a)   (b)   (c)

Fig. 3. Prototype product of the capacitive instrument measuring transducer: (a) – in empty state, (b) – filled with transmission oil ($\varepsilon = 2.01$), (c) – filled with mazut ($\varepsilon = 2.67$).

We can see experimental setup on Fig. 4. It has a container with instrument measuring transducer, secondary measuring transducer, oscilloscope Tektronix 2213A to control the shape and duration of pulses, taken from secondary measuring transducer's check point, variable air capacitor, digital multimeter UTM18803 to measure dc voltage on the output of a secondary transducer and accurate RLC-meter UTM 1612 to measure capacitance value of the variable air capacitor.

Fig. 4. Experimental setup for moisture measurement.

Process of measurement consisted of several steps. Container with instrument measuring transducer was filled with substance under research. Then capacitive sensors (three sections) of the instrument measuring transducer were connected one by one to the input of the secondary measuring transducer.

Secondary measuring transducer converts values of C_1, C_2 and C_3 electrical capacitance into corresponding dc voltages and consists of two 555 or 777 interval timers, where one of them is connected as a multivibrator and generates rectangular pulses. Other one, to which three sections of instrument measuring transducer are connected one by one, operates like biased multivibrator, and duration of its output pulses is in straight proportion with the capacitance of instrument transducer section. Low-pass filter on the biased multivibrators' output operates like pulse duration into dc voltage converter in a range from 0,000 to 2,000 V [9].

As it was written above, three capacitive sensors were connected one by one to the input of secondary measuring transducer and three values of dc voltage from its output were fixed. After that instrument measuring transducer was substituted by variable air capacitor. Using method of substitution electric capacitances of each instrument measuring transducers' section were defined as an average of 10 measurements.

5   Results/Discussion

Results of capacitance C_1, C_2, C_3, C_4, C_2', C_3', C_4' measurements in picofarads, fulfilled with a help of UTM 1612 RLC-meter on 10 kHz frequency, are given in Table 2.

Values of moisture content, received after substitution capacitance values from Table 2 into formula (1), are placed in Table 3.

Table 2. Results of capacitance measurement.

Substance	W, %	$\overline{C}_1$	$\overline{C}_2$	$\overline{C}_3$	$\overline{C}_4$	$\overline{C}'_2$	$\overline{C}'_3$	$\overline{C}'_4$
Transmission oil	0	30,35	39,35	60,70	78,70	51,61	121,40	206,44
	10	39,18	46,41	78,36	92,82	54,71	156,72	218,84
	20	50,50	59,82	101,00	119,64	70,82	202,00	283,28
	30	64,82	76,50	129,64	153,00	90,73	259,28	362,92
Mazut	0	40,56	52,57	81,12	105,14	67,41	162,24	269,64
	10	52,56	62,28	105,12	124,56	73,47	210,24	293,88
	20	67,42	79,72	134,84	159,44	94,30	269,68	377,20
	30	85,80	101,18	171,60	202,36	119,82	343,20	479,28

Table 3. Moisture content values.

Substance	Moisture content W, %			
	0	10	20	30
Transmission oil	0,049	9,207	20,370	30,022
Mazut	0,003	9,665	19,458	28,746

It was possible to estimate the method part of moisture measurement absolute error by using average absolute error $\overline{\Delta}_m$ as a measure of distance between standard and measured values of moisture content: $\overline{\Delta}_m = \frac{1}{8}\sum |W_{mi} - W_{si}|$.

$$\overline{\Delta}_m = \frac{|0.049-0|}{8} + \frac{|9.207-10|}{8} + \frac{|20.370-20|}{8} + \frac{|30.022-30|}{8}$$
$$+ \frac{|0.003-0|}{8} + \frac{|9.665-10|}{8} + \frac{|19.458-20|}{8} + \frac{|28.746-30|}{8} = 0.42 \%.$$

It makes possible to see that substitution of direct water additives with metallic plates of different thickness, which must be introduced into a gap between electrodes, provides good measurement accuracy of the moisture content measuring instrument.

6 Conclusions

In the described instrument measuring transducer we have metallic plates introduced into the gap between electrodes instead of capsules with water or direct water addition. It was divided into three sections. First section is free from testing influences on the material under research, in second section 10% of water addition is simulated by increasing its capacitance into 27,64%. Third section simulates 20% addition of water and, to provide it, electrical capacitance of this section should be increased in 59,74%. In accordance with that thickness of metallic plates and electrical capacitances of three empty (without a material) sections were calculated. In the process of experimental researches two oil products with different dielectric permittivity vales had been used: transmission oil ($\varepsilon = 2.01$) and mazut ($\varepsilon = 2.67$). Experimental setup was assembled using the substitution method of measurement to provide acceptable accuracy for the conditions of capacitance measurement in substances with significant dielectric loss. Results of moisture measurement have good accuracy (average absolute error has a value of $\overline{\Delta}_m = 0.42\,\%$).

References

1. Dryer, F.L.: Water addition to practical combustion systems – concepts and applications. Symp. (Int.) Combust. **16**(1), 279–295 (1977). https://doi.org/10.1016/S0082-0784(77)80332-9
2. Kadota, T., Yamasaki, H.: Recent advances in the combustion of water fuel emulsion. Prog. Energy Combust. Sci. **28**(5), 385–404 (2002). https://doi.org/10.1016/S0360-1285(02)00005-9
3. Shinjo, J., Xia, J., Ganippa, L.C., Megaritis, A.: Physics of puffing and microexplosion of emulsion fuel droplets. Phys. Fluids **26**(10), 103302 (2014). https://doi.org/10.1063/1.4897918
4. Zabolotny, A.V., Koshevoi, M.D.: Improving efficiency of the quality control of substances with dielectric properties. Telecommun. Radio Eng. **57**(2–3), 177–190 (2002). https://doi.org/10.1615/TelecomRadEng.v57.i2-3.200
5. Aints, M., Paris, P., Tufail, I., et al.: Determination of the calorific value and moisture content of crushed oil shale by LIBS. Oil Shale **35**(4), 339–355 (2018). https://doi.org/10.3176/oil.2018.4.04
6. Sharma, P., Yeung, H.: Recent advances in water cut sensing technology. In: Yurish, S.E. (ed.) Advances in Measurements and Instrumentation: Reviews, vol. 1, pp. 147–175. IFSA Publishing, Barcelona (2018)
7. Garvey, R., Fogel, G.: Estimating water content in oils: moisture in solution, emulsified water, and free water. In: Technology Showcase: Integrated Monitoring, Diagnostics and Failure Prevention. Proceedings of a Joint Conference, Mobile, Alabama, pp. 1–14, 22–26 April 1996
8. Zabolotnyi, O.V., Zabolotnyi, V.A., Koshevoi, M.D.: Conditionality examination of the new testing algorithms for coal-water slurries moisture measurement. Sci. Bull. Nat. Min. Univ. **1**, 51–59 (2018). https://doi.org/10.29202/nvngu/2018-1/21
9. Zabolotnyi, O.V.: Proximate testing method of moisture measurement for substances of dielectric nature. Radio Electron. Comput. Sci. Control **1**, 7–17 (2019). https://doi.org/10.15588/1607-3274-2019-1-1
10. Josh, M., Clennell, B.: Broadband electrical properties of clays and shales: comparative investigations of remolded and preserved samples. Geophysics **80**(2), 129–143 (2015). https://doi.org/10.1190/GEO2013-0458.1
11. Forejt, J.: Kapacitni merice neeekltckych velicin. SNTL, Czech Republic (1963)

Sorption-Capacitive Gas Humidity Sensor of Increased Sensitivity

Oleksandr Zabolotnyi$^{(\boxtimes)}$ and Maksym Sukhobrus

National Aerospace University "Kharkiv Aviation Institute",
17 Chkalova Street, Kharkiv 61070, Ukraine
pretorian14@ukr.net

Abstract. A new sorption-capacitive gas humidity sensor was suggested. It's technical composition provided better sensitivity in comparison with closest analogues. Calculations of initial capacitance and sensitivity of a new sensor and two closest analogues were done. Theoretical calculations confirm significant advantage in sensitivity for suggested humidity sensor. Three prototype products of capacitive humidity sensors were manufactured to fulfill experimental researches. Static generator of humid gas with salt solutions was used to reproduce values of relative air humidity in a range from 33% to 93%. Results of experiments confirmed the results of theoretical researches and proved that new humidity sensor has two times better sensitivity in comparison with two closest analogues.

Keywords: Relative humidity · Sorption-capacitive sensor · Silk insulation · Sensitivity · Static generator of humid gas

1 Introduction

Gas humidity is an important parameter of control during the process of natural gas extraction, processing and transportation. Natural gas that comes out from the well is usually saturated with water. To avoid pipelines' corrosion and gas hydrates formation, which destroy fittings and other gas network components, natural gas should be dehydrated to a certain level before its transportation. During the process of transportation humidity control is still necessary to prevent condensate appearance. At the same time final product, delivered to the consumer, should satisfy requirements of its specifications. So we can say that reliable and accurate gas humidity measurement is necessary for all stages of natural gas production.

Humidity control usually happens in strongly variable conditions of medium and faces with different hardships [1]. Among them we can emphasize measurements in a flow, in condition of pressure falls or low temperatures, etc., what requires high accuracy and sensitivity from the measuring instrument.

Humidity analyzers are in often use as a part of automated control systems, what forms complementary requirements to a high-speed performance and accuracy of modern humidity transducers and measuring instruments [2–8].

© The Author(s), under exclusive license to Springer Nature Switzerland AG 2021
M. Nechyporuk et al. (Eds.): ICTM 2020, LNNS 188, pp. 92–101, 2021.
https://doi.org/10.1007/978-3-030-66717-7_8

2 Problem Statement

When studying the problem of natural gas humidity measurement, some number of interesting technical solutions, connected with gas humidity instrument measuring transducers were detected [9, 10]. In that instrument measuring transducers single and stranded wires, covered with silk insulation, were applied as humidity sorbent. Silk, natural or rayon, used as an insulator, is a good sorbent of moist for different gaseous mediums. When relative humidity of natural gas changes from 0 to 100%, equilibrium moisture content of silk changes from 0 to 30% [11]. So, change of natural gas humidity causes proportional change of moisture content in silk insulation. Presence of moisture in silk changes it's dielectric permittivity. And if silk is used as dielectric between the electrodes of a capacitive sensor, presence of moisture will change its electric capacitance.

Mentioned above capacitive instrument measuring transducers seemed to be interesting because of their simple practical implementation and rather high sensitivity. Besides, it will be easy to provide good interchangeability and repeatability of multiple measurements. Technical composition of these capacitive gas humidity sensors is given on Fig. 1.

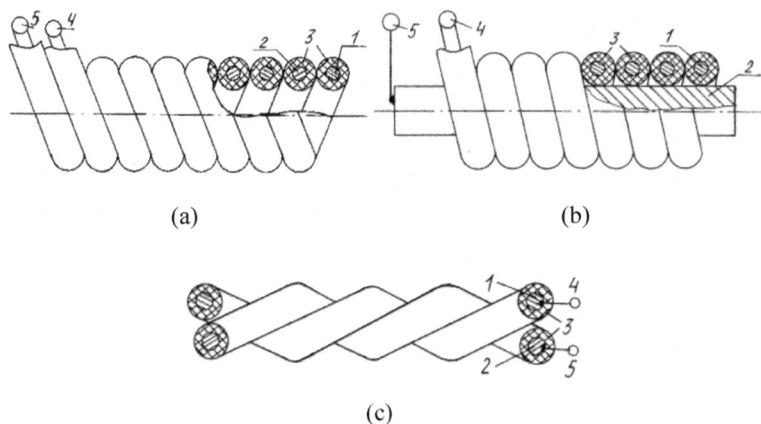

(a)   (b)

(c)

Fig. 1. Variants of technical composition for a capacitive gas humidity sensor: 1, 2 – electrodes, 3 – silk insulation, 4, 5 – clamps for connection with secondary transducer.

To provide theoretical calculations of electric capacitance for the variants of technical composition from Fig. 1, taken for identical linear dimensions, the method of calculating spatial characteristics of electric field had been applied. It allows relatively easy and accurate calculation of electric capacitance for the capacitive sensor with any configuration of electrodes.

Analyzing capacitive sensors from literary sources [9, 10] we managed to get an idea how to modify technical composition of present capacitive sensors and to provide approximately three times better sensitivity staying in current linear dimensions.

So, the main purpose of current research is to develop modified capacitive gas humidity sensor with better sensitivity, provide its calibration and fulfill experimental researches.

3 Materials and Methods

Suggested technical composition of a sorption-capacitive humidity sensor is given on Fig. 2.

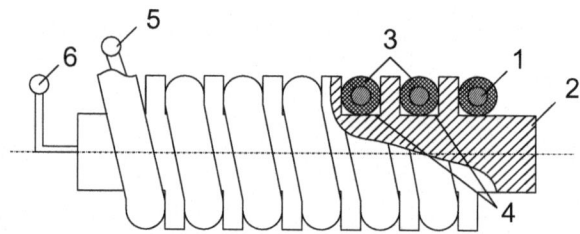

Fig. 2. New technical composition of a sorption-capacitive humidity sensor: 1 – first electrode; 2 – second electrode; 3 – silk insulation; 4 – rectangular slots; 5 – clamp of a first electrode; 6 – clamp of a second electrode.

Suggested sorption-capacitive humidity sensor consists of the first electrode 1 in a form of metallic wire and second electrode 2 in a form of metallic rod. Electrode 1 is covered with silk insulation 3 and placed in rectangular slots 4 of metallic rod 2. Clams 5 and 6 are necessary for connection with a secondary transducer of humidity measuring instrument. Placing the wire with silk insulation (electrode 1) in slots 4 of the second electrode increases the space between electrodes what in its turn provides significant increase of sensors' sensitivity without increasing linear dimensions [12].

For further theoretical and experimental researches a new sensors' prototype product was created together with prototype products of two analogues: number 1 (Fig. 1a) and number 2 (Fig. 1b). Images of all prototype products are given on Fig. 3.

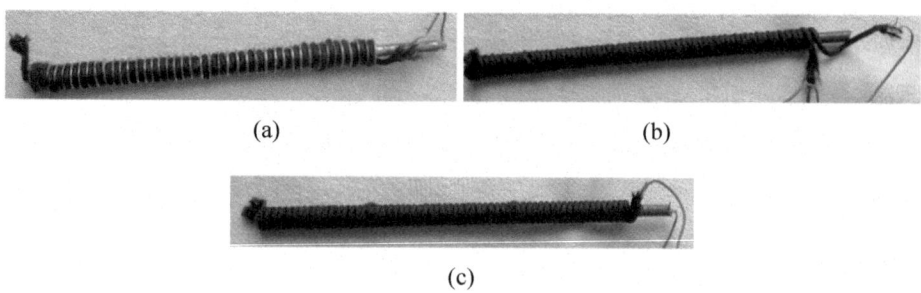

(a)   (b)

(c)

Fig. 3. Prototype products of three sorption-capacitive humidity sensors: (a) – prototype product of a new sensor; (b) – prototype product of the first analogue; (c) – prototype product of the second analogue.

Analogue c from Fig. 1 was excluded from further researches because of significantly smaller space between the electrodes and, therefore, evidently smaller sensitivity. Each prototype product was made from a standard wire with round cross-section, length $l = 500$ mm, diameter with insulation $d_1 = 1.6$ mm, without insulation $d = 0.6$ mm.

4 Theory/Calculation

In this section it was necessary to calculate electrical capacitance and sensitivity of modified sorption-capacitive humidity sensor. Sensors' electric capacitance can be calculated using formula (1):

$$C = \varepsilon_0 \varepsilon_n g_r,$$ (1)

where ε_0 is the dielectric constant ($\varepsilon_0 = 8,85 \cdot 10^{-12}$ F/m); ε_n is the dielectric permittivity of a medium in a space between electrodes; g_r is the spatial characteristic of electric field between two electrodes (basic electric field).

Mentioned above spatial characteristic of basic electric field would have a shape of parallelepiped without a half cylinder (Fig. 4a).

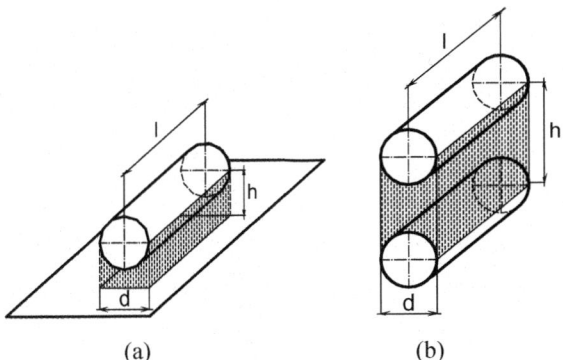

(a)   (b)

Fig. 4. Spatial characteristic of basic electric field: (a) – parallelepiped without a semicylinder; (b) – parallelepiped without two semicylinders.

We can calculate its value using formula:

$$g_r = 3 \cdot \frac{2\pi l}{ln\left(\frac{2h}{d} + \sqrt{\left(\frac{2h}{d}\right)^2 - 1}\right)} = 3 \cdot \frac{2 \cdot 3.14 \cdot 0.5}{ln\left(\frac{2 \cdot 0.0008}{0.0006} + \sqrt{\left(\frac{2 \cdot 0.0008}{0.0006}\right)^2 - 1}\right)} = 5.74 \text{ m},$$

where l is the length of the electrode; h is the distance between electrodes; d is the electrodes' diameter.

Dielectric permittivity of a medium silk – air in the gap between electrodes can be calculated with a help of general Lichterecker mixing rule [13]:

$$lg\varepsilon_{medium} = \beta_{silk} \cdot lg\varepsilon_{silk} + \beta_{air}lg\varepsilon_{air}, \tag{2}$$

where β_{silk} is the volume concentration of silk; ε_{silk} is the dielectric permittivity of silk; β_{air} is the volume concentration of air; ε_{air} is the dielectric permittivity of air.

Dielectric permittivity of a humid silk for maximal gas humidity (100% RH) was defined with a help of universal Wiener equation [14]:

$$\varepsilon_{humid\ silk} = \varepsilon_{silk} \cdot \left(1 + \frac{3W}{\frac{\varepsilon_{water} + 2\varepsilon_{silk}}{\varepsilon_{water} - \varepsilon_{silk}} - W}\right) = 4 \cdot \left(1 + \frac{3 \cdot 0.3}{\frac{80 + 2 \cdot 4}{80 - 4} - 0.3}\right) = 8.2. \tag{3}$$

Using formula (2) we can find dielectric permittivity of a medium silk – air in the gap between electrodes for dry and humid (100% RH) air:

$$\varepsilon_{medium}(0\%RH) = 10^{\beta_{silk} \cdot lg\varepsilon_{silk} + \beta_{air}lg\varepsilon_{air}} = 10^{0.75 \cdot lg4 + 0.25lg1} = 2.82, \tag{4}$$

$$\varepsilon_{medium}(100\%RH) = 10^{\beta_{silk} \cdot lg\varepsilon_{humid\ silk} + \beta_{air}lg\varepsilon_{air}} = 10^{0.75 \cdot lg8.2 + 0.25lg1} = 4.85. \tag{5}$$

By placing received values into formula (1), we can calculate electric capacitances of the gas humidity sensor from Fig. 2 or Fig. 3, a, for dry and humid (100% RH) air:

$$C(0\%RH) = \varepsilon_0\varepsilon_{medium}(0\%RH)g_r = 8.85 \cdot 10^{-12} \cdot 2.82 \cdot 4.78 = 143.25 \text{ pF},$$

$$C(100\%RH) = \varepsilon_0\varepsilon_{medium}(100\%RH)g_r = 8.85 \cdot 10^{-12} \cdot 4.85 \cdot 4.78 = 246.38 \text{ pF},$$

Sensitivity of a new gas humidity sensor has a value:

$$S = [C(100\%RH) - C(0\%RH)]/(W_{max} - W_{min})$$
$$= (246.38 - 143.25)/(100 - 0) = 1.03 \text{ pF}/\%,$$

After that it was necessary to fulfill similar calculations for two analogues from Fig. 1a, b. First of them (Fig. 1a or Fig. 3b) consists of two electrodes as two wires of round cross-section with silk insulation.

Spatial characteristic of basic electric field in a gap between two electrodes has a shape of a parallelepiped without two semicylinders (Fig. 4b). It was calculated using formula:

$$g_r = \pi l \left/ n\left(\frac{h}{d} + \sqrt{\left(\frac{h}{d}\right)^2 - 1}\right) \right. l = 3.14 \cdot 0.5 \left/ ln\left(\frac{0.0016}{0.0006} + \sqrt{\left(\frac{0.0016}{0.0006}\right)^2 - 1}\right)\right.$$
$$= 0.96 \text{ m}.$$

Using values, received in formulas (4), (5) we can calculate electric capacitances of the first analogue from Fig. 1a or Fig. 3b:

$$C(0\%\text{RH}) = \varepsilon_0 \varepsilon_{\text{medium}}(0\%\text{RH})g_r = 8.85 \cdot 10^{-12} \cdot 2.82 \cdot 0.96 = 23.96 \text{ pF},$$

$$C(100\%\text{RH}) = \varepsilon_0 \varepsilon_{\text{medium}}(100\%\text{RH})g_r = 8.85 \cdot 10^{-12} \cdot 4.85 \cdot 0.96 = 41.20 \text{ pF}.$$

Sensitivity of the first analogue has a value:

$$S = [C(100\%\text{RH}) - C(0\%\text{RH})]/(W_{max} - W_{min})$$
$$= (41.20 - 23.96)/(100 - 0) = 0.17 \text{ pF}/\%.$$

Image of the second analogue can be found on Fig. 1b and Fig. 3c. This humidity sensor is organized with first electrode as a simple metallic rod (position 2 on Fig. 1b) and second electrode (metallic wire 1 with silk insulation 3, winded up on a metallic rod 2). Length of metallic wire was 680 mm.

Spatial characteristic for the second analogue has a shape of parallelepiped without a half cylinder (Fig. 4a).

$$g_r = \frac{2\pi l}{ln\left(\frac{2h}{d} + \sqrt{\left(\frac{2h}{d}\right)^2 - 1}\right)} = \frac{2 \cdot 3.14 \cdot 0.68}{ln\left(\frac{2 \cdot 0.0008}{0.0006} + \sqrt{\left(\frac{2 \cdot 0.0008}{0.0006}\right)^2 - 1}\right)} = 2.60 \text{ m}.$$

Capacitance values for the second analogue:

$$C(0\%\text{RH}) = \varepsilon_0 \varepsilon_{\text{medium}}(0\%\text{RH})g_r = 8.85 \cdot 10^{-12} \cdot 2.82 \cdot 2.60 = 64.89 \text{ pF},$$

$$C(100\%\text{RH}) = \varepsilon_0 \varepsilon_{medium}(100\%\text{RH})g_r = 8.85 \cdot 10^{-12} \cdot 4.85 \cdot 2.60 = 111.60 \text{ pF}.$$

Sensitivity of the second analogue has a value:

$$S = [C(100\%\text{RH}) - C(0\%\text{RH})]/(W_{max} - W_{min})$$
$$= (111.60 - 64.89)/(100 - 0) = 0.47 \text{ pF}/\%.$$

5 Experiments

During the process of hygrometers' verification generators of humid gas are used as standards to reproduce values of relative humidity. All generators of humid gas can be classified into: static generators with salt solutions; dynamic two-temperature generators; dynamic generators of humid gas with two pressures; generators of humid gas with mixed flows of water steam and dehydrated gas.

Most simple of them are static generators of humid gas with salt solutions [15, 16]. Principle of their work uses a property of saturated solutions of different salts to decrease gas humidity in enclosed space. Static generator of humid gas with salt solutions can be easily reproduced in practice. For that purpose verified sensor should be placed inside hermetically sealed container and connected to the electric circuit of the measuring instrument. After that a cotton swab, soaked with saturated salt solution, should be placed into that container and its lead should be closed. Approximately 30 min later appropriate value of relative humidity will set inside hermetically sealed container. Using different saturated salt solutions it is possible to reproduce all values of sensors' static function in a range from 12 to 98% of relative humidity and to linearize it if necessary.

To carry out experimental researches authors prepared four types of saturated salt solutions that cover the range from 33 to 93% of relative humidity for a temperature of standard conditions (Table 1).

Table 1. Relative air humidity in a space with four saturated salt solutions, %.

Salt solution	Temperature, °C						
	10	15	20	25	30	35	40
Potassium nitrate KNO_3	95	94	93	92	91	89	88
Sodium chloride NaCl	76	76	76	75	75	75	75
Ammonium nitrate NH_4NO_3	73	69	65	62	59	55	53
Magnesium chloride $MgCl_2$	34	34	33	33	33	32	32

Prepared salt solutions were placed in plastic containers with hermetic leads. Inside this containers two fixtures were located what allowed placing prototype products of different humidity sensors (Fig. 5). To control variation of electric capacitance of a sensor under research multimeter M890G was used in a mode of measuring electric capacitance in a range from 0 to 200 pF.

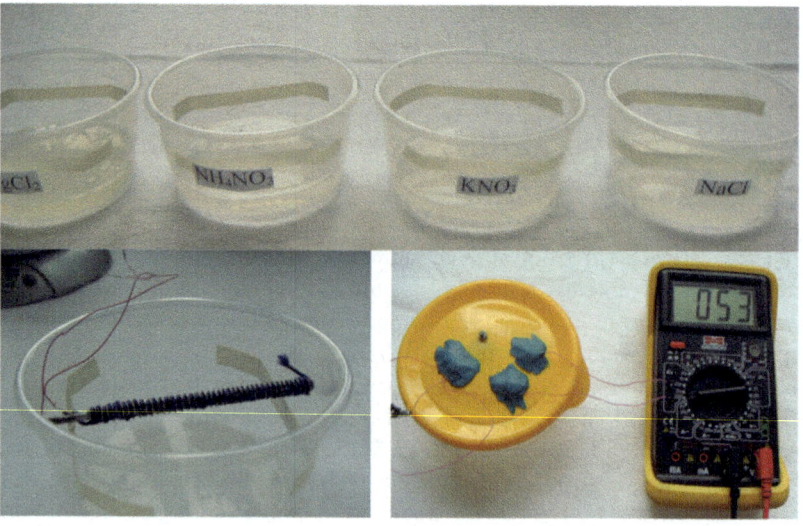

Fig. 5. Static generators of humid gas with fixtures for prototype products of capacitive sensors.

To get more information about the medium inside plastic containers with saturated salt solutions, module SY-HS-230B for air humidity measurements had been used as humidity comparator. It was placed on the inside surface of containers' lead (Fig. 6). It has temperature compensation, analogue dc voltage output and powered from 5.0 V power supply.

Fig. 6. Displacement of humidity sensor and humidity comparator, process of measuring electric capacitance of a prototype product.

Experimental researches continued four days, and during this time period all three prototype products were four times placed into each container with appropriate saturated salt solution.

At first workability of each prototype product was checked, then each prototype product was placed into static generator of humid gas for 30 min. Then results of measurements from the prototype product output and humidity comparator SY-HS-230B were taken. Because of rather long time of response between each salt solution change prototype products were placed in aired room for an hour to renovate their initial state. Authors used random order of changing salt solutions to reduce the probability of systematic error appearance. Experiments had been carried out at a stable temperature of +21 °C and 23% of air humidity, very close to standard conditions.

Results of experimental researches are given on Fig. 7 as an approximated static function for each prototype product.

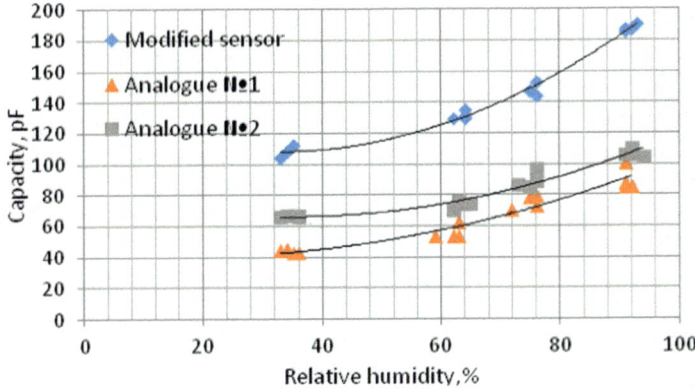

Fig. 7. Experimental static functions of three humidity sensors' prototype products.

As we can see, all prototype products have nonlinear dependence between dielectric permittivity of a system silk-water and water content in the humid air.

Presence of random variation in repeated measurements can be explained with sensitivity of static generator of humid gas to changes of temperature and presence of random error in a humidity measuring module SY-HS-230B.

After receiving the results of experimental researches, sensitivity of all three prototype products was calculated again (Table 2).

Table 2. Theoretical and experimental calculations and their comparison.

		Modified sensor	Analogue №1	Analogue №2
Theoretical values	Capacitance values for the range 0–100% RH, pF	119.29–205.17	23.96–41.2	64.89–111.6
	Sensitivity, pF/%	0.8588	0.1724	0.4671
Experimental values	Capacitance values for the range 33–93% RH, pF	107.5–187.5	43.5–87.75	66–105.5
	Sensitivity, pF/%	1.3	0.74	0.66

Analyzing data from Table 2 it is possible to say that both theoretical and experimental calculations of sensors' sensitivity prove that modified humidity sensor exceeds both analogues. It has almost two times better sensitivity having the same linear dimensions. There is a difference in theoretical and experimental calculations for the first analogue and it can be explained with some simplifications in theoretical calculations.

6   Conclusions

A new sorption-capacitive gas humidity sensor had been developed. Modified gas humidity sensor has relatively big initial capacitance in comparison with three closest analogues due to mutual displacement of two electrodes. Theoretical calculations, performed during the process of development, showed two times better sensitivity of new sensor in spite of that it has linear dimensions, same with closest analogues.

Prototype products of a new sensor and two closest analogues were produced. To get experimental static functions of three humidity sensors' prototype products static generator of humid gas with four salt solutions was used. Obtained experimental results confirmed that new sorption-capacitive humidity sensor has approximately two times better sensitivity (1.3 pF/%) relatively to the analogues (0.74 pF/% and 0.66 pF/%).

References

1. Zabolotny, A.V., Koshevoi, M.D.: Improving efficiency of the quality control of substances with dielectric properties. Telecommun. Radio Eng. **57**(2–3), 177–190 (2002). https://doi.org/10.1615/TelecomRadEng.v57.i2-3.200
2. Lee, C.Y., Lee, G.B.: Humidity sensors: a review. Sensor Lett. **3**(1–2), 1–15 (2005). https://doi.org/10.1166/sl.2005.001
3. Luijten, C.C.M., van Dongen, M.E.H., Stormbom, L.E.: Pressure influence in capacitive humidity measurement. Sens. Actuators B Chem. **49**(3), 279–282 (1998). https://doi.org/10.1016/S0925-4005(98)00148-8
4. Kotoh, K., Irube, M., Muta, M., Nishikawa, M.: Analytical method of calibration for moisture sensor of capacitor type. J. Nucl. Sci. Technol. **30**(8), 785–795 (1993). https://doi.org/10.1080/18811248.1993.9734549
5. Lorek, A.: Humidity measurement with capacitive humidity sensors between −70 °C and 25 °C in low vacuum. J. Sens. Sens. Syst. **3**(2), 177–185 (2014). https://doi.org/10.5194/jsss-3-177-2014
6. Farahani, H., Wagiran, R., Hamidon, M.N.: Humidity sensors principle, mechanism, and fabrication technologies: a comprehensive review. Sensors **14**(5), 7881–7939 (2014). https://doi.org/10.3390/s140507881
7. Islam, T., Mistry, K.K., Sengupta, K., Saha, H.: Measurement of gas moisture in the ppm range using porous silicon and porous alumina sensors. Sens. Mater. **16**(7), 345–356 (2004)
8. Majewski, J.: The dynamic behaviour of capacitive humidity sensors. Devices Methods Measur. **11**(1), 53–59 (2020). https://doi.org/10.21122/2220-9506-2020-11-1-53-59
9. Minaiev, I.G., Vostrukhin, A.V.: Capacitive sorption sensor of gas humidity. RU Patent 94030042A, 20 July 1996
10. Minaiev, I.G., Rebrova, O.V.: Capacitive pickup of gaseous and liquid media dielectric properties. SU Patent 1125530A1, 23 November 1984
11. Morton, W.E., Hearle, J.W.S.: Physical properties of textile fibres. In: Physical Properties of Textile Fibres, 4th edn., pp. 178–194. Woodhead Publishing Limited, Cambridge (2008). https://doi.org/10.1533/9781845694425.178
12. Sukhobrus, M.A., Zabolotnyi, A.V., Sukhobrus, A.A., Zabolotnyi, V.A.: Sorption-capacitor sensor of humidity of gases. UA Patent 103800C2, 25 November 2013
13. Wu, Y., Zhao, X., Li, F., Fan, Z.: Evaluation of mixing rules for dielectric constants of composite dielectrics by MC-FEM calculation on 3D cubic lattice. J. Electroceram. **11**(3), 227–239 (2003). https://doi.org/10.1023/B:JECR.0000026377.48598.4d
14. Josh, M., Clennell, B.: Broadband electrical properties of clays and shales: comparative investigations of remolded and preserved samples. Geophysics **80**(2), D129–D143 (2015). https://doi.org/10.1190/geo2013-0458.1
15. Gridnev, A.S., Mandrokhlebov, V.F.: Salt-solution humidity generators. Meas. Tech. **25**(9), 772–776 (1982). https://doi.org/10.1007/BF00827805
16. Young, J.F.: Humidity control in the laboratory using salt solutions – a review. J. Appl. Chem. **17**(9), 241–245 (1967). https://doi.org/10.1002/jctb.5010170901

Photoelectric Measurement and Control Methods of Angular Displacement of the Aircraft Control Surfaces

Nikolay Koshevoy[1] 🄳, Oleg Burlieiev[2] 🄳, Oleksandr Zabolotnyi[1],
Olena Kostenko[3] 🄳, Irina Koshevaya[1] 🄳,
and Oleksii Potylchak[1(✉)] 🄳

[1] National Aerospace University "Kharkiv Aviation Institute",
17 Chkalova Street, Kharkiv 61070, Ukraine
kafedraapi@ukr.net, o.potylchak@khai.edu
[2] Private Institution of Higher Education, Ukrainian Humanities Institute,
14 Instytutska Street, Bucha, Kiev Region 08292, Ukraine
[3] Poltava State Agrarian Academy,
1/3 Skovorody Street, Poltava 36003, Ukraine

Abstract. The measurement and control photoelectric methods of the angular displacement of the aircraft control surfaces have been considered. Their application makes it possible to automate these processes and implement them without changing the aircraft design. Their main advantages and disadvantages have been listed. The functional schemes implementing these methods have been shown.

Keywords: Automation · Angular displacement · Measurement · Photoelectric methods · Control surfaces · Functional scheme

1 Introduction

One of the most important characteristics that require monitoring at the maintenance of the aircraft is the angular displacement of its control surfaces [1–16]. Availability of error between the desired angular displacement of the control surface and the actual one will lead to higher fuel consumption, increase in flight time, as well as the origination of emergency during takeoff and landing. In doing so, this angular movement control process of the aircraft surfaces on domestic aircraft plants was reduced to manual measurement of this test parameter using bevel squares. In order to automate and improve the accuracy of the process there has been developed photoelectric methods of measuring the angular displacement of control surfaces.

In paper [17, 18] proposed was a photoelectric measurement method. There its radiation source was fixed on the end of the aircraft control surface and optical elements receiving its radiation were arranged along an arc outlining the surface.

Figure 1 shows the functional scheme of this device. In exercising the method the following components are used (Fig. 1): 1 – laser emitter; 2 – clamp; 3 – aircraft control; 4 – optical light guides; 5 – input ends of optical light guides; 6 – output ends of optical light guides; 7, 8 – photo-receivers.

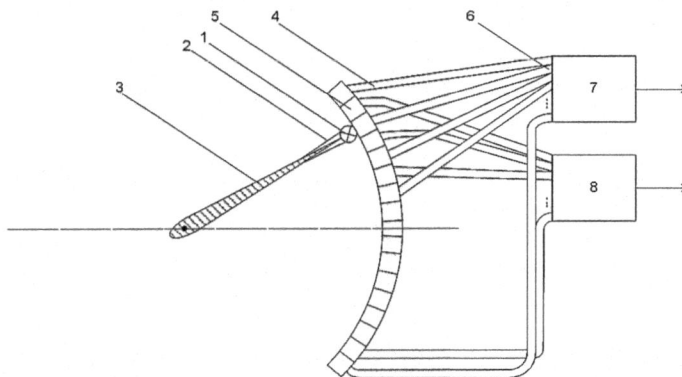

Fig. 1. A functional scheme of the device that implements the proposed optical measurement method.

To avoid making it possible to light several input ends of the light guides with a light source at the same time, there are two solutions to this problem:

- use an additional unit designed for data sharing;
- share the output ends of data channel light guides into odd and even groups, and optically couple them with photo-receivers.

Installation of the arc section with the input ends of the optical light guides was carried out using a metal shoe, which was mounted on the surface of the wing or stabilizer. To avoid possible sliding the shoe, a layer of rubber was used on its lower surface (see Fig. 2).

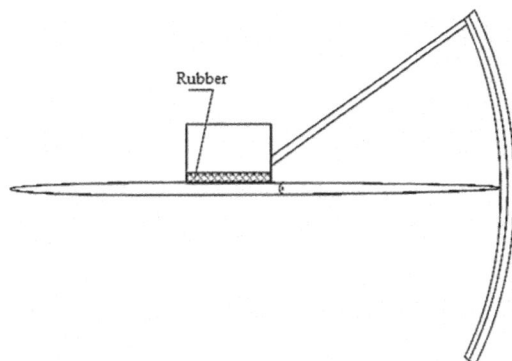

Fig. 2. Installation of arc section with input ends of optical light guides on wing surface or stabilizer.

The advantage of the proposed method of the optical angular displacement measurement is as follows:

- automation of measuring the angular displacement;
- measurement is made without changing the design of the control surface;
- higher workability compared with the capacitive method;
- high speed.

Disadvantages:

- a large number of optical elements;
- the use of the shoe, and the arc section of large radius, which increases the weight and the overall performances of devices for its implementation.

2 Main Results of Research

An improved measurement method is proposed (patent No. 33044 (Ukraine)). It is based on their photoelectric conversion using such a component as a mirror. The use of this method of measurement makes it possible to:

- automate the data acquisition process on the angular displacements;
- conduct highly accurate measurements without changing the design of control surfaces;
- reduce the overall dimensions of the measuring device.

The method merit is as follows. On the object studied a radiation source is fixed so that its major axis should be radially directed with respect to its rotation axis. When the object under study is in its initial position, at distance B from the axis of rotation up to touching the beam point in respect to the mirror at angle β to the horizontal line, passing through the rotation axis of the center of the object under study, a mirror is mounted on a support and fixed with a spring. At a distance h from the horizontal line passing through the beam touching point the mirror when the object under study is in its final position, the input ends of light guides are locked and the output ends of optical light guides thereof are connected to the light receiving unit. The input ends of optical light guides are arranged along a horizontal line so that each end position should meet the specific angle of rotation of the object under study, in doing so, angle β must be greater than the maximum angle of rotation of the object, and the quantity of the optical light guides is given by formula

$$n = \frac{\alpha_{max}}{\Delta \alpha}, \tag{1}$$

where α_{max} is maximum rotation angle of the object under study; $\Delta \alpha$ is resolution capability.

For multimode optofiber with shield diameter of $d = 125$ µm, measuring range of angular displacement $\alpha \in [0°; 30°]$ and parameter values of $B = 100$ mm, $h = 20$ mm, $\beta = 45°$ respectively, the accuracy levels up of 2.5'.

Figure 3 shows a functional scheme of a photoelectric converter implementing the improved method with the mirror in use.

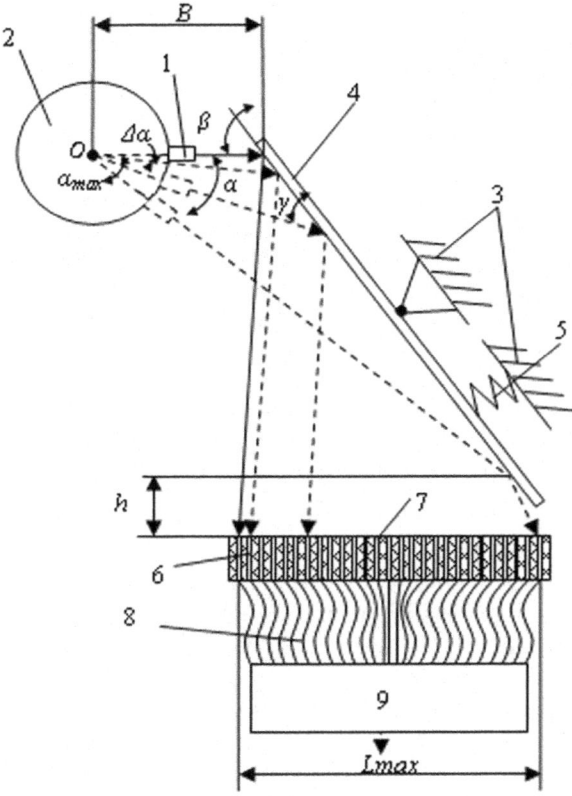

Fig. 3. Functional scheme of the photoelectric converter angular movements using a mirror.

The photoelectric transducer of the angular displacements includes light source 1, which is fixed to object 2 under study placed in one housing 3 of mirror 4, that via spring 5 is set at angle β in respect to the horizontal axis that passes through the center of rotation of object 2 under study, optical fibers 6, input ends 7 thereof are placed on the horizontal line so as the position of each end should meet the specific angle of rotation of the object, and the output ends 8 should be assembled into harnesses and optically connected with photo-receiving unit 9.

The dependence of the distance traveled by the beam along the horizontal line stowing input ends 7 of the optical fibers 6, on the angular movement of the object of study 2 is as follows:

$$L = \frac{H}{\text{tg}(2\beta - \alpha)} + \frac{B\sin\alpha}{\sin(\beta - \alpha)}\left(\cos\beta - \frac{\sin\beta}{\text{tg}(2\beta - \alpha)}\right),$$ (2)

where $H = \frac{B \cdot sin\alpha_{max}}{sin(\beta - \alpha_{max})} \cdot sin\beta + h$; h is the distance from the horizontal line passing through the beam touching point the mirror when the object under of study takes its final position to the input ends of the optical fibers; B is the distance from the axis of rotation of the object under study up to mirror beam touching point when the object of study is in its initial position; β is the angle of the mirror in relation to the horizontal axis that passes through the rotation center of the object under study ($\beta > \alpha_{max}$); α is the angular displacement of the object under study.

The disadvantage is a complex mathematical relationship (2), which affects the speed of data processing.

A further improvement of the optical method based on the use of such a component as a level support unit, which is mounted on the rotation object and allows you to retain over a beam perpendicular direction from the light source in relation to the input ends of optical fibers connected to the photo- receiving unit.

Application of the improved method (patent No. 61840 (Ukraine)) allows you to:

- simplify the mathematical relationship (2);
- conduct highly accurate measurements without changing the design of the aircraft control surfaces;
- reduce the overall parameters of the measuring transducer.

Method for measuring the angular displacement including the level support unit is as follows. On the object under study mounted is the level support unit. On it installed is a light source optically coupled with optical light-guides the input ends thereof are fit in a horizontal line so that the position of each end meets the specific angle of rotation of the object under study. The output ends of the optical light-guides (fibers) are connected to the photo receiving unit.

The dependence of the distance traveled by the beam along the stowing horizontal line of the input ends of optical light-guides (fibers) on the angular displacement α of the object under of study is as follows:

$$L = R(1 - cos\alpha), \tag{3}$$

where R is a distance from the axis of rotation of the object to the light source connection point with the object under study; α is an angular displacement of the object of study.

For multimode optic fiber with shield diameter of $d = 125$ μm, a measuring range of angular displacement $\alpha \in [0°; 30°]$ and the value $R = 150$ mm, the accuracy top up of 4.8 '.

Measuring transducer (see Fig. 4) for implementation of the improved method includes light source 1 mounted on level support unit 2, which is connected with object 3 under study. Light source 1 is optically coupled with optical light-guides 4, input ends 5 thereof are stowed along a horizontal line, in doing so, the position of each end should meet the specific angle of rotation of object 3 under study. Output ends 6 are optically coupled with light receiving unit 7 [19].

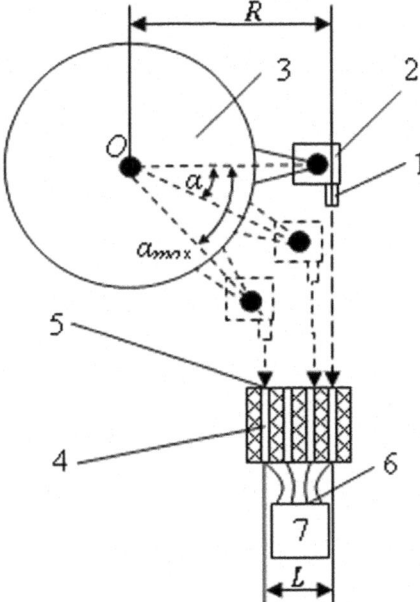

Fig. 4. The measuring transducer with level support unit.

Disadvantages of the method are as follows:

- use of a large number of optical light-guides;
- limited range of measurement of angular displacement.

To obviate the mentioned shortcomings there has been proposed an improvement of the optical method for measuring the angular displacement of control surfaces due to the use of a code disk with three coding tracks (patent No. 40489 (Ukraine)).

Due to the use of said component, the following advantages are achieved:

- expansion of measurement range up to 360°;
- the number of optical light-guides is reduced to six.

An improved method of measuring by using the code disk is as follows. The rotation shaft of the object is fastened with a case-placed shaft encoder disk, three coding tracks thereof are made with holes having resolution of q. The coding disk is installed between the output end of the first group of optical light-guides and the input end of the second group of optical light-guides. The number of optical light-guides in the first and second groups is equal to three. The input ends of the first group of optical light-guides fibers are placed into the cone concentrator that by the lock is attached to the outlet port of the illuminator. The input ends of the second group of optical light-guides are optically coupled to the output ends of the optical light-guides of the first group and code disk holes, as well as to the corresponding three photo receivers, combined in one unit. The outputs of the photo receivers' detectors are connected to the corresponding circuits of series-connected amplifiers and filters, which are connected to the microprocessor unit or a computer, and that in its turn is connected to the data display device.

The accuracy of the method depends on the accuracy of manufacture of the code disk. For single-mode optical fiber with core diameter of d = 10 μm [20], and the disk diameter of $D = 50$ mm accuracy tops of 1,4′.

For the improved method with the code disk there has been developed the interconnection technique of rotational object and measuring device (patent No. 61854 (Ukraine)). This angular displacement measurement process can be carried out without changing the design of aircraft control surfaces.

Figure 5 shows a functional scheme of the angular displacement photoelectric transducer, realizing the improved method by using the code disk.

Figure 6 shows code disk with three coded tracks.

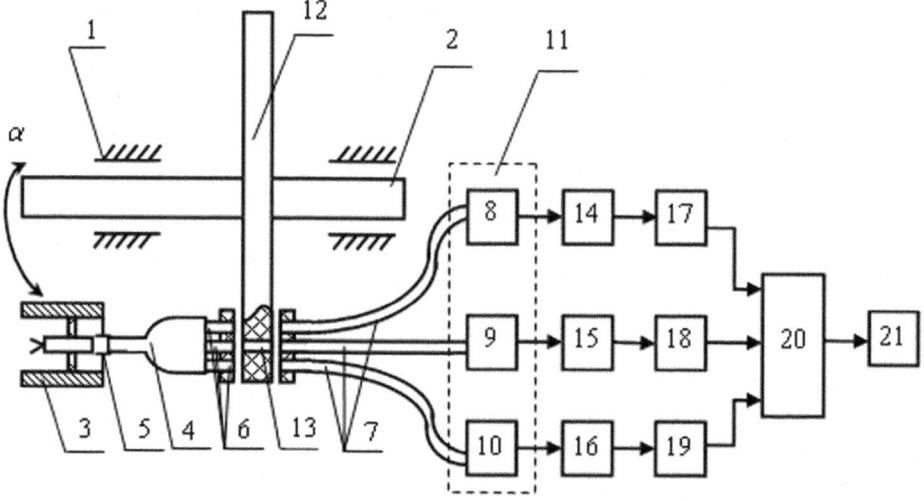

Fig. 5. Functional scheme of angular displacement photoelectric transducer with code disk.

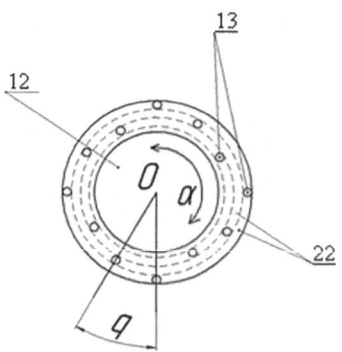

Fig. 6. Code disk with holes.

Notation in Fig. 6: 12 – shaft; 13 – holes filled with fiber-optics material; 22 – tracks.

The photoelectric transducer of the angular displacement is composed of housing 1, shaft 2 disposed in the housing and designed for coupling to an object, optically coupled illuminator 3, focon 4 (focusing concentrator), which is attached via lock 5 to the output hole of the illuminator with the ends of its first group of optical light guides (optical fibers) 6 arranged therein, the second group of optical light guides 7 with its output ends optically coupled with respective three photo receivers (photo detectors) 8, 9 and 10 of photo receiving unit 11, disk 12 with holes 13, mounted on shaft 2 and disposed between the output ends of the first group of optical light guides (fibers) 6 and the input ends of the second group of optical light guides 7, in doing so the outputs of photo receivers 8, 9, 10 are connected to the respective circuits of series-connected amplifiers (14, 15, 16) and filters (17, 18, 19), which with their outputs are connected to microprocessor unit 20 connected to display device 21 and holes 12 are formed on the disk with three tracks having resolution q.

Taking in view high manufacturing cost of the code disk and optical fiber components, used in the above-mentioned methods for measuring angular displacements, it is advisable to reduce the cost measuring technique of angular displacement of the aircraft control surfaces, by improving the optical method with no expensive components used, but at the same time one may retain opportunity to conduct highly accurate measurements without changing the design of rotation object.

This method of measurement uses two stands with holders, one of which is glued to the wing of the airplane, and the other one to the control surface respectively. In the initial position the distance between the holders is equal to L. The radiation source is connected to the power driver (PD) and mounted on the holder, and the stand thereof is fixed to the aircraft control surface. Two photo-receivers optically coupled to the radiation source are fixed at a distance R from each other on the other holder with the stand thereof mounted on the aircraft wing. Then the photo-receivers are connected to an electronic data-processing unit (EDPU). The distance values R and L are determined on the basis of the geometric dimensions of the control surface.

Functional diagram of a photoelectric transducer, implementing the proposed method of measuring the angular displacement of the aircraft control surfaces, is shown in Fig. 7.

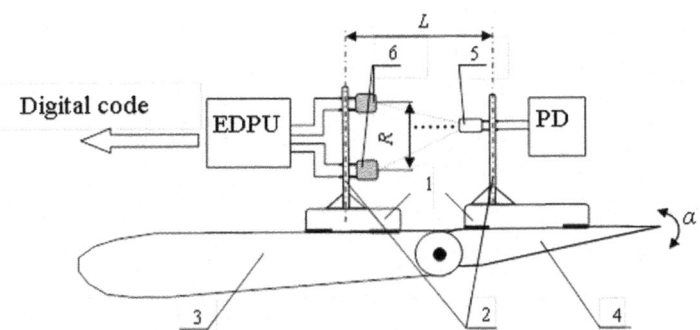

Fig. 7. Photoelectric transducer for measuring the rotation angle of the aircraft control surfaces.

The device comprises stand 1 with two holders 2, which are attached by adhesive to aircraft wing 3 and control surface 4. In the initial position holders 2 are at distance L from each other. Radiation source 5 is connected to the power driver (PD) and holder 2 is mounted on stand 1 installed on aircraft control surface 4. Two photo receivers (detectors) 6 located at distance R from each other in one plane, are connected to an electronic data processing unit (EDPU) at whose output a digital code is generated corresponding to the measured angular displacement of control surface 4.

3 Conclusion

Thus, the accuracy of processes for measurement and control of the angular displacements of the aircraft control surfaces due to the use of the proposed methods are automated and improved. These methods can also be used in industries where the rotation objects are under operation and their angular parameters cannot be changed in the process of measuring and monitoring.

References

1. Shimizu, Y., Matsukuma, H., Gao, W.: Optical sensors for multi-axis angle and displacement measurement using grating reflectors. Sensors **19**(23), 5289 (2019). https://doi.org/10.3390/s19235289
2. Jia, X., Wan, Q., Yu, H., et al.: Small high-resolution angular displacement measurement technology based on near-field image acquisition. IEEE Sens. J. **19**(15), 6141–6146 (2019). https://doi.org/10.1109/ISEN.2019.2911095
3. Yu, H., Wan, Q., Lu, X., et al.: Small-size, high-resolution angular displacement measurement technology based on an imaging detector. Appl. Opt. **3**(56), 755–760 (2017). https://doi.org/10.1364/AO.56.000755
4. Huang, Y., Yang, Y., Liang, J., et al.: An optical glass plane angle measuring system with photoelectric autocollimator. Nanotechnol. Precis. Eng. **2**(2), 71–76 (2019). https://doi.org/10.1016/j.npe.2019.06.001
5. Jia, B., He, L., Yan, G., Feng, Y.: A differential reflective intensity optical fiber angular displacement sensor. Sensors **16**(9), 1508 (2016). https://doi.org/10.3390/s16091508
6. Li, R., Konyakhin, I., Zhang, Q., et al.: Error compensation for long-distance measurements with a photoelectric autocollimator. Opt. Eng. **58**(10), 104112 (2019). https://doi.org/10.1117/1.OE.58.10.104112
7. Shl, H., Fu, M., Zhu, G., et al.: An absolute displacement measurement method of simple graphic decimal shift encoding. In: 2019 International Conference on Optical Instruments and Technology: Optoelectronic Measurement Technology and Systems, Beijing, vol. 11439, pp. 114390U. SPIE (2020). https://doi.org/10.1117/12.2543441
8. Zhang, W.Y., Zhang, J., Wu, L.Y.: Small-angle measurement of laser beam steering based on total Internal-reflection effect. J. Phys. Conf. Ser. **48**(1), 766–770 (2006). https://doi.org/10.1088/1742-6596/48/1/145
9. Gao, X., Li, S., Ma, Q.: Subdivided error correction method for photoelectric axis angular displacement encoder based on particle swarm optimization. IEEE Trans. Instrum. Meas. **69**(10), 8372–8382 (2020). https://doi.org/10.1109/TIM.2020.2986852

10. Shan, M., Min, R., Zhong, Z., et al.: Differential reflective fiber-optic angular displacement sensor. Opt. Laser Technol. **68**, 124–128 (2015). https://doi.org/10.1016/j.optlastec.2014.10.016

11. Matyushin, S.: Fiber optic sensors for moving robotic platforms. Sci. Educ. Transp. **2**, 185–189 (2018). (in Russian)

12. Palamar, M., Chaykovskyi, A.: Intelligent Optoelectronic Angle Sensors: Circuit and Software-Algorithmic Methods of Synthesis. Dzhura Publ., Ternopil (2015). (in Ukrainian)

13. Legayev, V.: Transmitters and Sensors: a Tutorial. Vladimir State University Publ., Vladimir (2019). (in Russian)

14. Steck, D.A.: Classical and modern optics (Revision 1.5.1). University of Oregon, Eugene (2013)

15. Kiryanov, V., Kiryanov, A.: Increasing the accuracy of angular measurements using photoelectric converters of combined type. Autometry **48**(6), 84–91 (2012). (in Russian)

16. Peatross, J., Ware, M.: Physics of Light and Optics. Brigham Young University, Provo (2013)

17. Koshevoi, N.D., Kostenko, E.M., Oganesyan, A.S., Tsekhovskoi, M.V.: Aircraft system for measuring the angular deflections of control surfaces. Russ. Aeronau. (Iz VUZ) **56**(4), 418–422 (2013). https://doi.org/10.3103/S1068799813040168

18. Oganesyan, A.S., Thehovsky, M.V., Koshevoy, N.D., Gordienko, V.A.: Investigation into optoelectronic aviation angle meter by the design-of-experiments method. Telecommun. Radio Eng. **69**(9), 841–847 (2010). https://doi.org/10.1615/TelecomRadEng.v69.i9.70

19. Koshevoy, N.D., Burleiev, O.L., Gordienko, V.A.: Photoelectric methods of measuring the steering angle in aviation. Telecommun. Radio Eng. **71**(8), 759–762 (2012). https://doi.org/10.1615/TelecomRadEng.v71.i8.80

20. FOA Online Reference Guide: Optical fiber. https://www.thefoa.org/tech/ref/basic/fiber.html. Accessed 05 July 2020

Method for Designing Low-Orbit Clusters of Small Satellites Under Stochastic Disturbances

Olena Tachinina[1]([✉]) [iD], Oleksandr Lysenko[2] [iD], Iryna Alekseeva[2] [iD], and Valeriy Novikov[2] [iD]

[1] National Aviation University,
1 Liobomyra Husara Avenue, Kiev 03058, Ukraine
tachinina5@gmail.com
[2] National Technical University of Ukraine "Igor Sikorsky Kyiv Polytechnic Institute", 37 Peremohy Avenue, Kiev 03056, Ukraine
lysenko.a.i.1952@gmail.com,
{alexirl,novikov1967}@ukr.net

Abstract. Under the new paradigm of exploration of space, both near and distant, continue using mini-, micro-, nano -, pico satellite armadas or fleets (compound dynamic systems – CDS). These CDS will be arranged in clusters in such a way that clusters made up of an excess of small satellites (mini-, micro-, nano-, pico-satellites) have increased functional stability and the quality of assigned functions. The efficiency of the CDS, consisting of thousands of small satellites (micro-, nano-, pico-satellites) will be determined by how efficiently the clusters of the required configuration are created, held (stabilized) in the created configuration, and reconfigured. Clusters for various practical purposes: communications; telecommunications and global Internet; navigation; monitoring of the Earth's surface, troposphere, stratosphere, ionosphere and exosphere have specific features. However, the common feature of all specialized clusters – they should be rebuilt (change the spatial topology) in a minimum time and stabilized in a new configuration at rational (preferably with minimal) energy costs. This article is devoted to the development of a method for designing (formation) low-orbit clusters of small satellites with a given spatial topology and stabilization of this topology in terms of rational energy costs under the action of stochastic disturbances. The method consists in the synthesis of control actions that are technically easy to implement using the small satellites' onboard hardware. Control actions allow to form a low-orbit cluster satellite (LCS) in a minimum amount of time and to keep the LCS in a configuration that differs little from the required.

Keywords: Compound dynamic systems · Low-orbit · Clusters · Small satellites · Cluster satellites

M. Nechyporuk et al. (Eds.): ICTM 2020, LNNS 188, pp. 112–121, 2021.
https://doi.org/10.1007/978-3-030-66717-7_10

1 Introduction

In articles [1–3], a concept was proposed for creating and maintaining the functioning of ultra-low orbit satellite systems of communication, navigation and monitoring of the earth's surface.

The implementation of this concept, which the authors called Clear Space, will almost completely prevent the further contamination of space.

The phenomenon of self-cleaning of space is well known and exists due to the deceleration of space objects during their free movement in a rarefied atmosphere.

If the objects will freely glide along the upper edge of atmosphere dense layers (at an altitude of 220–250 km), then after about a year they will slow down, drop off orbit and burn up in the dense layers of the Earth's atmosphere.

To implement the Clear Space concept, all satellites must have small masses and dimensions in order to completely burn out in the Earth's atmosphere and not pose a threat to all objects on the earth's surface and in the air up to an altitude of 80 km.

The small satellites (mini-, micro-, nano-, pico-satellites) which should be clustered in orbit (low-orbit cluster satellites – LCS or low-orbit distributed satellites LDS).

However, to solve common human problems, for example, the global coverage of the Earth by the Internet, the LCS group should consist of tens of thousands of small satellites (mini-, micro-, nano-, pico-satellites).

The Clear Space concept [4] takes into account only use of aerospace systems with retained stages (flying launching sites) will allow to solve economically both the problem of launching into space and the replenishment of thousands of functionally redundant low-orbit groups of small satellites.

It should be noted that the LCS that move to orbits at ultra-low altitudes (220–250 km) do not pose a threat for international space stations.

At the same time, there is the task of collecting tens of thousands of small satellites into thousands of clusters with a given spatial topology and stabilizing this topology for a given time interval under stochastic disturbances.

Stochastic disturbances are caused by the unpredictable effect of a rarefied atmosphere on small satellites that fly at altitudes of 200–250 km above the Earth's surface.

This article is devoted to the development of a method for designing (formation) low-orbit clusters of small satellites with a given spatial topology and stabilization of this topology rational in terms of energy costs under stochastic disturbances.

2 Problem Statement

The cluster (distributed) satellites for various practical purposes: communications; telecommunications and the global Internet; navigation; monitoring of the Earth's surface, troposphere, stratosphere, ionosphere and exosphere have specific features [5–21].

However, it is common for all specialized cluster satellites that they should be rearrange in a minimum time (change the spatial topology) and stabilized during a given time interval in a new configuration with rational (preferably minimal) energy costs.

For low-orbit cluster satellites, the feature is added that the restructuring of the spatial topology and the stabilization of the new topology occur under stochastic disturbances.

Schematically, the process of configuring or reconfiguring the LCS (designing low-orbit clusters of small satellites) corresponds to the "assembly in a beam" scheme (see Fig. 1).

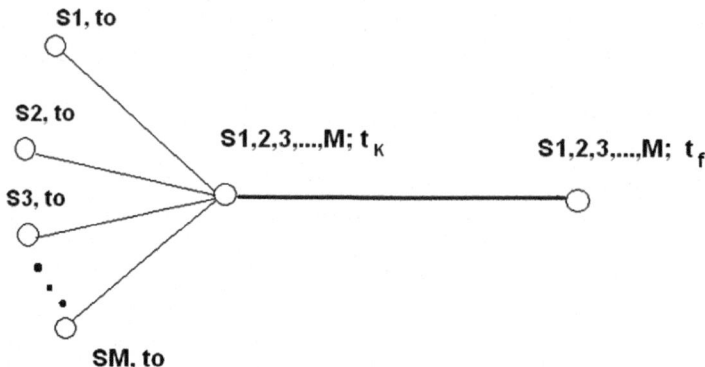

Fig. 1. The design scheme of low-orbital cluster satellite (LCS): $S1, 2, 3, ..., M$ – symbol designation of small satellites (mini-micro, micro-, nano-, pico-satellites), which are part of the LCS; M – total number of small satellites in the cluster; t_o - time point of start of the LCS formation; t_K – point in time when all small satellites will occupy their position in the cluster (time of completion of the LCS formation); t_f – time moment of the end of LCS current configuration.

The mathematical setting up of the problem of formation a low-orbit cluster of small satellites under stochastic disturbances can be formulated as follows.

$$dx_i(t) = (A \cdot x_i(t) + B \cdot u_i(t))dt + \sigma_i \cdot d\omega_i(t),$$
$$y_i(t) = C \cdot x_i(t) + D \cdot u_i(t), \ i = \overline{1, M}, \tag{1}$$

which describe the stochastic dynamics of the relative coplanar motion of small satellites in a force-free field [21], from a known initial state $x_i(t_0)$ to a given neighborhood X_i near the final state $x_i(t_{iK})$, $t_{iK} \in (t_0; t_K]$, followed by retention in this area for a period of time $[t_K; t_f]$ so that

$$W(u_1(t), u_2(t), \ldots, u_M(t)) = E[t_K - t_0] \to \min_{u_1(t) \in G_1, \ldots, u_M(t) \in G_M, \ t \in [t_0, t_f]}, \tag{2}$$

where $E[\ldots]$ is the mathematical expectation operation; t_0, t_f are known (deterministic, predetermined) time instants of the start of formation of the LCS and completion of its functioning in a given spatial topology (configuration); t_K is a random point in time when all small satellites will occupy their position in the cluster (the point in time of completion of the LCS formation); $x_i(t) = [x_{i1}(t) \ x_{i2}(t)]^T$ is the state vector;

$x_{i1}(t)$, $x_{i2}(t)$ are the relative coordinate of the center of mass position and relative velocity of the center of mass i-small satellite from the LCS group, which are counted along OX axis; positive OX axis direction coincides with the direction of motion of the main (or root [15, 16]) small satellite from the LCS group (any small satellite or some virtual point on OX axis can be chosen as the "reference item"); $u_i(t)$ is the control action (load factor) created by the jet engine along OX axis on board of i-small satellite (suppose that $u_i(t) \in \{-a,\ 0,\ a\} = G_i, a > 0,\ t \in [t_0; t_f]$); $\sigma_i \cdot d\omega_i(t)$ is the stochastic braking effect of the atmosphere; σ_i is the parameter that characterizes the intensity of the atmosphere stochastic effect; $A = \begin{bmatrix} 0 & 1 \\ 0 & 0 \end{bmatrix}$, $B = \begin{bmatrix} 0 \\ 1 \end{bmatrix}$, $C = \begin{bmatrix} 1 & 0 \\ 0 & 1 \end{bmatrix}$, $D = \begin{bmatrix} 0 & 0 \\ 0 & 0 \end{bmatrix}$.

3 Problem Solving

To solve the problem (1), (2), a method is proposed to form low-orbit clusters of small satellites under stochastic disturbances. The method consists of two stages.

The first stage is the alignment of the speeds of movement of small satellites from the LCS group with the optimum minimum time (after alignment, the relative speeds are approximately equal to 0). The jet engine is started up for maximum thrust and, with the appropriate orientation, either accelerates or brakes in a minimum time a small satellite from the LCS group.

The second stage is the reconfiguration of LCS space topology (rearrangement of the small satellites from the LCS group in the proper order) and maintaining the achieved configuration (spatial topology) of the LCS for a given time interval, i.e., up to moment t_f.

At the second stage, taking into account the physical meaning of the problem, we replace criterion (2) with a criterion equivalent in the mathematical sense

$$W(u_1(t), u_2(t), \ldots, u_M(t)) = E[t_K - t_0] \rightarrow \min_{u_1(t) \in G_1, \ldots, u_M(t) \in G_M,\ t \in [t_0, t_f]} \Leftrightarrow$$
$$\Leftrightarrow W(u_1(t), u_2(t), \ldots, u_M(t)) = E[t_K] - t_0 \rightarrow \min_{u_1(t) \in G_1, \ldots, u_M(t) \in G_M,\ t \in [t_0, t_f]} \tag{3}$$

where $E[t_K] = \max\{E[t_{1K}], E[t_{2K}], \ldots, E[t_{MK}]\}$, t_0 is the known (determined) value, t_{iK} is the point in time when i-small satellite will occupy a given position in cluster $(i = \overline{1, M})$, i.e. will go to a given neighborhood X_i near the final state $x_i(t_{iK})$, $t_{iK} \in (t_0; t_K]$.

Reduction criterion (2) to form (3) allows us to more clearly state the meaning of the algorithm for implementation of the second stage of LCS formation. To perform the second stage of LCS formation, it is proposed to use an algorithm that consists of three procedures.

Procedure 1 is intended to find the first approximation of solving the problem of the LCS reconfiguration in a stochastic condtition. Procedure 1 is based on the solution of reconfiguration problem in a deterministic formulation [17–20]. This means that there is no stochastic effect in the mathematical model (1), and in the criteria (2) and (3) the operation of calculating – there is no the mathematical expectation. The result of solving the LCS reconfiguration problem in a deterministic formulation is shown in Model 1 (see Fig. 2) and graphs (see Fig. 3a, b).

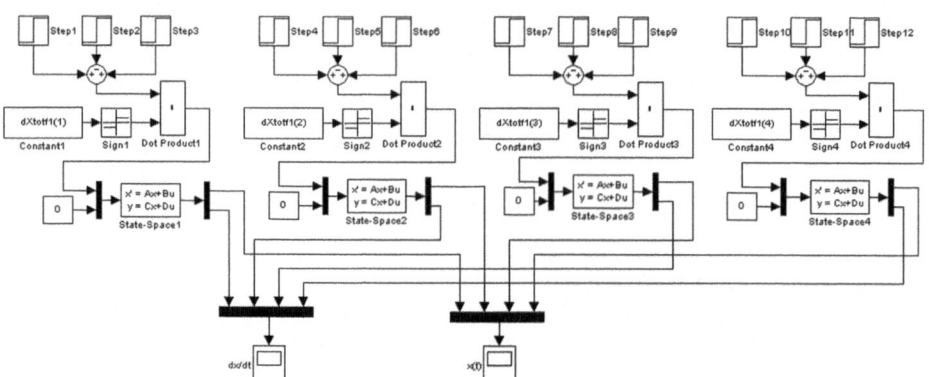

Fig. 2. Computer MATLAB+ Simulink mathematical model (model 1) of the deterministic reconfiguration of a low-orbit cluster satellite (LCS), which consists of four ($M = 4$) small satellites: State-Space 1, 2, 3, 4 – modules that simulate Eq. (1) (the motion dynamics of the small satellite with a zero signal acting on a disturbing input and the optimal "bang bang" control (forced control when the control takes one of two boundary values) [17–20].

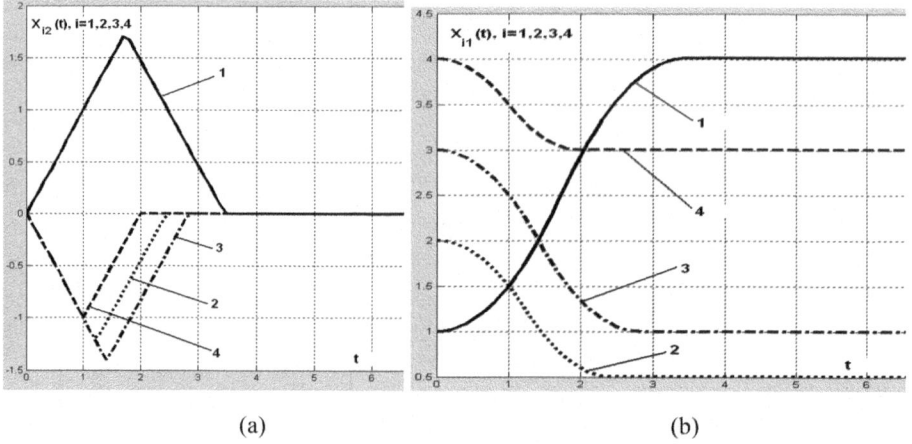

(a)   (b)

Fig. 3. Oscillograph indications of the computer deterministic model shown in Fig. 2: (a) graphs of time changes $x_{i2}(t)$; $i = 1, 2, 3, 4$ of relative velocity of small satellites included in the cluster, (b) graphs of time changes $x_{i1}(t)$; $i = 1, 2, 3, 4$ of relative coordinate of the motion of small satellites included in the cluster; on graphs (a), (b), numbers 1, 2, 3, 4 indicate the coordinates of state vector of the specific small satellite

It is assumed at modeling: $a = 1$; start and final relative coordinates of micro-(nano-) satellite are set by the vectors $X_{to1} = [1\ 2\ 3\ 4]$ and $X_{tf1} = [4\ 0.5\ 1\ 3]$; $dX_{totf1} = X_{tf1} - X_{to1}$; the time intervals of "bang bang" control switching were calculated according to the formula described in [20], using functions MATLAB sqrt(abs (dX_{totf1}/a)); all physical quantities are presented in dimensionless form.

Procedure 2 is designed to correct the control, which is calculated as a result of implementation of procedure 1, i.e. first approximation procedure. Procedure 2 (the second approximation procedure) is based on the use of numerical algorithms of optimizing the systems described by stochastic differential equations [17].

Procedure 3 is intended to design a rational control in terms of energy costs, which allows to maintain the LCS spatial topology achieved as a result of procedure 2 up to time t_f. Procedure 3 is based on the use of computer simulation model that allows modeling the system (2). This model allows using computer simulation methods to select quasi-optimally (rationally) the time of a control pulse that counteracts the atmosphere braking effect.

The result of applying procedures 1, 2, 3 to the solution of problem of formation LCS in a stochastic statement is presented by model 2 (see Fig. 4) and graphs (see Fig. 5a, b and Fig. 6a, b).

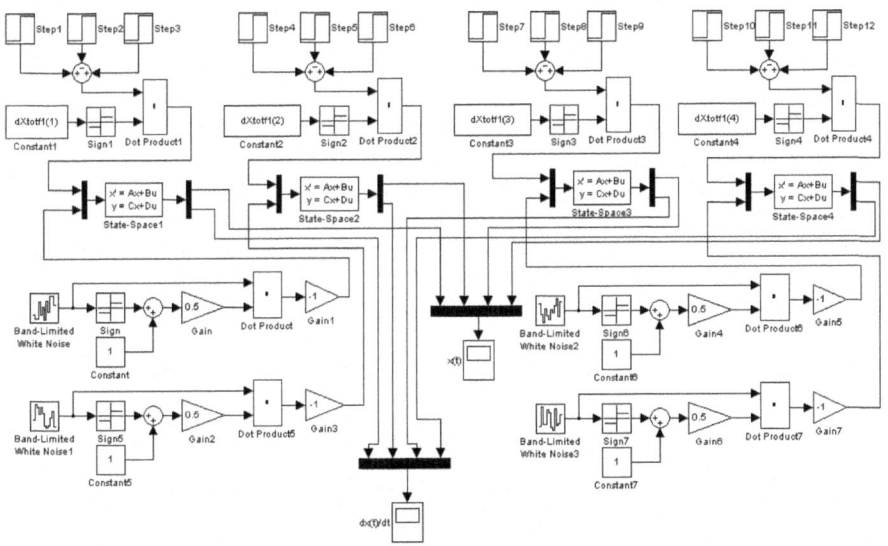

Fig. 4. Computer MATLAB+ Simulink mathematical model (model 2) of the stochastic reconfiguration of a low-orbit cluster satellite (LCS): State-Space 1, 2, 3, 4 – modules that simulate Eq. (1) (the motion dynamics of small satellite) with a stochastic signal acting on a disturbing input and the optimal control synthesized as a result of simulation experiment.

In this example, the intensity of stochastic braking effect is two orders of magnitude less than the limiting load factor a that can be created by the jet engine of each individual small satellite. This ratio corresponds to the physics of phenomenon [21].

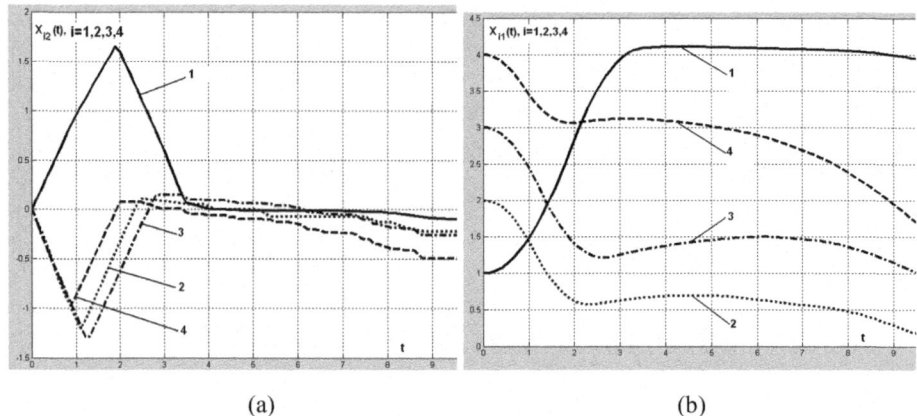

(a)   (b)

Fig. 5. Oscillograph indications of the computer stochastic model shown in Fig. 4: (a) graphs of time changes of relative speed of small satellites included in the cluster; (b) graphs of time changes of relative coordinate of motion of small satellites included in the cluster, in which the adjusted "bang bang" control operates: numbers 1, 2, 3, 4 on the graphs are coordinates of the state vector of small satellite.

As shown in Fig. 4, when reconfiguring the LCS in a deterministic formulation, the problem is decomposed into four autonomous tasks.

This is because after redeployment small satellites theoretically indefinitely keep the achieved position unchanged.

In the stochastic statement of reconfiguration problem of the LCS, this is not observed. Simulation using model 2 (stochastic model) showed that:

- when solving the problem of LCS reconfiguration in a stochastic formulation, the cluster does not retain the spatial topology achieved after procedures 1 and 2 (see Fig. 5a, b);
- in order to keep small satellites in an acceptable area (to maintain the order in a reconfigured cluster), it is necessary to synthesize corrective control (perform procedure 3);
- impulse corrective control is energetically more profitable than continuous corrective control. This means that with a certain interval it is necessary to apply corrective control actions (see Fig. 6a, b, c). The calculation results showed that in the example under consideration corrective impulse actions should be applied every 6–8 min for a duration of 12–15 s and an amplitude equal to a.

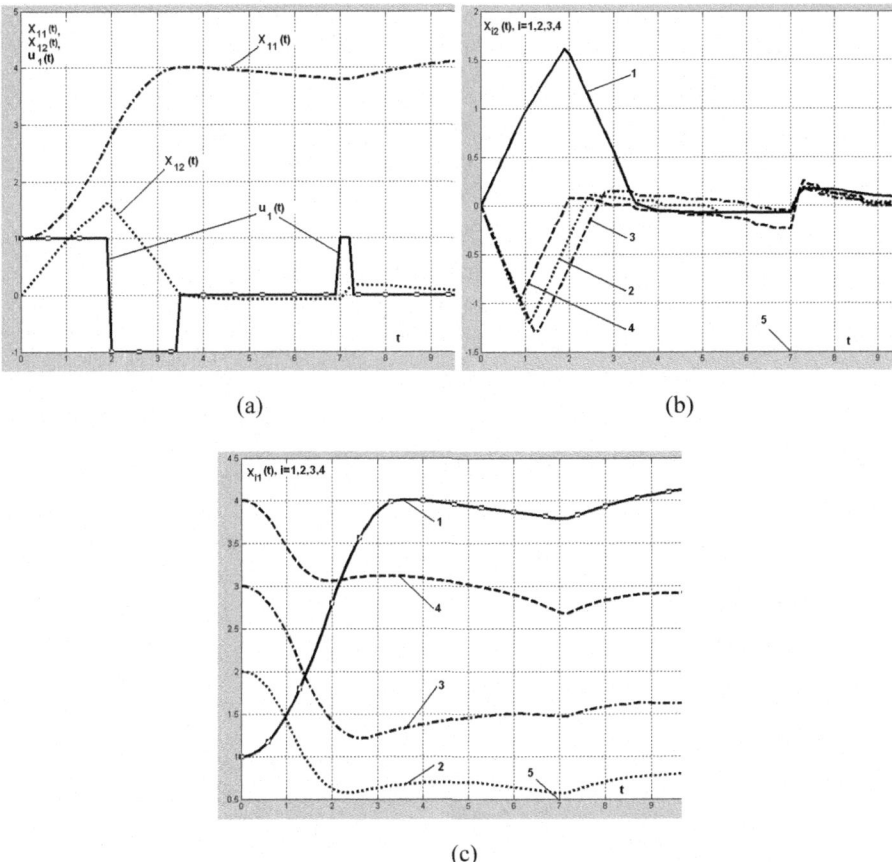

(a)   (b)

(c)

Fig. 6. Oscillograph indications of the computer stochastic model shown in Fig. 4: (a) example of modeling the motion of first small satellite under the action of average optimal control $u(t)$, which consists of "bang bang" – first-approximation control and periodic corrective impulse control; (b) graphs of time changes of relative velocity of small satellites included in the cluster; (c) graphs of time changes of relative coordinate of motion of small satellites included in the cluster; in graphs (b) and (c) the coordinates of state vector of the small satellite are indicated by numbers 1, 2, 3, 4; 5 – time point for corrective control

4 Conclusions

The article describes a method of designing (formation) the low-orbit clusters of small satellites under stochastic disturbances. This method can be implemented:

– during preliminary control synthesis, if there is sufficient information about the intensity of stochastic effect of the atmosphere on small satellites;
– during preliminary synthesis of the first approximation to rational control, which adapts to specific flight conditions based on the operational identification of the

intensity of atmosphere stochastic effect on small satellites (foggy and/or cloud computing technologies can be used for identification).

The method consists in the synthesis of control actions that are technically easy to implement using the small satellites' on board hardware. For the synthesis of control actions, information is needed on the initial and final positions of small satellites, as well the intensity of the atmosphere stochastic effect. Control actions allow:

– to form a low-orbit cluster satellite (LCS) in a minimum time;
– to keep the LCS in a configuration that differs little from the required, during a given time interval with a rational energy consumption.

The proposed method does not assume use of energy-intensive continuous monitoring of the position of mini satellites in a cluster and continuous control. This control is required to be carried out periodically: approximately once per circle and the control to keep the cluster in the required configuration with permissible deviations is quasi-optimal in terms of energy costs due to specially calculated time intervals for the supply of control pulses.

Subsequent studies should be carried out in the direction of refining the mathematical model of the relative motion of mini satellites in a cluster while keeping the method structure described in the article.

References

1. Lysenko, O., Sparavalo, M., Tachinina, O., et al.: Feasibility reasoning of creating ultra-low orbit communication systems based on small satellites and method of their orbits designing. Inf. Telecommun. Sci. **11**(1), 59–69, (2020). https://doi.org/10.20535/2411-2976.12020.59-70
2. Yavisya, V., Lysenko, O., Alekseeva, I., Tachinina, O.: Approach of construction of the space segment of the satellite communication system CLEAR SPASE. In: Proceedings of the International Scientific Conference "Modern Challenges in Telecommunications", Kyiv, pp. 22–26. NTUU "KPI" (2019). (in Ukrainian)
3. Yavisya, V., Lysenko, O., Alekseeva, I., Tachinina, O.: The method of creation and allocation of nanosatellite clusters of clear space satellite communication system. In: Proceedings of the International Scientific Conference "Modern Challenges in Telecommunications", pp. 285–287. NTUU "KPI", Kyiv (2019). (in Ukrainian)
4. Plokhikh, V., Karp, K.: Conceptual Studies and Synthesis of Reusable Aerospace Systems of Horizontal Launch. MAI Publ., Moscow (2006). (in Russian)
5. Qu, Z., Zhang, G., Cao, H., Xie, J.: LEO satellite constellation for Internet of Things. IEEE Access **5**, 18391–18401 (2017). https://doi.org/10.1109/ACCESS.2017.2735988
6. Kota, S., Giambene, G.: Satellite 5G: IoT use case for rural areas applications. In: The Eleventh International Conference on Advances in Satellite and Space Communications (SPACOMM), pp. 7–14. IARIA (2019)
7. OneWeb minisatellite constellation for global internet service. http://directory.eoportal.org/web/eoportal/satellite-missions/content/-/article/oneweb. Accessed 13 July 2020
8. Wang, B.: FCC will approve Kepler, LeoSat, Telesat and SpaceX satellites. https://www.nextbigfuture.com/2018/11/fcc-will-approve-kepler-leosat-telesat-and-spacex-satellites.html (2018). Accessed 13 July 2020

9. Welts, M.: SpaceBelt: Our first outer space partner (2018). https://wasabi.com/blog/spacebelt. Accessed 13 July 2020

10. Gunter's Space Page: LeoSat. https://space.skyrocket.de/doc_sdat/leosat.htm. Accessed 13 July 2020

11. Handley, M.: Delay is not an option: low latency routing in space. In: Proceedings of the 17th ACM Workshop on Hot Topics in Networks, Redmond, pp. 85–91. ACM (2018). https://doi.org/10.1145/3286062.3286075

12. Coldewey, D.: SpaceX reveals more Starlink info after launch of first 60 satellites (2019). https://techcrunch.com/2019/05/24/spacex-reveals-more-starlink-info-after-launch-of-first-60-satellites. Accessed 13 July 2020

13. Satnews: Ten million euros to fund the first demonstration of GLoT by NanoAvionics, KSAT and Antwerp Space (2019). http://www.satnews.com/story.php?number=1179815788. Accessed 13 July 2020

14. Narayanasamy, A., Ahmad, Y.A., Othman, M.: Nanosatellites constellation as an IoT communication platform for near equatorial countries. IOP Conf. Ser. Mater. Sci. Eng. **260** (1), 012028 (2007). https://doi.org/10.1088/1757-899X/260/1/012028

15. Narytnik, T., Rassamakin, B., Prisyazhny, V., Kapshtyk, S.: Coverage area formation for a low-orbit broadband access system with distributed satellites. In: 2018 International Conference on Information and Telecommunication Technologies and Radio Electronics (UkrMiCo), Odessa, pp. 1–4. IEEE (2018). https://doi.org/10.1109/UkrMiCo43733.2018.9047526

16. Ilchenko, M., Narytnik, T., Rassamakin, B., et al.: Creation of the architecture of "distributed satellite" for low-orbital information-telecommunication systems based on the grouping of micro and nano satellites. Aerosp. Tech. Technol. **2**, 33–43 (2018). https://doi.org/10.32620/aktt.2018.2.05. (in Russian)

17. Bodner, V., Rodnischev, N., Yurikov, E.: Optimization of Terminal Stochastic Systems. Mashynostroenie, Moscow (1987). (in Russian)

18. Goodwin, G.C., Graebe, S.F., Salgado, M.E.: Control System Design. Prentice Hall, Upper Saddle River (2001)

19. Dorf, R.C., Bishop, R.H.: Modern Control Systems. Pearson, Harlow (2017)

20. Sage, A.P., White, C.C.: Optimum System Control. Prentice-Hall, Englewood cliffs (1977)

21. Ferrario, A., Furlan, B.: CubeSat formation flying mission as Wi-Fi data transmission and GPS based relative navigation technology demonstrator. Master's thesis, Politecnico di Milano (2011)

Big Data and Data Science

Technology of Integrated Application of Classical Decision Making Criteria for Risk-Uncertainty Assessment of Group Systems of Preferences of Air Traffic Controllers on Error's Dangers

Oleksii Reva[1]([✉]) [iD], Andrii Nevynitsyn[2] [iD], Sergii Borsuk[3] [iD],
and Valerii Shulgin[2] [iD]

[1] Ukrainian Institute of Scientific and Technical Expertise and Information,
180 Antonovicha Str., Kiev 03150, Ukraine
ran54@meta.ua
[2] Flight Academy of the National Aviation University,
1 Dobrovolskogo Str., Kropyvnytskyi 25005, Ukraine
{nevatse, vashulgin}@ukr.net
[3] Wenzhou University, Wenzhou, Zhejiang, China
greyone.ff@gmail.com

Abstract. The study of individual systems of air traffic controllers' (ATC) preferences on the dangers of characteristic errors has a positive proactive character. Group systems of preferences (GSP) reveal features of functioning of separate societies – ATC's shifts. Individual systems of preferences $m = 37$ tested air traffic controllers were built. The implementation of a multi-step technology for detecting and rejecting marginal thoughts has led to a statistically consistent GSP: Kendall's concordance coefficient is $W = 0.700$ and is statistically significant at an unusually high level of significance for human factor research $\alpha = 1\%$. A decision matrix has been formed – a "cost matrix", for the solution of which the methodology of application of classical decision-making criteria by Wald (W), Savage (S), Bayes-Laplace (B-L), Hurwitz (HW) has been implemented. Empirical preferences coincide: the values of Spearman's rank correlation coefficients are equal to $R_S^{B-L-W/S} = 0.8922$, $R_S^{B-L-HW} = 0.9263$ and are statistically significant at the level of significance $\alpha = 1\%$. The values of the normalized risk index of indistinguishability of error risks in group systems of advantages are equal to: $R_{BL}^* = 0$, $R_{HW}^* = 0.19 \cdot 10^{-2}$, $R_{W/S}^* = 5.58 \cdot 10^{-2}$. For the group as a whole $R_g^* = 0.52 \cdot 10^{-2}$.

Keywords: Flight safety · Human factor · Classical decision making criteria · Air traffic controllers errors · Individual and group systems of preferences · Risk of indistinguishability of error's dangers

© The Author(s), under exclusive license to Springer Nature Switzerland AG 2021
M. Nechyporuk et al. (Eds.): ICTM 2020, LNNS 188, pp. 125–134, 2021.
https://doi.org/10.1007/978-3-030-66717-7_11

1   Introduction

Today, air traffic controllers (ATC), together with flight crew members, are considered "front-line" and "last frontier" aviation operators (AO), as they have a direct, both positive and, statistically, mostly negative impact on ensuring proper level of flight safety (FS) [1, 2]. Therefore, the study of HF problems, especially proactive, and the practical implementation of their results, is a more important factor in preventing accidents. After all, during the last 60–70 years at least 2/3–3/43 of these events arose due to the negative impact of the human factor (HF) [3].

Purposeful and complex polyergatic control system "Flight crew – aircraft – environment – ATC" is humanistic, according to one of the founders of fuzzy mathematics [4]. Thus aviators also have the right to make mistakes [1–4 etc.]. Moreover, these errors should be considered in the context of decision-making (DM), as the professional activity of "front line" aviation operators (AO) is usually considered as a continuous chain of decisions that are made and implemented in explicit and implicit forms and under the influence of many different factors. In addition, the vast majority of accidents are the result of wrong decisions.

Recalling the well-known Latin proverb "Praemonitus, praemunitus" (warned, therefore, armed), it would be expedient to form in JSC "leading edge" skills of recognition, assessment of dangers, memorization, and, consequently, prevention of erroneous actions and decisions in professional activity.

That's why, as the experience of research [5–7] shows, the identification of individual and group systems of advantages (SP) of "leading edge" AO on the indicators and characteristics of professional activity, in particular, on the dangers of characteristic errors that they may assume, performing operating procedures. It was found that the controllers of ATS, who were accidentally involved in the construction of such individual SP before training, made in its process a third fewer errors than those that were not covered by this procedure.

In the context of our research, the system of preferences (SP) will be understood as the representation of the ATS about the most and the least dangerous mistake, and hence – about the completely orderly series of mistakes that they can make in professional activities.

2   Analysis of Researches and Publications

Let us draw attention to the ICAO-recognized basic model of error management, proposed by Professor of the University of Texas, Dr. Robert Helmreich [8]. However, the model is focused on pilot's error management and does not take into account the specifics of the professional activities of air traffic controllers (ATC's).

Significant studies of ISP and GSP of air traffic controllers from Azerbaijan and Ukraine were conducted under the guidance of one of the co-authors. Based on world accidents and incidents statistics at ATC, ICAO recommendations, as well as personal experience of practical ATC, teaching and instructor work of the authors of this publication, a list of $n = 21$ characteristic errors was generated, which is currently the most complete and comprehensive coverage of the inappropriate actions of air traffic controllers: $Er._1$ is the Violation of radiotelephony phraseology; $Er._2$ is the Inconsistent entry of the aircraft into the zone of the adjacent ATC; $Er._3$ is the Violation of longitudinal course time separation; $Er._4$ is the Violation of time separation on reciprocal tracks; $Er._5$ is the Violation of separation between aircrafts on crossing tracks; $Er._6$ is the Address less ATC messaging; $Er._7$ is the Error in determining of aircraft call sign; $Er._8$ is the Error in aircraft identification; $Er._9$ is the Misuse of ATC schedule; $Er._{10}$ is the Absence of mark of the control transfer to the adjacent Air traffic control center in the strip; $Er._{11}$ is the Absence of mark of the coordination of the entrance of the aircraft to the adjacent ATC area in the strip; $Er._{12}$ is the Violation of coordinated geographic boundary by ATC; $Er._{13}$ is the Violation of coordinated time of control transfer at FIR – boundary by ATC; $Er._{14}$ is the Negligence in applying to the strip of the letter-digital information (the possibility of double interpretation); $Er._{15}$ is the Non-economical ATC; $Er._{16}$ is the Violation of shift handover procedures; $Er._{17}$ is the Issued commands to change the altitude or direction of flight are not reflected on the strip; $Er._{18}$ is the Attempt to control the aircraft under condition of TCAS system operation in the resolution advisory mode; $Er._{19}$ is the Errors in entering information about aircraft into an automated system; $Er._{20}$ is the Violation of emergency procedures; $Er._{21}$ is the Violations of airspace use.

A multi-step procedure for detecting and weeding out marginal thoughts was implemented, classical DM criteria were used to detect GSP, an indicator for assessing the degree of their risk (indistinguishability of alternatives-errors) was introduced, and the Kemeny's median was constructed as an optimized GSP that gives the most complete idea of true group wounds.

However, when establishing the ISP, the normative method of distribution of the total risk of errors was used, which led to a certain "coarsening" of rank assessments of the risk of errors in both the ISP and the GSP. The classical Hurwitz's criterion was not used in the construction of the GSP.

Previously a differential method of detecting part of the total risk of errors was developed and implemented by authors, which led to obtaining more accurate ISPs, and hence GSPs. Which contributed to an increase in the consistency of errors (Kendall's concordance coefficient) immediately by 1.92 times relative to its indicator calculated for GSP, which were obtained by generalization of ISP, obtained by the traditional method of normative distribution of the total risk of error. The multi-step technology of detection and elimination of marginal thoughts is realized. For the first time, the criteria of danger and frequency of adverse events proposed by ICAO were used for an integrative (holistic, aggregate) assessment of the level of undesirability of errors. However, neither the classical DM criteria nor the Kemeny's median was used to obtain the GSP based on these results.

Therefore, taking into account the impact of HF on FS, as well as the results of analysis of ATC's SP on the dangers of characteristic errors, the purpose of this publication is to build GSP using the classical criteria of DM, as well as assess their risk from the standpoint of distinguishing the dangers of orderly errors.

3 Forming and Solving a Matrix of Solutions

The methodology of applying the classical criteria of DM to solve applied technical and economic problems is well known. However, recommendations for taking into account the peculiarities of the impact of the HF on DM, especially in aviation systems with the use of classical criteria contains a limited number of works. Based on the above, consider the appropriate technology.

Thus, $m = 37$ of the respondents-ATCs arranged $n = 21$ of the characteristic errors, using the differential method of determining the comparative risk of errors. Using multi-step technology to detect and weed out marginal thoughts, this sample was reduced to $m_A = 27$ members who have a high level of intragroup agreement: the Kendall concordance coefficient is equal to $W_{m_A} = 0.7$ and is statistically plausible at an unusually high significance level for HF studies $\alpha = 1\%$. The aggregation of ISPs for members of this subgroup in GSP was carried out using a group decision strategy such as summation and averaging of ranks. The formal type is

$$E_{18} \underset{m_A}{\succ} E_{20} \underset{m_A}{\succ} E_5 \underset{m_A}{\succ} E_{21} \underset{m_A}{\succ} E_4 \underset{m_A}{\succ} E_3 \underset{m_A}{\succ} E_8 \underset{m_A}{\succ} E_{17} \underset{m_A}{\succ} E_{13} \underset{m_A}{\succ} E_2 \underset{m_A}{\succ} E_{16} \succ$$
$$\underset{m_A}{\succ} E_{19} \underset{m_A}{\succ} E_6 \underset{m_A}{\succ} E_{12} \underset{m_A}{\succ} E_7 \underset{m_A}{\succ} E_1 \underset{m_A}{\succ} E_{14} \underset{m_A}{\succ} E_{11} \underset{m_A}{\succ} E_9 \underset{m_A}{\succ} E_{10} \underset{m_A}{\succ} E_{15}, \tag{1}$$

where $\underset{m_A}{\succ}$ is the mark of the advantage of the danger of one error over another in the GSP (1), constructed by the differential method of comparing their dangers.

The ISPs of the members of the m_A subgroup form a decision matrix (Table 1), which is the so-called "loss matrix", because the smaller the absolute value of the rank of the i-th error in the ISP of the j-th expert (r_{ij}), the more dangerous it is. The solution of the decision matrix can be done with the help of classical DM criteria. Abraham Wald's criterion is considered a criterion of extreme pessimism (caution), because its application contributes to a guaranteed result. The approach based on the main principles of systems analysis, known as "removal of uncertainty". According to him, the best solution (the most dangerous mistake) is from the analysis of Table 1 as follows:

$$Z_W = \min_i r_{ik} = \min_i \max_j r_{ij}. \tag{2}$$

According to Table 1 and Eqs. (1)–(5), we conclude that the ATCs GSP on the dangers of characteristic errors were the same when applying the Wald criterion and the Savage criterion. Their formal description is as follows:

$$E_4 \underset{W}{\succ} E_{20} \underset{W}{\succ} E_5 \underset{W}{\approx} E_{18} \underset{W}{\succ} E_3 \underset{W}{\succ} E_{21} \underset{W}{\succ} E_2 \underset{W}{\succ} E_{12} \underset{W}{\approx} E_{17} \underset{W}{\succ} E_{13} \underset{W}{\succ} E_8 \underset{W}{\succ}$$
$$\underset{W}{\succ} E_1 \underset{W}{\succ} E_7 \underset{W}{\succ} E_6 \underset{W}{\approx} E_9 \underset{W}{\approx} E_{10} \underset{W}{\approx} E_{11} \underset{W}{\approx} E_{14} \underset{W}{\approx} E_{15} \underset{W}{\approx} E_{16} \underset{W}{\approx} E_{19}, \tag{3}$$

where $\underset{W/S}{\succ}$, $\underset{W/S}{\approx}$ are marks of comparative advantage and adequacy of hazard errors in GSP, constructed using the classical Wald's criterion.

The application of Wald's criterion can lead to the loss of a very successful solution (error of the first kind), when a significant error can get an inadequate rank for its

Table 1. Matrix of solutions for construction of group systems of advantages of air traffic controllers on dangers of characteristic errors (fragment).

ATC_j	E_i							
	E_1	E_2	E_3	E_4	...	E_{18}	...	E_{21}
1	19^{18}	10^9	5^4	5^4	...	2^1	...	1^0
2	14^{13}	7^6	6^5	1^0	...	2^1	...	4^3
3	10^9	8^7	5^4	5^4	...	1^0	...	2^1
4	12^{11}	10^9	6^5	4^3	...	1^0	...	3^2
5	9^8	10^9	5^4	6^5	...	1^0	...	2^1
6	17^{16}	7^6	6^5	3^2	...	2^1	...	1^0
7	11^{10}	7^6	6^5	5^4	...	1^0	...	3^2
8	18^{17}	17^{16}	6^5	3^2	...	1^0	...	2^1
9	19^{18}	7^6	5^4	2^1	...	1^0	...	6^5
10	20^{19}	13^{12}	5^4	1^0	...	8^7	...	3^2
11	7^6	16^{15}	5^4	2^1	...	1^0	...	6^5
12	14^{13}	12^{11}	2^1	$3.5^{2,5}$	...	1^0	...	5^4
13	16^{15}	9^8	6^5	5^4	...	$2.5^{1,5}$	...	1^0
14	17^{16}	12^{11}	10^9	2^1	...	1^0	...	3^2
15	$10.5^{6,5}$	17^{13}	4^0	4^0	...	4^0	...	4^0
16	15^{14}	13^{12}	7^6	$3.5^{2,5}$	...	1^0	...	14^{13}
17	7^6	9^8	5^4	5^4	...	2^1	...	3^2
18	18^{17}	11^{10}	5^4	3^2	...	1^0	...	8^7
19	19^{18}	15^{14}	7^6	3^2	...	5^4	...	2^1
20	20^{19}	$9,5^{8,5}$	4^3	$5,5^{4,5}$	...	1^0	...	3^2
21`	1^0	6^5	4^3	5^4	...	$2,5^{1,5}$	...	8^7
22	1^0	6^5	4^3	5^4	...	3^2	...	8^7
23	14^{13}	10^9	5^4	6^5	...	1^0	...	2^1
24	10^9	8^7	7^6	4^3	...	3^2	...	1^0
25	$17.5^{16,5}$	9^8	7^6	2^1	...	1^0	...	6^5
26	16^{15}	9^8	8^7	2^1	...	1^0	...	5^4
27	19^{18}	$7.5^{6,5}$	6^5	1^0	...	3^2	...	4^3
r_i^W	12	7	5	1	...	3,5	...	6
r_i^S	12	7	5	1	...	3,5	...	6
r_i^{B-L}	16	10	6	5	...	1	...	4
r_i^{HW}	10	8	5	1	...	3	...	6

danger. Therefore, in addition to it, other classical criteria of DM should be applied to the construction of ATC's GSP on the dangers of characteristic errors.

Savage's criterion is usually seen as the development and refinement of Wald's criterion. This criterion is considered democratic for group decision-making because it takes into account the views of both the majority and the minority of experts. According to this criterion, such a strategy (GSP) is considered optimal, which provides the minimum general deviation from it of the ICP of respondents in the most unfavorable situation. This deviation is traditionally called risk, regret, fine, sadness.

The best solution (the most dangerous error) when applying the Savage criterion to the data in Table 1 is by the following formula:

$$Z_S = \min_i \max_j a_{ij} = \min_i \max_j \left| \min_i r_{ij} - r_{ij} \right|. \tag{4}$$

Consistent application of time (4) to the data of Table 1 led to the following GSP:

$$E_4 \underset{S}{\succ} E_{20} \underset{S}{\succ} E_5 \underset{S}{\approx} E_{18} \underset{S}{\succ} E_3 \underset{S}{\succ} E_{21} \underset{S}{\succ} E_2 \underset{S}{\succ} E_{12} \underset{S}{\approx} E_{17} \underset{S}{\succ} E_{13} \underset{S}{\succ} E_8 \underset{S}{\succ}$$
$$\underset{S}{\succ} E_1 \underset{S}{\succ} E_7 \underset{S}{\succ} E_6 \underset{S}{\approx} E_9 \underset{S}{\approx} E_{10} \underset{S}{\approx} E_{11} \underset{S}{\approx} E_{14} \underset{S}{\approx} E_{15} \underset{S}{\approx} E_{16} \underset{S}{\approx} E_{19}, \tag{5}$$

where $\underset{W/S}{\succ}$, $\underset{W/S}{\approx}$ are marks of comparative advantage and adequacy of hazard errors in GSP, constructed using the classical Savage's criterion.

As we can see, in our specific case GSP (3) and (5), obtained using the Wald's and Savage's criterions, respectively, are identical.

The Bayes-Laplace criterion is unusually simple and comes down to finding the sum of the error ranks, their further averaging, and the ordering of the errors in ascending order of the calculated averages. Which corresponds to the application to the data of Table 1 of the following formula:

$$Z_{BL} = \min_i \bar{r}_i = \min_i \left(\frac{1}{n} \sum_{j=1}^{n} r_{ij} \right). \tag{6}$$

where $\bar{r}_i$ is the rank of the i-th error, obtained by summing and averaging the opinions (ranks) of all m respondents-ATC's.

Therefore, applying formula (6) to the data of Table 1, we obtain the following GSP:

$$Er_{\cdot 18} \underset{BL}{\succ} Er_{\cdot 20} \underset{BL}{\succ} Er_{\cdot 4} \underset{BL}{\succ} Er_{\cdot 21} \underset{BL}{\succ} Er_{\cdot 5} \underset{BL}{\succ} Er_{\cdot 3} \underset{BL}{\succ} Er_{\cdot 2} \underset{BL}{\succ} Er_{\cdot 8} \underset{BL}{\succ} Er_{\cdot 13} \underset{BL}{\succ} Er_{\cdot 16} \underset{BL}{\succ} Er_{\cdot 17} \underset{BL}{\succ}$$
$$\underset{BL}{\succ} Er_{\cdot 12} \underset{BL}{\succ} Er_{\cdot 19} \underset{BL}{\succ} Er_{\cdot 6} \underset{BL}{\succ} Er_{\cdot 1} \underset{BL}{\succ} Er_{\cdot 7} \underset{BL}{\succ} Er_{\cdot 14} \underset{BL}{\succ} Er_{\cdot 11} \underset{BL}{\succ} Er_{\cdot 9} \underset{BL}{\succ} Er_{\cdot 10} \underset{BL}{\succ} Er_{\cdot 15} \tag{7}$$

where $\underset{BL}{\succ}$ are marks of the advantage of the danger of one error over another in the GSP, constructed using the Bayes-Laplace criterion.

Note that the Bayes-Laplace criterion led to a GSP of the form (7), which duplicates such a strategy of group decisions as summation and averaging of ranks, illustrated by the GSP (1). Therefore, the GSP obtained with its help is checked for consistency using the Kendall concordance coefficient. This is important, because when it comes to averaging ranks, it contributes to a risky result, when the found average value of a certain rank does not correspond to any opinion (Fig. 1).

The initial position of ATC as a person being DM, when applying the Bayes-Laplace criterion, is more optimistic than in the case of Wald's criterion, but assumes a higher level of awareness and long and frequent implementation. Therefore, this criterion is also called the criterion of insufficient justification.

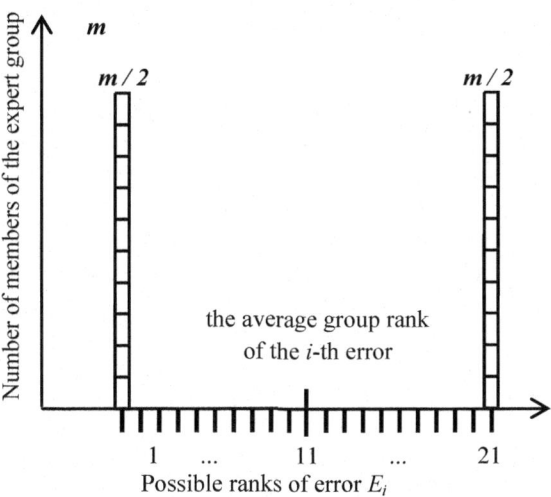

Fig. 1. Illustration of the dangers of simply averaging conflicting and opposing opinions about the dangers of E_i mistake.

The Hurwiz's criterion is based on the desire to take an equilibrium position when choosing an indicator (rank) that best characterizes the risk of error. To do this, enter the coefficient of optimism α ($0 \leq \alpha \leq$) and the corresponding coefficient of pessimism $1 - \alpha$. The value of the coefficient is determined based on the initial (more optimistic or pessimistic) position of the expert. It is assumed that the danger of each error is best characterized by the weighted sum of the highest and lowest rank given to it in the ISP:

$$Z_{HW} = \min_i c_i = \min_i \left[\alpha \cdot \min_j r_{ij} + (1 - \alpha) \cdot \max_j r_{ij} \right]. \tag{8}$$

We assume that the coefficient of optimism is equal to $\alpha = 0.3$, then, accordingly, the pessimism coefficient is equal to $1 - \alpha = 0.7$.

Applying the following values of the coefficients of optimism / pessimism and expression (8) to the data of Table 1, we obtain the following GSP:

$$E_4 \underset{HW}{\succ} E_{20} \underset{HW}{\succ} E_{18} \underset{HW}{\succ} E_5 \underset{HW}{\succ} E_3 \underset{HW}{\succ} E_{21} \underset{HW}{\succ} E_{13} \underset{HW}{\succ} E_2 \underset{HW}{\succ} E_1 \underset{HW}{\succ} E_{12} \underset{HW}{\succ} E_8 \underset{HW}{\succ}$$
$$\underset{HW}{\succ} E_6 \underset{HW}{\approx} E_{16} \underset{HW}{\approx} E_{19} \underset{HW}{\succ} E_7 \underset{HW}{\succ} E_{11} \underset{HW}{\succ} E_9 \underset{HW}{\succ} E_{15} \underset{HW}{\succ} E_{14} \underset{HW}{\succ} E_{10} \underset{HW}{\succ} E_{17} \tag{9}$$

where $\underset{HW}{\succ}$, $\underset{HW}{\approx}$ are the corresponding marks of comparative advantage and adequacy of the risk of errors in the GSP, obtained using the Hurwitz's criterion.

Note that the measuring properties of the ordering scale used to determine the ranks of error hazards impose certain restrictions on their mathematical processing. Therefore, when applying the Bayes-Laplace criterion, we used a simple sum of ranks. In addition, when applying the Hurwitz's criterion, it was assumed that the scale is linear and quantitative.

4 Determination of the Risk Index of Indistinguishability of Error's Danger

GSP (3), (5), (7), (9) were based on an additional factor of Spearman's rank correlation. Its empirical values equal: $R_S^{BL-W/S} = 0.8922$, $R_S^{BL-HW} = 0.9263$, $R_S^{W/S-HW} = 0.9477$, and is statistically plausible at an unusually high significance level for HF studies:

$$t_{emp.}^{BL-W/S} = 8.611 > > t_{1\%, k=19} = 2.861; t_{emp.}^{BL-HW} = 10.716 > > t_{1\%, k=19} 2.861$$
$$t_{emp.}^{W/S-HW} = 12.943 > > t_{1\%, k=19} = 2.861.$$

The above means that the coincidence of the ranks of the danger of errors in the GSP (3), (5), (7), (9) is a regularity, and not coincidence is an accident.

Based on the presence/absence of related ranks in the obtained GSPs (3), (5), (7), (9), it seems possible to calculate the following normalized indicators of the degree of fragmentation/indistinguishability of the danger of errors in them:

$$R^* = \frac{T}{T_{max}} = \frac{\sum_{\gamma=1}^{n} \left(t_\gamma^3 - t_\gamma \right)}{n^3 - n}, \tag{10}$$

where T is the indicator of the presence of related ranks in the GSP, which is determined from the formula for calculating the Kendall concordance coefficient. It makes sense of the correction factor, which is calculated in all k "cases" of indistinguishability of ordered objects-errors; t_γ is the number of indistinguishable errors of one "case"; $n = 21$ is the number of errors ranked; T_{max} is the indicator of maximum error indistinguishability, when all ordered errors are conditionally considered to be the same in terms of danger:

$$E_1 = E_2 = ... = E_n \Leftrightarrow r_{E_1} = r_{E_2} = ... r_{E_{21}} \Leftrightarrow T_{max} = n^3 - n = 21^3 - 21 = 9240. \tag{11}$$

If condition (11) is really fulfilled, then the error indistinguishability index in the GSP is the maximum and is equal to $R^* = R_{max}^* = 1$. If, on the contrary, all errors are strictly ordered, that is, there are no associated (middle) ranks in the GSP, this indicator is minimal, this figure is minimal $R^* = R_{min}^* = 0$.

Applying formulas (10), (11) to the data of Table. 1 and GSP (3), (5), (7), (9), we obtain $R^*_{BL} = 0$, which is quite natural, since in GSP (7), which was obtained using the Bayes-Laplace criterion, there are no related ranks.

The studied indicator reaches the maximum (among the obtained) values $R^*_{W/S} = 5.58 \cdot 10^{-2}$ for GSP (3), (5), constructed using Wald/Savage criteria. Which is 29 times more than for GSP (9), built using the Hurwitz criterion: $R^*_{HW} = 0.19 \cdot 10^{-2}$.

Note that although the absolute values of the established empirical indicators R^* are small, the results still give an idea of the comparative effectiveness of the applied classical criteria of DM for risk assessment – the uncertainty of the indistinguishability of the dangers of errors in them. To assess the degree of differentiation of the dangers of errors by the expert group as a whole, expression (10) is converted into the following:

$$R^*_g = \frac{1}{m}\sum_{j=1}^{m}R^*_j = \frac{1}{m}\sum_{j=1}^{m}\frac{\sum_{\gamma=1}^{n}\left(t^3_{\gamma j} - t_{\gamma j}\right)}{n^3 - n} = \frac{1}{m(n^3 - n)}\sum_{j=1}^{m}\sum_{\gamma=1}^{n}\left(t^3_{\gamma j} - t_{\gamma j}\right), \qquad (12)$$

where R^*_j is the indicator of the risk of indistinguishability of the dangers of errors in the ISP of the j-th expert-ATC; $t_{\gamma j}$ is the number of indistinguishable errors of one "case" in the ISP j-th ATCs.

Using formula (12) and the data in Table. 1, we establish that $R^*_g = 0.52 \cdot 10^{-2}$. As we can see, this indicator is almost identical to the result obtained for GSP (9), constructed using the Hurwitz criterion.

5 Conclusions and Prospects for Further Research

Based on the new scientific results obtained and presented in this publication. It is necessary to state the fact of a real solution to the problem of correct application of the spectrum of classical decision making (DM) criteria (Wald, Savage, Bayes-Laplace, Hurwitz) for construction of group system of preferences (GSP) of Ukrainian ATC's on characteristic errors, which they make in their professional activities. Some partial results include the following:

1. From the comparison of the obtained GSP follows the adequacy of the rankings the results of the application of the Wald and Savage criterion, as well as the Bayes-Laplace criterion and such a strategy of group decisions as summation and averaging of ranks.
2. All the obtained group systems of preferences (GSP) coincide, confirming the unusually high in absolute value positive values of Spearman's rank correlation coefficients, statistically significant at an unusually high level of significance $\alpha = 1\%$ for human factor (HF) studies.
3. The normalized risk factor is introduced – the uncertainty of the indistinguishability of alternatives-errors, based on one of the components of the formula for determining the Kendall concordance coefficient. The minimum risk of indistinguishability is

observed in the GSP obtained using the Bayes-Laplace criterion ($R^*_{BL} = 0$), the maximum – under the conditions of application of the Wald/Savage criterion ($R^*_{W/S} = 5.58 \cdot 10^{-2}$). Some intermediate place is occupied by the results of the application of the Hurwitz test ($R^*_{HW} = 0.19 \cdot 10^{-2}$). At the same time, the error indistinguishability index for the expert group as a whole reaches a value and is close to the indicator calculated for the GSP, determined using the Hurwitz criterion.

4. The given methodology of application of classical criteria of DM is universal and can be applied to construction of GSP for researches in any field of human activity.
5. Based on the above, it should be noted the fact of expanding the methodology of expert procedures in the study of the human factor (HF). Further research should be conducted in the following areas (without ranking):

- construction of the Kemeny's median as an optimization ATC's GSP on the spectrum of characteristic errors;
- carrying out of the comparative analysis of efficiency of methods of construction of ATC's GSP on dangers of a spectrum of characteristic errors;
- clarification of the possible influence of cross-cultural factors on the attitude of ATC's to the dangers of typical errors and etc.

References

1. ICAO Doc 9758-AN/966. Human factors guidelines for air traffic management (ATM) systems. International Civil Aviation Organization, Montreal (2000)
2. Virovac, D., Domitrović, A., Bazijanac, E.: The influence of human factor in aircraft maintenance. Promet Traffic Transp. 29(3), 257–266 (2017). https://doi.org/10.7307/ptt.v29i3.2068
3. ICAO Doc 9806 AN/763. Human factors guidelines for safety audits manual. International Civil Aviation Organization, Montreal (2002)
4. Leveson, N.: A new accident model for engineering safer systems. Saf. Sci. 42(4), 237–270 (2004). https://doi.org/10.1016/S0925-7535(03)00047-X
5. Zarei, E., Yazdi, M., Abbassi, R., Khan, F.: A hybrid model for human factor analysis in process accidents: FBN-HFACS. J. Loss Prev. Process Ind. 57, 142–155 (2019). https://doi.org/10.1016/j.jlp.2018.11.015
6. Kelly, D., Efthymiou, M.: An analysis of human factors in fifty controlled flight into terrain aviation accidents from 2007 to 2017. J. Saf. Res. 69, 155–165 (2019). https://doi.org/10.1016/j.jsr.2019.03.009
7. Reva, O., Kamyshyn, V., Nevynitsyn, A., et al.: Criteria indicators of the consistency of air traffic controllers' preferences on a set of characteristic errors. In: Stanton, N. (ed.) Advances in Human Aspects of Transportation. AHFE 2020. AISC, vol. 1212, pp. 617–623. Springer, Cham (2020). https://doi.org/10.1007/978-3-030-50943-9_79
8. Karanikas, N., Chionis, D., Plioutsias, A.: "Old" and "new" safety thinking: perspectives of aviation safety investigators. Saf. Sci. 125, 104632 (2020). https://doi.org/10.1016/j.ssci.2020.104632

Smartphone for Smart Physics Learning

Oksana Luchsheva$^{(\boxtimes)}$ ⑩, Ihor Turkin ⑩, Ihor Klymenko ⑩,
and Vitaliy Narozhnyy ⑩

National Aerospace University "Kharkiv Aviation Institute",
17 Chkalova Street, Kharkiv 61070, Ukraine
o.luchsheva@khai.edu

Abstract. The possibility of using microelectromechanical system (MEMS) of the mobile devices to physics learning is the main subject of this work. We have used the smartphone as a tool for physics measurements. The proposed solution illustrates the easing of smartphone use in experiments for physics learning. Other sensors can also be integrated into physics training experiments. The drawbacks of each embedding sensor can be suppressed by using the sensor data fusion. The evolution of the developed application should be carried out in the following directions: use of augmented reality technology in order to make the obtained results more demonstrative and understandability, algorithmic integration of measurement results by different sensors, application of modern methods of digital filtering in the application. The concepts discussed in this paper can prove useful to both students and teachers for the practical implementation of a virtual laboratory workshop to study the laws of mechanics at universities.

Keywords: Science education · Laboratory work · Smartphones · Gyroscope · Microelectromechanical system · Accelerometer · Physical pendulum

1 Introduction

This introductory section provides a brief overview of the advantages and current tendencies of the technology BYOD (Bring Your Own Devices) in the high school learning process. BYOD in the classroom is the principle of active use for the training of smartphones, laptops, tablets, and other digital devices, which are not provided by the university, and students use their tools. Mobile technology has recently been developed and made available to a large number of users in the world, and especially among young people. Modern smartphones and tablets are powerful and sophisticated computing devices with a variety of different data sensors.

Usually, students use smartphones for communication in social networks, video viewing, music listening, or in a better case, for information searching. Therefore not surprising that in Ukraine, the ban on the use of mobile phones during the educational process was introduced in May 2007, and seven years ago, in August 2014, it was abolished to spread the use of information and communication technologies. The French parliament in July 2018 has approved a far-reaching ban on mobile phones in schools following the result of the vote in the National Assembly. The law fundamentally bans the use of mobile phones in all preschools, primary, and middle schools.

M. Nechyporuk et al. (Eds.): ICTM 2020, LNNS 188, pp. 135–146, 2021.
https://doi.org/10.1007/978-3-030-66717-7_12

Nevertheless, due to the widespread use of these devices, they began to be introduced in education. Modern educational technologies offer many options for developing new and modernization of existing training courses – the search for an appropriate structure of educational strategies and relevant teaching aids, then for searching for their purposeful and meaningful interconnection [1], given the dissemination of mobile devices (more than 100 million smartphone owners in the United States already in 2012 [2]), especially among students. According to the statistics, the number of smartphones worldwide exceeded 2.8 billion in 2019. Smartphone capabilities and embedded sensors allow them to play a more significant role in health, identification, localization, tracking, and education [3].

Three-axis accelerometers embedded in mobile devices allow us to determine motion characteristics by measuring the instantaneous acceleration of an object compared to gravity at any given time. An alternative to the accelerometer is a gyroscope, which is capable of the rotation speed measuring. The gyroscope allows determining the orientation of the device in space and connects this data with the virtual world [4]. Smartphones, equipped with inertial sensors, provide an opportunity to solve a wide range of practical tasks. Authors of the tutorial [5] focus on the signal processing aspects of position and orientation estimation using inertial sensors - microelectromechanical system (MEMS). The article [6] discusses the capabilities of the diverse array of sensors and radios that modern smartphones contain, as well as the advantages and disadvantages of each of them for indoor localization. Models, methods, and technology for the simultaneous use of a magnetometer, accelerometer, and gyroscope for the joint assessment of both three-dimensional location and three-dimensional orientation can significantly improve the accuracy of the location estimate [7]. Heterogeneous tools and development languages make it difficult to develop multiplatform mobile applications that use embedded sensors. The work [8] examines the issues of using the Model Driven Architecture (MDA) approach to create mobile applications that use embedded sensors. Using MDA allows you to generate native code without re-developing all files and lines of code, which are sometimes redundant and difficult to customize.

The papers [9, 10] give examples of the smartphone's sensors using for investigation of the several types of one-dimensional motions, where acceleration plays an essential role in the characterization of systems. The primary kinematic relationships between angular velocity and centripetal acceleration and the coherence of measurements obtained with different sensors were verified using smartphone sensors in work [11]. In work [12], the real acceleration, velocity, and angle of rotation of the smartphone, which was measured made by the in-built acceleration and gyroscope sensors, were compared with independent results from video recordings. A comparison of the advantages and disadvantages of two different instrumentational technologies for their using in science learning experiments was given in the paper [13]. Each of these technologies can be of interest to both researchers and practitioners in advanced learning strategies. The work [14] investigated theoretically and experimentally using smartphone' sensors for a study of the dynamics of a ubiquitous, traditional, and straightforward toy, namely the yoyo. The work shows that the usage of these traditional toys to teach physics is a new approach to promote engagement and creativity.

The primary purpose of this study is to create an additional possibility of using visual mobile applications for the study of mechanical motion, for the formation of such learning outcomes, as understanding how moving, rotation, trajectory, speed, average speed, acceleration, and so on, connected. We suppose that the student's learning outcome is to their practical ability of mechanics laws using by solving the problem of determining the location of the constructive placement of an accelerometer in their smartphone.

2 Model and Method

Baseline data – oscillograms of measurement results of accelerations along the X–Y axis and the angular velocity along the Z-axis, obtained from the gyroscope and accelerometer of the smartphone fixed on a physical pendulum (see Fig. 1).

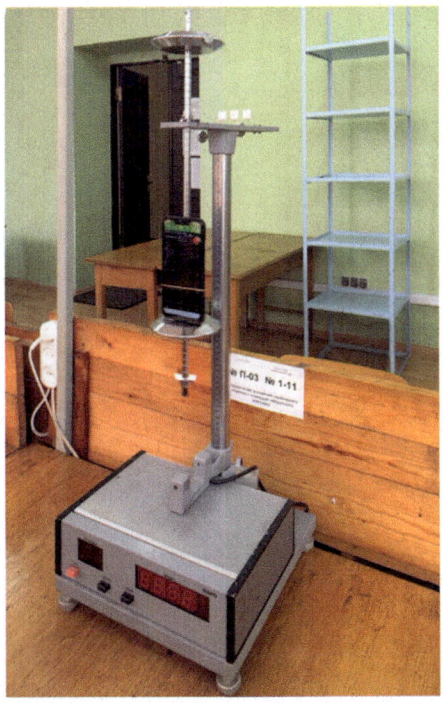

Fig. 1. Snapshot of the physical laboratory pendulum with a fixed smartphone.

The expected result in the student's report is the determined coordinates of the accelerometer position in the smartphone with an assessment of the accuracy of their determination. The initial laboratory setup allows us to do independent measurements of the oscillation period of the pendulum with an accuracy of 0.001 s using an optical sensor. In our design, the oscillation amplitude decreases by less than 1% within 30 s.

Therefore, we can neglect friction when considering a short time interval 15–20 s, i.e., of the order of 10–15 periods. Consider small oscillations of a pendulum with a smartphone fixed on it. Assume that (see Fig. 2): $\theta_Z(\tau)$ is the angular displacement of the physical pendulum's center of mass from the equilibrium position at the moment of time τ; $\Delta\theta$ is the unknown angular position of the accelerometer relative to the axis, passing through the fixed point O and the center of mass of the pendulum; ρ is the distance from the static point O to the unknown position of the accelerometer.

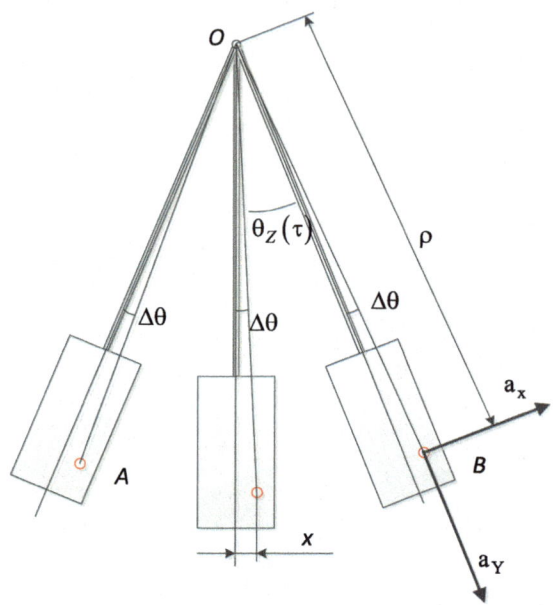

Fig. 2. Kinematic diagram of the physical laboratory pendulum.

A 3-axis accelerometer measures proper acceleration, i.e., acceleration of a body in its instantaneous rest frame. Proper acceleration differs from the coordinate acceleration in a fixed inertial system, then, as to an example, an accelerometer at rest on the surface of the Earth will measure the acceleration due to Earth's gravity $g \approx 9.81$ m/s². Vice versa, accelerometers in free fall will measure zero.

Consequently, a 3-axes accelerometer in a smartphone will measure the sum of two terms. The first one is the projection of acceleration of gravity on the axes of the accelerometer, and the second is the projection of the accelerations, which forced by the oscillatory motion are equal:

$$a_X(\tau) = -g \cdot \sin(\theta_Z(\tau) + \Delta\theta) - \theta_Z''(\tau) \cdot \rho,$$
$$a_Y(\tau) = g \cdot \cos(\theta_Z(\tau) + \Delta\theta) + (\theta_Z'(\tau))^2 \cdot \rho. \tag{1}$$

Considering that the accelerometer measures proper acceleration in gravitational acceleration unit's g and using the small-angle approximation:

$$\sin(\theta_Z + \Delta\theta) \approx \theta_Z + \Delta\theta,$$
$$\cos(\theta_Z + \Delta\theta) \approx 1 - (\theta_Z + \Delta\theta)^2 \big/ 2, \qquad (2)$$

we present Eq. (1) in the next view:

$$a_X(\tau) \approx -\theta_Z(\tau) - \Delta\theta - \theta_Z''(\tau) \cdot p/g,$$
$$a_Y(\tau) \approx 1 - (\theta_Z(\tau) + \Delta\theta)^2 \big/ 2 + (\theta_Z'(\tau))^2 \cdot p/g. \qquad (3)$$

If we can provide conditions of a physical experiment that will correspond to the definition of the "simple pendulum", following constraints of the physical pendulum must be respected – the start deviation angle from the pendulum vertical must satisfy the condition $\theta_0 \approx \sin\theta_0$.

Suppose that the above constraint is satisfied, then the solution of the differential equation of motion of the mathematical pendulum will be:

$$\theta_Z(\tau) \approx \theta_0 \cdot \cos(\omega_0 \cdot (\tau - \tau_0)),$$
$$\theta_Z'(\tau) \approx -\omega_0\theta_0 \cdot \sin(\omega_0 \cdot (\tau - \tau_0)), \qquad (4)$$
$$\theta_Z''(\tau) \approx -\omega_0^2\theta_0 \cdot \cos(\omega_0 \cdot (\tau - \tau_0)),$$

where τ_0 is a start time-point for measuring movement's characteristics, in reality, this is any time stamp for which angular velocity changes its sign from positive to negative (see Fig. 3), θ_0 is the start deviation angle from the pendulum vertical, $\omega_0 = 2\pi/T$ is the angular frequency of periodic motion with a period T, s.

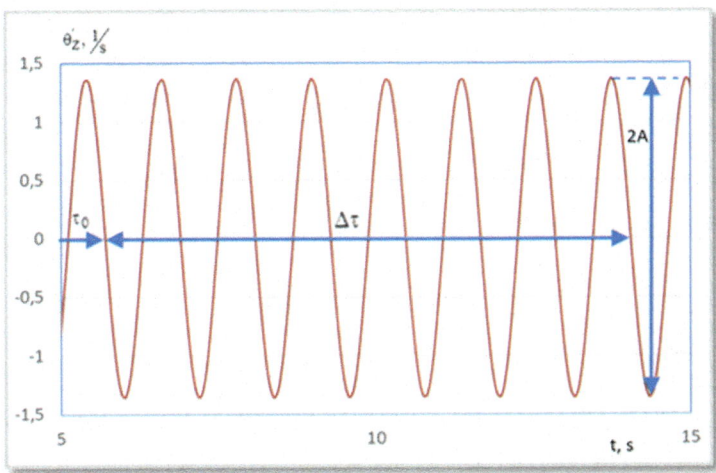

Fig. 3. The mathematical model of the time series $\theta_Z'(\tau)$ to determine ω_0 and θ_0.

The method of determining unknown parameters $\Delta\theta$ and ρ is adapted to the capabilities of first-year students at the university or high school. It means it does not require knowledge of modern methods for the identification of mathematical models. In this formulation of the problem, it is necessary to refrain from using the results of numerical differentiation and/or numerical integration of measurement results also because the mistakes of the first approach critically depend on random measurement errors, and the second, on systematic measurement errors. In the first case, these are quantization and discretization errors, in the second, the "0" drift of MEMS-sensors.

The method is doing by the following five consecutive steps:

Step 1. Fix the smartphone on the physical laboratory pendulum, run the program below for measuring and recording time series $a_X(\tau)$, $a_Y(\tau)$, $\theta'_Z(\tau)$. Deviate the pendulum from the equilibrium position by an angle θ_0 (the recommended value is 5–10°), release the pendulum, and allow it to do about 10–15 oscillations.

Step 2. Using the measured time series $\theta'_Z(\tau)$, select the starting point τ_0 on the chart and the time $\Delta\tau$ of n oscillations. Measurement results allow determining the cyclic frequency of oscillatory motion (see Fig. 3):

$$\omega_0 = 2\pi n/\Delta\tau, \quad \theta_0 = A/\omega_0. \tag{5}$$

Step 3. Determine the $a_Y(\tau)$ at the upper boundary points of the deviation of the pendulum (when the angular displacements of the center of the pendulum of mass from the equilibrium position are maximum $\theta_Z \approx \pm\theta_0$) and their difference:

$$\Delta a_Y \approx g \cdot (\cos(\theta_0 + \Delta\theta) - \cos(\theta_0 - \Delta\theta)) = -2 \cdot g \cdot \sin(\theta_0) \cdot \sin(\Delta\theta), \tag{6}$$

that allows us to calculate $\Delta\theta$:

$$|\Delta\theta| \approx \frac{\Delta a_Y}{2 \cdot g \cdot \theta_0}. \tag{7}$$

The sign of the $\Delta\theta$ indicates the direction of reference of the angular position of the accelerometer relative to the axis, which passes through the fixed point O and the center of mass of the pendulum.

Value Δa_Y may be obtained from the time series $a_Y(\tau)$ (see Fig. 4) or phase trajectories $a_Y(\theta'_Z)$ (see Fig. 5), $a_Y(a_X)$ (see Fig. 6).

Step 4. The ρ is calculated using an equation:

$$\rho \approx g \frac{a_{Y_{max}} - 1}{\theta'^2_{Z_{max}}} = g \frac{a_{Y_{max}} - 1}{(\omega_0 \theta_0)^2}, \tag{8}$$

in which we use the value $a_{Y_{max}}$ obtained from the measured time series $a_Y(\tau)$ ($a_{Y_{max}}$ corresponds to the pendulum position $\theta_Z \approx 0$ at which $|\theta'_Z(\tau)| \approx$ max). Value $a_{Y_{max}} - 1$ also may be obtained from the time series $a_Y(\tau)$ (see Fig. 4) or phase trajectories $a_Y(\theta'_Z)$ (see Fig. 5) and $a_Y(a_X)$ (see Fig. 6).

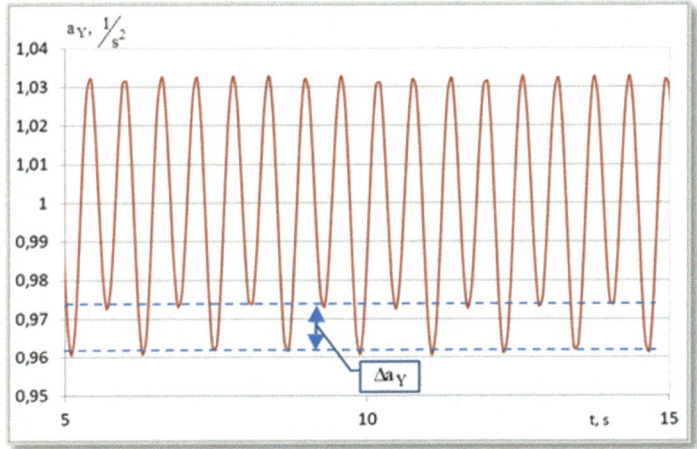

Fig. 4. Determination Δa_Y for (7) from the measured time series $a_Y(\tau)$.

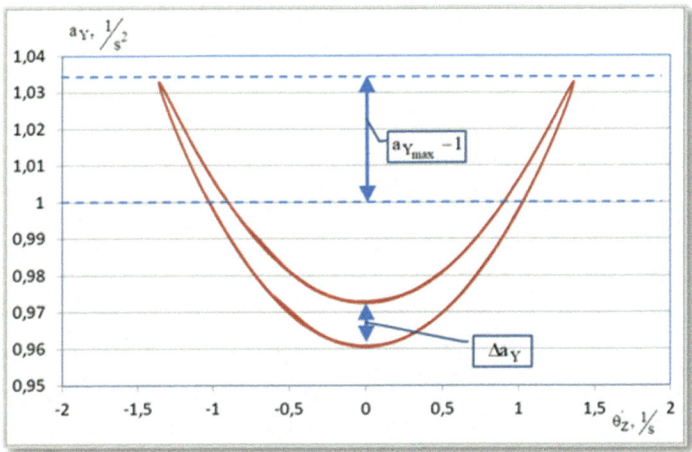

Fig. 5. Determination Δa_Y for (7) and $a_{Y_{max}} - 1$ for (8) from the phase trajectory $a_Y(\theta'_Z)$.

Step 5. Compare the measurement results with the calculated data and estimate the accuracy of the model obtained. To evaluate the measurement accuracy, we suggest repeating the experiment with the new position of the smartphone (for example, to turn it over 90° or 180°) and compare the positions of the accelerometer obtained from different measurements. Besides, having experimentally determined the accelerometer coordinates (the distance ρ from the fixed point O and the accelerometer position) and the oscillation period of the pendulum using an optical sensor, you can calculate the maximum values of the angular velocity of the pendulum and compare it with the measurement results of this value with the accelerometer.

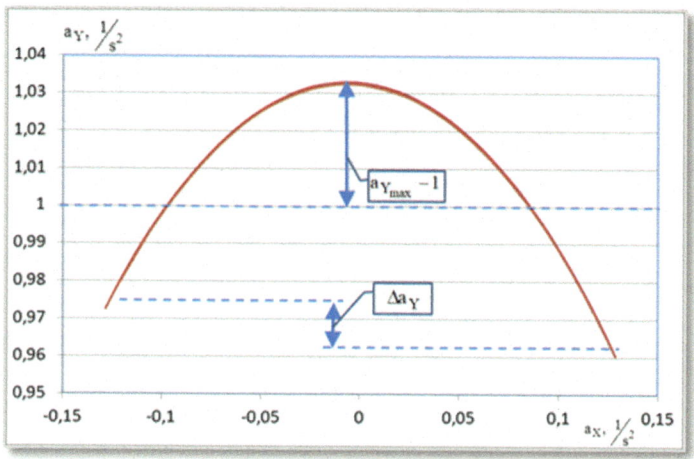

Fig. 6. Determination Δa_Y for (7) and $a_{Y_{max}} - 1$ for (8) from the phase trajectory $a_Y(a_X)$.

3   Results

The software complex implements the following functions to automate the educational process:

– visualization and recording of measurement results to a file;
– sending messages to students' mobile devices. Messages can contain lists of tasks for the current day. After completing assignments, students send messages to the controller, notifying that the assignment is complete;
– formation of reporting on statistics;
– view the results of the task in real-time.

The project is realized in the Java programming language in the Android Studio IDE for use in the Java virtual machine of the Android operating system. The project uses the SQLite relational database management system. SQLite gives a significant advantage for all interested parties, as there is the possibility of remote access to information and the ability to control in real-time. In general, a specific software package has three interacting parts:

– the server part implements mechanisms for accessing the database;
– Android-application for students provides inertial sensors measurement, visualization of the results of these measurements, result saving to a file, interaction with the user through the menu system (see Fig. 7);
– Android-application app for the teacher allows you to manage assignments, view learning outcomes, etc.

The volume of data obtained allows us to "visualize" the equations of motion of the pendulum (see Fig. 8 and 9).

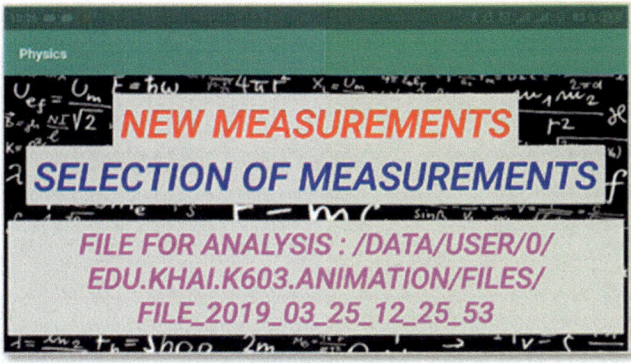

Fig. 7. Menu of the software package.

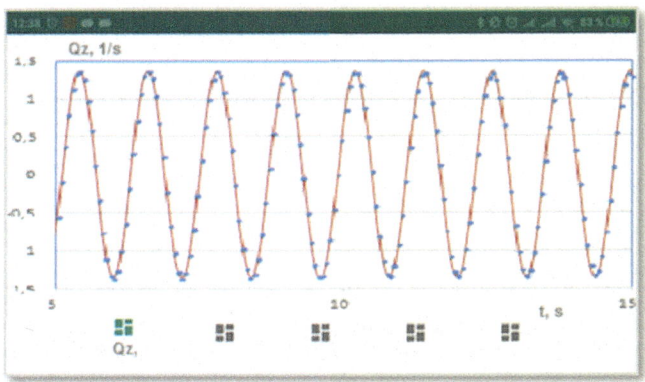

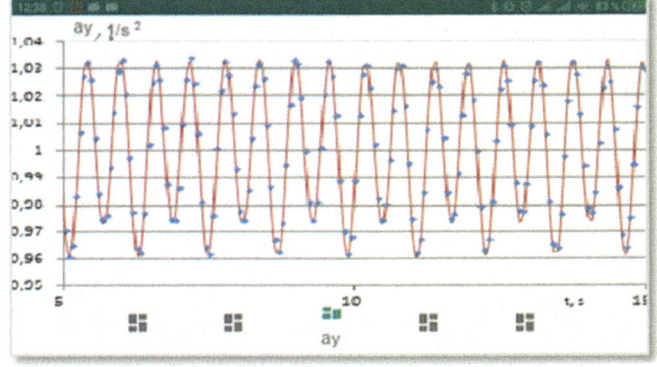

Fig. 8. The measured time series $\theta'_Z(\tau)$, $a_Y(\tau)$.

The simplicity of the construction of parametric diagrams makes the process of data processing and extraction of important information readily available (see Fig. 3, 4, 5 and 6).

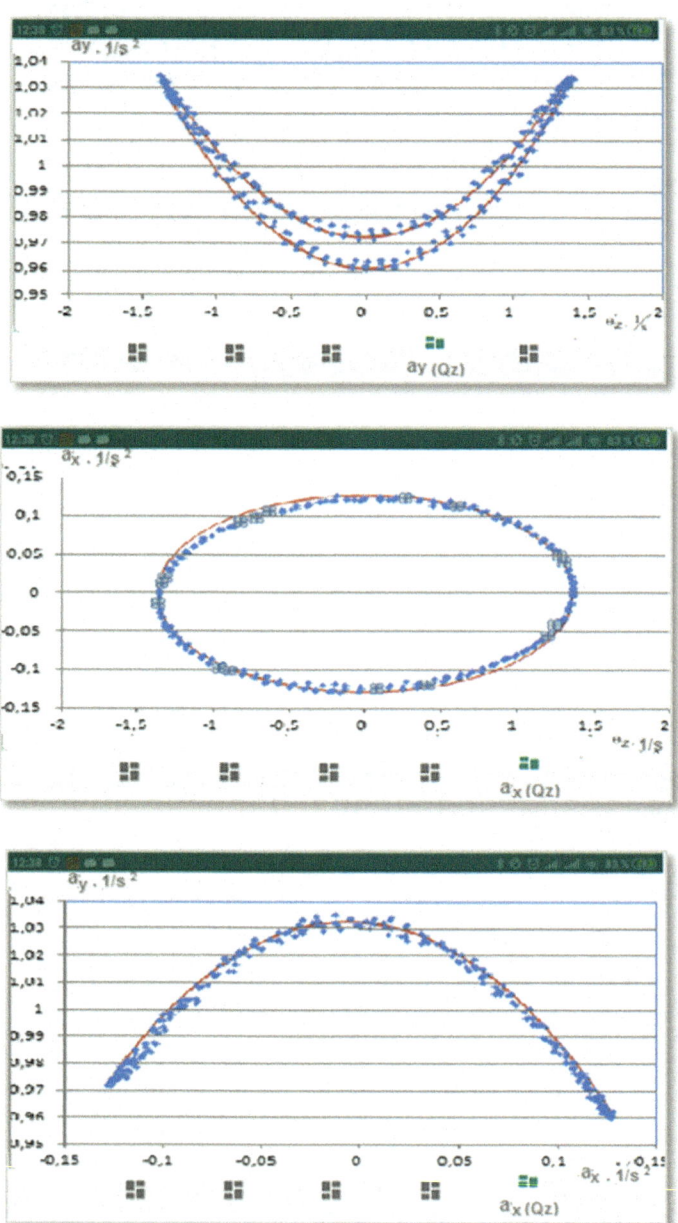

Fig. 9. The phase trajectories $a_X(\theta'_Z)$, $a_Y(\theta'_Z)$, $a_Y(a_X)$.

A comparison of accelerometer measurements of angular velocity and linear accelerations with indirect measurements of these parameters by the traditional method showed a discrepancy within 10%. The error correlates with the accuracy of an accelerometer declared by their manufacturers. The use of phase trajectories not only provides clarity in the processing of experimental data but also introduces students to the basics of parametric modeling. Compared with the recently popular virtual modeling, the proposed approach is based on the receipt and processing of experimental data of a real object under study. In this case, the student's attention is primarily focused not on the process of fixing the motion parameters, but on the processing and interpretation of the data used by parametric models, the reproducibility of the experiment.

4   Conclusion and Perspectives

This study proposes a simple Android-based solution for undergraduate learning outcome achievement, namely a practical ability of mechanics laws using by solving the problem of determining the location of the constructive placement of an accelerometer in their smartphone.

We have used the smartphone as a tool for physics measurements. The proliferation of smartphones, almost ubiquitous, and the new sensors that incorporate them make them almost essential in a modern physics laboratory. The proposed solution illustrates the easing of smartphone use in experiments for physics learning. Other sensors, such as a light sensor and a magnetic field sensor, can also be integrated into physics training experiments. A fusion of data from all three position sensors (Accelerometer, Magnetometer, and Gyroscope) is the best solution for robust and spatial rotation estimation. Non-gravitational acceleration effect and magnetic interference can be suppressed by using data from the gyroscope. On the contrary, the drift of the gyroscope can be compensated by data from the accelerometer and magnetometer. Data from an accelerometer and magnetometer can also be used to calculate the reference rotation instead of the starting position.

We believe that the evolution of the developed application should be carried out in the following directions:

– use of augmented reality technology to make the obtained results more demonstrative and understandability;
– algorithmic integration of measurement results by different sensors, which will allow to mutually compensate for the errors of each of them and improve measurement accuracy;
– application of modern methods of digital filtering in the application, the priority of which we believe is the use of the Kalman filter.

The finished goal of its improvements is the development and practical implementation of a virtual laboratory workshop to study the laws of mechanics at universities. We believe that the concepts discussed in this paper can prove useful to both students and teachers.

References

1. Strecker, S., Kundisch, D., Lehner, F., et al.: Higher education and the opportunities and challenges of educational technology. Bus. Inf. Syst. Eng. **60**(2), 181–189 (2018). https://doi.org/10.1007/s12599-018-0522-8
2. Oprea, M., Miron, C.: The study of the free and damped harmonic oscillations using the accelerometer of a smartphone. In: Vlada, M., et al. (eds.) Proceedings of the 12th International Conference on Virtual Learning (ICVL), pp. 448–454. Thomson Reuters (2017)
3. Masoud, M., Jaradat, Y., Manasrah, A., Jannoud, I.: Sensors of smart devices in the Internet of Everything (IoE) era: big opportunities and massive doubts. J. Sens. **2019**, 6514520 (2019). https://doi.org/10.1155/2019/6514520
4. Passaro, V., Cuccovillo, A., Vaiani, L., et al.: Gyroscope technology and applications: a review in the industrial perspective. Sensors **17**(10), 2284 (2017). https://doi.org/10.3390/s17102284
5. Kok, M., Hol, J.D., Schön, T.B.: Using inertial sensors for position and orientation estimation. Found. Trends Signal Process. **11**(1–2), 1–53 (2017). https://doi.org/10.1561/2000000094
6. Langlois, C., Tiku, S., Pasricha, S.: Indoor localization with smartphones: harnessing the sensor suite in your pocket. IEEE Consum. Electron. Mag. **6**(4), 70–80 (2017). https://doi.org/10.1109/MCE.2017.2714719
7. Shen, S., Gowda, M., Choudhury, R.R.: Closing the gaps in inertial motion tracking. In: Proceedings of the 24th Annual International Conference on Mobile Computing and Networking, pp. 429–444. ACM, New York (2018). https://doi.org/10.1145/3241539.3241582
8. Lachgar, M., Lamhaddab, K., Abdelmounaim, A.: Mobile phone sensors meta-model. Int. J. Ad hoc Sens. Ubiquit. Comput. **9**, 15–32 (2018). https://doi.org/10.5121/ijasuc.2018.9202
9. Monsoriu, J.A., Giménez, M.H., Ballester, E., et al.: Smartphone acceleration sensors in undergraduate physics experiments. In: Proceedings of the Joint International Conference on Engineering Education & International Conference on Information Technology (ICEE/ICIT), Riga, pp. 109–116 (2014)
10. Ballester, E., Castro-Palacio, J.C., Velázquez-Abad, L., et al.: Smart physics with smartphone sensors. In: 2014 IEEE Frontiers in Education Conference (FIE) Proceedings, pp. 1–4). IEEE, Madrid (2014). https://doi.org/10.1109/FIE.2014.7044031
11. Monteiro, M., Cabeza, C., Marti, A.C., et al.: Angular velocity and centripetal acceleration relationship. Phys. Teach. **52**(5), 312 (2014). https://doi.org/10.1119/1.4872422
12. Monteiro, M., Cabeza, C., Marti, A.C.: Acceleration measurements using smartphone sensors: dealing with the equivalence principle. Revista Brasileira de Ensino de Física **37**(1), 1303 (2015). https://doi.org/10.1590/S1806-11173711639
13. Hochberg, K., Kuhn, J., Müller, A.: Science education with handheld devices: a comparison of Nintendo WiiMote and iPod touch for kinematics learning. Perspect. Sci. **10**, 13–18 (2016). https://doi.org/10.1016/j.pisc.2016.01.008
14. Salinas, I., Monteiro, M., Marti, A.C., Monsoriu, J.A.: Dynamics of a yoyo using a smartphone gyroscope sensor. arXiv preprint, arXiv:1903.01343 (2019)

Classification of Diabetes Disease Using Logistic Regression Method

Andrew Hrimov[1] , Ievgen Meniailov[1] ,
Dmytro Chumachenko[1(✉)] , Kseniia Bazilevych[1] ,
and Tetyana Chumachenko[2]

[1] National Aerospace University "Kharkiv Aviation Institute",
17 Chkalova Str., Kharkiv 61070, Ukraine
andrew.hrimov@gmail.com, evgenii.menyailov@gmail.com,
dichumachenko@gmail.com, ksenia.bazilevich@gmail.com
[2] Kharkiv National Medical University, 4 Nauky Av., Kharkiv 61022, Ukraine
tatlchum@gmail.com

Abstract. At the moment, there are many methods of analysis and classification aimed at building the most accurate and effective mathematical models that are widely used in medicine as a decision-making tool. Existing methods make it possible to identify the relationships between input and output variables in the sample, build models reflecting these relationships, compare them in terms of accuracy, profitability and costs, and choose the most effective model. The increase in the incidence of diabetes not only in the world, but also in Ukraine, dictates the need to introduce a mathematical apparatus for automatic diagnosis of the disease. Within the framework of the study, the classification of patients with diabetes by the logistic regression method was implemented. Python is used for software implementation.

Keywords: Machine learning · Diagnostics · Classification · Diabetes · Logistic regression

1 Introduction

The global coronavirus pandemic has once again demonstrated the need to introduce digital tools into healthcare [1]. The digitalization of medicine is one of the most urgent tasks in the modern world, and already created products and solutions are used in various areas of public health: in the management of medical institutions [2–5], diagnostics of various diseases [6–8], modeling epidemic processes [9, 10] and predicting morbidity [11], surgical treatment [12, 13] and even in training of medical personnel [14–17].

The modern development of information technologies makes it possible to develop not just shells for automating the work of medical institutions [18], but also complex systems based on methods and means of artificial intelligence [19], multi-agent modeling [20], game theory [21], decision theory [22], machine learning [23, 24], computer vision [25, 26] and other modern methods and approaches.

M. Nechyporuk et al. (Eds.): ICTM 2020, LNNS 188, pp. 147–157, 2021.
https://doi.org/10.1007/978-3-030-66717-7_13

The worldwide attention to the incidence of Covid-19 not only does not diminish the importance of global epidemics of other diseases, but often exacerbates them. One of these diseases is diabetes.

Diabetes is a chronic disease that develops when the pancreas does not produce enough insulin, or when the body cannot use the insulin it makes efficiently. Insulin is a hormone that regulates blood sugar levels. A common result of uncontrolled diabetes is hyperglycemia, or elevated blood sugar levels, which over time leads to severe damage to many systems in the body, especially nerves and blood vessels.

According to the latest official data from the Ministry of Health, there were 1.27 million people with diabetes in Ukraine. Among them, almost 200,000 patients require daily insulin intake. From 2010 to 2017, the total number of patients increased by 4%, and the rate per 100 thousand population – by 12%. The specific weight of diabetes mellitus cases among all diseases during this period increased by 0.3% (from 1.4% to 1.7%). According to the Public Health Center, in Ukraine, almost half of the patients with diabetes are not diagnosed.

One of the tools in solving the problem of diabetes diagnosis is the use of a machine learning apparatus to identify infected based on test data. Thus, the **aim of the research** is the automated classification of patients with suspected diabetes based on the logistic regression method.

2 Materials and Methods

At the moment, there are many methods of analysis and classification aimed at building the most accurate and effective mathematical models that are widely used in medicine as a decision-making tool [27]. Existing methods make it possible to identify the relationships between input and output variables in the sample, build models reflecting these relationships, compare them in terms of accuracy, profitability and costs, and choose the most effective model [28, 29]. In our case, this may be the presence or absence of diabetes.

Linear regression is used to model linear relationships between a continuous output variable and a set of input variables [30]. Under certain conditions, the linear regression equation serves as an irreplaceable and very high-quality tool for analysis and forecasting. The linear regression model is the most common and simplest equation for the relationship between input and output variables. In addition, the constructed linear regression equation can be the starting point for data analysis.

When analyzing data, there are often problems where the output variable is categorical, and then the use of linear regression is difficult. Therefore, when looking for relationships between a set of input variables and a categorical output variable, logistic regression has become widespread. Logistic regression is a binary classification method. It allows you to estimate the probability of realization (or non-realization) of an event depending on the values of some independent variables. The logistic regression line, unlike the linear one, is not straight.

ROC curve (Receiver Operator Characteristic) is a curve used to represent the results of binary classification in machine learning. Since there are two classes, one of them is called a class with positive outcomes, the other – with negative outcomes.

The ROC curve shows the dependence of the number of correctly classified positive examples on the number of incorrectly classified negative examples.

In the terminology of ROC analysis, the former are called true positive, and the latter, false negative. In this case, it is assumed that the classifier has a certain parameter, by varying which, we will obtain one or another division into two classes. This parameter is often called the threshold, or cut-off value. Depending on it, different values of type I and II errors will be obtained.

In logistic regression, the cut-off threshold ranges from 0 to 1 – this is the calculated value of the regression equation. Let's call it a rating.

To understand the essence of type I and II errors, consider a four-field confusion matrix (Table 1), which is built on the basis of the results of classification by the model and the actual (objective) belonging of the examples to classes.

Table 1. Confusion matrix.

Model	Actually positive	Actually negative
Positive	*True Positives* are correctly classified positive examples (the so-called true positive cases)	*False Positives* are negative examples classified as positive (type II error). This is a false detection because in the absence of an event, a decision is mistakenly made about its presence (false positive cases)
Negative	*False Negatives* are positive examples classified as negative (Type I error). This is the so-called "false pass" – when the event of interest to us is mistakenly not detected (false negative examples)	*True Negatives* are correctly classified negative examples (true negative cases)

What is positive and what is negative depends on the specific task.

When analyzing, they often operate not with absolute indicators, but with relative shares (rates), expressed as a percentage:

– the proportion of True Positive Rate:

$$TPR = \frac{TP}{TP + FN} \cdot 100\%;$$

– the proportion of False Positive Rate:

$$FPR = \frac{FP}{TN + FP} \cdot 100\%.$$

Let us introduce two more definitions: sensitivity and specificity of the model. They determine the objective value of any binary classifier.

Sensitivity is the proportion of truly positive cases:

$$S_e = TPR = \frac{TP}{TP + FN} \cdot 100\%.$$

Specificity is the proportion of true negative cases that were correctly identified by the model:

$$S_p = \frac{TN}{TN + FP} \cdot 100\%.$$

A high-sensitivity model often gives a true result if there is a positive outcome. Conversely, a model with high specificity is more likely to give a true result when there is a negative outcome (it detects negative examples). If we talk in terms of medicine - the problem of diagnosing a disease, where the model for classifying patients into sick and healthy is called a diagnostic test, then we get the following:

- a sensitive diagnostic test manifests itself in overdiagnosis – the maximum prevention of missing patients;
- a specific diagnostic test only diagnoses patients with certainty. This is important in the case when, for example, the treatment of a patient is associated with serious side effects and overdiagnosis of patients is not desirable.

3   Results

For implementation the classification method we have used open dataset of Diabetes patients: PIMA Indians Diabetes Database. Each instance represents individual patients and their various medical attributes along with diabetes classification. Database has 768 instances and 9 attributes (Table 2).

Table 2. Full list of parameters.

Parameter name	Description	Data type
Pregnancies	Number	Decimal
PG Concentration	Count	Integer
Diastolic BP	Count	Integer
Tri Fold Thick	Count	Integer
Serum Ins	Count	Integer
BMI	Count	Integer
DP Function	Count	Integer
Age	Years	Decimal
Diabetes	Present or not	0/1

First step of data analysis is data preprocessing. For analysis and program real-ization we have used Python language. First of all, we need to import necessary modules and upload data from database (Fig. 1).

```python
import pandas as pd
import numpy as np
import matplotlib.pyplot as plt
from sklearn.linear_model import LogisticRegression
from sklearn.preprocessing import StandardScaler
from sklearn.model_selection import train_test_split
from sklearn.metrics import confusion_matrix

df1 = pd.read_excel('Diabetes.xls')
df1.head()
```

	Pregnancies	PG Concentration	Diastolic BP	Tri Fold Thick	Serum Ins	BMI	DP Function	Age	Diabetes
0	6	148	72	35	0	33.6	0.627	50	Sick
1	1	85	66	29	0	26.6	0.351	31	Healthy
2	8	183	64	0	0	23.3	0.672	32	Sick
3	1	89	66	23	94	28.1	0.167	21	Healthy
4	0	137	40	35	168	43.1	2.288	33	Sick

Fig. 1. Data import.

For correct analysis we have to change format of "infected" and "healthy" values to "0" and "1". Let's form data frames of characteristics and Boolean values of the disease. Next, we will set the data for training and validation and build a logistic regression model. Next, we will check it on test data and find the accuracy of its classification.

Figure 2 shows the error matrix for the constructed model. Here 39 is the number of correctly predicted healthy people, 35 are incorrectly predicted healthy people, 16 are incorrectly predicted patients and 141 people were correctly identified as sick, in other words, the model correctly predicted 39 + 141 = 180 people, and was mistaken in the case of 35 + 16 = 52 persons.

```python
confusion_matrix = confusion_matrix(y_test, y_pred)
print(confusion_matrix)

[[ 39  35]
 [ 16 141]]
```

Fig. 2. Error matrix.

To improve the accuracy of the model, it is necessary to analyze the characteristics that were used for the classification (Fig. 3). Here, you can see the differences in mean values for sick and healthy patients, in order to better understand the influence of each characteristic, we visualize their values (Fig. 4).

Next, let's build several models, taking into accounts the factors Age, Pregnancies, Serum Ins, DP Function, and the second – PG Concentration, Diastolic BP, Tri Fold Thick and BMI, similarly creating data frames and setting test and data for training the model.

```
print(df1.groupby('Diabetes').mean())
```

```
          Pregnancies  PG Concentration  Diastolic BP  Tri Fold Thick  \
Diabetes
0            4.865672        141.257463      70.824627       22.164179
1            3.298000        109.980000      68.184000       19.664000

          Serum Ins       BMI  DP Function        Age
Diabetes
0        100.335821  35.142537     0.550500  37.067164
1         68.792000  30.304200     0.429734  31.190000
```

Fig. 3. Characteristics used for classification.

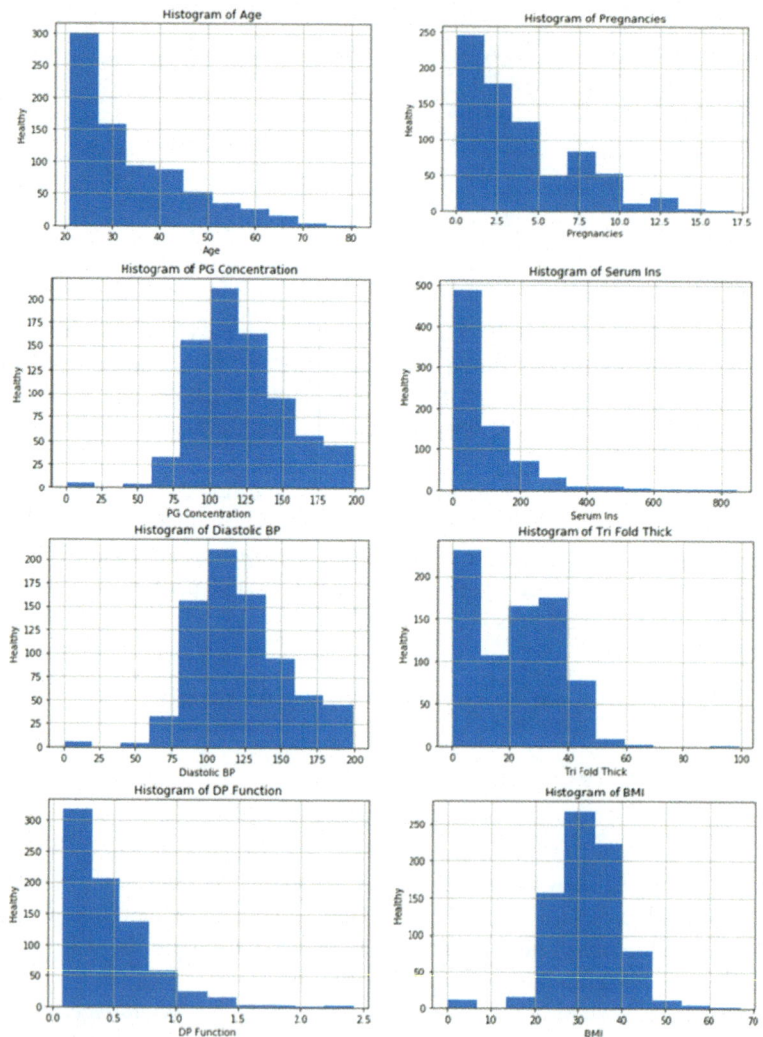

Fig. 4. Data visualization.

The first model classifies with an accuracy of 0.6926. Second model with an accuracy of 0.7706. ROC-curves of the general model, the model of factors 1 and factors 2 are presented in Fig. 5, 6 and 7, respectively.

Analysis shows that the characteristics of the second group have a greater impact on the classification accuracy. The dashed line represents the ROC curve of a completely random classifier. A good classifier remains as far from it as possible (towards the upper left corner). In this case, the optimal classifiers can be called those presented in Fig. 5 and 7.

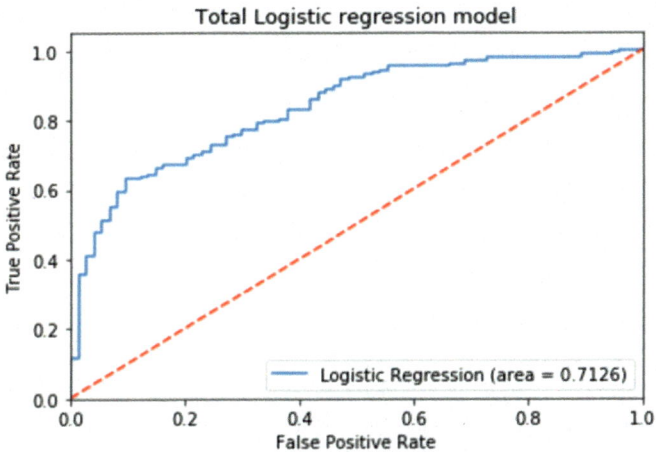

Fig. 5. Model of all factors.

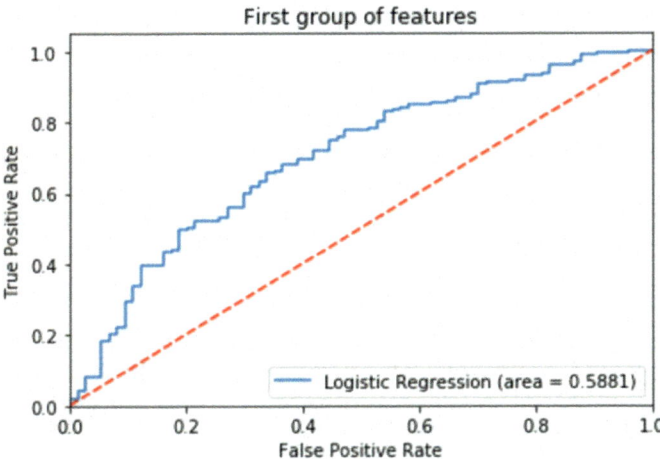

Fig. 6. Model of first group of factors (Age, Pregnancies, Serum Ins, DP Function).

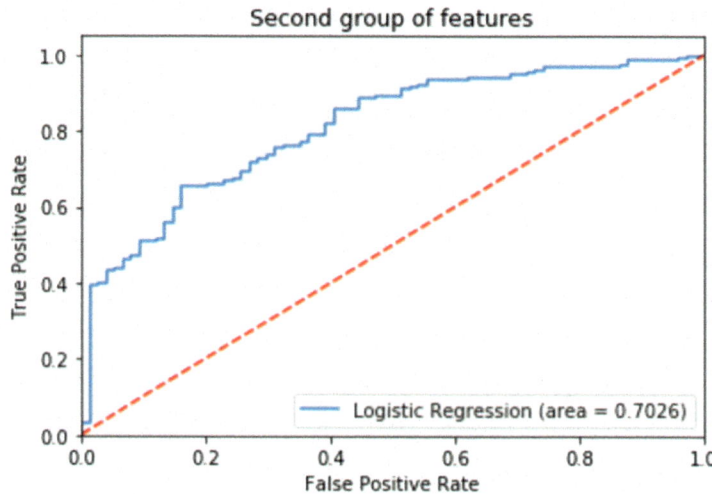

Fig. 7. Model of second group of factors (PG Concentration, Diastolic BP, Tri Fold Thick and BMI).

4 Conclusions

The article presents a logistic regression method as a tool for developing a mathematical-statistical model for predicting the probability of an event of interest to the researcher in the presence of two possible outcomes. The ROC analysis method was selected and described in detail as a tool for assessing the quality of the model. The capabilities of these methods are demonstrated by a real example of creating and evaluating the effectiveness (sensitivity and specificity) of a model for predicting the likelihood of diabetes incidence.

The analysis showed that the factors PG Concentration, Diastolic BP, Tri Fold Thick and BMI most affect the accuracy of the disease detection.

The accuracy of the model was 77%. At first glance, the accuracy of the model is relatively low for the problem of diagnosing morbidity. However, in our case, the option with the maximum sensitivity and specificity of the tests was chosen, which indicates overdiagnosis of patients. In the task of diagnosing diabetes, this is the best option, because a false-positive result can threaten, for example, only an additional visit to the doctor, and a false-negative result can not reveal a dangerous, but curable disease.

Acknowledgment. The study was funded by the Ministry of Health of Ukraine for the state budget in the framework of the research work on the theme "To develop a scientifically substantiated strategy of prevention antibiotic resistance of the bacteria causing of healthcare-associated infections in healthcare facilities" (State registration number 0118U000944).

References

1. Ohannessian, R., Duong, T.A., Odone, A.: Global telemedicine implementation and integration within health systems to fight the COVID-19 pandemic: a call to action. JMIR Public Health Surveill. **6**(2), e18810 (2020). https://doi.org/10.2196/18810
2. Dotsenko, N., Chumachenko, D., Chumachenko, I.: Modeling of the process of critical competencies management in the multi-project environment. In: 2019 IEEE 14th International Conference on Computer Sciences and Information Technologies (CSIT), vol. 3, pp. 89–93. IEEE, Lviv (2019). https://doi.org/10.1109/STC-CSIT.2019.8929765
3. Dotsenko, N., Chumachenko, D., Chumachenko, I.: Management of critical competencies in a multi-project environment. In: CEUR Workshop Proceedings, vol. 2387, pp. 495–500 (2019)
4. Dotsenko, N., Chumachenko, D., Chumachenko, I.: Project-oriented management of adaptive teams' formation resources in multi-project environment. In: CEUR Workshop Proceedings, vol. 2353, pp. 911–920 (2019)
5. Dotsenko, N., Chumachenko, D., Chumachenko, I.: Modeling of the processes of stakeholder involvement in command management in a multi-project environment. In: 2018 IEEE 13th International Scientific and Technical Conference on Computer Sciences and Information Technologies (CSIT), vol. 1, pp. 29–32. IEEE, Lviv (2018). https://doi.org/10.1109/STC-CSIT.2018.8526613
6. Bazilevych, K., Meniailov, I., Fedulov, K., et al.: Determining the probability of heart disease using data mining methods. In: CEUR Workshop Proceedings, vol. 2488, pp. 383–394 (2019)
7. Strilets, V., Bakumenko, N., Chernysh, S., et al.: Application of artificial neural networks in the problems of the patient's condition diagnosis in medical monitoring systems. . In: Nechyporuk, M., et al. (eds.) Integrated Computer Technologies in Mechanical Engineering. AISC, vol. 1113, pp. 173–185. Springer, Cham (2020). https://doi.org/10.1007/978-3-030-37618-5_16
8. Bakumenko, N., Strilets, V., Ugryumov, M.: Application of the C-means fuzzy clustering method for the patient's state recognition problems in the medical monitoring systems. In: CEUR Workshop Proceedings, vol. 2362, pp. 218–227 (2019)
9. Chumachenko, D., Meniailov, I., Bazilevych, K., et al.: Development of an intelligent agent-based model of the epidemic process of syphilis. In: 2019 IEEE 14th International Conference on Computer Sciences and Information Technologies (CSIT), vol. 1, pp. 42–45. IEEE, Lviv (2019). https://doi.org/10.1109/STC-CSIT.2019.8929749
10. Chumachenko, D., Chumachenko, T.: Intelligent agent-based simulation of HIV epidemic process. In: Lytvynenko, V., et al. (eds.) Lecture Notes in Computational Intelligence and Decision Making. ISDMCI 2019. AISC, vol. 1020, pp. 175–188. Springer, Cham (2020). https://doi.org/10.1007/978-3-030-26474-1_13
11. Polyvianna, Y., Chumachenko, D., Chumachenko, T.: Computer aided system of time series analysis methods for forecasting the epidemics outbreaks. In: 2019 IEEE 15th International Conference on the Experience of Designing and Application of CAD Systems (CADSM), pp. 1–4. IEEE, Polyana (2019). https://doi.org/10.1109/CADSM.2019.8779344
12. Nechyporenko, A.S., Krivenko, S.S., Alekseeva, V., et al.: Uncertainty of measurement results for anatomical structures of paranasal sinuses. In: 2019 8th Mediterranean Conference on Embedded Computing (MECO), pp. 1–4. IEEE, Budva (2019). https://doi.org/10.1109/MECO.2019.8760032

13. Alekseeva, V., Lupyr, A., Urevich, N., et al.: Significance of anatomical variations of maxillary sinus and ostiomeatal components complex in surgical treatment of sinusitis. Novosti Khirurgii **27**(2), 168–176 (2019). https://doi.org/10.18484/2305-0047.2019.2.168
14. Herasymova, A., Chumachenko, D., Padalko, H.: Development of intelligent information technology of computer processing of pedagogical tests open tasks based on machine learning approach. In: CEUR Workshop Proceedings, vol. 2631, pp. 121–130 (2020)
15. Zabolotnyi, O.V., Zabolotnyi, V.A., Koshevoi, M.D.: Conditionality examination of the new testing algorithms for coal-water slurries moisture measurement. Naukovyi Visnyk Natsionalnoho Hirnychoho Universytetu **1**, 51–59 (2018)
16. Chumachenko, D., Balitskii, V., Chumachenko, T., et al.: Intelligent expert system of knowledge examination of medical staff regarding infections associated with the provision of medical care. In: CEUR Workshop Proceedings, vol. 2386, 321–330 (2019)
17. Zabolotnyj, A.V., Koshevoj, N.D.: Development, examination and optimization of the device for quality control of dielectric materials. Pribory i Sistemy Upravleniya **1**, 39–42 (2004)
18. Mazorchuck, M., Dobriak, V., Chumachenko, D.: Web-application development for tasks of prediction in medical domain. In: 2018 IEEE 13th International Scientific and Technical Conference on Computer Sciences and Information Technologies (CSIT), vol. 1, pp. 5–8. IEEE, Lviv (2018). https://doi.org/10.1109/STC-CSIT.2018.8526684
19. Mashtalir, V.P., Shlyakhov, V.V., Yakovlev, S.V.: Group structures on quotient sets in classification problems. Cybern. Syst. Anal. **50**(4), 507–518 (2014). https://doi.org/10.1007/s10559-014-9639-z
20. Chumachenko, D., Chumachenko, K., Yakovlev, S.: Intelligent simulation of network worm propagation using the code red as an example. Telecommun. Radio Eng. **78**(5), 443–464 (2019). https://doi.org/10.1615/TelecomRadEng.v78.i5.60
21. Chumachenko, D., Meniailov, I., Bazilevych, K., Chumachenko, T.: On intelligent decision making in multiagent systems in conditions of uncertainty. In: 2019 XIth International Scientific and Practical Conference on Electronics and Information Technologies (ELIT), pp. 150–153. IEEE, Lviv (2019). https://doi.org/10.1109/ELIT.2019.8892307
22. Mashtalir, V.P., Yakovlev, S.V.: Point-set methods of clusterization of standard information. Cybern. Syst. Anal. **37**(3), 295–307 (2001). https://doi.org/10.1023/A:1011985908177
23. Karatekin, T., Sancak, S., Celik, G., et al.: Interpretable machine learning in healthcare through generalized additive model with pairwise interactions (GA2M): predicting severe retinopathy of prematurity. In: 2019 International Conference on Deep Learning and Machine Learning in Emerging Applications (Deep-ML), pp. 61–66. IEEE, Istanbul (2019). https://doi.org/10.1109/Deep-ML.2019.00020
24. Piletskiy, P., Chumachenko, D., Meniailov, I.: Development and analysis of intelligent recommendation system using machine learning approach. In: Nechyporuk, M., et al. (eds.) Integrated Computer Technologies in Mechanical Engineering. AISC, vol. 1113, pp. 186–197. Springer, Cham (2020). https://doi.org/10.1007/978-3-030-37618-5_17
25. Gargin, V., Radutny, R., Titova, G., et al.: Application of the computer vision system for evaluation of pathomorphological images. In: 2020 IEEE 40th International Conference on Electronics and Nanotechnology (ELNANO), pp. 469–473. IEEE, Kyiv (2020). https://doi.org/10.1109/ELNANO50318.2020.9088898
26. Nechyporenko, A., Reshetnik, V., Alekseeva, V., et al.: Assessment of measurement uncertainty of the uncinated process and middle nasal concha in spiral computed tomography data. In: 2019 IEEE International Scientific-Practical Conference Problems of Infocommunications, Science and Technology (PIC S&T), pp. 585–588. IEEE, Kyiv (2019). https://doi.org/10.1109/PICST47496.2019.9061557

27. Bondarenko, A.V., Pokhil, S.I., Lytvynenko, M.V., et al.: Anaplasmosis: experimental immunodeficient state model. Wiadomosci Lekarskie **72**(9), 1761–1764 (2019)
28. Kozko, V.M., Bondarenko, A.V., Gavrylov, A.V., et al.: Pathomorphological peculiarities of tuberculous meningoencephalitis associated with HIV infection. Intervent. Med. Appl. Sci. **9** (3), 144–149 (2017). https://doi.org/10.1556/1646.9.2017.31
29. Lytvynenko, M., Shkolnikov, V., Bocharova, T., et al.: Peculiarities of proliferative activity of cervical squamous cancer in HIV infection. Georgian Med. News **270**, 10–15 (2017)
30. Kim, K., Seol, J.A., Kim, D.B., et al.: Health management based on history of personalized physiological data using linear regression analysis. In: 2018 Global Medical Engineering Physics Exchanges/Pan American Health Care Exchanges (GMEPE/PAHCE), p. 1. IEEE, Porto (2018). https://doi.org/10.1109/GMEPE-PAHCE.2018.8400736

Conceptual Model of Information Security

Vladimir Pevnev⬤, Mikhail Tsuranov$^{(\boxtimes)}$⬤, Heorhii Zemlianko⬤,
and Olena Amelina

National Aerospace University "Kharkiv Aviation Institute",
17 Chkalova Str., Kharkiv 61070, Ukraine
{v.pevnev,m.tsuranov,g.zemlynko}@csn.khai.edu,
o.amelina@student.csn.khai.edu

Abstract. The presented paper deals to create a conceptual model of information security. The purpose of the paper is to investigate existing models and
develop a first-level conceptual model and lower-level models. The above
analysis of existing models showed their main disadvantage. This is the absence
of any dialectical connections between model elements. Such schemes are
primitive in terms of their use in the organization of protection systems. The
proposed model allows you to look at the task of creating information protection
systems taking into account the impact of vulnerabilities on possible scenarios
of attacks by the intruder, the impact of detected vulnerabilities on the protection
system, in terms of its improvement, the possibility of passive counteraction to
the actions of the intruder. The second-level models examined the elements of
the first-level model in more detail, which made it possible to penetrate more
deeply into the ideology of creating protection systems, including proactively
depriving some of the capabilities of a potential violator.

Keywords: Information security · Conceptual model · Vulnerabilities ·
Threats · Violator · Protection system

1 Introduction

The application of information technologies today determines the level of development
of industry, medicine, science, agriculture and all other segments of the industrial
complex, the market for services and all the vital activities of mankind. Information
technology controls our roads, critical production life support systems and the supply
of water to our homes, allows you to carry out complex operations and explore space.
Their use has changed the world today. At the same time, all systems become
increasingly vulnerable to aggressive influences. Having posed a new problem for
consumers of information technologies – the problem of information security (IS).

There are many options for solving this problem, in each case they should be
determined on the basis of specific conditions and the availability of material resources.
Any information security system was costly. This is the main argument of opponents of
the construction of such systems, until there are any incidents that cause loss of money,
in one form or another.

Many factors must be taken into account when building an information security
system. It is extremely difficult to do this, even to a specialist with extensive experience

© The Author(s), under exclusive license to Springer Nature Switzerland AG 2021
M. Nechyporuk et al. (Eds.): ICTM 2020, LNNS 188, pp. 158–168, 2021.
https://doi.org/10.1007/978-3-030-66717-7_14

in creating such systems. The conceptual model of information security can be used as a keynote for solving such a problem.

The main property of conceptual models (CM) is abstraction. The conceptual model allows us to look at the problem of information lack as if from the outside. This allows you to abstract from the effects independent of the researcher, while the possibility of their neutralization remains. High-quality conceptual models for describing information security of the first level are suitable for building any security systems. Further, these models can be developed by second-level models, in which you can detail the main elements of the first-level models.

The purpose of this article is to analyze existing ones and develop a conceptual model of information security materials.

2 Analysis Models of Information Security

2.1 Conceptual Model Requirements

Conceptual models are subject to a number of requirements. The most important of which is the abstraction of the model. What is understood in the work behind abstraction? This is a reflection of the main patterns of the model, links within the model in order to highlight the most common features. The next requirement is the versatility of the model. Universality should be understood as the possibility of using the model in question when building an information security protection system for any objects and organizations, from personal security to country information security. The requirement of comprehensiveness is based on the possibility of using the proposed model to create harmonized requirements for the information security system.

Completeness involves taking into account all critical elements that are taken into account in the construction of the protection system. Specificity implies the absence of redundant elements that complicate the process of building protection systems. The last requirement for systems to be analyzed is the requirement of informatively. The model should be understandable and digestible for the specialist who works with it.

Meeting these requirements, in our opinion, allows us to create an effective information security system using a particular conceptual model.

2.2 CIA Model (Confidentiality, Integrity, Availability)

We will begin the analysis of information security models with the simplest of them - the CIA model [1]. The diagram of this model is presented in Fig. 1.

This model is based on the principle of building information security. This is to ensure accessibility and confidentiality and to monitor integrity. This model is best seen as a way of thinking and reasoning about how to best protect data in your network. Modern standards add authorization, authentication, and non-compliance requirements to privacy, integrity, and availability requirements.

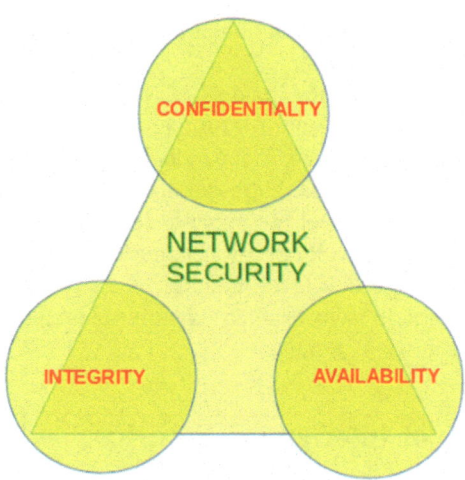

Fig. 1. CIA model.

Advantages: high level of abstraction, definition of specific criteria.

Disadvantages: The model is not a ready-made approach that can ensure information security in your system.

2.3 DREAD Model

The model in question allows you to identify and evaluate the threats that a potential intruder can use. By taking preventive action, these threats can be repaired or minimized.

Based on the proposed approach, it is possible to create a safety profile that can include various categories, ranging from input verification to system audit [2]. The model of this system is shown in Fig. 2.

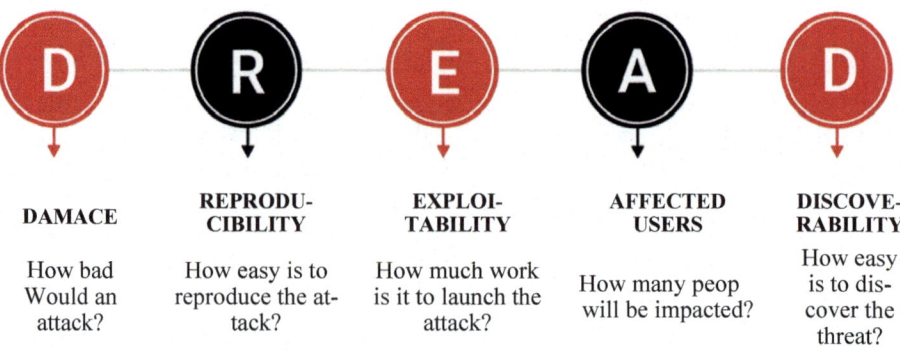

Fig. 2. DREAD model.

The idea of this model is based on the assumption of future attacks. This is achieved by answering the questions presented in the model. Depending on the qualifications and knowledge of the alleged offender, the protection specialist can build a strategy for creating a protection system with varying degrees of detail.

The advantages of the DREAD model include a high level of abstraction.

2.4   STRIDE Model

A fairly popular threat model that allows you to group the six most common computer security violations (Fig. 3). Among violations considered [3]:

- Spoofing.
- Tampering.
- Repudiation.
- Information disclosure.
- Denial of service.
- Escalation of privileges.

Most often used with a system model.

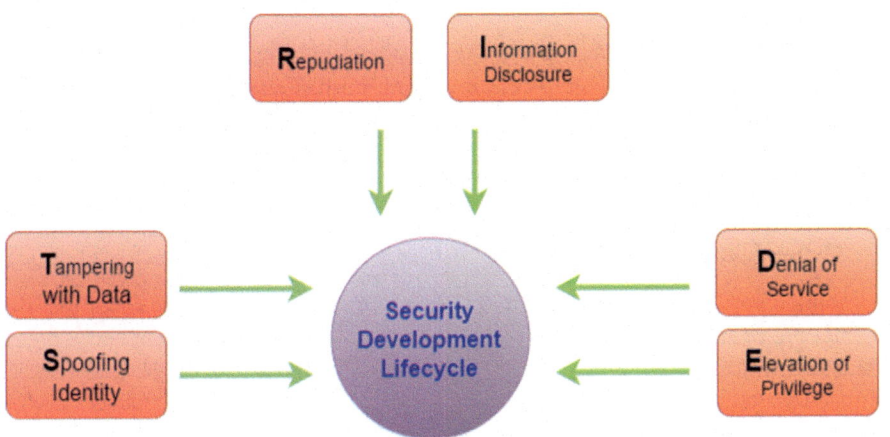

Fig. 3. STRIDE model.

The goal of STRIDE threat modeling is to ensure that the application meets such security properties as privacy, integrity, availability, authorization, authentication, and non-provability.

The merit of the model in question should include a high level of abstraction.

2.5   «Pyramid of Pain» Model

The proposed model allows controlling possible attacks of the intruder by detecting the impact and counteracting this effect depending on the pyramid level (Fig. 4) [4].

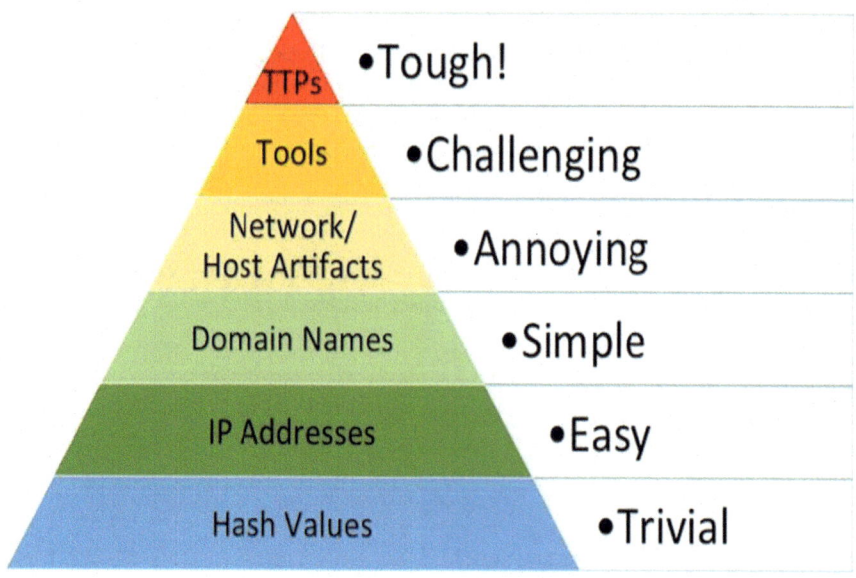

Fig. 4. «Pyramid of Pain» model.

It should be noted that all actions in the organization of protection are passive. The lower levels of the pyramid control the incoming hash files of the function, -address and domain names. Changing network and host artifacts causes the intruder to reconfigure the attack system. The upper levels change the protection system, forcing the intruder to look for new approaches and methods to organize attacks.

2.6   "Cyber Kill Chain" Model

The presented model (Fig. 5) [5] is a description of the actions of the intruder that they take when organizing attacks. This model is based on the consideration of each stage of the attack as a mandatory component of the success of the attack [6].

The proposed actions, according to the model, of the defending side are a chain break anywhere. The probability of repelling the attack increases. If this happens in the early stages. If it is possible to control the attack, then there is the possibility of its reflection in later stages.

In this case, you can gather more information about the goals, awareness and methods of organizing attacks, which will allow you to purposefully change the structure of the protection system.

2.7   Conceptual Information Security

A large number of models presented a set of threats, sources of threats, access calls and other information that really complicated the perception of business. One such diagram is shown in Fig. 6 [7].

Fig. 5. "Cyber Kill Chain" model.

The use of such schemes is good at the training stage, where you can explain to students the basics of building information security systems and show explanations for each element of the scheme.

In addition to the graphical representation of information security models, there is a table view. This view is most often used to develop information security models for specific systems that take into account local conditions, territorial location, the presence of a protection loop and much more.

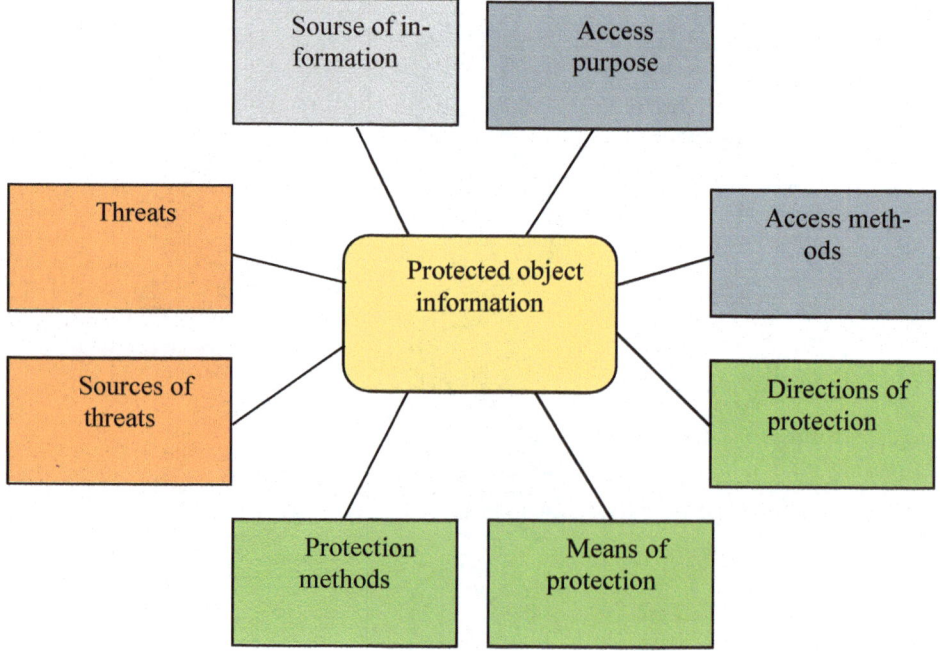

Fig. 6. Conceptual information security scheme.

2.8   Comparative Model Analysis

We will compare the models according to the set of requirements set out in Sect. 2.1. This is abstraction, complexity, completeness, specificity, informativity of the models considered.

Each model is rated from 0 to 4 points for each requirement. "0" – complete absence of this requirement, "4" – complete compliance.

The results of the analysis show that none of the presented models is conceptual.

3   Development of a Conceptual Model of Information Security

The main disadvantage of existing models is that they present in great detail composite elements that are difficult to understand. One of the main requirements for CM is a high level of abstraction. When constructing such models, it is necessary to answer the most general questions of providing the IS of a particular object, and the complexity of the system should not affect the CM. The necessary high level of abstraction allows you to solve the tasks set for these models (Table 1).

Table 1. The updated SAM fragment using the Erlang phase method.

Model	Abstractness	Complexity	Completeness	Concreteness	Informativeness	Average score
CIA	4	2	1	1	2	2
DREAD	4	1	2	2	2	2.2
STRIDE	4	1	1	2	1	1.8
Pyramid of Pain	3	2	2	3	2	2.4
Cyber Kill Chain	3	2	2	3	3	2.6
Conceptual information security scheme	3	2	2	2	2	2.2

3.1   Assumptions to the Model Development

The central element of the conceptual model should be the security object. In our case, such an object is information. Before building a security system, it is necessary to determine what information will circulate in this system. Based on the obtained definition, protection is built. Any security system cannot meet all information security needs. There is a rule that the cost of the system should not exceed the cost of information. In other words, any system has some vulnerabilities. The developer of the protection system must understand this and lay down the capabilities to modernize the created system. Any vulnerability is a threat to the offender. It should be noted that the offender can act both alone and in a preliminary criminal conspiracy with other violators [8]. Based on these discussions, a conceptual model of information security of the first level is proposed (see Fig. 7).

The advantage of the model that is presented is, firstly, a high level of abstraction, because this model fits any system, secondly, it presents all dialectical links between information that needs to be protected, a protection system, possible vulnerabilities that were not taken into account during the creation process or arose during operation, Threats that may exploit existing vulnerabilities and violators themselves that may exploit existing vulnerabilities due to threats to affect information.

3.2   Model of Intruder

The intruder model is the main second-level model in the construction of an integrated IS model. What should be emphasized? The first is the location of the intruder. It can be either inside the protection loop or outside. Secondly, the violator carries out possible actions based on awareness of the object, the organization of the protection system, the available rights and the level of access to the object, acquaintance with staff or the possibility of blackmail or bribery to these persons.

The intruder's goal may be to familiarize himself with information, modify information (up to its destruction), impose information, as well as violate the technical component of information processing, transmission and storage systems and protection

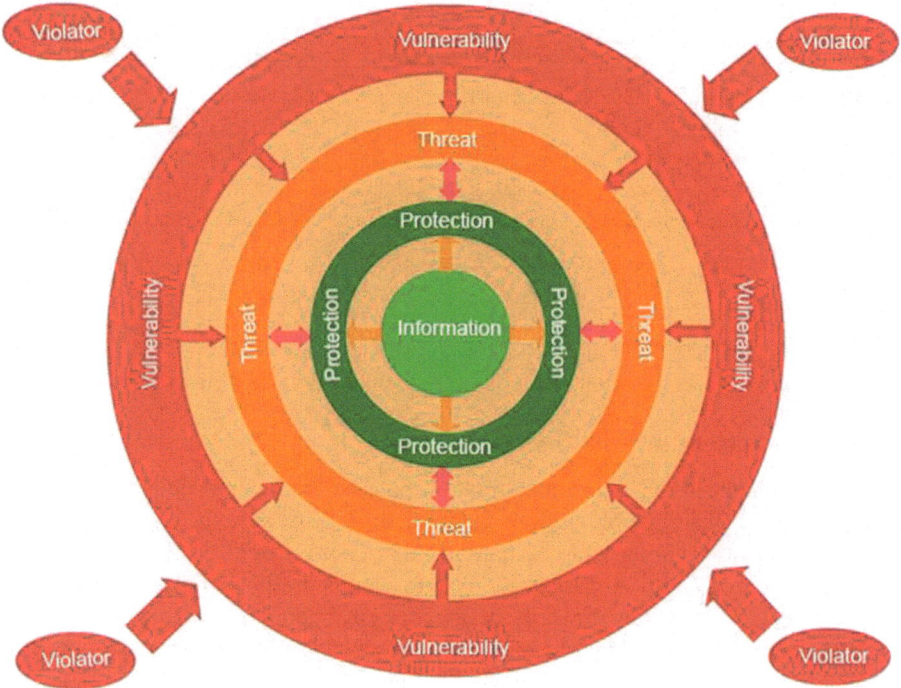

Fig. 7. Conceptual model of information security.

systems. The target of the offender may be the technical system itself, the life of which is provided by the ICS, including the protection system. The model of the intruder is shown in Fig. 8.

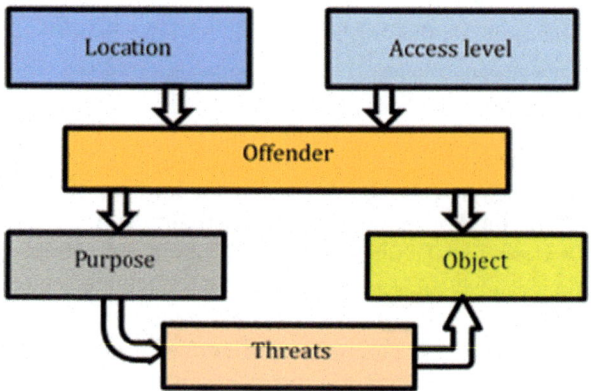

Fig. 8. Model of Intruder.

3.3 Threat Model of Information Security

Threat analysis and detection is an important component for the provision of IS. Therefore, the threat, according to [9], refers to the leak, the possibility of blocking or violation of the integrity of information. A clearer definition of threats is contained in [10]. They are defined as a set of conditions and factors that create a potential violation of the IS. [11] treats the threat as a possible cause of an unwanted incident that could harm the system or organization.

The identification of IS threats should be systematic and should be carried out both at the stage of creating and creating requirements for its protection, and during its operation. A systematic approach to IS threat identification is needed to identify the need for specific IP requirements and to establish an adequate and effective IP system in the ICS. IP measures taken at the design and operation stages should ensure effective and timely detection and blocking of IS threats, as a result of which unacceptable negative consequences may occur [12]. The process of determining IS threats is carried out indirectly through identified vulnerabilities and covers all protection objects and segments within the logical and physical boundaries of the ICS, in which the operator accepts and monitors IP activities.

Most often, the threat is due to the existence of vulnerabilities in the protection of ICS, for example, uncontrolled access to personal computers or unlicensed software (unfortunately, even licensed software for security is not devoid of vulnerabilities) [13, 14]. Highlights the main types of information security threats (Fig. 9).

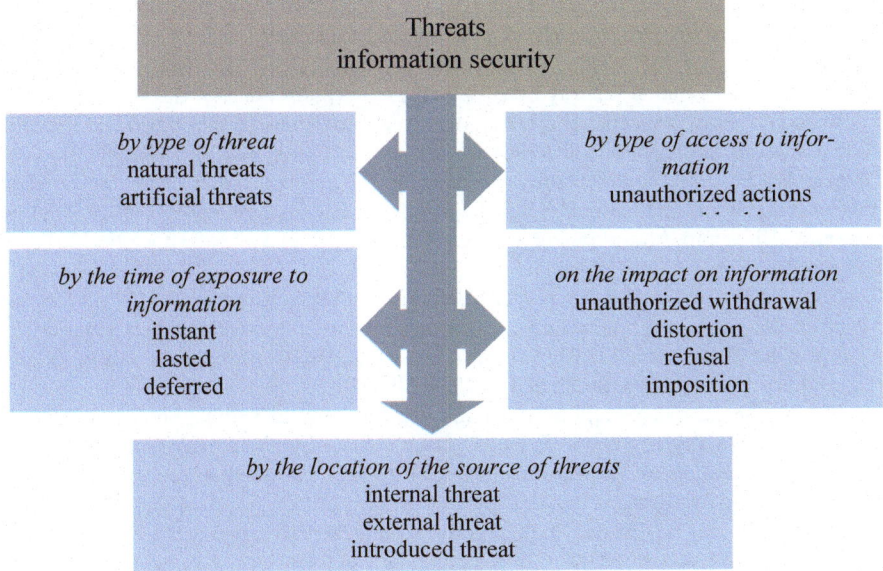

Fig. 9. Model of information security threats.

4 Conclusions

The analysis of some information security models and their comparative characteristics revealed the absence of conceptual models that would meet the needs of specialists in building information security systems.

The proposed conceptual model of information security has a high level of abstraction and is suitable for working with any object. The most important advantage of this model is the well-traced dialectical links between the information to be protected, the protection system that directly depends on the information, the vulnerabilities that affect the protection system, the threats that exploit existing vulnerabilities, and the violator who uses the latter to solve his tasks.

The second-level models presented made it possible to predict the possible actions of the violator through various organizational and technical measures.

References

1. Rouse, M.: Confidentiality, integrity, and availability (CIA triad). https://whatis.techtarget.com/definition/Confidentiality-integrity-and-availability-CIA. Accessed 10 June 2020
2. Howard, M., LeBlanc, D.: Writing Secure Code. Microsoft Press, Redmond (2002)
3. Future Learn: STRIDE. https://futurelearn.com/courses/cyber-security/0/steps/19631. Accessed 10 June 2020
4. Chenette, S.: Emulating attacker activities and the pyramid of pain. https://attackiq.com/2019/06/26/emulating-attacker-activities-and-the-pyramid-of-pain. Accessed 15 Aug 2020
5. Lockheed Martin: The Cyber Kill Chain®. https://www.lockheedmartin.com/en-us/capabilities/cyber/cyber-kill-chain.html. Accessed 30 July 2020
6. Cassetto, O.: Cyber Kill Chain: Understanding and mitigating advanced threats. https://exabeam.com/information-security/cyber-kill-chain. Accessed 30 July 2020
7. Information security concept. The main conceptual provisions of the information security system. https://studfile.net/preview/3904693. Accessed 10 Aug 2020 (in Ukrainian)
8. Pevnev, V.: Ensuring confidentiality when using infocommunication technologies. Paper presented at the 3rd International Scientific and Technical Conference IPST-2014, National Technical University "Kharkiv Polytechnic Institute", Kharkiv, 21–23 October 2014. (in Russian)
9. Rogozin, D.: War and peace in terms and definitions. PoRog, Moscow (2004). (in Russian)
10. Musman, S.: Assessing prescriptive improvements to a system's cyber security and resilience. In: 2016 Annual IEEE Systems Conference (SysCon), pp. 1–6. IEEE, Orlando (2016). https://doi.org/10.1109/SYSCON.2016.7490660
11. Pevnev, V.: Threat models and information integrity. Systems and Technologies 2, 80–95 (2018). https://doi.org/10.32836/2521-6643-2018.2-.56.6. (in Ukrainian)
12. Stepanov, Y., Korneyev, I.: Information security and information protection. Infra-M, Moscow (2001). (in Russian)
13. Choras, M., Kozik, R., Bruna, M.P.T., et al.: Comprehensive approach to increase cyber security and resilience. In: 2015 10th International Conference on Availability, Reliability and Security, pp. 686–692. IEEE, Toulouse (2015). https://doi.org/10.1109/ARES.2015.30
14. Galinec, D., Steingartner, W.: Combining cybersecurity and cyber defense to achieve cyber resilience. In: 2017 IEEE 14th International Scientific Conference on Informatics, pp. 87–93. IEEE, Poprad (2017). https://doi.org/10.1109/INFORMATICS.2017.8327227

Serverless and Containerization Models and Methods in Challenger Banks Software

Yuliia Kuznetsova$^{(\boxtimes)}$ ⓘ, Artem Kolomytsev, Maksym Somochkin,
and Oleksandr Vdovitchenko ⓘ

National Aerospace University "Kharkiv Aviation Institute",
17 Chkalova Str., Kharkiv 61070, Ukraine
y.kuznetsova@khai.edu

Abstract. The article describes the perks of the serverless and containerization models and methods in challenger banks software. Data sovereignty in the context of microservices means that the stored model of the domain may differ same between subsystems. This approach to data modeling creates an easier way to write a query that merges data from various sources. Microservices intercommunication methods shows that typical microservices-based application is a distributed backend running as multiple processes, in most cases on multiple hosts in the same network. Service in the simplest case is just a process which can interact with other parts of the system. Methods of identification of the domain boundaries for each service imply the goal which should be to achieve the most meaningful separation leaded by your (or some expert's in your team) domain knowledge. The main idea here is it's not the size, that matters, but business capabilities and concerts separation. Methods for handling transactions in asynchronous distributed system show that each service is usually an isolated system apart with its own dedicated database. This means you can no longer take advantage of the 2PC to maintain the consistency of the whole system. Dealing with transient errors, eventual (in)consistency between microservices, isolations, and version rollbacks are scenarios that should be considered during the design phase.

Keywords: Challenger bank · Cloud software · Containerization · Serverless · Microservices · Distributer systems

1 Introduction

Formulation of the Problem. People use banks' services and make payments with cards every day. In the UK only there were approximately 39 million of cards' transactions per day at 2016. There is a new type of banking that become popular in 2017/2018 years: "challenger banks".

Challenger banks are usually small retail banks. They compete with some older banks that exist in the country. These banks try to differentiate themselves from the existing banks by modern IT practices: almost all services of the bank may be used via mobile app that save the costs and resolve some complexities of traditional banking.

To understand how popular this trend is become we can review investment data: "Over the 2017 year, challenger banks have been making headlines by attracting big venture

M. Nechyporuk et al. (Eds.): ICTM 2020, LNNS 188, pp. 169–185, 2021.
https://doi.org/10.1007/978-3-030-66717-7_15

capital investments. Biggest three UK's challengers are: Monzo (raised US$93 million in 2017 and built on a $27.5-million financing round during the first quarter), Atom Bank ($140 million) and Starling Bank ($54 million in latest funding) [1].

Challenger banks are usually established by a people with good experience in banking. Companies that act as challenger banks mostly small or midsize and it allow them to be more agile in competition with larger institutions. These banks don't bury under the weight of legacy technology. They can bounce over traditional infrastructure and disrupt the status quo. They are more agile, faster adapt to user needs and more personal than older banks. Biggest perk of new challengers is that they are building everything from scratch with a digital offering and the use of the latest technology available. For example it's much easier to new bank to introduce 3D-secure payment validation, provide ApplePay and GooglePay integration – than for older banks. Modern banks can even use messengers as a communication method for their support staff. Traditional banks respond much slower to market demands and unable to keep up with technological developments.

Because challenger banks' backends often are created from scratch – they can use lots of modern cloud technologies for faster TTM in new features development, increases scalability and reliability. Two most important and growing trends in cloud application's development in 2018/2019 are serverless and containerization [2–4].

Containerization is an isolation method on an OS-level used to push and run applications without launching a dedicated server or VM for each app. Multiple isolated applications or services run on one host and access the same OS kernel [5]. Containers are a streamlined way to develop, test, push, and redeploy applications on several environments from a software engineer local machine to an on-premises hardware and even the cloud [6]. Containerization benefits:

- containers require fewer system resources in comparison to traditional VM environments because they don't come with OS (operating system) images;
- applications in containers may be crossdeployed easily to various OS's and hardware platforms;
- applications in containers have the same behavior, regardless of host where they are being deployed;
- containers improve the speed of application deployment, upgrades and scaling.

Containers can become really efficient if they are used together with serverless computing technologies. "Serverless" term means applications where backend logic is still written by the app developer, but, unlike older architectures, it's run in managed by a third party, stateless isolated containers that are triggered by event, ephemeral (exist only for single invocation). One way to name this is FaaS – "Functions as a Service". AWS Lambda is now the most popular managed implementation of a FaaS platform at present, but there are some others, too [7]. Serverless perks:

- reduced operational cost;
- faster TTM;
- reduced deployment complexity;
- ease of continuous experimentation;
- Green computing [8, 9].

Therefore, there is a need to build challenger bank software using serverless and containerization. **The aim of the project** is to increase the reliability, scalability, efficiency, and maintainability of challenger bank backend software. **The object of study** is serverless and containerization technologies. **The subject of the study** is cloud software design patterns and approaches, modern trends in cloud software development that leads to increasing the resilience and efficiency of the software. **The study methods** are:

- modern cloud applications design patterns were used for increasing the efficiency, scalability, and maintainability of challenger bank backend software;
- serverless and containerization is the subset of modern cloud application design that was used in the study;
- OpenFaaS was used as FaaS (serverless) technology and Docker as containerization technology;
- in this software is being used such design patterns as FaaS, containerization, dependency injection containers, OOP, OOD, distributed cloud systems approach, CQRS and Mediator as DB access patterns, queue messaging as service inter-communication system.

2 Model of Data Sovereignty per Microservice

An essential rule for microservices/SOA architecture is that each service must own its domain data and logic. Just as a whole monolithic application owns its data and logic, so must each service own its data and logic under an isolated lifecycle and isolated deployment cycle per service [10]. This means that the stored model of the domain may not be the same between subsystems. This principle is similar in DDD, where each BC [11] or autonomous subsystem or service must own its domain model (business logic and data model), see Fig. 1.

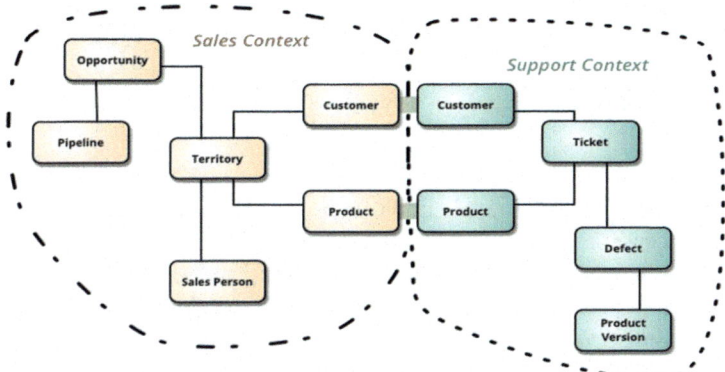

Fig. 1. Bounded contexts.

Bounded Context is a fundamental pattern in DDD. It is the core of DDD's strategic design section. That section is all about dealing with extensive models and teams. DDD deals with extensive models by breaking them down into distinct Bounded Contexts and being explicit about their interconnection. DDD is about designing software architecture based on the model of the underlying domain area. The model is a "Ubiquitous Language" that aids in communication between domain experts and software engineers.

It also acts as the conceptual foundation for the architectural design of the software itself – how it's splitted into objects or functions. To be efficient, the model needs to be internally consistent so that there are no contradictions within it.

Each Bounded Context has several self-sufficient services. In a decent implementation, they would be loosely coupled but would have high cohesion. In monolithic architecture applications often have a single PoF (point-of-failure), and all application components are tightly coupled with it. This is often a regular RDBMS with one or multiple databases that are being used for the whole application backend and all its internal subsystems, as shown in Fig. 2.

Data in Traditional approach

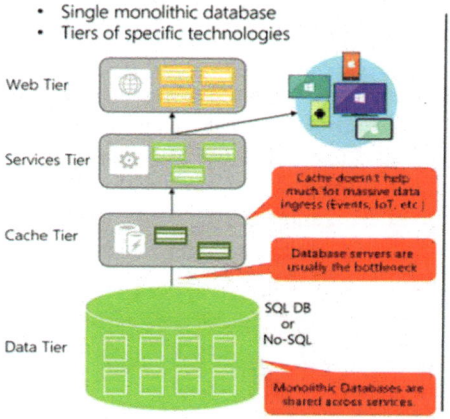

Data in Microservices approach

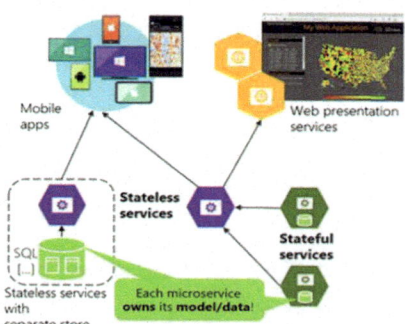

Fig. 2. Data sovereignty in microservices/Service oriented architectures versus monolithic applications.

In a traditional monolithic application with a single relational DB there are two most important perks: ease use of fully ACID transactions and option to write queries that are working across all the tables and data in the database. This is very useful in analytics, reporting and audit tasks, which are hard problems to solve in typical microservice architecture.

Even if we have microservice system, that uses same DB engine for all databases and supports cross-database queries (like postgres_FDW) – we should try not to use it. It breaks the isolation between services, we leak implementation details and our solution becomes very fragile. If we want our services to be able to evolve independently of one another we must encapsulate the data. DB schema management would also become an arduous task: it's hard to do it properly and without any downtime even for monolith, but when multiple services always accessing the same data we must test all of them with the new schema, before deploying it. This would break the microservice lifecycle and deployment autonomy. This distributed data storages situation mean that you can't make an ACID transaction across your data store. It leads to embracing eventual consistency as an approach to system state storage. Eventual consistency makes transactions guarantees harder to achieve, even makes them impossible in many cases. But it grants us an opportunity to improve reliability and availability of the software system as a whole – because now, when data store is distributed, a problem in a datacenter with "the main database" doesn't lead to downtime – you do not have "the main database" now. Now your system consists of many small units, and, even when some of them fail – other parts work correctly. A lot of systems have non-critical parts and if they do not work – it will not be a huge problem.

As a plus side, the distributed nature of microservices storage allows us to use various database types for the single system: like store financial records in RDBMS, but use document DB for less structured data.

3 Microservices Intercommunication Methods

API microservices must sometimes talk to each other. There are a few potential issues with services inter-communication that need to be discussed.

In a single monolithic application, all your components are parts of the same program. It means intercommunication between components is local – it's just a method call. It can be synchronous or asynchronous, you can use local synchronization primitives (like mutexes) – but it's all simple and in the boundaries of a single application. [12]. The most direct approach for moving from local in-process calls to a distributed system would be just replacing all methods that are out of the boundaries of a single service with RPC calls. But it would cause a lot of extra network traffic and would be really inefficient and unstable method of communication.

The challenges in design of the distributed system properly are well studied and there's even a set of assertions made by Laurence Peter Deutsch known as the "Fallacies of distributed computing". It lists false hypotheses that developers with small experience in distributed applications always make.

Agile teams are expected to regularly and rapidly push new features and fixes. Those teams are typically in constant search for ways to decrease their organizational complexity. For example: if you have 3 backend and 3 frontend teams in your project, there is more overhead (in comparison to 1 backend team and 1 frontend team, or even single combined full-stack team) such as meetings, discussions, planning, tech stack and processes.

Microservices is an approach to software architecture design and software engineering that is based on isolated, separately deployable services. Those services act as bulkheads that improve operational flexibility and resilience as they can be deployed in isolation of all other existing systems. Microservices also give an opportunity to a single agile team to build a solution alone in a small office somewhere without requiring huge bureaucratic complexity and without huge communication overhead.

Complex platforms tend to be developed by a dedicated team. An enterprise that runs all service internal messaging through a single solution/system would need to call members of that team in most projects. This is unappealing to microservice-oriented teams who need to develop, own and deploy their own isolated services.

Microservices-oriented teams may wish to put transient error handling such as timeout/circuit-breaker into communication endpoints, in this case services, rather than operate integration platforms that need dedicated team members to customize and support. Communication technologies (like message bus, queues etc.) can be a potential single PoF that can bring down all microservices at once. As such, well-tested and straightforward asynchronous communication reinforces the reliability of microservices architecture.

There is no silver bullet or the only right template on how to design distributed systems, but several. You can then use async communication between the microservices and replace tightly coupled and unreliable synchronous communication that's typical in intra-process communication between objects with hard-grained communication.

Typical microservices-based application is a distributed backend running on multiple processes, in most cases on multiple hosts in the same network. Service in the simplest case is just a process which can interact with other parts of the system. It can do so using a high-level communication protocol such as HTTP, AMQP, GRPC (over HTTP/2) or more "closer to metal" binary protocol like TCP or UDP, depending on the nature of each service. Services do not even need to communicate directly, they could have some middle-man for communication, like message broker (Kafka, RabbitMQ). The microservice community (represented by different speakers or book authors) promotes the methodology of "smart endpoints & dumb pipes". It's a design principle that facilitate simple, and well-tested, async communication mechanisms over complex platforms (like relying your whole application on some-thing like Akka or Hadoop, which may be crucial in some cases – but would be a burden to support in smaller teams and would bring more downsides than benefits.

This encourages a design that has low coupling as possible between services, and high cohesion as possible inside of a single service. As explained above, each microservice owns its data and its domain logic.

Two most common approaches to interservices communication are messaging and bare HTTP calls. But client and services can talk through many diverse types of communication, where each one targets a distinct scenario and objective.

The most crucial things that must be taken into consideration, when choosing communication strategies are the answer to such questions:

- Do we need to provide guaranteed and immediate responses?
- Do we need to send the data to single or multiple receivers?
- How large is our data payload?

And based on the answers to those questions, we may choose one of (but not limited to) the following.

Use Simple (and Blocking) Synchronous Communication. In this type of communication services communicates directly, most often via HTTP or gRPC over HTTP. HTTP is a synchronous protocol. The client sends a request and waits for a response from the server. An implementation may be synchronous (thread is blocked) or async (with epoll in Linux, as an example) – but this is just an implementation detail. The main idea when using this approach is that the client's routine (go routine/task or other high-level abstraction) would be blocked until a response is received.

Use Non-blocking Asynchronous Communication. Services would communicate using some message broker and messages would be delivered eventually. They may even be delivered as early as possible, but in most cases, the sender's routine would not care and would not receive acknowledgment if the message was delivered. Communication between service and broker would still (obviously) be synchronous, but now we do not need wait for receiver to process and create some response. Response may be sent later via the same broker. This communication pipeline allows service to work even if the receiver is not running. We may spin up the receiver when we have N messages in the queue, process them – and put it to sleep. It would help us to more efficiently manage our costs in a cloud environment. And we do not need to do it manually – serverless AWS lambdas already doing it for quite some time and used by many developers.

If we need to have only single receiver for our message synchronous communication (with some load balancer between caller and receiver) may be sufficient for us. But the implementation of multiple receivers with synchronous communication is not an easy task to solve. And sometimes it's required, for example when we want to do A/B testing on multiple versions of backends. With synchronous communication, we must have them all running simultaneously when with asynchronous queue we could just duplicate those messages across consumer groups or store for later use – and test in any way we want.

A microservice-based software will often use a mix of these communication methods. For example API for web/mobile client would be in almost any case simple HTTP(S) backend. It can be hidden under a load balancer, implementation details underneath it may be various – but for the client, it's just a simple HTTP endpoint, no difference if it's running on the server in the same office, or on a server-less environment behind AWS Elastic Load Balancer.

But for intercommunication between services most developers would prefer asynchronous messaging: if your service doesn't depend on another service to be running and may just push messages into queue – your service would be much more reliable.

Obviously the most reliable method of communication would be not to have any inter-service communication at all. So it's really important to draw borders between services – but not split some stuff, that may comfortably live in the same microservice and would be easily maintained with much less effort.

If possible, service should almost never rely on a synchronous communication between multiple microservices, even for simple queries. The objective of each microservice is to be isolated and available to the user, even if the other microservices that are part of the backend are down or not responding.

When it is crucial to obtain communication with other microservices from the selected microservice (for example, to receive data through the execution of an HTTP request) in order to return a response to the client application, there is only an unstable architecture in which some of the services may not work.

Any dependency can be unreliable, but the aim is to implement a solution that can work (at least partially) even if underlying systems are down. If servers of company that print plastic cards for the bank are down – it doesn't mean that the whole bank must stop working. Also, if there are connections between microservices, for example, when you create loops of requests or responses with chains of HTTP requests, and these loops can be both nested and long (see Fig. 3, its first part), then this the current situation, first of all, makes your services not isolated. It also leads to a decrease in performance when one of the running services from this chain stops working normally.

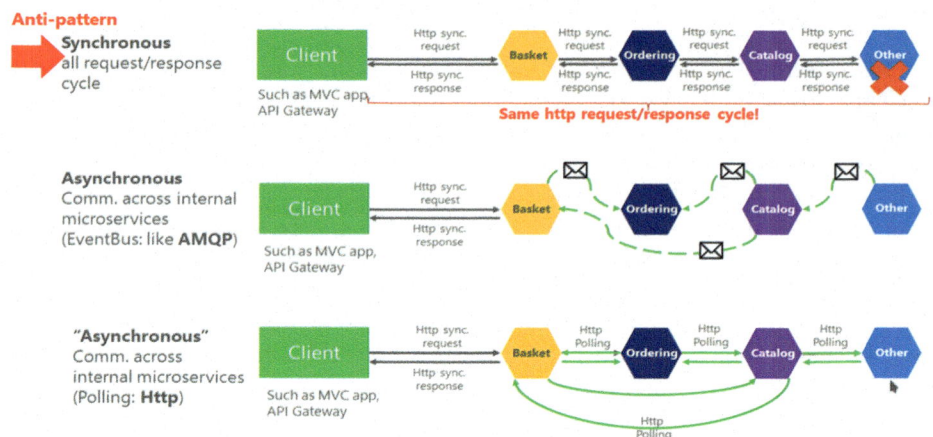

Fig. 3. Microservices communication anti-patterns and patterns.

However, if several of synchronous interdependencies between services, such as request requests, increases, then the total response time for applications running on the client side decreases.

In the case that you need to perform another operation in another microservice using your service, you should not do any action synchronously or as part of a microservice request and response operation. It is best to perform these actions asynchronously (using async messaging or events bus or queues, etc.). Also, you should not execute (invoke) these actions synchronously as part of the original synchronous request and response operation.

And most importantly (this is where most of the issues arise in the process of building microservices), if a microservice requires data that originally belonged to other microservices, then there is a need to perform synchronous requests for this data. Instead, copy or propagate this data (only the required attributes) to the original service database using eventual consistency (usually using events bus or queue as detailed in the following pages).

If duplicate data is encountered in some microservices, this is not wrong or flawed design. But it becomes achievable to convert the data into a special language or terms of this additional domain or limited context. Also you can use various protocols for communication. When using a synchronous communication protocols like HTTP are most appropriate, especially when publishing services outside of the Docker host or your microservice cluster.

When communicating internally between services (within your microservices cluster Docker node), it is achievable to use binary communication mechanisms (for example, .Net Remoting or Windows Communication Foundation over TCP). Also you can use async messaging communication mechanisms such as SQS or AMQP.

There are some message formats (text-based like JSON or XML, or like messagePack) that may be more efficient. In the event where the option you have chosen is not standard, most likely you should not publish your microservices using the chosen format. A non-standard format can be used for intercommunication between microservices. Similar actions can be done when communicating between containers on same Docker host or orchestrated cluster (Docker Swarm or K8s), or for custom client apps that interact with microservices.

Request/Response Communication via HTTP and REST. If a client uses a communication request, it sends a request to the response service, after which the service processes the request and sends a response. The request/response is most suitable for requesting real-time UI (live UI) from client applications. Therefore, first of all, the reactive communication mechanism for messages is needed (see Fig. 4).

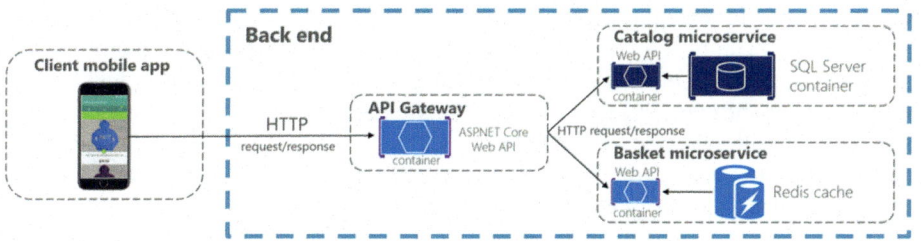

Fig. 4. Using HTTP request/response communication.

When using a request/response communication, it is assumed that the response will be delivered within a short amount of time, usually less than a 500 ms, or a few seconds. Deferred responses require an implementation of async communication based on message queues or event bus, which is a distinct approach.

A common architectural style for how requests and responses interact is the REST approach, which is based on and utilizes HTTP protocol, including HTTP verbs such as POST, PUT, GET, etc. REST is the most frequently used approach when creating an external or internal API for a service.

The Representational State Transfer (REST) introduced and described by R. Fielding in his PhD dissertation in 2003. It's is an abstraction of the architectural elements within a distributed hypermedia system [13].

REST ignores the implementation details and underlying protocol (most often it's just HTTP over TCP) in order to describe the roles of components. It means that it's not limited via HTTP and synchronous communication; it can be done on top of any protocol. Limits the interaction between the rest of the components, as well as the interpretation of the most significant data items. Introduces major constraints on components, connectors, and data that define the foundation of the web architecture and therefore the behavior of that architecture as a network software application.

Async Messaging. Asynchronous messaging or event based communication are crucial when propagating data changes and system events across several services and their domain models. When utilizing messaging, services communicate by sending and receiving messages asynchronously directly or via broker.

A client creates a command or a request to some service by sending a message or event to it. If the receiver needs to make a reply, it sends another message to the original sender. Since it's an async communication, the original sender doesn't expect that response would be received immediately and can also handle situation where there would be no response at all.

A message consists of some metadata defined by protocol and a body. Body can contain any arbitrary binary or text data. Messages are usually sent via async protocols as SQS or AMQP.

For most microservices applications it would be better to use a small and lightweight message broker instead of some enterprise buses, so often used in SOA. Lightweight queue/topic-based messaging is closer to microservices philosophy, because it allows to follows the "dumb pipe" paradigm and service doesn't depend indirectly on any other services that may or may not publish something into the bus. Broker may be self-hosted (like nsq) or managed (like SQS). In this scenario, most of the "smart" thinking still lives within the borders of a single service.

It's preferred to use async methods for intercommunication inside the distributed system, but create synchronous API for external clients (website or mobile app) for ease of integration in the future. It may be SQS for communication inside the system and HTTPS for public API as an example.

There are two types of async messaging: message-based single-recipient communication and message-based multi-recipient communication. The meaning of these types of asynchronous exchange is disclosed in more detail below.

Single-Receiver. It's a type of async and message-based intercommunication with a single sender and a single receiver. A good example of it would be ZeroMQ – it's not a broker, it doesn't need to have something hosted outside the borders of a service. It's just an external library that can be used in any project and those projects can interact between themselves [14].

Exceptional situations can also arise. For example, if the backend application would be killed by node failure or the orchestrator tries to reschedule it – we may not know if the receiver processed a message, so we may resend it. It's called at-least-once delivery semantics. It's more resilient, but it leads to messages duplication. Due to network or other failures, the client must be able to resend the message, and the server must implement the operation in order or use some message-ID for reduplication algorithm to remain idempotent and process a particular message only once.

Message-based communication with one sink is particularly well suited for send-ing asynchronous commands from one microservice to another (see Fig. 5 for an illus-tration of this approach).

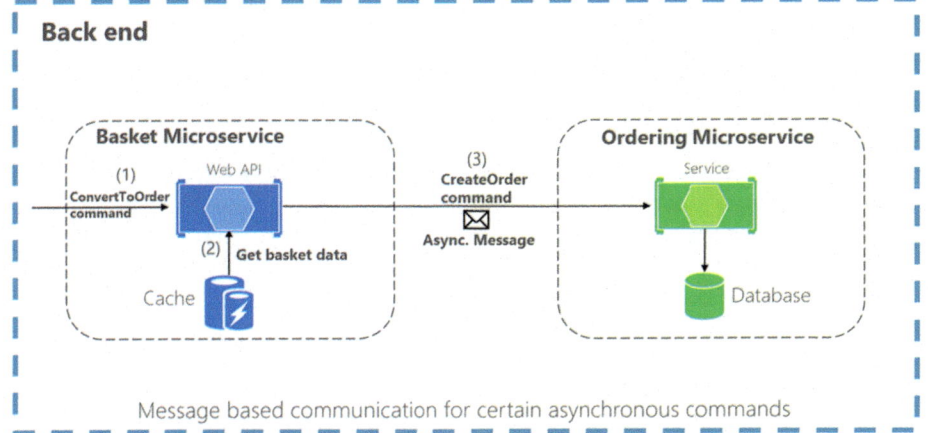

Fig. 5. A single microservice receiving an async message.

Once you start sending messages based on messages (using commands or events), you should avoid mixing messages based on messages with synchronous HTTP communication.

Multiple-Receivers. As a more flexible approach, a pub/sub mechanism can be used to make the sender's message available to multiple internal services or external applications. So it helps to follow the open/close principles in the dispatch service. Therefore, in the future, you can add additional subscribers without having to change the sender service.

While using pub/sub communication, messages are usually published into some topic and there may be $0...N$ readers from it. They may load-balance messages between themselves, or the broker may duplicate messages for all of them. All these approaches are used, it just depends on the use-case.

In the case of event-driven async communication, the microservice publishes an integration event, and when something happens in its domain, the other microservice needs to know about it, for example, a price change in the product catalog microser-vice. Additional microservices subscribe to events in order to receive them asyn-chronously. When this happens, recipients can update their own domain entities, which can result in more integration events being posted. The pub / sub may be implemented via event bus or topics with one or multiple consumer groups The event bus can be designed as an abstraction or as an API that is required to subscribe to or opt out of events, and to publish events. The event bus can be implemented in several ways: for example, based on any interprocess broker and proxy for messaging (messaging queue), or as a service bus that supports asynchronous communication and a publish/subscribe model.

If the system uses event-driven end-to-end consistency, the approach needs to be completely transparent to the end user. The system should not use an approach that mimics integration events such as SignalR or client polling systems. The end user and business owner must explicitly accept the possible consistency in the system and understand that in many cases the business has no problem with this approach if it is explicit. This is important because users may expect to get some immediate results that may not ultimately happen.

Integration events can be used to implement business goals that span multiple microservices. In this way, consistency can be achieved between these services. As a result, a consistent transaction will consist of a set of distributed actions. With each action, the associated microservice updates the domain object and publishes a different integration event that triggers the next action under the same cross-cutting business problem.

In case you want to communicate with multiple microservices subscribed to the same event, you can use event-driven publish / subscribe messaging (see Fig. 6).

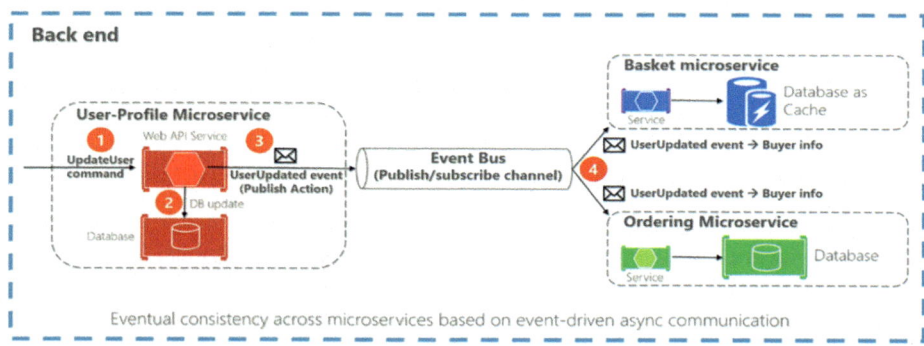

Fig. 6. Asynchronous event-driven message communication.

Pub/sub is not something, that was invented by microservice developers can only be used there. BC in DDD should communicate using same or similar approach, or to the way you propagate updates from the write database to the read database in the CQRS architecture pattern. We do not need to have multiple data sources in the system to be in sync, we just want them to be eventually consistent.

4 Methods of Identification of the Domain-Model Borders for Each Service

Achieving the largest possible granularity is not considered as a goal for defining the boundaries and size of the model of each microservice. Despite this, if possible, you should aim for small microservices [15]. The goal is to get to the most meaningful division while guided by domain knowledge. The main thing here is not size, but business opportunities and division of concerts. In addition, when there is a need for a high degree of consistency (for a specific area of the application) based on a large number of dependencies, this also indicates the need for one microservice. Cohesion is a way of determining how microservices can be split or grouped.

Microservices development is best done by building on the BC pattern we introduced earlier. There may be cases where a BC consists of multiple physical services, but not vice versa (a service cannot include multiple BCs). A domain model consisting of specific domain objects can be applied to a specific BC as well as a service. The BC limits the applicability of the domain model and also provides the development team members with a clear or shared understanding of what can be coherent and what needs to be developed independently. Identical goals are set for microservices.

To identify context constraints, you can use the DDD template, commonly referred to as the context display template. In terms of domain-oriented design, context can be defined as "the setting in which a word or statement appears that determines its meaning" [16]. By using a display context, it is possible to identify the various contexts in an application, as well as the boundaries of those applications.

With context mapping, you identify the various contexts in your application and their boundaries. More often than not, there are different boundaries and contexts for any small subsystem. A context map is a way to define and make explicit boundaries between domains.

BC is stand-alone and contains details of one domain, such as domain objects. It also defines contracts for integration with other BCs. Similar to the definition of a microservice: it is self-contained, implements specific domain capabilities, and also provides interfaces. This is why context mapping and the bounded context pattern are both acceptable approaches for defining the domain model of the microservices that you develop.

5 Methods for Handling Transactions in Asynchronous Distributed System

Transactions are the main part of software applications. Without them, maintaining data consistency would be impossible. The most important type of transaction is two-stage commit, where committing the first transaction depends on the completion of the second.

If you need to update multiple objects at the same time (for example, order confirmation and instant inventory update) this can be extremely useful. In the case of working with microservices, everything is much more complicated. Each service is a separate system with its own database, so there is no way to exploit the simplicity of local two-phase commits to maintain the consistency of the entire system.

Dealing with transients, possible consistency between services, and rollbacks and isolation are scenarios needed to be considered during the design phase.

The most famous distributed transaction pattern is the Saga pattern. Saga is a sequence of local transactions, each of which updates data within a single service. The very first transaction is usually initiated by an external request that corresponds to a system operation. After that, any next step is started by completing the previous one. There are several ways to implement a saga transaction. However, let's look at two of them that are the most popular:

- events/choreography: when there is no centralized coordination, then any service can produce and listen to the events of another service, and then decide if there is a need to take any action (Fig. 7);
- command/orchestration: the case where the saga coordinator service is responsible for centralizing decision making as well as streamlining business logic (Fig. 8).

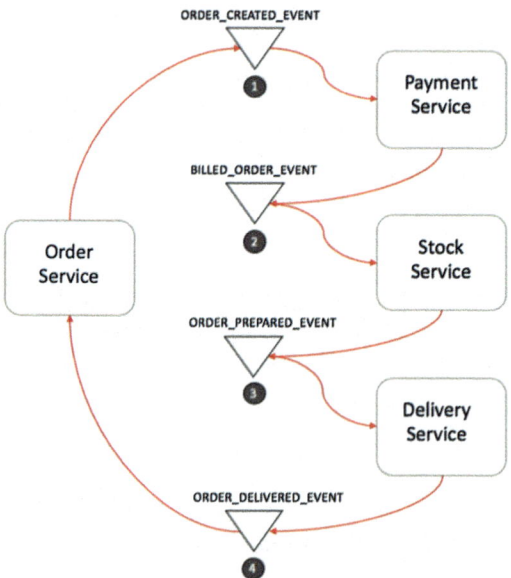

Fig. 7. Data sovereignty comparison: monolithic database versus microservices.

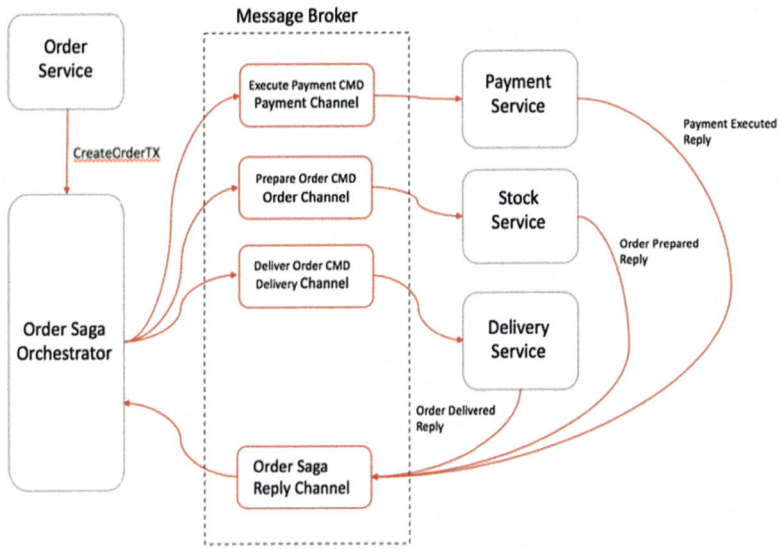

Fig. 8. Data sovereignty comparison: monolithic database versus microservices.

6 Conclusions

A bond is an instrument of debt of the bond issuer to its holders. Stocks and bonds are securities, however, there is a major difference between them, which is that shareholders have a stake in the company's capital, while bondholders have a creditor's stake in the company. As creditors, bondholders have priority over shareholders. This suggests that they can be paid off before the shareholders, although in the event of bankruptcy they are inferior to secured creditors. There is also another difference, which is that bonds usually have a specific maturity or, in other words, a maturity date, after which the bond is redeemed. Typically, bonds are issued by government agencies as well as credit companies or organizations. The issuance of bonds on primary markets by supranational organizations is quite acceptable. The most common way of issuing bonds is underwriting. A fixed rate bond in finance is a type of bond with a fixed interest (coupon) rate debt instrument, as opposed to a floating rate bond. A bond that has a fixed interest rate is a long-term debt paper that has a predetermined interest rate.

Before accessing any app features, user must to pass phone number verification, email verification, KYC which is the process of a business verifying the identity of its clients and assessing potential risks of illegal intentions for the business relationship and PEP/sanctions checks. Backend APIs use JWT signed with RS256. Different methods of services communications are being used in the backend – from synchronous REST requests, to asynchronous messaging. Microservices designed according to model of data sovereignty per microservice. This means that each microservice must own its domain data and logic. Each service is separated from each other and contains bounded context.

Saga is a pattern that being used to handle distributed transactions in the system. Developed models and methods are approbated in the software of challenger bank Dozens. Based on results from competitiveness indicators calculation dozens can be seen as a competitor for existing challenger banks on UK market. The project can be enhanced with the use of actor-model computing using Ak-ka.Net or Orleans frameworks, moving from RDBMS to NoSQL and NewSQL DB engines.

Scientific Novelty. Information technology of challenger banking in terms of algorithms and software has been further developed through the integrated use of innovative technologies like serverless computing and containerization linked together with message queues, using Docker software, OpenFaaS as serverless framework, SQS as message broker.

Practical significance of the obtained results is approbation that use of server-less computing and containerization in challenger bank backend development allows faster time-to-market in new features development, increases scalability and reliability. It allows using cloud hosting with better price/efficient coefficient than traditionally systems and also more eco-friendly [17].

References

1. Ozcan, P., Dinçkol, D., Zachariadis, M.: Monzo, Revolut and other challenger banks are shaking up the industry. The Conversation (2018). https://theconversation.com/monzo-revolut-and-other-challenger-banks-are-shaking-up-the-industry-99564. Accessed 2 Sept 2020
2. Baldini, I., Castro, P., Chang, K., et al.: Serverless computing: current trends and open problems. In: Chaudhary, S., et al. (eds.) Research Advances in Cloud Computing, pp. 1–20. Springer, Singapore (2017). https://doi.org/10.1007/978-981-10-5026-8_1
3. da Silva, V.G., Kirikova, M., Alksnis, G.: Containers for virtualization: an overview. Appl. Comput. Syst. **23**(1), 21–27 (2018). https://doi.org/10.2478/acss-2018-0003
4. Hardin, T.: What is serverless computing? Advantages and predictions. Learning Hub (2018). https://learn.g2.com/trends/serverless-computing. Accessed 2 Sept 2020
5. Pahl, C., Brogi, A., Soldani, J., Jamshidi, P.: Cloud container technologies: a state-of-the-art review. IEEE Trans. Cloud Comput. **7**(3), 677–692 (2019). https://doi.org/10.1109/TCC.2017.2702586
6. Bachiega, N.G., Souza, P.S., Bruschi, S.M., De Souza, S.D.R.: Container-based performance evaluation: a survey and challenges. In: 2018 IEEE International Conference on Cloud Engineering (IC2E), pp. 398–403. IEEE, Orlando (2018). https://doi.org/10.1109/IC2E.2018.00075
7. Soltani, B., Ghenai, A., Zeghib, N.: Towards distributed containerized serverless architecture in multi cloud environment. Procedia Comput. Sci. **134**, 121–128 (2018). https://doi.org/10.1016/j.procs.2018.07.152
8. Lloyd, W., Ramesh, S., Chinthalapati, S., et al.: Serverless computing: an investigation of factors influencing microservice performance. In: 2018 IEEE International Conference on Cloud Engineering (IC2E), pp. 159–169. IEEE, Orlando (2018). https://doi.org/10.1109/IC2E.2018.00039

9. Shostak, I., Matyushenko, I., Romanenkov, Y., et al.: Computer support for decision-making on defining the strategy of Green IT development at the state level. In: Kharchenko, V., et al. (eds.) Green IT Engineering: Social, Business and Industrial Applications. SSDC, vol. 171, pp. 533–559. Springer, Cham (2019). https://doi.org/10.1007/978-3-030-00253-4_23

10. Microsoft: Data sovereignty per microservice (2018). https://docs.microsoft.com/en-us/dotnet/standard/microservices-architecture/architect-microservice-container-applications/data-sovereignty-per-microservice. Accessed 2 Sept 2020

11. Fowler, M.: BoundedContext (2014). https://martinfowler.com/bliki/BoundedContext.html. Accessed 2 Sept 2020

12. Microsoft: Communication in a microservice architecture (2020). https://docs.microsoft.com/en-us/dotnet/standard/microservices-architecture/architect-microservice-container-applications/communication-in-microservice-architecture. Accessed 2 Sept 2020

13. Fielding, R.T.: Architectural styles and the design of network-based software architectures. Dissertation, University of California (2000)

14. Microsoft: Asynchronous message-based communication (2018). https://docs.microsoft.com/en-us/dotnet/standard/microservices-architecture/architect-microservice-container-applications/asynchronous-message-based-communication. Accessed 2 Sept 2020

15. Microsoft: Identify domain-model boundaries for each microservice (2018). https://docs.microsoft.com/en-us/dotnet/standard/microservices-architecture/architect-microservice-container-applications/identify-microservice-domain-model-boundaries. Accessed 2 Sept 2020

16. Kapferer, S., Zimmermann, O.: Domain-specific language and tools for strategic domain-driven design, context mapping and bounded context modeling. In: Proceedings of the 8th International Conference on Model-Driven Engineering and Software Development (MODELSWARD), vol. 1, pp. 299–306. SciTePress, Valletta (2020). https://doi.org/10.5220/0008910502990306

17. Shostak, I., Danova, M., Romanenkov, Y., Kuznetsova, Y.: A retrospective analysis technology of the Green Software Ecosystems development on the parametric identification of the Brown's model. In: 2018 IEEE 9th International Conference on Dependable Systems, Services and Technologies (DESSERT), pp. 572–577. IEEE, Kyiv (2018). https://doi.org/10.1109/DESSERT.2018.8409197

A Graphical Environment for Algorithms Training

Sergiy Markovych[(⊠)] ⓘ, Andrey Chukhray ⓘ,
Vladislav Lukashov ⓘ, Olena Havrylenko ⓘ, and Olena Novytska ⓘ

National Aerospace University "Kharkiv Aviation Institute", 17 Chkalova Street,
Kharkiv 61070, Ukraine
{s.markovych, o.havrylenko, o.novytska}@khai.edu,
achukhray@gmail.com, vladislav.lukashov.120@gmail.com

Abstract. Currently, there are a lot of training systems for acquiring programming skills, and at the same time there is an acute shortage of tools that would allow to acquire and improve the skills of various user groups in the field of algorithmization. The aim of the work is to develop an Intelligent Tutoring System that supports various user groups in the study of algorithms and their application in various practical fields. In the work we used the method of intellectual computer training in algorithmic thinking of various user groups through deep diagnosis and adaptive prompts. The proposed models and method are implemented in the ITS prototype, the work of which is demonstrated on the example of training medical staff, aircraft pilots and students in three different subject areas. The experimental operation of the system has shown its effectiveness in the areas considered. From a scientific point of view the information technology of training in the compilation of algorithms in the form of flowcharts based on new models and methods is one of major interest. In practical terms the developed prototype of a web system for adaptive training in algorithms with the potential to expand the circle of users and problem areas is of great value.

Keywords: Algorithm · Flowchart · Reference model · Diagnostic model · Intelligent tutoring system

1 Problem Statement

An important component of training the competencies of various specialists is the focus on making effective decisions. Therefore it is necessary to teach algorithmic thinking and reference decision-making algorithms in various fields of human activity.

The subject of the current research is to develop an intelligent tutoring system (ITS) for compiling algorithms in the form of flowcharts. The system should work in three modes: demo, training and test.

The first mode of operation is a demo one. The task is formulated and performed by a computer program. A student can independently adjust the pace of learning, i.e., the speed of the system demonstrating how the task is performed. In the training mode of operation, the task is formulated by the program, and the student must complete it. In the case of errors or student's hesitations, the system can provide him with tips. In the test mode the task is formulated by the program, and the student must complete it. Unlike the previous mode, tooltips are not available.

For implementation it is necessary to create models for representing algorithms in the form of flowcharts, diagnostic models [1] for training, and a method of intelligent computer training [2] for compiling algorithms in the form of flowcharts.

2 Review of Existing Solutions

Currently there are products suitable for acquiring programming skills [3], but we also need tools that would allow to acquire and improve a learner's skills in the field of algorithmization. In existing systems of intellectual learning, programming languages lack a more flexible approach to organizing the learning process.

If we consider systems for teaching algorithms without an emphasis on specific programming languages, we will find out that their amount is much smaller. Among such products we can name the Solution-based Intelligent Tutoring System (SITS), also known as the Flowchart-based Intelligent Tutoring System (FITS) [4]. This program does not allow to form the optimal solution since the reference solution is fixed. The program focuses on the study of basic algorithmic structures (Fig. 1). Comparison with reference solutions is only possible because each task provides fixed locations for block distribution. Moreover, the content of the blocks is known in advance.

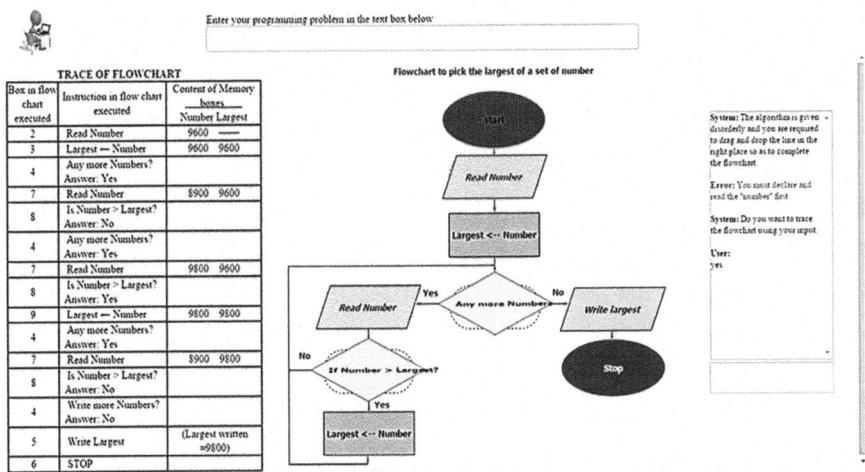

Fig. 1. FITS system interface.

Thus, it is necessary to provide more flexible approaches for training users with the attraction of tools, which are more likely to be applicable for different programming languages.

The current ITS in the field of testing skills in the implementation of algorithms and writing final programs in any programming language that solve certain algorithmic problems are limited in their ability to isolate students' errors in the program code that were found when comparing the student's solution with the standard one. Some errors in the student's code can be detected using static analysis [5].

In [6–8], the examples of testing systems are considered. However, they determine the correctness of the program based only on the output data for the given input tests and do not allow to build the learning process if the student is not able to independently develop the correct algorithm.

There are several ways to recognize and classify the learner's mistakes when compiling a program. One of them is a structural diagnosis, the other is a lexical analysis. They can be used both for analyzing the student's code for the presence or absence of keywords [6], and for finding the "problem" part of the code based on the reference solution to the task proposed to the student. Structural analysis can be carried out at the level of code representation using abstract syntax trees [9].

A promising tool for analyzing learner's algorithms are diagnostic models (DM). This is a new class of mathematical models linking indirect and direct signs of errors [1, 2].

Let us consider in more detail the analysis of algorithms using an example.

3   The Modeling of the Subject Area

Consider the following algorithm:

1. Start.
2. Press the "Start" button.
3. Turn on the toggle switch 1.
4. If the red light is off, then read the ammeter.
5. Turn off the toggle switch 1.
6. Press the Stop button.
7. The end.

Having a graphic reference model of the algorithm (Fig. 2), we will create its mathematical model in the form of an ordered four $\hat{G} = (\hat{N}, \hat{E}, \hat{\Gamma}, \hat{C})$; Where $\hat{N}$ is the name of the algorithm; $\hat{E}$ is a set of vertices; $\hat{\Gamma}$ is a multivalued mapping of the set $\hat{E}$ in itself, i.e. a set of connections; $\hat{C}$ is a set of branches. Suppose we have a functional mapping $f(e_i) = A_i$, i.e., each action of the algorithm is characterized by a unique identifier.

Then for this algorithm we have: $\hat{N} = "A \lg 1"$, $\hat{E} = \{S, e1, e2, e3, e4, e5, e6, F\}$, $\hat{\Gamma} = \{(S, e1), (e1, e2), (e2, e3), (e3, e4), (e3, e5), (e5, e4), (e4, e6), (e6, F)\}$, $\hat{C} = \{(e3, 1, e4), (e3, 0, e5)\}$.

Let us analyze all the errors that a student can make during the construction of such an algorithm: missing vertices; missed connections; extra vertices; extra connections; skipped branches; extra branches. Obviously, incorrect vertices in the general case are the missing correct vertices and extra vertices. The same can be said about incorrect connections and incorrect branches. It is also obvious that skipping vertices entails skipping connections, and using extra vertices means using extra connections.

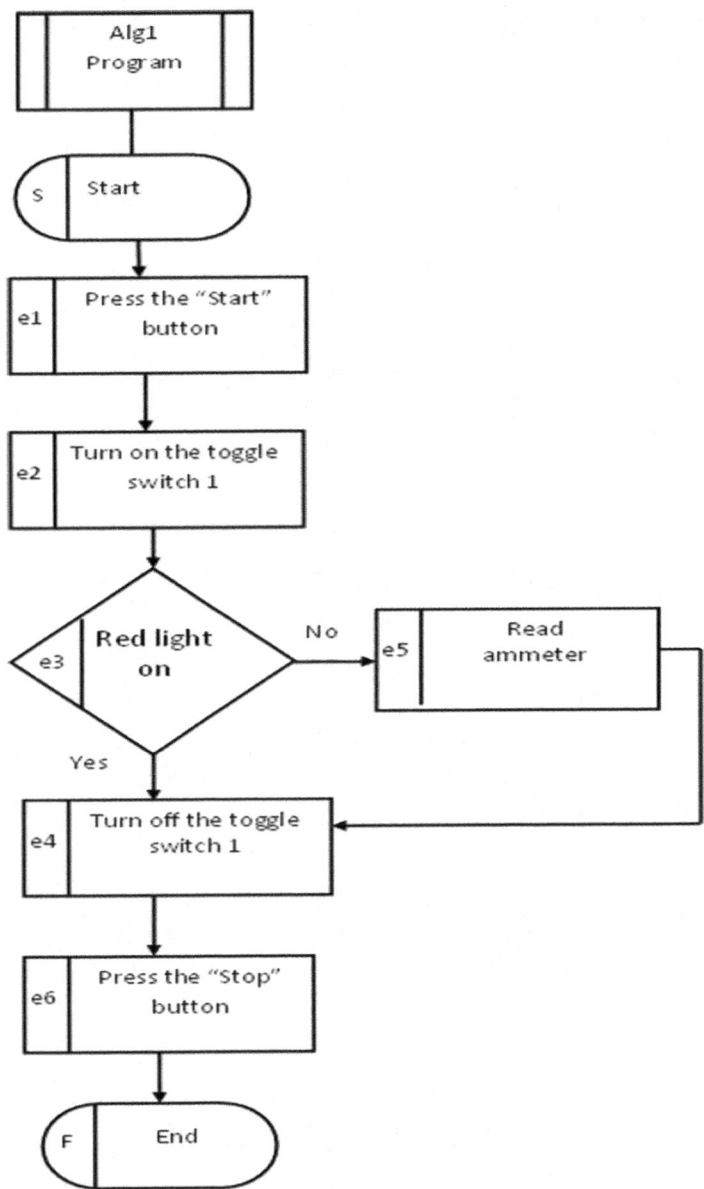

Fig. 2. The graphical reference model of the algorithm.

In addition, in the general case, the student can make several mistakes at the same time, for example, skip the correct vertices (and, therefore, the correct connections), as well as form incorrect branches.

Suppose the student build the algorithm shown in Fig. 3.

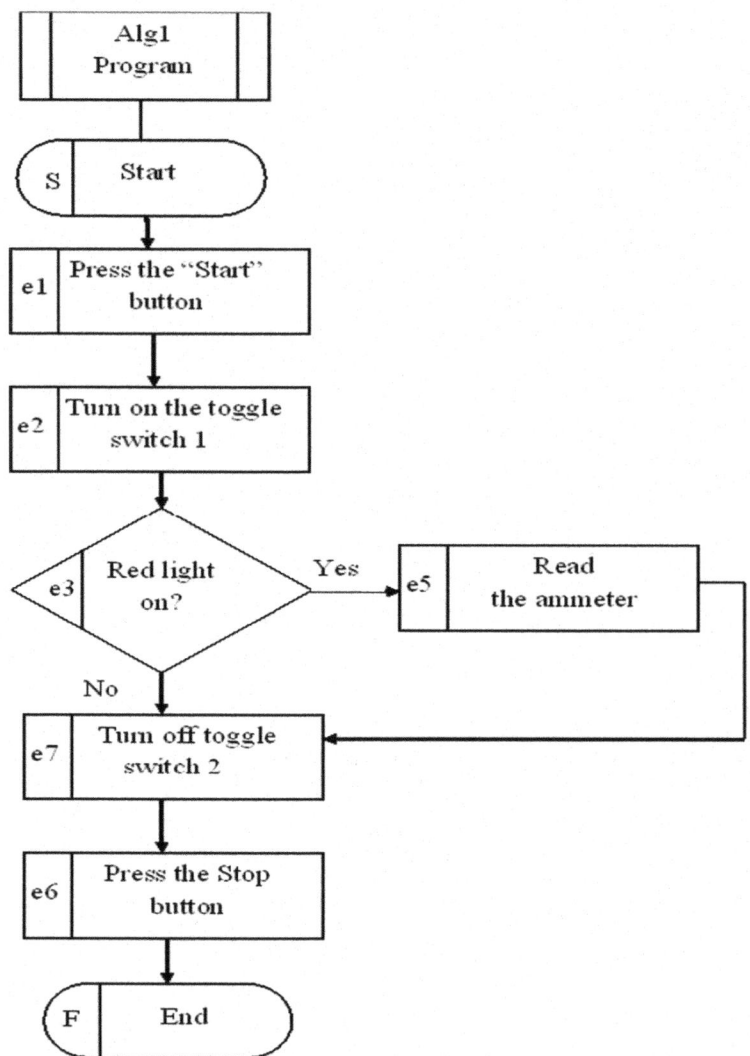

Fig. 3. The graphical model of the algorithm built by a learner.

According to this graphical representation, it is easy to automatically switch to the mathematical model in the form $\tilde{G} = (\tilde{N}, \tilde{E}, \tilde{\Gamma}, \tilde{C})$, where $\hat{N}$ is the name of the algorithm, $\tilde{E} = \{S, e1, e2, e3, e5, e6, e7, F\}$, $\tilde{\Gamma} = \{(S, e1), (e1, e2), (e2, e3), (e3, e5),$ $(e3, e7), (e5, e7), (e7, e6), (e6, F)\}$, $\tilde{C} = \{(e3, 1, e5), (e3, 0, e7)\}$.

To detect missing vertices DM we have $\Delta_{aE} = \hat{E}/\tilde{E} \neq \emptyset$. In this case $\Delta_{aE} = \{e4\}$. To search for extra vertices, use DM $\emptyset(\Delta_{rE} = \{e7\})$. Missed connections can be detected using DM $\Delta_a = \hat{\Gamma}/\tilde{\Gamma} \neq \emptyset$ $(\Delta_a = \{(e3, e4), (e5, e4), (e4, e6)\})$. Finding unnecessary connections is provided for by DM $\Delta_r = \tilde{\Gamma}/\hat{\Gamma} \neq \emptyset$ ($\Delta_r = \{(e3, e7), (e5, e7), (e7, e6)\}$). Similarly DMs are constructed to search for missing branches $\Delta_{aC} = \hat{C}/\tilde{C} \neq \emptyset$ ($\Delta_{aC} = \{(e3, 1, e4), (e3, 0, e5)\}$) and identification of unnecessary branches $\Delta_{rC} = \tilde{C}/\hat{C} \neq \emptyset$ ($\Delta_{rC} = \{(e3, 1, e5), (e3, 0, e7)\}$).

Summarizing the given diagnostic models, one can get more abstract diagnoses of the form "errors at the vertices" $\Delta_{arE} = (\hat{E}/\tilde{E}) \cup (\tilde{E}/\hat{E}) = \tilde{E} - \hat{E} \neq \emptyset$, where the "–" sign denotes the symmetric difference of the sets, "link errors" $\Delta_{ar} = (\hat{\Gamma}/\tilde{\Gamma}) \cup (\tilde{\Gamma}/\hat{\Gamma}) = \hat{\Gamma} - \tilde{\Gamma} \neq \emptyset$ and "branches errors" $\Delta_{ar} = \hat{\Gamma} - \tilde{\Gamma} \neq \emptyset$.

In addition, you can build models designed to identify learning abilities. We denote $Err(t) = |\Delta_{arE}(t)| + |\Delta_{ar}(t)| + |\Delta_{ar}(t)|$– the total number of errors made by the student at time t. Then there is a positive learning dynamic when $Err(t + \Delta t) < Err(t)$. Accordingly, if $Err(t + \Delta t) \geq Err(t)$, then the learner does not improve his result.

Study a more complex algorithm when executed on real equipment, it will be necessary to respond to various, including random, events:

1. Start.
2. Set the timer for sound signals "Take ammeter readings" after 20, 40 and 60 s and "Stop" after 70 s.
3. Press the "Start" button.
4. Turn on the toggle switch 1.

If the green light is on, turn on the timer and enter standby mode. If the light does not light, then turn off the toggle switch 1 and press the Stop button.

Events and reactions (Table. 1):

– when there is the sound signal "Read the ammeter readings", record the ammeter readings in table One;
– when the sound signal is "Stop", turn off the switch 1 and press the "Stop" button;
– with the sound signal "Problems", turn off the toggle switch 1 and press the "Stop" button;
– when there is the sound signal "Insufficient input voltage", turn on the switch 2, if the voltmeter readings are from 11 to 12 V.

Graphic models of such an algorithm are shown in Fig. 4–5.

Table 1. Events and Actions.

Events	Actions
Sound signal "To take readings of the ammeter"	Perform the "Commit" procedure
Sound signal "Stop"	Perform the Stop procedure
Sound signal "Problems"	Perform the Stop procedure
Sound signal "Insufficient input voltage"	Perform the "Voltage Boost" procedure

Now the mathematical model of this algorithm can be represented by an ordered six $CG = (N, V, S, E, , C)$, where N is the name of the algorithm; V is a set of events that

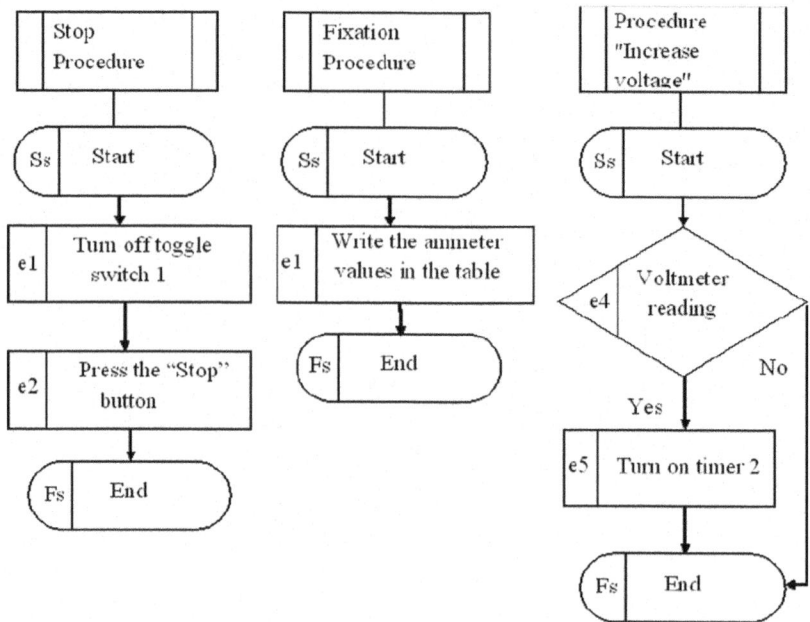

Fig. 4. The graphical reference model of the algorithm (events and reactions).

you need to respond to, and each event is an ordered two $v = (ev, A)$; S is a set of subroutines:

$$N = "A \lg 2";$$
$$V = \{(onFix, e15), (onStop, e16), (on \Pr oblems, e16), (onNotEnouhVoltage, e17)\};$$
$$S = \{S_{Stop}, S_{Fix}, S_{Inc}\};$$
$$S_{Stop} = (N_{S_{Stop}}, E_{S_{Stop}, S_{Stop}}, C_{S_{Stop}})$$
$$= ("Stop", \{S_s, e_1, e_2, F_s\}, \{(S_s, e_1), (e_1, e_2), (e_2, F_s)\}, \{\});$$
$$S_{Fix} = (N_{S_{Fix}}, E_{S_{Fix}, S_{Fix}}, C_{S_{Fix}}) = ("Fix", \{S_s, e_3, F_s\}, \{(S_s, e_3), (e_3, F_s)\}, \{\});$$
$$S_{Inc} = (N_{S_{Inc}}, E_{S_{Inc}, S_{Inc}}, C_{S_{Inc}})$$
$$= ("Inc", \{S_s, e_4, e_5, F_s\}, \{(S_s, e_4), (e_4, e_5), (e_4, F_s)(e_5, F_s)\}, \{(e_4, 1, e_5), (e_4, 0, F_s)\});$$
$$E = \{S, e_6, e_7, e_8, e_9, e_{10}, e_{11}, e_{12}, F\};$$
$$= \{(S, e_6), (e_6, e_7), (e_7, e_8), (e_8, e_9), (e_9, e_{10}), (e_{10}, e_{11}), (e_9, e_{12}), (e_{12}, F)\};$$
$$C = \{(e_9, 1, e_{10}), (e_9, 0, e_{12})\}, f(e_{15}) = \prod_N S_{Fix}, f(e_{16}) = \prod_N S_{Stop}, f(e_{17}) = \prod_N S_{Inc},$$

where Π is the projection operator which selects one of the components out of the ordered n, in this case with the value of the subroutine name.

In addition to the mistakes made when compiling simple algorithms (in this case they can occur both when compiling the main program and individual subroutines), the following errors can also be made here:

– there are no events and reactions to them, $\tilde{V} = \emptyset$;
– incomplete event processing $\hat{V}/\tilde{V} \neq \emptyset$;

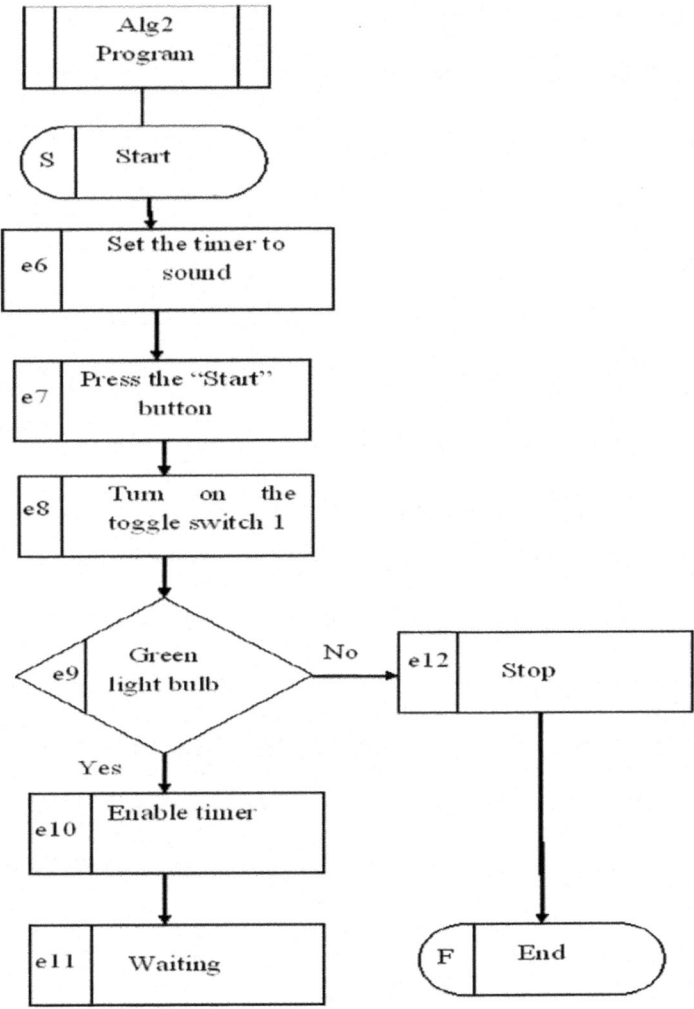

Fig. 5. Graphical reference model of the algorithm (main program).

- wrong reaction to events
- $\exists \hat{v} \in \hat{V}, \exists \tilde{v} \in \tilde{V} : (\Pi_{ev}\hat{v} = \Pi_{ev}\tilde{v}) \wedge (\Pi_A \hat{v} \neq \Pi_A \tilde{v})$;
- no subroutines $\tilde{S} = \emptyset$;
- incomplete set of subroutines $\hat{S}/\tilde{S} \neq \emptyset$;
- excessive number of subroutines $\tilde{S}/\hat{S} \neq \emptyset$.

Therefore, the method of analysis of the algorithm designed by the student in a graphical environment should be expanded to check the correctness of events and their processing, as well as the correctness of the subroutines.

To formalize actions with parameters we define the ParamActions class with each action $A_i = (Text, Params) \in ParamActions$ is a pair consisting of text and many parameter values inside this text.

Then, if the student makes a mistake of the type "Set the timer for sound signals", "Read the ammeter after 20, 30 (instead of 40!) And 60 s" and "Stop after 70 s", then a similar diagnostic model can be used to identify it $\exists \hat{A}_i, \tilde{A}_j \in ParamActions$:

$$(\hat{\Gamma}^1(f^{-1}(\hat{A}_i)) = \tilde{\Gamma}^1(f^{-1}(\tilde{A}_j))) \wedge (\hat{\Gamma}^{-1}(f^{-1}(\hat{A}_i))$$
$$= {}^{\sim 1}(f^{-1}(\tilde{A}_j))) \wedge (f^{-1}(\hat{A}_i) \neq f^{-1}(\tilde{A}_j)) \wedge (\Pi_{Text}\hat{A}_i = \Pi_{Text}\tilde{A}_j) \wedge (\Pi_{Params}\hat{A}_i \neq \Pi_{Params}\tilde{A}_j).$$

Despite the apparent simplicity of the examples, this mathematical tool can be used for a wide range of tasks: training operators of complex technical objects, training medical personnel, managers, etc.

4   The Development of a Prototype System

Since the development is carried out according to the three-level architecture model, we consider a set of tools for each of the layers of architecture starting from the bottom.

The database level is represented by the MySQL server version 8.0.12. If the version of the remote server is not lower than 5.7, then there should not be problems with compatibility between it and the local server on which the development is underway. Node.JS version 12.0.10 with built-in Node Package Manager is used as the application server for the logic level. To work with the client level JavaScript is now used in accordance with the ES6 specification in conjunction with HTML5 and CSS3.

Regarding the storage level, the following data handling models were considered:

- relational;
- document-oriented;
- graph based.

Let us analyze them starting with the last.

1. The graph-based data presentation model has its own specific context where abstract entities are interpreted as vertices of the graph, and the relationships between them are defined as connections. This form of data presentation is suitable, for example, for designing a repository of some kind of social network where tens of millions of users and related data are required to be stored. But to solve this problem this approach is not the best choice, since the product being developed conceived in the form of one of the modules of the web system has a different development scale. This module works with several entities and, at maximum load, several hundred objects.
2. The document-oriented data model defines entities, as well as their attributes, in the form of abstract trees in the context of this model called documents. The leaf nodes of the tree contain data that, when a document is added to the repository, is entered into indexes which allows flexible retrieval of individual pieces of information

without the need to load the entire document into RAM. To work with data the NoSQL approach is used whereas the name implies the SQL syntax is completely absent. That is information in the form of a collection created by the means of the selected programming language is directly recorded in the storage as a continuous block which due to some hardware features gives a significant increase in performance. Another advantage is the direct work with collections and, as a result, the absence of strict size restrictions which are determined only by the amount of RAM. Each time you make changes to the collection you can overwrite the previously created record in the database without having to manually delete the old one and create a new field. The main disadvantage is the absence of strict typing in some cases. That is each record in the repository will contain a line that upon the request will need to be divided into substrings and carried out other conversions thus increasing the application operating time. But the rejection of this approach is justified, first of all, by the impossibility of creating a strict data structure which is critical for this project.

3. Therefore, excluding the alternatives discussed above, a relational data model was chosen. Although it is clearly inferior in performance to a document-oriented project on the current scale, it has a clearly traceable structure and mathematical apparatus of relational algebra implemented using SQL syntax. To work with relational databases, four server implementations were considered: MySQL, PostgreSQL, Oracle DB, MSSQL Server. The last two options were excluded due to cumbersomeness, which does not preclude the use of Oracle DB in the future when other project modules are implemented and it becomes possible to create a single repository of the web system. It turned out to be preferable to use MySQL as the optimal solution for the developed module in terms of scale, cost, prevalence and testability.

The choice of implementation of the application server was carried out based on the requirements for using the Java, JavaScript languages and related technologies. For performance reasons, Node.JS was chosen as the application server. JavaScript code executed on the basis of this platform is directly translated into machine code. And the presence of an integrated npm package manager in the platform makes it easy to download the required components and connect them to the developed system.

At this stage of development the amount of code on the server is ten times less than on the client side. This is due to the specificities of the module being developed, where most of the code is responsible for the graphical interface and manipulation of graphic objects, and on the server the main logic consists in processing requests that are responsible for working with the database and the process of diagnosing the learner algorithm model based on the reference model. Another reason is the inability of the platform to work with the BOM and DOM elements of the browser, that is, access to the graphical tools of the browser is impossible.

Code is written on the client side in order to achieve increased productivity, as well as the ease of maintenance and system modifiability.

To work with the flowchart, two panels are provided in the user interface. Vertical, containing many specialized blocks and horizontal, responsible for operations on these blocks. To add an item from the vertical panel you need to perform a single click on it with the left button of the touch mouse and repeat clicking on the canvas to the left of the panel in the place required for displaying (Fig. 6). There are also functions for adding text, moving and deleting one element and a group, canceling and repeating the last actions.

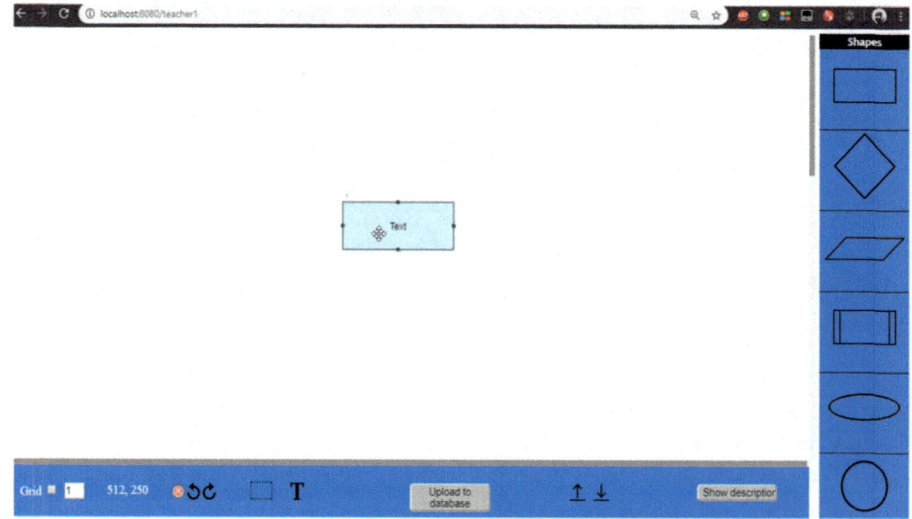

Fig. 6. The system interface view.

Each node has a set of connection ports between which you can build connections. Any port belongs to one of three types:

- *mixed* – in this case the port can connect to any type of port on the remote host;
- *output* – in this case the port can connect either to the mixed or input type ports;
- *input* – in this case the port can connect either to ports of a mixed type or an output one.

Any node has a predefined input port on top, an output port on the bottom, and a mixed one on either side, if provided.

In fact, creating a concept such as a port is needed to introduce restrictions on the process of building flowchart connections. So, you cannot connect a pair of ports of the output type defined at the bottom of the node or that of the input type at the top. In addition, it is not permissible to construct two connections for which the nodes of the beginning and the end coincide.

To build a connection you need to move the cursor to an arbitrary port, perform a single click and move the cursor in the direction of another node and its port (Fig. 7). During the movement the path of future connection with the "pivot points" or "knees" of the connection is built.

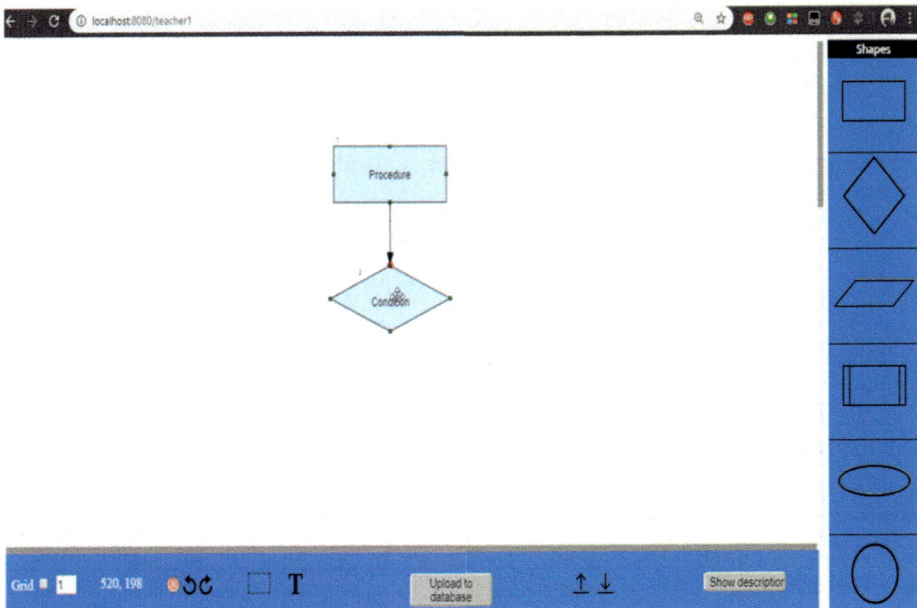

Fig. 7. Building elements connections.

In this case, the state of the canvas is updated and saved each time the left mouse button is returned to its original position or the deletion is performed, i.e. the save arrow is provided for convenience only.

The mode described above is the compilation of a reference flowchart by a teacher. The learning mode interface is similar, except for the vertical panel, which is not required here. When the system starts in this mode, a lot of blocks are saved from the database stored in the reference flowchart. The learner's task is to select the correct sequence of blocks and establish connections between them. After clicking the "Check solution" button the result of the construction is sent to the server and compared with the reference solution after which the page of verification results and the type of the correct solution are loaded from the server. In the process of loading the page the verification results are processed and the corresponding messages are displayed.

5   Results

As the examples that demonstrate the effectiveness of the proposed models and method we consider several subject areas of training. The first is the algorithm of actions of medical personnel to destroy narcotic drugs and psychotropic substances.

The protocol for the destruction of narcotic drugs and psychotropic substances is as follows:

1) liquid dosage forms (DF) in glass ampoules, vials are destroyed by crushing the primary packaging, liquid DF in plastic ampoules, syringe tubes are destroyed by crushing the primary packaging with subsequent dilution of the resulting contents with water in a ratio of 1: 100 and draining the resulting solution into the sewer;

2) solid pharmaceutical preparations containing water-soluble pharmaceutical substances of narcotic drugs and psychotropic substances after crushing to a powder state must be diluted with water in a ratio of 1: 100 and drained of the resulting suspension (solution) into the sewer;

3) water-soluble pharmaceutical substances are destroyed by diluting with water in a ratio of 1: 100 and draining the resulting solution into the sewer;

4) solid DF containing water-insoluble pharmaceutical substances of narcotic drugs and psychotropic substances, soft DF, transdermal dosage forms are destroyed by burning;

5) water-insoluble pharmaceutical substances are destroyed by incineration.

After analyzing the textual description of the algorithm, we can distinguish that the actions to destroy narcotic drugs and psychotropic substances depend on:

– type of DF (liquid, hard, soft or transdermal);
– type of pharmaceutical substance (soluble or insoluble in water);
– type of primary packaging (glass ampoules, vials, plastic ampoules, syringe tubes);
– the method of destruction (discharge into the sewer or burning).

There is redundancy in the text of the algorithm – repetition of the same type action description which is not allowed in the flowcharts. Thus the action "dilution with water in a ratio of 1:100 and the discharge of the resulting solution into the sewer" is repeated three times, and "destruction by burning" – two times.

We transform the textual description of the algorithm into a flowchart. For clarity we use identifiers – conditional numbers of nodes in the diagram (Fig. 8). As a starting point determining the next steps we take the form of DF as it was in the document.

So, we add the selection block 1 with the options "Liquid", "Solid", "Soft or transdermal" to the first element of the flowchart after the start node. The actions with liquid drugs are described in paragraph 1 of the protocol.

Using the selection node 2 we will separate the methods for destroying DF packaged in glass ampoules and vials (operation 3 – "Crushing") and DF produced in plastic ampoules and syringe tubes (operation 4 – "Crushing"). Further actions with the obtained contents are the same: 5 – "Dilution with water in a ratio of 1: 100" and 6 – "Drain into the sewer". The left branch of the flowchart is completed so we put the "End" block.

We deal with actions in relation to solid DF. Actions vary depending on what type of pharmaceutical substance these forms contain: water-soluble (paragraph 2) or insoluble in water (paragraph 4). We add selection block 7 to the flowchart – branching depending on the type of pharmaceutical substance.

The actions with solid pharmaceutical preparations containing water-soluble substances do not differ from operations with plastic ampoules and syringe tubes which we

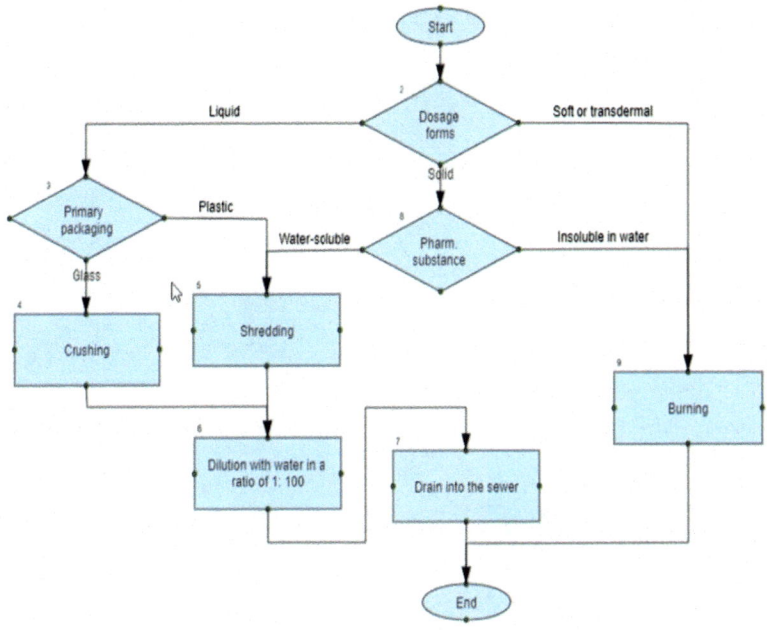

Fig. 8. A reference solution for the algorithm of the medical staff actions.

have indicated in the flowchart earlier: 4, 5, 6. Therefore, we will connect the output of block 7 corresponding to the option "Soluble in water" to the block 4 already available in the diagram.

Returning to paragraph 4 and noting paragraph 5 of the document we add block 8 of the "Burning" operation to the flowchart and connect the outputs of the blocks of choice 1 – "Soft or transdermal" and 7 – "Insoluble in water" with block 8 which, in turn, is connected with the terminator "End". After analyzing the remaining paragraph 3 we connect the output of block 3 with block 5.

Next you need to add the nodes of the reference solution to the block area by alternately selecting elements and using the key combination "Ctrl + c". It is also necessary to add several "erroneous" blocks, and then click the "Upload to database" button.

Now we proceed directly to learning how to build this flowchart (the "Go to the learner module" button). At the first call the student's page loads many blocks previously set by the teacher (Fig. 9) from the repository, and provides the ability to select the necessary blocks and sequentially transfer them to the main canvas adding connections. After clicking on the "Check solution" button the student's decision is sent for verification. The verification takes place in several stages. At the first stage, a comparison is made with the reference solution in terms of the number of nodes and connections. And if it does not match, then there is no point making the other checks. At the second stage, the comparison is based on the type of node, its contents, as well as the index that corresponds to the solution step.

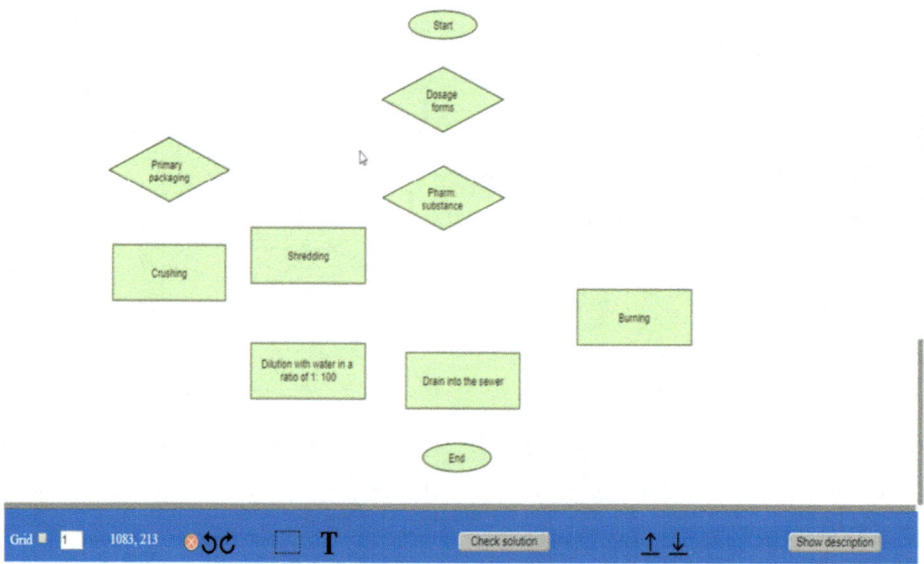

Fig. 9. Building the algorithm by a student.

After performing all the checks the corresponding messages will be displayed on the page (Fig. 10). If the solution is correct, the corresponding message is displayed and the training mode is completed. Other situations shown in Fig. 10, namely: the absence of a block and / or connection, incorrect contents and/or type of block, imply further error correction until the solution matches the reference one.

Type and content messages are displayed for each node separately. If the node has both errors, then type information is prioritized. To remove error messages you need to perform a single click on the main canvas. The backlighting of erroneous nodes is removed after a double click. Also, it is possible to view the reference solution by clicking on the "Show reference solution" link where viewing is available within one minute.

Now let's look at another example – the algorithm of the actions of the crew of the An-24 aircraft to eliminate the fire inside one of the AN-24 engines.

When a fire occurs inside an An-24 engine the red light "FIRE INSIDE RIGHT ENG"or" FIRE INSIDE THE LEFT ENG" appears on the fire extinguishing system control panel and the siren is buzzing.

In this case, it is necessary to:

1) turn off the autopilot (if it was turned on);
2) perform an emergency decline;
3) report the incident to the ATC service;
4) turn on the signal "DISTRESS";
5) feather the blades of the engine on fire by pressing the KFL-37 button duplicating the feathering by emergency hydro-stop;

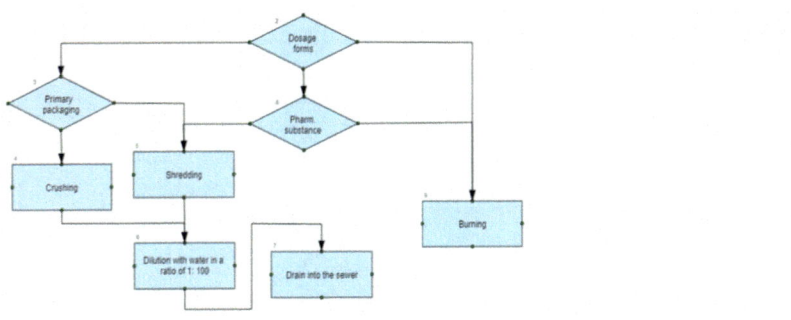

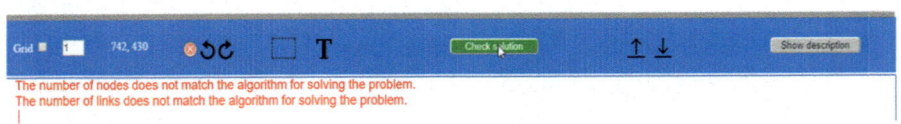

Fig. 10. Diagnostic messages in medical staff training mode.

6) press the button for turning on the discharge of fire extinguishers inside the engine on fire while the corresponding yellow warning lights I and II of the fire extinguishers discharge control should go out;
7) turn off the fuel priming pumps on the half-wing on which the engine on fire is located, and on stalling the engine: close the shut-off valve; turn off alternating current and alternating current generators; close the air intake; set the engine control lever to the "O" position according to the control gear and the SHUTOFF switch to the "CLOSED" position;
8) not earlier than after 15 s check the elimination of the fire for which the main switch should be put in the neutral position, and then in the "FIRE FIGHTING" position.

If the fire is extinguished (the light does not light up), proceed to the nearest airfield.

Note. If the alarm system fails and the fire is visually detected, instruct the flight engineer to turn on the fire extinguishing system manually by pressing the appropriate button. In the future, all necessary actions should be performed as if a fire alarm is triggered.

If the flasher lights are on again, which indicates the continuation of the fire, it is urgent to make an emergency landing at the nearest airfield or at any suitable site.

Now, having a clear sequence of steps in place we set up many blocks and build a reference flowchart, which the student must reproduce having the correct and incorrect blocks. As a result, we should obtain the flowchart identical to the reference one in the number of connections and nodes, as well as their index, type and content. After checking either a success message is displayed (Fig. 11) or it is necessary to correct the blocks highlighted in red and pass the check again.

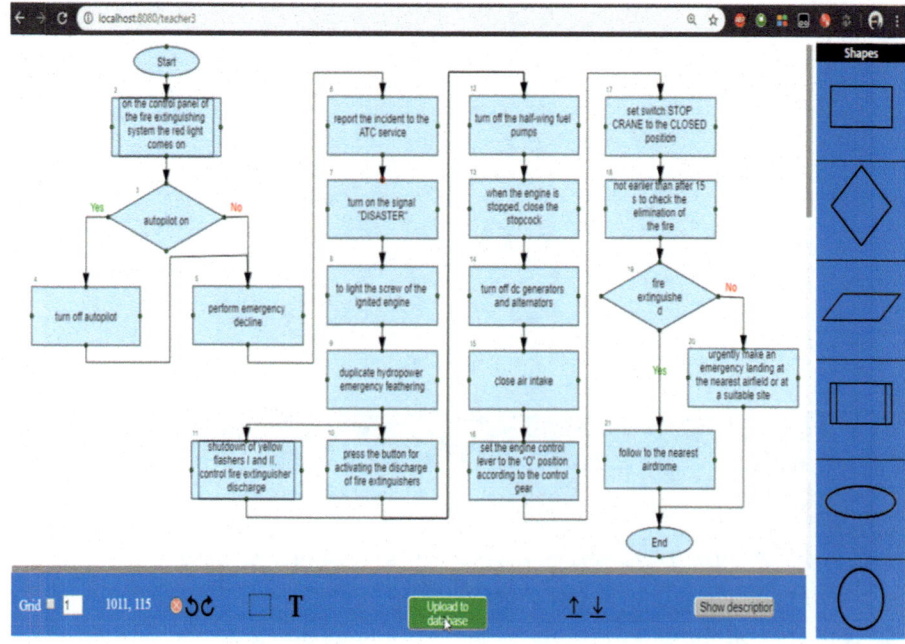

Fig. 11. Reference flowchart for training aircraft pilots.

Another example is not an algorithmic problem but it is reducible to a certain linear algorithm: checking for the correctness of the compilation of terms and definitions within the discipline "Theory of automatic control" from the educational plan of 3rd year students of the specialty "Avionics".

Validation begins by choosing a term or domain-specific definition. For example: "The analysis of the types and consequences of the malfunctions is a qualitative reliability analysis which consists of studying the types of possible malfunctions of each component part of an object and determining the effect of the consequences of malfunctions of each type on other component parts and on the required functions of the object."

Such a definition can be divided into syntactic or semantic blocks of arbitrary length:

1) analysis of the types and consequences of the malfunctions;
2) qualitative reliability analysis which consists of;
3) studying the types of possible malfunctions of each component of an object;
4) and determining the effect of the consequences of malfunctions of each type;
5) on other component parts;
6) and on the required functions of the object.

This will be the initial set of flowchart nodes connected in series in the reference model with the addition of terminal nodes of the start and the end (Fig. 12).

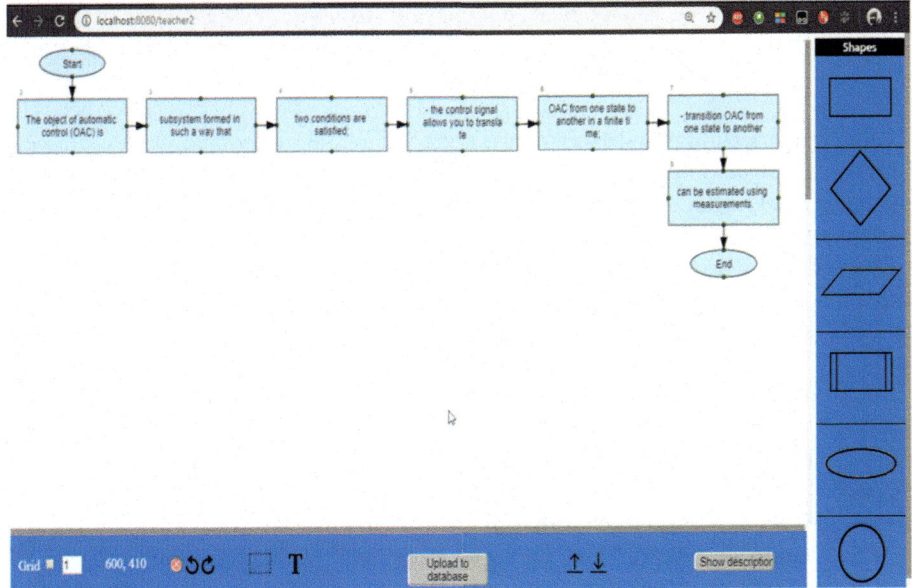

Fig. 12. The reference flowchart for teaching students the definition of a concept.

Also, for the training mode you need to add several blocks taking them from the definitions of other terms that are close in meaning, for example:

1) qualitative reliability analysis method which includes;
2) identification and analysis of the nature of occurrence.

The compilation of the algorithm by the trainees and the corresponding diagnostic messages are similar to the previous examples, and the physical movement and connection of the definition blocks allow you to use various storage mechanisms.

Thus, the considered software module can act as a prototype for testing complex solutions with improved intellectual support. Further development directs to covering other scientific and engineering spheres [10, 11] and requires clarification of the subject area and the expansion of the resource base for research in the field of intelligent training programs.

6 Conclusion

Even at the current stage of development we can confidently say that there are no analogues of the above-described web system under open access. The last of the considered examples demonstrates the range of possibilities of the proposed approach covering an extensible class of problems the solution of which is a subject of algorithmization.

Considering the principles laid in the basis of the development it can be concluded that through the system it is possible to teach the compilation of algorithms of any complexity using various pedagogical strategies, for example: 1) the student is shown a reference flowchart in the demonstration mode; 2) an incomplete flowchart is shown in the demo mode, i.e. some elements and links are missing; 3) the reference flowchart is not shown.

Then, in the training mode the trainee goes through an individual way of drawing up the correct algorithm while diagnostic adaptive messages are sent to him.

The test mode without prompts is advisable to use for the current or module control of the trainee and the final assessment of his competencies within a specific course.

The scientific novelty of the research lies in the fact that for the first time an information technology teaching the development of algorithms in the form of flowcharts based on new models and the method of intelligent computer learning has been created.

The practical significance of the work which is a continuation of research [12–14] lies in the fact that through the developed web-system it is possible to effectively and adaptively train a wide range of users in algorithmic techniques and making rational tactical decisions.

The prospects for further research include training doctors in algorithms for diagnosing complex diseases and choosing drug therapy, training students and IT specialists in software development and management algorithms for this process [10, 11], training pilots in emergency situations, training marketers and managers in making decisions in dynamically changing conditions, also training engineers in innovative technical algorithms [15, 16], etc.

References

1. Kulik, A.: Rational intellectualization of the aircraft control: resources-saving safety improvement. In: Kharchenko, V., et al. (eds.) Green IT Engineering: Components, Networks and Systems Implementation. SSDC, vol. 105, pp. 173–192. Springer, Cham (2017). https://doi.org/10.1007/978-3-319-55595-9_9
2. Chukhray, A.: Methodology for learning algorithms. National Aerospace University "Kharkiv Aviation Institute", Kharkiv (2017). (in Russian)
3. Rattan, D., Bhatia, R., Singh, M.: Software clone detection: a systematic review. Inf. Softw. Technol. 55(7), 1165–1199 (2013). https://doi.org/10.1016/j.infsof.2013.01.008
4. Hooshyar, D., Ahmad, R.B., Yousefi, M., et al.: A flowchart-based intelligent tutoring system for improving problem-solving skills of novice programmers. J. Comput. Assist. Learn. 31(4), 345–361 (2015). https://doi.org/10.1111/jcal.12099
5. Kim, M., Notkin, D., Grossman, D., Wilson, G.: Identifying and summarizing systematic code changes via rule inference. IEEE Trans. Softw. Eng. 39(1), 45–62 (2013). https://doi.org/10.1109/TSE.2012.16
6. Zhang, K., Shasha, D.: Simple fast algorithms for the editing distance between trees and related problems. SIAM J. Comput. 18(6), 1245–1262 (1989). https://doi.org/10.1137/0218082
7. Tai, K.C.: The tree-to-tree correction problem. J. ACM 26(3), 422–433 (1979). https://doi.org/10.1145/322139.322143
8. Wagner, R.A., Fischer, M.J.: The string-to-string correction problem. J. ACM 21(1), 168–173 (1974). https://doi.org/10.1145/321796.321811

9. Yang, W.: Identifying syntactic differences between two programs. Softw. Pract. Exp. **21**(7), 739–755 (1991). https://doi.org/10.1002/spe.4380210706
10. Zabolotnyi, O.: Proximate testing method of moisture measurement for substances of dielectric nature. Radio Electron. Comput. Sci. Control **1**, 7–17 (2019). https://doi.org/10. 15588/1607-3274-2019-1-1
11. Zabolotnyi, O.V., Zabolotnyi, V.A., Koshevoi, M.D.: Conditionality examination of the new testing algorithms for coal-water slurries moisture measurement. Naukovyi Visnyk Natsionalnoho Hirnychoho Universytetu **1**, 51–59 (2018). https://doi.org/10.29202/nvngu/ 2018-1/21
12. Gaydachuk, D., Havrylenko, O., Bastida, J.P.M., Chukhray, A.: Structural diagnosis method for computer programs developed by trainees. CEUR Workshop Proc. **2387**, 485–490 (2019)
13. Bastida, J.P.M., Havrykenko, O., Chukhray, A.: Developing a self-regulation environment in an open learning model with higher fidelity assessment. In: Bassiliades, N., et al. (eds.) Information and Communication Technologies in Education, Research, and Industrial Applications. ICTERI 2017. CCIS, vol. 826, pp. 112–131. Springer, Cham (2018). https:// doi.org/10.1007/978-3-319-76168-8_6
14. Chukhray, A., Havrylenko, O.: Proximate objects probabilistic searching method. In: Nechyporuk, M., et al. (eds.) Integrated Computer Technologies in Mechanical Engineering. AISC, vol. 1113, pp. 219–227. Springer, Cham (2020). https://doi.org/10.1007/978-3-030- 37618-5_20
15. Fevralev, D.V., Ponomarenko, N.N., Lukin, V.V., et al.: Efficiency analysis of color image filtering. EURASIP J. Adv. Signal Process. **2011**, 41 (2011). https://doi.org/10.1186/1687- 6180-2011-41
16. Kharchenko, V., Gorbenko, A., Sklyar, V., Phillips, C.: Green computing and communi- cations in critical application domains: challenges and solutions. In: The International Conference on Digital Technologies 2013, Zilina, pp. 191–197. IEEE (2013). https://doi.org/ 10.1109/DT.2013.6566310

Cyberterrorism Attacks on Critical Infrastructure and Aviation: Criminal and Legal Policy of Countering

Mykola Nechyporuk⑩, Volodymyr Pavlikov⑩,
Nataliia Filipenko$^{(\boxtimes)}$⑩, Hanna Spitsyna⑩, and Ihor Shynkarenko⑩

National Aerospace University "Kharkiv Aviation Institute",
17 Chkalova Street, Kharkiv 61070, Ukraine
n.filipenko@khai.edu

Abstract. Analysis of scientific opinions on the content of criminal law policy for countering cyberterrorist attacks on critical infrastructure and aviation. Scientifically substantiated recommendations concerning the directions of counteraction to cyber incidents are given and the innovation development algorithm for the needs of law enforcement agencies is offered.

Keywords: Cyberthreats · Information security · Malware · Crime counteraction · Interference in the work of information and telecommunications systems · Innovations · Crime investigation

1 Introduction

Rapid humankind development in recent decades is due to the development and implementation of the latest scientific developments that can accelerate the pace of vital processes and cover all areas of public life. Constant use of computer technology, latest means and methods of information delivery marked the countdown to new reality, namely: global informatization and the emergence of cyberspace (cyberreality). According to its technical characteristics and purpose, cyberspace is a transnational area of information technology infrastructures and interdependent networks including Internet, telecommunications networks, computer systems and embedded processors in critical industries. Globally interconnected and interdependent cyberspace is the basis for development of modern society providing important support for the world economy, flawless operation of military, civilian and critical infrastructure. Therefore, protection and safeguarding of critical infrastructure and aviation from cyberterrorism and cyberattacks is a priority for all law enforcement and law enforcement agencies.

Despite the significant contribution a lot of scientists to the doctrine development of the basics of countering attacks on critical infrastructure and aviation their investigations have not exhausted this problem but on the contrary, have raised a number of new issues that need to be resolved. Nevertheless, in recent years, several opinions have emerged on the organization of preventive activities in the field of computer technology, priority areas of law enforcement agencies, the process of reforming the regulatory framework for such activities, and so on. Therefore, the issues of criminal law

M. Nechyporuk et al. (Eds.): ICTM 2020, LNNS 188, pp. 206–217, 2021.
https://doi.org/10.1007/978-3-030-66717-7_17

policy for countering cyber-terrorist attacks on critical infrastructure and aviation are relevant and require separate research.

2   Main Content Presentation

An important part of Ukrainian modern criminal law policy is countering various terrorist threats. The most dangerous and destructive consequences are terrorist acts related to the computer technology use, as they pose a real threat to the stable operation of critical infrastructure and aviation that threatens human life and health, disrupts work of enterprises, institutions and organizations, industrial and economic facilities, etc. In recent years, criminals have been increasingly using scientific and technological advances and as a result, the total number of new types of crimes using computer technology and software is growing. Traditionally, most cyber-attacks have been carried out by criminal organizations, with the majority of incidents failing to register on an enterprise risk scale of businesses that faced significant setbacks. In 2017, this dynamic changed with the WannaCry and NotPetya incidents. These two attacks affected organizations in more than 150 countries, prompted business interruption and other losses estimated at well over USD 300 million by some companies, brought reputational damage, and resulted in loss of customer data [1].

According to the Security Service of Ukraine, in 2019, information security specialists neutralized more than 480 cyber incidents and cyberattacks on public authorities and critical infrastructure (Fig. 1). During this period, more than 1,000 Web resources used for criminal purposes also ceased to function [2]. This is particularly damaging for financial information sites such as a stock market. They have a legal duty to give equal access to different users. They would normally shut down and stop trading for a while rather than allow some people to get information before others. These attacks are not designed to steal data or do insider trading. They are generally set up to demand ransom from the victims, usually asking for thousands of dollars paid in bitcoin or another cryptocurrency which is effectively untraceable. Governments, terrorist organisations, political groups and even pranksters have also been known to use these attacks [3].

Various indicators of implemented measures in the field of protection of computer and telecommunication networks from cyber-attacks and creating conditions for safe operation of cyberspace are evaluated and used to monitor and compare the state of cybersecurity in different countries in annual international rankings, the most authoritative of which is the Global Cybersecurity Index (GCI). According to the GCI-2018 ranking, Ukraine took 54th place among 193 countries, rising by 5 positions over the last year. At the same time, experts noted: progressive steps in building a legal framework to guarantee the cybersecurity of the state; sustainability of government initiatives to increase cybersecurity in the field of ICT; a significant improvement in the cyber resilience of organizations over the past year, despite more than doubling targeted cyberattacks. However, if we compare Ukraine's performance in this ranking with, for example, post-Soviet countries (Table 1), it becomes clear that many of them did a much better job of building cyber resilience, as they are significantly ahead of us in this ranking. Thus, Lithuania took 4th place in the overall ranking, Estonia – 5,

Georgia – 18, the Russian Federation – 26, Kazakhstan – 40, Latvia – 44, Moldova – 53 and bypassed us in the GCI-2018 ranking [4].

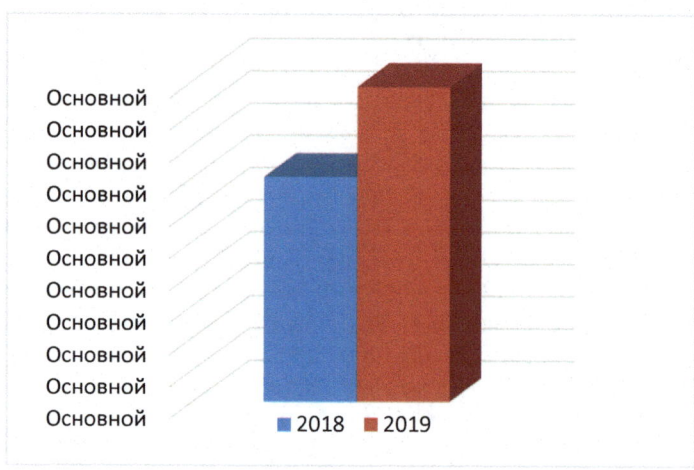

Fig. 1. Comparative number of cyberattacks on public authorities and critical infrastructure in 2018/2019 years.

However, this is just the tip of the "information iceberg". The lion's share of terrorist cyberattacks remains in "shadow" causing significant damage, both material and moral. The complexity of prevention, detection and investigation of these cyber incidents is due to the following:

– rapid emergence of the latest developments in the field of computer technology for civil and military purposes;
– weakness of system of countering terrorist cyberattacks. Unfortunately, we should state that level of protection of domestic critical infrastructure and aviation does not always meet modern world safety requirements. This is very often due to corruption or shortcomings in organizational and economic activities at critical infrastructure and aviation facilities;
– lack of a sufficient number of top-quality experts working on critical infrastructure, aviation and have the necessary level of knowledge in the field of computer technology;
– gaps in the patriotic upbringing of the population leading to betrayal of the national interests of Ukraine. This is especially relevant in the realities Ukrainian present when the state is forced to resist foreign aggression on its borders;
– increasing stress and psychological burden on society, the emergence of panic caused by the presence of negative military, economic and social factors [5];
– weak coordination of actions of law enforcement agencies during anti-terrorist operations;
– significant prevalence on Internet of sites with outright extremist ideology [6];

Table 1. Rating of Ukraine in GCI-2018.

Country	Place among 193 countries
Lithuania	4th place
Estonia	5th place
Georgia	18th place
Russian Federation	26th place
Kazakhstan	40th place
Latvia	44th place
Moldova	53th place
Ukraine	54th place

- lack of operational and investigative information regarding persons who create a hacker group of criminal orientation;
- weakness of domestic anti-terrorist legislation containing many legal conflicts and contradictions;
- high level of secrecy and latency of cyber incidents, etc.

Therefore, the measures of criminal law policy to counter cyberterrorist attacks on critical infrastructure and aviation can be grouped into several large blocks.

The first large block includes the problems of application in domestic legislation of various scope and level of subordination of regulations.

According to experts, the criminal law policy of the state in the field of crime prevention is a strategy developed by the state, the basic concepts, directions, goals and means of influencing crime [7]. Legal policy in the field of countering terrorist activities is part of the domestic policy of the state. Thus, it can be considered in certain, independent of each other, planes, namely: criminal law policy, criminal procedure policy and criminological policy to counter terrorist activities including cyberterrorist attacks on critical infrastructure and aviation.

At the same time, the criminal law policy of countering cyberterrorist attacks on critical infrastructure and air aviation cannot be opposed to other areas of the unified state policy, as the stabilization of criminogenic situation is possible only due to the whole set of technical, economic, social, ideological, educational and legal measures. However, given the object of criminal law policy and the means of influencing cyberterrorism, there is reason to believe that it is relatively independent. In other words, the criminal law policy to combat cyberterrorist attacks on critical infrastructure and aviation is an independent area of domestic legal policy of Ukraine taking into account the general theory of crime, develops strategy and tactics, formulates basic tasks, principles, directions, purpose of influence and the means for their achieving.

We propose to consider the more specific issue of criminal law policy to counter cyberterrorist attacks on critical infrastructure and aviation in the penal area. The penal term (in Latin: poena) is understood as punishable, criminally punishable; the one who is at a disadvantage, under pain of punishment. Penology is a science of punishment and means of criminal legal influence [8]. The purpose of the penal policy is reflected in Part 2 of Art. 50 of the Criminal Code of Ukraine "The concept of punishment and its purpose". However, we believe that the provisions set out in this article determine the purpose only in the field of penitentiary policy, i.e. policy in the field of execution of punishments. Although it is obvious that the scope of criminal law policy to combat cyberterrorist attacks on critical infrastructure and air transport includes the full range of emerging issues.

The first decisive steps of the penal policy to counter cyberterrorist attacks on critical infrastructure and aviation in Ukraine have been taken for a long time. The provisions of the Law of Ukraine: On combating terrorism contain the following definition: Terrorism is a socially dangerous activity consisting in deliberate use of violence by taking hostages, arson, murder, torture, intimidation of the population and the authorities, or committing other encroachments on the life or health of innocent people or threatening to commit criminal actions in order to achieve criminal goals [9]. We consider the concept lack of cyberterrorism in its test to be a significant shortcoming of this normative legal act, as it reflects the challenges posed by the present. Instead, the Law contains the concept of technological terrorism: crimes committed for terrorist purposes with the use of nuclear, chemical, bacteriological (biological) and other weapons of mass destruction or their components, other harmful substances, electromagnetic means, computer systems and communication networks, including the seizure, decommissioning and destruction of potentially dangerous objects that directly or indirectly create or threaten the threat of an emergency as a result of these actions and pose a danger to personnel, the public and the environment; create conditions for accidents and man-made disasters [10].

The United States Criminal Law (Title VIII: Strengthening the criminal laws against terrorism is the eighth of ten titles which comprise the USA PATRIOT Act, an anti-terrorism bill passed in the United States one month after the September 11, 2001 attacks. Title VIII contains 17 sections and creates definitions of terrorism, and establishes or re-defines rules with which to deal with it) states that … Several aspects of cyberterrorism are dealt with in title VIII. Under Sect. 814 of the Patriot Act, it is clarified that punishments apply to those who either damage or gain unauthorized access to a protected computer and thus cause a person an aggregate loss greater than $5,000; adversely affects someone's medical examination, diagnosis or treatment; causes a person to be injured; causes a threat to public health or safety; or causes damage to a governmental computer that is used as a tool to administer justice, national defence, or national security. It is only through these specific actions that civil action may be taken against an offender. Section 814 also prohibits any extortion via a protected computer, and not just extortion against a "firm, association, educational institution, financial institution, government entity, or other legal entity". Punishments were expanded to include attempted illegal use or access of protected computers. The punishment for attempting to damage protected computers through the use of viruses or other software mechanism is now imprisonment for not more than 10 years, while the punishment for unauthorized access and subsequent damage to a protected computer is now more than five years' imprisonment. Should the offense occur a second time, the penalty increases to no more than 20 years' imprisonment. The Federal sentencing guidelines were amended to allow any individual convicted of computer fraud and abuse to be subjected to appropriate penalties, without regard to any mandatory minimum term of imprisonment.

Section 816 specifies the development and support of cybersecurity forensic capabilities. It directs the Attorney General to establish regional computer forensic laboratories that have the capability of performing forensic examinations of intercepted computer evidence relating to criminal activity and cyberterrorism, and that have the capability of training and educating Federal, State, and local law enforcement personnel and prosecutors in computer crime, and to "facilitate and promote the sharing of Federal law enforcement expertise and information about the investigation, analysis, and prosecution of computer-related crime with State and local law enforcement personnel and prosecutors, including the use of multijurisdictional task forces". US $50,000,000 was authorized for establishing such labs [11].

In provisions of the Law of Ukraine: "On the Basic Principles of Cyber Security of Ukraine", cyberterrorism means terrorist activity carried out in cyberspace or with its use [10]. In the wake of cyber terrorism, the United States and the United Kingdom have passed legislation to specifically combat cyber terrorists. According to the US and the UK laws, cyber terrorism is treated as an act of terrorism, a special kind of cybercrime [12].

Cyberattack means deliberate actions in cyberspace carried out by means of electronic communications (including information and communication technologies, software, software and hardware, other technical and technological means and equipment) and aimed at achieving one or a combination of the following objectives: violation of confidentiality, integrity, availability of electronic information resources processed (transmitted, stored) in communication and/or technological systems, obtaining unauthorized access to such resources; violation of security, stable, reliable and regular operation of communication and/or technological systems; the communication system use, its resources and means of electronic communications to carry out cyberattacks on other cyber protection objects.

According to domestic scientists, the cyberterrorism is the use of terrorist methods in cyberspace (for example, imposition of false coordinates on a moving object). Cyberattack means measures used to undermine the security of systems or realize the threat to the security characteristics of the resources of the information aviation system (e.g., aviation) through the use of their vulnerabilities [13]. That cyberterrorism is defined by its location or the medium through which it is executed can be criticized to some extent. To address such criticisms, a comparison can be made to aircraft hijacking terrorist acts, such as the 9/11 terrorist attacks on the World Trade Centre; or vehicle-based terrorist attacks, such as when a truck deliberately drove into a crowd of people on the Nice promenade in 2016. In reality, the scope of cyberterrorism appears to follow the general tendency for many "real world" phenomena to be replicated online. Thus, it is common to talk about "cyber activism" as a type of activism carried out online; or "cyberbullying" being a type of bullying which also occurs online. Similarly, it's not difficult to imagine that, with the rise of terrorism, there has also emerged its virtual strain: cyberterrorism [14].

The concept of terrorist activity is proposed in Art. 258 of the Criminal Code of Ukraine, namely: terrorist act, i.e. the use of weapons, explosion, arson or other acts that endanger human life or health or cause significant property damage or other serious consequences, if such acts were committed to violate public safety, intimidate the population, provocation of military conflict, international complication or for the purpose of influencing decisions or committing or failing to act by public authorities or local self-government bodies, officials of these bodies, associations of citizens, legal entities, or drawing public attention to certain political, religious or other views of the perpetrator (terrorist), as well as the threat of committing these actions for the same purpose [15]. However, in our opinion, this is not enough. Taking into account social and political situation in the country, as well as the unfinished armed aggression against Ukraine and in order to increase the effectiveness of countering cyberterrorist attacks on critical infrastructure and aviation, we propose to take the following measures:

Firstly, to strengthen social rehabilitation of victims of a terrorist act. Historically, criminal justice systems have been largely focused on the apprehension, prosecution and punishment of perpetrators of crime, while the role of victims of crime has often been limited to that of witnesses or forgotten altogether. However, the adoption by the United Nations General Assembly in 1985 of the Declaration of Basic Principles of Justice for Victims of Crime and Abuse of Power, which contains 21 recommended measures aimed at securing access to justice and fair treatment, and ensures restitution, compensation and social assistance for victims, represented a landmark change of approach towards more victim-centred criminal justice responses. The Declaration provided the basis for the subsequent development and implementation of international standards and norms concerning the fair treatment of victims of crime within legal and criminal justice systems, in accordance with the rule of law, human rights and fundamental freedoms [16].

For this purpose, adopt the Law of Ukraine: On Rehabilitation of Victims of a Terrorist Act. Today, only the Resolution of the Cabinet of Ministers of Ukraine of July 28, 2004 is valid in Ukraine. № 982: On approval of the Procedure for social rehabilitation of persons affected by a terrorist act. This resolution was adopted by the Government of Ukraine in 2004 and does not take into account the social and political today realities: Ukraine has been in a state of permanent military aggression for six years; According to the Office of the United Nations High Commissioner for Human Rights (OHCHR) for the entire period of the conflict, from April 14, 2014 to March 31, 2020, the victims of the military conflict were about 41–44 thousand persons! During the 6 years of the war, from 13 to 13.2 thousand persons died, including about 4 thousand 100 Ukrainian servicemen. 3 thousand 55 cases of death of civilians: 1 thousand 814 men, 1 thousand 57 women, 98 boys, 49 girls and 37 adults which gender is unknown [17]. Therefore, we believe that the adoption of the latest legal act will implement the following principles of criminal law policy for combating cyberterrorist attacks on critical infrastructure and aviation: specify the responsibilities of the state for the security of its citizens, citizens of other states and stateless persons who are legally on the territory of Ukraine; strengthen assistance to victims of a terrorist act in compensating for material and moral damage, ensuring their return to a normal way of life; to improve the social protection of persons who as a result of a terrorist act cannot return to a normal way of life; provide social protection for the families of persons who died as a result of a terrorist act; develop a system of rehabilitation of persons who have suffered as a result of anti-terrorist operations or other actions of law enforcement agencies. It is necessary to amend some laws of Ukraine in order to improve the legal regulation of international cooperation in the field of countering terrorist activities, in particular, the prevention of terrorist financing. When drawing up such acts, you can use international experience. Federal anti-money laundering and combating the financing of terrorism (AML/CFT) laws in the United States are highly developed, mature and complex. They originated with passage of the Bank Secrecy Act 1970 (31 USC Sections 5311-67), which imposed extensive record-keeping, reporting and procedural compliance obligations on US financial institutions in order to prevent and detect money laundering and terrorist financing. In 1986 the United States enacted the Money Laundering Control Act (18 USC Sections 1956-57) and became the first nation to make money laundering a criminal offence. Since then, the scope of US AML/CFT law and regulation has continuously expanded – most significantly with the passage of the USA PATRIOT Act 2001 (31 USC Sections 5301 and following). Title III of the USA PATRIOT Act, captioned "International Money Laundering Abatement and Antiterrorist Financing", amended the Bank Secrecy Act by: enquiring financial institutions to adopt and implement risk-based policies and procedures for the detection, prevention and reporting of money laundering or terrorist financing activities; increasing the extraterritorial scope of US AML/CFT laws; and expanding the powers of law enforcement authorities with respect to AML/CFT measures. Most terrorism financing prosecutions in the United States are brought under 18 USC Section 2339B, which was enacted in 1996 and prohibits the knowing provision of "material support or resources" to a foreign terrorist organisation. In addition, individual states have enacted and enforce money laundering statutes. The most prominent is New York State, which prohibits money laundering under Article 470 of the New York Penal Law [18].

The second block of problems related to the implementation of criminal law policy to combat cyberterrorist attacks on critical infrastructure and aviation includes the difference between cyber incidents and so-called "ordinary" terrorism (for example, use of weapons, explosives and substances). Given the international experience [19] and practical side of the implementation of the act of illegal interference in the activities of civil aviation, the following categories of threats can be identified (Fig. 2): sabotage against aircraft and airports; acts of illegal aircraft seizure; attempts to illegally seize aircraft; use of means of communication as a weapon; attacks on communications facilities in flight; attacks on airport buildings and facilities; other acts directed against the safety of civil aviation (threats not included in the previous categories) [20].

In order to counter these incidents, it is necessary to: immediate upgrade of existing navigation and tracking systems for aircraft and airports; adoption of the state program for the development of the domestic instrument industry in order to gradually abandon foreign components and raise the domestic economy; improving the efficiency of training and staffing of critical infrastructure and aviation by specialists who have in-depth skills in working with communications, computer technology and meet basic requirements of professional competence; purchase and implementation of critical infrastructure and aviation of the latest models of equipment, including foreign equipment; increasing the overall security of such facilities (an example of the protection of air transport systems and facilities is the transition from the Aeronautical Fixed Telecommunications Network (AFTN), to the Aeronautical Telecommunication Network (ATN) [21]; the use of innovative technology (for example, use unmanned aerial vehicles (UAV). In June 2015 EasyJet began testing UAVs in the maintenance of their Airbus A320s [22] and in July 2016 at the Farnborough Airshow, Airbus (manufacturer of the A320), demonstrated the use of UAVs for the visual inspection of an aircraft [23]. However, some aircraft maintenance professionals remain wary of the technology and its ability to properly catch potential dangers.

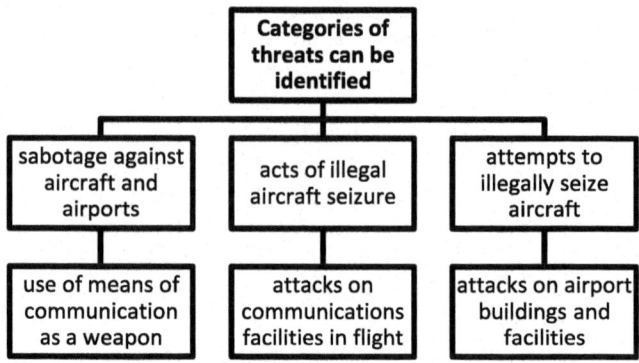

Fig. 2. Categories of threats can be identified.

In particular, with their help it is possible to carry out various preventive measures: round-the-clock surveillance of the territory adjacent to critical infrastructure facilities and airports; monitoring the activities of ground-based security devices and protecting the data they integrate and transmit. They can be used in conjunction with traditional methods of shooting and the latest complexes using laser scanning equipment, GNSS receivers, inertial sensors, with the help of which geospatial data of the highest quality are obtained.

Unfortunately, there are still many unresolved issues in the Ukrainian legislation on the use of such devices (for example, issues of using information obtained with the help of UAV still remain unresolved). There are also no uniform safety requirements for certain classes of drones. This situation requires scientific development and implementation of methods for using UAV to counter cyberterrorist attacks on critical infrastructure and aviation; training of professional staff of operators and users of unmanned aerial vehicles, etc.

Another example of a cyberattack on critical infrastructure can be considered the incidents that occurred in December 2015. The world witnessed the first known power outage caused by a malicious cyber-attack. Three utilities companies in Ukraine were hit by BlackEnergy malware, leaving hundreds of thousands of homes without electricity for six hours. According to cyber security firm Trend Micro, the malware targeted the utility firms' SCADA (supervisory control and data acquisition) systems and probably began with a phishing attack. The blackout was followed two months later by the news that the Israel National Electricity Authority had suffered a major cyber-attack, although damage was mitigated after the Israel Electricity Corporation shut down systems to prevent the spread of a virus [24].

The third block includes the development and introduction of modern innovative technologies in Ukraine which will increase the overall innovation potential of the state. As scholars rightly point out [25], not only financial success depends on this, but also security and sovereignty of the state, its competitiveness in modern world. According to the Innovation Index provided by Bloomberg in 2018, Ukraine ranked 53rd place among 60 countries surveyed. At the same time, our country ranked 58th place in terms of technological capabilities, and 54th place in terms of research and development expenditures in gross domestic product. The amount of funding for innovation in Ukraine is declining every year and the main source of its support are developers of innovative products. Therefore, development and all kinds of support from the state are needed. Much has already been done in this regard. Thus, the Cabinet of Ministers of Ukraine approved the Strategy for the Development of Innovation for the period up to 2030 analysing in detail the problems that need to be addressed through public policy instruments. In particular, the document states there is insufficient budget funding and lack of established communications between scientists-developers and representatives of the customer of innovative products at the stage of creating innovations [26]. However much remains to be done. For example, it is necessary to increase the interaction and coordination level and joint efforts of Ukrainian and foreign experts. The rapid and unrelenting pace of changes and challenges in cybersecurity and NATO's increased emphasis on improving cybersecurity awareness, preparedness and resilience are the driving forces to develop new courses on cybersecurity. As part of the NATO Defence Education Enhancement Programme for Ukraine, experts from allied visited the to assist with the development of a new course on cybersecurity. Ukraine is one of the first NATO partners (together with Tunisia) to develop such a course [27]. Contacts between scientists from Ukrainian universities and forensic and legal research institutions play an important role while such interaction.

Priorities in the context of coordination of joint efforts in this direction are: implementation of efforts aimed at adapting the legislation of Ukraine on countering cyberterrorism to NATO security standards; continuing to implement measures to conclude cooperation agreements in the field of combating cyberterrorism with EU and NATO member states.

3   Conclusions

Summing up, firstly we note that improvement of the entire criminal law policy of the state is impossible without in-depth legislative work that is currently being conducted intensively in Ukraine. Therefore, it is necessary to obtain new legally defined tools for countering cyber and information influences. This requires development and implementation in the legislative field of the latest legal act aimed at improving the system of countering terrorism in cyberspace. After all, this document will regulate the issues of criminal law policy to counter cyberterrorist attacks including at critical infrastructure and aviation. It should provide for the creation of effective mechanisms aimed at rapid detection, response, prevention and neutralization of cyber threats, cyberattacks and cybercrimes, elimination of their consequences and restoration of stability and reliability of communication, technological systems, audit and creation of a mechanism to update cybersecurity requirements and recommendations regarding state information systems, information and telecommunication networks of critical infrastructure and aviation.

Secondly, it is necessary to create a unified classification and formal model of cyber threats that will greatly facilitate the development of effective measures for their countering and investigating.

Thirdly, an important element of criminal law policy to counter cyberterrorist attacks on critical infrastructure and aviation is the introduction of public law partnerships. Most often, partnerships are established between public law enforcement agencies and private IT companies that are developers several the latest electronic products, regional and international companies. When establishing such partnerships, the facts of cyber threats are established more effectively, exchange of operational information is accelerated, analytical work is carried out and preventive measures of a comprehensive nature are applied. One of the directions of such partnership is creation at the state level of a professional, creative and patriotic team of professionals who ensure the cyber and information security of Ukraine. The state needs to increase funding for information and cyber security units for the purchase of the latest software and equipment, as well as material incentives for high level professionals.

Fourthly, comprehensive nature of criminal law policy measures to combat cyberterrorist attacks on critical infrastructure and aviation requires the definition of innovative approaches to protection formation and development of information space in the globalization context.

Fifthly, develop state measures for information society development that involves strengthening moral and patriotic principles in the public consciousness, promoting a culture of information security and raising the level of digital literacy of Ukrainian citizens that is essential in the realities of the Ukrainian present when the state is forced to resist foreign aggression on its borders.

References

1. Patel, A., Tailor, J.: A malicious activity monitoring mechanism to detect and prevent ransomware. Comput. Fraud Secur. **2020**(1), 14–19 (2020). https://doi.org/10.1016/S1361-3723(20)30009-9
2. Katerynchuk, P.: Challenges for Ukraine's cyber security: national dimensions. Eastern Rev. **8**, 137–147 (2019). https://doi.org/10.18778/1427-9657.08.05
3. Taplin, R.: Cyber Risk, Intellectual Property Theft and Cyberwarfare: Asia, Europe and the USA. Routledge, London (2020). https://doi.org/10.4324/9780429453199
4. ITU: Global cybersecurity index 2018. International Telecommunication Union, Geneva (2019)
5. Hamden, R.H.: Psychology of Terrorists: Profiling and Counteraction. CRC Press, Boca Raton (2018). https://doi.org/10.4324/9781315156750
6. Scrivens, R., Gill, P., Conway, M.: The role of the internet in facilitating violent extremism and terrorism: suggestions for progressing research. In: Holt, T., Bossler, A. (eds.) The Palgrave Handbook of International Cybercrime and Cyberdeviance, pp. 1417–1435. Palgrave Macmillan, Cham (2020). https://doi.org/10.1007/978-3-319-78440-3_61
7. Weisburd, D., Farrington, D.P., Gill, C.: What works in crime prevention and rehabilitation: an assessment of systematic reviews. Criminol. Public Policy **16**(2), 415–449 (2017). https://doi.org/10.1111/1745-9133.12298
8. Harrison, K.: Penology: Theory, Policy and Practice. Red Globe Press, London (2020)

9. Verkhovna Rada of Ukraine: On fight against terrorism. The law of Ukraine. https://zakon.rada.gov.ua/laws/show/638-15. Accessed 20 Sep 2020. (in Ukrainian)
10. Verkhovna Rada of Ukraine: On basic principles of cyber security support in Ukraine. The law of Ukraine. https://zakon.rada.gov.ua/laws/show/2163-19. Accessed 20 Sep 2020. (in Ukrainian)
11. Stefoff, R.: The Patriot Act. Marshall Cavendish Benchmark, New York (2011)
12. Pitaksantayothin, J.: Cyber terrorism laws in the United States, the United Kingdom and Thailand: a comparative study. Chulalongkorn Law J. **32**(2), 169–185 (2014)
13. Fox, S.J.: Flying challenges for the future: aviation preparedness – in the face of cyber-terrorism. J. Transp. Secur. **9**(3–4), 191–218 (2016). https://doi.org/10.1007/s12198-016-0174-1
14. Mayer Lux, L.: Defining cyberterrorism. Revista Chilena de Derecho y Tecnología **7**(2), 5–25 (2018). https://doi.org/10.5354/0719-2584.2018.51028
15. Verkhovna Rada of Ukraine: Criminal code of Ukraine. The law of Ukraine. https://zakon.rada.gov.ua/laws/show/2341-14. Accessed 20 Sep 2020. (in Ukrainian)
16. UNODC: Good practices in supporting victims of terrorism within the criminal justice framework. United Nations, New York (2015)
17. Tronc, E., Nahikian, A.: Ukraine-conflict in the Donbas: civilians hostage to adversarial geopolitics. Harvard Humanitarian Initiative (2020). https://doi.org/10.2139/ssrn.3657394
18. Basha, L.A.: Overview on recent developments for the legislation, regulatory and anti-money laundering US update. Trusts Trustees **25**(1), 138–148 (2019). https://doi.org/10.1093/tandt/tty173
19. Thomas, C.S., Kirby, M.J.: The Convention for the suppression of unlawful acts against the safety of civil aviation. Int. Comp. Law Q. **22**(1), 163–172 (1973). https://doi.org/10.1093/iclqaj/22.1.163
20. Oster, C.V., Jr., Strong, J.S., Zorn, C.K.: Analyzing aviation safety: problems, challenges, opportunities. Res. Transp. Econ. **43**(1), 148–164 (2013). https://doi.org/10.1016/j.retrec.2012.12.001
21. ICAO Doc 9896. Manual on the Aeronautical Telecommunication Network (ATN) using Internet Protocol Suite (IPS) Standards and Protocol. International Civil Aviation Organization, Montreal (2015)
22. Stevenson, B.: EasyJet tests UAV with A320 inspection. FlightGlobal (2015). https://www.flightglobal.com/easyjet-tests-uav-with-a320-inspection/117108.article. Accessed 20 Sep 2020
23. Airbus: Airbus demonstrates aircraft inspection by drone at Farnborough (2016). https://www.airbus.com/newsroom/press-releases/en/2016/07/airbus-demonstrates-aircraft-inspection-by-drone-at-farnborough.html. Accessed 20 Sep 2020
24. Lee, J., Mo, J.: Analysis of technological innovation and environmental performance improvement in aviation sector. Int. J. Environ. Res. Public Health **8**(9), 3777–3795 (2011). https://doi.org/10.3390/ijerph8093777
25. Harust, Y.V., Melnyk, V.I.: Law Enforcement Agencies on Protection of Economic Security of Ukraine Administrative and Legal Aspect. Sumy State University, Sumy (2019)
26. Verkhovna Rada of Ukraine: Strategy for the development of innovative activity for the period up to the year 2030. Order of the Cabinet of Ministers of Ukraine. https://zakon.rada.gov.ua/laws/show/526-2019-%D1%80. Accessed 20 Sep 2020. (in Ukrainian)
27. Shypovskyi, V., Cherneha, V., Marchenkov, S.: Analysis of the ways of improvement of Ukraine–NATO cooperation on cybersecurity issues. Soc. Dev. Secur. **10**(2), 11–15 (2020). https://doi.org/10.33445/sds.2020.10.2.2

Information Technology in Creation of Rocket Space Systems

Effect of Parameters of Adhesive Application by Intaglio Printing on Honeycomb Core Bonding Strength

Andrii Kondratiev[1]([✉]) [iD], Sergiy Melnikov[2] [iD],
Tetyana Nabokina[2] [iD], and Anton Tsaritsynskyi[2] [iD]

[1] O. M. Beketov National University of Urban Economy in Kharkiv,
17 Marshala Bazhanova Street, Kharkiv 61002, Ukraine
a.kondratiev@khai.edu
[2] National Aerospace University "Kharkiv Aviation Institute",
17 Chkalova Street, Kharkiv 61070, Ukraine

Abstract. Currently, the use of intaglio printing for applying the adhesive on the foil is a promising way to obtain the honeycomb core with high geometric characteristics of a cell. In this method, adhesive strips are featuring the structure with a number of individual points of specified thickness and size, ensuring high accuracy of resulting strips and, accordingly, honeycombs' characteristics. The paper specifies the field of tolerances for the technological parameters of the process of adhesive application using the intaglio printing method. Due to the fact, that adhesive strips in the intaglio printing are applied on the foil as discrete microzones, the geometrical mathematical models for the relationship of adhesive amount applied and its thickness are proposed and implemented. These models and the method implementing the same allowed establishing the tolerance field for the adhesive application depending on its final thickness formed in the honeycomb pack molding operation. The mathematical model and method for determining the relationship of uneven tearing strength of the adhesive layer as an integral characteristic of the adhesive tensile strength, elastic moduli of the foil and adhesive, as well as thickness of the foil and adhesive layer, is proposed and introduced. The resulting analytical dependence allows finding the tolerance field for the uneven tearing strength to be realized in the function of the nominal adhesive thickness, which is implemented in a specific technological process. The results, in the aggregate, allow significantly reducing the amount of technological preparation of considered stages of the technological process for the manufacturing of honeycomb cores by means of reduction of the experimental research.

Keywords: Adhesive joint · Honeycomb core · Foil · Uneven tearing strength · Tolerance limit · Process deviations

M. Nechyporuk et al. (Eds.): ICTM 2020, LNNS 188, pp. 221–233, 2021.
https://doi.org/10.1007/978-3-030-66717-7_18

1 Introduction

At present time, in space, aviation and transport industries there is a class of problems connected with the need to increase the specific strength of products obtained by structural bonding [1, 2]. Honeycomb cores are among the industrial materials, which are the most widely used for the structural bonding [3, 4]. Usage of the honeycomb cores in sandwich aggregates of various applications together with advanced polymeric composite materials allows significantly reducing the weight of structures providing high specific strength and stiffness at the different types of external impacts [5, 6]. Among many types of honeycomb cores that differ in cell configuration and manufacturing technology, hexagonal honeycombs are the most widespread ones [6, 7]. Honeycombs on the base of the aluminium foil represent the most common filling material used in the structures operating under various conditions [8, 9]. During manufacturing of honeycomb cores, the main parameters of the adhesive joint influencing the honeycomb characteristics are the thickness of the adhesive strip, mass application of the adhesive, graphic accuracy of the adhesive strip location, and strength of the adhesive bond at delamination [10, 11]. The most advanced technology which provides obtaining high-quality honeycomb core is considered to be the technology where the adhesive is applied by the intaglio method derived from the printing industry [3, 10, 12]. One of the ways to improve the adhesive application procedure by this method is the validation and selection of the conditions of printing process, which provide the minimum mass application of the adhesive with the necessary tolerances for the width of adhesive strip and delamination strength of the adhesive joint of honeycomb faces.

2 Literature Review

The papers [11, 12] include the comparative characteristics of various methods for applying adhesive strips on the foil. Comparison was made between the methods of intaglio printing, relief printing, screen printing and flow of the adhesive. These papers show a decrease in the adhesive thickness to 3–5 mcm achieved by intaglio printing compared with that obtained by flow method (10–50 mcm) at the reduction of tolerance for the width of the adhesive strip to 0.1 mcm and high application rate.

The regression mathematical model was developed in the papers [13, 14] with the use of experimental data. This model is based on the scheme of planning of the complete factorial experiment 2^3. The proposed model of the process under study, as a result of evaluation according to a number of criteria, allowed obtaining the minimum applied amount of adhesive with guaranteed uneven tearing strength. In this case, the connection between the thickness of the adhesive strip and the adhesive amount applied, as well as the width and pitch of the adhesive strips remained unspecified.

The quality of bonding of the honeycomb cores is characterized by its reliability, provided by geometric parameters of the adhesive joint, and continuity. It terms of statement of such problems, the papers [15, 16] solving the task about permissible dimensions of disbonds in the honeycomb structures should be noted. In these studies, the issues related to the specification of the manufacturing tolerances were not reflected. In the number of works [17] a conclusion is made that analytical models can be abandoned in favor of the finite element modeling. The complexity of such approaches is obviously justified only in cases where the checking calculations are necessary. Some works deal with certain aspects of the influence of the temperature-time conditions and bonding pressure of ultra-lightweight honeycomb structures of aerospace engineering [18]. However, these studies are not sufficient for specification of the technological tolerance field for the process of application of adhesive strips by intaglio printing. The paper [19] deals with the systematic study of a number of typical process deviations of the honeycomb cores. However, the proposed approach to standardization of tolerances on honeycomb cells' bonding parameters is not focused on the technological capabilities of the honeycomb core manufacturing; instead, it is oriented on ensuring the regulated deviation of the bearing capacity of a particular panel within the object under consideration. The paper [20] represents the results of mathematical modeling of technological methods for correcting the physico-mechanical characteristics of honeycomb cores according to the results of implementation of the method of superposition of matrices constructed with due account for restrictions on the honeycomb geometric parameters. This work deals with the ranges of possible changes in the pitch of adhesive strips depending on the method of adhesive application. However, the method developed in [20] does not allow establishing for the intaglio printing method the rational parameters, which ensures implementation of the desired admissible region of the regulated physical and mechanical characteristics of the honeycomb core.

It follows from the brief analysis of the above publications, that the problem of improving the quality of bonding of honeycomb cores is already solved to a greater or lesser degree. However, for one of the most advanced technologies of adhesive application by intaglio printing, a number of issues were not taken into account. Solving of these issues can have significant effect on the results and requires a separate study.

The objective of this work consists in reducing the amount of technological preparation of the process of adhesive application by the intaglio printing method for the honeycomb cores by means of reduction of the experimental research. To achieve this objective, the following tasks were solved:

- determination of the relationship of the adhesive amount applied and its thickness, taking into account the micro-relief of its structure in the intaglio printing method;
- determination of the relationship of the uneven tearing strength of the adhesive layer as an integral characteristic of the adhesive tensile strength, elastic moduli of the foil and adhesive, as well as thickness of the foil and adhesive layer.

3   Determination of Relationship of Amount and Thickness of Adhesive Taking into Account Micro-relief of Its Structure

Intaglio printing differs from the other printing method in that the printing elements on the form cylinder are located below spacing elements. As can be seen in Fig. 1, adhesive in the form cylinder 1, on the surface of which the system of raster base lines and deep printing elements 7 is located, is removed by the blade 6, remaining only in the deep printing elements. At the time of printing, the printing material 4, passing between the form cylinder 1 and printing (rubberized) cylinder 2, comes into contact with the plate 1, with which the adhesive 5 is transferred onto printing material.

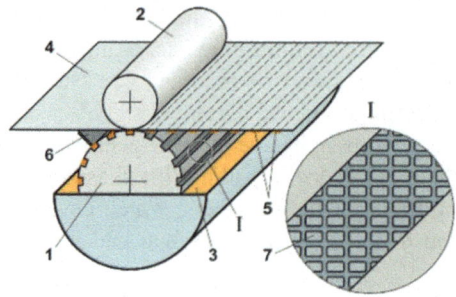

Fig. 1. Scheme of applying adhesive strips by intaglio printing: 1 – form cylinder; 2 – printing cylinder; 3 – container with adhesive; 4 – printing material; 5 – adhesive strips; 6 – blade; 7 – deep printing elements.

The papers [13, 14] establishes the relationship between the adhesive amount applied on the surface y_η and technological parameters of the process of applying adhesive strips – adhesive viscosity μ_η, foil roll speed V_p, depth of printing elements of the applicator cylinder h_p.

$$y_\eta = 1.732 \cdot 10^{-4} \left(3.07 \cdot 10^{-2} h_p + 1.18 \cdot 10^{-4} \mu_\eta V_p - 1\right) \qquad (1)$$

We establish the relationship between the amount of applied adhesive y_η, thickness of adhesive layer η, its density ρ_η and geometrical parameters of the honeycomb core formation – width of adhesive strips a_s and pitch of application t (Fig. 2). Then we consider the process of forming the honeycomb core cell by applying onto the roll foil using the form cylinder and further combining the roll material into a pack with staggered offset of adhesive strips by half a pitch, i.e. $t/2$.

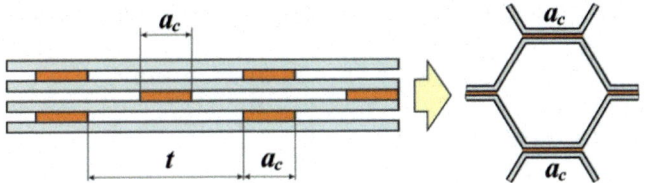

Fig. 2. Geometric parameters of honeycomb pack formation.

The papers [13, 14] establish the relationship between the amount of adhesive applied y_η, reduced thickness of the adhesive layer $\tilde{\eta}$, adhesive density ρ_η and geometric parameters a_c, t

$$y_\eta = \frac{a_c \tilde{\eta} \rho_\eta}{(a_c + t)}. \tag{2}$$

In the intaglio printing method, the adhesive is applied on the foil in discrete zones, instead of the continuous layer; such zones represent rhombuses (see Fig. 1) with the areas (Fig. 3) not filled with the adhesive ("spaces"). During the intaglio printing, spaces of the printing element are filled with the adhesive partially transferred onto the foil. Therefore, all adhesive of $\tilde{\eta}$ thick on the representative element of the foil features the mesh, instead of continuous, structure. In order to determine the reduced (true) adhesive density, we consider a representative element of the printing element.

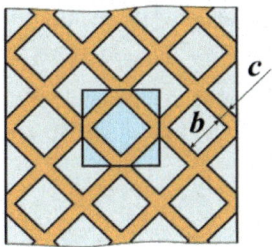

Fig. 3. Representative element of the structure of adhesive and spacing zones in the process of adhesive application by intaglio printing.

The area occupied by the adhesive when it is transferred onto the foil is the area of spaces in the representative element of the printing element $F^{gap} = 2c(2b + c)$. The area on which the adhesive is not present during transfer onto the foil is equal to $F^{rhom} = 2b^2$.

The total area of the representative element of the printing element is the sum of

$$F_\Sigma = F^{gap} + F^{rhom} = 2\left[b^2 + c(2b+c)\right].$$

Then reduced adhesive density $\tilde{\rho}_\eta$ will be.

$$\tilde{\rho}_\eta = \rho_\eta \theta = \rho_\eta \frac{F^{rhom}}{F_\Sigma} = \rho_\eta \frac{(2b+c)}{b^2 + c(2b+c)}, \tag{3}$$

where θ is the relative proportion of adhesive in the adhesive layer. Taking into account (3), amount of adhesive applied will be equal to.

$$y_\eta = \frac{a_c(2b+c)}{(a_c+t)[b^2 + c(2b+c)]} \rho_\eta \tilde{\eta} = K_y \tilde{\eta}. \tag{4}$$

When modeling the microstructure of the adhesive, we proceed from the fact that the real thickness $\eta < \tilde{\eta}$, since in the above model the adhesive zone is represented in the form of a rectangle. In fact, the adhesive zone will blur, taking the shape of a second-order surface segment (Fig. 4).

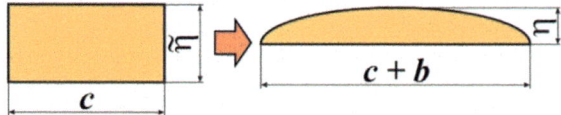

Fig. 4. Transformation of shape of adhesive micro-element applied onto the adhesive strip.

Assuming that the sides of the adhesive zone close at the boundaries of spaces, and approximately assuming the area of the new adhesive zone to be equal to the area of the parabolic segment.

$$S_\eta = \frac{2}{3}\eta(c+b), \tag{5}$$

and the area of the rectangular zone of adhesive equal to

$$S_{\tilde{\eta}} = \tilde{\eta} c, \tag{6}$$

and equating these areas as the volumes of unit width, we obtain

$$\eta = \frac{3}{2(c+b)} \tilde{\eta}. \tag{7}$$

In the model analyzed above, the adhesive, when applied onto the foil, has a microrough surface with the peaks of η high. However, in the course of formation of the honeycomb pack, non-polymerized adhesive softens at the elevated temperature and molding pressure, and the peaks η are smoothed, and the adhesive structure takes the rectangular shape with the height of η^* (Fig. 5) again. In this case, the volume of the unit of adhesive surface equivalent to the cross-sectional area remains unchanged, i.e. S_η is determined by dependencies (5) and (6); $S_{\eta^*} = \eta^*(c+b)$.

Fig. 5. Transformation of adhesive microelement shape in the process of honeycomb block molding.

From this equality of areas, it follows that

$$\eta^* = \frac{c}{c+b}\tilde{\eta}. \tag{8}$$

Substituting in (4) instead of $\tilde{\eta}$ its final value η^*, from (8) we obtain the final value of the amount of adhesive applied y_η.

$$y_\eta = \frac{a_c(c+b)(2b+c)}{(a_c+t)[b^2+c(2b+c)]}\rho_\eta\eta^* = K_y^*\eta^*. \tag{9}$$

With the values adopted in the planned experiment of papers [13, 14] and included in (9) parameters of adhesive thickness lie in the range of $\eta^* = 2.5...7.75$ mcm.

When standardizing the tolerance range for application of adhesive Δy_η by intaglio printing we use the dependence of

$$y_\eta \pm \Delta y_\eta = \frac{(a_c \pm \Delta a_c)[(c \pm \Delta c) + (b \pm \Delta b)][2(b \pm \Delta b) + (c \pm \Delta c)](\rho \pm \Delta\rho)(\eta^* \pm \Delta\eta^*)}{[(a_c \pm \Delta a_c) + (t \pm \Delta t)][(b \pm \Delta b)^2 + 2(c \pm \Delta c)(b \pm \Delta b) + (c \pm \Delta c)^2]} \tag{10}$$

This dependence with the tolerances adopted in the printing industry [14] $\Delta c = \pm 0.05c$, $\Delta b = \pm 0.05b$, $\Delta h = \pm 5$ mcm and tolerance field substantiated in [19, 20] $\Delta a_c = \pm 0.05a_c$, $\Delta t = \pm 0.05t$ using the results of multifactorial experiment 2^3 of [13] allowed obtaining the following asymmetric tolerance limit with optimal printing parameters $h_{popt} = 50$ mcm, $\mu_{\eta opt} = 50$ c and $V_{popt} = 30$ m/min:

$$y_\eta = 3.65^{+0.92}_{-0.42} \cdot 10^{-5}\eta^*, \text{ g/cm}^2 \tag{11}$$

4   Relationship of Uneven Tearing Strength of the Adhesive and Its Application

There is a description of a large number of test methods for determining the adhesive strength at uneven tearing S, but the most common is the "tear" method using the drum-type clamp [21]. Testing of uneven tearing strength is carried out also with the use of simplified method, with the direct tearing of samples of a certain width of two bonded foil strips (Fig. 6) [10].

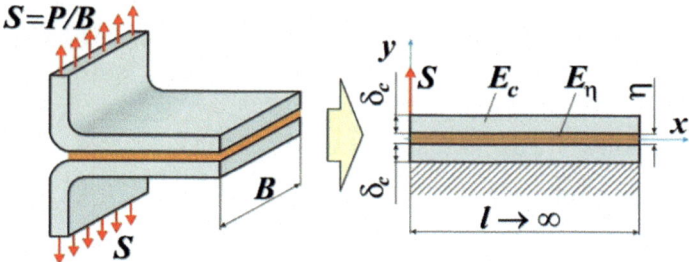

Fig. 6. Simplified method for determining the adhesive uneven tearing strength and the relevant design test setup.

Uneven tearing strength of the adhesive is equal to $S = P/B$, where P is breaking strength with sample width of B.

As a mathematical model of the adhesive joint, a beam of infinite length on the elastic base (a foil tightly wound on a drum) is adopted. The differential equation of bending of this beam is written as [22]:

$$E_c I_c \frac{d^4 f}{dx^4} + rf - q(x) = 0, \tag{12}$$

where E_c is the elastic modulus of foil; I_c is the moment of inertia of the strip of foil of B wide and δ_c thick; f is the foil bending in section; $q(x)$ is the intensity of external distributed load; $r = E_\eta/\eta$ is the coefficient of elasticity of the base, representing the reactive force per unit of the beam length during deflection equal to unity; E_η is the elastic modulus of adhesive.

Introducing the designations $4\alpha^4 = r/E_c I_c$, $I_c = \delta_c^3/12$, $r = E_\eta/\eta^*$ and assuming $q = 0$, we rewrite Eq. (12) as.

$$\frac{d^4 f}{dx^4} + 4\alpha^4 f = 0. \tag{13}$$

General solution of Eq. (13) can be written as

$$f = e^{\alpha x}(C_1 \sin \alpha x + C_2 \cos \alpha x) + e^{-\alpha x}(C_3 \sin \alpha x + C_4 \cos \alpha x)$$

Assuming the top layer of foil being torn off the drum infinitely long and taking into account that at $x \to \infty$ $f \to 0$, we get $C_1 = C_2 = 0$. Constants C_3 and C_4 are determined from the boundary conditions for $x = 0$. According to Fig. 6 for the adhesive joint there is no bending moment M at the edge $x = 0$, and the shearing force Q is equal to the current line force S i.e.

$$M = E_c I_c \frac{d^2 f}{dx^2} = 0; \quad Q = \frac{dM}{dx} = E_c I_c \frac{d^3 f}{dx^3} = S. \tag{14}$$

Having determined the constants C_3, C_4 from conditions (14), we finally obtain.

$$f = \frac{S}{2\alpha^3 E_c I_c} e^{-\alpha x} \cos \alpha x.$$

Deformation of the adhesive layer is equal to $\varepsilon = f/\eta$, and stresses in it are determined from Hooke's law $\sigma_\eta = E_\eta \varepsilon$. Finally, we obtain the following expression for the normal stresses in the adhesive layer

$$\sigma_\eta = 2S\sqrt[4]{\frac{3E_\eta}{\eta^* E_c \delta_c^3}} e^{-\alpha x} \cos \alpha x \tag{15}$$

From equalities (15) it follows that the maximum value of stresses is realized for $x = 0$, and at the moment of foil tearing it will be equal to the tensile strength of the adhesive $\sigma_{e\eta}$. It follows:

$$S = \sigma_{e\eta} \sqrt[4]{\frac{E_c \delta_c^3 \eta^*}{3E_\eta}} \tag{16}$$

It should be noted that the theoretical dependence (16) due to the approximation of the mathematical model can give significant errors at very small $\eta^* \to 0$ and rather high $\eta^* > \delta_c$. In the remaining range $0 < \eta^* < \delta_c$ the model should accurately describe the phenomenon.

For the adhesive with tolerance field specified in [13, 19, 20], i.e. $\Delta E_c \approx 0$, $\Delta \delta_c = \pm 0.1 \delta_c$, $\Delta E_\eta = \pm 0.05 E_\eta$, assuming $\Delta \sigma_\eta = \pm 0.05 \sigma_\eta$ with the use (16), we obtain.

$$S = 6.3^{+1.26}_{-0.88} \cdot 10^{-5} \sqrt[4]{\eta^*}, \text{ N/cm.} \tag{17}$$

Based on dependences (17) and (11) Fig. 7 represents the graph $S = f(y_\eta)$ with tolerance field and experimental points obtained by the method of planning the experiment in [13, 14].

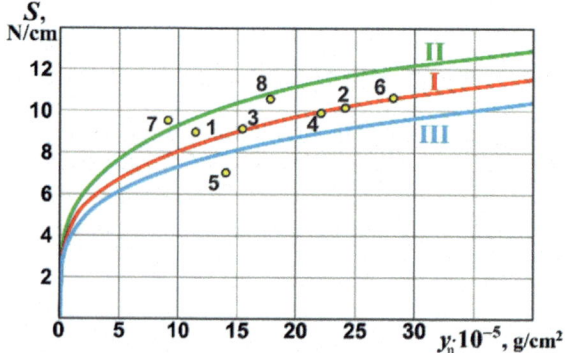

Fig. 7. Dependence of uneven tearing strength of the adhesive on the amount of adhesive applied: ⦿ – experimental points; theoretical curves: I – on the nominal values; II – on the upper limit of tolerances; III – on the lower limit of tolerances.

As can be seen in Fig. 7, one experimental point is higher than the upper limit and one point is significantly lower than the lower limit of the tolerance field. The remaining points are within the tolerance field S.

5   Results and Discussion

A peculiar feature of the intaglio printing method is that the adhesive is applied onto the foil in the form of discrete rhombic microzones, between which there are sections not filled with the adhesive. To account for this phenomenon, a new model of a representative foil element is proposed. The proposed model of the process under study allows obtaining the dependence (4), which reflects the previously unknown relationship between the thickness of adhesive strips and their pitch, as well as density of adhesive. The developed model unlike [13, 14] allowed taking into account the microrelief of the adhesive layer. Thus, the approximate model (2) proposed in these papers, which is obtained by considering a foil element with the continuous adhesive layer, gives significantly overestimated results for the thickness of the adhesive layer. The proposed refined model of the adhesive micro-relief (4) taking into account the transformation of the shape of the adhesive micro-element at the time of application and in the process of honeycomb pack molding, made it possible to obtain the expression (9) for adhesive amount applied. Applied adhesive y_η obtained using the formula (9) taking into account the accepted tolerance on the thickness of the foil is fully consistent with the range of 3…5 mcm defined in the experiments of [13, 14]. This may indicate the adequacy of the proposed refined model to the real process of applying adhesive strips by intaglio printing and forming a honeycomb pack. As it follows from (9), a linear relationship exists between the adhesive amount applied and the conditional thickness of the adhesive layer on the foil (which represents microprotrusions of discrete droplets of adhesive columns). This dependence allows determining the adhesive amount applied by weighing a foil sample with applied adhesive strips containing n representative elements of the foil on its surface, and further determining the adhesive thickness. This method of determination of thickness of the adhesive layer on foil compared to its measurements with the use of measuring instruments is seems more accurate, since η at the intaglio printing is 3...5 mcm, and the maximum accuracy (at measuring sensitivity of 2 μ) with the use of caliper gage is ± 1 mcm [10, 21]. At the same time, analytical scales for weighing the sample have accuracy of $\pm 2 \cdot 10^{-4}$ g [21].

To establish the relationship between the identified tolerance fields for adhesive uneven tearing strength and amount of adhesive applied, a model is proposed that allows establishing for the first time the analytical dependence of the adhesive uneven tearing strength as an integral characteristic of the tensile strength of the adhesive, the elastic moduli of the foil and adhesive, and also the thickness of foil δ_c and adhesive. With the known tolerance fields for these parameters, the obtained dependence (16) allowed finding the tolerance field for uneven tearing strength of the adhesive in the function of the nominal adhesive thickness as (17). As can be seen from the graphs (Fig. 7), the dependences obtained by formula (17) for the adhesive uneven tearing strength on the amount of adhesive applied within the tolerance field are in good agreement with the experiments of [13, 14].

It should be noted that the resulting formula (16) can give errors for very thin and very thick adhesive strips. However, there can also be another important source of errors of this formula associated with neglecting the discrete (mesh) structure of the adhesive strip at the micro-level, which requires the additional research.

6 Conclusions

Mathematical models for the relationship of the adhesive amount applied and its thickness are proposed and implemented. These models and the method implementing the same allowed establishing the tolerance field for the adhesive application by intaglio printing depending on its final thickness formed in the honeycomb pack molding operation.

The mathematical model and method for determining the relationship of uneven tearing strength of the adhesive layer as an integral characteristic of the adhesive tensile strength, elastic moduli of the foil and adhesive, as well as thickness of the foil and adhesive layer, is proposed and introduced. The resulting analytical dependence allows finding the tolerance fields for the uneven tearing strength to be realized in the function of the nominal adhesive thickness, giving a possibility to predict the adhesive uneven tearing strength implemented in a specific technological process.

The results, in the aggregate, allow significantly reducing the amount of technological preparation of considered stages of the technological process for the manufacturing of honeycomb cores by means of reduction of the experimental research.

References

1. Rodichev, Y.M., Smetankina, N.V., Shupikov, O.M., Ugrimov, S.V.: Stress-strain assessment for laminated aircraft cockpit windows at static and dynamic loads. Strength Mater. **50**(6), 868–873 (2018). https://doi.org/10.1007/s11223-019-00033-4
2. Fomin, O., Lovska, A., Kulbovskyi, I., et al.: Determining the dynamic loading on a semi-wagon when fixing it with a viscous coupling to a ferry deck. Eastern-European J. Enterp. Technol. (2019). https://doi.org/10.15587/1729-4061.2019.160456
3. Nunes, J.P., Silva, J.F.: Sandwiched composites in aerospace engineering. In: Rana, S., Fangueiro, R. (eds.) Advanced Composite Materials for Aerospace Engineering, pp. 129–174. Woodhead Publishing, Duxford (2016). https://doi.org/10.1016/B978-0-08-100037-3.00005-5
4. Slyvyns'kyy, V., Gajdachuk, V., Kirichenko, V., Kondratiev, A.: Basic parameters' optimization concept for composite nose fairings of launchers. In: Proceedings of the 62nd International Astronautical Congress (IAC 2011), vol. 7, pp. 5701–5710 (2011)
5. Slyvynskyi, V.I., Sanin, A.F., Kharchenko, M.E., Kondratyev, A.V.: Thermally and dimensionally stable structures of carbon-carbon laminated composites for space applications. In: Proceedings of the International Astronautical Congress (IAC), vol. 8, pp. 5739–5751 (2014)
6. Kondratiev, A., Gaidachuk, V.: Weight-based optimization of sandwich shelled composite structures with a honeycomb filler. Eastern-European J. Enterp. Technol. **1**(1), 24–33 (2019). https://doi.org/10.15587/1729-4061.2019.154928
7. Herrmann, A.S.: Design and manufacture of monolithic sandwich structures with cellular cores. In: Sandwich Construction 4, vol. 2, pp. 719–728 (1998)
8. AMS-A-81596A. Aluminum foil for sandwich construction. SAE International (2018). https://doi.org/10.4271/AMSA81596A
9. Sypeck, D.J.: Wrought aluminum truss core sandwich structures. Metall. Mater. Trans. B **36**(1), 125–131 (2005). https://doi.org/10.1007/s11663-005-0012-5
10. Bitzer, T.: Honeycomb Technology: Materials, Design, Manufacturing, Applications and Testing. Springer, Dordrecht (1997). https://doi.org/10.1007/978-94-011-5856-5
11. Slivinsky, M., Slivinsky, V., Gajdachuk, V., et al.: New possibilities of creating efficient honeycomb structures for rockets and spacrafts. In: Proceedings of the 55th International Astronautical Congress, vol. 3, pp. 1923–1932 (2004). https://doi.org/10.2514/6.IAC-04-I.3.A.10
12. Slyvyns'kyy, V., Gajdachuk, V., Gajdachuk, A., Slyvyns'ka, N.: Weight optimization of honeycomb structures for space applications. In: Proceedings of the 56th International Astronautical Congress, vol. 6, pp. 3611–3620 (2005). https://doi.org/10.2514/6.IAC-05-C2.3.07
13. Slyvyns'kyy, V., Slyvyns'kyy, M., Gajdachuk, A., et al.: Technological possibilities for increasing quality of honeycomb cores used in aerospace engineering. In: Proceedings of the 58th International Astronautical Congress, vol. 8, pp. 5434–5447 (2007)
14. Melnikov, S.: The relationship of the tolerance fields on the strength of the adhesive with uneven tearing and its deposition on the foil in the production of honeycomb core. Issues Des. Manuf. Aircr. Struct. **44**(1), 114–119 (2006). (in Russian)
15. Huang, S.J., Chiu, L.W.: Modeling of structural sandwich plates with 'through-the-thickness' inserts: five-layer theory. Comput. Model. Eng. Sci. **34**(1), 1–32 (2008). https://doi.org/10.3970/cmes.2008.034.001

16. Okada, R., Kortschot, M.T.: The role of the resin fillet in the delamination of honeycomb sandwich structures. Compos. Sci. Technol. **62**(14), 1811–1819 (2002). https://doi.org/10.1016/S0266-3538(02)00099-4

17. Mackerle, J.: Finite element analyses of sandwich structures: a bibliography (1980–2001). Eng. Comput. **19**(2), 206–245 (2002). https://doi.org/10.2514/2.991

18. Rion, J., Stutz, S., Leterrier, Y., Månson, J.A.E.: Influence of process pressure on local facesheet instability for ultra-light sandwich structures. J. Sandwich Struct. Mater. **11**(4), 293–311 (2009). https://doi.org/10.1177/1099636209104513

19 Gaydachuk, V., Koloskova, G.: Mathematical modeling of strength of honeycomb panel for packing and packaging with regard to deviations in the filler parameters. Eastern-European J. Enterp. Technol. **6**(1), 37–43 (2016). https://doi.org/10.15587/1729-4061.2016.85853

20 Kondratiev, A.V., Prontsevych, O.O.: Stabilization of physical-mechanical characteristics of honeycomb filler based on the adjustment of technological techniques for its fabrication. Eastern-European J. Enterp. Technol. **5**(1), 71–77 (2018). https://doi.org/10.15587/1729-4061.2018.143674

21. Shah, V.: Handbook of Plastics Testing and Failure Analysis. Wiley, Hoboken (2007)

22. Beer, F.P.: Mechanics of Materials. McGraw-Hill, New York (2012)

Analysis of Laminated Composites Subjected to Impact

Sergey Ugrimov$^{(\boxtimes)}$, Natalia Smetankina , Oleg Kravchenko ,
and Vladimir Yareshchenko

A. Pidgorny Institute of Mechanical Engineering Problems of the National
Academy of Sciences of Ukraine, 2/10 Pozharskogo Street, Kharkiv 61046,
Ukraine
sugrimov@ipmach.kharkov.ua

Abstract. The paper proposes both theoretical and experimental approaches to the analysis of laminated composite response to impact loading. For theoretical modelling of dynamic behavior of a composite, the generalized model is used that takes into account the spatial character of deformation on near to the impact point. This model is based on a power series expansion of the displacement vector component in each layer for the transverse coordinate. The results of calculations are compared with the data obtained by other researchers for the case of low-velocity impact, as well as with the experimental data obtained by ourselves at medium-velocity impacts on composite panels. In the experimental study, maximum deflections of composite samples during the impact of an indenter were investigated. A pneumatic gun was used to launch the indenter, and a crusher was used to register the maximum deflections. An experimental study of the response of an eleven-layer fiber-glass composite to indenter impacts at different velocities was performed. For launching, the 600 g indenter was used. It is established that the calculation results and experimental data are in good agreement.

Keywords: Layered composite · Impact · Stress-strained state · Experiment

1 Introduction

Thin-walled composite shell structures are widely used in aerospace engineering, shipbuilding, chemical industry, and automobile industry [1, 2]. They work under conditions of both stationary and non-stationary force loads. Dynamic loads are observed in operating modes, and also in some emergency situations. Such loads occur during various explosions, when an aircraft collides with a bird, when an aircraft tire fragment hits the composite wing of the aircraft, when the composite cabin of an electric locomotive is hit by solid objects from the oncoming train, and so on [3–5]. These are practically important tasks that are considered in the design of composite structural elements in various fields of technology.

Particularly dangerous to thin layered structures is the transverse impact of a rigid projectile [6, 7]. In composites, such a localized intense load can lead to matrix destruction, fiber damage, and structure delamination. This is why special attention is paid to both theoretical and experimental methods of analyzing the stress-strained state of thin-walled composite structures under impact loading [8].

When solving the problem of impact on layered anisotropic structures, the known mathematical difficulties associated with an adequate description of the multilayer structure of composite [9–13] and the need to solve a dynamic problem [10–13], are aggravated by the presence of a localized load [10], which in general acts on a previously unknown area. This area must be determined in the process of solving the problem itself. In addition, the impact in the composite excites a wide range of oscillations with different frequencies, which imposes additional requirements for the structural model being applied.

All these problems result in a very limited number of works on the analysis of the response of laminated composite structures to the impact of elastic bodies [6, 8, 12, 13]. Typically, such problems are solved by numerical methods based on the discretization of a complex domain and its boundary, such as the finite element method (FEM). There are very few works devoted to analytical methods for solving such problems, especially for layered structures having a noncanonical shape. In this case, the immersion method [14], and the method of R-functions [15] can be used. With this, the behavior of a multilayer structure is usually described by low-order shell theories, and the process of impact interaction is described by Hertz's law or its modifications. Some problems of modeling the response of a composite to the low-velocity impact were considered in [6, 12, 13, 16]. In [6, 12, 13], refined first-order theories were used to model the behavior of a composite, which do not take into account the compression and nonlinear nature of stress distribution over the composite thickness. In [13], the solution was obtained by an analytical method, and in [6, 16], with the use of the FEM.

Nosier *et al.* [12] applied a refined discrete-structural theory of layered plates and investigated six models to describe the loading model. In the first five, the domain of interaction of the indenters and plates is proposed to be known. As a result, the problem is reduced to a nonlinear integral equation similar to Tymoshenko's equation. The impact force is interpolated by the Legendre and Hermite polynomials. In the sixth one, based on Hertz's law, the time dependence of the contact area was taken into account, which also led to the need to solve a nonlinear integral equation.

Pierson and Vaziri [13] obtained an analytical solution for the problem of analyzing the response of simply supported composite plates to the low-velocity impact. The equations of motion for the plates were based on the Whitney and Pagano first-order theory. Local indentation was taken into account on the basis of Hertz's law, and the coefficients that enter into this dependence were determined experimentally. The solution to the problem was based on the decomposition of displacements into Fourier series. The domain of integration was divided over time into equal segments, at each of which the impact force was assumed to be constant.

Tan and Sun [17] proposed a modified Hertz's law to improved description of impact on composite targets. The dependence was obtained experimentally in the study of the response of graphite-epoxy structures subjected to impact.

Choi and Hong [18] presented the results of theoretical and experimental studies of the response of layered composite plates to the impact. They applied both a high-order theory and the FEM. The impact force is described by modified Hertz's law. It is established that for a more accurate description of the behavior of composite plates at impact, it is necessary to apply a high-order theory.

In [10], it was proved that with localized impact loading, in the analysis of stresses in even thin composites, it is necessary to use high-order theories, and first-order theories can be used only for the approximate calculation of forces and displacements.

The presence of various assumptions in the models of layered composites, as well as simplifications in the description of interaction of indenter and composite structure requires, at the final stage of design of critical structures, conducting an experimental test of their response to impact. This requires using special equipment, namely launch devices, as well as equipment for registering deflections and strains. Low-velocity impact tests are usually performed by simply dropping the impactor onto a target [19, 20]. When studying the response of structures to medium- or high-speed impacts, it is already necessary to use special acceleration devices [4, 8, 20]. To register the behavior of structures during impact, high-speed video recording can be used, which allows one to assess the behavior of the entire structure, a variety of crushers to register the maximum structural displacement. One of the most common methods of registering strains is the method of dynamic wide-range strain gauging, which allows registering the time-dependent change in plane strains at a point [4]. But the rapidity of the deformation process requires the use of equipment with a high clock frequency. All this makes it difficult to perform the experiment.

The more detailed review of simulation and experimental study of composite structures subjected to low-velocity impact can be found, for example, in works Abrate [8], Patil *et al.* [19], Cantwell and Morton [20], Panettieri *et al.* [21].

This paper proposes a theoretical approach to the analysis of the response of a composite subjected to the impact loading. The approach is based on the hypotheses of the generalized theory of multilayer structures, which allows one to investigate laminated structures under localized loads.

2　Theoretical Modeling

2.1　Mathematical Model of Laminated Composite

A laminated composite consists of I layers of constant thickness, h_i is the thickness of the i-th layer (Fig. 1).

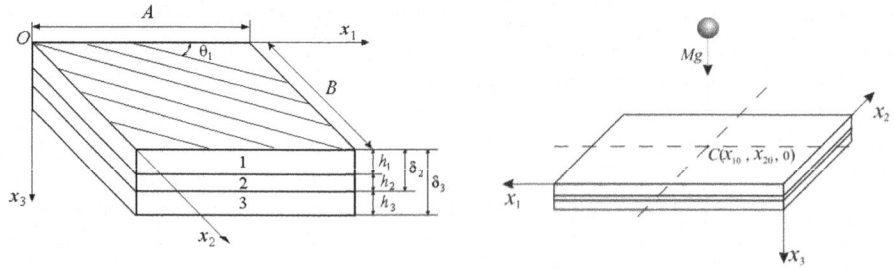

Fig. 1. Laminated composite. Impact problem.

The layers are made of orthotropic materials, and θ_i is the reinforcing angle in the i-th layer. The reinforcement directions in each layer are assumed to be parallel to the coordinate axes Ox_1, Ox_2. It is assumed that the contact between the layers excludes their delamination and mutual slipping.

The behavior of the layered plate is described by equations of the generalized theory of multilayer plates [10]. The displacement of a point on the i-th layer is described by the following kinematic relationships:

$$u_\alpha^i(x_1,\ x_2,\ x_3, t) = u_\alpha + \sum_{k=1}^{K_\alpha} \left[\sum_{j=1}^{i-1} h_j^k u_{\alpha k}^j + (x_3 - \delta_{i-1})^k u_{\alpha k}^i \right], \qquad (1)$$

where $h_j^k = (h_j)^k$, $\delta_i = \sum_{j=1}^{i} h_j$, $\delta_{i-1} \le x_3 \le \delta_i$, $i = \overline{1, I}$; u_α^i $(\alpha = \overline{1,\ 3})$ are the displace-ments of a point in the i-th layer in the direction of the coordinate axes Ox_α; u_α, $u_{\alpha k}^i$ are the coefficients of expanding displacements into power series, which are functions of the arguments x_1, x_2, t; K_α are the maximum powers of the transverse coordinate for plane ($\alpha = 1, 2$) and transverse ($\alpha = 3$) displacements of the i-th layer. The parameters K_1 and K_2, which describe the number of retained power series terms for plane dis-placements, will be the same and equal to K, while the parameter K_3, which describes the number of retained power series terms for transverse displacements, shall be equal to L. The generalized theory shall be designated by the number of retained terms in power series (1) for plane and transverse displacements, viz., theory $\{K, L\}$.

By varying the number of retained terms in power series (1), it is possible to obtain two-dimensional approximations of the stress-strained state with different accuracy. Particular cases of the generalized model are Grigoliuk's model [22], the refined first-order theory, which takes into account the influence of the transverse normal and shear strains in each layers [23], as well as the high-order theory [24].

The strains in each layer are supposed to be small and are described by the linear relationships. With account of the accepted hypotheses (1), the the strain tensor of a point in the i-th layer ($\varepsilon_{\alpha\beta}^i$) take the form

$$\varepsilon_{vv}^i = u_{v,v} + \sum_{k=1}^{K} \left[\sum_{j=1}^{i-1} h_j^k u_{vk,v}^j + (x_3 - \delta_{i-1})^k u_{vk,v}^i \right],$$

$$\varepsilon_{33}^i = \sum_{\ell=1}^{L} \ell(x_3 - \delta_{i-1})^{\ell-1} u_{3\ell}^i,$$

$$\varepsilon_{12}^i = \frac{1}{2} \left(u_{1,2} + u_{2,1} + \sum_{k=1}^{K} \left[\sum_{j=1}^{i-1} h_j^k (u_{1k,2}^j + u_{2k,1}^j) + (x_3 - \delta_{i-1})^k (u_{1k,2}^i + u_{2k,1}^i) \right] \right),$$

$$\varepsilon_{v3}^i = \frac{1}{2}\left(\sum_{k=1}^{K} k\,(x_3 - \delta_{i-1})^{k-1} u_{vk}^i + u_{3,v} + \sum_{\ell=1}^{L}\left[\sum_{j=1}^{i-1} h_j^\ell u_{3\ell,v}^j + (x_3 - \delta_{i-1})^\ell u_{3\ell,v}^i\right]\right).$$

(2)

Applying hypotheses (1) yields a displacement field, which is continuous over the pack thickness, and ensures the continuity of the strains ε_{11}^i, ε_{22}^i and the piecewise continuity of the transverse strain ε_{33}^i, ε_{v3}^i (2) over the pack thickness. Therefore, within the theory being suggested, it is possible in principle to satisfy the conditions of contact between the layers with a specified accuracy.

The relation between the components of strain tensors and the stresses for the orthotropic case being considered has the form [10]

$$\begin{pmatrix} \varepsilon_{11}^i \\ \varepsilon_{22}^i \\ \varepsilon_{33}^i \end{pmatrix} = \begin{pmatrix} 1/E_1^i & -v_{21}^i/E_2^i & -v_{31}^i/E_3^i \\ -v_{12}^i/E_1^i & 1/E_2^i & -v_{32}^i/E_3^i \\ -v_{13}^i/E_1^i & -v_{23}^i/E_2^i & 1/E_3^i \end{pmatrix} \cdot \begin{pmatrix} p_{11}^i \\ p_{22}^i \\ p_{33}^i \end{pmatrix}, \quad \begin{array}{l} \varepsilon_{12}^i = p_{12}^i/2G_{12}^i \\ \varepsilon_{13}^i = p_{13}^i/2G_{13}^i \\ \varepsilon_{23}^i = p_{23}^i/2G_{23}^i. \end{array}$$

where E_α^i, $v_{\alpha\beta}^i$ are Young's modulus and Poisson ratios, G_{12}^i, G_{13}^i, G_{23}^i are shear moduli, and $p_{\alpha\beta}^i$ is the stress tensor for the i-th layer.

Forces and moments in the i-th layer are determined by the formulas

$$N_{\alpha\beta}^{ik} = N_{\beta\alpha}^{ik} = \int_{\delta_{i-1}}^{\delta_i} (x_3 - \delta_{i-1})^k p_{\alpha\beta}^i dx_3, \quad \alpha, \beta = \overline{1,3}, \ k = \overline{1,K}, \quad i = \overline{1,I}.$$

The equations of motion for the forces and moments have the form [10]

$$\sum_{i=1}^{I}\left[L_\alpha^i - I_{\alpha 1}^i\right] + q_\alpha^1 = 0,$$

$$N_{1\alpha,1}^{ik_\alpha} + N_{\alpha2,2}^{ik_\alpha} - k_\alpha N_{\alpha3}^{ik_\alpha - 1} + h_i^{k_\alpha}\sum_{j=i}^{I-1}\left[L_\alpha^{j+1} - I_{\alpha 1}^{j+1}\right] - I_{\alpha k_\alpha + 1}^i = 0,$$

(3)

where $L_1^i = N_{11,1}^{i0} + N_{12,2}^{i0}$, $L_2^i = N_{22,2}^{i0} + N_{12,1}^{i0}$, $L_3^i = N_{13,1}^{i0} + N_{23,2}^{i0}$;

$$I_{\alpha r}^i = \frac{\rho_i h_i^r}{r}\left(u_{\alpha 0,tt} + \sum_{k=1}^{K_\alpha}\left[\sum_{j=1}^{i-1} h_j^k u_{\alpha k,tt}^j + \frac{rh_i^k}{k+r} u_{\alpha k,tt}^i\right]\right), \ k_\alpha = \overline{1,K_\alpha}, \ i = \overline{1,I}, \ \alpha = \overline{1,3}.$$

The boundary conditions on the support contour for a simply supported rectangular plate are given below at $x_1 = 0$, $x_1 = A$,

$$\sum_{i=1}^{I} N_{11}^{i0} = 0, \quad u_2 = 0, \quad u_3 = 0, \quad N_{11}^{ik_1} + h_i^{k_1} \sum_{j=i}^{I-1} N_{11}^{j+10} = 0, \quad u_{2k_2}^i = 0, \quad u_{3k_3}^i = 0;$$

at $x_2 = 0$, $x_2 = B$,

$$u_1 = 0, \quad \sum_{i=1}^{I} N_{22}^{i0} = 0, \quad u_3 = 0, \quad u_{1k_1}^i = 0, \quad N_{22}^{ik_2} + h_i^{k_2} \sum_{j=i}^{I-1} N_{22}^{j+10} = 0, \quad u_{3k_3}^i = 0. \quad (4)$$

Equations of motion (3) can be written in terms of displacements

$$\Omega \cdot \overline{U}_{,tt} - \Lambda \cdot \overline{U} = \overline{Q}, \tag{5}$$

where $\overline{U}$ is a vector whose components are the sought for functions $\overline{U}^T = \left(u_\alpha, \ u_{\alpha k_\alpha}^i \right)$, $\alpha = \overline{1, 3}$, $i = \overline{1, I}$, $k_\alpha = \overline{1, K_\alpha}$, Λ, Ω are the symmetrical matrixes of stiffness and mass of order $(2K + K_\alpha)I + 3$ [10]; $\overline{Q}$ is a vector whose components are a function of an external force applied to the external surface of the layered plate $\overline{Q}^T = (q_1, \ q_2, \ q_3, \ 0, ..., 0)$.

Equations of motion (5) and boundary conditions (4) are supplemented with zero initial conditions

$$u_\alpha = u_{\alpha k_\alpha}^i = 0, \quad u_{\alpha,t} = u_{\alpha k_\alpha,t}^i = 0, \quad \text{at } t = 0. \tag{6}$$

Hence, the dynamic behavior of a layered composite is described by the system of Eqs. (5), as well as boundary conditions (4) and initial conditions (6). The method of solving the obtained system of equations is based on the expansion of the sought functions u_α, $u_{\alpha k_\alpha}^i$ ($\alpha = \overline{1, 3}$, $k_\alpha = \overline{1, K_\alpha}$, $i = \overline{1, I}$) and the external load q_α into trigonometric series by functions $B_{\alpha mn}(x_1, x_2)$ satisfying the boundary conditions.

$$\left[u_\alpha, u_{\alpha k}^i, q_\alpha \right] = \sum_{m=1}^{m1} \sum_{n=1}^{n1} \left[\Phi_{\alpha mn}(t), \ \Phi_{\alpha kmn}^i(t), \ q_{\alpha mn}(t) \right] B_{\alpha mn}(x_1, x_2),$$

where $m1$, $n1$ is the number of the terms retained in the series.

For a simply supported rectangular plate, function $B_{\alpha mn}(x_1, x_2)$ has the form

$$B_{1mn} = \cos c \sin d, \ B_{2mn} = \sin c \cos d, \ B_{3mn} = \sin c \sin d, \ c = m \pi x_1/A, \ d = n \pi x_2/B.$$

As a result, the problem on non-stationary deformation of a laminated composite for each pair of values m and n is reduced to integrating a system of ordinary second-order differential equations with constant coefficients and zero initial conditions.

2.2   Impact on a Laminated Composite

Let us investigate the process of non-stationary deformation of a horizontally located simply supported rectangular layered plate under low-velocity transverse impact (Fig. 1) [5, 10]. The impact is delivered in the middle of the outer surface of the first plate layer by a ball of radius R and mass M. At the moment of collision with the plate, the ball has a velocity V_0. The impact force and the area of contact are unknown in advance and must be determined in the process of solving the problem itself, so it is necessary to consider the mutual displacement of the plate and the indenter. The system of equations that describes the behavior of the plate is supplemented by the equations of motion for the indenter, as well as the condition of joint displacements of indenter and composite plate.

With this, it is convenient to specify the shape of the area and the nature of load distribution, and to determine their parameters while solving the problem. There are several classic models of the load area: a point load area, a rectangular area, and a circular area [12]. The most realistic picture can be obtained by using a circular area, especially for isotropic bodies. During impact on orthotropic plates, the load area has an elliptical character, but as a first approximation, a circular area is allowed to be used, which gives good results [12]. Therefore, we assume that the area of interaction between the indenter and the orthotropic plate is a circle of radius $r(t)$.

The character of load distribution over the contact area is unknown. To study it in the contact area, different mathematical models are used, for example, uniform and sinusoidal distributions [12]. However, a more accurate model is the ellipsoidal stress distribution over the contact area [12, 25]. In the study of impact, it will be used to model the distribution of stresses in the contact area.

Thus, the area of interaction between the indenter and plate is assumed to be a circle of radius $r(t)$, with the contact pressure being distributed over the contact area according to the law

$$q_3(x_1, x_2, t) = P_0(t) \left[1 - \frac{(x_1 - x_{10})^2 + (x_2 - x_{20})^2}{r^2} \right]^{0.5}, \tag{7}$$

$$q_1(x_1, x_2, t) = q_2(x_1, x_2, t) = 0,$$

where x_{10} and x_{20} are the coordinates of the ball and plate contact point at the initial instant of time.

The contact force with account of (7) is.

$$P(t) = \iint\limits_S q_3 dS = (2/3) P_0 \pi a^2,$$

where S is the contact interaction area.

The equation of motion for centre of mass of the indenter and the initial conditions have the form [10]

$$Mz_{,tt} = Mg - P, \; z(0) = 0, \; z_{,t}(0) = V_0, \tag{8}$$

where $z = z(t)$ is the indenter displacement, P is the contact force, and g is the gravity factor.

The condition of the joint displacement of the indenter and composite plate is

$$u_3^1(x_{10}, \, x_{20}, \, 0, \, t) + a(t) - z(t) \geq 0, \tag{9}$$

where $a(t)$ is the contact approach of the ball and plate in the contact point.

The indenter and plate come into contact when inequality (13) becomes equality. Contact approach is found using Hertz's law $a = kP^{2/3}$ [10, 12, 13, 26]. The coefficient k, which depends on materials and shapes of interacting bodies, for a contact of a ball with an isotropic plate has a rather simple expression obtained by Dinnik [25]. During impact on an orthotropic half-space, a similar expression for the coefficient k is absent [12]. In practice, the coefficient k is determined from an experiment or formulas for the isotropic case and averaged mechanical characteristics for the orthotropic half-space are used, for example $v_a = (v_{12} + v_{21})/2$, $E_a = (E_1 + E_2)/2$, where E_a, v_a are the averaged values of Young's modulus and Poisson's ratio for the first layer [12].

The radius of the contact area $r(t)$ is computed using the formulas [10, 12]

$$r(t) = [(3/16) \cdot P(t) \cdot R \cdot (\theta + \theta_1)]^{1/3}, \; \theta_1 = 4(1 - v_a^2)/E_a, \quad \theta = 4(1 - v^2)/E,$$

where E, v are Young's modulus and the Poisson ratio for the ball material.

Thus, the non-stationary deformation problem of a laminated composite at impact is reduced to the integration of a system of equations describing the behavior of the composite plate (4)–(6), together with the equation of motion for the indenter (8) and the condition of the joint displacements (9). The method of solving the obtained system is described in detail in [10].

2.3 Numerical Results

The response of a symmetric ten-layer composite $(0°/90°/0°/90°/0°)_s$ to the impact of a steel ball that has a diameter of 12.7 mm, a mass of 8.5 g and an initial velocity of 3 m/s is considered [13]. The geometric parameters of the composite are $A = B = 0.2$ m, and $\delta_I = 2.69$ mm. The properties of the material are $E_{11} = 120$ GPa, $E_{22} = E_{33} = 7.9$ GPa, $G_{12} = G_{23} = G_{13} = 5.5$ GPa, $v_{12} = v_{13} = v_{23} = 0.3$, $\rho = 1580$ kg/m^3. The coefficient k that is used in Hertz's law was the same as in [10, 12, 13] for this problem.

Fig. 2 shows both the change in time for the contact force and deflections under the point of impact. The solid line shows the analytical solution to the two-dimensional problem [13], the dots show the solution by the FEM [12, 13], and the dotted line shows the solution according to the generalized model {7, 6}. The figure shows that in the time interval being investigated, the rebound is followed by a re-collision, and the results of calculations according to the proposed theory are in good agreement with the known solutions.

In the calculation, the generalized model {7, 6} was used, but for the general analysis of displacements and the contact force, lower-order models can be used [10]. High-order theories are necessary for a detailed analysis of stresses and strains in composite layers [10].

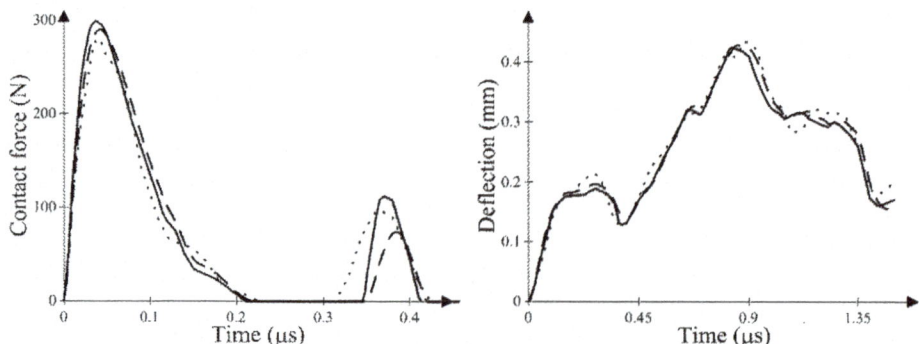

Fig. 2. Change of the contact force and deflections in time.

3   Experimental Setup and Results

For the experimental research, the test stand of the A. Pidgorny Institute of Mechanical Engineering Problems of the NAS of Ukraine was used. A pneumatic gun was used for launching (Fig. 3). This gun consists of a bore with a length of 4,000 mm and a diameter of 125 mm; a tank filled with compressed air; and a special membrane for rapid air intake [5].

A system for measuring the velocity of the object being launched is installed at the muzzle of the bore (Fig. 4). Velocity measurements were performed by registering the time between the rupture of two wires of diameter 0.3 mm, spaced at a distance of 100 mm from each other. Time measurement was performed using an E 20–10 analog-to-digital converter.

Fig. 3. Pneumatic gun.

Fig. 4. Speed measurement system.

A 600 g indenter with a cylinder nose of diameter 20 mm was used for launching. Its main part is made of polyfoam, which together with two disks plays the role of a wad. The impactor part of the indenter is made of aluminum; and its weight, radius, and shape can be chosen depending on the objectives of the study.

Fig. 5 shows the scheme of an experimental setup (1 − gun bore, 2 – indenter, 3 – composite plate, 4 − crusher, 5 – support frame, 6 − clamping bar, 7 − rubber gaskets, 8 − base plate). A plasticine crusher was installed between the test sample and the base plate (Fig. 6). The difference between the length of the crusher before and after the test gives the maximum value of deflection.

The target for testing was fixed in a special box (Fig. 6), perpendicular to the direction of indenter movement (Fig. 6). The impact was delivered to the center of the outer surface of the composite plate.

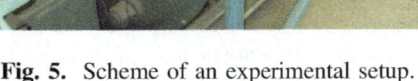
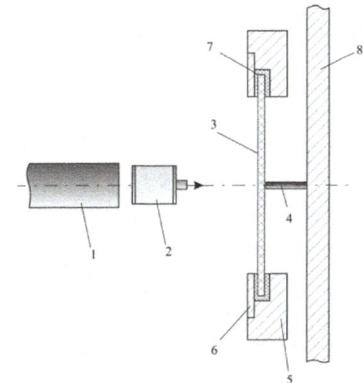

Fig. 5. Scheme of an experimental setup.   **Fig. 6.** A target and a box for tests.

In the experiment, the maximum deflection of a simple-supported flat fiberglass sample with a plan size of 500×500 mm was investigated. The fiberglass plastic consists of 11 layers: one front layer, two layers of glass-fiber mat, 8 layers of glass-fiber fabrics. For impregnation, the Crystic 1355 PA resin is used. The orientation of the layers is longitudinal-transverse. Layer thicknesses are $h_1 = h_2 = h_i = 0.6$ mm ($i = \overline{4, 11}$) and $h_3 = 0.9$ mm. Physical and mechanical characteristics of the front layer are $E_1^1 = E_2^1 = E_3^1 = 6.4$ GPa, $v_{12}^1 = v_{13}^1 = v_{23}^1 = 0.44$, $G_{12}^1 = G_{13}^1 = G_{23}^1 = 2.286$ GPa, $\rho^1 = 1600$ g/cm^3; of the glass-fiber mats, $E_1^2 = E_2^2 = E_3^2 = 6.4$ GPa, $v_{12}^2 = v_{13}^2 = v_{23}^2 = 0.44$, $G_{12}^2 = G_{13}^2 = G_{23}^2 = 2.286$ GPa, $\rho^2 = 100$ g/cm^3, $E_1^3 = E_2^3 = 15$ GPa, $E_3^3 = 7.689$ GPa, $v_{12}^3 = 0.12$, $v_{13}^3 = v_{23}^3 = 0.41$, $G_{12}^3 = 2.554$ GPa, $G_{13}^3 = G_{23}^3 = 2.184$ GPa, $\rho^3 = 450$ g/cm^3; of the glass-fiber fabrics, $E_1^i = E_2^i = 12.1$ GPa, $E_3^i = 6.331$ GPa, $v_{12}^i = 0.13$, $v_{13}^i = v_{23}^i = 0.41$, $G_{12}^i = 2.176$ GPa, $G_{13}^i = G_{23}^i = 2.06$ GPa, $\rho^i = 1580$ g/cm^3.

Deflections of three composite samples upon impact with velocities of 106, 97 and 84 m/s are studied. Table 1 shows the results of numerical and experimental studies of the maximum deflections of the fiberglass plastic in the middle of its rear side under the point of impact. The calculation assumes that the area of interaction between the indenter and the plate is a circle of the same diameter as the diameter of the impact cylinder. It can be seen that the results of calculations are in good agreement with the experimental data, which confirms the efficiency of the proposed method.

Table 1. Maximum deflections of composite at impact.

Velocity, m/s	Deflection, mm	
	Theory	Experiment
84	83.1	80
97	97.2	102
106	104.9	107

4   Conclusions

A theoretical and experimental approach to modeling the response of a laminated composite to the impact is proposed. To theoretically model the behavior of a composite, equations of the generalized model of layered structures are used, which allows taking into account the spatial nature of the deformation near the point of impact.

The possibilities of the proposed theoretical approach are shown using examples of a number of problems of calculating the stress-strained state of laminated composites with different sets of properties of layers. The probability of the obtained results is illustrated by their comparison with the calculation data of other authors, obtained using different two-dimensional theories.

An experimental study of the response of an eleven-layer glass-fiber composite to the impact by indenter was performed. For launching, the 600 g indenter was used. A pneumatic gun was used for launching the indenter, and crushers were used to register the maximum deflections. It is established that the calculation results and the experimental data are in good agreement.

The theory proposed has a wide field of application and allows for a valid description of the impact response of layered structures having a practically any composition of layers and pack thickness.

Acknowledgment. The work was supported in part by the budget program of the NAS of Ukraine KPKVK 6541230 "Supporting the development of priority areas of scientific research".

References

1. Nicolais, L., Meo, M., Milella, E. (eds.): Composite Materials: A Vision for the Future. Springer, London (2011)
2. Slyvynskyi, V.I., Sanin, A.F., Kharchenko, M.E., Kondratyev, A.V.: Thermally and dimensionally stable structures of carbon-carbon laminated composites for space applications. In: Proceedings of the International Astronautical Congress (IAC), vol. 8, pp. 5739–5751 (2014)

3. EASA/AV CS-25. Certification specifications and acceptable means of compliance for large aeroplanes. European Aviation Safety Agency (2020)
4. Shupikov, A.N., Ugrimov, S.V., Smetankina, N.V., et al.: Bird dummy for investigating the bird-strike resistance of aircraft components. J. Aircr. **50**(3), 817–826 (2013). https://doi.org/10.2514/1.C032008
5. Smetankina, N., Kravchenko, O., Ugrimov, S., Yareshchenko, V.: Calculation and experimental study of the stress-strained state of laminated composites at impact loading. Paper presented at the 7th International Conference Space Technologies: Present and Future, Yuzhnoe State Design Office, Dnipro, 21–24 May 2019
6. Tiberkak, R., Bachene, M., Rechak, S., Necib, B.: Damage prediction in composite plates subjected to low velocity impact. Compos. Struct. **83**(1), 73–82 (2008). https://doi.org/10.1016/j.compstruct.2007.03.007
7. Richardson, M.O.W., Wisheart, M.J.: Review of low-velocity impact properties of composite materials. Compos. Part A Appl. Sci. Manuf. **27**(12), 1123–1131 (1996). https://doi.org/10.1016/1359-835X(96)00074-7
8. Abrate, S.: Impact on laminated composite materials. Appl. Mech. Rev. **44**(4), 155–190 (1991). https://doi.org/10.1115/1.3119500
9. Reddy, J.N.: On the generalization of displacement-based laminate theories. Appl. Mech. Rev. **42**(11S), S213–S222 (1989). https://doi.org/10.1115/1.3152393
10. Ugrimov, S.V., Shupikov, A.N.: Layered orthotropic plates. Generalized theory. Compos. Struct. **129**, 224–235 (2015). https://doi.org/10.1016/j.compstruct.2015.04.004
11 Kurennov, S.S.: Longitudinal-flexural vibrations of a three-layer rod. An improved model. J. Math. Sci. **215**(2), 159–169 (2016). https://doi.org/10.1007/s10958-016-2829-7
12. Nosier, A., Kapania, R.K., Reddy, J.N.: Low-velocity impact of laminated composites using a layerwise theory. Comput. Mech. **13**(5), 360–379 (1994). https://doi.org/10.1007/BF00512589
13. Pierson, M.O., Vaziri, R.: Analytical solution for low-velocity impact response of composite plates. AIAA J. **34**(8), 1633–1640 (1996). https://doi.org/10.2514/3.13282
14 Shupikov, A.N., Smetankina, N.V.: Non-stationary vibration of multilayer plates of an uncanonical form. The elastic immersion method. Int. J. Solids Struct. **38**(14), 2271–2290 (2001). https://doi.org/10.1016/S0020-7683(00)00166-9
15. Suvorova, I.G., Kravchenko, O.V., Baranov, I.A.: Mathematical and computer modeling of axisymmetric flows of an incompressible viscous fluid by the method of R-functions. J. Math. Sci. **184**(2), 165–180 (2012). https://doi.org/10.1007/s10958-012-0861-9
16. Sun, C.T., Chen, J.K.: On the impact of initially stressed composite laminates. J. Compos. Mater. **19**(6), 490–504 (1985). https://doi.org/10.1177/002199838501900601
17. Tan, T.M., Sun, C.T.: Wave propagation in graphite/epoxy laminates due to impact. Report No. NASA CR-168057. Purdue University, West Lafayette (1982)
18. Choi, I.H., Hong, C.S.: Low-velocity impact response of composite laminates considering higher-order shear deformation and large deflection. Mech. Compos. Mater. Struct. **1**(2), 157–170 (1994). https://doi.org/10.1080/10759419408945825
19. Patil, S., Reddy, D.M., Reddy, M.: Low velocity impact analysis on composite structures – a review. AIP Conf. Proc. **1943**(1), 020009 (2018). https://doi.org/10.1063/1.5029585
20. Cantwell, W.J., Morton, J.: The impact resistance of composite materials – a review. Composites **22**(5), 347–362 (1991). https://doi.org/10.1016/0010-4361(91)90549-V
21. Panettieri, E., Fanteria, D., Montemurro, M., Froustey, C.: Low-velocity impact tests on carbon/epoxy composite laminates: a benchmark study. Compos. Part B Eng. **107**, 9–21 (2016). https://doi.org/10.1016/j.compositesb.2016.09.057
22. Grigoliuk, E.I., Chulkov, P.P.: The theory viscoelasticity of multilayer shells with rigid filler at finite deflections. J. Appl. Mech. Tech. Phys. **5**, 109–117 (1964). (in Russian)

23. Smetankina, N.V., Shupikov, A.N., Sotrikhin, S.Y., Yareshchenko, V.G.: A noncanonically shape laminated plate subjected to impact loading: theory and experiment. J. Appl. Mech. **75** (5), 051004 (2008). https://doi.org/10.1115/1.2936925

24 Shupikov, A.N., Ugrimov, S.V., Kolodiazhny, A.V., Yareschenko, V.G.: High-order theory of multilayer plates. The impact problem. Int. J. Solids Struct. **35**(25), 3391–3403 (1998). https://doi.org/10.1016/S0020-7683(98)00020-1

25. Dinnik, A.N.: Selected Works. Academy of Sciences of the Ukrainian SSR Publ, Kiev (1952).(in Russian)

26. Zukas, J.A., Nicholas, T., Swift, H.F., et al.: Impact Dynamics. Wiley, New York (1982)

A Method of Rapid Measurement of Vessels Volume with Complex Shape by Critical Nozzles

Sergiy Plankovskyy[1] ⓘ, Olga Shypul[2] ⓘ, Sergiy Zaklinskyy[2] ⓘ,
Yevgen Tsegelnyk[2](✉) ⓘ, and Volodymyr Kombarov[2] ⓘ

[1] O. M. Beketov National University of Urban Economy in Kharkiv,
17 Marshala Bazhanova Street, Kharkiv 61002, Ukraine
[2] National Aerospace University "Kharkiv Aviation Institute",
17 Chkalova Street, Kharkiv 61070, Ukraine
y.tsegelnyk@khai.edu

Abstract. A subject of the study is ways to determine the volume of vessels with complex shapes. The work aims to develop and scientifically substantiate a rapid method of measuring the volume of the vessel's inner cavity with complex geometric shapes. The following results were obtained. Invented method of rapid measurement of the vessel volume with complex shape bases on a critical outflowing filled gas and its drainage through a nozzle with a predetermined discharge coefficient. The dynamic pressure change in a vessel is a determining parameter of its volume. Two conditions must be met for this method to be accurate. The first one is the calming of transients in the measured vessel after the beginning of gas critical outflow and the second is providing an adiabatic flow. The proposed method of measuring the vessel volume was checked by simulating. Obtained results show the accuracy of determining the volume of about 0.06% compared with CAD system data. In practice, the measurement by the method should be performed in two stages. At the first stage, the value of the nozzle discharge coefficient must be found based on the results of the control measurement when flowing out of the etalon vessel with a known volume. At the second stage, the required volume is determined using the found nozzle discharge coefficient. We expect the duration of the direct measurement by the invented method up to 1 s and its accuracy by about 0.1%.

Keywords: Method of vessels volume measurement · Critical outflowing · Numerical simulation · Critical nozzles

1 Introduction

The task of measuring the volume of complex shape vessels is relevant in many practical cases. Such as problems associated with the formation of gas mixtures for using in laser [1], plasma [2] and detonation processing [3]. For many of these processes, high dosing accuracy of the gas mixture components must be ensured while at the same time high mixing productivity. A promising way to solve this problem is using the critical holes method with the effect of supercritical gas outflow, providing

© The Author(s), under exclusive license to Springer Nature Switzerland AG 2021
M. Nechyporuk et al. (Eds.): ICTM 2020, LNNS 188, pp. 247–255, 2021.
https://doi.org/10.1007/978-3-030-66717-7_20

automatic stabilization of volume flow through the nozzles. In [4], a modification of the critical holes method due to using outflow of mixture components from intermediate vessels is proposed. The application of this method requires accurate determination of the volumes of the gas path elements of the mixture generator, which can have a rather complex shape.

A more difficult task is to measure the volume of rockets tanks. Their feature is the presence of in-tank structures and equipment, which makes it difficult to determine the volume. It is important to take into account the change in the volume of the tanks caused by their deformation under the influence of boost pressure and inertial forces. Moreover monitoring of the remaining fuel in tanks requires ascertainment of the tanks volumes especially by the Pressure Volume Temperature (PVT) method [5].

One of the most recognized ways to determine the volume of vessels with complex shape is filling it with a liquid [6] and finding the volume of liquid with a high-precision measurer or weighing [7]. If the vessel has a complex shape it is difficult to ensure filling full measure volume with the liquid due to the formation of gas pockets. The disadvantages of this technology for determining the volume of rocket tanks are the high labor intensity and long-time duration of measurements. Besides that, this method does not allow testing the volumes corresponding to the rocket flight conditions. Particularly neither the deformation of the tank body under the influence of boost gas pressure nor the influence of the overload caused by acceleration is not taken into account.

It is known the methods of volume determination based on the measurement of gas outflow parameters. In this the pressure is measured while direct and reverse flow of gas from a measuring vessel with a fixed volume to the measured one. After finding the pressure in the vessels, the volume of the measured vessel is calculated by the formula, based on the equation of state of the ideal gas [8]. The disadvantage of this measuring is the poor accuracy because of changing the gas parameters inside the vessel during the test. Also this method is characterized by the same drawbacks as for the previous considered one.

Modern methods of laser scanning can be used to determine the volumes of tanks [9]. The advantage of these methods is the ability to measure the volume of large tanks (up to 5000 m^3) [10]. However, during the 3D scanning process, data scatter is inevitable and affects the measurement. Mentioned methods have limited application for measuring volumes of small vessels or having a complex shape.

A study [11] considers the possibility of determining the volume of satellite fuel systems in microgravity using Electrical Capacitance Tomography (ECT). This method can directly measure the mass of fuel in the tank. This method requires profound selection of the number and size of electrodes to provide a high spatial resolution of tomography images. In addition, the expertise of obtained numerical data found that the method does not provide reliable results of image error estimation. These shortcomings are planned to eliminate in the next version of the ECT system. At that rate, this method has a significant promise.

The patent [12] proposes a combined method for determining the volume of the rocket tank, taking into account deformations from gas pressurization and acceleration in flight. The method involves the combined use of laser scanning and liquid measurement of volume when filling the tank to specified levels. However, volume measuring according to this patent takes still time-consuming.

The way to solve the mentioned above disadvantages can be applying a test method based on the supercritical gas outflow from the measured vessel. Taking into account that the temperature in the tank changes significantly during control and introduces an error in the calculation of the measured volume, the proposed method needs to be improved. Thus the aim of the study is a development and scientifically substantiate a rapid method of measuring the volume of the vessel's inner cavity with complex geometric shapes bases on a critical outflowing filled gas.

2   Rapid Measurement of Volume at Critical Gas Draining

At supercritical pressure drop, the instantaneous value of the mass flow through the critical hole is determined by the Eq. [13]

$$G = \frac{\mu F P}{\sqrt{RT}} \psi, \tag{1}$$

where $\psi = \sqrt{k(2/(k+1))^{\frac{k+1}{k-1}}}$; μ is the nozzle discharge coefficient; F is the cross-section area of the nozzle critical hole; k, R is the adiabatic exponent and the gas constant of outflowing gas.

The current values of gas temperature and pressure in the vessel under the condition of adiabatic flow are determined by the equations

$$T = T_0 (P/P_0)^{\frac{k-1}{k}}, \tag{2}$$

$$P = P_0 (1 + Bt)^{\frac{-2k}{k-1}}, \tag{3}$$

where T_0, P_0 is the initial gas temperature and pressure; V is the volume of the vessel and

$$B = \frac{(k-1) F \sqrt{RT_0}}{2V} \psi.$$

In paper [14] the method of nozzle discharge coefficient finding is proposed. According to it, the researched nozzle is connected to an etalon vessel with a known volume. The vessel is filled with air to pressure at 2.0 MPa. The cross-sectional area of the pipelines is 10 times bigger than the area of the nozzle critical hole. The pressure in the vessel is measured during the critical air leakage with the time range at 0.5...1.0 s. At this time range, the thickness of the heated layer near the walls δ does not exceed 2...3 mm. The heated layer near the walls is calculated by the equation $\delta \cong \sqrt{at}$, where a is the temperature conductivity. Because of the thickness of this air layer does not exceed 1.5% of the vessel diameter when the outflow takes place at high-pressure drops, it is justified the outflow process is adiabatic. Then using (1)–(3) the nozzle discharge coefficient can be calculated as

$$\mu = \frac{1}{N_{nozzle}} \sum_{i=1}^{N_{nozzle}} \frac{\left(P_i^{nozzle}/P_0\right)^{-(k-1)/2k} - 1}{t_i^{nozzle}} \frac{2V_{et}}{\psi(k-1)F\sqrt{RT_0}}, \tag{4}$$

where N_{nozzle} is the number of pressure measurements in the time range; V_{et} is the volume of etalon vessel.

The measured vessel should be filled with compressed gas and connected to the ascertained nozzle by means of a normally closed solenoid valve. The pressure in the vessel is set at a level that ensures a critical flow mode. The valve connecting the measuring vessel and the nozzle then opens and the gas from the measured vessel flows to the environment. Given that the operating time of solenoid valves used in gas paths can reach 0.1 s, as well as the presence of transients in gas paths which can last up to 0.4 s, for high accuracy the measurement should be carried out under the condition of their calming. In view of this, it is advisable to set the time with the beginning for measurements $t \geq 0.5$ s. To eliminate random errors a series of measurements should be performed, and the required volume should be determined by the average value

$$V_{vessel} = \frac{1}{N_{vessel}} \sum_{i=1}^{N_{vessel}} \frac{t_i^{vessel}}{\left(P_i^{vessel}/P_0\right)^{-(k-1)/2k} - 1} \frac{\mu\psi(k-1)F\sqrt{RT_0}}{2}. \tag{5}$$

Due to defining the pressure in the vessel retaining the adiabatic flow conditions and eliminating of random errors the accuracy of determining vessel volume increases. Thus, the proposed method provides high accuracy while significantly reducing the measurement volume time.

The method has some limitations. To ensure the critical outflow mode in the entire range of pressure measurement time, the condition must be met

$$\beta = \frac{P_{env}}{P} = \frac{P_{env}}{P_0(1 + Bt)^{\frac{-2k}{k-1}}} \geq \left(\frac{2}{k+1}\right)^{\frac{k}{k-1}}, \tag{6}$$

where P_{env} is the environmental pressure.

On the other hand, the pressure change must be larger at least ten times over the outflowing time than the pressure sensor accuracy to be reliably measured with it. When using pressure sensors with an accuracy class of 0.01, this requirement can be written as

$$\frac{P_t}{P_0} = (1 + Bt)^{\frac{-2k}{k-1}} \leq 0.9. \tag{7}$$

Expressions (6) and (7) make it possible to reasonably choose the characteristics of the critical nozzle, the initial pressure and the range of the expiration time for a known value of the reference volume. Another important parameter for ensuring the accuracy according to dependence (5) is the cutoff frequency of the sensor f_c. For reliable compensation of random errors with outflowing times $t \sim 1$ s it is need to ensure $f_c \sim 10^2 \ldots 10^3$.

For the proposed method, the best accuracy in determining the volume will be achieved if the nozzle is calibrated before each measurement using the same gas that will pressurize the measured vessel. In this case, dependencies (4) and (5) can be combined, which gives the formula for calculating the volume:

$$V_{vessel} = \frac{V_{et}}{N_{vessel} \cdot N_{nozzle}} \sum_{i=1}^{N_{vessel}} \frac{t_i^{vessel}}{\left(P_i^{vessel}/P_0^{vessel}\right)^{-(k-1)/2k} - 1}$$
$$\times \sum_{i=1}^{N_{nozzle}} \frac{\left(P_i^{nozzle}/P_0^{nozzle}\right)^{-(k-1)/2k} - 1}{t_i^{nozzle}}. \tag{8}$$

The volume value calculated by the Eq. (8) is average over the gas flow time. When measuring the volume of a deformable vessel, such as, for example, a rocket tank, it must be used taking into account the elastic deformation of the tank. In this case, as shown in [12], the volume of the tank is linearly dependent on the value of the boost pressure. Therefore, to determine it by Eq. (8), it is necessary to calculate the current values of the tank volume by the follow

$$V_i = \frac{V_{et}}{N_{nozzle}} \cdot \frac{t_i^{vessel}}{\left(P_i^{vessel}/P_0^{vessel}\right)^{-(k-1)/2k} - 1} \sum_{i=1}^{N_{nozzle}} \frac{\left(P_i^{nozzle}/P_0^{nozzle}\right)^{-(k-1)/2k} - 1}{t_i^{nozzle}},$$

then approximate them by the dependence

$$V(P) = V_0 + bP,$$

and determine the coefficients of approximation by the method of least squares, for example.

3 Numerical Simulation of the Proposed Method

In order to investigate the possibility of the proposed method a two-stage modelling procedure was adopted. First, the nozzle discharge coefficient was calculated with the described method. For this stage, a model of the reference vessel was constructed using the CAD system (Fig. 1), and the volume was amounted to $1.575 \cdot 10^{-3}$ m^3.

Adjacent to the vessel is a short pipe with a nozzle, which is a conical diffuser that passed into a cylindrical part. The cross-sectional diameter of the cylindrical channel was 1 mm. To determine the parameters of supercritical leakage for the created CAD model, a calculated grid is constructed, and in the pipeline and nozzle the grid has 7 wall layers, the total thickness of which was chosen based on calculations of the expected thickness of the boundary layer. To correctly model the flow of gas from the co-plate to the calculation area was added a cylindrical part (not shown in Fig. 1), at the boundary of which the ambient conditions are set $P = 1$ bar, $T = 300$ K. Given the short flow time on the walls the condition of constant temperature $T = 300$ K is given, and the no slip condition is set for speed. The initial conditions in the vessel are set to pressure $P = 5$ bar, temperature $T = 300$ K. It was assumed that the whole vessel is filled with nitrogen $k = 1.404$, $R = 297$ J/kg·K.

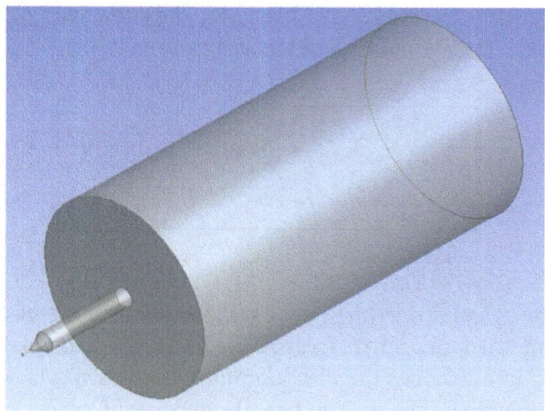

Fig. 1. Model of the etalon vessel for calculation nozzle discharge coefficient.

The following settings were used for simulation. Turbulence model is the SAS SST. The time solution step is adaptive to the Courant number ($C < 5$). Scheme for time – second order backward Euler. Other solver settings – high resolution. In Fig. 2 shows a graph of the change in pressure in the reference vessel in the process of critical outflowing. Transients lasted up to 0.1 s. The calculation of the nozzle discharge coefficient in the time range of 0.1…0.3 s gave the mean value of $\mu = 0.753$ with a standard deviation of $\delta = 0.002$.

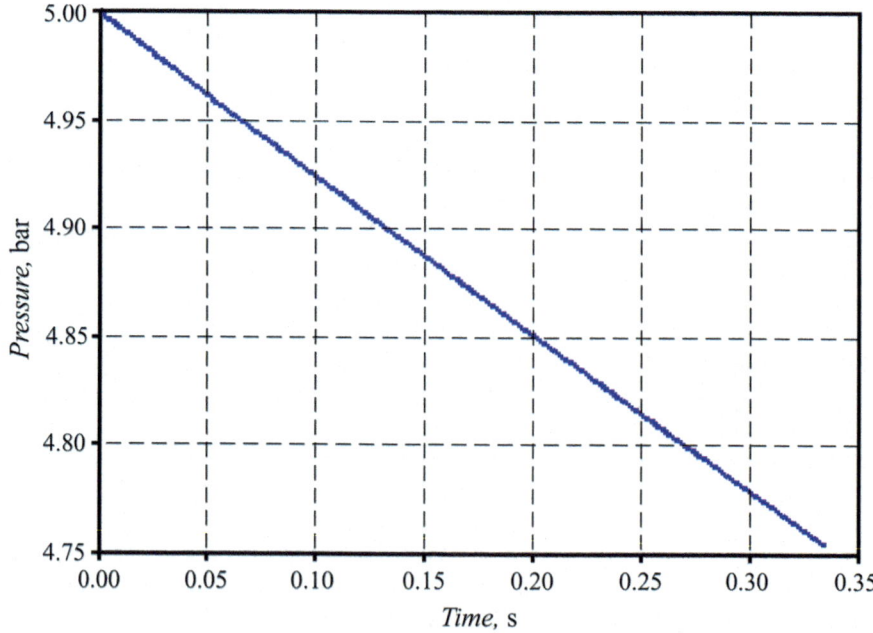

Fig. 2. The dependence of the pressure in the etalon vessel on time at critical outflowing.

After that, in the second stage a determined nozzle discharge coefficient is used to calculate the volume of the complex shape vessel (Fig. 3). The boundary and initial conditions, as well as the settings of the solver are identical to those used in the first stage of simulation. Taking into account the transient processes, pressure values in the time range of 0.1…0.3 s are used to determine the volume of the vessel.

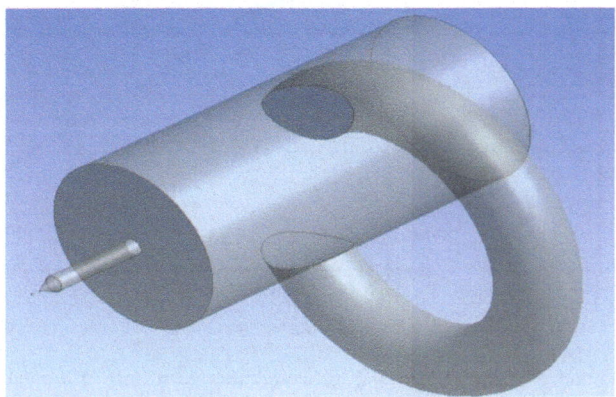

Fig. 3. Model of a complex shape vessel for determination volume by critical outflowing parameters.

In Fig. 4 shows a graph of the pressure change in the measured vessel in the process of supercritical gas leakage. According to the determined data of pressure and flow coefficient, the calculation of the volume according to the above dependence gave the value $V = 2.0771 \cdot 10^{-3}$ m^3 at the volume value defined in CAD $2.0784 \cdot 10^{-3}$ m^3. Thus, the accuracy of volume determination relative to CAD system data was 0.0625%.

Such high accuracy is due to the complete identity of the conditions of critical outflowing the cases under consideration. Therefore, to use the proposed method in practice, the measuring instruments should include a reference vessel with a known volume. Volume measurement should be performed in two stages, using in both cases the same gas for supercharging, the simplest – prepared (purified and dried) air.

In the first stage, according to the results of the control measurement at the outflow from the etalon vessel, the value of the nozzle discharge coefficient should be specified, and at the second stage, at the outflow from the measured vessel – the required volume. Since the same gas and the same nozzle will be used in both cases, the accuracy of the volume determination should be expected to be close to the calculated one with an error of $\Delta \approx 0.1\%$.

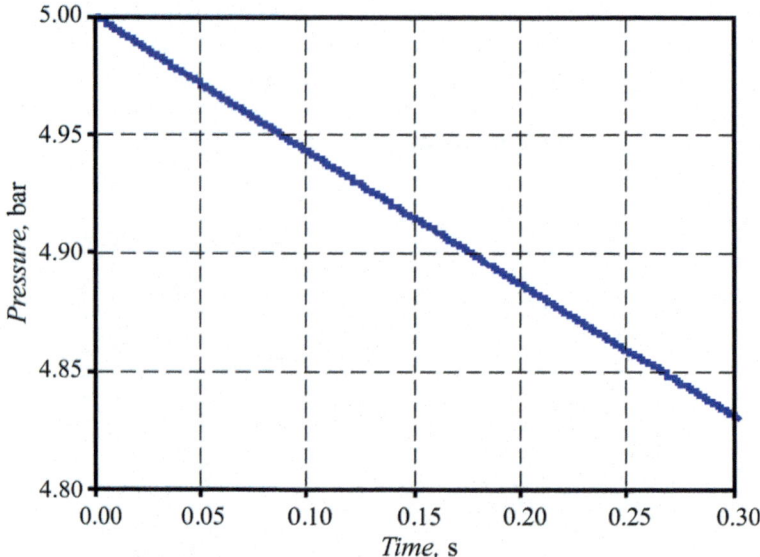

Fig. 4. Dependence of pressure in the measured vessel on time at critical outflowing.

4   Conclusion

A method of rapid measurement of vessel volume is proposed. The method is based on the measured vessel filling with gas and its drainage at critical outflow from the nozzle while measuring the pressure in the vessel. The volume of the measured vessel can be determined by the proposed equation. The method allows to take into account the elastic deformation under the action of supercharging pressure when measuring the volumes of thin-walled tanks.

During the simulation of the process of measuring the volume of a complex shape vessel on the basis of the proposed method, the accuracy of determining the volume of the vessel is 0.0625% relative to the CAD system data. To implement the proposed method in practice in the composition of the measuring instruments should include an etalon vessel and realized measuring in two stages, using in both cases for supercharging the same gas. However, we expect the accuracy of the volume determination to be in error $\Delta \approx 0.1\%$.

References

1. Yilbas, B.S., Ali, H., Karatas, C.: Laser gas assisted treatment of Ti-alloy: analysis of surface characteristics. Opt. Laser Technol. **78**, 159–166 (2016). https://doi.org/10.1016/j.optlastec. 2015.11.002
2. Samadi, M., Eshaghi, A., Bakhshi, S.R., Aghaei, A.A.: The influence of gas flow rate on the structural, mechanical, optical and wettability of diamond-like carbon thin films. Opt. Quant. Electron. **50**(4), 1–14 (2018). https://doi.org/10.1007/s11082-018-1456-6

3. Plankovskyy, S., Popov, V., Shypul, O., et al.: Advanced thermal energy method for finishing precision parts. In: Pramanik, A., Gupta, K. (eds.) Advanced Machining and Finishing. Elsevier, Amsterdam (2021). https://doi.org/10.1016/B978-0-12-817452-4.00014-2
4. Plankovskyy, S., Shypul, O., Zaklinskyy, S., Tryfonov, O.: Dynamic method of gas mixtures creation for plasma technologies. Probl. At. Sci. Technol. 6(118), 189–193 (2018)
5. Seo, M., Jeong, S., Jung, Y.S., et al.: Improved pressure–volume–temperature method for estimation of cryogenic liquid volume. Cryogenics 52(4–6), 290–295 (2012). https://doi.org/10.1016/j.cryogenics.2012.01.012
6. BPM 8871101-96/1:2005. Liquid tank calibration methodic (3–200 m^3). KTU Metrology Institute, Lithuania (2005)
7. Youhuan, X., Jinfeng, S., Liqun, L., et al.: Accuracy analysis on measuring model of rocket propellant filling based on weight measurement. In: 2014 IEEE International Conference on Signal Processing, Communications and Computing (ICSPCC), Guilin, pp. 281–286. IEEE (2014). https://doi.org/10.1109/ICSPCC.2014.6986199
8. Lal, A., Raghunandan, B.N.: Uncertainty analysis of propellant gauging system for spacecraft. J. Spacecr. Rockets 42(5), 943–946 (2005). https://doi.org/10.2514/1.9511
9. Knyva, M., Knyva, V., Meškuotiene, A., et al.: 3D laser scanning pointcloud processing uncertainty estimation for fuel tank volume calibration. MAPAN 35(3), 333–341 (2020). https://doi.org/10.1007/s12647-020-00367-4
10. Huadong, H., Xianlei, C., Haolei, S., et al.: The automatic measurement system of large vertical storage tank volume based on 3D laser scanning principle. In: 2017 13th IEEE International Conference on Electronic Measurement & Instruments (ICEMI), Yangzhou, pp. 211–216. IEEE (2017). https://doi.org/10.1109/ICEMI.2017.8265768
11. Gut, Z.: Using electrical capacitance tomography system for determination of liquids in rocket and satellite tanks. Trans. Aerosp. Res. 2020(1), 18–33 (2020). https://doi.org/10.2478/tar-2020-0002
12. Morozov, V.S., Romanets, N.S., Valov, O.A.: Method of measuring internal volume of liquid rocket fuel tank and calibration of volume of tank on levels. RU Patent 2577090C1, 10 March 2016
13. Zvegintsev, V.I.: Gas-Dynamic Installations of Short-Term Action. Part 1. Installations for Scientific Research. Parallel' Publ., Novosibirsk (2014). (in Russian)
14. Bykovskii, F.A., Vedernikov, E.F.: Flow rates of nozzles and their combinations for forward and reverse flow. J. Appl. Mech. Tech. Phys. 37(4), 98–104 (1996). (in Russian)

Optimal Design of the Cyclically Symmetrical Structure Under Static Load

Serhii Misura[1,2(✉)] ⓘ, Natalia Smetankina[2] ⓘ,
and Ievgeniia Misiura[3] ⓘ

[1] National Technical University "Kharkiv Polytechnic Institute",
2 Kyrpychova Street, Kharkiv 61002, Ukraine
misurasy@gmail.com

[2] A. Pidgorny Institute of Mechanical Engineering Problems of the National
Academy of Sciences of Ukraine, 2/10 Pozharskogo Street, Kharkiv 61046,
Ukraine
nsmetankina@ukr.net

[3] Simon Kuznets Kharkiv National University of Economics,
9a Nauky Avenue, Kharkiv 61166, Ukraine
misuraeu@gmail.com

Abstract. A method is proposed for solving the problems of optimal design of cyclically symmetric structures under static loading, which has been tested on critical structural elements of hydraulic turbines. One of the basic problems in the design of hydraulic turbines is considered, namely, ensuring their strength and reliability under continuous operation under the influence of a static loading. The problem of optimal design of the initial and modified covers of a rotary-blade hydraulic turbine operating in the normal mode has been solved. A Kaplan turbine cover is a complex spatial structure consisting of thin-walled elements. Therefore, the finite element method is used for the calculation to most fully take into account the design features and the spectrum of external influences acting during operation. As the initial design, covers with an initial and modified hole in the rib were selected. The geometric parameters of the cover are modified to minimize the cover weight. The thicknesses of structural elements are taken as design variables. The minimum and maximum thicknesses, as well as maximum stress intensity values are limited. The objective function is the cover weight. The problem of optimal design is solved with the help of the gradient method using a finite-difference analogue of a gradient of the objective function. The distribution of axial displacements and stress intensity in the original and modified cover design during normal operation was obtained. It was found that the mass of the cover structure was reduced by 30%, and the rolled stock thickness range was downsized by five positions, which is significant in the manufacture of a new design. In this case, the stress values in the optimal structure during the modification of the hole in the ribs did not exceed the admissible values. The proposed approach will subsequently be applied to the analysis of elements of aircraft structures.

Keywords: Optimization · Kaplan turbine · Strain-stress state · Gradient method

M. Nechyporuk et al. (Eds.): ICTM 2020, LNNS 188, pp. 256–266, 2021.
https://doi.org/10.1007/978-3-030-66717-7_21

1   Introduction

When designing hydraulic turbines, one of the main problems is to ensure their strength and reliability during continuous operation under the influence of static and dynamic loadings [1–3].

The specific feature of a hydroelectric power station workflow requires special design solutions that ensure reliable operation of units and structures, one of which is the cover of a hydraulic turbine. It is a large-size welded fixed ring part that limits the flow part from above and serves as a base for accommodating guide apparatus parts.

Despite significant achievements in the study of the strength of cyclically symmetric metal structures, the study of their reliability remains relevant. That is why, design of hydropower turbines for hydroelectric power plants requires methods for determining their strain-stress state, allowing to create design models with sufficient accuracy [4].

Typically, the shape of impellers of a hydraulic turbine is optimized [5–7]. The problems of optimal cover design are less studied [8, 9].

In this work, the strain-stress state of a Kaplan turbine cover is analyzed using advanced effective methods and programs for calculating the strength and characteristics of welded load-bearing structures. The methods and programs are based on the elasticity theory the finite element method and the theory of thin plates and shells [10–13]. The optimization problem is solved using the gradient method [14, 15].

The aim of the paper is optimal design of the initial and modified Kaplan turbine cover configurations. The objective function is the cover weight. The optimization parameters are the thicknesses of the structural elements.

2   Model of a Kaplan Turbine Cover

A finite-element cover model of a Kaplan turbine under static axisymmetric load is offered. The cover is a spatial cyclically symmetric structure consisting of thin-wall shells of revolution joined by n ribs. The ribs are meridional plates of complex configuration. Thus, the cover consists of sectors, on whose boundaries the conditions of cyclic symmetry are satisfied. The development of a model with such structures begins with developing a sector model.

When constructing a sector model, the key points in the plane of the rib, along which the lines are drawn, are first defined, and then a rib model is created. To obtain the shell parts of the structure and a complete sector model, the lines of intersection of ribs and shell surfaces are rotated clockwise and counterclockwise through an angle of $360/(2n)$, where n is the number of sectors.

Since the cover is a spatial structure consisting of thin-wall elements, for which the ratio of the thickness of the structural elements and the characteristic size does not exceed 1/10, the theory of thin plates and shells is used. The system of governing equations is.

$$[K]\{u\} = \{F\},$$

where $[K]$ is stiffness matrix; $\{u\}$ is vector of nodal displacements; $\{F\}$ is vector of forces determining the influence of external loads.

To solve the problem, a triangular elastic shell finite element with three nodes is used. An element in each node has six degrees of freedom, namely displacements in the direction of the coordinate X, Y and Z axes and rotations about them.

The model is divided into finite elements, after which the conditions of cyclic symmetry, as well as the conditions of structure fixing and loading are introduced at the boundaries with neighboring sectors.

Figure 1 shows the scheme of the cover. To place the mechanisms and reduce the cover weight, round holes are provided in the ribs. Curved holes are created in the annular plates in the form of a blade profile, which are designed for dismantling and repairing individual blades without completely disassembling the guide apparatus. The cover has such overall dimensions: diameter 3.44 m; height, 1.05 m.

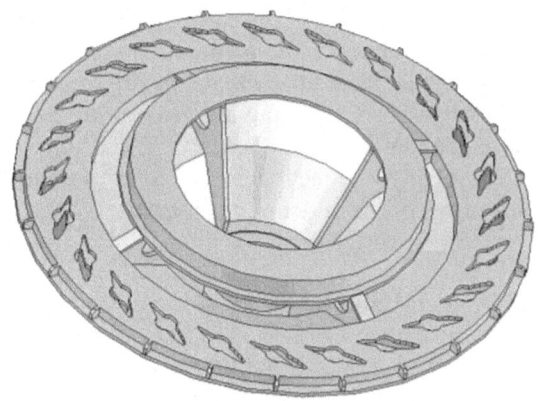

Fig. 1. Kaplan turbine cover.

The cover is made of sheet steel St20 or its analog ASTM A516 Gr.60. The mechanical properties of the material are as follows: $E = 2.1 \cdot 10^5$ MPa is Young's module; $v = 0.3$ is Poisson's ratio; $\rho = 7850$ kg/m^3 is material density; $\sigma_\tau = 215$ MPa is yield strength; $\sigma_s = 430$ MPa is ultimate strength; $[\sigma] = 0.5 \cdot \sigma_\tau = 107.5$ MPa is admissible stress.

The design scheme is adopted as a cover sector with a solution angle of 90° and symmetry conditions at the boundaries (see Fig. 2).

The conditions are introduced for fixing the cover on the supporting surface of the flange connecting it to the stator ring, which is considered absolutely rigid, along the circumference on which the studs of the flange connection are located.

Figure 3 shows the scheme of cover loading and fastening. The weight of the generator and impeller is taken into account in the form of equivalent pressure $P = 2.45 \cdot 10^5$ N applied to the surface of the upper ring (see Fig. 3).

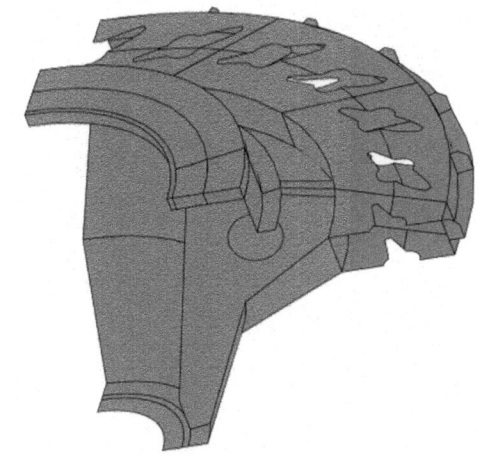

Fig. 2. Cover sector.

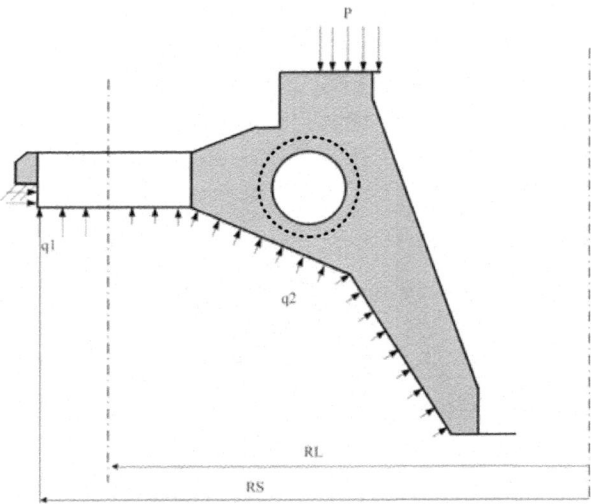

Fig. 3. Scheme of cover loading and fastening.

In the normal mode, the hydraulic pressure $q2 = 0.0965$ MPa is applied to the bottom. During an emergency shutdown of the turbine unit, the pressure in the supply pipe from the radius of the circle RL, on which the guide apparatus vanes are located to the circle radius on which the studs of the flange connection RS are located, rises sharply from $q2$ to $q1 = 0.1254$ MPa. Therefore, numerical results are presented precisely for this case.

In Fig. 3, the dotted line shows the contour of a modified hole in the cover rib to place equipment and reduce the weight. The radius of this hole is increased 1.5 times relative to the original one. Figure 4 shows the finite element model of the cover sector.

Fig. 4. Finite element model of the cover sector.

3   Stress Analysis of the Cover

First, we obtain the distribution of axial displacements (Fig. 5) and stress intensity (Fig. 6) in the cover in the normal mode.

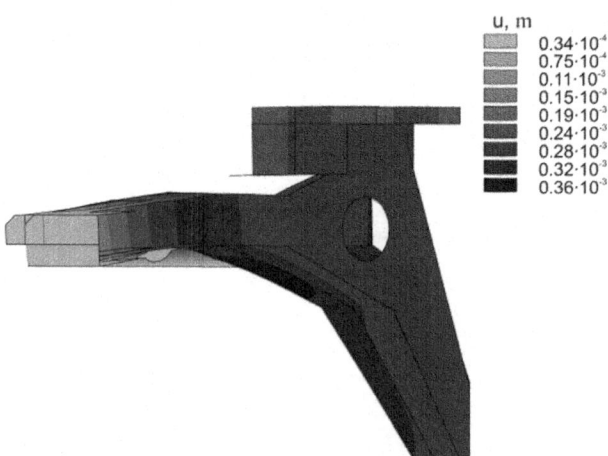

Fig. 5.   Distribution of axial displacements in the cover sector.

Maximum stresses occur in the ribs, which are located in the duct where the vanes of the guide apparatus pass. In Fig. 6, the arrow ($\rightarrow$) shows the zone of highest stress concentration. Calculation yields zones where maximum displacements occur. They are located in the duct on the guide apparatus side. The maximum displacement and

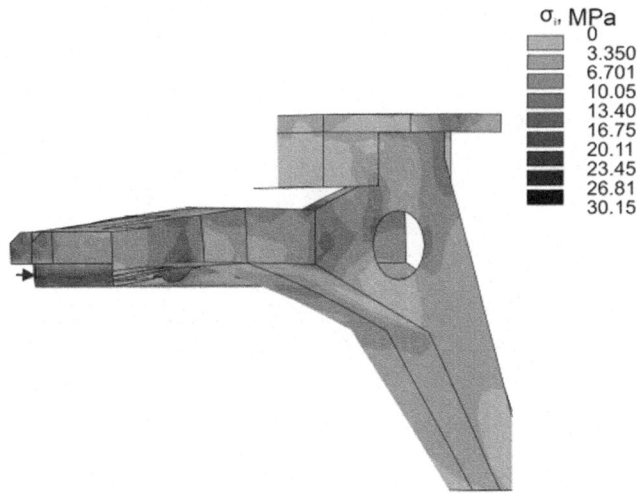

Fig. 6. Distribution of stress intensity in the cover sector.

stresses have the following values: $u_{max} = 3.6 \cdot 10^{-4}$ m and $\sigma_{max} = 30.15$ MPa, respectively.

Similar distributions of axial displacements and stress intensities in the cover in the normal mode were obtained when the hole was modified (see Fig. 3). The calculation results are shown in Fig. 7 and Fig. 8.

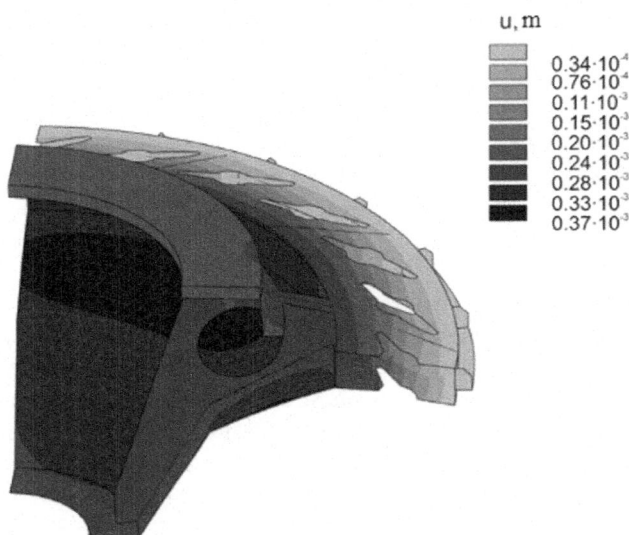

Fig. 7. Distribution of axial displacements in the cover sector with the modified hole.

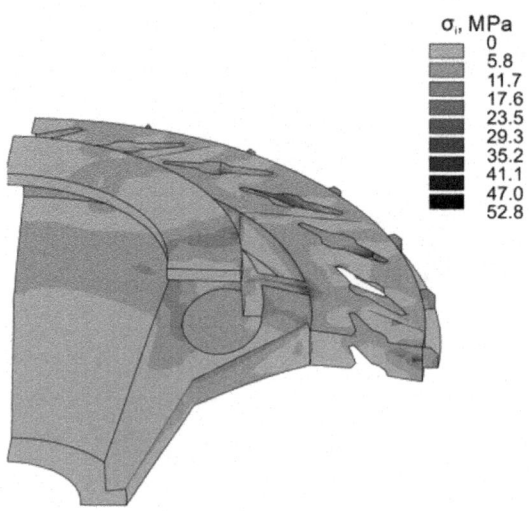

Fig. 8. Distribution of stress intensity in the cover sector with the modified hole.

4   Statement of the Optimal Design Problem

In a broad sense, the general problem of nonlinear programming consists in finding the extreme point.

$$\mathbf{C} = \mathbf{C}^*, \ \mathbf{C} \in E_m,$$

where E_m is the space of design variables at which the objective function reaches a minimum value.

$$F^* = F(\mathbf{C}^*) = \min F(\mathbf{C}),$$

and constraints are met.

$$G_j(\mathbf{C}^*) \geq 0, \ j = \overline{1, J}.$$

Here, $\mathbf{C}$ is the vector of the space of design variables. The objective function is the cover weight. The design variables were the thicknesses of the cover structural elements, namely shells, plates and ribs. Constraints are imposed on the minimum and maximum values of thicknesses. This is most often the case because of manufacturing and operational requirements. The minimum possible thickness for all elements is 0.016 m. The maximum thicknesses are the initial values of thicknesses of the not modified cover. The maximum stress intensity values are limited by the admissible value $[\sigma] = 107$ MPa.

The problem of optimal design is solved with the gradient method using a finite-difference analogue of a gradient for the objective function described previously [14].

5   Optimization Results

The numerical solution of the optimization problem yields the optimal thicknesses of shells, plates and ribs of the cover. As a start design, the covers with initial and modified holes were selected in turn.

Figure 9 and Fig. 10 show the distributions of axial displacements and stress intensity in the optimal cover, respectively. Maximum stresses arise in the area of ribs as indicated by the arrow ($\rightarrow$).

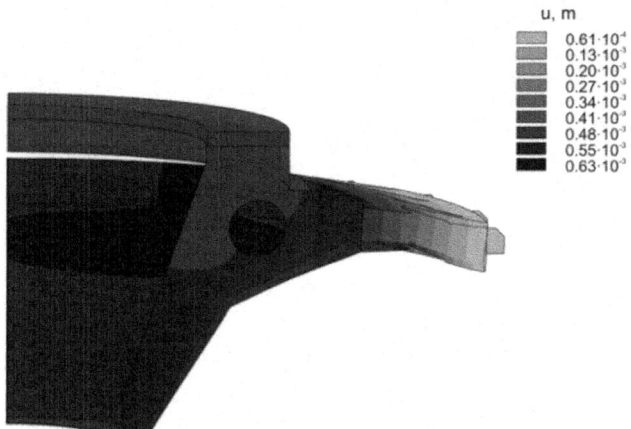

Fig. 9. Distribution of axial displacements in the optimal cover.

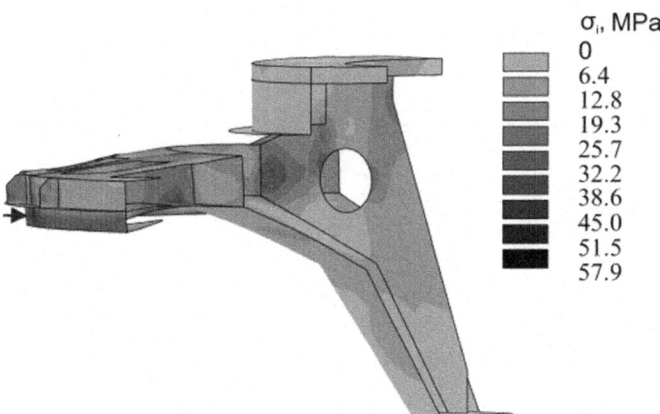

Fig. 10. Distribution of stress intensity in the optimal cover.

Next, the cover with the modified hole was optimized. The distribution of axial displacements and stress intensity in the optimal cover in the normal mode with the modified hole is shown in Fig. 11 and Fig. 12, respectively.

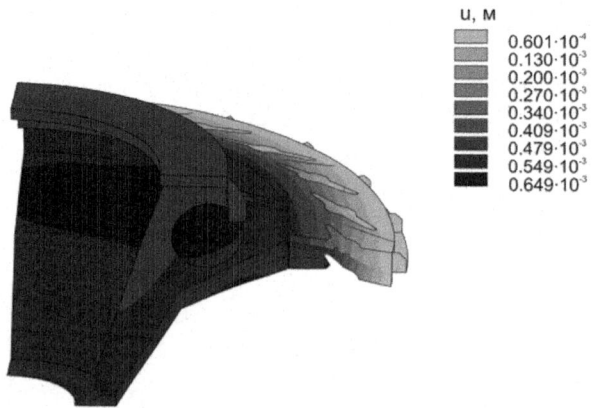

Fig. 11. Distribution of axial displacements in the optimal cover.

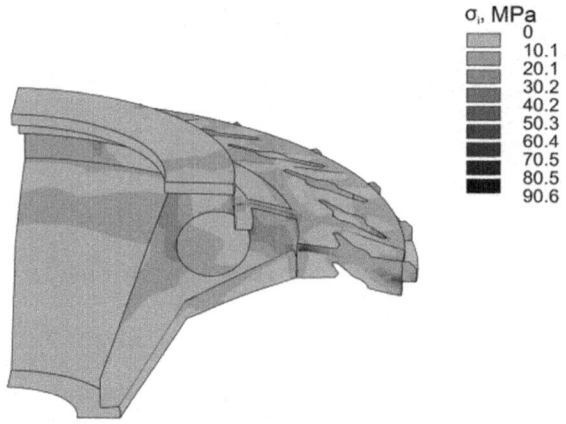

Fig. 12. Distribution of stress intensity in the optimal cover with the modified hole.

Table 1 presents the maximum values of stress intensity σ_{max}, axial displacements u_{max} and the cover weight obtained by design of original and optimal structures. The optimal design is characterized by the following range of structural thicknesses (shells, plates and ribs) 0.018 m, 0.02 m, 0.022 m, 0.03 m, 0.032 m, and 0.04 m.

Table 1. Parameters of optimal design of the cover.

Cover structure	Stress intensity σ_{max} [MPa]	Displacement $u_{max} \cdot 10^{-3}$ [m]	Weight [kg]
Original cover	32.2	0.371	4080.7
Optimal cover with the hole	57.9	0.630	2578.5
Optimal cover with the modified hole	90.6	0.649	2569.2

During optimization, only constraints on the minimum thicknesses were applied. Hence, the thicknesses of all elements of the optimal cover turned out to be the same and equal to 0.16 m. The thickness of the flange of the cover attached to the stator remained equal to 0.03 m.

Note that the optimal cover weight is 1510 kg less than that of the initial cover. In addition, the nomenclature of optimal design elements was reduced by five positions, which is a clear advantage.

The stresses in the optimal cover for all structures do not exceed admissible values.

6 Conclusions

The paper suggests an approach to minimum weight design of Kaplan covers subject to geometrical and strength constraints. Optimal design parameters are the structural thicknesses of covers. Constraints are imposed on the minimum and maximum values of thicknesses of shells, plates and ribs in the cover. Constraints are also imposed on maximum stress intensity values.

A special technique based on the finite element method was developed to analyze the strain-stress state of covers under a static axisymmetric load. The optimization problem is formulated in terms of nonlinear programming, and then solved using the gradient method.

Several numerical examples are given that allow following the variation of the optimal design depending on the type of covers. Covers with holes and without holes in ribs were considered.

In all the cases, the weight of the optimal cover is one third less than that of the original structures, and the stresses in the optimal covers do not exceed the admissible values. In addition, the rolled stock thickness range is downsized by five positions. That is crucial for manufacturing turbines.

In practice, the execution of an optimal design is limited by the thicknesses of structural elements. However, optimal design makes it possible for a designer to see how close it fits an optimal one. Therefore, optimal designs similar to those given here may be useful when designing real turbines and aircraft structures.

Acknowledgment. The work was supported in part by the budget program of the NAS of Ukraine KPKVK 6541230 "Supporting the development of priority areas of scientific research".

References

1. Yershov, S., Rusanov, A., Gardzilewicz, A., Lampart, P.: Calculation of 3D viscous compressible turbomachinery flows. In: 2nd Symposium on Computational Technologies for Fluid/Thermal/Chemical Systems with Industrial Applications, vol. 397.2, pp. 143–154. ASME PVP (1999)
2. Suvorova, I.G., Kravchenko, O.V., Baranov, I.A.: Mathematical and computer modeling of axisymmetric flows of an incompressible viscous fluid by the method of R-functions. J. Math. Sci. **184**(2), 165–180 (2012). https://doi.org/10.1007/s10958-012-0861-9

3. Strelnikova, E., Kriutchenko, D., Gnitko, V., Degtyarev, K.: Boundary element method in nonlinear sloshing analysis for shells of revolution under longitudinal excitations. Eng. Anal. Boundary Elem. **111**, 78–87 (2020). https://doi.org/10.1016/j.enganabound.2019.10.008
4. Strelnikova, E.A., Medvedovskaya, T.F., Medvedeva, E.L., et al.: Use of computer technologies in modernization of head covers for PL 20-B-500 Kaplan turbines. J. Mech. Eng. **21**(1), 35–44 (2018)
5. Chehouri, A., Younes, R., Ilinca, A., et al.: Optimal design for a composite wind turbine blade with fatigue and failure constraints. Trans. Canad. Soc. Mech. Eng. **39**(2), 171–186 (2015). https://doi.org/10.1139/tcsme-2015-0013
6. Avdyushenko, A.Y., Cherny, S.G., Chirkov, D.V., et al.: Numerical simulation of transient processes in hydroturbines. Thermophys. Aeromech. **20**(5), 577–593 (2013). https://doi.org/10.1134/S0869864313050059
7. Skotak, A., Obrovsky, J.: Shape optimization of a Kaplan turbine blade. In: Proceedings of the 23rd IAHR Symposium on Hydraulic Machinery and Systems, Yokohama, Japan (2006)
8. Lipej, A., Poloni, C.: Design of Kaplan runner using multiobjective genetic algorithm optimization. J. Hydraul. Res. **38**(1), 73–79 (2000). https://doi.org/10.1080/00221680009498361
9. Shupikov, A.N., Misyura, S.Y.: Minimizing stresses in the stiffeners of a turbine cover. Prob. Mech. Eng. Mach. Reliab. **5**, 79–84 (2014). [in Russian]
10. Misiura, S.Y.: Hydroelastic vibrations of the covers on water turbines with the upper ring of the guide vanes. East. Eur. J. Enterp. Technol. **78**(7), 4–10 (2015). [in Russian]. https://doi.org/10.15587/1729-4061.2015.55664
11. Shupikov, A.N., Misyura, S.Y.: Minimization of stresses in stiffening ribs of the cover of a water turbine. J. Mach. Manuf. Reliab. **43**(5), 416–421 (2014). https://doi.org/10.3103/S1052618814050173
12. Misura, S.Y., Smetankina, N.V., Misiura, I.I.: Rational modelling of hydroturbine cover for a strength's analysis. Bull. Nat. Tech. Univ. "KhPI". Ser. Dyn. Str. Mach. **1**, 34–39 (2019). [in Ukrainian]. https://doi.org/10.20998/2078-9130.2019.1.187415
13. Plankovskyy, S., Myntiuk, V., Tsegelnyk, Y., et al.: Analytical methods for determining the static and dynamic behavior of thin-walled structures during machining. In: Shkarlet, S., et al. (eds.) Mathematical Modeling and Simulation of Systems (MODS 2020). AISC, vol. 1265, pp. 82–91. Springer, Cham (2021). https://doi.org/10.1007/978-3-030-58124-4_8
14. Shupikov, A.N., Smetankina, N.V., Sheludko, H.A.: Selection of optimal parameters of multilayer plates at nonstationary loading. Meccanica **33**(6), 553–564 (1998). https://doi.org/10.1023/A:1004311229316
15. Smetankina, N.V.: Nonstationary deformation, thermoelasticity, and optimization of laminated plates and cylindrical shells. Mis'kdruk Publ, Kharkiv (2011).[in Russian]

Dynamic Response of Laminate Composite Shells with Complex Shape Under Low-Velocity Impact

Natalia Smetankina[(✉)] [iD], Alyona Merkulova[iD],
Dmytro Merkulov[iD], and Oleksii Postnyi[iD]

A. Pidgorny Institute of Mechanical Engineering Problems of the National
Academy of Sciences of Ukraine,
2/10 Pozharskogo Str., Kharkiv 61046, Ukraine
nsmetankina@ukr.net

Abstract. Investigating dynamic response parameters for impact loading is a key effort in analyzing vibrations of laminated composite structures. The work presents an analytical approach to vibration analysis of laminated orthotropic shells with a complex plan shape under low-velocity impact. The dynamic behavior of shells is described by the first-order theory. The equations of motion of shells and boundary conditions are obtained from the Hamilton's variational principle. The motion equations are added by the indenter equation of motion and the condition of joint displacement of the indenter and shell. The analytical solution of the problem is derived by the immersion method. The system of motion equations of shells is integrated by expansion into Taylor series. The method potentialities are demonstrated by calculating deflections and stresses in orthotropic shells with different boundary conditions. A good match of results obtained by different methods confirms the feasibility and effectiveness of the method offered.

Keywords: Composite shell · Impact loading · Mathematical modelling

1 Introduction

The development and application of new structural materials, namely composite materials, is a characteristic trend in the development of modern technology [1, 2]. The operational properties of composite materials fully satisfy the needs of the main industries due to the fact that the properties of the composite material can significantly differ from the properties of its constituent materials. The use of composites allows expanding the capabilities of the created structures and obtaining significant improvement in a number of important parameters: weight reduction, increase in strength, wear resistance, increased resistance to various influences [3, 4]. In mechanical engineering, shipbuilding, production of aircraft and space technologies, military equipment, the latest composite materials are gradually replacing traditional ones. Further widespread adoption of composites in the practice of designing engineering objects directly depends on the level of knowledge of the processes of their

M. Nechyporuk et al. (Eds.): ICTM 2020, LNNS 188, pp. 267–276, 2021.
https://doi.org/10.1007/978-3-030-66717-7_22

deformation. Of particular relevance is the problem of modelling composite structural elements under the influence of shock loads, which often cause destruction and delamination of the structural material [5–7]. This indicates the importance and relevance of studies of composites, as well as the need to develop specialized methods for assessing their strength during operation.

Laminated composite shells are one of the main elements of various aircraft structures that can be subjected to intense dynamic loads [7–10]. The most common research methods for the dynamic behaviour of non-canonical laminated shells are numerical methods, namely, the finite element method [11–14].

Theoretical methods have not been developed enough, which is associated with the complexity of mathematical models that describe the process of deformation of such shells under intense short-term effects [15, 16].

In an analytical form, solutions to problems of non-stationary dynamics obtained only for laminated plates and shells of canonical shape in plan using R-functions [17] and B-splines [18].

An analysis of the cited works allows us to conclude that despite the fact that in our time there are many numerical methods for calculating structural elements, there is a need to develop analytical methods that allow us to analyse the influence of individual factors on the stress-strain state and optimize the parameters of composite elements. At the same time, the issue of unsteady dynamics of laminated composite shells with a complex plan shape remains insufficiently studied, which requires further development and improvement of methods for calculating such shells.

The aim of the work is to develop a method for solving the problem with unsteady vibrations of laminated orthotropic open cylindrical shells of complex shape in plan at impact load, which allows to obtain a solution to the problem in an analytical form.

2   Problem Statement

Consider an open laminated cylindrical shell of radius R. The shell consists of I orthotropic layers of constant thickness h_i and occupies on the coordinate surface (the outer surface of the first layer) the region Ω bounded by the contour Γ: $x_\Gamma = x(s)$, $y_\Gamma = y(s)$ (s is the current arc length). Impulse loads $\mathbf{P} = \{p_j(x, y, t)\}$ $(j = \overline{1, 3I + 3})$ act on the shell. The x coordinate varies along the generatrix, the y coordinate varies along the arc of the shell cross section. The positive direction of the Z-axis coincides with the direction of the external normal to the coordinate surface.

The impact is delivered by the indenter in the form of a ball with mass M and radius r, which is dropped from a height H onto the outer surface of the first layer of the shell. The velocity of its collision with the shell is determined by the formula $V_z = \sqrt{2gH}$, where g is gravity acceleration. A low-velocity impact is considered when the shell deforms elastically.

According to studies [7, 19], the interaction region of the indenter and the shell has a weakly elliptical shape, which can be replaced by a circular contact pad. It was shown that such an assumption does not affect the results of further calculations. Thus, it is assumed that the region of interaction of the indenter and the shell is a circle of radius

$a(t)$ centered at a point with coordinates (x_0, y_0). The radius of the contact area is calculated by the formula

$$a(t) = \left[\frac{3}{16}F(t)(\theta_1 + \theta)\right]^{1/3},$$

where $\theta_1 = \frac{4(1-v_1^2)}{E_1}$, $\theta = \frac{4(1-v^2)}{E}$, $F(t)$ is the indenter's contact force of action on the shell, t is time, E_1 and v_1 are the average values of the elastic modulus and Poisson's ratio of the material of the first layer of the shell, E and v are the corresponding characteristics of the indenter material.

The indenter equation of motion is

$$Mz_{,tt} = Mg - F(t), \quad z(0) = 0, z_{,t}(0) = V_z, \tag{1}$$

where $z = z(t)$ is the displacement of the indenter.

The condition of joint displacement of the indenter and shell is written as follows [20]:

$$w_0 + \alpha_c - z = 0. \tag{2}$$

Here α_c is the contact approach of the indenter and the shell at the point of contact (x_0, y_0), $w_0 = w(x_0, y_0, t)$ is deflection of the outer surface of the first layer of the shell at the point (x_0, y_0).

Contact approach α_c is determined by solving the Hertzian problem of ball indentation into an elastic semispace

$$\alpha_c = kF^{2/3}. \tag{3}$$

The coefficient k depending on the material and shape of the interacting bodies and it is selected on the basis of the experiment.

The dynamic behavior of the shell is described by kinematic hypotheses that take into account transverse shear strains, thickness reduction and normal element rotation inertia in each layer

$$u_k^i = u_k + \sum_{j=1}^{i-1} h_j u_{3+I(k-1)+j} + (z - \delta_{i-1})u_{3+I(k-1)+i}, \quad k = 1, 2, 3, \quad i = \overline{1, I} \tag{4}$$

where $\delta_i = \sum_{j=1}^{i} h_j$, $\delta_{i-1} \leq z \leq \delta_i$; $u_k = u_k(x, y, t)$, $(k = 1, 2, 3)$ is displacement of point of the coordinate surface in the direction of the coordinate axes; $u_{3+I(k-1)+i} = u_{3+I(k-1)+i}(x, y, t)$, $(k = 1, 2)$ are angles of rotation of a normal element in the i-th layer around the coordinate axes X-axis and Y-axis; $u_{3+2I+i} = u_{3+2I+i}(x, y, t)$ is compression of the normal element in the i-th layer; t is time.

Shell strains have the form

$$\varepsilon_x^i = u_{1,x}^i, \varepsilon_y^i = \frac{1}{1+z/R}\left(u_{2,y}^i + \frac{1}{R}u_3^i\right), \varepsilon_z^i = u_{3,z}^i, \gamma_{xy} = \frac{1}{1+z/R}u_{1,y}^i + u_{2,x}^i,$$

$$\gamma_{xz}^i = u_{1,z}^i + u_{3,x}^i, \gamma_{yz} = u_{2,z}^i + \frac{1}{1+z/R}u_{3,y}^i - \frac{1}{R(1+z/R)}u_2^i, i = \overline{1, I}. \qquad (5)$$

Stresses and strains in the i-th layer are related by Hooke's law for an orthotropic body [21]

$$\sigma_x^i = B_{11}^i \varepsilon_x^i + B_{12}^i \varepsilon_y^i + B_{13}^i \varepsilon_z^i, \sigma_y^i = B_{12}^i \varepsilon_x^i + B_{22}^i \varepsilon_y^i + B_{23}^i \varepsilon_z^i$$

$$\sigma_z^i = B_{13}^i \varepsilon_x^i + B_{23}^i \varepsilon_y^i + B_{33}^i \varepsilon_z^i, \tau_{yz}^i = B_{44}^i \gamma_{yz}^i, \tau_{xz}^i = B_{55}^i \gamma_{xz}^i, \tau_{xy}^i = B_{66}^i \gamma_{xy}^i, \quad i = \overline{1, I} \quad (6)$$

Taking into account relations (4)–(6) with Hamilton's variational principle [20] we obtain the equation of motion of the shell under the influence of shock loads P

$$[\Omega^p]U_{,tt} - [A]U = P, (x, y) \in \Omega, U = U_{,t} = 0, \quad t = 0, \qquad (7)$$

and the system of boundary conditions on the contour Γ

$$[B^\Gamma]U = P^\Gamma, \quad (x, y) \in \Gamma, \qquad (8)$$

where $[\Omega^p]$ and $[A]$ are symmetric matrices, $U = \{u_j(x, y, t)\}$, $B_{ij}^\Gamma = \chi_i^1 B_{ij}^u + \chi_i^2 B_{ij}^\sigma$, $i, j = \overline{1, 3I + 3}$.

The form of the matrix elements $[B^\Gamma]$ and the ultimate load vector P^Γ depends on the boundary conditions on the shell contour. By providing various values for the coefficients χ_i^1 and χ_i^2, we can simulate the necessary boundary conditions on the shell contour.

3   Solution Method

The method for solving the problem (1)–(8) is based on the method of expanding a given region [22, 23]. The initial laminated shell expands to an auxiliary laminated shell with the same stack of layers. The shape and boundary conditions of the auxiliary shell are chosen so that a solution to the problem could be obtained quite simply. The simplest type of solution can be obtained by choosing a simply supported shell rectangular in plan as an auxiliary one. This allows us to find a solution to the original problem in the form of expansions into trigonometric series.

To ensure the fulfillment of the initial boundary conditions (8), additional compensating loads $Q^{comp} = \{q_j^{comp}(x, y, t)\}$, $(j = \overline{1, 3I + 3})$ are added to the auxiliary shell, which are continuously distributed along the contour. Thus, the problem of vibrations of the shell with complex shape and arbitrary boundary conditions is reduced to the problem of vibrations of the simply supported rectangular in plan shell.

Compensating loads are included in the equations of motion of the auxiliary shell in the form of such integral relations

$$p_j^{comp}(x,y,t) = \sum_{k=1}^{3I+3} \oint_\Gamma \varsigma_{jk} q_k^{comp}(s,t)\delta(x - x_\Gamma, y - y_\Gamma)ds, \ j,k = \overline{1,3I+3}, \quad (9)$$

where $\delta(x - x_\Gamma, y - y_\Gamma)$ is the two-dimensional δ-function.

To satisfy boundary conditions (8), a system of integral equations for finding the unknown intensities of compensating loads

$$[B^\Gamma] U[Q^{comp}(x,y,t)] = P^\Gamma, (x,y) \in \Gamma. \quad (10)$$

The method for solving system (10) is that the displacement functions (4), given and compensating loads (9) are expanded into trigonometric series for functions satisfying simply supported conditions

$$u_j(x,y,t) = \sum_{m=1}^\infty \sum_{n=1}^\infty \Phi_{jmn}(t)C_{jmn}(x,y),$$

$$p_j(x,y,t) = \sum_{m=1}^\infty \sum_{n=1}^\infty p_{jmn}(t)C_{jmn}(x,y),$$

$$p_j^{comp}(x,y,t) = \sum_{m=1}^\infty \sum_{n=1}^\infty p_{jmn}^{comp}(t)C_{jmn}(x,y), j = \overline{1,3I+3},$$

where

$$C_{1mn} = \cos\frac{m\pi x}{A}\sin\frac{n\pi y}{B}, C_{2mn} = \sin\frac{m\pi x}{A}\cos\frac{n\pi y}{B}, C_{3mn} = \sin\frac{m\pi x}{A}\sin\frac{n\pi y}{B},$$

$$C_{3+i\ mn} = C_{1mn}, C_{3+I+i\ mn} = C_{2mn}, C_{3+2I+i\ mn} = C_{3mn}, i = \overline{1,I},$$

$$p_{jmn}(t) = \frac{4}{AB}\int_0^A\int_0^B p_j(t)C_{jmn}(x,y)dxdy,$$

$$p_{jmn}^{comp}(t) = \frac{4}{AB}\sum_{k=1}^{3I+3}\oint_\Gamma L_{jk}q_k^{comp}(s,t)C_{jmn}(x_\Gamma,y_\Gamma)ds, j = \overline{1,3I+3};$$

A is the length of the generatrix, B is the length of the arc of the auxiliary shell.

The functions of compensating loads are expanding into a series along the contour Γ

$$q_j^{comp}(s,t) = \sum_{\alpha=1,2}\sum_{\mu=0}^\infty q_{j\alpha\mu}(t)b_{\alpha\mu}(s), j = \overline{1,3I+3}, \quad (11)$$

where $b_{1\mu} = \sin[\mu\gamma(s)]$, $b_{2\mu} = \cos[\mu\gamma(s)]$, $\gamma(s) = 2\pi\int_0^s d\tilde{s}/\oint_\Gamma d\tilde{s}$, $0 \leq \gamma(s) \leq 2\pi$, $\mu = \overline{0,\mu^*}$.

The limit functions entering into the outgoing boundary conditions on the contour (8) also expand in a series along the contour Γ. As a result, system (10) at each time step turns into a system of linear algebraic equations with respect to expansion coefficients of compensating loads in a series along the contour Γ. The order of the resulting system depends on the number of layers in the shell and the number of members in the developed series (11). The system of equations of motion (7) turns into a system of ordinary differential equations of the second order, which is integrated by expanding the solution into Taylor series [22] taking into account condition (2). Thus, after calculating the compensating loads, the solution to the problem takes the form

$$u_j(x, y, t) = \sum_{m=1}^{\infty} \sum_{n=1}^{\infty} (\frac{4}{AB} \sum_{k=1}^{3l+3} \pi_{jk}^{mn} \sum_{l=1}^{3l+3} \sum_{\alpha=1,2} \sum_{\mu=0}^{\infty} \theta_{kl\alpha\mu}^{mn} q_{l\alpha\mu}(t)$$
$$+ \varepsilon_{jmn}(t)) C_{jmn}(x, y),$$

where π_{jk}^{mn}, $\theta_{kl\alpha\mu}^{mn}$, ε_{jmn} are elements of matrices obtained as a result of numerical transformations.

We obtain the solution of the equation of motion of the indenter (1) using the Laplace integral transform. After calculating the compensating loads, displacements (4), strains (5) and stresses (6) in the layers of the initial shell are determined.

4 Results

In order to confirm the reliability of the numerical results, we considered the deformation of a simply supported three-layer orthotropic shell under impact loading and compared the calculation results with the results found in the paper [24]. The shell contour is described by the Lame curve equations

$$\Gamma: \quad \left(\frac{x}{\alpha}\right)^c + \left(\frac{y}{\beta}\right)^c = 1,$$

where $\alpha = \beta = 0.127$ m, $c = 10$.

The shell has the following geometric parameters and mechanical properties: $R = 5$ m (shell radius), $h_i = 2.1$ mm (layer thickness), $E_1^i = 173.06$ GPa, $E_2^i = 33.1$ GPa, $E_3^i = 5.17$ GPa (elastic moduli), $G_{12}^i = 9.38$ GPa, $G_{13}^i = 8.27$ GPa, $G_{23}^i = 3.24$ GPa (shear moduli), $v_{12}^i = 0.36$, $v_{13}^i = 0.25$, $v_{23}^i = 0.17$ (Poisson's ratios), $\rho_i = 1568$ kg/m^3 (density of the material of the layers), $i = \overline{1,3}$, $\theta_1 = \theta_3 = 0°$, $\theta_2 = 90°$ (angles of reinforcing of the layers).

The indenter in the form of a steel ball is dropped from a height $H = 0.33$ m onto the outer surface of the first layer of the shell at a point with coordinates $x_0 = y_0 = 0.127$ m, $z_0 = 0$. Its characteristics are as follows: $M = 8.4$ g, $r = 6.35$ mm, $E = 2.1 \cdot 10^5$ MPa (elastic modulus), $v = 0.3$ (Poisson's ratio), $\rho = 7.85 \cdot 10^3$ kg/m^3 (material density).

The coefficient in the Hertzian law (3), as in [24], is equal $k = 0.4736 10^{-6} \frac{m}{N^{2/3}}$.

Figure 1 shows the change of the shell deflection in time on the outer surface of the third layer at the point $x_r = y_r = 0.127$ m, $z_r = \delta_3$. The figure also shows the design scheme of the shell. The solid line corresponds to the results obtained by the finite element method in [24], and the dashed line corresponds to the calculation results by the proposed method. It can be seen that these two dependences are in good agreement with each other, which confirms the reliability of the calculation results and the operability of the developed method.

Also, we consider the vibrations of a three-layer clamped shell with a complex plan shape, which is shown in Fig. 2. The shell has the following geometric parameters: $R = 2$ m, $h_i = 1.27$ mm $(i = \overline{1,3})$, $s_1 = 0.48$ m, $s_2 = 0.31$ m, $s_3 = 0.36$ m, $s_4 = 0.24$ m; $R_1 = 0.12$ m, $R_2 = 0.06$ m, $R_3 = 0.2$ m, $R_4 = 0.07$ m. The layers are made of T300/5208 epoxy carbon fiber [21] with the following characteristics: $E_1^i = 132.5$ GPa, $E_2^i = 10.8$ GPa, $E_3^i = 10.8$ GPa, $G_{12}^i = G_{13}^i = 5.7$ GPa, $G_{13}^i = 8.27$ GPa, $G_{23}^i = 3.4$ GPa, $v_{12}^i = v_{13}^i = 0.24$, $v_{23}^i = 0.49$, $\theta_1 = \theta_3 = 0^o$, $\theta_2 = 90^o$, $\rho_i = 1500$ kg/m³. The geometric parameters of the indenter are $M = 14.17$ g and $r = 4.8$ mm.

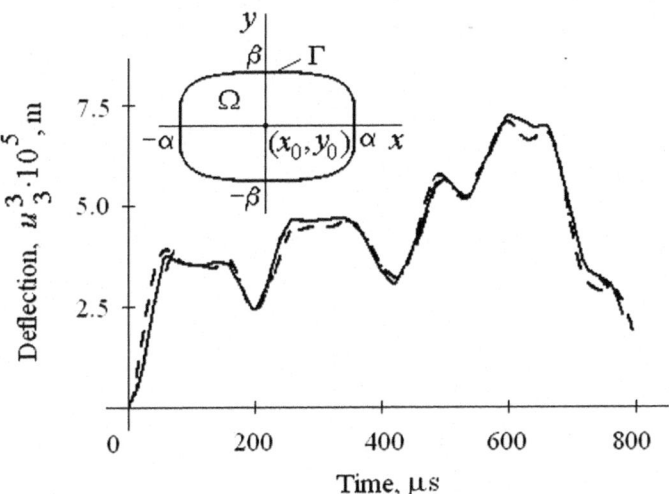

Fig. 1. Change of the deflection of the shell over time.

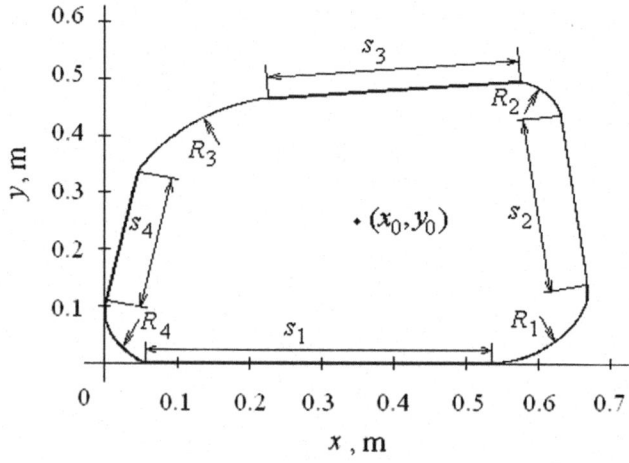

Fig. 2. Shape of the shell plan.

The impact is applied at a speed of 22.6 m/s on the outer surface of the first layer to a point with coordinates $x_0 = 0.35$ m, $y_0 = 0.25$ m, $z_0 = 0$. The coefficient in the law (3) is equal $k = 10^{-8} \frac{m}{N^{2/3}}$.

The stresses in the shell layers are calculated. In Fig. 3, the solid line shows the time dependence of the stress σ_x^3, and the dashed line shows the stress σ_y^3 on the outer surface of the third layer at the point with coordinates $x_r = 0.35$ m, $y_r = 0.25$ m, $z_r = \delta_3$.

Despite the high level of load intensity, the stresses do not exceed their permissible values.

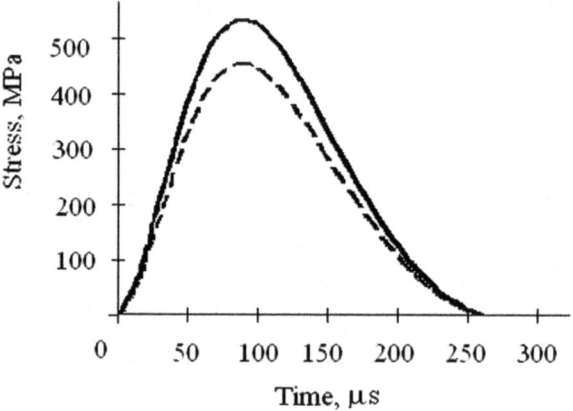

Fig. 3. The change of stresses σ_x^3 and σ_y^3 over time.

5 Conclusions

A method for studying unsteady vibrations of orthotropic laminated open cylindrical shells with a complex plan shape under impact loading has been developed. The method allows us to obtain the problem solution in the form of expansions in trigonometric series.

The capabilities of the method are illustrated by calculating the deflections and stresses of orthotropic shells of various shapes in plan and with different boundary conditions upon impact by a ball indenter. Good agreement of the calculation results with the data obtained by the finite element method, confirmed the reliability of the results obtained using the proposed method.

The method can be used to calculate laminated shells of various geometries in a wide range of changes in the mechanical properties of the layers. The results can be used in the design process of shell elements of aircraft structures under the action of intense fast loads.

References

1. Kassapoglou, C.: Design and analysis of composite structures with applications to aerospace structures. Wiley, Chichester (2013)
2. Sarasini, F., Tirillò, J., Ferrante, L., et al.: Drop-weight impact behaviour of woven hybrid basalt–carbon/epoxy composites. Compos. Part B Eng. **59**, 204–220 (2014). https://doi.org/10.1016/j.compositesb.2013.12.006
3. Shah, S.Z.H., Karuppanan, S., Megat-Yusoff, P.S.M., Sajid, Z.: Impact resistance and damage tolerance of fiber reinforced composites: a review. Compos. Struct. **217**, 100–121 (2019). https://doi.org/10.1016/j.compstruct.2019.03.021
4. Danek, W., Katunin, A., Wronkowicz-Katunin, A.: Analysis of selected parameters in numerical modeling of low-velocity impact damage in composite structures. Procedia Struct. Integr. **25**, 19–26 (2020). https://doi.org/10.1016/j.prostr.2020.04.005
5. Kreculj, D., Rašuo, B.: Review of impact damages modelling in laminated composite aircraft structures. Tehnički vjesnik **20**(3), 485–495 (2013)
6. Jalón, E., Hoang, T., Rubio-López, A., Santiuste, C.: Analysis of low-velocity impact on flax/PLA composites using a strain rate sensitive model. Compos. Struct. **202**, 511–517 (2018). https://doi.org/10.1016/j.compstruct.2018.02.080
7. Gholizadeh, S.: A review of impact behaviour in composite materials. Int. J. Mech. Prod. Eng. **7**(3), 35–46 (2019)
8. Kazancı, Z.: A review on the response of blast loaded laminated composite plates. Prog. Aerosp. Sci. **81**, 49–59 (2016). https://doi.org/10.1016/j.paerosci.2015.12.004
9. Ansari, M.M., Chakrabarti, A.: Impact behavior of FRP composite plate under low to hyper velocity impact. Compos. Part B Eng. **95**, 462–474 (2016). https://doi.org/10.1016/j.compositesb.2016.04.021
10. Şahan, M.F.: Viscoelastic damped response of cross-ply laminated shallow spherical shells subjected to various impulsive loads. Mech. Time-Depend. Mater. **21**(4), 499–518 (2017). https://doi.org/10.1007/s11043-017-9339-y
11. Chai, G.B., Manikandan, P.: Low velocity impact response of fibre-metal laminates – a review. Compos. Struct. **107**, 363–381 (2014). https://doi.org/10.1016/j.compstruct.2013.08.003

12. Ahmad, F., Abbassi, F., Park, M.K., et al.: Finite element analysis for the evaluation of the low-velocity impact response of a composite plate. Adv. Compos. Mater. **28**(3), 271–285 (2019). https://doi.org/10.1080/09243046.2018.1510589

13. Zhi, J., Tay, T.E.: Explicit modeling of matrix cracking and delamination in laminated composites with discontinuous solid-shell elements. Comput. Methods Appl. Mech. Eng. **351**, 60–84 (2019). https://doi.org/10.1016/j.cma.2019.03.041

14. Chaker, A., Koubaa, S., Mars, J., et al.: An ABAQUS implementation of a solid-shell element: application to low velocity impact. In: Aifaoui, N., et al. (eds.) Design and Modeling of Mechanical Systems – IV. CMSM 2019. LNME, pp. 770–777. Springer, Cham (2020). https://doi.org/10.1007/978-3-030-27146-6_84

15. Rodichev, Y.M., Smetankina, N.V., Shupikov, O.M., Ugrimov, S.V.: Stress-Strain assessment for laminated aircraft cockpit windows at static and dynamic loads. Strength Mater. **50**(6), 868–873 (2019). https://doi.org/10.1007/s11223-019-00033-4

16. Plankovskyy, S., Myntiuk, V., Tsegelnyk, Y., et al.: Analytical methods for determining the static and dynamic behavior of thin-walled structures during machining. In: Shkarlet, S., et al. (eds.) Mathematical Modeling and Simulation of Systems (MODS'2020). AISC, vol. 1265, pp. 82–91. Springer, Cham (2021). https://doi.org/10.1007/978-3-030-58124-4_8

17. Sklepus, S.N.: Numerical-Analytical Method of Studying Creep and Sustained Strength Characteristics of a Multilayer Shell. Strength Mater. **49**(2), 313–319 (2017). https://doi.org/10.1007/s11223-017-9871-7

18. Baştürk, S., Uyanık, H., Kazancı, Z.: An analytical model for predicting the deflection of laminated basalt composite plates under dynamic loads. Compos. Struct. **116**, 273–285 (2014). https://doi.org/10.1016/j.compstruct.2014.05.018

19. Her, S.C., Liao, C.C.: Impact analysis of composite laminate shell structures. Appl. Mech. Mater. **764**, 1185–1188 (2015). https://doi.org/10.4028/www.scientific.net/AMM.764-765.1185

20. Smetankina, N.V., Shupikov, A.N., Sotrikhin, S.Y., Yareschenko, V.G.: Dynamic response of an elliptic plate to impact loading: theory and experiment. Int. J. Impact Eng. **34**(2), 264–276 (2007). https://doi.org/10.1016/j.ijimpeng.2005.07.016

21. Reddy, J.N.: Mechanics of Laminated Composite Plates: Theory and Analysis. CRC Press, Boca Raton (1997)

22. Shupikov, A.N., Smetankina, N.V.: Non-stationary vibration of multilayer plates of an uncanonical form. The elastic immersion method. Int. J. Solids Struct. **38**(14), 2271–2290 (2001). https://doi.org/10.1016/S0020-7683(00)00166-9

23. Smetankina, N.V., Shupikov, A.N., Sotrikhin, S.Y., Yareshchenko, V.G.: A noncanonically shape laminated plate subjected to impact loading: theory and experiment. J. Appl. Mech. **75**(5), 051004 (2008). https://doi.org/10.1115/1.2936925

24. Nosier, A., Kapania, R.K., Reddy, J.N.: Low-velocity impact of laminated composites using a layerwise theory. Comput. Mech. **13**(5), 360–379 (1994). https://doi.org/10.1007/BF00512589

Improving the Noise Immunity
of the Measuring and Computing
Coherent-Optical Vibrodiagnostic Complex

Mykhaylo Tkach[1(✉)] , Yuri Zolotoy[1(✉)] , Yurii Halynkin[1(✉)] ,
Arkadii Proskurin[1] , Irina Zhuk[2] , Volodymyr Kluchnyk[1] ,
and Igor Bobylev[1]

[1] Admiral Makarov National University of Shipbuilding,
9 Heroes of Ukraine Avenue, Mykolayiv 54025, Ukraine
mykhaylo.tkach@gmail.com,
vladimir.kluchnyk@gmail.com,
igor.bobylev.nuk@gmail.com, goldspekl@ukr.net,
{yurii.galynkin, arkadii.proskurn}@nuos.edu.ua
[2] Petro Mohyla Black Sea National University,
10 68-Desantnikov Street, Mykolaiv 54003, Ukraine
iryna.zhuk@chmnu.edu.ua

Abstract. The paper shows the possibilities of using the measuring and computing complex of electronic speckle pattern interferometry (ESPI) to determine the vibrational dynamic parameters of structural elements. The characteristics of the methods for obtaining patterns of bands by the method of difference speckle correlation and the proposed method for determining the contrast of the dynamic speckle pattern are compared. A comparative analysis shows their equivalence in the quality of interferograms and the information content of the obtained data. At the same time, varying the volume of the processed specklegram buffer makes it possible to reduce the influence of the instability of the optical parameters of the interferometer and the electrical control signals of the experiment on the test results. A slight increase in the inertia of the response to a change in the oscillation regime results in a significant decrease in the sensitivity of the installation to external disturbances, which, in combination with the noise-resistant optical circuit of the interferometer, allows the complex to be operated in off-stand conditions.

Keywords: Interferometry · Speckle · Correlation fringes · Vibrational analysis · Natural frequencies · Free boundary conditions · Turbine blade vibration · Shell vibration

1 Introduction

The experimental determination of vibration parameters is a prerequisite for the effective solution of the problems of dynamic structural strength at the design and development stage of machines, using a computational-experimental research method for vibration-resonant non-destructive testing of critical elements [1].

M. Nechyporuk et al. (Eds.): ICTM 2020, LNNS 188, pp. 277–289, 2021.
https://doi.org/10.1007/978-3-030-66717-7_23

In experimental mechanics, a special place belongs to non-contact coherent-optical methods for studying vibration parameters. The most informative in the study of the dynamic parameters of machine elements is the method of holographic interferometry [2]. But the technique for obtaining holographic interferograms is quite complicated, and involves research in laboratory conditions with effective vibration isolation of the interferometric stand.

Sufficient for vibrometry the completeness of information about the oscillatory process with significantly less labor and less stringent requirements for the experimental technique has the method of electron speckle interferometry (ESPI) [3–6].

The sensitivity of a digital speckle interferometer (DSI) to mechanical and temperature disturbances can be significantly lower than holographic, and do not require special vibration protection measures. Moreover, the noise immunity of the stand and the level of requirements for the technical parameters of the video system used depend both on the optical design of the interferometer and on the method for obtaining speckle interferograms. The development of such methods, that contributes to increasing the stability of the stand to external influences and reducing the requirements for the hardware of the experiment is relevant and in demand.

When determining waveforms using the ESPI method, digital photographic recording of two or more patterns of interference of the reference and subject waves is assumed for different phase relationships between them. The optical setup schemes provide the formation of the subject and reference beams with their subsequent combination in the plane of the video matrix and obtaining an adjustable phase difference of the interfering waves (Fig. 1) [7].

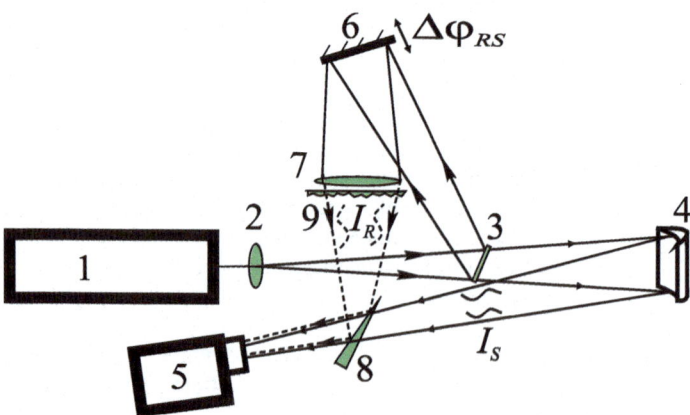

Fig. 1. Optical design of speckle interferometer. 1 – laser; 2 – micro lens; 3 – flat beam splitter; 4 – test object; 5 – video camera; 6 – phase-shifting mirror; 7 – lens; 8 – optical wedge; 9 – diffuser.

In the work [8] laser beam splitting and combining of interfering waves occurs on a single beam splitter.

During research, camcorder 5 operates in the webcam mode and in real time transfers a video stream to the computer with a brightness division into 256 gradations. Speckle sizes are controlled by the iris of the camera lens. Brightness B_j of each point of the reproduced image is proportional to the intensity I_i i-th speckle of the resulting speckle field corresponding to the j-th pixel of the matrix sensor.

$$B_j = KI_i, i = 1, 2 \ldots N, j = 1, 2 \ldots m, \tag{1}$$

where K is the proportionality coefficient determined by the parameters of the video system, N is the number of speckles of the resulting speckle field, m is the number of pixels displaying them.

2 Analysis of Speckle Correlation Methods for Vibration Studies

The resulting speckle structure of the vibrating object in explicit form does not contain the image of the vibrational mode. Indeed, the image of the specklegram of a resting object has the form [3]

$$B_j = K(I_{R_i} + I_{S_i} + 2\sqrt{I_{R_i}I_{S_i}} \cos(\varphi_{R_i} - \varphi_{S_i})), \tag{2}$$

where I_{R_i} and I_{S_i} are local intensities, and $(\varphi_{R_i} - \varphi_{S_i}) = (\varphi_{RS_i})$ is the initial phase difference of the reference and subject light waves in a given speckle.

The brightness of the image B_{j_τ} of the same speckles, but fluctuating during the vibration of the object and averaged by an inert video sensor over the time τ of recording the frame, is determined by the relation [7]

$$B_{j_\tau} = K(I_{R_i} + I_{S_i} + 2\sqrt{I_{R_i}I_{S_i}} J_0\left(\frac{4\pi}{\lambda}U_{0_i}\right)\cos(\varphi_{RS_i})), \tag{3}$$

where U_{0_i} is the amplitude of the i-th point oscillations occurring in the direction of illumination and observation of the object, J_0 is the Bessel function of the first kind of zero order.

It follows that the regular interferogram of the vibrational mode is encoded in the specklegram by a random distribution of the intensities I_{R_i}, I_{S_i} and the initial phases φ_{R_i} and φ_{S_i} of the interfering waves. In experimental practice, such interferograms are predominantly obtained by the method of correlation comparison of speckle fields for two structural states, which in ESPI is realized by a simple procedure of element-by-element subtraction of their brightness [3]. The non-informative component of the $I_{R_i} + I_{S_i}$ signal is thereby eliminated, and the given difference module in the form of a strip pattern displays the distribution of amplitudes and the shape of the part.

If, when scanning the resonance spectrum, a video stream of time-averaged specklegrams (3) is entered into a computer and the measure of their correlation with a previously registered reference frame (2) of a stationary object is determined in real time, then we obtain an interferogram whose brightness B_{I_j} corresponds to the expression [9]

$$B_{I_j} = 2K \left| \sqrt{I_{R_i} I_{S_i}} \cos\varphi_{RS_i} \right| \left(1 - J_0 \left(\frac{4\pi}{\lambda} U_{0_i} \right) \right). \tag{4}$$

The resulting picture of the bands with average brightness normalized to unity (Fig. 2a, curve 1) is similar to the holographic interferogram of a vibrating object observed in real time [2]. Unlike the usual interferogram, it has an inverted brightness, the darkest stripes correspond to the nodal lines, and the price of the strip can approximately be considered equal $\lambda/2$. Moreover, relation (4) is valid if the phase difference φ_{RS_i} in each speckle of subtracted images (2) and (3) is constant.

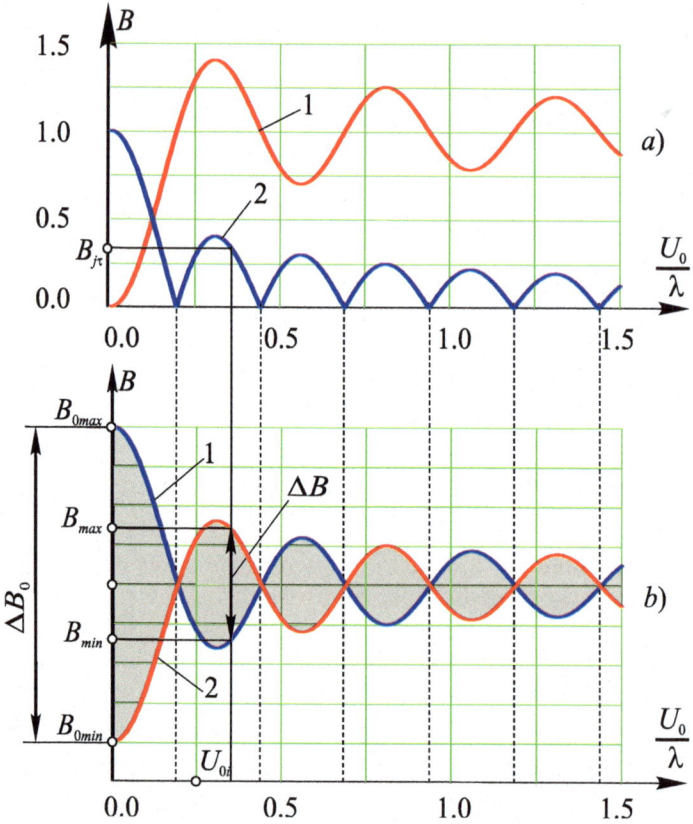

Fig. 2. Matching brightness changes: (a) brightness distribution of speckle correlation bands; (b) a diagram of the dependence of the range of variation of the speckle brightness on the amplitude of oscillations on time-averaged interferograms.

But with continuous scanning of the resonance spectrum, the time interval between recordings of the reference (2) and current (3) images increases, and the cumulative effect of the influence of thermal and mechanical disturbances leads to an uncontrolled change in the value of φ_{RS_i}. In this case, the shape of the bands is distorted and the noise immunity of the interferometer is affected, since it is necessary to interrupt the spectrum scan periodically to update the reference image and take additional measures to heat the stand. The method is advisable to use in the case of studies of oscillations with large amplitudes and the need to reduce the sensitivity of the interferometer. The listed problems are practically absent in the variant of correlation comparison of time-averaged specklegrams (3) of a vibrating object after a preliminary phase shift of the reference wave on one of them [7]. To do this, the mirror 6 makes a reciprocating motion in the normal direction with a frequency of 2...4 Hz and amplitude that provides a change in the phase difference of the reference and subject beams $\Delta\varphi_{RS} = \pi$ rad. Then the module of the difference in images corresponding to the extreme values of the phase shift gives a contrast system of bands on the computer monitor according to [9].

$$B_{I_j} = 4K\sqrt{I_{R_i}I_{S_i}}\left|J_0\left(\frac{4\pi}{\lambda}U_{0_i}\right)cos(\varphi_{RS_i})\right|. \tag{5}$$

In this case, the reference frame is updated automatically every half-period of a change in $\Delta\varphi_{RS}$, which increases the noise immunity of the interferometer and does not interrupt the research process.

According to (5), the dependence of the brightness b_{I_j} of the image points on the oscillation amplitude, up to a constant factor, is determined by the expression.

$$b_{I_j} = \left|J_0\left(\frac{4\pi}{\lambda}U_{0_i}\right)\right|. \tag{6}$$

The contrast of such a pattern of bands (Fig. 2a, curve 2) tends to unity, their period is half that of the interferogram (4), and the sensitivity is twice as high.

But the high quality of such speckle interferograms is ensured with the phase shift stability $\Delta\varphi_{RS}$ and stable synchronization of its amplitude values with the moments of image registration. Otherwise, random oscillations of the position and visibility of speckle correlation bands arise. Therefore, when implementing the method, it is advisable to use a specialized video camera with an external synchronization function.

3   The Principle and Features of Speckle Decoding by Speckle Contrast Determination

The contrast V of the speckle structure characterizes the depth of spatial modulation of a homogeneous scattered field as the relative mean-square dispersion of speckle intensity [10], or, by virtue of equality (1), as the relative mean-square dispersion of the brightness of the pixels of their digital image.

$$V = {}^{\sigma_B}/_{\langle B \rangle}'$$

(7)

where $\langle B \rangle$ is the average brightness value of the digital image of the speckle structure; σ_B is its standard deviation.

The contrast of the inhomogeneous field N of fluctuating speckles displayed by m pixels (3) of the inertial photodetector varies from point to point, and its variations also contain information on the distribution of the vibration amplitudes of the object's points [11]. Taking into account the unbiased estimation of variance, the local contrast of a small image area of $n \times n$ pixels in the vicinity of this point is determined by the expression [12]

$$V = \sigma_B/\langle B \rangle = \sqrt{\sum\nolimits_{j=1}^{n^2} (B_j - \langle B \rangle)^2 / n^2 - 1} / \frac{1}{n^2} \sum\nolimits_{j=1}^{n^2} B_j.$$

(8)

Dependencies (3) and (8) allow the lines of equal contrast of the resulting dynamic speckle pattern to be considered as lines of equal amplitudes of oscillation of the points of the object. Such an interpretation of isocontrast lines has long been used as an auxiliary method for express analysis of the resonance spectrum of products [2]. But the depth of modulation of the contrast of the specklegram is small, and no more than two consecutive bands are detected by visual observation. The result of pointwise calculation of the local contrast of the specklegram according to algorithm (8) also has an unacceptably high level of random errors associated with the complexity of choosing the size of the calculation area, the non-optimal ratio and macroinhomogeneities of the interfering beams, and depolarization of the object wave.

The influence of these factors is eliminated if, for the synthesis of the interferogram, we use not the integral value of the local contrast of the specklegram, but only that dynamic part of it V_d, which is determined solely by the ripple of the speckle intensity. In expression (3), the brightness of the specklegram points depends on the amplitude of vibration only in the third term. Therefore, imagine the brightness of the image B_{j_τ} of the i-th speckle by the sum of the stationary B_{st} and dynamic B_d components

$$B_{j_\tau} = B_{st} + B_d = B_{st} + 2K\sqrt{I_{R_i}I_{S_i}}\left(1 + J_0\left(\frac{4\pi}{\lambda}U_{0_i}\right)\cos\left(\varphi_{RS_i}\right)\right).$$

(9)

Then the dynamic contrast of the V_d image of a given single speckle is estimated by the contrast of a hypothetical set of the same speckles with brightness $B_d = 2K\sqrt{I_{R_i}I_{S_i}}(1 + J_0(\frac{4\pi}{\lambda}U_{0_i})\cos(\varphi_{RS_i}))$, which differs only in randomly distributed φ_{RS_i} values. Since the possible values of φ_{RS_i} are equally probable, the average brightness $\langle BD \rangle$ of this speckle field will be equal to half the range of variation of possible values of B_{j_τ} at rest (at $U_{0_i} = 0$), and its standard deviation σ_B at a given vibration amplitude is directly proportional to the range of variation of possible values of B_{j_τ}. The range of variation of the brightness function (3) of the digital speckle image, determined only by their fluctuation, is shown in Fig. 2b by the diagram bounded by the two extreme positions 1 and 2 of the curve $B_{j_\tau}(U_0)$ at extreme values $cos\varphi_{RS_i} = \pm1$. Then

$$\langle B_d \rangle = \frac{\Delta B_0}{2} = 2K\sqrt{I_{R_i}I_{S_i}}; \sigma_B \sim \Delta B = 2K\sqrt{I_{R_i}I_{S_i}}\left|J_0\left(\frac{4\pi}{\lambda}U_{0_i}\right)\right|,$$

$$V_d \sim \frac{\Delta B}{\Delta B_0} = \left|J_0\left(\frac{4\pi}{\lambda}U_{0_i}\right)\right|,$$

where $\Delta B = (B_{max} - B_{min})$ is the range of brightness of the vibrating point, $\Delta B_0 = (B_{max_0} - B_{min_0})$ is the brightness variation range of the same point at rest (Fig. 2b). As a result, the interferogram of the vibrational mode is formed according to the expression

$$B_{I_j} = \frac{\beta\Delta B}{\Delta B_0} = \beta\left|J_0\left(\frac{4\pi}{\lambda}U_{0_i}\right)\right|, \tag{10}$$

where β is the luminance factor selected by visual observation.

From relation (10) and comparison of curve 2 (Fig. 2a) and the brightness diagram (Fig. 2b), it follows that the obtained pattern of the bands is similar to the interferogram (6). The oscillation amplitudes of the dark line points are determined by the equality $\frac{4\pi}{\lambda}(U_0)_k = n_k$, where $k = 1, 2 \ldots$ is the zero number of the Bessel function, i.e. dark band number (starting from the node); $(U_0)_k$ is the amplitude of the oscillations of the points of this band, n_k is the value of the argument of the Bessel function at this zero. Approximately the amplitude at these points can be estimated from the relation $(U_0)_k = k\lambda/4$, where $k = 0, 1, 2 \ldots$ is the number of the bright band [3], however, with the amplitudes $U_0 > (2 \div 3)\lambda$ the determination of the orders of the bands becomes problematic due to a significant decrease in their brightness.

For the experimental implementation of this algorithm, oscillations of the mirror 6 with a frequency of 2–4 Hz change the phase difference between the interfering beams. Moreover, a confident but single inversion of speckle brightness (a change in $\Delta\varphi_{RS} \approx 2\pi$ rad) [13] should occur on the amplitude of the oscillations.

But applying this algorithm to pixels with a very low or zero speckle intensity (to speckle field dislocations), where the denominator of expression (10) tends to zero, and the level of electronic noise prevails in the photosensor signal, leads to a high noise level of the synthesized interferogram. Therefore, preliminary computer filtering of the pixels used in the calculation is required and the brightness values at the missing points are determined by two-dimensional approximation methods, which is problematic when scanning the resonance spectrum in real time.

In addition, expression (10) does not contain information about the relief of the brightness of the object; therefore, difficulties arise in linking the pattern of strips to the surface under study.

The positioning of the interferogram relative to the image of the object is manifested if in relation (10) for all pixels we take $\Delta B_0 = 256$ (maximum brightness value). Then the reconstruction formula for the interferogram takes the form

$$B_{I_j} = \beta \Delta B / 256 = \beta 4K \sqrt{I_{R_i} I_{S_i}} \left| J_0 \left(\frac{4\pi}{\lambda} U_{0_i} \right) \right| / 256. \tag{11}$$

Thus, the system of interference bands $\left| J_0 \left(\frac{4\pi}{\lambda} U_{0_i} \right) \right|$ is modulated by the factor $\sqrt{I_{S_i}}$ at a level sufficient to determine the position of the vibrational forms relative to the surface under study.

Note that the brightness zeros in distributions (10) and (11) exactly coincide, but only the brightness of the bright bands differ, therefore these interferograms can be considered equivalent. The advantages of the method for producing interferograms according to algorithm (11) are that it:

- does not require periodic registration of specklegrams of a resting object;
- satisfactorily solves the issues of visualization of the investigated surface and binding interference fringes to it;
- free from problems in speckle field dislocations and related computational problems.

The use of algorithm (10) is justified only if it is necessary to digitize interferograms with photometric fractional parts of the bands.

Note some features of the bench control algorithm. The phase shift modulation frequency of 2–4 Hz suggests an optimal buffer size of 6–8 frames, and its update can be performed in steps or sequentially. When the buffer is updated in stages after processing the set of frames and reproducing the picture of the bands, the buffer is completely replaced. The procedure is cyclically repeated, updating the image of the waveform 3–4 times per second. In a sequential update, the arrival of the next frame in the buffer is accompanied by discarding the last one. In this case, the change in the pattern of the vibrational form occurs with the frame rate of the video camera, but requires increased computing resources of the computer.

It is important that in the correlation ESPI, interferograms are obtained by comparing only two specklegrams, and in the contrast determination method, by calculating relation (11) according to computer processing of all the images contained in the buffer. An increase in buffer size reduces the effect of multiplicative perturbations as a result of their averaging over a large number of frames. Therefore, the sensitivity of installations to external influences is significantly lower than in the correlation ESI mode, and in combination with noise-resistant optical circuits [8] allows operation without the use of an interferometric table. But the inertia of the response of the interferogram to a change in the amplitude of the oscillations also increases.

In addition, the use of the speckle contrast determination method eliminates the requirements of strict observance of the phase shift and its strict synchronization with the moments of frame recording, which allows using a simple video camera to record specklegrams. All interferograms presented in this work were obtained using a household camera (Panasonic NV-GS47) with resolution 720×580, not having an external synchronization function.

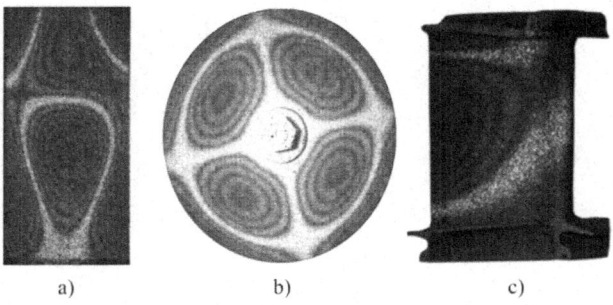

a)   b)   c)

Fig. 3. Time-averaged speckle interferograms waveforms obtained by the determination method speckle dynamic picture contrast: (a) – cantilevered rectangular plate; (b) – a cutting disc fixed to the shaft; (c) – a turbine blade cantilevered in a lock.

The application of the proposed method does not require the creation of a special optical scheme. If the software is available without any reconfiguration, it can be used in parallel with any differential DSI, which provides for the implementation of a periodic controlled phase shift of interfering waves.

For example, Fig. 3 shows the time-averaged speckle interferograms. We note some features of the application of the method.

4   The Study of Oscillations Under Free Boundary Conditions of the Object

The low sensitivity of the proposed method to low-frequency mechanical and convective perturbations in combination with a noise-resistant optical circuit allows us to study the modal spectrum of the product without mandatory rigid mounting on the stand (if this is not suggested by the test conditions) up to the boundary conditions close to free. Such conditions, as a rule, are used in determining the properties of materials by resonance methods [14], when the elastic constants are calculated from the parameters of the resonance spectrum of the simplest objects such as rods or plates.

A similar situation arises in vibroresonant flaw detection [1], which requires repeated accurate reproduction of the conditions for vibration excitation of parts. This requirement is met as correctly as possible under free boundary conditions for specimen fixation. Several interferograms of resonant vibrations obtained by contrast determination under these difficult boundary conditions for interferometry are illustrated in Fig. 4. The vibration is excited by a miniature bimorph piezoelectric element.

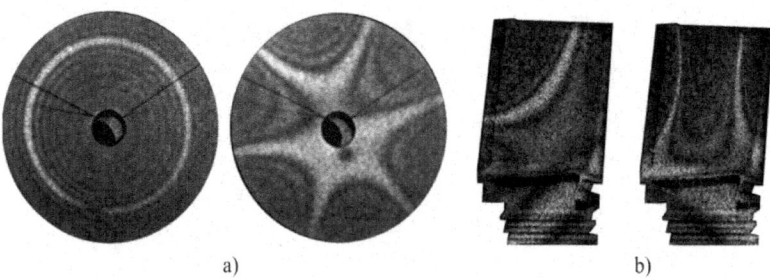

a)   b)

Fig. 4. Interferograms of vibrational forms under free boundary conditions: (a) – a free disk suspended on thin threads; (b) – turbine blades on soft silicone racks.

5   Refined Positioning of the Interferogram

When solving the problems of vibration resistance of products with a complex profile, the presence of stress concentrators in the form of holes and cracks, it is important to accurately determine the position of the interferogram relative to small surface elements as much as possible. In the method for determining contrast, the problem is solved quite simply by modulating the image of the surface under study by interference fringes. For this purpose, before starting the research, the basic image of the object in white light is pre-registered with the brightness of the display pixels B_{b_j}, $j = 1, 2 \ldots m$. Moreover, in the process of testing it is important not to disturb the relative position of the structure 4 and the video camera 5 (Fig. 1). Computer processing of the speckle-gram buffer set is carried out in real time, as in the implementation of algorithm (12). But now the interference image modulates the basic image of the object according to the expression

$$B_{I_j} = \beta B_{b_j} \Delta B / 256, \ j = 1, 2, \ldots m. \tag{12}$$

The clarity of the image of the surface details and the positioning of the strips is significantly improved. As an example, Fig. 5 a,b shows speckle interferograms of the vibrational modes of a composite shell with a perforated surface obtained by direct reproduction of the contrast distribution (11). In Fig. 5 d,e – interferograms of the same vibration modes of a similar design, synthesized by modulating the basic image by the interference signal (Fig. 5c) according to algorithm (12). The advantages of the second option in more accurate positioning of the waveforms relative to the shell are obvious.

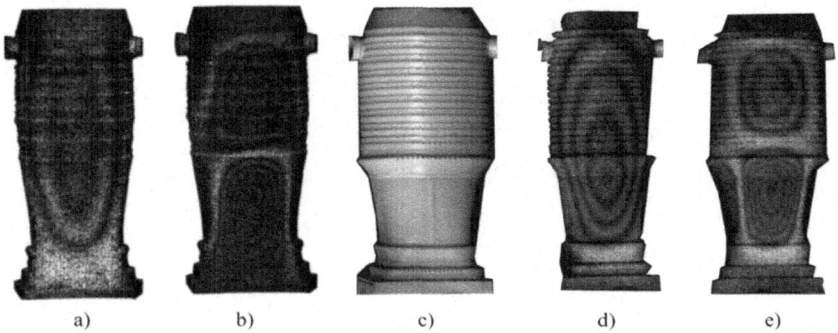

a)   b)   c)   d)   e)

Fig. 5. Improving the visualization of vibration modes of the shell structure by modulating the base image: (a), (b) – interferograms of vibration modes synthesized according to the interference signal (11); (c) – the basic image in white light; (d), (e) – is the basic image modulated by the interference signal (12).

6 Realization of Stroboscopic Speckle Interferometry

The study of the problems of vibration strength of machine elements, in addition to elucidating the dislocations of nodal lines, often involves determining the amplitudes of points in the vibrational mode to evaluate strains and stresses [14]. But the quantitative interpretation of speckle interferograms is difficult due to a sharp decrease in the brightness and visibility of the bands of time-averaged interferograms with an increase in the amplitude of oscillations. The problem of reducing the brightness of the bands is removed by applying the method of stroboscopic ESPI [2, 15]. The laser radiation is modulated in amplitude by rectangular pulses of duration Δt, synchronized with the amplitude moments of the object's vibrations. At $T \gg \Delta t$ during time τ, the object is repeatedly recorded on the specklegram in turn in 2 amplitude states. The contrast V_d of the virtual speckle system of the averaged specklegram will be high only at those points on the surface for which speckles with maximum and minimum intensities will retain their brightness in all amplitude states. The amplitude of oscillations at these points must meet the condition

$$U_0 = k\lambda/4, k = 0, 1, 2\ldots \tag{13}$$

If in neighboring amplitude positions the speckle intensity has opposite extremes, then it will be averaged on the specklegram, and the contrast will become minimal. These considerations are ideally valid for very high gating rates. But even with a duty cycle of 7–10 units, it is possible to study oscillations whose amplitudes amount to tens of wavelengths [16].

To reduce speckle noise, it is advisable to use the method proposed in [6] for multi-frame averaging of images of an identical pattern of bands with uncorrelated noise.

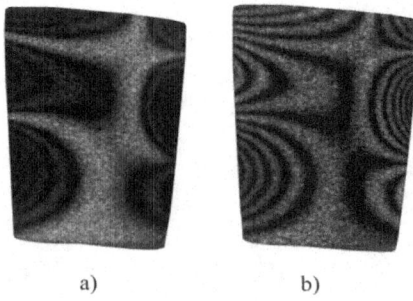

a)   b)

Fig. 6. Interferograms of the third torsional vibration mode of the compressor blade: (a) – averaged over time; (b) – with gated lighting.

In Fig. 6 the images of the third torsional vibration mode of the compressor blade are shown, obtained by the method for determining the speckle contrast: (a) – time-averaged interferogram; (b) – stroboscopic speckle interferogram with a duty cycle of gating 7 units. An increase in the resolution and visibility of the strip pattern under gated illumination greatly facilitates the estimation of the amplitude relief of the vibrational mode. It is important that the test procedure, in addition to connecting the elements of the strobe complex, remains similar to the time averaging method.

7   Conclusions

The method for determining the contrast of the dynamic speckle pattern is an effective method for reconstructing the images of vibrational forms in speckle interferometry and a full-fledged alternative to difference correlation vibrometry. At a commensurate cost of computing resources, the interferograms obtained by this method are equivalent in quality to the difference speckle interferograms, both time-averaged and gated. But the elimination of the strictly regulated magnitude of the adjustable phase shift and the need for its synchronization with the recording of specklegrams, contribute to the stability of the installation, eliminate the need for frequent monitoring of its operation modes, which is important when automated scanning of the resonance spectrum of products or serial control of samples. Moreover, for video recording, you can use a simple video camera, without the function of external synchronization.

Improved positioning of the bands and visualization on the interferogram of small relief elements of the investigated surface allow the presence of stress concentrators. Therefore, in combination with the ability to study objects under various boundary conditions, the method is effective in resonant flaw detection, and high-resolution interferograms make it possible to estimate the distribution of vibration displacements and deformations in vibrational form.

If the reconstruction of speckle interferograms is carried out with an increased buffer volume of the processed specklegrams, the sensitivity of the installation to mechanical and thermal convective disturbances is reduced, and when implementing a noise-resistant optical circuit, it allows operation without special measures of isolation from interference.

References

1. Tsareva, A.M., Makaeva, R.K., Safina, D.M., Galimova, R.K.: Diagnostics of fracture of the blower impeller of gas turbine engine core by using the holographic interferometry. Russ. Aeronaut. **63**(2), 362–365 (2020). https://doi.org/10.3103/S1068799820020257
2. Vest, C.M.: Holographic Interferometry. Wiley, New York (1982)
3. Jones, R., Wykes, C.: Holographic and Speckle Interferometry, 2nd edn. Cambridge University Press, Cambridge (1989). https://doi.org/10.1017/CBO9780511622465
4. Qin, J., Gao, Z., Wang, X., Yang, S.: Three-dimensional continuous displacement measurement with temporal speckle pattern interferometry. Sensors **16**(12), 2020 (2016). https://doi.org/10.3390/s16122020
5. Volkov, I.V.: Use of the speckle holography technique in experimental mechanics. Meas. Tech. **60**(2), 161–165 (2017). https://doi.org/10.1007/s11018-017-1167-6
6. Halama, R., Horňáček, L., Pečenka, L., et al.: 3-D ESPI measurements applied to selected engineering problems. Appl. Mech. Mater. **827**, 65–68 (2016). https://doi.org/10.4028/www.scientific.net/AMM.827.65
7. Bukshtab, M.: Spectroscopic interferometry and laser-excitation spectroscopy. In: Photometry, Radiometry, and Measurements of Optical Losses. SSOS, vol. 209, pp. 655–717. Springer, Singapore (2019). https://doi.org/10.1007/978-981-10-7745-6_12
8. Bystrov, N.D., Zhuzhukin, A.I.: Speckle-interferometry in the investigation of large-size gas turbine engine structures vibration. Procedia Eng. **176**, 471–475 (2017). https://doi.org/10.1016/j.proeng.2017.02.346
9. Blevins, R.D.: Formulas for Dynamics, Acoustics and Vibration. Wiley, Chichester (2016)
10. Goodman, J.W.: Statistical Optics, 2nd edn. Wiley, Somerset (2015)
11. Rossing, T.D., Chiaverina, C.J.: Holography. In: Light Science, pp. 279–303. Springer, Cham (2019). https://doi.org/10.1007/978-3-030-27103-9_11
12. Gusev, M.E., Alekseenko, I.V.: A study of multicomponent mechanical oscillations by the method of digital holographic vibrometry. Radiophys. Quantum Electron. **57**(8–9), 543–550 (2015). https://doi.org/10.1007/s11141-015-9537-x
13. Tkach, M.R., Zolotiy, Y.G., Dovgan, D.V., Guk, I.Y.: Determination of the natural vibration forms of gas turbine engine elements in real time by electron speckle interferometry. Aerosp. Tech. Technol. **8**, 203–207 (2012). [in Russian]
14. Mrozek, P., Mrozek, E., Werner, A.: Electronic speckle pattern interferometry for vibrational analysis of cutting tools. Acta Mechanica et Automatica **12**(2), 135–140 (2018). https://doi.org/10.2478/ama-2018-0021
15. Demtröder, W.: New techniques in optics. In: Electrodynamics and Optics. ULNP, pp. 353–387. Springer, Cham (2019). https://doi.org/10.1007/978-3-030-02291-4_12
16. Tkach, M., Morhun, S., Zolotoy, Y., Zhuk, I.: Modal analysis of the axial compressor blade: advanced time-dependent electronic interferometry and finite element method. Int. J. Turbo Jet-Eng. (2020). https://doi.org/10.1515/tjj-2020-0014

Determination of the Acoustic Strength of Solar Battery Panel for Space Applications

Maksym Nesterenko[1]($\boxtimes$) (iD) and Andrii Kondratiev[2] (iD)

[1] National Aerospace University "Kharkiv Aviation Institute", 17 Chkalova Street, Kharkiv 61070, Ukraine
maksim_nest@yahoo.co.jp
[2] National Technical University of Ukraine "Igor Sikorsky Kyiv Polytechnic Institute", 37 Peremohy Avenue, Kyiv 03056, Ukraine

Abstract. High requirements are imposed on rapidly developing rocket and space technology. One of the most important requirement is the acoustic influence on the structure. In this regard, there is a need to search for methods for predicting and modeling the reaction of individual structural elements of rocket and space technology. Electric energy sources play a prominent role in creation and ensuring the operability of space and rocket technology. These sources are usually solar batteries. The frames of solar panels in the form of ultra-lightweight rigid sandwich panels made of polymer composite materials with honeycomb are most widely used. The paper presents an approach to determining the stress-strain state of solar panel during acoustic loading. The implementing approach is presented in the form of a block diagram with a detailed description of each of the four blocks highlighted below. «Input data» is the definition of the acoustic loading spectrum, material characteristics and panel shape. "Preliminary design" is the definition of the reduced characteristics of the solar panel and the definition of the method of representation the loading. «Finite-element method» is the analysis of either random vibrations or harmonics of a steady state in the package of the finite element method. "Output data" is the selection of interest data obtained from the designing by the finite element method and their subsequent analysis. The developed approach will be useful for enterprises to introduce into the structure analysis process of rocket and space technology for strength under acoustic influence.

Keywords: Acoustic influence · Sandwich panel · Composite materials

1 Introduction

During the period of scientific and technological progress, rocket and space technology is increasingly rapidly picking up steam in its development. A feature of rocket and space technology is its high reliability and operational safety. One of the important indicators that determine reliability and guarantee the safety of a product of rocket and space technology is the strength of its structures [1].

Electric energy sources play a prominent role in creation and ensuring the operability of space and rocket technology. These sources are usually solar batteries (SB), photovoltaic converters (PC) of which are glued to frames of SB panels. The frames of

SB panels in the form of ultra-lightweight rigid sandwich panels made of polymer composite materials with honeycomb core (HC) are most widely used [2–8]. SB panels experience large overloads during the launch-vehicle lift-off and thermal cycling in specific space conditions. One of the poorly studied types of loading of space SB panels during their operation is acoustic influence. The acoustic effect is created by the power plant of the launch vehicle at lift-off and by the high-speed flow acting on the structure of the head unit at the launch phase. Ensuring the minimum weight of SB panels under the specified operating conditions is associated with the need to develop special methods for optimal design.

2 Literature Review

Currently, close attention is being paid to the creation of cellular structures of rocket and space technology of minimum surface mass, which can be confirmed by a series of contributions, for example [3, 8, 9]. An integrated approach to solving the minimization of the mass of SB panels is described in [10, 11]. In these papers developed the concept, the essence of which consists in the analytical prediction of maximum possible reduction in the surface mass of SB panels, taking into account the modern technological capabilities of their manufacture. For this, the authors developed a comprehensive optimization technique that harmonizes a specific structural layout of the panel with a library of analytical models based on various mathematical models. However, the authors used the analytical model of the beam-strip when analyzing the stress-strain state (SSS) of the honeycomb panels under consideration, as well as when optimizing their parameters. This was the reason for the obviously approximate obtained results, since this did not take into account the interference of the structure zones outside the beam-strips under consideration. This simplification is also used in other contributions, for example [12]. State-of-the-art of integrated computer engineering design technologies has allowed the authors of [13] to solve the problem of increasing the mass efficiency of cell panels taking into account their current level of production by developing a new conceptual approach to the synthesis of rational parameters of composite units of the class under consideration. However, the complexity of this approach is obviously justified only in cases where it is necessary to conduct re-analysis of SSS. One of the types of loading of space SB panels during their operation, which cannot be directly taken into account during optimization as a design case, is acoustic influence. A number of studies present the results of various research on acoustic influence. So in [14–18], the nature of occurrence and the principles of predicting the acoustic load acting on structural elements of rocket and space technology are described. In [8, 19, 20], the results of determining the stress-strain properties (SSP) of sandwich (composite) structures exposed to acoustic effect are presented. Contributions [21–26] are devoted to predicting the damping properties (DP) of composite structures. Contributions [17, 27] are devoted to numerical simulation of structures exposed to acoustic load. Currently, two main approaches have been formed in acoustic design: the use of finite-element modeling and statistical energy analysis [28]. Finite-element modeling is most often used for acoustic analysis in the low-frequency range. To solve acoustic problems in the medium and high frequencies, statistical energy analysis is used. In [28], a simplified approach to the calculation of SB panels when exposed to sound pressure was implemented, based on finite-element modeling of the appearance of resonant acoustic vibrations in them.

Today, the only reliable method for assessing acoustic strength is a full-scale experiment on a finished structure or laboratory research on its models, which are one of the most expensive types of tests, and their preparation and implementation take a lot of time [16].

3 Determination of the Stress-Strain State of the Panel Under the Influence of Acoustic Loading

Figure 1 shows a block diagram of the algorithm for determining the SSS of a structure under the influence of acoustic loading. The algorithm consists of the following main blocks: input and output data (Block № 1 and Block № 4, respectively), preliminary design (Block № 2) and the finite-element method (FEM) (Block № 3).

The output follows from the objective itself: to determine the SSS structure under acoustic influence. The resonance ranges are more interesting, and we will be interested in determining the SSS at the resonance ranges, i.e. the SSS structure, when the exciting frequencies coincide with the natural frequencies of the structure under consideration, and therefore with the predicted resonant effect.

In the input data (Block № 1), first of all it is necessary to specify the characteristics of the structure itself, the SSS of which we want to determine. Also important is the assignment of the exciting factor, i.e. assignment of data on the acoustic influence itself. As with any other technique or algorithm relating to the design of any engineering equipment or object, we have an idea of the structure as a whole, as something complicated. And to solve design problems, the object under consideration is simplified to the design case and to the mathematical model. In other words, the structure under consideration must be defined as a certain shape with its dimensions, with a certain configuration. It is also necessary to divide the structure into several elements (if necessary), for example, in the case of composite structures or structures consisting of different parts having distinct characteristics from each other. Therefore, from the above, such data as the shape and dimensions of the structure under consideration, as well as the characteristics of the materials of which the structure itself consists, follow (Block № 1.2). As for specify of the exciting factor: the acoustic effect is a complex process and usually has a random nature. To obtain the most reliable results, it is advisable to present the acoustic load as a spectrum of sound pressure levels (SPL) in accordance with a certain frequency. It is difficult to obtain these quantities by analytical methods and completely obtained by experimental ones. As a result, for the further implementation of the algorithm, it is necessary to obtain a table of the spectrum of SPL depending on the frequency (Block № 1.1).

The range of the spectrum can be represented in the form of 1-octave, 1/3-octave and other ranges. Type of partitioning into a third octave range has proved to be better. For a better understanding, the 1/3 octave range is the frequency range where between any two taken frequencies of the spectrum there will be a 1/3 octave frequency difference (the frequency of the first taken octave is 2 times less than the frequency of the

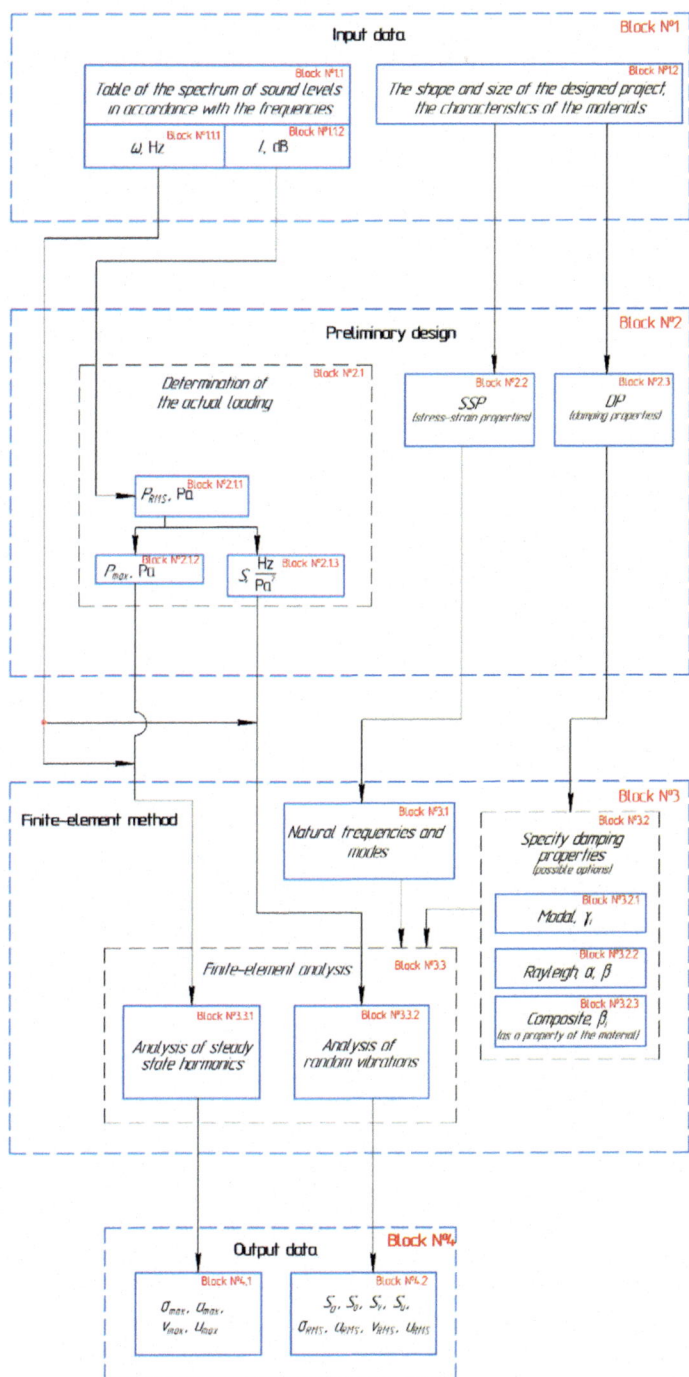

Fig. 1. The block diagram of the algorithm for determining the SSS of a structure under the influence of acoustic loading.

subsequent octave) [29]. Having determined the partitioning of the frequency spectrum (Block № 1.1.1), it is necessary to determine the SPL on the surface of the object under consideration for each frequency from the range of the previously determined frequency spectrum (Block № 1.1.2). To do this, it is necessary to conduct experiments by installing sensors near the payload and directly operating or simulating the flight of the launch vehicle, conduct analytical estimations or take an available spectrum from the requirements document of previously conducted tests of such complexes (complex is a booster plus payload). In this issue, papers [14–18] will be useful.

After all the data of block № 1 is collected, you should proceed to the preliminary design (Block № 2). At this stage, you should get the missing data that is directly the input data for determining the SSS of the object under consideration in the resonant ranges using the FEM (Block № 3). It should be noted that further two ways of solving the problem will be presented (Block № 3.3), depending on the type of FEM analysis, which in turn depends on the type of loading (Block № 2.1).

The first way is that the analysis will be performed on harmonic loading, i.e. the load will be set by the spectrum of harmonic oscillations depending on the frequency. For this way, it is necessary to determine the amplitude values of the pressure of each harmonic depending on the frequency (Block № 2.1.2).

The second way is that the acoustic excitation of the structure is random and therefore the loading will be set in the form of spectral power densities (PSD) of pressure depending on the frequency (Block № 2.1.3).

For these two ways, it is first necessary to determine the root-mean-square (RMS) sound pressure (Block № 2.1.1) for each frequency from the known values of SPL from block № 1.1.2. To do this, you can use the well-known equation:

$$L = 20 \cdot \lg \frac{p_{RMS}}{p_0}, \tag{1}$$

where L is the SPL, dB; p_{RMS} is the RMS sound pressure, created by all components of its spectrum, Pa; $p_0 = 2 \cdot 10^{-5}$ Pa is the reference sound pressure.

Expressing the RMS sound pressure (Block № 2.1.1), we obtain:

$$p_{RMS} = p_0 \cdot 10^{\frac{L}{20}}. \tag{2}$$

The next step is to determine the amplitude values of pressure (Block № 2.1.2) or PSD of pressure (Block № 2.1.3) depending on the chosen solution way. The obtained values of this calculation (Blocks № 2.1.1, № 2.1.2, № 2.1.3) together with the initial data (Blocks № 1.1.1, № 1.1.2) are conveniently written in one table.

In addition to determining the actual loading data, it is necessary to obtain data on the reduced mass, stiffness and dissipative properties of the structure under consideration or its individual components (for example, in the case of two-, three- or multilayer structures). It is convenient to distinguish two separate prediction here: prediction of the reduced SSP (Block № 2.2) [8, 19, 20] and prediction of DP (Block № 2.3) [21–26].

The reduced SSP should include such as: elasticity modulus E, modulus of elasticity in shear G, Poisson ratio μ, density ρ. The first three, as a rule, must be determined by three coordinates x, y, z.

For example, take a three-layer panel consisting of honeycomb core and composite skins with two monolayers. The input data will contain such parameters as: elasticity modulus, modulus of elasticity in shear, Poisson ratio and density of the honeycomb foil and fibers for monolayers, or monolayers themselves. This may also include the density of secondary elements, such as coating composition and adhesive substances. And through preliminary design, we obtain the reduced characteristics given separately for each of the three layers, i.e. the aforementioned SSP of the honeycomb core (its density and other characteristics differ from the characteristics of the foil itself), and two skins (their SSP may be different depending on the coordinate x, y, z).

Summarizing, we can say that at this stage, there is a transformation of the real structure with its characteristics into a analytical model with selected parts (or layers) with their own characteristics (Block № 2.2). And this analytical model is used when modeling it in the FEM packages (Block № 3).

As mentioned above, a separate prediction distinguishes the prediction of UDC (Block № 2.3). Defining them is a complex process. Currently, full-scale experiments or laboratory research are a more reliable method for determining the DP of a structure, its components or the materials themselves. However, there are some simplified methods of analytical predictions [21–26]. It should be noted that already at this stage it is necessary to decide on the option of specify the DP (Block № 3.2) in the FEM package (Block № 3). The method for determining DP will depend on this option.

Let us single out the main three options for defining damping properties in FEM packets (Block 3.2).

Modal damping (Block № 3.2.1). Here it is necessary to set the dissipation factor ξ_i depending on the frequency, where i is the mode number (number of natural frequency). In other words, for each mode, it is necessary to set your own damping factor.

Rayleigh damping (Block № 3.2.2). Here you need to set two factors α and β, satisfying equation:

$$[C] = \alpha[M] + \beta[K], \tag{3}$$

where $[C]$ is a damping matrix, $[M]$ is a mass matrix, $[K]$ is a stiffness matrix.

Composite modal damping (Block № 3.2.3). In this case, damping is specified as a property of the material, or rather, as a property of a reduced element or layer of the considered analytical model. For an example with a three-layer panel, it is necessary to set the dissipation factor β_i for the honeycomb core as a whole and for composite skin, where i is the order number of the element or layer of the analytical model.

For any of the options for setting damping, the input data (Block № 1.2) must contain the values of the damping properties of materials (determined experimentally or taken from existing requirements documents) so that they can be reduced to design factors.

For modal damping, there are complex methods of analytical designs provided that the standard restraint cases are observed in the analytical model.

For Rayleigh damping, the dissipation factors are determined either by numerical investigation or full-scale experiments.

For composite damping, there are less reliable methods (more simplified methods) for determining damping factors, since their analytical determination does not take into account deviations in the values for each frequency of natural modes. However, even for some types of structure, the factors are taken from the statistical and experimental data of tests and research already carried out due to the lack of methods for their analytical determination.

Also, in some FEM packages, there are recommendations for setting the damping properties of various types of structures and materials.

The following steps are followed directly by analysis in the FEM package (Block № 3). At this stage, the analytical model is generate (the shape and configuration are set), mesh generation is performed with the characteristics of the elements of the analytical model specified (defined in blocks № 2.2, № 2.3, taking into account the choice in block № 3.2), the restraint (displacements) is set, the necessary settings are made (see user manual of the selected FEM package). Note that a detailed description of generation of the analytical model in the FEM package is not given in this paper.

The next step is the eigenvalues prediction (modal analysis) in the FEM package (Block № 3.1). Here you can configure this type of analysis, indicate the number of frequencies and modes taken into account. Increasing this value increases the accuracy of subsequent predictions, but decreases computer capacity. The most important are the first 5–10 modes, depending on the complexity of the structure under study (for very complex structures, you may need the first 100 modes). Next, a direct analysis is carried out for natural frequencies and modes. At this stage, it is possible to carry out more reliable analytical predictions to determine the eigenvalues and forms of natural vibrations, but only for comparison with the FEM analysis and their refinement by editing the analytical model for FEM. However, such a comparison is more appropriate to do with previous tests, if any. This stage is directly important for subsequent predictions (analyzes) in the FEM packages, because the subsequent prediction is based on this analysis in the FEM packages.

The next step of block № 3 is the choice of a solution to the problem (Block № 3.3). The first way is the harmonic analysis (Block № 3.3.1), the second is the prediction of random vibrations (Block № 3.3.2). Depending on the selected analysis (generically indicated by block № 3.3), a subsequent analysis is set up. Here the type of dynamic analysis is set (the chosen way to solve the problem). The loading spectrum is set depending on the frequency (Block № 1.1.1): the spectrum of the amplitude values of harmonics (Block № 2.1.2) in the case of the first path, or PSD of pressure (Block № 2.1.3) in the case of the second path. The values of the factors of the selected damping are read from block № 3.2. Settings are made for predicting the response in the form of time steps for which it is necessary to obtain the required plots. The analysis is carried out directly and the postprocessing of the response is performed (settings are made to display the data of interest in the form of tables, plots and a visual display of deformations). Here (Block № 3.3), the SSS of the structure is determined, which must be analyzed and the results of interest selected in the output of block № 4.

When you select the first path in the input data, the maximum values of displacements u_{max}, velocities v_{max}, accelerations a_{max} and stresses σ_{max} are obtained; as well as interesting plots showing the nature of the stress state of the analytical model (Block № 4.1).

When choosing the second path, the output should contain the PSD of displacements S_u, velocities S_v, accelerations S_a, stresses S_σ and RMS of displacements u_{RMS}, velocities v_{RMS}, accelerations a_{RMS}, stresses σ_{RMS}; as well as interesting plots showing the nature of the stress state of the analytical model (Block № 4.2).

4 Conclusions

The paper presents a detailed algorithm for determining the structure response of rocket and space technology under acoustic influence. As an example, a sandwich SB panel for space applications was taken. It is noted that the study of acoustic loads and their impact on the structure is a difficult task. The only reliable method for assessing acoustic strength is a full-scale experiment on a finished structure or laboratory studies on its models, which are an expensive type of test, and their preparation and conduct take a lot of time.

The algorithm implementation is based on the finite element modeling method. The algorithm is presented in the form of four blocks: input and output data, preliminary design and the actual analysis in the FEM package. The proposed algorithm makes it possible to assess the stress-strain state of SB panels and other composite structures of space-rocket technology when resonant acoustic vibrations occur in them using almost any finite-element modeling package, even in the absence of specialized package, which makes it quite attractive.

References

1. Slyvynskyi, V.I., Sanin, A.F., Kharchenko, M.E., Kondratyev, A.V.: Thermally and dimensionally stable structures of carbon-carbon laminated composites for space applications. Proc. Int. Astronaut. Cong. (IAC) **8**, 5739–5751 (2014)
2. Slyvynskyi, V.I., Alyamovskyi, A.I., Kondratjev, A.V., Kharchenko, M.E.: Carbon honeycomb plastic as light-weight and durable structural material. In: Proceedings of the 63th International Astronautical Congress 2012, vol. 8, pp. 6519–6529 (2012)
3. Adams, D.O., Webb, N.J., Yarger, C.B., et al.: Multi-functional sandwich composites for spacecraft applications: an initial assessment. Report No. NASA/CR-2007–214880. NASA, Hampton (2007)
4. Jet Propulsion Laboratory: Solar cell and array technology for future space science missions. NASA, Pasadena (2002)
5. Jet Propulsion Laboratory: Solar power technologies for future planetary science missions. NASA, Pasadena (2017)
6. Roibás-Millán, E., Alonso-Moragón, A., Jiménez-Mateos, A.G., Pindado, S.: Testing solar panels for small-size satellites: the UPMSAT-2 mission. Meas. Sci. Technol. **28**(11), 115801 (2017). https://doi.org/10.1088/1361-6501/aa85fc
7. Kodiyalam, S., Nagendra, S., DeStefano, J.: Composite sandwich structure optimization with application to satellite components. AIAA J. **34**(3), 614–621 (1996). https://doi.org/10.2514/3.13112

8. Zenkert, D.: The Handbook of Sandwich Construction. Engineering Materials Advisory Services, West Midlands (1997)
9. Kondratiev, A.: Improving the mass efficiency of a composite launch vehicle head fairing with a sandwich structure. East. Eur. J. Enterp. Technol. 6(7), 6–18 (2019). https://doi.org/10.15587/1729-4061.2019.184551
10. Slyvyns'kyy, V., Gajdachuk, V., Gajdachuk, A., Slyvyns'ka, N.: Weight optimization of honeycomb structures for space applications. In: Proceedings of the 56th International Astronautical Congress, vol. 6, pp. 3611–3620 (2005). https://doi.org/10.2514/6.IAC-05-C2.3.07
11. Slyvyns'kyy, V., Slyvyns'kyy, M., Polyakov, N., et al.: Scientific fundamentals of efficient adhesive joint in honeycomb structures for aerospace applications. In: Proceedings of the 59th International Astronautical Congress 2008, vol. 8, pp. 5307–5314 (2008)
12. Slivinsky, M., Slivinsky, V., Gajdachuk, V., et al.: New possibilities of creating efficient honeycomb structures for rockets and spacrafts. In: Proceedings of the 55th International Astronautical Congress, vol. 3, pp. 1923–1932 (2004). https://doi.org/10.2514/6.IAC-04-I.3.A.10
13. Kondratiev, A., Gaidachuk, V.: Weight-based optimization of sandwich shelled composite structures with a honeycomb filler. East. Eur. J. Enterp. Technol. 1(1), 24–33 (2019). https://doi.org/10.15587/1729-4061.2019.154928
14. Clarkson, B.L.: Structural Aspects of Acoustic Loads. AGARD, Neuilly sur Seine, France (1960)
15. Morshed, M.M., Hansen, C.H., Zander, A.C.: Prediction of acoustic loads on a launch vehicle: non-unique source allocation method. J. Spacecraft Rock. 52(5), 1–22 (2015). https://doi.org/10.2514/1.A33204
16. Panda, J.: Aeroacoustics of Space Vehicles. Paper presented at Applied Modeling & Simulation (AMS) Seminar Series, NASA Ames Research Center, CA, 8 April 2014. https://www.nas.nasa.gov/assets/pdf/ams/2014/AMS_20140408_Panda.pdf. Accessed 10 July 2020
17. Space engineering: Spacecraft mechanical loads analysis handbook (ECSS-E-HB-32–26A). European Space Agency, Noordwijk (2013)
18. James, M.M., Salton, A.R., Gee, K.L., et al.: Full-scale rocket motor acoustic tests and comparisons with empirical source models. Proc. Meet. Acoust. 18(1), 040007 (2012). https://doi.org/10.1121/1.4870984
19. Petras, A.: Design of sandwich structures. Dissertation, Cambridge University (1998)
20. Jones, R.M.: Mechanics of Composite Materials, 2nd edn. Taylor & Francis, Philadelphia (1999)
21. Lopes, J.P.: Effects of design parameters on damping of composite materials for aeronautical applications. Dissertation, University of Beira Interior (2013)
22. Abbasloo, A., Maheri, M.R.: Prediction of modal damping of FRP-honeycomb sandwich panels with arbitrary geometries. Latin Am. J. Solids Struct. 14(1), 17–35 (2017). https://doi.org/10.1590/1679-78252537
23. Rydberg, S.: Prediction of vibrational amplitude in composite sandwich structures: Prediction and implementation of the orthotropic damping in carbon-fibre-reinforced epoxy. Master's thesis, Chalmers University of Technology (2013)
24. Panigrahi, S.K.: Damping of composite material structures with bolted joints. Bachelor's thesis, National Institute of Technology (2012)
25. Lavanya, K., Krishna, P.V., Sarcar, M.M.M., Sankar, H.R.: Analysis of the damping characteristics of glass fibre reinforced composite with different orientations and viscoelastic layers. Int. J. Concept. Mech. Civil Eng. 1(1), 88–92 (2013)

26. Butaud, P., Foltete, E., Ouisse, M.: Sandwich structures with tunable damping properties: on the use of shape memory polymer as viscoelastic core. Compos. Struct. **153**, 401–408 (2016). https://doi.org/10.1016/j.compstruct.2016.06.040
27. Rassaian, M., Huang, Y., Lee, J., Arakawa, T.T.: Structural analysis with vibro-acoustic loads in LS-DYNA®. In: Proceedings of the 10th International LS-DYNA® Users' Conference, pp. 45–60 (2008)
28. Kondratiev, A., Gaidachuk, V., Nabokina, T., Tsaritsynskyi, A.: New possibilities in creating of effective composite size-stable honeycomb structures designed for space applications. In: Nechyporuk, M., et al. (eds.) Integrated Computer Technologies in Mechanical Engineering. AISC, vol. 1113, pp. 45–59. Springer, Cham (2020). https://doi.org/10.1007/978-3-030-37618-5_5.
29. Blevins, R.D.: Formulas for Dynamics, Acoustics and Vibration. Wiley, Chichester (2015)

Investigation of Condensing Heating Surfaces with Reduced Corrosion of Boilers with Water-Fuel Emulsion Combustion

Victoria Kornienko[1]([⊠]) ⓘ, Roman Radchenko[2] ⓘ,
Łukash Bohdal[3] ⓘ, Leon Kukiełka[3] ⓘ, and Stanisław Legutko[3] ⓘ

[1] Kherson Branch of Admiral Makarov National University of Shipbuilding,
44 Ushakova Avenue, Kherson 73000, Ukraine
kornienkovikal987@gmail.com
[2] Admiral Makarov National University of Shipbuilding,
9 Heroes of Ukraine Avenue, Mykolayiv 54025, Ukraine
[3] Koszalin University of Technology,
2 Śniadeckich Street, Koszalin 75-453, Poland

Abstract. Using of condensing low-temperature heating surfaces in exhaust gas boilers allows to increase the economic efficiency of boilers and thermal power plants. Analysis of literary sources showed, that there were no quantitative data of the low-temperature corrosion intensity of exhaust gas boiler low-temperature heating surfaces while water-fuel emulsion combustion. Experimental investigations of corrosion processes of low-temperature heating surfaces in exhaust gas boilers with excess air factor in the range of 1.5…3.0 were carried out. The empirical correlations for dependence of specific metal mass loss on water content of water-fuel emulsion, sulfur content in output fuel and excess air factor α at wall temperatures below the dew point temperature of sulfuric acid vapor, which characterize the low-temperature corrosion intensity of condensation surfaces at different operating modes of exhaust gas boilers were received. For estimation of the influence of the quality of combustible fuel and its combustion regimes on the corrosion intensity, a computer simulation was conducted by using the statistical program package Statgraphics Centurion XV. These correlations show that the smallest values of corrosion intensity are observed at large values of water content in water-fuel emulsion of about 30%. The minimum values of exhaust gas temperature at the exit from exhaust gas boiler and of wall temperature are determined, at which the permissible speed of low-temperature corrosion at a level of 0.25 mm/year is ensured.

Keywords: Exhaust heat utilization · Low-temperature heating surface · Low-temperature corrosion · Exhaust gases

1 Introduction

Applications of condensing heating surfaces increases the efficiency of any thermal power plant based on cogenerative internal combustion engines [1, 2] and gas turbines [3], trigeneration plants generating heat, electricity power and cooling [4, 5] including

M. Nechyporuk et al. (Eds.): ICTM 2020, LNNS 188, pp. 300–309, 2021.
https://doi.org/10.1007/978-3-030-66717-7_25

refrigeration [6] and air conditioning [7, 8], i.e. practically in all waste heat recovery technics in particular in ship power plants [9] with internal combustion engines (ICE) as the main engines. Application of condensing low-temperature heating surfaces (LTHS) in exhaust gas boilers (EGB) allows to increase the economic efficiency [10, 11] and environmental performance of boilers [12] and thermal power plants (TPP) in the whole [13].

When fuel oils are burnt in boilers, combustion chambers of ICE [14, 15] and gas turbine engines [16], the formation of the main toxic ingredients of exhaust gases NOx and SOx, as well as solid particles of ash and soot, occurs. When the combustion products formed in the exhaust gases move to the chimney cutoff, precipitation of solid ash and soot particles occurs, H_2SO_4 vapors are formed and they condense on LTHS at wall temperatures below the dew point temperature, which causes low-temperature corrosion (LTC) of metal [17] and their pollution. Decreasing of heat transfer intensity in the heating surfaces, EGB steam capacity and depth of gas heat utilization has taken place.

Therefore, lowering the LTC intensity at the surface temperature below a dew point temperature of sulfuric acid vapor H_2SO_4 is practically the only possibility of reducing the EGB exhaust gas temperature and improving ecology and economy of TPP.

2 Literature Review

According to literature data, a heat loss of exhaust gases represents a high proportion of the total heat loss of combustion engines [18] turbines and boilers in TPP. Nevertheless, the wall temperature of heating surface of EGB and exhaust gases must be no lower than dew point temperature of sulfuric acid vapor in exhaust gas. Because, the acid vapor are condensed on the heating surface, and corrode exhaust heat economizers, which will shorten their life [19]. The condensed acid vapor also glues the ash in exhaust gas, and adheres on heating surface, which increases the resistance of exhaust gas flow and heat transfer, affecting the units' reliable and economical operation [20, 21].

A promising method of suppressing the formation of these substances in furnace processes is the burning of fuel oil in the form of a water-fuel emulsion (WFE), which differs from fuel oil in both physicochemical properties and the characteristics of combustion, mass transfer, and heat transfer [22, 23]. The experience of using WFE in boilers and diesel engines indicates the undeniable advantages of applying this type of fuel: the effective specific fuel consumption decreases by about 8% [24], the concentration of nitrogen oxides in the exhaust gases is significantly reduced in 1.4...3.1 times [25], the concentration of CO – in 1.3...1.5 times [26], smoke – in 1.3...2.4 times [27].

In spite of application of modern methods of modelling and experimental analysis of metal alloys properties and technologies and of waste heat recovery [28] and cooling processes in ICE [29] and condensation processes in heat exchangers [30], there is almost no specific information about the course of corrosion processes on the heating surfaces of boilers using WFE, and the published few data are qualitative. This needs quantitative data on the LTC intensity of LTHS during combustion of WFE, because they determine not only the reliability of boilers in deep utilization of exhaust gases, but also the temperature of exhaust gases, economic performance and, consequently, efficient use of fuel and secondary energy sources.

The aim of research is to obtain the correlations for dependence of specific mass loss of metal ΔG_c of condensing heating surfaces of boilers with WFE combustion on the excess air factor α, water W^r and sulfur content S^r.

The research tasks: experimental research of LTHS corrosion processes of EGB at excess air factor $\alpha = 1.5\ldots3.0$, that influence a depth of gas heat utilization, and EGB working reliability; study of influence of excess air factor α, water content of emulsion W^r and sulfur content of fuel S^r on corrosion intensity; obtaining the wall temperature range of condensing heating surfaces that provides their safe operation.

3   Research Methodology

Experimental researches of pollution intensity at wall temperature values below dew point temperature of sulfuric acid vapors were carried out at the experimental setup with combustion of fuel oil and WFE (Fig. 1). WFE preparation for combustion in experimental setup was carried out in special installation.

Fig. 1. General view of experimental setup.

During the tests in the combustion chamber of the ICE of 13200 kW power output the mixture of fuels was combusted at $S^r = 1.5\%$, $\alpha = 2.9$, $W^r = 2\%$, ash content was at the level of 0.01%. In the gas pipeline of EGB, four probes were installed between the packets, which provided measurements of mass loss of metal in 2, 4, 8 and 12 h. The tube samples were cooled by water or oil. The temperature of exhaust gases and tube samples were measured by thermocouples, which were wound in metal spacers between the tubes. Determination of specific mass loss of metal ΔG_c of tube samples caused by LTC was carried out by the gravitation method. The masses were defined by using the laboratory weights.

The processing of measurement results for study of corrosion process kinetic was carried out after 2, 4, 8 and 12 h, based on which the approximation equations were obtained. The verification of approximation equations reliability was carried out on the results of experiment for 100 h.

For obtaining the regression equation for multifactorial corrosion process, estimating the joint effect of three factors considered of specific mass loss of metal ΔG_c, the constants and weights of components in multifactorial regression equation, evaluating the importance of factors on ΔG_c, the statistical processing of experimental data by Statgraphics system was used. This system provides access to a full set of statistical methods, provides an opportunity to conduct an extended regression analysis. The module for experiments planning of this system makes it possible to take into account the interaction of the analyzed factors by an amount ΔG_c.

Method of surface construction for three influence factors was chosen for analysis. Full factor rotatable plan of type 2^3 was chosen (that is the number of experiments in the plan should be $2^3 = 16$, twelve of which have specific observational data, and four missing values are estimated taking into account the experimental design method and recommendations used). It should be noted that knowing the levels of factors variation in preparation of the planning matrix, it is possible to introduce additional data within certain change limits of factors being studied. Thus, the calculation of all major effects impact as well as their two-factor combinations is available for calculation.

4 Results

Based on short-term experimental research data, approximation equations of corrosion kinetic in the form dependencies $\Delta G_c = f(\tau)$ were obtained (Fig. 2).

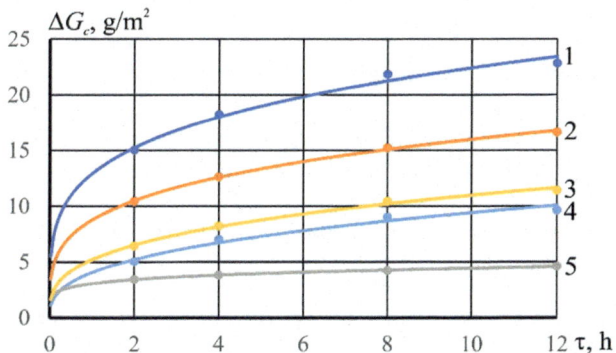

Fig. 2. Investigation of corrosion processes kinetics: $1 - (S^r = 1.5\%; \alpha = 2.9; W^r = 2\%)$, $\Delta G_c = 14.0884\tau^{0.1866}$, $R^2 = 0.9877$; $2 - (S^r = 1.5\%; \alpha = 1.25; W^r = 2\%)$, $\Delta G_c = 8.7075\tau^{0.2633}$, $R^2 = 0.9983$; $3 - (S^r = 0.98\%; \alpha = 1.35; W^r = 15\%)$, $\Delta G_c = 5.1568\tau^{0.3273}$; $R^2 = 0.9947$; $4 - (S^r = 2\%; \alpha = 1.01; W^r = 2\%)$, $\Delta G_c = 4.2884\tau^{0.4573}$, $R^2 = 0.9855$; $5 - (S^r = 1.5\%; \alpha = 1.15; W^r = 30\%)$; $\Delta G_c = 3.0246\tau^{0.1645}$, $R^2 = 0.9939$.

As the results of experimental research have shown, the time for stabilization of corrosion process is 2…3 h, especially when WFE is burnt. This allows to assess the level of corrosion process and to predict process for a long time from the results of kinetic studies for 2…12 h. In first 1…3 h, processes are accelerated, and then proceed with deceleration. After 4…5 h, dynamic balance is almost attained and the corrosion process proceeds at an almost constant rate. This allows, under the same working conditions to determine (predict) the value of ΔG_c by calculation during any period of operation, but while ensuring stable operating modes.

Empirical correlations of specific mass loss of metal ΔG_c from water content W^r in WFE (Fig. 3), sulfur content in fuel S^r (Fig. 4) and air excess factor α (Fig. 5) were obtained based on experimental data by Statgraphics. Analysis of correlations showed that the lowest value is observed when $W^r = 30\%$, $\alpha = 1.5$, $S^r = 1\%$, and the largest - $W^r = 2\%$, $\alpha = 3$, $S^r = 2\%$.

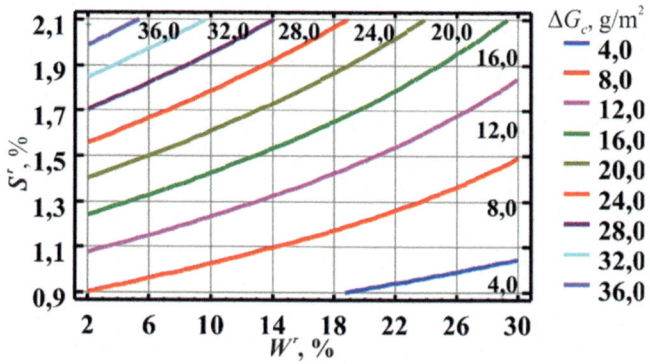

Fig. 3. Correlations of specific mass loss of metal $\Delta G_c = f\,(S^r,\ W^r)$ while $\alpha = 2.9$.

The R^2 statistic indicates that the model as fitted explains 98% of the variability in ΔG_c.

The Pareto chart obviously shows that sulphur content S^r (factor C), water content in emulsion W^r (factor B), α (factor A) and quadratic term BC have statistically significant effects (Fig. 6).

As result of multivariate regression analysis, the average value of ΔG_c was obtained as function of other parameters (W^r, α, S^r):

$$\Delta G_c = -14.8326 + 7.3776\alpha + 0.4531\,W^r + 6.0669S^r - 0.5593\,W^r S^r. \qquad (1)$$

The equation gives acceptable values of ΔG_c in the range of values $\alpha = 1.5…2.9$, $S^r = 0.9…2\%$ and $W^r = 2…30\%$.

Based on experimental studies the minimum value of wall temperature is defined (Fig. 7), at which the permissible speed of LTC at a level of 0.25 mm/year and the minimum value of exhaust gas temperature at the exit from EGB are ensured (Fig. 8). Figure 8 show that the smallest values of corrosion intensity are observed for large values of water content in water-fuel emulsion of about 30%.

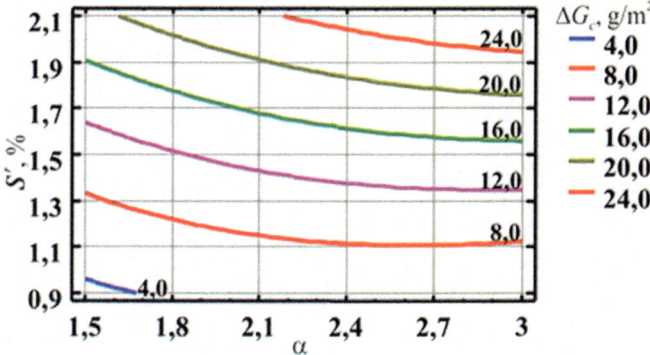

Fig. 4. Correlations of specific mass loss of metal $\Delta G_c = f(S^r, \alpha)$ while: $W^r = 15\%$.

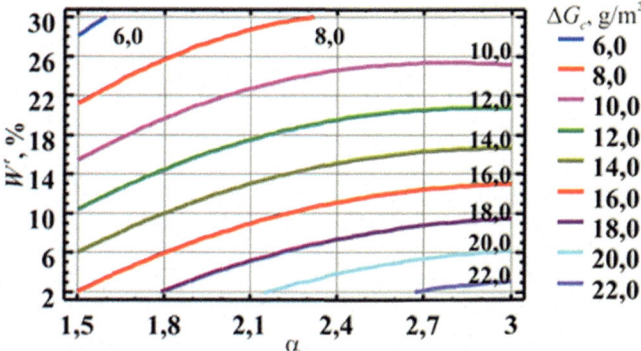

Fig. 5. Correlations of specific mass loss of metal $\Delta G_c = f(W^r, \alpha)$ while $S^r = 1.5\%$.

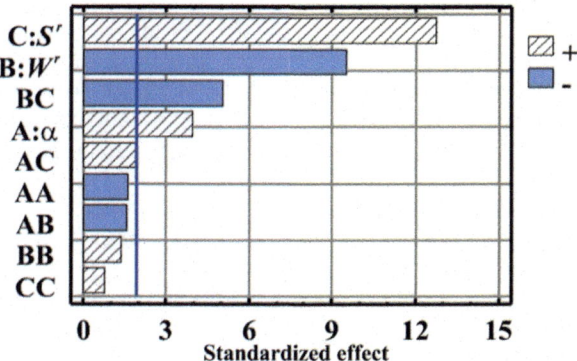

Fig. 6. Pareto chart for specific mass loss of metal ΔG_c.

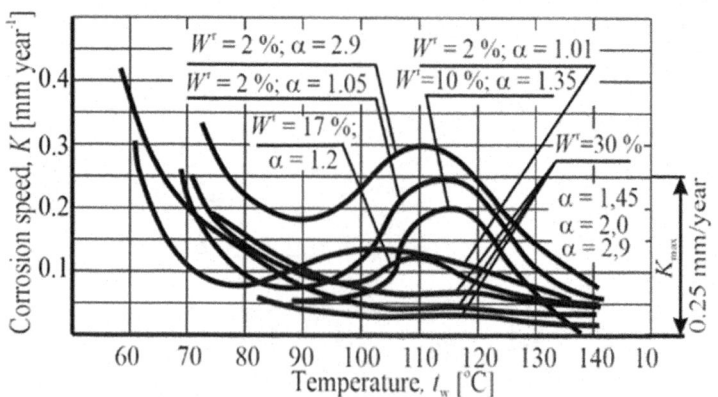

Fig. 7. Correlation of corrosion rate K from wall temperature.

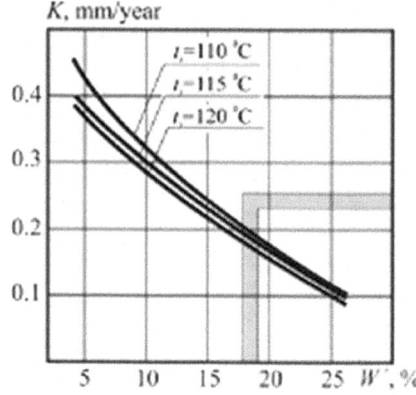

Fig. 8. Correlation of corrosion rate K from water content W^r in emulsion.

The obtained regression equation allows calculating corrosion intensity of condensing heating surfaces of EGB, which is necessary to obtain exhaust gas temperature value, a depth of gas heat utilization, and EGB working reliability.

5 Conclusions

The kinetics of LTC on the EGB condensing surfaces with WFE combustion are investigated to obtain approximation equations for predicting processes development.

The semi-empirical correlations for dependences of the specific mass loss of metal on the water content of WFE, sulfur content in fuel oil and excess air factor with wall temperatures below dew point temperature of sulfuric acid pair have been obtained on the base of experimental data.

The regression equations have been obtained that make it possible to estimate the influence of various factors such as water content of the WFE, sulfur content in fuel oil and excess air factor on corrosion intensity.

The minimum value of wall temperature is determined, at which the permissible speed of low-temperature corrosion at a level of 0.25 mm/year is ensured and the minimum value of exhaust gas temperature at the exit from exhaust gas boiler. The wall temperature range of condensing heating surfaces that provides their safe operation is determined, which opens opportunities for deep exhaust gas heat utilization.

In the case of WFE combusted with water content of $W^r = 30\%$ there is a significant decrease of LTC intensity to the level 0.25 mm/year. That makes it possible to apply condensing LTHS at a wall temperature tw below the dew point temperature of sulfuric acid vapors within high working reliability of these condensing LTHS.

References

1. Radchenko, A., Mikielewicz, D., Forduy, S., Radchenko, M., Zubarev, A.: Monitoring the fuel efficiency of gas engine in integrated energy system. In: Nechyporuk, M., Pavlikov, V., Kritskiy, D. (eds.) Integrated Computer Technologies in Mechanical Engineering. AISC, vol. 1113, pp. 361–370. Springer, Cham (2020). https://doi.org/10.1007/978-3-030-37618-5_31

2. Radchenko, M., Portnoi, B., Kantor, S., et al.: Rational thermal loading the engine inlet air chilling complex with cooling towers. In: Tonkonogyi, V., et al. (eds.) Advanced Manufacturing Processes II. InterPartner 2020. LNME. Springer, Cham (2021)

3. Radchenko, A., Bohdal, L., Zongming, Y., Portnoi, B., Tkachenko, V.: Rational designing of gas turbine inlet air cooling system. In: Tonkonogyi, V., Ivanov, V., Trojanowska, J., Oborskyi, G., Edl, M., Kuric, I., Pavlenko, I., Dasic, P. (eds.) InterPartner 2019. LNME, pp. 591–599. Springer, Cham (2020). https://doi.org/10.1007/978-3-030-40724-7_60

4. Forduy, S., Radchenko, A., Kuczynski, W., Zubarev, A., Konovalov, D.: Enhancing the gas engines fuel efficiency in integrated energy system by chilling cyclic air. In: Tonkonogyi, V., Ivanov, V., Trojanowska, J., Oborskyi, G., Edl, M., Kuric, I., Pavlenko, I., Dasic, P. (eds.) InterPartner 2019. LNME, pp. 500–509. Springer, Cham (2020). https://doi.org/10.1007/978-3-030-40724-7_51

5. Radchenko, A., Stachel, A., Forduy, S., Portnoi, B., Rizun, O.: Analysis of the efficiency of engine inlet air chilling unit with cooling towers. In: Ivanov, V., Pavlenko, I., Liaposhchenko, O., Machado, J., Edl, M. (eds.) DSMIE 2020. LNME, pp. 322–331. Springer, Cham (2020). https://doi.org/10.1007/978-3-030-50491-5_31

6. Radchenko, A., Trushliakov, E., Tkachenko, V., et al.: Improvement of the refrigeration capacity utilization for the ambient air conditioning system. In: Tonkonogyi, V., et al. (eds.) Advanced Manufacturing Processes II. InterPartner 2020. LNME. Springer, Cham (2021)

7. Trushliakov, E., Radchenko, A., Radchenko, M., Kantor, S., Zielikov, O.: The efficiency of refrigeration capacity regulation in the ambient air conditioning systems. In: Ivanov, V., Pavlenko, I., Liaposhchenko, O., Machado, J., Edl, M. (eds.) DSMIE 2020. LNME, pp. 343–353. Springer, Cham (2020). https://doi.org/10.1007/978-3-030-50491-5_33

8. Radchenko, M., Mikielewicz, D., Tkachenko, V., Klugmann, M., Andreev, A.: Enhancement of the operation efficiency of the transport air conditioning system. In: Ivanov, V., Pavlenko, I., Liaposhchenko, O., Machado, J., Edl, M. (eds.) DSMIE 2020. LNME, pp. 332–342. Springer, Cham (2020). https://doi.org/10.1007/978-3-030-50491-5_32

9. Trushliakov, E., Radchenko, A., Forduy, S., Zubarev, A., Hrych, A.: Increasing the operation efficiency of air conditioning system for integrated power plant on the base of its monitoring. In: Nechyporuk, M., Pavlikov, V., Kritskiy, D. (eds.) Integrated Computer Technologies in Mechanical Engineering. AISC, vol. 1113, pp. 351–360. Springer, Cham (2020). https://doi.org/10.1007/978-3-030-37618-5_30

10. Huang, S., Li, C., Tan, T., et al.: An improved system for utilizing low-temperature waste heat of flue gas from coal-fired power plants. Entropy 19(8), 423 (2017). https://doi.org/10.3390/e19080423

11. Liu, J., Sun, F.: Experimental study on operation regulation of a coupled high–low energy flue gas waste heat recovery system based on exhaust gas temperature control. Energies 12(4), 706 (2019). https://doi.org/10.3390/en12040706

12. Chen, H., Pan, P., Wang, Y., Zhao, Q.: Field study on the corrosion and ash deposition of low–temperature heating surface in a large–scale coal–fired power plant. Fuel 208, 149–159 (2017). https://doi.org/10.1016/j.fuel.2017.06.120

13. Radchenko, R., Kornienko, V., Pyrysunko, M., Bogdanov, M., Andreev, A.: Enhancing the efficiency of marine diesel engine by deep waste heat recovery on the base of its simulation along the route line. In: Nechyporuk, M., Pavlikov, V., Kritskiy, D. (eds.) Integrated Computer Technologies in Mechanical Engineering. AISC, vol. 1113, pp. 337–350. Springer, Cham (2020). https://doi.org/10.1007/978-3-030-37618-5_29

14. Chen, T., Zhuge, W., Zhang, Y., Zhang, L.: A novel cascade organic Rankine cycle (ORC) system for waste heat recovery of truck diesel engines. Energy Convers. Manage. 138, 210–223 (2017). https://doi.org/10.1016/j.enconman.2017.01.056

15. Hoang, A.T.: Waste heat recovery from diesel engines based on Organic Rankine Cycle. Appl. Energy 231, 138–166 (2018). https://doi.org/10.1016/j.apenergy.2018.09.022

16. Mikielewicz, D., Kosowski, K., Tucki, K., et al.: Gas turbine cycle with external combustion chamber for prosumer and distributed energy systems. Energies 12(18), 3501 (2019). https://doi.org/10.3390/en12183501

17. Gruber, T., Schulze, K., Scharler, R., Obernberger, I.: Investigation of the corrosion behaviour of 13CrMo4–5 for biomass fired boilers with coupled online corrosion and deposit probe measurements. Fuel 144, 15–24 (2015). https://doi.org/10.1016/j.fuel.2014.11.071

18. Radchenko, R., Pyrysunko, M., Radchenko, A., et al.: Ship engine intake air cooling by ejector chiller using recirculation gas heat. In: Tonkonogyi, V., et al. (eds.) Advanced Manufacturing Processes II. InterPartner 2020. LNME. Springer, Cham (2021)

19. Tee, K.F., Khan, L.R., Li, H.: Reliability analysis of underground pipelines using subset simulation. Int. J. Civ. Environ. Eng. 7(11), 843–849 (2013). https://doi.org/10.5281/zenodo.1088836

20. Moustabchir, H., Pruncu, C.I., Azari, Z., Hariri, S., Dmytrakh, I.: Fracture mechanics defect assessment diagram on pipe from steel P264GH with a notch. Int. J. Mech. Mater. Des. 12(2), 273–284 (2015). https://doi.org/10.1007/s10999-015-9296-z

21. Karami, M.: Review of corrosion role in gas pipeline and some methods for preventing it. J. Pressure Vessel Technol. 134(5), 054501 (2012). https://doi.org/10.1115/1.4006124

22. Kornienko, V., Radchenko, R., Konovalov, D., Andreev, A., Pyrysunko, M.: Characteristics of the rotary cup atomizer used as afterburning installation in exhaust gas boiler flue. In: Ivanov, V., Pavlenko, I., Liaposhchenko, O., Machado, J., Edl, M. (eds.) DSMIE 2020. LNME, pp. 302–311. Springer, Cham (2020). https://doi.org/10.1007/978-3-030-50491-5_29

23. Kornienko, V., Radchenko, R., Mikielewicz, D., et al.: Improvement of characteristics of water-fuel rotary cup atomizer in a boiler. In: Tonkonogyi, V., et al. (eds.) Advanced Manufacturing Processes II. InterPartner 2020. LNME. Springer, Cham (2021)

24. Baskar, P., Kumar, A.S.: Experimental investigation on performance characteristics of a diesel engine using diesel-water emulsion with oxygen enriched air. Alexandria Eng. J. **56** (1), 137–146 (2017). https://doi.org/10.1016/j.aej.2016.09.014

25. Gupta, R.K., Sankeerth, K.A., Sharma, T.K., et al.: Effects of water-diesel emulsion on the emission characteristics of single cylinder direct injection diesel engine – a review. Appl. Mech. Mater. **592**, 1526–1533 (2014). https://doi.org/10.4028/www.scientific.net/AMM.592-594.1526

26. Patel, K.R., Dhiman, V.: Research study of water- diesel emulsion as alternative fuel in diesel engine – an overview. Int. J. Latest Eng. Res. Appl. **2**(9), 37–41 (2017)

27. Wojs, M.K., Orliński, P., Kamela, W., Kruczyński, P.: Research on the influence of ozone dissolved in the fuel-water emulsion on the parameters of the CI engine. In: IOP Conference Series: Materials Science and Engineering, vol. 148, p. 012089 (2016). https://doi.org/10.1088/1757-899X/148/1/012089

28. Konovalov, D., Trushliakov, E., Radchenko, M., Kobalava, H., Maksymov, V.: Research of the aerothermopressor cooling system of charge air of a marine internal combustion engine under variable climatic conditions of operation. In: Tonkonogyi, V., Ivanov, V., Trojanowska, J., Oborskyi, G., Edl, M., Kuric, I., Pavlenko, I., Dasic, P. (eds.) InterPartner 2019. LNME, pp. 520–529. Springer, Cham (2020). https://doi.org/10.1007/978-3-030-40724-7_53

29. Konovalov, D., Kobalava, H.: Efficiency analysis of gas turbine plant cycles with water injection by the aerothermopressor. In: Ivanov, V., Trojanowska, J., Machado, J., Liaposhchenko, O., Zajac, J., Pavlenko, I., Edl, M., Perakovic, D. (eds.) DSMIE 2019. LNME, pp. 581–591. Springer, Cham (2020). https://doi.org/10.1007/978-3-030-22365-6_58

30. Bohdal, T., Sikora, M., Widomska, K., Radchenko, A.M.: Investigation of flow structures during HFE-7100 refrigerant condensation. Arch. Thermodyn. **36**(4), 25–34 (2015). https://doi.org/10.1515/aoter-2015-0030

One-Dimensional Axisymmetric Model of the Stress State of the Adhesive Joint

Kostiantyn Barakhov$^{(\boxtimes)}$ ⓘD, Daria Dvoretska ⓘD,
and Oleksandr Poliakov ⓘD

National Aerospace University "Kharkiv Aviation Institute",
17 Chkalova Str., Kharkiv 61070, Ukraine
kpbarakhov@gmail.com, o.poliakov@khai.edu

Abstract. Local damage repair of modern aircraft structures can be done by creating patchs that are glued to the main structure. The patch takes on part of the load, unloading the damaged area. This method of repair provides tightness and aerodynamic efficiency of the structure. The stress state of such glued structures is calculated, as a rule, using the finite element method. Classical models of the lap joint stress state are one-dimensional. I.e. the change in the stress state is considered only along one coordinate. In this case, the joint is considered rectangular. The aim of this work is to create a mathematical model of the stress state of circular axisymmetric adhesive joints, and to build an appropriate analytical solution to the problem. It is assumed that there is no bending and the deformations of the plates are uniform in thickness. The adhesive layer only works on shear. Both plates – the main plate and the patch are assumed to be isotropic. The solution is built in the polar coordinate system, in which the stress state of the joint depends only on the radial coordinate, i.e. is one-dimensional. The solution is obtained in an analytical form. This mathematical model is a generalization of the Wolkersen classical adhesive joint model for a circular or annular domain and is considered for the first time. The model problem is solved. The calculation results are compared with the calculations performed using the finite element method.

Keywords: Adhesive joint · Axisymmetric model · Analytical solution · Circular plate

1 Introduction

Considerable attention is paid to the local repair problem of composite and metal structures in aerospace engineering [1–4]. In this case, as a rule, two types of crippling of the main panel or plate are considered – in the form of a circular hole or in the form of a crack. In the overwhelming number of works devoted to the study of the stress state of the plates and patchs joints, the calculations are performed using the finite element method [5–8]. There are analytical methods for the stress state calculation of patched plates with holes [9–11], where it is assumed that the patch is fixed with the main plate along the line, while in the real design the patch is often fixed over the lining area. A plate with a glued circular patch in the adhesive area can be considered as a

© The Author(s), under exclusive license to Springer Nature Switzerland AG 2021
M. Nechyporuk et al. (Eds.): ICTM 2020, LNNS 188, pp. 310–319, 2021.
https://doi.org/10.1007/978-3-030-66717-7_26

three-layer plate with a soft joining layer. To build a solution, it is necessary to use models of three-layer plates, which have been used in the study of the adhesive joint stress state [12]. Well-known analytical solutions to the problem of the reinforcing patch and lap joints assume the rectangular geometry of the patch and plates to be joined, as well as a uniform distribution of stresses by the joint width [13–15]. There are also several different approximate two-dimensional models and corresponding methods for solving the problems of the adhesive joint stress state [16–22], where also the shape of the patch is assumed to be rectangular. Therefore, these approaches do not allow to obtain an analytical solution to the problem for a plate with a circular hole and a circular patch. It is obviously, that the axial symmetry of the structure under consideration requires the use of a polar coordinate system. This allows us to reduce the problem to a one-dimensional problem, as far as there is no dependence of stresses on the angular coordinate.

One-dimensional models of lap joints are used not only for glued beams calculation, but also for the axisymmetric coaxial tubes joint stress state calculation [23–25]. In this case, axial symmetry actually provides a uniform distribution of stresses along the circumferential coordinate. However, the stress state problem for the axisymmetric joint of a circular patch and plate is solved for the first time.

The aim of this work is to build an analytical solution and study the stress state of the lapped adhesive joint of a circular plate which has a circular hole with a coaxial circular patch. In the future, the proposed model can be developed for dynamic problems that arise, for example, during impacts on aerospace multilayer structures, including glazing [26] or attachment knots of honeycomb panels [27].

2   Formulation of the Problem

Consider the adhesive joint of two circular plates of the same thickness, shown in Fig. 1. The main plate is loaded with symmetrical biaxial tension. The radius of the hole in the main plate is R_1 patch radius is R_2. The main plate has a thickness δ_1 the patch has a thickness δ_2. The base and patch are made of isotropic materials, the elastic modulus of which, respectively E_1 and E_2, Poisson's ratio μ_1 and μ_2. The plates are joined using a joint layer, the thickness of which is δ_0, and the shear modulus is G_0.

Note that in practice, most often repairs are made using the same sheet-material the main structure is made, i.e. the thickness and properties of the patch coincide with the thickness and properties of the base. In addition, to reduce the bending of the structure, repairs can be performed using two identical patchs glued to the base from two sides. In this case, the thickness of the patch is two times less than the thickness of the base, and due to the symmetry, the problem can be reduced to the one solved in this paper.

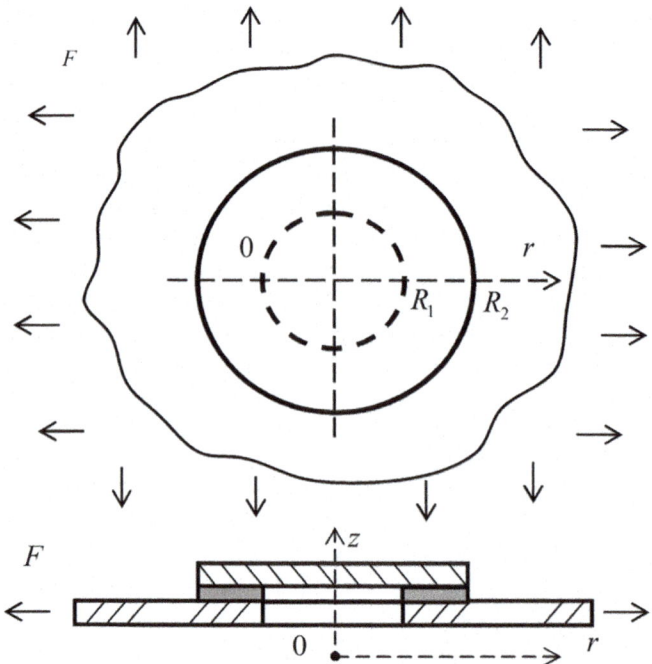

Fig. 1. Diagram of the adhesive joint of the plate with the patch.

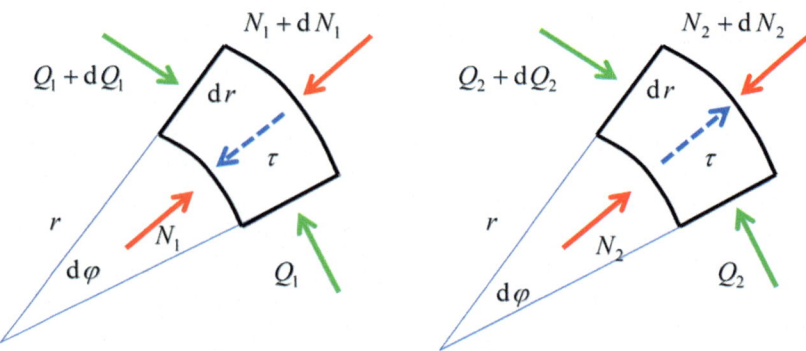

Fig. 2. Differential elements of the joint.

Due to axial symmetry, the tangential forces in the base layers Q_1 and Q_2 are independent on the angular coordinate, tangential forces in the base layers are absent. The subscript "1" corresponds to the main plate, and the subscript "2" corresponds to a circular plate within the gluing area $x \in [R_1; R_2]$.

Setting the direction of tangential stresses in the adhesive layer, we can write down the equilibrium equations of the base layers elements in the following form

$$\frac{N_1 - Q_1}{r} + \frac{dN_1}{dr} - \tau = 0, \quad \frac{N_2 - Q_2}{r} + \frac{dN_2}{dr} + \tau = 0, \tag{1}$$

where N_k, Q_k are radial and tangential forces in the base layer k, $k = 1, 2$; τ is the tangential stresses in the adhesive layer in the radial direction.

We assume that the stresses in the adhesive are proportional to the difference of the shifts sided to the adhesive layer of the both plates.

$$\tau = P(U_1 - U_2), \tag{2}$$

where P is a shear rigidity of the adhesive layer, $P = G_0/\delta_0$; U_k is a radial shifts of the layers, $k = 1, 2$.

The physical law equations for the plates have the form:

$$N_k = B_k\left(\varepsilon_{k,r} + \mu_k \varepsilon_{k,\varphi}\right), \quad Q_k = B_k\left(\varepsilon_{k,\varphi} + \mu_k \varepsilon_{k,r}\right), \tag{3}$$

where $B_k = \delta_k E_k/(1 - \mu_k^2)$ is a membrane rigidity of plates; $\varepsilon_{k,r}$ and $\varepsilon_{k,\varphi}$ are radial and tangential deformation of the layer k.

Kinematic relations of the elasticity theory are

$$\varepsilon_{k,r} = \frac{dU_k}{dr}, \quad \varepsilon_{k,\varphi} = \frac{U_k}{r}, \tag{4}$$

3 Constructing the Solution

Equations (1), using (4) and (2), can be represented as:

$$\frac{\tau}{B_1} + \frac{d^2 U_1}{dr^2} + \frac{1}{r}\frac{dU_1}{dr} - \frac{U_1}{r^2} = 0, \quad -\frac{\tau}{B_2} + \frac{d^2 U_2}{dr^2} + \frac{1}{r}\frac{dU_2}{dr} - \frac{U_2}{r^2} = 0. \tag{5}$$

Differentiating (2) and using the equations obtained above, we can obtain:

$$\frac{d^2\tau}{dr^2} + \frac{1}{r}\frac{d\tau}{dr} - \left(\frac{P}{B_1} + \frac{P}{B_2} + \frac{1}{r^2}\right)\tau = 0. \tag{6}$$

This equation has an analytical solution.

$$\tau = C_1 I_1(\lambda r) + C_2 K_1(\lambda r). \tag{7}$$

where $\lambda = \sqrt{\frac{P}{B_1} + \frac{P}{B_2}}$; I_1, K_1 are modified Bessel functions; C_1, C_2 an arbitrary constants.

It can be noted that in the adhesive joint stress state problem of rectangular plates, the tangential stresses in the adhesive are described by a linear combination of exponential functions [12, 15, 17]. In the simplest formulation of the problem, the so-called Wolkersen models [12], tangential stresses in the adhesive can be represented as a superposition of the hyperbolic sine and hyperbolic cosine. In the axisymmetric circular patch problem which is under consideration, the unlimited and non-periodic modified Bessel functions act as an analog of these hyperbolic and exponential functions.

Substituting tangential stresses (7) into one of Eqs. (5), then solving the nonhomogeneous Euler differential equations that was early obtained, and using relation (2), we get

$$
\begin{aligned}
U_1 &= -\frac{C_1}{\lambda^2 B_1} I_1(\lambda r) - \frac{C_2}{\lambda^2 B_1} K(\lambda r) + C_3 r + \frac{C_4}{r}, \\
U_2 &= \frac{C_1}{\lambda^2 B_2} I_1(\lambda r) + \frac{C_2}{\lambda^2 B_2} K(\lambda r) + C_3 r + \frac{C_4}{r}.
\end{aligned} \tag{8}
$$

4   Shifts Outside the Adhesive Area and Boundary Conditions

Movements in the inner area $(r < R_1)$ and in the external area $(r > R_2)$, i.e. outside of adhesive domain, are described by the well-known deformation equations for round plates

$$
\frac{d^2 U}{dr^2} + \frac{1}{r}\frac{dU}{dr} - \frac{U}{r^2} = 0.
$$

We denote the radial shifts of the patch in the inner part of the joint as U_3 and shifts of the main plate outside the joint as U_4. The equations which was given above have solutions

$$
U_3 = c_1 r + \frac{c_2}{r}, \quad U_4 = c_3 r + \frac{c_4}{r}. \tag{9}
$$

Shifts (8) and (9), as well as relations (4) and (3), make it possible to find forces in the main plate and patch both in the adhesive area and beyond.

Constants C_1, C_2, C_3, C_4, and c_1, $\ldots$, c_4 we find from the boundary conditions and the shift conjugation conditions with boundary forces of the domains.

Suppose that the main plate has a radius R_3. Boundary conditions at the outer boundary of the main plate are:

$$
N_4(R_3) = F;
$$

Conditions at the adhesive area border for the main plate are:

$$
U_1(R_2) - U_4(R_2) = 0; \quad N_1(R_2) - N_4(R_2) = 0; \quad N_2(R_2) = 0.
$$

Conditions at the adhesive area border for the hole are:

$$U_2(R_1) - U_3(R_1) = 0; \quad N_2(R_1) - N_3(R_1) = 0; \quad N_1(R_1) = 0.$$

And we find two other constants from the zero longitudinal shift conditions for the patch and the finite value of its transverse shifts at the origin point $(r = 0)$:

$$c_2 = 0.$$

Thus, we obtain a system of eight linear equations with respect to eight unknown constants.

5 Model Problem

When defining the geometry of the domain, we assume that the main plate has a very large radius R_3.

Model parameters are: $R_1 = 30$ mm, $R_2 = 50$ mm, $\delta_1 = \delta_2 = 3$ mm, $\delta_0 = 0,1$ mm, $E = 70$ GPa (aluminum alloy), $\mu = 0.28$, $E_0 = 0.8$ GPa, $G_0 = 0.3125$ GPa. Linear tensile forces F applied around the perimeter of the main plate, the outer radius of which is R_3 we will consider infinitely large $(R_3 = \infty)$.

In Fig. 3 it is shown the graphs of tangential stresses in the adhesive layer, calculated according to the proposed model (a), and also calculated using the finite element method (b). The stresses in the presented graphs are shown in dimensionless form.

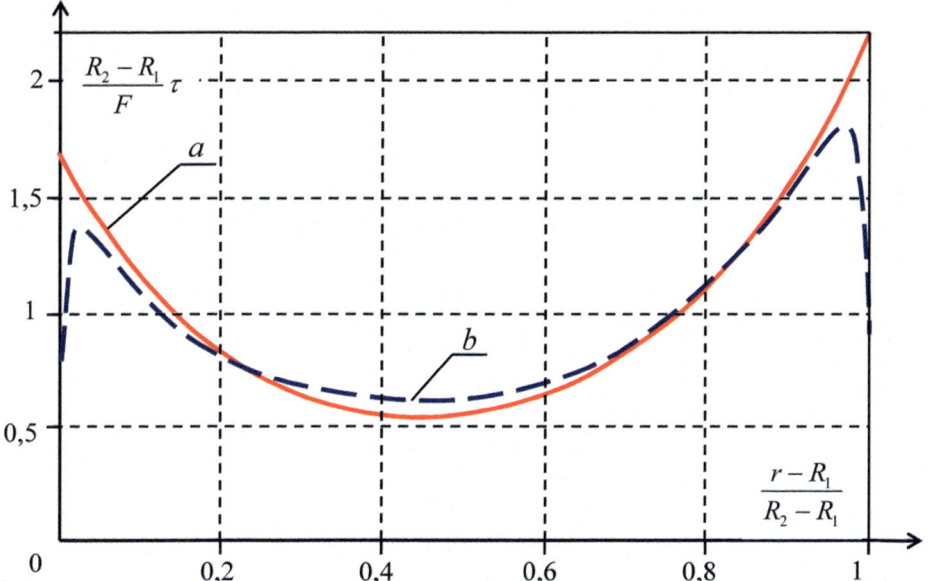

Fig. 3. Stresses in adhesive.

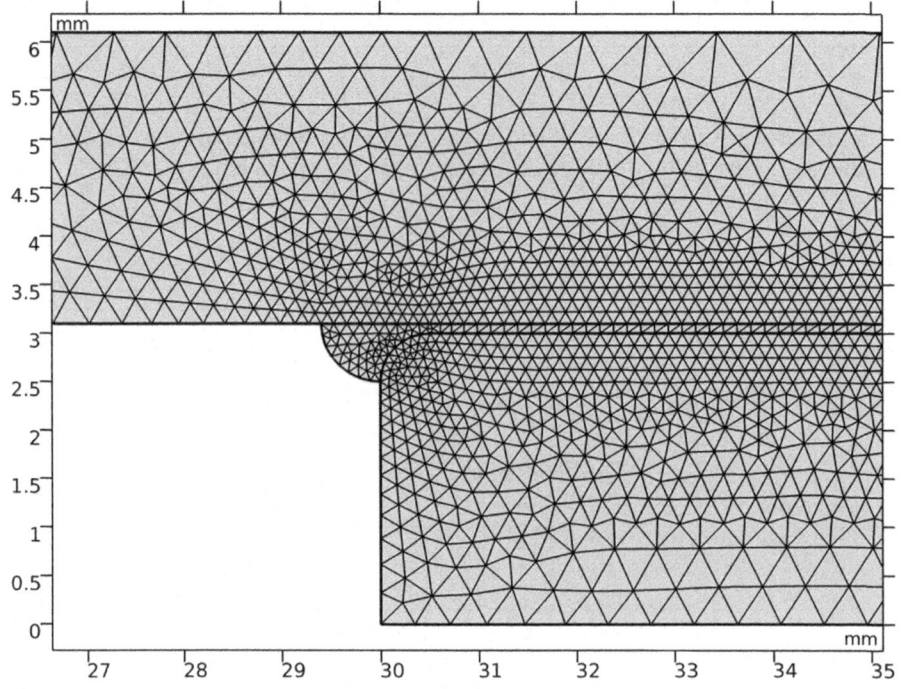

Fig. 4. Finite element model fragment.

The presented calculations were verified using finite element modeling in COMSOL Multiphysics 5.3a. For comparison, stresses were taken in the middle plane of the adhesive layer. Chamfers and squeezed glue excess are added to the finite element joint model. The neighborhood of the adhesive line edge and the grid of the finite elements are shown in Fig. 4.

Calculations showed that stresses calculated using the proposed model and using finite element modeling coincide in almost the entire adhesive area. Differences are observed only in small areas at the edges of the glue line, the length of this area being on the order of the thickness of the adhesive layer. The differences found do not exceed a few percent, and the graphs almost coincide in the results. Moreover, the approach proposed in this paper gives somewhat overrated results, which is acceptable for design problems. The described small differences between the results can be explained by the fact that the outer edge of the adhesive joint has a load-free border, as a result of which the tangential stresses at the edge of the adhesive layer should be zero, this fact cannot be effected by the proposed model. This feature of modeling the adhesive stress state by (4) is well known [12] and can be overcome using more accurate approaches to the description of the adhesive layer stress state [28].

To illustrate how the patch unloads the hole, we consider the graphs of the radial and tangential force relations in the main plate. In Fig. 5, radial and tangential forces are shown in dimensionless form in a certain neighborhood of the hole. It is obviously, that at infinity we have a uniform stress state $N_4 = Q_4 = F$. The ratio of radial forces to F is denoted as "a" on the graph, and the ratio of tangential forces to F as "b".

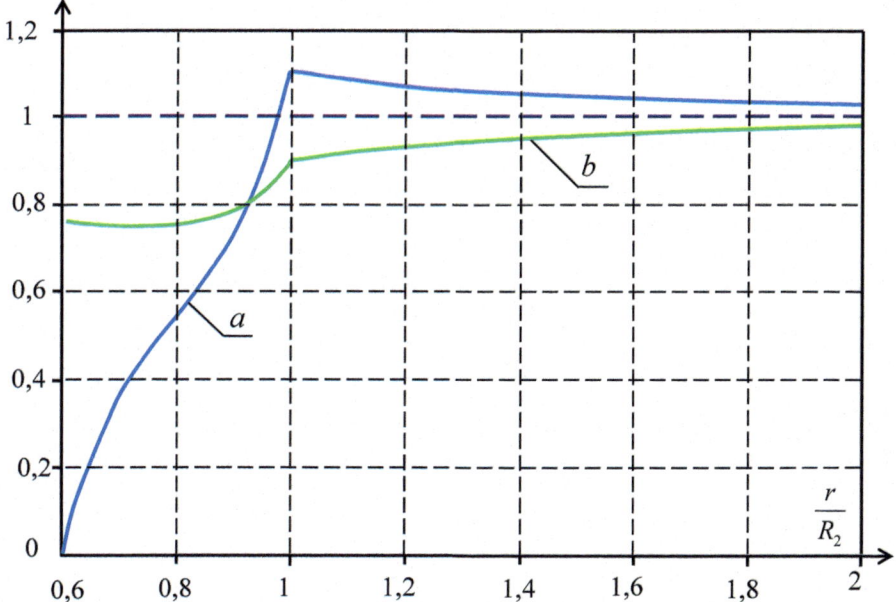

Fig. 5. Stresses in the main plate in the neighborhood of the hole: (a) – N/F; (b) – Q/F

The sharp bend in the graphs corresponds to the patch border.

In the absence of a patch at the hole edge, normal forces would be equal to zero, and tangential forces would equal $2F$. In this case, we have only a small increase in radial forces to $1.1\,F$. That is, as expected, the patch significantly unloads the hole. Calculations show that in the patch over the hole $Q_3 = N_3 < 0.9F$.

6 Conclusion

A stress state mathematical model for the axisymmetric plate with a circular hole, which is overlapped by an adherent coaxial circular patch, is proposed. The problem is reduced to a linear differential equation with respect to tangential stresses in the adhesive layer. This equation has an analytical solution within Bessel functions.

A feature of the problem is that, in contrast to beam joints under unidirectional loading, not all forces from the main plate are transferred to the patch by the adhesive layer. The patch a little bit unloads the hole, reducing the tangential stresses in the main plate in the neighborhood of the hole.

Possible directions for the development of the proposed model are

– taking into account the plates bending;
– inclusion into the model the temperature deformations;
– inclusion into the model the inertia forces and study the dynamic stresses;
– generalization to the case of uniaxial stress state of the base at infinity.

References

1. Tomblin, J.S., Salah, L., Welch, J.M., Borgman, M.D.: Bonded repair of aircraft composite sandwich structures. Final Report DOT/FAA/AR-03/74. Federal Aviation Administration, Office of Aviation Research, Washington (2004)
2. Baker, A.A., Rose, L.R.F., Jones, R. (eds.): Advances in the Bonded Composite Repair of Metallic Aircraft Structures, vol. 1. Elsevier, Amsterdam (2002)
3. Fedotov, A.A., Tsipenko, A.V.: An analytical model of the damage adhesive repair of an aircraft skin with taking into account the degradation of material properties. Scientific Bulletin MSTU CA **19**(6), 118–126 (2016). (in Russian)
4. Bakuckas, J.G., Chadha, R., Swindell, P., Fleming, M., Lin, J.Z., Ihn, J.B., Desai, N., Espinar-Mick, E., Freisthler, M.: Bonded repairs of composite panels representative of wing structure. In: Niepokolczycki, A., Komorowski, J. (eds.) ICAF 2019. LNME, pp. 565–580. Springer, Cham (2020). https://doi.org/10.1007/978-3-030-21503-3_45
5. Okafor, A.C., Singh, N., Enemuoh, U.E., Rao, S.V.: Design, analysis and performance of adhesively bonded composite patch repair of cracked aluminum aircraft panels. Compos. Struct. **71**(2), 258–270 (2005). https://doi.org/10.1016/j.compstruct.2005.02.023
6. Tsouvalis, N.G., Mirisiotis, L.S., Dimou, D.N.: Experimental and numerical study of the fatigue behaviour of composite patch reinforced cracked steel plates. Int. J. Fatigue **31**(10), 1613–1627 (2009). https://doi.org/10.1016/j.ijfatigue.2009.04.006
7. Sabelkin, V., Mall, S., Hansen, M.A., et al.: Investigation into cracked aluminum plate repaired with bonded composite patch. Compos. Struct. **79**(1), 55–66 (2007). https://doi.org/10.1016/j.compstruct.2005.11.028
8. Fedotov, A.A., Tsipenko, A.V., Lebedev, A.I.: Numerical modeling of adhesive repair joint. Sci. Bull. MSTU CA **21**(3), 125–138 (2018). https://doi.org/10.26467/2079-0619-2018-21-3-125-138. (in Russian)
9. Zemlyanova, A.Y., Sil'vestrov, V.V.: The problem of the reinforcement of a plate with a cutout by a two-dimensional patch. J. Appl. Math. Mech. **71**(1), 40–51 (2007). https://doi.org/10.1016/j.jappmathmech.2007.03.012
10. Sil'vestrov, V.V., Zemlianova, A.Y.: Patch repair of a plate with a circular cut. J. Appl. Mech. Tech. Phys. **45**(4), 176–183 (2004)
11. Zemlyanova, A.Y.: Reinforcement of a plate weakened by multiple holes with several patches for different types of plate-patch attachment. Math. Mech. Solids **21**(3), 281–294 (2016). https://doi.org/10.1177/1081286513519812
12. Da Silva, L.F., das Neves, P.J., Adams, R.D., Spelt, J.K.: Analytical models of adhesively bonded joints – part i: literature survey. Int. J. Adhes. Adhes. **29**(3), 319–330 (2009). https://doi.org/10.1016/j.ijadhadh.2008.06.005
13. Zhang, X., Wu, J., Fan, Z., et al.: Cohesive shear stress and strength prediction of composite patch bonded to metal reinforcement. Int. J. Adhes. Adhes. **90**, 144–153 (2019). https://doi.org/10.1016/j.ijadhadh.2019.02.008
14. Kurennov, S.S., Koshevoi, A.G., Polyakov, A.G.: Through-thickness stress distribution in the adhesive joint for the multilayer composite material. Russ. Aeronaut. (Iz VUZ) **58**(2), 145–151 (2015). https://doi.org/10.3103/S1068799815020026
15. Lee, J., Cho, M., Kim, H.S.: Bending analysis of a laminated composite patch considering the free-edge effect using a stress-based equivalent single-layer composite model. Int. J. Mech. Sci. **53**(8), 606–616 (2011). https://doi.org/10.1016/j.ijmecsci.2011.05.007
16. Rapp, P.: Mechanics of adhesive joints as a plane problem of the theory of elasticity. Part II: displacement formulation for orthotropic adherends. Arch. Civ. Mech. Eng. **15**(2), 603–619 (2015). https://doi.org/10.1016/j.acme.2014.06.004

17. Kurennov, S.S.: An approximate two-dimensional model of adhesive joints. analytical solution. Mech. Compos. Mater. **50**(1), 105–114 (2014). https://doi.org/10.1007/s11029-014-9397-z

18. Kurennov, S.S.: Determining stresses in an adhesive joint with a longitudinal unadhered region using a simplified two-dimensional theory. J. Appl. Mech. Tech. Phys. **60**(4), 740–747 (2019). https://doi.org/10.1134/S0021894419040199

19. Kurennov, S.S.: A simplified two-dimensional model of adhesive joints. nonuniform load. Mech. Compos. Mater. **51**(4), 479–488 (2015). https://doi.org/10.1007/s11029-015-9519-2

20. Kurennov, S.S., Barakhov, K.P.: The stressed state of the double-layer rectangular plate under shift. The simplified two-dimensional model. PNRPU Mech. Bull. **3**, 166–174 (2019). https://doi.org/10.15593/perm.mech/2019.3.16. (in Russian)

21. Kim, H.S., Cho, M., Lee, J., et al.: Three dimensional stress analysis of a composite patch using stress functions. Int. J. Mech. Sci. **52**(12), 1646–1659 (2010). https://doi.org/10.1016/j.ijmecsci.2010.08.006

22. Kessentini, R., Klinkova, O., Tawfiq, I., Haddar, M.: Transient hygro-thermo-mechanical stresses analysis in multi-layers bonded structure with coupled bidirectional model. Int. J. Mech. Sci. **150**, 188–201 (2019). https://doi.org/10.1016/j.ijmecsci.2018.10.004

23. Lubkin, J.L., Reissner, E.: Stress distribution and design data for adhesive lap joints between circular tubes. Trans. ASME **78**, 1213–1221 (1956)

24. Selahi, E.: Elasticity solution of adhesive tubular joints in laminated composites with axial symmetry. Arch. Mech. Eng. **65**(3), 441–456 (2018). https://doi.org/10.24425/124491

25. Kurennov, S.S., Barakhov, K.P., Poliakov, A.G.: Stressed state of the axisymmetric adhesive joint of two cylindrical shells under axial tension. Mater. Sci. Forum **968**, 519–527 (2019). https://doi.org/10.4028/www.scientific.net/MSF.968.519

26. Rodichev, Y.M., Smetankina, N.V., Shupikov, O.M., Ugrimov, S.V.: Stress-strain assessment for laminated aircraft cockpit windows at static and dynamic loads. Strength Mater. **50**(6), 868–873 (2019). https://doi.org/10.1007/s11223-019-00033-4

27. Kondratiev, A., Gaidachuk, V., Nabokina, T., Tsaritsynskyi, A.: New possibilities of creating the efficient dimensionally stable composite honeycomb structures for space applications. In: Nechyporuk, M., Pavlikov, V., Kritskiy, D. (eds.) Integrated Computer Technologies in Mechanical Engineering. AISC, vol. 1113, pp. 45–59. Springer, Cham (2020). https://doi.org/10.1007/978-3-030-37618-5_5

28. Amidi, S., Wang, J.: Three-parameter viscoelastic foundation model of adhesively bonded single-lap joints with functionally graded adherends. Eng. Struct. **170**, 118–134 (2018). https://doi.org/10.1016/j.engstruct.2018.05.076

Improving of Energy Efficiency of Cruise Ships by Applying of Thermochemical Recuperation

Oleksandr Cherednichenko◉, Mykhaylo Tkach◉, and Vira Mitienkova$^{(\boxtimes)}$◉

Admiral Makarov National University of Shipbuilding, 9 Heroes of Ukraine Avenue, Mykolayiv 54025, Ukraine
oleksandr.cherednichenko@nuos.edu.ua,
mykhaylo.tkach@gmail.com, vera.mitenkova@gmail.com

Abstract. The growth of the cruise industry leads to the proliferation of the fleet of cruise ships, but it will increment the atmospheric emission. The International Maritime Organization (IMO) has imposed restriction on the various pollution components of exhausted gases from ships, including carbon dioxide that contributes greatly to the greenhouse effect and is an indicator of energy efficiency for maritime transport. In this paper, ten cruise ships built in 2013–2019 with basic diesel-electric power plants are considered. The schematic diagrams of the alternative combined gas-turbine-electric and diesel-electric power plants applying thermochemical regenerators fed by exhaust gases have been developed for vessels under consideration. Energy efficiency design index for the cruise ships with both the basic and alternative power plants operating on marine fuel oils, liquefied natural gas and syngas has been estimated and compared with the IMO guideline values for this type of vessels. It has been revealed that the application of the thermochemical technologies of waste heat recovery increases the energy efficiency of cruise ships by 15–25%.

Keywords: Cruise ships · Energy efficiency design index · Gas-turbine engine · Syngas · Thermochemical heat recovery

1 Introduction

A review of development trends of the marine cruise business indicates a significant contribution of this industry to the global economy. For instance, the number of passengers on cruise ships in recent years is more than 25 million people annually. The financial impact of the cruise industry on the global economy is $126 billion, and it ensures more than 1 million jobs paying $41 billion in wages and salaries. Moreover, the cruise ship order book from 2018–2025 includes 50 new ocean-going vessels with an investment value of $51 billion [1].

Practice design of the propulsion complexes for passenger ships shows that ship power plants (SPP) with electric propulsion are nowadays the most widely used [2]. At the same time, single power plants with rudder propeller units like podded propulsors dominate on the modern cruise ships [3]. The application of these design solutions allows generating electrical energy in the same generating sets for both the propulsion units and general ship demands [3]. A feature of SPP of cruise ships is the surplus of total plant power with respect to propulsion power. It has to do with the redundancy of power to ensure reliable and safe ship operation and, most importantly, with large power needs to serve numerous household consumers.

Medium-speed diesel engines are the most common generator set engines on the cruise ships. It can be from four to eight auxiliary engines and from two to four propulsion motors in single ship power plants that ensures the necessary reliability and survivability of SPP [2]. The disadvantage of such design solutions is the complication of the entire plant, including an increase of the mass-to-power ratio and volumetric and dimensional parameters.

In the early 2000s, at relatively low oil prices, combined gas-turbine-electric and diesel-electric power plants (COGED) found application in a number of cruise ships with electric propulsion and rudder propeller units [4]. The cruise ships with combined steam-turbine and gas-turbine power plants were also put into operation, and such plants included two gas-turbine engines (GTE) and a heat recovery steam turbine generator [5]. With rising oil prices, the application of gas turbine engines became economically inefficient, becoming unprofitable with the price of marine gas oil of $75 [6].

Nevertheless, GTE-based plants have several advantages compared to diesel ones such as a mass-to-power ratio (measured in kg/kW), the possibility of obtaining a large amount of heat energy without additional fuel consumption by using waste heat of GTE exhaust gases, and reduction of harmful exhaust emission components. Smaller mass-dimensions parameters of combined plants allow accommodating up to 45 additional cabins on board [6].

The new IMO (International Maritime Organization) requirements relating to the maximum sulfur content in marine fuels became effective on January 1, 2020. According to these, the sulfur content in fuels should not exceed 0.5% outside emission control areas (SECA) and 0.1% in SECA. The new requirements apply to marine fuels for both main and auxiliary engines and boilers [7]. Also since January 1, 2020, EEDI Phase 2 for reduction of the carbon dioxide emission became effective as a part of the global strategy to decrease greenhouse gas emissions. CO_2 emission at Phase 2 for the new transport vessels covered by the energy efficiency design index (EEDI) is expected to reduce by 15%–20% of 2013 reference value dependent on the deadweight and ship types [8]. Regarding the emission of nitrogen oxides, Tier II and Tier III limits govern this component of exhaust gases of marine engines, both of these limits were introduced in 2011 and 2016, respectively, and applied outside the special emission control areas (NECA) and in such areas [7].

The operation of cruise ships is most frequently carried out in inshore traffic zones, bays and inland seas. In these areas, the IMO requirements for regulated emission including nitrogen oxides, sulfur and carbon dioxide are harder than in high seas [9]. Hence, the issue of ecological efficiency improving of cruise ships is sufficiently relevant. While the emission of nitrogen oxides and sulfur can be significantly reduced by using the secondary cleaning methods, for example, scrubbers and catalytic converters, CO_2 as a greenhouse gas can be decreased mainly by lowering the specific fuel consumption or by burning low carbon fuels.

Cruise ships ranks seventh among all the type of ships in terms of CO_2 emission (Fig. 1) [10]. An effective way to reduce carbon dioxide pollution for transport ships is to run engines on natural gas (NG) or other alternative fuels, for instance, methanol [11]. Another promising option can be the application of thermochemical regenerators of waste heat recovery (TCRs), as confirmed by the efficiency of using this technical solution on gas carriers transporting liquefied natural gas (LNG) [12]. There are also known the solutions to reduce specific fuel consumption by cooling the air at the inlet of the gas turbine compressor [13–17]. However, while operating on natural gas, the emission of another greenhouse gas such as methane grows simultaneously (so called methane slip) [18]. Although nowadays the level of methane emission is not regulated, the growth in the number of LNG-fueled ships may lead to a revision of this issue. It is worth noting that in contrast to diesel engines, GTE operating on natural gas have near zero methane slip [6].

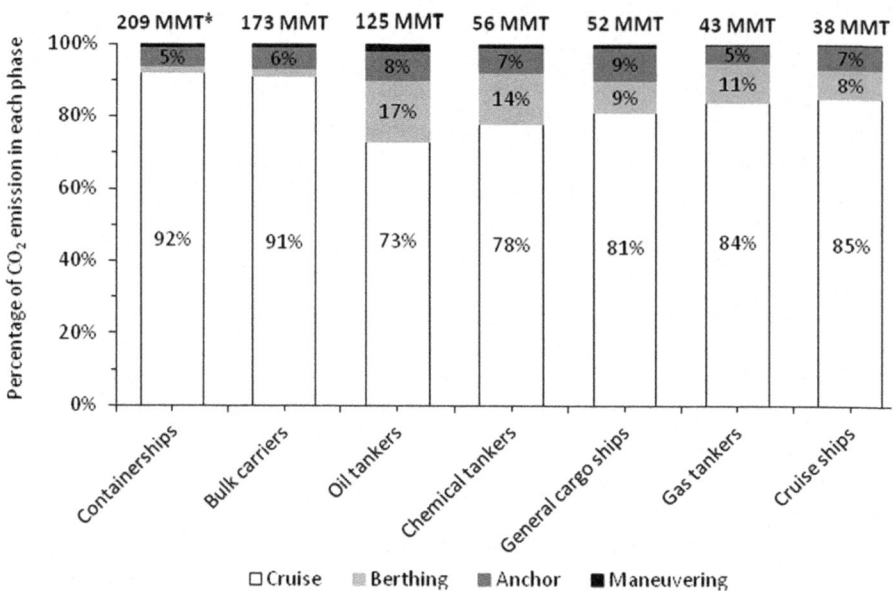

Fig. 1. CO_2 emissions by phase for top-emitting ship classes, 2015 (*MMT – million metric tons).

To address the mentioned above issue associated with improving the energy efficiency of cruise ships, the research objectives have been formulated:

- to develop the schematic diagrams and layout arrangements of alternative power plants for modern cruise ships;
- to evaluate the EEDI for cruise vessels with a basic single diesel-electric power plant;
- to estimate and analyze the impact of the application of thermochemical technologies on the ship energy efficiency.

The paper is organized as follows: in Sect. 2 the schemes of alternative SPP applying thermochemical regenerators for cruise ships are developed, in Sect. 3 the energy efficiency design index for modern cruise vessels with basic and alternative power plants including ones with TCRs are estimated.

2 Development of Schematic Diagrams of Alternative Power Plants with TCRs for Cruise Ships

The concept of the application of thermochemical recuperation of waste heat of marine heat engines follows the main trends of SPP development given in the analytical review *Global Marine Technology Trends* 2030 prepared by Lloyd's Register [19–21].

According to the review, one of the promising directions is the application of alternative fuels. Thermochemical technologies make it possible to convert a basic fuel using waste heat of exhaust gases. Ethanol, methanol, natural gas, petroleum gas of different compositions, or gaseous hydrocarbon fuels (ethane, propane, butane and others) can be used as the main energy source.

The second direction follows the trend in the application of TCRs in COGED. It is envisaged in such power plants to apply thermochemical recuperation of exhaust gases heat of GTE via steam conversion of a hydrocarbon fuel. Conversion products are used as a fuel.

The efficiency of thermochemical technologies of waste heat recovery requires that the temperature and power potential of exhaust gases are sufficient to produce syngas at the required volume and composition from a basic fuel.

In the article, the design study of modernization of cruise ship power plants has been conducted to determine whether the proposed diagrams and parameters of power plants meet the IMO energy efficiency requirements. The application of system analysis methods has been allowed for a multidimensional classification of the main diagrams of cruise ship power plants.

The type of the main engine or power plant has been taken as the major parameter of classification. Depending on this parameter, engines can be two-stroke slow-speed diesel (D2S), four-stroke medium-speed diesel (D4S), gas-turbine (GT) of a simple cycle (E) or a combined cycle (C). The additional parameters of classification are the number of fuels in use (SF – single-fueled and DF – dual-fueled) and power transmission (M – mechanical and E – electrical).

Ten cruise ships falling under the EEDI have been considered, all of the ships are equipped with single power plants with four-stroke medium-speed diesel engines and electrical power transmission (D4S/E). Ship designs have been selected from the *Significant Ships* issues published by the Royal Institution of Naval Architects in 2014–2020. The designs comprise a reasonably representative sample, which allows determining the major trends in use of various shipbuilding technologies.

In previous studies, the authors identified promising schemes of ship power plants using TCRs [22]. In accordance with these studies, the alternative power plant for the cruise ships was selected according to the scheme DF/GTE/E + DF/D4S/E + TCRs, i.e. a combined gas-turbine-electric and diesel-electric power plant with thermochemical regenerators of waste heat recovery. As an intermediate option, the SPP of

DF/GTE/E + DF/D4S/E type without TCRs operating on LNG has been considered. Only a gas-turbine engine operates on natural gas in the diagram with TCRs, while diesel engines burn syngas produced from NG. An example of such a diagram for one of the cruise ships under study is shown in Fig. 2. Depending on the required power, three configurations of alternative SPP have been considered:

- 3 generator set engines (one GTE and two diesel engines of equal power);
- 4 generator set engines (one GTE, two diesel engines of equal power and one diesel engine of different power);
- 5 generator set engines (one GTE and two pairs of diesel engines of equal power).

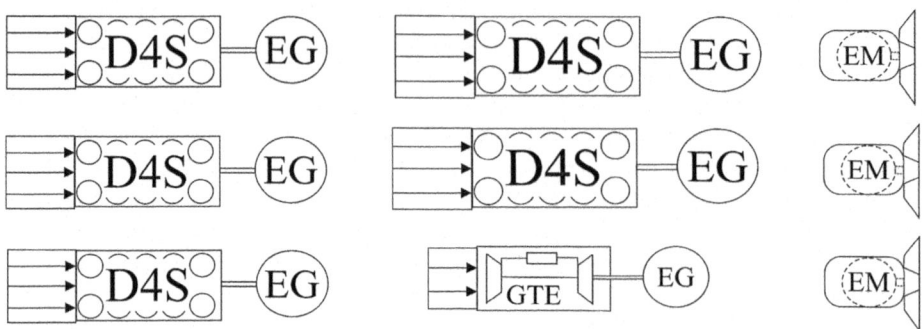

Fig. 2. The alternative schematic diagram of the power plant for the cruise ship "MSC SEASIDE".

Table 1 provides the parameters of basic and alternative power plants for the cruise ships under discussion.

As indicated by numerous studies, the application of TCRs (including the plasma-chemical method) ensures not only a decrease in specific fuel consumption in the propulsion system but also a significant reduction in carbon dioxide generated at fuel combustion (its amount is defined by the carbon content in the fuel) [12].

Carbon content can be calculated using the following methodology [23]:

$$CC_{Mixture} = \frac{\sum_{i=1}^{\#of\_Components} (n \times AW_{Carbon} \times Xm_i)}{\sum_{i=1}^{\#of\_Components} (MW_i \times Xm_i)}, \tag{1}$$

where $CC_{Mixture}$ is the carbon content weight fraction of the mixture; n is the number of carbon atoms in the component; AW_{Carbon} is the atomic weight of carbon; MW_i is the molecular weight of component i; Xm_i is the mole fraction.

The main components of syngas formed during steam conversion of hydrocarbon fuel are H_2, H_2O, CH_4, CO_2, CO.

Table 1. Parameters of the cruise ships and SPP configuration.

	Ship names (Gross tonnage, gt)	Year of built	Power, MW		Type of basic SPP, (number × power of main engines, MW)	Type of alternative SPP, (number × power of main engines, MW)
			Main engines	Power plant		
1	EUROPA 2 (42830)	2013	24	24	D4S/E (4 × 6,0)	(DF/GTE/E) + (DF/D4S/E) (1 × 5,1 + 2 × 9,1)
2	COSTA DIADEMA (133000)	2014	52	52	D4S/E (4 × 12,6 + 2 × 0,8)	(DF/GTE/E) + (DF/D4S/E) (1 × 24,8 + 2 × 11,4 + 1 × 5,7)
3	MEIN SCHIFF 3 (99526)	2014	48	48	D4S/E (2 × 14,4 + 2 × 9,6)	(DF/GTE/E) + (DF/D4S/E) (1 × 24,8 + 2 × 11,4)
4	VIKING STAR (47842)	2015	23,52	23,52	D4S/E (2 × 6,72 + 2 × 5,04)	(DF/GTE/E) + (DF/D4S/E) (1 × 5,1 + 2 × 9,1)
5	AID APRIMA (125572)	2016	46,8	46,8	D4S/E (3 × 12 + 1 × 10,8 DF)	(DF/GTE/E) + (DF/D4S/E) (1 × 24,8 + 2 × 11,4)
6	KONNIGSDAM (99836)	2016	50,4	50,4	D4S/E (4 × 12,6)	(DF/GTE/E) + (DF/D4S/E) (1 × 24,8 + 2 × 10,3 + 1 × 6,9)
7	SEABOURN (41865)	2016	23,04	23,04	D4S/E (4 × 5,76)	(DF/GTE/E) + (DF/D4S/E) (1 × 5,1 + 1 × 7,6 + 2 × 5,7)
8	SEVEN SEAS EXPLORER (55254)	2016	36	36	D4S/E (4 × 8,0)	(DF/GTE/E) + (DF/D4S/E) (1 × 24,8 + 2 × 5,7)
9	MSC SEASIDE (153516)	2017	62,4	62,4	D4S/E (2 × 14,4 + 2 × 16,8)	(DF/GTE/E) + (DF/D4S/E) (1 × 24,8 + 2 × 11,4 + 2 × 8.6)
10	SPIRIT OF DISCOVERY (58119)	2019	21,6	21,6	D4S/E (4 × 5,4)	DF/GTE/E) + (DF/D4S/E) (1 × 5,1 + 2 × 5,7)

The carbon content in syngas produced via steam conversion of natural gas by thermochemical regenerators of waste heat depends on the temperature of exhaust gases and composition of a basic hydrocarbon fuel made by dry components. The values of the carbon content in syngas needed for the calculation of EEDI have been determined for every ship from Table 1 when conducting the design study of alternative ship power plants.

3 Analysis of the Impact of the Application of Thermochemical Technologies on the Energy Efficiency of Cruise Ships

The EEDI for new ships has been adopted in Annex VI of MARPOL 73/78. The attained EEDI for new ships is a measure of their energy efficiency estimated as carbon dioxide emission per ship transport work and measured in g/(t·nm). The EEDI is applicable also to cruise ships and calculated by the following formula [24]:

$$
EEDI = \frac{\left(\prod_{j=1}^{M} f_j\right)\left(\sum_{i=1}^{nME} P_{ME(i)} \cdot C_{FME(i)} \cdot SFC_{ME(i)}\right) + (P_{AE} \cdot C_{FAE} \cdot SFC_{AE}*)}{f_i \cdot f_c \cdot f_l \cdot Capacity \cdot V_{ref} \cdot f_w}
$$
$$
+ \frac{\left(\left(\prod_{j=1}^{M} f_j \cdot \sum_{i=1}^{nPTI} P_{PTI(i)} - \sum_{i=1}^{neff} f_{eff(i)} \cdot P_{AEeff(i)}\right) C_{FAE} \cdot SFC_{AE}\right) - \left(\sum_{i=1}^{neff} f_{eff(i)} \cdot P_{eff(i)} \cdot C_{FME} \cdot SFC_{ME}\right)}{f_i \cdot f_c \cdot f_l \cdot Capacity \cdot V_{ref} \cdot f_w}.
$$

$$(2)$$

*If a part of the normal maximum sea load is provided by shaft generators, SFC_{ME} and C_{FME} replace SFC_{AE} and C_{FAE} for that part of the power.

Here, SFC is the specific fuel consumption for main engines (ME) and auxiliary engines (AE), measured in g/kWh.

For cruise ships, gross tonnage should be used as $Capacity$. Other elements of the formula include the following:

C_F is a non-dimensional conversion factor between fuel consumption and CO_2 emission measured in g, which is based on carbon content in the fuel; V_{ref} is the ship speed measured in knots; $P_{ME(i)}$ is the power of the ME measured in kW; $P_{PTI(i)}$ is 75% of the rated power consumption of each shaft motor divided by the weighted average efficiency of the generator(s), kW; $P_{eff(i)}$ is the output of the innovative mechanical energy-efficient technology for propulsion at 75% main engine power; $P_{AEeff(i)}$ is the auxiliary power reduction due to innovative electrical energy-efficient technology; P_{AE} is the auxiliary engine power required to supply the normal maximum sea load; f_j is a correction factor to account for a ship's specific design elements; f_w is a non-dimensional coefficient indicating the decrease of speed in representative sea conditions; $f_{eff(i)}$ is the availability factor of each innovative energy efficiency technology; f_i is the capacity factor for any technical/regulatory limitation on capacity; f_c is the cubic capacity correction factor; and f_l is the factor for general cargo ships equipped with cranes and other cargo-related gear to compensate for deadweight losses [24].

For cruise ships, the formula of the EEDI reference lines is as follows:

$$Reference\,EEDI = 170,84 * GT^{-0,214},\qquad(3)$$

where *GT* denotes the gross tonnage in accordance with the International Convention of Tonnage Measurement of Ships 1969, annex I, regulation 3. It is expected to reduce permissible CO_2 emission calculated by the formula (3) by 5% by the end of 2019 (Phase 1), by 20% by the end of 2024 (Phase 2) and by 30% since 2025 (Phase 3) [8].

Here, we will briefly cover some features of the attained EEDI calculation for cruise ships. It includes the estimation of the main and auxiliary engines power for diesel-electric power plants taking into account the distribution of produced energy between propulsion and general ship demands.

Currently, the design procedure of the EEDI for ship combined power plants has not been developed yet, but, the appropriate IMO regulations permit extending the EEDI to combined diesel-gas-turbine power plants for cruise ships [8, 24]. The standard techniques have been applied to EEDI calculation for alternative power plants (Table 1). The proportion between the rated power of installed GTE and diesel engines has been talking into consideration, while the estimation of propulsion power input to propulsion motors and general ship consumers. In addition, specific fuel consumption for every engine type based on the share of output power is taken into account. In power plants DF/GTE/E + DF/D4S/E type with TCRs, C_F from formula (2) for GTE has been taken for natural gas, but for diesel engines, it has been calculated in accordance with the formula (1).

The results of the determined EEDI for cruise ships under study with basic and alternative power plants without and with TCRs are shown in Fig. 3 and Fig. 4.

It has been calculated the following EEDI for every ship from Table 1:

- *EEDI-ref* is a reference energy efficiency index calculated in accordance with the formula (3);
- *EEDI-base* is the attained EEDI for a basic ship power plant operating on marine fuel oils;
- *EEDI-LNG* is the attained EEDI for an alternative ship power plant without TCRs operating on natural gas;
- *EEDI-SG-LNG* is the attained EEDI for an alternative ship power plant with TCRs operating on syngas (conversion product of natural gas). The last three values have been calculated in accordance with the formula (2).

As seen in Fig. 3, that even with basic SPP, the EEDI for cruise ships under study does not exceed the reference value. Despite the higher specific fuel consumption in GTE in comparison with diesel engines, the application of COGED operating on LNG allows further reducing CO_2 emission by 3.5–13.5% for eight ships out of ten ones. The application of TCRs will reduce specific emission by 14.5–24% compared with basis power plants for all ten cruise ships. It is worth noting that the attained EEDI for all ships under consideration with power plants DF/GTE/E + DF/D4S/E type meets the most stringent IMO requirements for CO_2 emissions. Therefore, cruise ships with such power plants will remain environmentally friendly in terms of CO_2 emissions, even with further tightening of the requirements for the emission of this greenhouse gas.

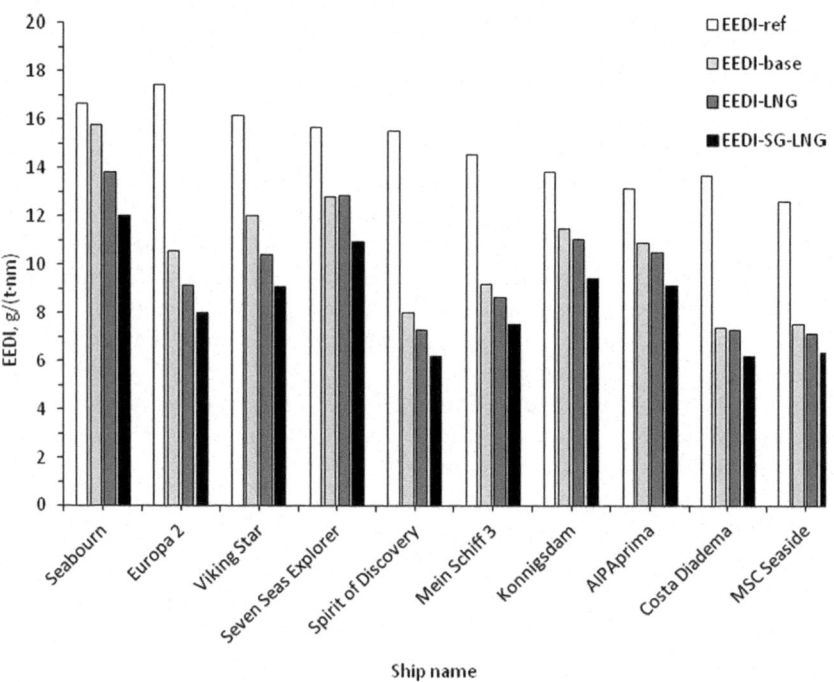

Fig. 3. Comparison of the EEDI for basic and alternative power plants of cruise ships operating on marine fuel fuels, LNG and syngas.

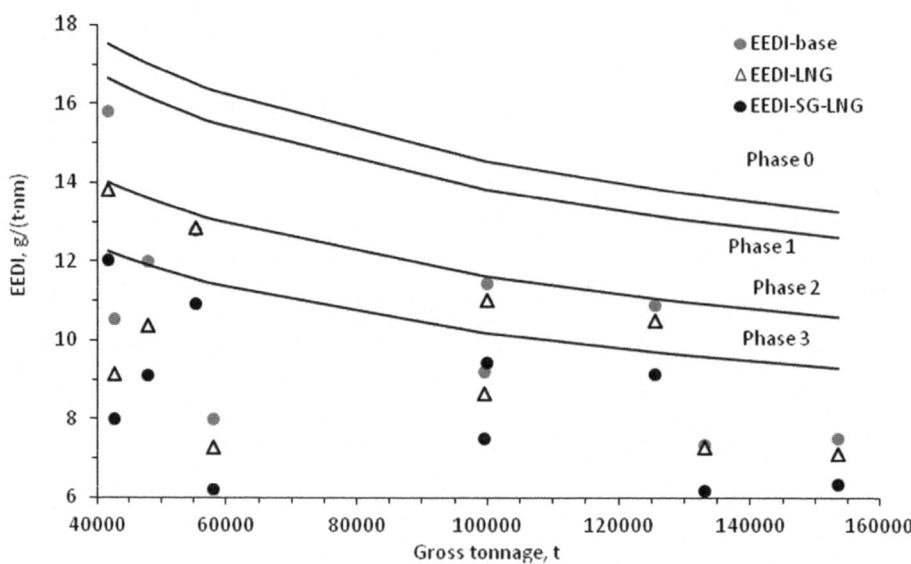

Fig. 4. Comparison of the attained EEDI and the EEDI reference lines for cruise ships.

Furthermore, as mentioned above, while GTE operating on natural gas and diesel engines operating on syngas, the emission of nitrogen oxides, sulfur and methane slip are reduced significantly. Consequently, the comprehensive environmental efficiency of cruise ships with power plants DF/GTE/E + DF/D4S/E type with TCRs will be higher than for vessels with diesel electric power plants. If smaller mass-dimensions parameters and other advantages of GTE application described above are taken into account, the operation of cruise ships will be economically feasible at the current oil prices.

4 Conclusion

The IMO stricter requirements for exhaust gas emissions from ships in conjunction with falling oil prices, and as a result, oil products, make it promising and economically feasible to use combined plants DF/GTE/E + DF/D4S/E type on cruise ships.

The application of equipment for thermochemical recuperation of waste heat recovery in ship power plants enables to obtain syngas without additional energy consumption and provides flexibility in choosing of hydrocarbon fuels that are feedstock for syngas generation.

The operation of diesel engines on syngas will significantly reduce IMO regulated emission and ensure the energy and ecological efficiency of cruise ships at the level of the world's stringiest standards.

Replacing single diesel-electric power plants by COGED allows reducing specific carbon dioxide emissions by 0.5–13.5% for eight cruise ships out of ten ones under study, thereby improving the ship energy efficiency.

The application of the thermochemical recuperation in plants DF/GTE/E + DF/D4S/E type allows cutting specific emission of CO_2 by 14.5–24% compared to basic power plants for all cruise ships under discussion.

References

1. Cruise Industry Overview. The Florida-Caribbean Cruise Association, Florida (2018)
2. Bond, M.: ABB to power all three new Virgin Voyages cruise ships. Seatrade Cruise News. https://www.seatrade-cruise.com/news-headlines/abb-power-all-three-new-virgin-voyages-cruise-ships (2018). Accessed 14 July 2020
3. ABB: Azipod® electric propulsion: The driving force behind safe, efficient and sustainable operation. https://new.abb.com/marine/systems-and-solutions/azipod. Accessed 14 July 2020
4. Armellini, A., Daniotti, S., Pinamonti, P.: Gas Turbines for power generation on board of cruise ships: A possible solution to meet the new IMO regulations? Energy Procedia **81**, 540–547 (2015). https://doi.org/10.1016/j.egypro.2015.12.127
5. ShipTechnology: Celebrity millennium. https://www.ship-technology.com/projects/millennium/. Accessed 14 July 2020
6. Knight, S.: Could turbines turn again for cruise?. TheMotorShip. https://www.motorshipcom/news101/engines-and-propulsion/could-turbines-turn-again-for-cruise. Accessed 14 July 2020

7. IMO: Prevention of air pollution from ships. International Maritime Organization. http://www.imo.org/en/OurWork/Environment/PollutionPrevention/AirPollution/Pages/Air-Pollution.aspx. Accessed 14 July 2020

8. Train the trainer (TTT) course on energy efficient ship operation. Module 2 – Ship energy efficiency regulations and related guidelines. International Maritime Organization, London (2016)

9. MARPOL Consolidated Edition. International Maritime Organization, London (2017)

10. Olmer, N., Comer, B., Roy, B., et al.: Greenhouse gas emissions from global shipping, 2013–2015. International Council on Clean Transportation, Washington (2017)

11. Methanol as a marine fuel. Naval Architects, January, 32–34 (2019)

12. Cherednichenko, O., Mitienkova, V.: Analysis of the impact of thermochemical recuperation of waste heat on the energy efficiency of gas carriers. J. Marine Sci. Appl. **19**(1), 72–82 (2020). https://doi.org/10.1007/s11804-020-00127-5

13. Radchenko, A., Mikielewicz, D., Forduy, S., Radchenko, M., Zubarev, A.: Monitoring the fuel efficiency of gas engine in integrated energy system. In: Nechyporuk, M., Pavlikov, V., Kritskiy, D. (eds.) Integrated Computer Technologies in Mechanical Engineering. AISC, vol. 1113, pp. 361–370. Springer, Cham (2020). https://doi.org/10.1007/978-3-030-37618-5_31

14. Trushliakov, E., Radchenko, M., Bohdal, T., Radchenko, R., Kantor, S.: An innovative air conditioning system for changeable heat loads. In: Tonkonogyi, V., Ivanov, V., Trojanowska, J., Oborskyi, G., Edl, M., Kuric, I., Pavlenko, I., Dasic, P. (eds.) InterPartner 2019. LNME, pp. 616–625. Springer, Cham (2020). https://doi.org/10.1007/978-3-030-40724-7_63

15. Konovalov, D., Trushliakov, E., Radchenko, M., Kobalava, H., Maksymov, V.: Research of the aerothermopressor cooling system of charge air of a marine internal combustion engine under variable climatic conditions of operation. In: Tonkonogyi, V., Ivanov, V., Trojanowska, J., Oborskyi, G., Edl, M., Kuric, I., Pavlenko, I., Dasic, P. (eds.) InterPartner 2019. LNME, pp. 520–529. Springer, Cham (2020). https://doi.org/10.1007/978-3-030-40724-7_53

16. Forduy, S., Radchenko, A., Kuczynski, W., Zubarev, A., Konovalov, D.: Enhancing the gas engines fuel efficiency in integrated energy system by chilling cyclic air. In: Tonkonogyi, V., Ivanov, V., Trojanowska, J., Oborskyi, G., Edl, M., Kuric, I., Pavlenko, I., Dasic, P. (eds.) InterPartner 2019. LNME, pp. 500–509. Springer, Cham (2020). https://doi.org/10.1007/978-3-030-40724-7_51

17. Radchenko, A., Bohdal, L., Zongming, Y., Portnoi, B., Tkachenko, V.: Rational designing of gas turbine inlet air cooling system. In: Tonkonogyi, V., Ivanov, V., Trojanowska, J., Oborskyi, G., Edl, M., Kuric, I., Pavlenko, I., Dasic, P. (eds.) InterPartner 2019. LNME, pp. 591–599. Springer, Cham (2020). https://doi.org/10.1007/978-3-030-40724-7_60

18. Ghadikolaei, M.A., Cheung, C.S., Yung, K.F.: Study of performance and emissions of marine engines fueled with liquefied natural gas (LNG). In: Proceedings of 7th PAAMES and AMEC2016, pp. 1–6. Hong Kong (2016)

19. Cherednichenko, O., Tkach, M., Dotsenko, S.: The usage of a waste heat recovery metal-hydride unit of continuous operation in the maritime energy. In: 2019 IEEE International Conference on Modern Electrical and Energy Systems (MEES), pp. 510–513. IEEE, Kremenchuk (2019). https://doi.org/10.1109/MEES.2019.8896386

20. Cherednichenko, O., Tkach, M., Timoshevskiy, B., et al.: Improving the efficiency of a gas-fueled ship power plant using a waste heat recovery metal hydride system. Sci. J. Maritime Univ. Szczecin **59**, 9–15 (2019). https://doi.org/10.17402/346

21. Global Marine Technology Trends 2030. Lloyd's Register (2015)

22. Cherednichenko, O., Serbin, S.: Analysis of efficiency of the ship propulsion system with thermochemical recuperation of waste heat. J. Marine Sci. Appl. **17**(1), 122–130 (2018). https://doi.org/10.1007/s11804-018-0012-x
23. API TR 2572: Carbon content, sampling, and calculation. American Petroleum Institute (2013)
24. MEPC 245(66): Guidelines on the method of calculation of the Attained Energy Efficiency Design Index (EEDI) for new ships. International Maritime Organization (2014)

Review of Methods for Obtaining Hardening Coatings

Sergiy Plankovskyy[2] ⓘ, Viktoriia Breus[1](✉) ⓘ, Vitalii Voronko[2] ⓘ,
Oleksandr Karatanov[1] ⓘ, and Olha Chubukina[1] ⓘ

[1] National Aerospace University "Kharkiv Aviation Institute",
17 Chkalova Street, Kharkiv 61070, Ukraine
v.breus@khai.edu
[2] O. M. Beketov National University of Urban Economy in Kharkiv,
17 Marshala Bazhanova Street, Kharkiv 61002, Ukraine

Abstract. The paper is devoted to the consideration of existing methods of deposition of hardening coatings. The paper discusses the relationship between the coating deposition process and the control circuit for the ion current density and energy control. The analysis of deposition methods is carried out and on the basis of this analysis the CIB deposition method of hardening coatings is selected. The authors studied the features of each of the considered methods, which made it possible to identify the main advantages and disadvantages. Based on the data obtained, it иs revealed that for equal-thickness coatings, it is necessary to use control circuits for the ion current density and energy.

Keywords: Nanocoatings · Hardening coatings · Coating deposition · Physical vapor deposition

1 Introduction

The production of ultrathin metal films is the most urgent problem of modern materials science. At the moment, modern enterprises are raising the issue of the need for nanocoatings in various industries of their application [1].

The rapidly developing technologies and the improvement of already existing techniques for applying nanocoatings lead to the question of studying the quality, structure, and wear resistance of films for their further application in more complex technologies and aggregates [2].

The development of modern technology is associated with an increase in the productivity of equipment, its durability, which requires an increase in the wear resistance of machine parts and tools. Ensuring stable performance characteristics of products can be achieved both by creating new construction materials and by applying functional coatings to machine parts and tools [3, 4].

Hardening coatings are widely used in the production of parts that make up the fuel equipment of aircraft engines, gearboxes, and actuators [5, 6]. Of particular note is the effectiveness of hardening coatings on the surface of gas turbine engine blades and parts of highly loaded friction pairs [7].

M. Nechyporuk et al. (Eds.): ICTM 2020, LNNS 188, pp. 332–343, 2021.
https://doi.org/10.1007/978-3-030-66717-7_28

At the present stage of technological development, it is necessary to develop such methods for obtaining coatings on cutting tools and machine parts in electronics, which can provide the required complex of surface characteristics [8]. There is a need to create coatings with special properties such as porosity, adhesion, oxidizability, electrical conductivity, roughness, microhardness, etc.

Particular interest in technologies for the deposition of coatings using ion and plasma flows is due to the possibility of obtaining a wide range of non-equilibrium chemical compounds [9]. In addition, the advantages of deposition of coatings from ion and plasma flows should be considered the purity of the process and the possibility of implementing the process at low temperatures, when the bulk of the material retains its structure obtained at the previous stages of processing [10].

The composition and properties of wear-resistant coatings largely depend on the technique and technology of their application; therefore, an important task in the development of coated tool material is the selection or improvement of existing coating methods. At the moment, in the world practice, there are many methods of coating. Vacuum ion-plasma technologies occupy a special place. According to the method of forming streams of deposited particles, they are divided into Chemical Vapor Deposition (CVD) and Physical Vapor Deposition (PVD) methods [11].

In order to understand how it is possible to achieve uniform thickness of the coating on the metal surface, this paper reviewed the most suitable methods for deposition of hardening coatings. The aim of the investigation is the determination of the relationship among the methods of coating deposition with the processes of controlling the ionic current density and energy.

2   Overview of Deposition Methods

This paper refers to the analysis of the technology of the second group. Therefore, CVD methods will not be considered in this paper. The existing PVD methods of applying surface layers can be divided into three groups: the method of thermal vacuum evaporation; ion sputtering method; CIB method – condensation with ion bombardment.

2.1   Thermovacuum Evaporation Method

The essence of this method consists in heating a substance placed in an evaporator to a temperature at which the evaporation flux becomes large enough, and vapor condensation on the substrate surface [12]. It should be noted that the most significant features of the condensation process are the low energy of the particles being sprayed and the practically zero degree of flow ionization. Low energy values primarily affect the value of adhesion. Indeed, to obtain an acceptable value of adhesion, it is necessary that the particle energy be at least 40–70 eV, whereas in the method of thermal vacuum evaporation it is only a fraction of an electron-volt. The scheme of thermal vacuum evaporation is shown in Fig. 1.

In some cases, the zero ionization degree of the flow can be considered an advantage of the method; in particular, upon condensation of a neutral coating, there

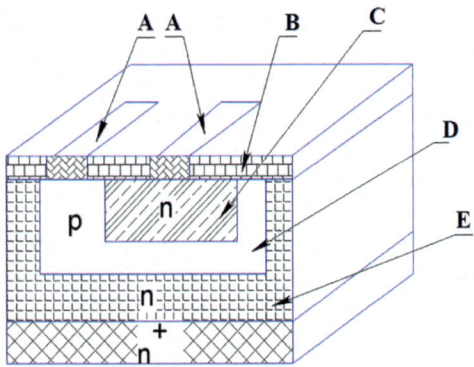

Fig. 1. Thermal vacuum evaporation method.

are no difficulties with the deposition of films on dielectrics, since compensation of the excess charge is not required.

2.2   Ion Sputtering Method

There are many varieties of this method, the main ones of which are: ion-plasma sputtering (diode and triode systems), magnetron sputtering (planar and cylindrical MRS), high-frequency sputtering, and reactive sputtering.

The ion sputtering scheme is shown in Fig. 2.

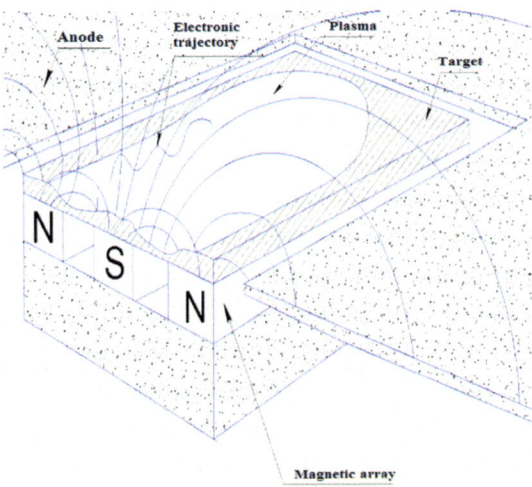

Fig. 2. Ion sputtering method.

When applying films by ion-plasma sputtering methods, the effect of cathode sputtering is used. The main advantage of this method is a higher adhesion than in the method of thermal vacuum evaporation, which is due to the higher energy of atoms of the condensing substance. The disadvantage is the need to maintain a relatively high gas pressure and, as a consequence, contamination of the film. To maintain a glow discharge at a pressure of less than 1 Pa, special devices are required.

Magnetron sputtering systems allow the process to be carried out in a high vacuum. [12, 13]. Their main difference from diode and triode spray systems is the presence of crossed electric and magnetic fields in the cathode region. The electrons diffusing in this region ensure effective ionization of the reaction gas, since they move along closed trajectories and, therefore, undergo numerous collisions with atoms. As a result, a high concentration of ions is created in the cathode region (despite the low gas pressure in the chamber) and, as a result, the avalanche is subjected to intense sputtering. Thus, the use of magnetron systems made it possible to significantly increase the rate of coating deposition without increasing the pressure of the working gas. Essentially, having crossed fields is analogous to an increase in pressure. In addition, the magnetron sputtering method does not require the use of high-voltage equipment, which is also its essential advantage.

In RF sputtering systems, plasma is generated and maintained using a high frequency alternating electromagnetic field (typically 13.56 MHz). The main feature of this method is the ability to sputter dielectric targets. The equipment is very complex and expensive, creates intense interference at radio frequencies and requires the use of special measures to protect against radiation when using gigahertz radiation sources.

Reactive spraying systems use reactive gases, most often a mixture of argon with oxygen and nitrogen. On the surface of the substrate, compounds of the deposited substance with active gases are formed. Thus, the reactive sputtering method is used to obtain films with a complex chemical composition. Most often, this method is used to obtain films of oxides and nitrides. The disadvantages of the reactive sputtering method are the low deposition rate, the complexity of the control of the sputtering and condensation processes, and high requirements for the purity of reactive gases.

2.3   Method of Condensation with Ion Bombardment (CIB)

The methods described above for obtaining wear-resistant coatings are characterized by a relatively low energy of particles deposited on the substrate, which does not allow the formation of films with a high adhesion value. In fact, the maximum attainable energy in magnetron systems does not exceed 50–100 eV. In addition, in all ion-plasma sputtering systems, the control of the condensation process is difficult. These features are due to the fact that in the systems mentioned above, the processes of generating a stream of condensed matter are closely related to the process of condensation. In installations using the CIB method, independent sources are used, which provides the ability to control the characteristics of the condensing matter flow over a wide range [14].

Thus, the main feature of the CIB method is the presence of independent sources of condensing matter and separate control loops for the generation and transportation of the flow. Most often, arc evaporators are used, providing a high rate of coating application (up to 50 μ/h), but in principle, other types of sources can be used in CIB installations, including thermal evaporators with subsequent ionization, for example, in a glow or RF discharge.

The use of independent sources made it possible, firstly, to carry out preliminary ionic cleaning of the surface of products, and secondly, to control the energy of the condensing substance regardless of the operating mode of the sources [15]. Thanks to these features, CIB installations provide the formation of high quality coatings with a high adhesion value.

In a number of cases, in CIB installations, it is effective to use pulsed plasma sources, providing flows of matter of very high density and temperature. In addition, pulsed sources make it possible to create streams of non-metallic plasma, for example, carbon plasma, to obtain diamond-like films. One of the features of the CIB method is the acceleration of ions of a substance due to the creation of a negative charge on the substrate (cutting tool). Also, the characteristic features of the CIB method include the high chemical activity of the cathode material, which is explained by the formation of condensate during the electric arc evaporation of the cathode material, which makes it possible to convert the condensate into a highly ionized flow of low-temperature plasma.

CIB method can be divided into two sequential processes:

– ion bombardment, activates thermomechanical processes, reduces the number of surface defects, as well as cleaning the surface with ions of the evaporated electrode, accelerated to an energy of (1–3) keV;
– condensation of the coating under the conditions of a plasma-chemical reaction between the cathode substance and the working gas.

A number of special devices are used to control the physical characteristics and velocity of the plasma flow, as well as to achieve high values of the ion flow density. In order to screen out the droplet phase of the ion flow, plasma flow separators are used. In the case of using the CIB method, it is possible to maintain a low temperature during the deposition process, which makes it possible to deposit coatings both on various grades of hard alloy and on high-speed steels.

By changing the technological parameters of the deposition process, it is possible to change the properties of the resulting coatings, such as microhardness and adhesion. An example of a multi-layer coating is shown in Fig. 3.

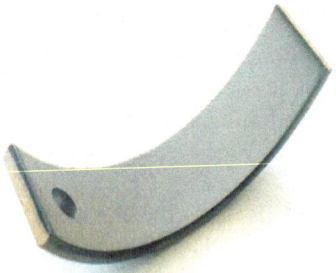

Fig. 3. Appearance of a multi-layer coated part.

For the deposition of coatings of various compositions, it is possible to manufacture cathodes with the required ratio of coating materials. So in the papers [16, 17] the manufacture of cathodes for covering a given composition is considered, which allows providing the necessary physical and mathematical properties.

Recently, single and multi-layer coatings based on nitrides, carbides and carbonitrides, but also based on various metals (Hf, Nb, Ta, Cr, Zr, Mo), have been widely used. Ensuring the uniformity of the coating thickness by controlling the energy of the ion flux density. Using the installation shown in Fig. 4, we can spray on any material and any substrate.

Fig. 4. Vacuum spraying unit.

In the paper [18] the problems of control of the energy and density of the ion flux along the surface of the processed substrate in technological devices of plasma-ion processing are considered. The analysis of modern technological systems and methods for controlling the parameters of the ion flow has been carried out.

A classification of schemes of interaction of plasma sources with a substrate is proposed, where the decisive function is performed by the substrate in the process of plasma generation.

The possibility of using ion and plasma flows to change the surface properties of materials and the formation of the necessary structures on their surface was discovered in the second half of the 19th century [19]. However, the intensive development of plasma-ion technologies has been noted since the second half of the XX century – the main reasons were the increased requirements for the resource of machine parts and the economy of expensive alloying materials (70–80s), the production of semiconductor electronics (90s.) [20, 21], as well as the planned transition to the level of nanotechnology [22]. Generally accepted technologies that do not use plasma depend significantly on such an equilibrium parameter of the system as temperature. However, the possibility of raising the temperature is limited by the phase transition temperatures of the substrate material, which are usually several hundred degrees Celsius, the requirements for the pre-hardened layer, and a number of other reasons. The most suitable way to solve the arising problems is to transfer matter into a state of plasma, which is relatively easy to obtain in technological devices as a rule, by ionization by electron impact [13]. A substance in a plasma state has two main advantages over the neutral phase: significantly higher reactivity and better controllability of substance flows. The latter property is explained by the ability of charged particles to react to applied electric and magnetic fields. In addition, plasma-ion technologies, as a rule, are implemented under conditions of reduced pressure, which makes it possible to reduce the level of undesirable impurities in the formed surface layers. This explains the presence of a wide range of technological processes of etching, cleaning, modification, nitriding and alloying of surface layers, as well as coating deposition.

Schemes of modern technological systems for plasma-ion treatment, energy control and typical distributions of ion current density are shown in Fig. 5. The simplest method of ionic surface treatment is plasma generation in a glow discharge and the associated process of sputtering a surface exposed to plasma [23]. In the ion sputtering method, the cathode surface is a source of metal vapor.

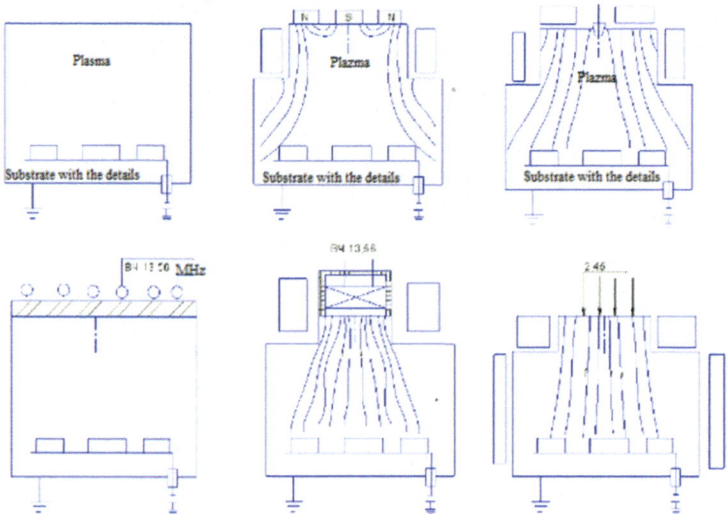

Fig. 5. Scheme of energy control and ion current density distribution.

When applying coatings by ion-plasma sputtering methods, a target fixed on the cathode is sputtered with high-energy ions, the particles of the sputtered substance condense on a substrate fixed on the anode. The use of a cathode as a substrate has the advantage that with a limited range of values of the interelectrode potential difference, it is possible to obtain a wide range of values of the current density, which can be controlled by changing the parameters of the external electric circuit.

The disadvantages include, first of all, the impossibility of separate control of the ion current density and ion energy, since the interelectrode potential drop.

The determining energy is determined by the current-voltage characteristic. This disadvantage is eliminated by separating the functions of the cathode and the substrate – in this case, the substrate ceases to be part of the plasma source (cathode). Then, when describing the current density and ion energy on the substrate, not volt-ampere characteristics are used, but the dependences of the current extracted to the substrate on the cathode bias potential – the process describes the cutoff of the electronic component from the total flow of charged particles to the surface.

This idea is implemented in the method of condensation from plasma streams with ion bombardment (CIB) [24], also widely known abroad as Plasma Immersed Ion Implantation and Deposition (PIII & D) [25]. In this case, it is possible to control the ion energy by varying the bias potential of the substrate in a wide range, although the dependence of the current on the bias potential tends to reach the saturation mode. Thus, when separating the functions of the cathode and the substrate, the ion current density is controlled by changing the operating mode of the plasma source, and the ion energy by the CIB method. The advantages and disadvantages of the considered methods are shown in Table 1.

Table 1. Advantages and disadvantages of ion-plasma sputtering methods

Method	Advantages	Disadvantages
Method of condensation with ion bombardment (CIB)	• Ability to use independent sources; • Ability to control the characteristics of the flow of the condensable substance within a wide range; • Ability to ensure the formation of high quality coatings with a high value of adhesion	• Low sedimentation rate
Ion sputtering method	• High adhesion	• The need to maintain a relatively high gas pressure and, as a consequence, film contamination occurs
Thermal vacuum evaporation method	• High efficiency; • Low cost of equipment; • Safety at work (low voltage on the electrodes); • Small overall dimensions	• Ability of contamination of the applied film with the heater material; • Short service life due to aging (destruction) of the heater, which requires periodic replacement

3 Methods for Controlling Metal Coatings

The problem of obtaining uniform thickness of coatings at the moment is the most relevant, in this regard, the following question arises regarding the methods of measuring the uniform thickness of coatings.

The suitability of metalized products for a particular purpose is determined by the quality of the coating and the strength of its adhesion to the base. Tests of coatings allow characterizing the quality of work and the economy of the metallization process. To test coatings, their thickness, gloss degree, electrical conductivity, adhesion value, as well as porosity and density, on which the products wear resistance, appearance, mechanical and electrical properties depend, are determined.

The properties of coatings depend not only on the type of metal, but also on the method of application. Thus, during the evaporation of metals in vacuum, cathodic sputtering and the classical method of metal reduction from a solution of silver salts, the coatings have a fine-crystalline structure and a relatively small thickness. The methods for determining the properties of such coatings naturally differ from the methods used for testing relatively thick metallization coatings, which are composed of solidified metal droplets (flakes). The thickness of the coatings depends on the application method and operating conditions and in practice ranges from $10^{-3} \ldots 10^{-4}$ mm (vacuum coatings) to 1 mm or more.

The coating thickness control is carried out by various methods. The simplest of them is to measure the thickness of products with a micrometer before and after coating. The method is applicable only for coatings with a thickness of more than 0.01 mm, moreover, for products of a simple profile with precisely ground surfaces. When assessing the thickness of metal coatings, a distinction should be made between the average and local thickness of the coating and, on this basis, the appropriate test method should be selected.

One of the methods for measuring the thickness of electroplated and metallized coatings is based on the principle of grinding the coating with a grinding wheel. The coating is ground with a strictly cylindrical circle with a diameter of 100–200 mm and a width of 4–10 mm until the base material is exposed. The thickness of the coating is calculated by the equation

$$h = \frac{b}{8r},$$

where b is the circle length track in the coating; r is the circle radius.

Physical and chemical methods of coating thickness control are also used:

- removal method;
- weighing before and after coating;
- jet method;
- drop method;
- optical methods.

Removal method is characterized by a relatively high accuracy and is applicable for coatings of various thicknesses, applied in various ways, with the exception of very thin vacuum ones. The method consists in the fact that all metal coating is removed with a solution that does not act on the base material. The thickness of the coating is determined by the difference in the weight of the article before and after removing the coating. The pre-weighed product is immersed in a solution and kept in it until the coating is completely dissolved. Then the product is thoroughly washed with a stream of water, dried and weighed again.

Jet and drop methods, in contrast to the stripping and weighing methods before and after coating, allow determining the local thickness of the metal layer. They apply primarily to galvanic coatings: zinc, cadmium, copper, nickel, silver, etc. The essence of the drop method lies in the fact that the coating area is dissolved by successively applied drops of solvent that are kept for a certain time. The thickness of the coating is calculated from the number of drops consumed. The accuracy of this method is ± 20%, it is associated with a relatively large investment of time. For quicker determination of local coating thickness, the jet method is recommended. The method consists in the fact that the coating area is dissolved by a solution flowing out at a certain speed, falling on the metal surface in the form of a jet. The calculation of the thickness of the coating is carried out according to the time taken to dissolve the coating in the test area.

Optical Method. The thickness of thin coatings obtained by sputtering in vacuum can be determined by the Fizeau method [26]. The method is based on the chemical interaction of iodine and silver. Optical control methods also include measuring the thickness of the coating by its light transmittance or light reflection [27].

4   Conclusions

This paper deals with the problem of uneven coating. To solve this problem, the most relevant methods are considered and analyzed. Based on the analysis performed, the CIB (condensation with ion bombardment) method is selected. This method is most suitable for depositing coatings on parts, provided that high heating temperatures are unacceptable.

Based on the analysis of the works of other authors, the importance of controlling the plasma fluxes in the process of deposition of hardening coatings is shown, which in turn provides an acceptable uniform thickness of the deposited coatings when processing large substrates.

References

1. Torabinejad, V., Aliofkhazraei, M., Assareh, S., et al.: Electrodeposition of Ni-Fe alloys, composites, and nano coatings – a review. J. Alloys Compd. **691**, 841–859 (2017). https://doi.org/10.1016/j.jallcom.2016.08.329
2. Popov, V., Kostyuk, G., Tymofyeyev, O., et al.: Design of new nanocoatings based on hard alloy. In: Ivanov, V., et al. (eds.) Advances in Design, Simulation and Manufacturing III. DSMIE 2020. LNME, pp. 522–531. Springer, Cham (2020). https://doi.org/10.1007/978-3-030-50794-7_51

3. Toboła, D., Kalisz, J., Czechowski, K., et al.: Surface treatment for improving selected physical and functional properties of tools and machine parts – a review. J. Appl. Mater. Eng. **60**(1), 23–36 (2020). https://doi.org/10.35995/jame60010003

4. Duriagina, Z.A., Lemishka, I.A., Trostianchyn, A.M., et al.: The effect of morphology and particle-size distribution of VT20 titanium alloy powders on the mechanical properties of deposited coatings. Powder Metall. Metal Ceram. **57**, 697–702 (2019). https://doi.org/10.1007/s11106-019-00033-8

5. Dolmatov, A.I., Sergeev, S.V., Kurin, M.O., et al.: Kinematics of the solid particle accelerated by a flow of gas in a supersonic nozzle and work hardening of the processed surface. Metallofizika i Noveishie Tekhnologii **37**(7), 871–885 (2015). https://doi.org/10.15407/mfint.37.07.0871

6. Denisov, L.V., Boitsov, A.G., Siluyanova, M.V.: Surface hardening in hydraulic cylinders for airplane engines. Russ. Eng. Res. **38**(12), 1080–1083 (2018). https://doi.org/10.3103/S1068798X18120237

7. Ziaei-Asl, A., Ramezanlou, M.T.: Thermo-mechanical behavior of gas turbine blade equipped with cooling ducts and protective coating with different thicknesses. Int. J. Mech. Sci. **150**, 656–664 (2019). https://doi.org/10.1016/j.ijmecsci.2018.10.070

8. Fotovvati, B., Namdari, N., Dehghanghadikolaei, A.: On coating techniques for surface protection: a review. J. Manufact. Mater. Process. **3**(1), 28 (2019). https://doi.org/10.3390/jmmp3010028

9. Kolesnyk, V.P., Chuhai, O.M., Slyusar, D.V., et al.: Structure and properties of ionic-plasma WC coatings. Mater. Sci. **55**(2), 220–224 (2019). https://doi.org/10.1007/s11003-019-00292-1

10. Plankovskyy, S., Shypul, O., Tsegelnyk, Y., et al.: Simulation of surface heating for arbitrary shape's moving bodies/sources by using R-functions. Acta Polytechnica **56**(6), 472–477 (2016). https://doi.org/10.14311/AP.2016.56.0472

11. Habig, K.H.: Chemical vapor deposition and physical vapor deposition coatings: properties, tribological behavior, and applications. J. Vac. Sci. Technol. Vac. Surf. Films **4**(6), 2832–2843 (1986). https://doi.org/10.1116/1.573687

12. Jehn, H.A., Thiergarten, F., Ebersbach, E., Fabian, D.: Characterization of PVD (Ti, Cr)N$_x$ hard coatings. Surf. Coat. Technol. **50**(1), 45–52 (1991). https://doi.org/10.1016/0257-8972(91)90191-X

13. Braithwaite, N.S.J.: Introduction to gas discharges. Plasma Sources Sci. Technol. **9**(4), 517 (2000). https://doi.org/10.1088/0963-0252/9/4/307

14. Baranov, O., Romanov, M.: Current distribution on the substrate in a vacuum arc deposition setup. Plasma Process. Polym. **5**(3), 256–262 (2008). https://doi.org/10.1002/ppap.200700160

15. Taran, V.S., Garkusha, I.E., Krasnyj, V.V., et al.: Recent developments of plasma-based technologies for medicine and industry. Nukleonika **57**, 277–282 (2012)

16. Taran, A., Plankovskyy, S., Ostrovsky, E., Ordanjan, S.: High-current-density cathodes based on Barium Hafnate with Tungsten. In: 2007 IEEE International Vacuum Electronics Conference, pp. 1–2. IEEE, Kitakyushu (2007). https://doi.org/10.1109/IVELEC.2007.4283309

17. Taran, A., Plankovskyy, S., Voronovich, D., Abashin, S.: Emission properties of Re-W dispenser cathodes. In: 2009 IEEE International Vacuum Electronics Conference, pp. 407–408. IEEE, Rome (2009). https://doi.org/10.1109/IVELEC.2009.5193581

18. Baranov, O., Romanov, M., Fang, J., et al.: Control of ion density distribution by magnetic traps for plasma electrons. J. Appl. Phys. **112**(7), 073302 (2012). https://doi.org/10.1063/1.4757022

19. Anders, A.: From plasma immersion ion implantation to deposition: a historical perspective on principles and trends. Surf. Coat. Technol. **156**(1–3), 3–12 (2002). https://doi.org/10. 1016/S0257-8972(02)00066-X

20. Perry, A.J., Matossian, J.N.: An overview of some advanced surface technology in Russia. Metall. Mater. Trans. A **29**(2), 593–610 (1998). https://doi.org/10.1007/s11661-998-0141-y

21. Bogaerts, A., Neyts, E., Gijbels, R., Van der Mullen, J.: Gas discharge plasmas and their applications. Spectrochim. Acta, Part B **57**(4), 609–658 (2002). https://doi.org/10.1016/ S0584-8547(01)00406-2

22. Ostrikov, K.: Colloquium: reactive plasmas as a versatile nanofabrication tool. Rev. Modern Phys. **77**(2), 489 (2005). https://doi.org/10.1103/RevModPhys.77.489

23. Grigoriev, S., Metel, A., Volosova, M., et al.: Surface hardening of massive steel products in the low-pressure glow discharge plasma. Technologies **7**(3), 62 (2019). https://doi.org/10. 3390/technologies7030062

24. Korusenko, P.M., Nesov, S.N., Povoroznyuk, S.N., et al.: Chemical composition and mechanical properties of coatings based on TiN formed using a condensation with ion bombardment. Protect. Metals Phys. Chem. Surf. **56**(3), 539–548 (2020). https://doi.org/10. 1134/S2070205120030193

25. Ueda, M., Silva, C., Santos, N.M., Souza, G.B.: Plasma immersion ion implantation (and deposition) inside metallic tubes of different dimensions and configurations. Nucl. Instrum. Methods Phys. Res., Sect. B **409**, 202–208 (2017). https://doi.org/10.1016/j.nimb.2017.03. 073

26. Fizeau, M.H.: On the effect of the motion of a body upon the velocity with which it is traversed by light. London, Edinburgh Dublin Philos. Mag. J. Sci. **19**(127), 245–260 (1860). https://doi.org/10.1080/14786446008642856

27. Li, Z., Palacios, E., Butun, S., et al.: Omnidirectional, broadband light absorption using large-area, ultrathin lossy metallic film coatings. Sci. Rep. **5**(1), 15137 (2015). https://doi. org/10.1038/srep15137

Stressed State of an Infinite Plate with a Circular Opening and a Concentric Cover Plate

Sergey Kurennov[1]([✉]) and Natalia Smetankina[2] [iD]

[1] National Aerospace University "Kharkiv Aviation Institute", 17 Chkalova Street, Kharkiv 61070, Ukraine
kurennov.ss@gmail.com

[2] A. Pidgorny Institute of Mechanical Engineering Problems of the National Academy of Sciences of Ukraine, 2/10 Pozharskogo Street, Kharkiv 61046, Ukraine
nsmetankina@ukr.net

Abstract. The paper offers an analytical solution of the problem of the stressed state of an infinite plate with a circular opening reinforced with a concentric round cover plate. The cover plate is assumed to be elastically attached to the main plate along its perimeter. The structure is loaded at infinity with uniform tension. The solution is obtained by expanding the components of the stress-strain state into a Fourier series about the angle coordinate. After satisfying the edge conditions, the solution retains only the first series terms. The model problem is solved. The cover plate was shown to reduce the stresses near the opening. The solution was verified by comparing computational results with calculations performed using the finite element method. The model suggested is highly accurate. A parametric study was performed to examine the impact of cover plate thickness and plate thickness ratios, and cover plate radius and opening radius ratios on the stress in the most loaded section.

Keywords: Repair patch · Analytical solution · Round hole

1 Introduction

Thin-wall structures, including thin elastic plates, can have defects in the form of holes and cracks that occur during operation, for instance, due to mechanical damage. The presence of holes in a plate results in concentration of stresses on the edges of the holes and, eventually, to premature failure of the structural element. Holes are often reinforced by using so-called repair patches attached to the main plate. The patch can be attached to the main plate over the entire patch surface (overlapping adhesive joints), along lines (weld joints) or in a system of points (riveted joints). Studying the stressed state of a plate with holes or cracks of different form is a classical mechanics problem and it has a longstanding history [1, 2]. The holes can be reinforced with internal elastic inserts [3, 4] or patches [5–10], which can be additionally secured with rivets [11]. To study the stressed state of reinforced damaged plates, experimental and numerical methods are used among others [12–16].

The concentration of stresses is reduced by providing holes of any form during repair work. As a rule, they have a round form. Therefore, of utmost interest is studying the stressed state of reinforced round holes. In known studies dedicated to this problem [7, 8] the solution is constructed as an assembly of complex Muskhelishvili potentials, and the patch attachment with the main plate is assumed absolutely rigid. This study offers an analytical solution constructed in the form of an expansion of forces in the plates into a Fourier series about the angle coordinate, which was developed in papers [17, 18] and seen as most convenient for analysis. In so doing, the most general case is considered – the attachment of the patch with the main plate is assumed elastic.

2 Formulation of the Problem

Let us consider an infinite plate with thickness δ_1, which has a circular opening with radius r_1. The opening is covered with a coaxial round patch with thickness δ_2 and radius r_2. The centre of the polar system of coordinates is aligned with the centres of both circles, Fig. 1.

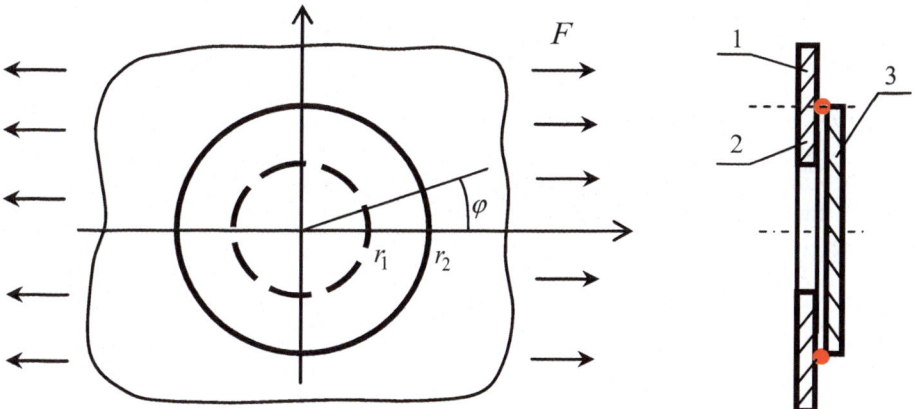

Fig. 1. Structure schematic diagram.

Over the perimeter, the patch is attached to the main plate with a weld joint. The main plate is loaded at infinity with tensile force F. Both plates are made of the same material with elasticity modulus E and Poisson's ratio μ.

Let us distinguish three domains: the main plate beyond the patch $\rho > r_2$, the main plate under the patch $r_1 > \rho > r_2$ and the patch proper $\rho < r_2$. The designation of domains is shown in Fig. 1.

Let us introduce radial R_i, tangential T_i and tangential forces Q_i in plates, which are equal to the product of respective stresses in the plate thickness. Subscript i designates the number of the respective domain. In the polar coordinate system, the forces are related by equilibrium equations

$$\frac{\partial R_i}{\partial \rho} + \frac{1}{\rho}\frac{\partial Q_i}{\partial \varphi} + \frac{R_i - T_i}{\rho} = 0, \quad \frac{\partial Q_i}{\partial \rho} + \frac{1}{\rho}\frac{\partial T_i}{\partial \varphi} + \frac{2Q_i}{\rho} = 0, \, i = 1,2,3, \tag{1}$$

and joint strain equations

$$\frac{\partial^2}{\partial \rho^2}(R_i + T_i) + \frac{1}{\rho}\frac{\partial}{\partial \rho}(R_i + T_i) + \frac{1}{\rho}\frac{\partial^2}{\partial \varphi^2}(R_i + T_i) = 0. \tag{2}$$

The forces in the plates relate to strain by physical law equations

$$Q_i = \frac{E}{2 + 2\mu}\delta_i\gamma_i, \, R_i = \frac{\delta_i E}{1 - \mu^2}\left(\varepsilon_{i,r} + \mu\varepsilon_{i,\varphi}\right), T_i = \frac{\delta_i E}{1 - \mu^2}\left(\varepsilon_{i,\varphi} + \mu\varepsilon_{i,r}\right), \tag{3}$$

where γ_i is shear strain; $\varepsilon_{i,r}$ and $\varepsilon_{i,\varphi}$ are radial and tangential strains in domain i. Strains are related to displacements by Cauchy relationships:

$$\varepsilon_{i,r} = \frac{\partial u_i}{\partial \rho}, \, \varepsilon_{i,\varphi} = \frac{u_i}{\rho} + \frac{1}{\rho}\frac{\partial v_i}{\partial \varphi}, \, \gamma_i = \frac{\partial v_i}{\partial \rho} - \frac{v_i}{\rho} + \frac{1}{\rho}\frac{\partial u_i}{\partial \varphi}, \tag{4}$$

where u_i are radial displacements; v_i are tangential displacements.

3   Constructing the Solution

From the conditions of symmetry of the structure's stressed state, it follows that the forces in the plates can be presented as Fourier series expansions in terms of even harmonics.

$$R_i = N_{i,0}(r) + \sum_{n=1}^{\infty} N_{i,n}(r)\cos 2n\varphi,$$

$$T_i = P_{i,0}(r) + \sum_{n=1}^{\infty} P_{i,n}(r)\cos 2n\varphi, \tag{5}$$

$$Q_i = \sum_{n=1}^{\infty} S_{i,n}(r)\sin 2n\varphi.$$

Substituting these expressions into equilibrium Eqs. (1) and joint strain Eqs. (2), we obtain ordinary differential equations for respective functions of ρ. By solving the differential equations and accounting for physical constraints imposed on the forces, and for the edge conditions at infinity, we obtain the interior and exterior solutions:

$$N_{1,0} = \frac{F}{2} + \frac{A_{1,0}}{\rho^2}, \ P_{1,0} = \frac{F}{2} - \frac{A_{1,0}}{\rho^2}, \ N_{3,0} = P_{3,0} = a_{3,0}, \ N_{1,1} = \frac{F}{2} - \frac{A_{1,1}}{\rho^4} + \frac{B_{1,1}}{\rho^2},$$

$$N_{1,n} = -\frac{A_{1,n}}{\rho^{2n+2}} - \frac{n+1}{n-1}\frac{B_{1,n}}{\rho^{2n}}, \ (n > 1), \ P_{1,1} = -\frac{F}{2} + \frac{B_{1,1}}{\rho^4}, \ P_{1,n} = \frac{A_{1,n}}{\rho^{2n+2}} + \frac{B_{1,n}}{\rho^{2n}},$$

$$S_{1,1} = -\frac{F}{2} - \frac{B_{1,1}}{\rho^4} + \frac{1}{2}\frac{A_{1,1}}{\rho^2}, \ N_{3,n} = -a_{3,n}\rho^{2n+2} - \frac{n-1}{n+1}b_{3,n}\rho^{2n},$$

$$S_{1,n} = -\frac{A_{1,n}}{\rho^{2n+2}} - \frac{n}{n-1}\frac{B_{1,n}}{\rho^{2n}}, \ P_{3,n} = a_{3,n}\rho^{2n-2} + b_{3,n}\rho^{2n},$$

$$S_{3,n} = a_{3,n}\rho^{2n+2} + \frac{n}{n+1}b_{3,n}\rho^{2n}.$$

Here, $A_{1,n}$, $B_{1,n}$ are arbitrary constants in the exterior solution; $a_{3,n}$, $b_{3,n}$ are constants in the interior solution; summands $F/2$ in the first terms of the exterior solution ensure the satisfying of conditions at infinity:

$$R_1^\infty = \frac{F}{2} + \frac{F}{2}\cos 2\varphi, \ T_1^\infty = \frac{F}{2} - \frac{F}{2}\cos 2\varphi, \ Q_1^\infty = -\frac{F}{2}\sin 2\varphi.$$

The above relationships describe the stressed state in the main plate beyond the patch (exterior solution, $i = 1$) and in the plate (interior solution, $i = 3$). Domain number two – the part of the main plate under the patch has an annular form. In this case, the general solution is a superposition of the interior and exterior solutions, i.e.

$$N_{2,0} = \frac{A_{2,0}}{\rho^2} + a_{2,0}, \ P_{2,0} = -\frac{A_{2,0}}{\rho^2} + a_{2,0}, \ N_{2,1} = \frac{B_{2,1}}{\rho^2} - \frac{A_{2,1}}{\rho^4} - a_{2,1},$$

$$P_{2,1} = \frac{B_{2,1}}{\rho^2} + a_{2,1} + b_{2,1}\rho^2, \ N_{2,n} = -\frac{A_{2,n}}{\rho^{2n+2}} - \frac{n+1}{n-1}\frac{B_{2,n}}{\rho^{2n}} - a_{2,n}\rho^{2n+2} - \frac{n-1}{n+1}b_{2,n}\rho^{2n},$$

$$P_{2,n} = \frac{A_{2,n}}{\rho^{2n+2}} + \frac{B_{2,n}}{\rho^{2n}} + a_{3,n}\rho^{2n-2} + b_{3,n}\rho^{2n},$$

$$S_{2,1} = \frac{1}{2}\frac{A_{2,1}}{\rho^2} - \frac{B_{2,1}}{\rho^4} + a_{2,n}\rho^{2n+2} + \frac{n}{n+1}b_{2,n}\rho^{2n},$$

$$S_{2,n} = -\frac{A_{2,n}}{\rho^{2n+2}} - \frac{n}{n-1}\frac{B_{2,n}}{\rho^{2n}} + a_{2,n}\rho^{2n+2} + \frac{n}{n+1}b_{2,n}\rho^{2n}.$$

Using the physical law Eqs. (3) and integrating the Cauchy relationships (4), we find the displacements

$$u_i = \frac{1}{\delta_i E}\left[U_{i,0}(\rho) + \sum_{n=1}^{\infty} U_{i,n}(\rho)\cos 2n\varphi\right], \ v_i = \frac{1}{\delta_i E}\sum_{n=1}^{\infty} V_{i,n}(\rho)\sin 2n\varphi$$

$$U_{1,0} = \frac{F}{2}(1-\mu)\rho - \frac{1+\mu}{\rho}A_{1,0}, \ U_{1,1} = \frac{F}{2}\alpha\rho + \frac{\alpha A_{1,1}}{3}\frac{}{\rho^3} - \frac{B_{1,1}}{\rho}, \ \alpha = 1+\mu,$$

$$U_{3,0} = a_{3,0}(1-\mu)\rho, \ U_{1,n} = \frac{\alpha}{2n+1}\frac{A_{1,n}}{\rho^{2n+1}} - \frac{n\alpha+1-\mu}{(n-1)(2n-1)}\frac{B_{1,n}}{\rho^{2n-1}},$$

$$V_{1,1} = -\frac{F}{2}\alpha\rho + \frac{\alpha A_{1,1}}{3}\frac{}{\rho^3} + \frac{1-\mu}{2}\frac{B_{1,1}}{\rho}, \ U_{3,n} = -\frac{\alpha b_{1,n}\rho^{2n-1}}{2n-1} - \frac{n\alpha-1+\mu}{(n-1)(2n+1)}a_{1,n}\rho^{2n+1},$$

$$U_{2,0} = a_{2,0}(1-\mu)\rho - \frac{a}{\rho}A_{2,0}, \ V_{2,1} = \frac{\alpha A_{2,1}}{3}\frac{}{\rho^3} + \frac{1-\mu}{2}\frac{B_{2,1}}{\rho},$$

$$V_{3,n} = \frac{n\alpha+2}{(n-1)(2n+1)}a_{1,n}\rho^{2n+1} + \frac{\alpha b_{1,n}\rho^{2n-1}}{2n-1},$$

$$U_{2,n} = \frac{\alpha}{2n+1}\frac{A_{2,n}}{\rho^{2n+1}} - \frac{n\alpha+1-\mu}{(n-1)(2n-1)}\frac{B_{2,n}}{\rho^{2n-1}} - \frac{\alpha b_{2,n}\rho^{2n-1}}{2n-1} - \frac{n\alpha-1+\mu}{(n-1)(2n+1)}a_{2,n}\rho^{2n+1},$$

$$V_{2,n} = \frac{\alpha}{2n+1}\frac{A_{2,n}}{\rho^{2n+1}} + \frac{n\alpha-2}{(n-1)(2n-1)}\frac{B_{2,n}}{\rho^{2n-1}} + \frac{\alpha b_{2,n}\rho^{2n-1}}{2n-1} + \frac{(n\alpha+2)a_{2,n}\rho^{2n+1}}{(n-1)(2n+1)}.$$

The edge conditions:

- no forces on the hole edge:

$$R_2(r_1,\varphi) = 0, \ Q_2(r_1,\varphi) = 0,$$

- the displacements of the main plate are continuous in the place of the patch attachment weld:

$$u_2(r_2,\varphi) = u_1(r_2,\varphi), \ v_2(r_2,\varphi) = v_1(r_2,\varphi),$$

- equilibrium conditions are satisfied in the weld area

$$R_1(r_2,\varphi) - R_2(r_2,\varphi) - R_3(r_2,\varphi) = 0, \ Q_1(r_2,\varphi) - Q_2(r_2,\varphi) - Q_3(r_2,\varphi) = 0.$$

- forces transmitted from the main plate to the patch via the attachment weld are proportional to the difference of displacements of attached layers [19]:

$$R_3(r_2,\varphi) = \frac{G_0 L_0}{\delta_0}[u_1(r_2,\varphi) - u_3(r_2,\varphi)], \ Q_3(r_2,\varphi) = \frac{G_0 L_0}{\delta_0}[v_1(r_2,\varphi) - v_3(r_2,\varphi)],$$

where δ_0 and L_0 are thickness and width of the attachment weld; G_0 is shear modulus of the attachment weld material. The attachment weld is assumed sufficiently narrow $L_0 << r_2$, i.e. commeasurable with the thicknesses of the joined plates. If $\delta_0 \to 0$ or $G_0 \to \infty$, then the conditions

$$u_1(r_2, \varphi) = u_3(r_2, \varphi), \ v_1(r_2, \varphi) = v_3(r_2, \varphi),$$

will be commeasurable. The conditions at infinity are not described because they are satisfied automatically.

Having satisfied the edge conditions and the matching conditions, and using the orthogonality of the system of functions $\{\cos 2n\varphi\}$ and $\{\sin 2n\varphi\}$ in the interval $(0, \ 2\pi)$, we obtain an infinite system of linear equations for unknown coefficients. This system can be decomposed into independent systems of eight equations for $A_{1,n}$, $B_{1,n}$, $A_{2,n}$, $B_{2,n}$, $a_{2,n}$, $b_{2,n}$, and $a_{3,n}$, $b_{3,n}$. For $n \geq 2$, all these systems are homogeneous. It follows that all the unknown coefficients in the equations are equal to zero. Hence, the forces in the plates are described by simple relationships in the following form:

$$R_i = N_{i,0}(\rho) + N_{i,1}(\rho)\cos 2\varphi, \ T_i = P_{i,0}(\rho) + P_{i,1}(\rho)\cos 2\varphi, \ Q_i = S_{i,n}(\rho)\sin 2\varphi.$$

4   Calculation Example

Let us consider an attachment of two plates with thicknesses $\delta_1 = \delta_2 = 1$ mm, elasticity modulus $E = 70$ GPa (an aluminium alloy), and $\mu = 0.28$. The hole radius is $r_1 = 50$ mm, and the patch radius is $r_2 = 100$ mm. The patch is attached over its perimeter to the base with a weld having the width $L_0 = 2$ mm, thickness $\delta_0 = 0.5$ mm, and it is also an aluminium alloy, $G_0 = \frac{E}{2(1+\mu)} = 27.34$ GPa. A uniform tensile load force F is specified at infinity as shown in Fig. 1.

A 3D finite element model was created to verify the computations. For reasons of symmetry, one quarter of the reinforced plate with an opening is considered, i.e. a fourth part of the opening and patch, Fig. 2.

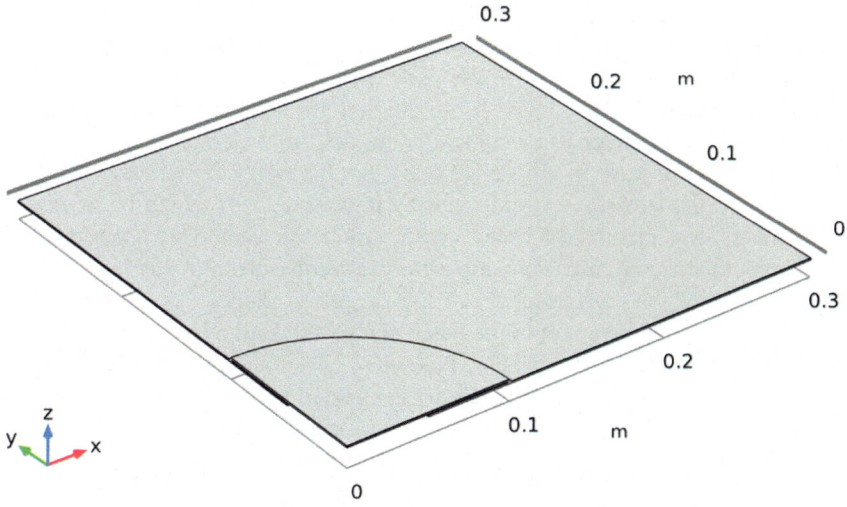

Fig. 2. Finite element model.

The infinite plate in the base is replaced in the finite element model with a square plate whose dimensions exceed the radii of the opening and patch by several-fold. Roller type boundary conditions are specified on the side faces adjoining the opening.

Figure 3 shows the graphs of radial (*a*) and tangential (*b*) forces in the weld in the sector $\varphi \in [0; \pi/2]$, which are calculated using the suggested model. A quarter circumference is also considered for stressed state symmetry.

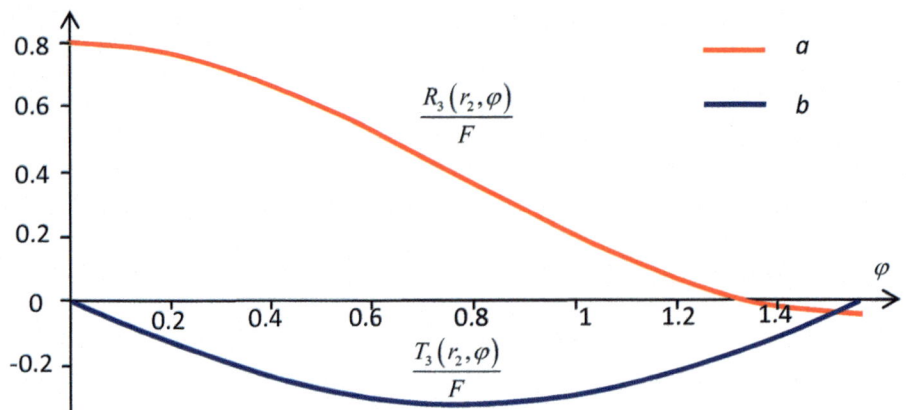

Fig. 3. Forces in the weld over its quarter length.

To illustrate to what extent the patch unloads the opening, let us consider the distribution of tangential stresses in the plate under the patch (T_2) along the radius in the direction $\varphi = \pi/2$. This section was chosen because it is well known that, with no patch, the tangential stresses on the opening boundary will exceed the stresses at infinity (Kirsch equations) by three-fold.

Figure 4 shows the graphs of tangential stresses (*a*) and radial stresses (*b*) calculated using the suggested model, as well as the tangential stresses found with the finite element method (*c*) and computed in the plate median plane under the patch (T_2^*).

Figure 5 shows, in dimensionless form, the first principle stresses in the main plate (beyond the patch and under it) calculated using the finite element method.

As evident, in this case in the dangerous zone the tangential stresses on the hole boundary exceed the stresses at infinity not by three-fold, as it would be in case of no patch, but merely less than by 1.5 times, i.e. the patch unloads the hole. Besides, we see that the computations virtually coincide with the results of finite element simulation. This is indicative of the high accuracy of the suggested model.

The functioning effectiveness of the patch can be assessed by evaluating the biggest stresses in the given structure, namely, the above-described tangential stresses in the hole $T_2(r_1, \pi/2)$. The structure stressed state is affected by different parameters. Let us consider in more detail the patch radius and the hole radius ratio (r_2/r_1), and the patch thickness and the base thickness ratio (δ_3/δ_1).

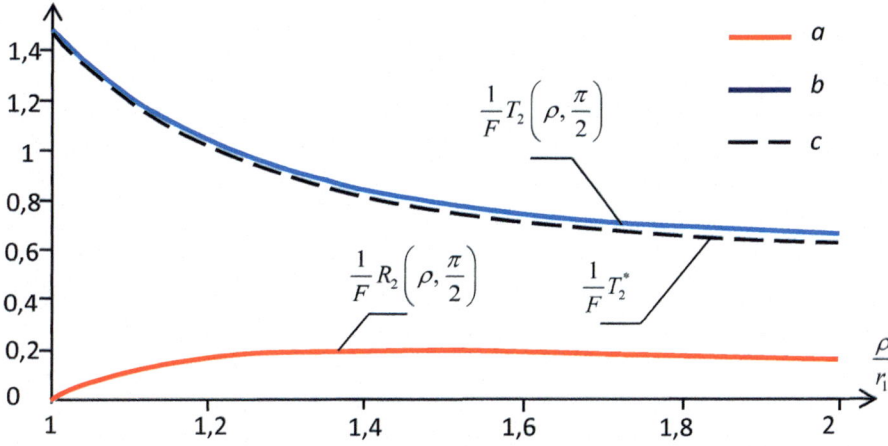

Fig. 4. Stresses under the patch along the straight line $\varphi = \pi/2$.

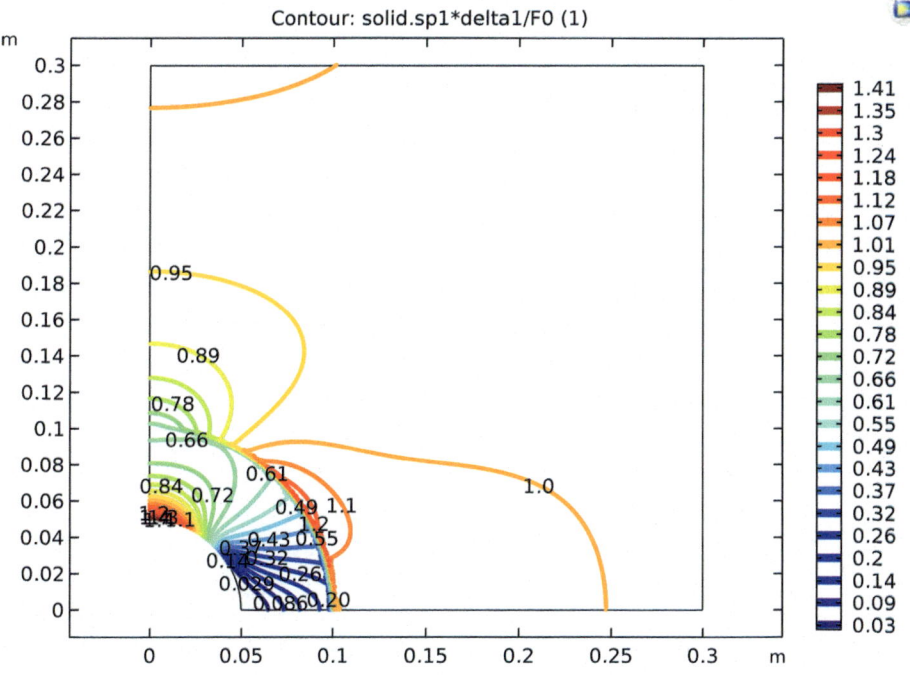

Fig. 5. First principle stresses in the main plate.

Figure 6 shows the graph $T_2(r_1, \pi/2)$ vs. the patch radius and main plate radius ratio. The thicknesses of the patch and main plate are assumed equal.

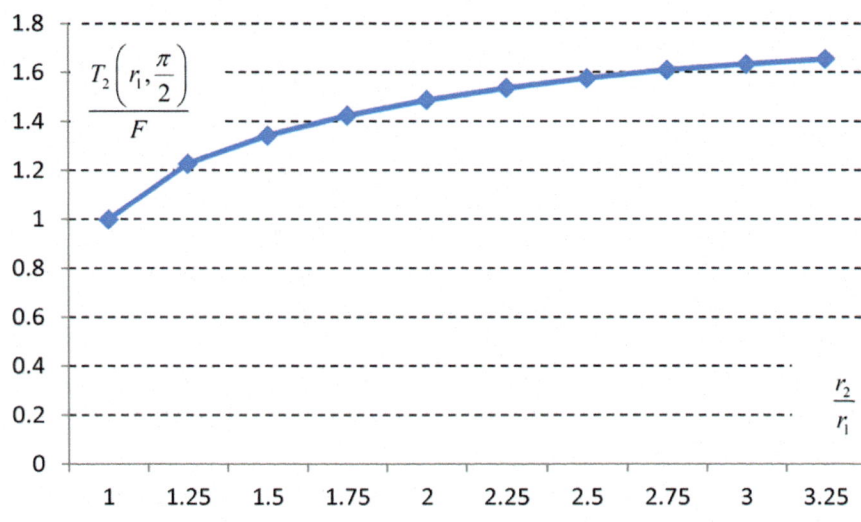

Fig. 6. Maximum stresses vs. patch radius.

Figure 7 shows the graph $T_2(r_1, \pi/2)$ vs. the patch thickness and main plate thickness ratio. The patch radius is assumed to exceed that of the opening one by two-fold.

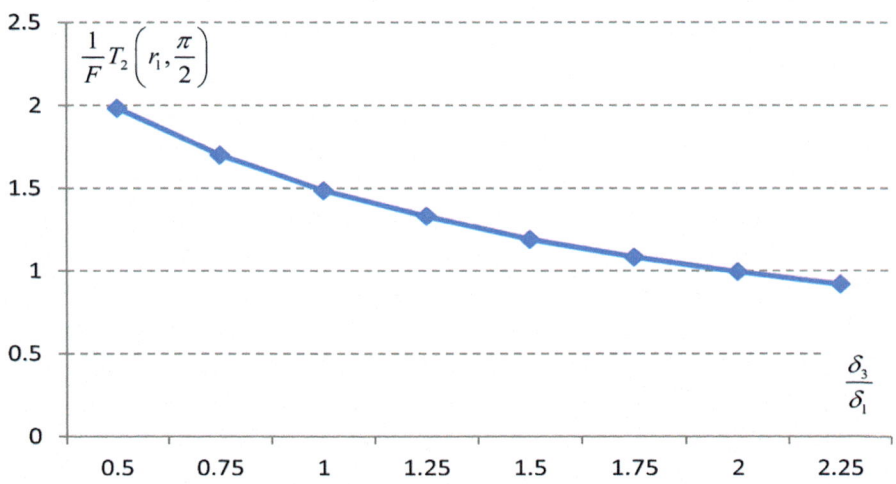

Fig. 7. Maximum stresses vs. the patch thickness and main plate thickness ratio.

5 Conclusion

The paper demonstrates an analytical solution of the problem of the stressed state of an infinite plate with a circular opening, which is reinforced with a circular patch under uniaxial tension. Comparison with numerical computations has shown that the obtained solution is highly accurate. Therefore, the suggested model can be used in the design of reinforced structures.

Further development of the model can be along the lines of creating a numerical-analytical technique of solving the problem for a circular patch attached to the opening not along a line, but over the entire surface. Besides, a problem can solved in optimal design of a minimal weight patch [20] with specified strength parameters of structure materials.

References

1. Savin, G.N.: Distribution of stresses near holes. Naukova Dumka Publ, Kyiv (1968). (in Russian)
2. Khoma, I.Y., Dashko, O.G.: Stress state of a nonthin transversely isotropic plate with a curved hole. Int. Appl. Mech. **51**(4), 461–473 (2015). https://doi.org/10.1007/s10778-015-0707-5
3. Smetankina, N.V., Sotrikhin, S.Y., Shupikov, A.N.: Theoretical and experimental investigation of vibration of multilayer plates under the action of impulse and impact loads. Int. J. Solids Struct. **32**(8–9), 1247–1258 (1995). https://doi.org/10.1016/0020-7683(94)00132-G
4. Smetankina, N.V., Shupikov, A.N., Sotrikhin, S.Y., Yareschenko, V.G.: Dynamic response of an elliptic plate to impact loading: Theory and experiment. Int. J. Impact Eng **34**(2), 264–276 (2007). https://doi.org/10.1016/j.ijimpeng.2005.07.016
5. Gui-fang, W.: Stress analysis of plates with a circular hole reinforced by flange reinforcing member. Appl. Math. Mech. **8**(6), 569–588 (1987). https://doi.org/10.1007/bf02017406
6. Maksimyuk, V.A., Storozhuk, E.À., Chernyshenko, I.S.: Stress–strain state of flexible orthotropic cylindrical shells with a reinforced circular hole. Int. Appl. Mech. **51**(4), 425–433 (2015). https://doi.org/10.1007/s10778-015-0703-9
7. Silvestrov, V.V., Zemlyanova, A.Y.: Repairing a plate with a circular opening by using a patch. Appl. Mech. Tech. Phys. **45**(4), 176–183 (2004). (in Russian)
8. Zemlyanova, A.Y.: Reinforcement of a plate weakened by multiple holes with several patches for different types of plate-patch attachment. Math. Mech. Solids **21**(3), 281–294 (2016). https://doi.org/10.1177/1081286513519812
9. Khan, M.A., Kumar, S.: Interfacial stresses in single-side composite patch-repairs with material tailored bondline. Mech. Adv. Mater. Struct. **25**(4), 304–318 (2018). https://doi.org/10.1080/15376494.2016.1255824
10. Engels, H., Zakharov, D., Becker, W.: The plane problem of an elliptically reinforced circular hole in an anisotropic plate or laminate. Arch. Appl. Math. **71**(9), 601–612 (2001). https://doi.org/10.1007/s004190100167
11. Maksimenko, V.N., Tiagnii, A.V.: Stressed state design for glued-riveted laminated plates with a crack. TsAGI Trans. **5**, 92–101 (1990). [in Russian]
12. Okafor, A.C., Singh, N., Enemuoh, U.E., Rao, S.V.: Design, analysis and performance of adhesively bonded composite patch repair of cracked aluminum aircraft panels. Compos. Struct. **71**(2), 258–270 (2005). https://doi.org/10.1016/j.compstruct.2005.02.023

13. Oudad, W., Belhadri, D.E., Fekirini, H., Khodja, M.: Analysis of the plastic zone under mixed mode fracture in bonded composite repair of aircraft structures. Aerosp. Sci. Technol. **69**, 404–411 (2017). https://doi.org/10.1016/j.ast.2017.07.001

14. Makwana, A.H., Shaikh, A.A.: The role of patch hybridization on tensile response of cracked panel repaired with hybrid composite patch: Experimental and numerical investigation. J. Adhes. (2019). https://doi.org/10.1080/00218464.2019.1629911

15. Ergün, R.K., Adin, H., Şişman, A., Temiz, Ş.: Repair of an aluminum plate with an elliptical hole using a composite patch. Mater. Test. **60**(11), 1104–1110 (2018). https://doi.org/10.3139/120.111255

16. Mohammadi, S., Yousefi, M., Khazaei, M.: A review on composite patch repairs and the most important parameters affecting its efficiency and durability. J. Reinf. Plast. Compos. (2020). https://doi.org/10.1177/0731684420941602

17. Tokovyy, Y.V., Hung, K.M., Ma, C.C.: Determination of stresses and displacements in a thin annular disk subjected to diametral compression. J. Math. Sci. **165**(3), 342–354 (2010). https://doi.org/10.1007/s10958-010-9803-6

18. Tokovyy, Y.V., Huang, Y.H., Yen, C.Y., Ma, C.C.: Analytical and experimental evaluation of stresses in elastic annuli subjected to three-point loading on the outer surface. Appl. Math. Model. **73**, 442–458 (2019). https://doi.org/10.1016/j.apm.2019.04.027

19. Kurennov, S.S.: A simplified two-dimensional model of adhesive joints. nonuniform load. Mech. Compos. Mater. **51**(4), 479–488 (2015). https://doi.org/10.1007/s11029-015-9519-2

20. Kondratiev, A.: Improving the mass efficiency of a composite launch vehicle head fairing with a sandwich structure. Eastern-Eur. J. Enterp. Technol. **6**(7), 6–18 (2019). https://doi.org/10.15587/1729-4061.2019.184551

Kelvin-Voigt Model of Dynamic Stress in the Conveyors' Belt

Oleh Pihnastyi🄳 and Georgii Kozhevnikov$^{(\boxtimes)}$ 🄳

National Technical University "Kharkiv Polytechnic Institute",
2 Kyrpychova Str., Kharkiv 61002, Ukraine
pihnastyi@gmail.com, kozhevnikov.gk@gmail.com

Abstract. The paper considers the reasons for the occurrence of dynamic stresses in a conveyor belt made of composite materials, the properties of which can be described by the Kelvin-Voigt model of an elastic element. A boundary value problem for the equation of elastic vibrations in a lightly loaded conveyor-type transport system is formulated. The dependencies determining the forces of resistance to the movement of the belt correspond to the recommendations of DIN 22101:2011-12. An expression is presented that determines the speed of propagation of disturbances in the conveyor belt, taking into account the uneven distribution of material along the transport route. The reasons for the appearance of elastic stresses in the conveyor belt associated with the start, acceleration, deceleration and stop of the transport conveyor are demonstrated. Dimensionless parameters were introduced to analyze the causes of dynamic stresses in the conveyor belt. Similarity criteria in the dimensionless transport conveyor model make it possible to use the results obtained for similar transport systems. It is shown that the value of static stresses for lightly loaded transport systems is characterized by a linear dependence on the distance between the belt element and the drive shaft of the conveyor section. The solution of the boundary value problem defining the expression for dynamic stresses in the tape is presented. It is shown that the decay speed of the arising disturbances is inversely proportional to the value of the viscosity coefficient of the composite material, which characterizes the mechanical properties of the conveyor belt. A qualitative assessment has been made for the characteristic decay time of dynamic disturbances in a conveyor belt.

Keywords: Transport conveyor · Belt stress · Distributed transport system

1 Introduction

Conveyor transport is the main element of mining enterprises. The length of modern multi-section conveyor systems has exceeded one hundred kilometers [1, 2]. Moreover, the length of a separate section of such transport systems is ten kilometers [3]. The costs of conveyor transport up to ten kilometers long for a mining enterprise account for more than 20% of the total cost of extracting material [4]. With an increase in the length of the transport route, the cost of transporting material at a constant belt speed increases non-linearly. With ineffective material flow control, transportation costs can increase several times. This is due to the factor of uneven material flow from the

© The Author(s), under exclusive license to Springer Nature Switzerland AG 2021
M. Nechyporuk et al. (Eds.): ICTM 2020, LNNS 188, pp. 355–365, 2021.
https://doi.org/10.1007/978-3-030-66717-7_30

mining site to the entrance of the transport system [5], which leads to an uneven specific density of material distribution along the transport route. An effective way to reduce transport costs is the use of belt speed control systems [6, 7]. The use of this method of reducing the specific energy consumption leads to the fact that the torque of the motor of the conveyor section periodically changes over time, causing dynamic stresses in the conveyor belt [8, 9]. There is a danger that the magnitude of the dynamic stresses will exceed the permissible value. This excess leads to the destruction of the conveyor belt and stopping the transportation process. Thus, the problem of designing effective control systems for the speed of the conveyor belt is closely related to the analysis of the dynamic stresses in the conveyor belt caused by the regulation of the speed of the belt. To a large extent, the process of dynamic stress propagation depends both on the mechanical properties of the material from which the conveyor belt is made and on the factor of uneven loading of the material on the conveyor belt along the transport route. The topicality of the use of belt speed control systems determined the purpose of this study: analysis of the propagation of dynamic stresses in a conveyor belt, the mechanical properties of the material of which correspond to the Kelvin-Voigt model of an elastic element.

2   Elastic Element Model

An overview of the models of elastic elements for materials of various properties is presented in [10, 11]. The characteristics of the elastic element models are considered: Hookean element, Newtonian element, Maxwell element, Kelvin element, Venant element, CDI geometric beam element and CDI five-element. The mechanical properties of composite materials for the manufacture of a rubber conveyor belt with polyester and polyamide cartridges are given in [12]. For rubber conveyor belts, experimental studies were carried out, the purpose of which was to determine the values of the tensile strength, elastic modulus, Poisson's coefficient. Indicators characterizing the properties of the material of the conveyor belt are used to select a model of an elastic element.

The choice of the elastic element model for the Kelvin-Voigt study is due to the presence in the model of terms with viscous properties. The presence of such terms affects the process of damping of dynamic disturbances in the conveyor belt. The Kelvin-Voigt element model equation (Fig. 1) provides a relationship between stress and deformation in the following form:

$$\sigma(t, S) = E\varepsilon(t, S) + \eta \frac{d\varepsilon(t, S)}{dt}, \tag{1}$$

where E is the elastic modulus of the element; η is element viscosity.

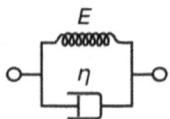

Fig. 1. Kelvin-Voigt element.

For a constant stress value in the element $\sigma(t, S) = \sigma_0 = const$ Eq. (1) admits the solution

$$\varepsilon(t, S) = \frac{\sigma_0}{E}\left(1 - e^{-t/t_\eta}\right), \quad t_\eta = \frac{\eta}{E}. \tag{2}$$

For long values of time, dependence (2), which determines the relative deformation $\varepsilon(t, S)$ at given stress σ_0 corresponds to Hooke's law

$$\sigma(t, S) \approx E\varepsilon(t, S). \tag{3}$$

The analysis of the arising dynamic stresses in accordance with Eq. (3) is studied in detail in [14].

3 Conveyor Line Model

To calculate the forces acting on the section dS of the density conveyor belt $[\chi]_{0C} = const$, on which the material with density $[\chi]_0(t, S)$ is located, we use the expression [14]

$$\frac{d\langle\mu\rangle(t, S)}{dt} dm = \sigma((t, S + dS)Bh - \sigma(t, S)Bh - dF_W, \tag{4}$$

where $\langle\mu\rangle(t, S)$, dm is the speed of the element dS and the mass of the belt element dS with the transported material

$$dm = \left([\chi]_0(t, S) + [\chi]_{0C}\right) dS,$$

B, h is width and thickness of the conveyor belt; $\sigma(t, S)$ is stress arising in the conveyor belt in accordance with the law (1).

F_W is the force of resistance to the movement of the conveyor belt:

$$F_W = F_H + F_N + F_{St} + F_S,$$

consisting of primary resistances associated with frictional resistance along the conveyor belt F_H, secondary movement resistances F_N, gradient movement resistances F_{St} and special resistances in the transport system F_S. In accordance with the recommendations of DIN 22101:2011-12 [13], the indicated resistance to the movement of the tape can be represented by the following expressions

$$dF_H = dS \cdot f_C \cdot g_m\left([\chi]_{0R} + \left([\chi]_0(t, S) + [\chi]_{0C}\right)\cos\delta_C\right),$$

$$F_N = (C - 1)F_H, \quad F_S \ll F_H,$$

$$dF_{St} = dS \cdot \sin \delta_C \cdot g_m \big([\chi]_0(t, S) + [\chi]_{0C}\big),$$

where f_C is the coefficient of rolling resistance of the driving rollers and the resistance of the belt indentation; $g_m = 9.81$ (m/sec^2); $C \approx 1,05$; $[\chi]_{OR}$ is linear load from rotating parts; δ_C is the angle of inclination of the section of the conveyor section. For a horizontally located conveyor section $\delta_C = 0$.

The absolute elongation of the conveyor belt $W(t, S)$ and the relative deformation of the element at the moment of time t for the technological position S are related by the differential relation

$$\varepsilon(t, S) = \frac{\partial W(t, S)}{\partial S}, \quad \varepsilon(t, S) \approx 10^{-2}. \tag{5}$$

The speed of movement $\langle \mu \rangle (t, S)$ of the element dS of the conveyor belt, on which the material is located, consists of the speed of the belt in the equilibrium $\mu_\psi(t, S)$ and vibrational part of the speed of the belt $dW(t, S)/dt$:

$$\langle \mu \rangle (t, S) = \mu_\psi(t, S) + \frac{dW(t, S)}{dt}. \tag{6}$$

Due to the fact that the relative deformation $\varepsilon(t, S)$ of the elastic element is small (5), it follows

$$\frac{dW(t, S)}{dt} = \frac{\partial W(t, S)}{\partial t} + \langle \mu \rangle \frac{\partial W(t, S)}{\partial S} = \frac{\partial W(t, S)}{\partial t} + \langle \mu \rangle \varepsilon(t, S) \approx \frac{\partial W(t, S)}{\partial t}. \tag{7}$$

Relations (5), (7) are the assumption that dynamic disturbances have a small gradient

$$|\langle \mu \rangle \varepsilon(t, S)| << \left| \frac{\partial W(t, S)}{\partial t} \right|. \tag{8}$$

We believe that for the case of large gradients of functions $W(t, S)$, $\langle \mu \rangle$ the conveyor belt is destroyed. Due to the smallness of the gradient, taking into account the inequality (8), it follows that

$$\frac{d \langle \mu \rangle}{dt} \approx \frac{d \mu_\psi}{dt} + \frac{\partial^2 W(t, S)}{\partial t^2} \tag{9}$$

Let us divide both sides of equality (4) by and using transformations (5)–(9), we obtain the equation of longitudinal vibrations in the conveyor belt

$$\frac{\partial^2 W(t, S)}{\partial t^2} = C_\psi^2(t, S) \left(\frac{\partial^2 W(t, S)}{\partial S^2} + \frac{\eta}{E} \frac{\partial^3 W(t, S)}{\partial t \partial S^2} - \frac{1}{BhE} \frac{\partial F_W}{\partial S} \right) - f_\psi(t). \tag{10}$$

where $f_\psi(t) = d\mu_\psi/dt$ is the acceleration of the conveyor belt for the steady state. The index "ψ" in the variables of the equation denotes that the value is determined for the unperturbed state of the parameter. Function $C_\psi^2(t, S)$

$$C_\psi^2(t, S) = \frac{BhE}{[\chi]_{0\psi}(t, S) + [\chi]_{0C}} \tag{11}$$

is the speed of propagation of longitudinal disturbances along the conveyor belt [14]. The speed of propagation of longitudinal vibrations depends on the density of distribution of the material along the transport route $[\chi]_{0\psi}(t, S)$. When deriving expression (11), it is assumed that there is no material sliding along the conveyor belt. In fact, the function $C_\psi^2(t, S)$ has a more complex dependence, taking into account the sliding and transformation of the shape of the bulk material when moving along the transport route [9]. Let us supplement the equation of longitudinal vibrations (10) with the boundary conditions:

$$\sigma(t, S_d) = \frac{T_1}{Bh} = \frac{1}{Bh}\left(k_s F_{W(2-3)} + k_s F_{\psi(2-3)} + F_{W(4-1)} + F_{\psi(4-1)}\right)\frac{\exp(k_b\alpha)}{\exp(k_b\alpha) - k_s}, \tag{12}$$

$$\sigma(t, 0) = \frac{T_4}{Bh} = \frac{T_1 - F_{W(4-1)} - F_{\psi(4-1)}}{Bh} = E\frac{\partial W(t, S)}{\partial S}\bigg|_{S=S_d} - \frac{F_{W(4-1)} + F_{\psi(4-1)}}{Bh}, \tag{13}$$

and initial conditions:

$$\frac{\partial W(t, S)}{\partial t}\bigg|_{t=0} = 0, \tag{14}$$

$$\sigma(0, S) = E\frac{\partial W(t, S)}{\partial S}\bigg|_{t=0} = \sigma(0, 0) + F_{W\Psi(4-1)}(0, S) + F_{\psi\Psi(4-1)}(0, S). \tag{15}$$

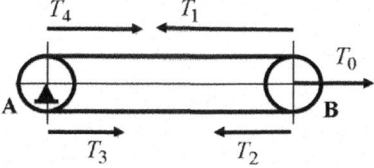

Fig. 2. Diagram of the acting forces of the conveyor belt.

The acting forces at the characteristic points of the conveyor belt are shown in Fig. 2 and can be determined as follows:

$$T_1 = T_2 \exp(k_b \alpha), \; T_3 = T_2 + F_{W23} + F_{\psi23}, \; T_4 = k_s T_3, \; T_1 = T_4 + F_{W41} + F_{\psi41} \quad (16)$$

$$F_{H23} = S_d f_C g_m \big([\chi]_{0R} + [\chi]_{0C}\big), \; F_{H41} = F_{H23} + f_C g_m \int_0^{S_d} [\chi]_0(t,S)dS, \quad (17)$$

$$F_{N23} = (C-1)F_{H23}, \; F_{N41} = (C-1)F_{N41}, \; F_{St23} = 0, \; F_{St41} = 0, \quad (18)$$

$$F_{\psi23} = F_\psi(t)[\chi]_{0C}S_d, \; F_{\psi41} = F_{\psi23} + F_\psi(t) \int_0^{S_d} [\chi]_0(t,S)dS, \quad (19)$$

where $F_{\psi23}$, $F_{\psi41}$ are forces of acceleration and deceleration of the conveyor belt; k_b is the coefficient of adhesion between the drum and the belt with a wrap angle α; k_s is the coefficient of losses on drum "A".

The boundary value problem (10), (12)–(15) allows studying the process of propagation of longitudinal vibrations on a conveyor belt with an uneven distribution of material along the transport route.

4 Analysis of Results

To solve the stated boundary value problem, we introduce dimensionless parameters:

$$\tau = \frac{t}{T_d}, \; \xi = \frac{S}{S_d}, \; \theta_0(\tau,\xi) = \frac{[\chi]_0(t,S)}{[\chi]_{0max}}, \; \psi(\xi) = \frac{\Psi(S)}{[\chi]_{0max}}, \quad (20)$$

$$W_0(\tau,\xi) = \frac{W(t,S)}{W_{max}}, \; W_{max} = \frac{\sigma_b S_d}{E}, \; \theta_C = \frac{[\chi]_{0C}}{[\chi]_{0max}}, \; \theta_R = \frac{[\chi]_{0R}}{[\chi]_{0max}}, \quad (21)$$

$$v_b = \frac{C f_C g_m \big([\chi]_{0max} + [\chi]_{0C}\big) S_d}{\sigma_b B h}, \; v_f(\tau) = \frac{\big([\chi]_{0max} + [\chi]_{0C}\big) f_\psi(t) S_d}{\sigma_b B h}, \; v_\eta = \frac{\eta}{E T_d}, \quad (22)$$

$$v_g^2 = \frac{B h E}{[\chi]_{0max} + [\chi]_{0C}} \left(\frac{T_d}{S_d}\right), \; K_{12} = \frac{(k_s+1)\exp(k_b\alpha)}{\exp(k_b\alpha) - k_s}, \; v_1^2 = v_g^2 \frac{1 + \theta_C}{\theta_C}, \quad (23)$$

$$\alpha_{12}(\tau) = v_b \frac{\theta_C}{1 + \theta_C}\left(1 + \frac{\theta_R}{\theta_C}\right) + v_f(\tau)\frac{\theta_C}{1 + \theta_C}, \; \alpha_1(\tau) = K_{12}\alpha_{12}(\tau). \quad (24)$$

The use of dimensionless parameters makes it possible to extend the research results to a wide class of similarly transport systems. Let's consider the solution of the problem for the case of movement of a lightly loaded conveyor line. The specific density of the material along such conveyor lines is low compared to the specific gravity of the conveyor belt

$$\theta_0(\tau, \xi) \ll \theta_C, \quad \psi(\xi) = \theta_0(0, \xi) \ll \theta_C.$$

Let's also introduce a simplification assuming a linear dependence of the belt acceleration on the drive shaft on time for the belt speed control modes [15, 16]:

$$v_f(\tau) = v_{f0} + v_{f1}\tau. \tag{25}$$

Taking these assumptions into account, Eq. (10) takes the form

$$\frac{\partial^2 W_0(\tau, \xi)}{\partial \tau^2} = v_1^2 \frac{\partial^2 W_0(\tau, \xi)}{\partial \xi^2} + v_\eta v_1^2 \frac{\partial^3 W_0(\tau, \xi)}{\partial \xi^2 \partial \tau} - v_1^2 \alpha_{12}(\tau), \tag{26}$$

with boundary conditions

$$\left.\frac{\partial W_0(t, \xi)}{\partial \xi}\right|_{\xi=1} = \alpha_1(\tau), \quad \left.\frac{\partial W_0(\tau, \xi)}{\partial \xi}\right|_{\xi=0} = \alpha_1(\tau) - \alpha_{12}(\tau), \tag{27}$$

and initial conditions

$$\left.\frac{\partial W_0(\tau, \xi)}{\partial t}\right|_{t=0} = 0, \quad \frac{\partial W_0(0, \xi)}{\partial \xi} = \alpha_1(\tau) - \alpha_{12}(\tau)(1 - \xi). \tag{28}$$

The solution to Eq. (26) is represented by the following form

$$W_0(\tau, \xi) = W_{00}(\tau, \xi) + A(\tau)\xi^2 + B(\tau)\xi. \tag{29}$$

Let us define the coefficients $A(\tau)$, $B(\tau)$ in such a way as to ensure the presence of boundary conditions for the function $W_{00}(\tau, \xi)$ in the form

$$\left.\frac{\partial W_{00}(\tau, \xi)}{\partial \xi}\right|_{\xi=1} = 0, \quad \left.\frac{\partial W_{00}(t, \xi)}{\partial \xi}\right|_{\xi=0} = 0. \tag{30}$$

From the solution of the system of Eqs. (30) it follows:

$$W_0(\tau, \xi) = W_{00}(\tau, \xi) + \frac{\alpha_{21}(\tau)}{2}\xi^2 + (\alpha_1(\tau) - \alpha_{12}(\tau))\xi. \tag{31}$$

Substituting the obtained expression (31) into the equation of longitudinal vibrations (26) under the assumption (25), we obtain

$$\frac{\partial^2 W_{00}(\tau, \xi)}{\partial \tau^2} = v_1^2 \frac{\partial^2 W_{00}(\tau, \xi)}{\partial \xi^2} + v_\eta v_1^2 \frac{\partial^3 W_{00}(\tau, \xi)}{\partial \xi^2 \partial \tau} + v_\eta v_1^2 v_{f1} \frac{\theta_C}{1 + \theta_C}, \tag{32}$$

$$\left.\frac{\partial W_{00}(\tau, \xi)}{\partial \xi}\right|_{\xi=1} = 0, \quad \left.\frac{\partial W_{00}(\tau, \xi)}{\partial \xi}\right|_{\xi=0} = 0 \tag{33}$$

$$\frac{\partial W_{00}(\tau, \xi)}{\partial \xi}\bigg|_{\xi=1} = -v_{f1}\frac{\theta_C}{1+\theta_C}\left(\frac{\xi^2}{2} - \xi + K_{12}\xi\right), \quad W_{00}(\tau, \xi) = 0. \qquad (34)$$

The solution of Eq. (32) due to boundary conditions (33) will be searched in the form

$$W_{00}(\tau, \xi) = T_0(\tau) + \sum_{n=1}^{\infty} T_n(\tau)\cos(\pi n\xi) \qquad (35)$$

Let's represent the initial conditions in the form of an expansion in a Fourier series:

$$\frac{\partial W_{00}(\tau, \xi)}{\partial \tau}\bigg|_{\tau=0} = \frac{a_0}{2} + \sum_{n=1}^{\infty} a_n\cos(\pi n\xi), \qquad (36)$$

$$a_0 = -2v_{f1}\frac{\theta_C}{1+\theta_C}\left(\frac{K_{12}}{2} - \frac{1}{3}\right), \quad a_n = -2v_{f1}\frac{\theta_C}{1+\theta_C}\left(\frac{1}{(\pi n)^2} + K_{12}\frac{(-1)^n-1}{(\pi n)^2}\right).$$

This will allow writing the system of equations

$$\frac{d^2 T_0(\tau)}{d\tau^2} = v_\eta v_1^2 v_{f1}\frac{\theta_C}{1+\theta_C} \qquad (37)$$

$$T_0(0) = 0, \quad \frac{dT_0(0)}{d\tau} = -v_{f1}\frac{\theta_C}{1+\theta_C}\left(\frac{K_{12}}{2} - \frac{1}{3}\right),$$

$$\frac{d^2 T_n(\tau)}{d\tau^2} + v_\eta v_1^2 (\pi n)^2\frac{dT_n(\tau)}{d\tau} + v_1^2(\pi n)^2 T_n(\tau) = 0, \qquad (38)$$

$$T_n(0) = 0, \quad \frac{dT_n(0)}{d\tau} = -2v_{f1}\frac{\theta_C}{1+\theta_C}\left(\frac{1}{(\pi n)^2} + K_{12}\frac{(-1)^n-1}{(\pi n)^2}\right),$$

for determining unknown functions $T_0(\tau)$, $T_n(\tau)$ in the expression for $W_{00}(\tau, \xi)$ (35). The solution this system of equations has the form:

$$T_0(\tau) = v_{f1}\frac{\theta_C}{1+\theta_C}\left(v_\eta v_1^2\frac{\tau^2}{2} - \left(\frac{K_{12}}{2} - \frac{1}{3}\right)\tau\right), \qquad (39)$$

$$T_n(\tau) = e^{-\gamma_n\tau}\sin(\omega_{pn}\tau)\frac{v_{f1}}{\omega_{pn}}\frac{\theta_C}{(1+\theta_C)}\left(\frac{1}{(\pi n)^2} + K_{12}\left(\frac{(-1)^2-1}{(\pi n)^2}\right)\right), \qquad (40)$$

$$\gamma_n = \frac{v_\eta}{2}\omega_n^2, \quad \omega_n^2 = (\pi n v_1)^2, \quad \omega_{pn} = \sqrt{\omega_n^2 - \gamma_n^2}, \text{ for } \omega_n^2 > \gamma_n^2.$$

Let us substitute the obtained solution (36), (40) in (35), perform differentiation (31) to the variable ξ, obtain an expression for determining the stress in the conveyor belt at the time τ for an arbitrary point of the transport route ξ

$$\frac{\partial W_0(\tau, \xi)}{\partial \xi} = \alpha_{12}(\tau)\xi + \alpha_1(\tau) - \alpha_{12}(\tau) - \sum_{n=1}^{\infty} T_n(\tau)\sin(\pi n\xi). \tag{41}$$

The stress in the conveyor belt can be represented by the sum of two components. The first component of dynamic stresses

$$\frac{\partial W_{0Slow}(\tau, \xi)}{\partial \xi} = \alpha_{12}(\tau)\xi + \alpha_1(\tau) - \alpha_{12}(\tau),$$

determines the slow change in stress over time as a result of acceleration or deceleration of the conveyor belt. The second component of dynamic stresses

$$\frac{\partial W_{0Fast}(\tau, \xi)}{\partial \xi} = -\sum_{n=1}^{\infty} T_n(\tau)\sin(\pi n\xi)$$

characterizes fast processes with the oscillation amplitude decaying over time

$$A_n(\tau) = e^{-\gamma_{n\tau}} \frac{v_{f1}}{\omega_{pn}} \frac{\theta_C}{1+\theta_C} \left(\frac{1}{(\pi n)^2} + K_{12}\frac{(-1)^n - 1}{(\pi n)^2} \right). \tag{42}$$

Let us introduce the notation

$$a_\eta = v_g v_\eta = \sqrt{\frac{BhE}{[\chi]_{0max} + [\chi]_{0C}}} \left(\frac{T_d}{S_d}\right) \frac{\eta}{ET_d} = \frac{t_\eta v_s}{S_d}, \quad v_s = \sqrt{\frac{BhE}{[\chi]_{0max} + [\chi]_{0C}}}. \tag{43}$$

Then the time variation of the damping amplitude can be represented by the expression

$$A_n(\tau) = \frac{e^{-\pi^2 n^2 a_\eta^2 n \frac{\tau}{v_\eta}}}{(\pi n)^3 \sqrt{4 - \pi^2 n^2 \frac{1+\theta_C}{\theta_C} a_\eta^2}} \frac{v_{f1}(1 + K_{12}(-1)^n - K_{12})}{v_1}, \tag{44}$$

$$\gamma_n \tau = \frac{v_\eta}{2} \omega_n^2 \tau = \frac{v_\eta}{2} (\pi n)^2 v_g^2 \frac{1+\theta_C}{\theta_C} \tau \approx v_\eta (\pi n)^2 v_g^2 \tau \approx \pi^2 n^2 a_\eta^2 \frac{\tau}{v_\eta}.$$

The coefficient a_η is the ratio of the distance that the wave can travel with the propagation speed v_s in time $t_\eta(2)$ to the total length of the conveyor section S_d. Figure 3 shows the damping of the oscillation amplitude for $n = 1.2$, depending on the value of the coefficient a_η.

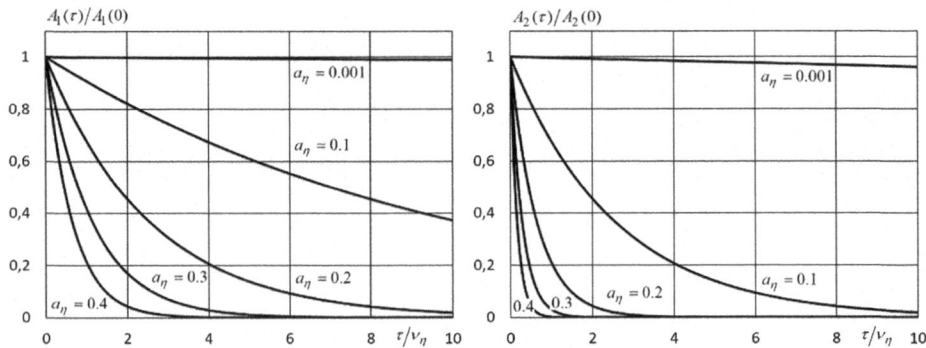

Fig. 3. Dependence of the relative oscillation amplitude $A_n(\tau)/A_n(0)$ on relative time τ/v_η for n = 1.2

The exponential decay of the dynamic stresses arising as a result of a change in the speed of the conveyor belt is observed. At small values of the coefficient a_η the vibration amplitude $A_n(\tau)$ is quasi-stationary.

5   Conclusions

The use of control systems for the speed of movement of the conveyor belt is a source of dynamic vibrations of the conveyor belt of the transport system. The properties of composite materials significantly affect the propagation of disturbances along the transportation route. The use of materials whose mechanical properties correspond to the Kelvin-Voigt elastic element model leads to damping of dynamic stresses. The decay rate of the oscillations exponentially depends on the viscosity coefficient and the harmonic of the oscillations. When constructing a model, the calculation of the resistance to the movement was carried out in accordance with DIN 22101:2011-12. An expression is given that determines the speed of propagation of disturbances in a conveyor belt, taking into account the uneven distribution of material along the transportation route. To analyze the transport system, similarity criteria were introduced.

References

1. Pihnastyi, O., Kozhevnikov, G., Khodusov, V.: Conveyor model with input and output accumulating bunker. In: 2020 IEEE 11th International Conference on Dependable Systems, Services and Technologies (DESSERT), pp. 253–258. IEEE, Kyiv (2020). https://doi.org/10.1109/DESSERT50317.2020.9124996
2. Siemens: Innovative solutions for the mining industry. https://www.siemens.com/mining. Accessed 08 Aug 2020
3. Pihnastyi, O., Khodusov, V.: Model of a composite magistral conveyor line. In: 2018 IEEE First International Conference on System Analysis & Intelligent Computing (SAIC), pp. 1–4. IEEE, Kiev (2018). https://doi.org/10.1109/SAIC.2018.8516739

4. Razumnyi, Y., Rukhlov, A., Kozar, A.: Improving the energy efficiency of conveyor transport of coal mines. Min. Electromech. Autom. **76**, 24–28 (2006). [in Russian]
5. Semenchenko, A., Stadnik, M., Belitsky, P., et al.: The impact of an uneven loading of a belt conveyor on the loading of drive motors and energy consumption in transportation. East. Eur. J. Enterp. Technol. **4**(1), 42–51 (2016). https://doi.org/10.15587/1729-4061.2016. 75936
6. Ristić, L.B., Bebić, M.Z., Jevtić, D.S., et al.: Fuzzy speed control of belt conveyor system to improve energy efficiency. In: 2012 15th International Power Electronics and Motion Control Conference (EPE/PEMC), pp. DS2a-9. IEEE, Novi Sad (2012). https://doi.org/10. 1109/EPEPEMC.2012.6397260
7. Halepoto, I.A., Shaikh, M.Z., Chowdhry, B.S., Uqaili, M.A.: Design and implementation of intelligent energy efficient conveyor system model based on variable speed drive control and physical modeling. Int. J. Control Autom. **9**(6), 379–388 (2016). https://doi.org/10.14257/ ijca.2016.9.6.36
8. Bebic, M., Ristic, B.: Speed controlled belt conveyors: drives and mechanical considerations. Adv. Electric. Comput. Eng. **18**(1), 51–60 (2018). https://doi.org/10.4316/AECE. 2018.01007
9. Zeng, F., Yan, C., Wu, Q., Wang, T.: Dynamic behaviour of a conveyor belt considering non-uniform bulk material distribution for speed control. Appl. Sci. **10**(13), 4436 (2020). https://doi.org/10.3390/app10134436
10. Yang, G.: Dynamics analysis and modeling of rubber belt in large mine belt conveyors. Sens. Transd. **181**(10), 210–218 (2014)
11. Nordell, L.K., Ciozda, Z.P.: Transient belt stresses during starting and stopping: elastic response simulated by finite element methods. Bulk Solids Handling **4**(1), 93–98 (1984)
12. Lu, Y., Lin, F.Y., Wang, Y.C.: Investigation on influence of speed on rolling resistance of belt conveyor based on viscoelastic properties. J. Theoret. Appl. Mech. **45**(3), 53–68 (2015). https://doi.org/10.1515/jtam-2015-0017
13. DIN 22101:2011-12. Continuous conveyors. Belt conveyors for loose bulk materials. Basis for calculation and dimensioning. Beuth-Vertrieb GmbH, Berlin (2011). https://doi.org/10. 31030/1821227
14. Pascual, R., Meruane, V., Barrientos, G.: Analysis of transient loads on cable-reinforced conveyor belts with damping consideration. In: Proceedings of the XXVI Iberian Latin-American Congress on Computational Methods in Engineering CILAMCE 2005, paper CIL0620 (2005)
15. He, D., Pang, Y., Lodewijks, G.: Determination of acceleration for belt conveyor speed control in transient operation. Int. J. Eng. Technol. **8**(3), 206–211 (2016)
16. Karolewski, B., Ligocki, P.: Modelling of long belt conveyors. Maint. Reliab. **16**(2), 179–187 (2014)

Effective Conveyor Belt Control Based on the Time-Of-Use Tariffs

Oleh Pihnastyi$^{(\boxtimes)}$ and Georgii Kozhevnikov

National Technical University "Kharkiv Polytechnic Institute",
2 Kyrpychova Str., Kharkiv 61002, Ukraine
pihnastyi@gmail.com, kozhevnikov.gk@gmail.com

Abstract. The paper proposes a method for constructing an algorithm for the speed control of a conveyor belt, based on the change in the price of electricity during the day. The analysis of methods for improving the energy efficiency of conveyor-type distributed transport systems is carried out. The influence of the uneven distribution of material along the transportation route on the cost of transportation of a unit weight of the material is demonstrated. The advantages of using Time-Of-Use (TOU) tariffs when designing belt speed control systems for long conveyor systems are considered. The TOU periods with peak, standard and low energy consumption depending on time are presented in detail, as well as the values of the tariff coefficients for the TOU periods. The dependence of the value of the tariff coefficient on time is an essential factor that must be taken into account when designing control algorithms. When developing the control algorithm, it was assumed that the resistance to motion in accordance with DIN 22101 is determined on the basis of the primary friction coefficients. To describe a separate section, an analytical model of the conveyor in a dimensionless form was used. The problem of constructing an optimal algorithm for controlling the speed of a conveyor belt for a steady-state is formulated. The criterion of the quality of the control process in the conditions of using a constant amount of electricity during the day has been determined. The Pontryagin function and the conjugate system of equations are written, taking into account the uneven distribution of material along the transport route.

Keywords: Transport conveyor · Distributed transport system · Energy management · Conveyor belt speed control · Transport delay · The uneven distribution of material

1 Introduction

The conveyor belt is widely used in the mining industry, used to transport material from the place of extraction to the place of loading or processing of material. The prevalence of conveyor systems is explained by the relatively small proportion of the unit cost of material transportation in comparison with other types of transportation systems. The cost of transporting material in a standard mode using a conveyor system with a length of several tens of kilometers is 20% of the total cost of extracting material [1]. Uneven incoming of material at the entrance of the transport system leads to an uneven distribution of material along the transport route, which causes a nonlinear

© The Author(s), under exclusive license to Springer Nature Switzerland AG 2021
M. Nechyporuk et al. (Eds.): ICTM 2020, LNNS 188, pp. 366–376, 2021.
https://doi.org/10.1007/978-3-030-66717-7_31

increase in transport costs depending on the filling factor of the transport system [2, 3]. For low loaded conveyor sections, transport costs can increase several times. This fact is of particular importance in the operation of extended and high-performance conveyor systems [4–6]. To increase the reliability of the conveyor system and increase the energy efficiency of material transportation, long transport systems are divided into sections [7–9]. Reducing energy consumption within a separate section is ensured by the belt speed control [10, 11], the flow of material from the accumulating hopper to the input of the conveyor section [12, 13] or combined control. Additional cost savings are achieved through the use of multiple drive conveyor systems [14, 15]. The listed methods of reducing the cost of material transportation are in constant development.

2 Literature Review

Significant method of reducing unit energy costs is the use of effective energy management methodologies [16]. The essence of this approach is to change the capacity consumed by the transport system in accordance with a given schedule based on a specific tariff structure. Electricity suppliers provide a differentiated tariff to large users to effectively manage the supply of electricity, which is beneficial for both the electricity supplier and the electricity consumer. Customers using the Time-Of-Use (TOU) tariff benefit from energy savings by scheduling the electrical equipment of the transport system during periods of time with low energy costs [17]. The longest transport conveyors are in Africa: Sasol–Shondoni Overland [18] (20.5 km single flight overland conveyor with multiple horizontal curves); Western Sahara [19] (the conveyor from the Bu Craa mine to the coast at El Aaiún, 128,7 km with 11 section). Therefore, let's will focus our attention on the analysis of the energy management methodology for South Africa. The periods Eskom – TOU (South Africa) are shown in Fig. 1 [20].

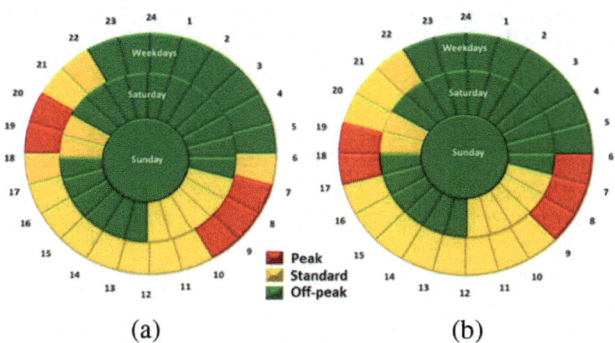

(a)   (b)

Fig. 1. Eskom–TOU periods: (a) – high demand season; (b) – low demand season [20].

TOU periods are represented by time intervals with peak demand, standard and low demand depending on the time of day and day of the week for high demand season

(June-August) and low demand season (September-September-May). Night save Eskom–TOU periods are shown in Fig. 2.

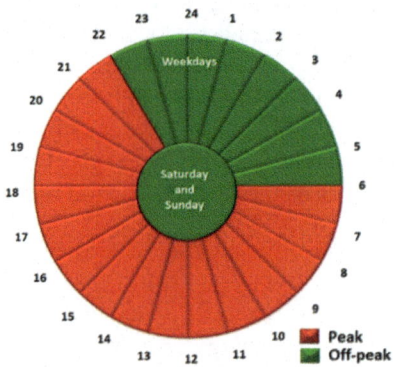

Fig.2. Night saves Eskom–TOU periods [20].

The price for electricity consumption depending on the tariff and recommendations for using tariffs are given in the articles [20–22]. In accordance with the Eskom–TOU pricing policy, coefficients were calculated that characterizes the ratio of the price of electricity consumption in different periods of the day (Fig. 3).

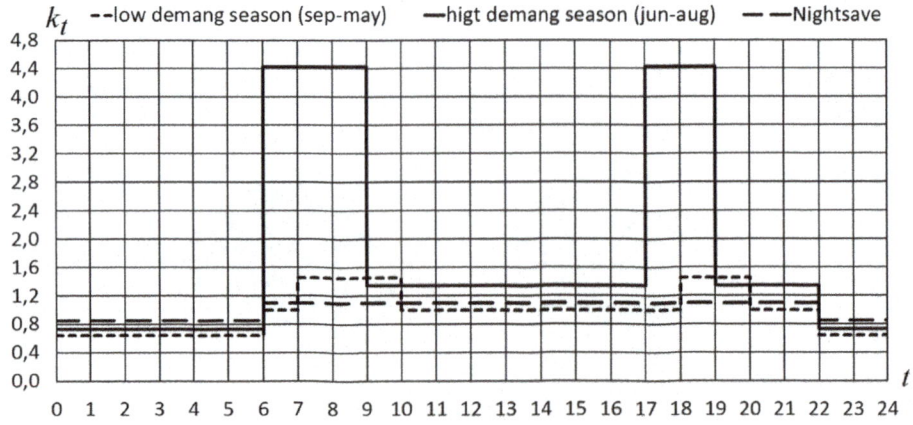

Fig. 3. Periods of tariff coefficients Eskom–TOU.

High peak loads match the price of electricity in winter (high demand season) [23]. Long-term strategy and principles for constructing tariffs for consumed electricity are given in the article [23]. The same document presents the profiles of electricity consumption during the day, which was the basis for the formation of Eskom-TOU tariffs [20]. Figure 3 shows a manifold increase in the price of consumed electricity during the period of peak loads, especially for the high demand season tariff. The two tariffs high demand season and low demand season can be considered as the main ones for the formation of highly efficient conveyor-type transport systems management systems. Figure 4 shows an example of a TOU strategy that was used by industrial enterprises in Ukraine [24].

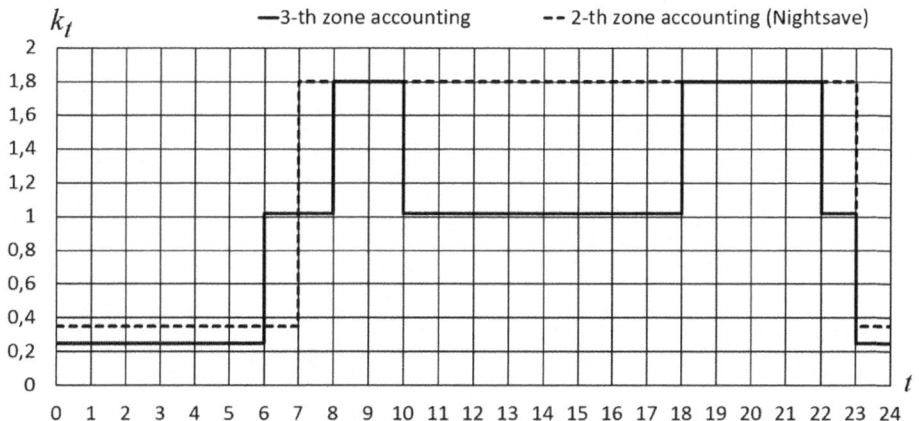

Fig. 4. Periods of tariff coefficients Ukraine–TOU.

TOU's electricity pricing strategy is typical of many countries and aims to efficiently distribute electricity consumption throughout the day.

3 Formal Problem Statement

The problem of energy management is actual for modern industrial production [25]. In [26], a statistical analysis of the data set on the electricity consumption of 12000 enterprises in 44 industrial sectors was carried out. Long conveyor transport system operating in continuous mode is a major consumer of electricity [5]. Continuous operation and a significant proportion of electricity in the cost of the extracted material determines the relevance of designing conveyor control algorithms, taking into account the use of the TOU strategy. The main attention will be paid to the design of algorithms for controlling the speed of the belt movement, which ensure the minimum cost of electricity consumed during the day (24 h). Moreover, let introduce the assumptions: a) a constant amount of electricity is consumed every day for the entire day period (24 h); b) the friction model in accordance with DIN 22101 is determined on the basis of the primary friction coefficient [27]; c) the effects of propagation of disturbances associated with instantaneous switching of the speed modes are not taken into account; d) the duration of the acceleration and deceleration modes of the conveyor belt is negligible with the characteristic time of the transportation process.

4 Conveyor Line Model

To describe the parameters of the conveyor line, let use the analytical PiKh – conveyor model [10]. The model that determines the state of the parameters of a separate section of the transport system has the form:

$$\frac{\partial [\chi]_0(t, S)}{\partial t} + \frac{\partial [\chi]_1(t, S)}{\partial S} = \delta(S)\lambda_1(t), \tag{1}$$

$$[\chi]_0(0, S) = \Psi(S), \tag{2}$$

$$[\chi]_1(t, S) = a(t)[\chi]_0(t, S), \tag{3}$$

$$H(S) = \begin{cases} 0, S < 0, \\ 1, S \geq 0, \end{cases} \quad \int\limits_{-\infty}^{\infty} \delta(S)dS = 1, \tag{4}$$

where $[\chi]_0(t, S)$, $[\chi]_1(t, S)$ are the linear density of the material and the material flow at the moment in time t at the point of the transport route with the coordinate $S \in [0, S_d]$; $a(t)$ is the speed of the conveyor belt; $\lambda_1(t)$ is material flow incoming the conveyor section; $\Psi(S)$ is the initial distribution of material along the transport route; $\delta(S)$ is Dirac function; $H(S)$ is Heaviside function.

Let us introduce dimensionless parameters for modelling the transport conveyor [10]:

$$\tau = \frac{t}{T_d}, \xi = \frac{S}{S_d}, H(S_d\xi) = H(S), \delta(\xi) = S_d\delta(S). \tag{5}$$

$$\psi(\xi) = \frac{\Psi(S)}{[\chi]_{0max}}, \quad \gamma_1(\tau) = \lambda_1(t)\frac{T_d}{S_d[\chi]_{0max}},$$

$$\theta_0(\tau, \xi) = \frac{[\chi]_0(t, S)}{[\chi]_{0max}} \theta_0(\tau, \xi) = \frac{[\chi]_0(t, S)}{[\chi]_{0max}},$$

where $[\chi]_{0max}$ is maximum permissible material density for the conveyor belt; T_d is the characteristic time of the transportation process. S_d is length of a single conveyor section. Taking into account the dimensionless parameters (5), the solution to Eq. (1)–(4) has the form

$$\theta_0(\tau, \xi) = (H(\xi) - H(\xi - G(\tau)))\frac{\gamma_1(\tau - \Delta\tau_\xi)}{g(\tau - \Delta\tau_\xi)} + H(\xi - G(\tau))\psi(\xi - G(\tau)), \tag{6}$$

$$\tau_\xi = G^{-1}(G(\tau) - \xi), G(\tau) = \int\limits_0^\tau g(\alpha)d\alpha, \Delta\tau_\xi(\tau) = \Delta\tau_\xi = \tau - \tau_\xi,$$

where $\Delta\tau_\xi(\tau)$ is transport delay of the incoming of material at the point of the transport route with the coordinate ξ at the time τ. The function $G^{-1}(\tau)$ is the inverse of $G(\tau)$. The transport delay between the input and output of material from the conveyor section $\Delta\tau_1(\tau) = \tau - G^{-1}(G(\tau) - 1)$ is determined $\theta_0(\tau, \xi)$ at the output from the conveyor Sect. (6)

$$\theta_0(\tau, 1) = (1 - H(1 - G(\tau)))\frac{\gamma_1(\tau - \Delta\tau_1)}{g(\tau - \Delta\tau_1)} + H(1 - G(\tau))\psi(1 - G(\tau)). \qquad (7)$$

For the steady mode $\tau \geq G^{-1}(1)$, in which the initial condition (2) does not affect the value of the dimensionless density at the output from the conveyor section, density $\theta_0(\tau, 1)$ is determined by the expression:

$$\theta_0(\tau, 1) = (1 - H(1 - G(\tau)))\frac{\gamma_1(\tau - \Delta\tau_1)}{g(\tau - \Delta\tau_1)}, \tau \geq G^{-1}(1), \qquad (8)$$

Taking into account the model of primary resistances [27], the expression for the force of movement of the conveyor belt can be represented as follows:

$$T = Cf_c g_m \int_0^{S_d} \left(2([\chi]_{OR} + [\chi]_{OC}) + [\chi]_0(t, S)\right)dS, \qquad (9)$$

where f_C is the coefficient of resistance to movement, includes the rolling resistance of the driving rollers and the resistance of the belt indentation [27, p. 13]; $g_m = 9.81(\text{m/sec}^2)$; $[\chi]_{OC}$ is the specific weight of the conveyor belt; $[\chi]_{OR}$ is linear load from rotating parts [27, p. 8]; C is the coefficient that takes into account the influence on the movement of secondary resistances [27, p. 17]. Taking into account the designations (5), the electrical power N_E, required to move the conveyor belt with the material is determined by the expression:

$$n_e(\tau) = g(\tau)m(\tau), m(\tau) = \int_0^1 (2\theta_{0R} + 2\theta_{0C} + \theta_0(\tau, \xi))d\xi, \qquad (10)$$

where

$$\theta_{0R} = \frac{[\chi]_{OR}}{[\chi]_{0max}}, \theta_{0C} = \frac{[\chi]_{OC}}{[\chi]_{0max}}, n_e = N_E\frac{T_d}{Cf_c g_m \eta_c [\chi]_{0max} S_d^2},$$

where η_c is efficiency; $m(\tau)$ is the mass of the transported material. The doubled value of the parameters θ_{0R}, θ_{0C} takes into account the upper and lower length of the conveyor S_d.

5 Method for Constructing a Control Algorithm

Let us formulate the problem of constructing an optimal program for controlling the speed of a conveyor belt for a steady mode of operation of a conveyor line (8): to determine the modes of switching the speed of the conveyor belt during a time interval $\tau = [0, \tau_{24}]$ with the value of the price coefficients of the cost of electricity $z(\tau)$ (Fig. 3, Fig. 4) with stepwise control of the speed of the conveyor belt $u(\tau) = (u_1, u_2)$, $0 < u_1 < u_2 < \infty$, $u_j = const$, which minimizes the functional:

$$\int_0^{\tau_{24}} z(\tau)g(\tau)m(\tau)d\tau \to \min, \tag{11}$$

with differential relations

$$\frac{dm(\tau)}{d\tau} = \gamma_1(\tau) - \theta_1(1, \tau), \tag{12}$$

and limitation on the total amount of energy consumed per day

$$\int_0^{\tau_{24}} g(\tau)m(\tau)d\tau = b = const. \tag{13}$$

The choice of a stepped control mode for analyzing the construction of an optimal control algorithm is due to its widespread use in control for transport systems [10, 28–30] and simplicity in the implementation of the control algorithm. Equation (12) determines the change in the amount of material moved along the conveyor section. The output flow $\theta_1(1, \tau)$ can be expressed in terms of the input flow in the material incoming the input of section $\gamma_1(\tau) = \theta_1(0, \tau)$ (7). To simplify the demonstration of the method for constructing an optimal algorithm for controlling the speed of the conveyor belt $g(\tau) = u(\tau)$ let's assume that the input flow is constant $\gamma_1(\tau) = 1$. Taking these remarks into account, let reformulate problem (11)–(13) in the following form:

$$\int_0^{\tau_{24}} z(\tau)u(\tau)m(\tau)d\tau \to \min, \tag{14}$$

with differential relations

$$\frac{dm(\tau)}{d\tau} = 1 - \frac{u(\tau)}{u(\tau - \Delta\tau_1)}, \tag{15}$$

$$\frac{dx_b}{d\tau} = u(\tau)m(\tau), \quad x_b(0) = 0, \quad x_b(\tau_{24}) = b, \tag{16}$$

The Pontryagin function and the conjugate system of equations can be represented in the form:

$$H = -z(\tau)u(\tau)m(\tau) + \psi_b u(\tau)m(\tau) + \psi_m \left(1 - \frac{u(\tau)}{u(\tau - \Delta\tau_1)}\right), \tag{17}$$

$$\frac{d\psi_b}{d\tau} = -\frac{\partial H}{\partial x_b} = 0, \tag{18}$$

$$\frac{d\psi_m}{d\tau} = (z(\tau) - \psi_b)u(\tau), \quad \psi_m(0) = 0, \quad \psi_m(\tau_{24}) = 0 \tag{19}$$

From Eq. (18) follows $\psi_b = C_b = const$. In the presence of continuous control of the belt speed, the optimal value of the belt speed is determined by the expression:

$$\frac{\partial H}{\partial u} = (\psi_b - z(\tau))m(\tau) - \psi_m \frac{\partial}{\partial u}\left(\frac{u}{u_1}\right) = 0, \tag{20}$$

$$\frac{\partial}{\partial u}\left(\frac{u}{u_1}\right) = \frac{1}{u_1}\left(1 - \frac{u^2 \dot{u}_1}{u_1^2 \dot{u}}\right), \quad u_1 = u(\tau - \Delta\tau_1),$$

If Eq. (20) has a solution, the optimal control is within the interval $[u_1, u_2]$. In the absence of a solution for Eq. (20) within the interval, the optimal control takes on a value on one of the boundaries of the interval.

In the case of a two-stage control mode [10] $0 < u_1 < u_2 < \infty$ the value of the optimal belt speed corresponds to the maximum value for the Pontryagin function (17). The points of switching control modes during the daily time interval are determined by the joint solution of Eqs. (15), (16), (19). The solution to the system of Eqs. (14)–(20) is presented in Fig. 5 for periods of tariff coefficients Ukraine – TOU (2-th zone, $k_\tau(\tau) = (0.35, 1.8)$).

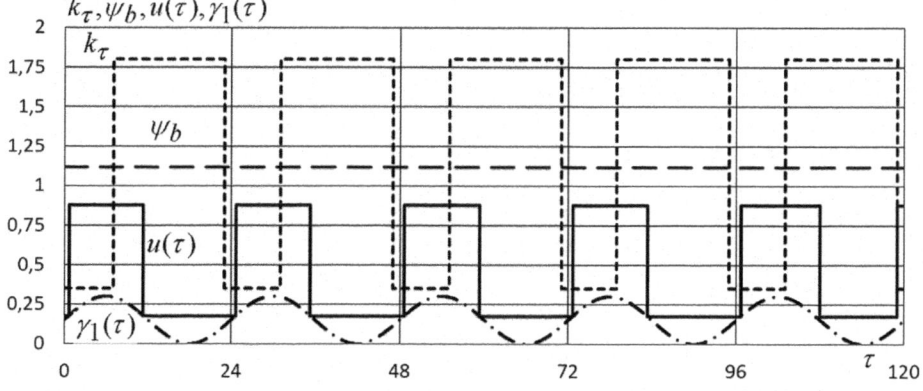

Fig. 5. The speed modes $u(\tau) = (u_1, u_2)$.

The switching points of the speed control mode $u(\tau) = (u_1, u_2)$ for the time-varying input flow intensity $\gamma_1(\tau) = 0.15 + 0.15 \sin(\pi\tau/12)$ are determined by solving the system of Eqs. (15)–(20) for the time interval with the value $\tau_{24} = 120$. The influence of the initial distribution of material $\psi(\xi) = 0.8523$ on subsequent cyclically repeating periods is smoothed out. The switching points are determined from a condition that limits the maximum permissible material density $\gamma_1(\tau) \leq u(\tau)$. For the considered regime, the energy consumption was at cost (14) equal to $x_b(\tau_{24}) = 20.34$ at cost (14), equal to 20.94. The ratio of the cost of energy consumed to the amount consumed is 1.03, which indicates an ineffective transition from a fixed tariff to a TOU tariff.

6   Conclusions

The paper proposes a method for constructing a conveyor belt speed control algorithm based on TOU tariffs. When constructing a control algorithm, the uneven distribution of material along the transport route and transport delay are taken into account. Simulation of the conveyor section is carried out with the assumption that the resistance to movement according to DIN 22101 is determined based on the coefficients of primary friction. To implement the method for constructing optimal control algorithms, the Pontryagin maximum principle is used.

The prospect for further research is the analysis of transport system control algorithms developed for various TOU tariffs.

References

1. Razumnyi, Y., Rukhlov, A., Kozar, A.: Improving the energy efficiency of conveyor transport of coal mines. Min. Electromech. Autom. **76**, 24–28 (2006). [in Russian]
2. Pihnastyi, O.: Control of the belt speed at unbalanced loading of the conveyor. Naukovyi Visnyk Natsionalnoho Hirnychoho Universytetu **6**, 122–129 (2019). https://doi.org/10.29202/nvngu/2019-6/18
3. Semenchenko, A., Stadnik, M., Belitsky, P., et al.: The impact of an uneven loading of a belt conveyor on the loading of drive motors and energy consumption in transportation. East.-Eur. J. Enterp. Technol. **4**(1), 42–51 (2016). https://doi.org/10.15587/1729-4061.2016.75936
4. Alspaugh, M.A.: Longer overland conveyors with distributed power. In: Paper presented at the Rockwell Automation Fair, St Louis, MO, 15 November 2005 (2005)
5. Pihnastyi, O., Kozhevnikov, G., Khodusov, V.: Conveyor model with input and output accumulating bunker. In: 2020 IEEE 11th International Conference on Dependable Systems, Services and Technologies (DESSERT), pp. 253–258. IEEE, Kyiv (2020). https://doi.org/10.1109/DESSERT50317.2020.9124996
6. Siemens: Innovative solutions for the mining industry. https://www.siemens.com/mining, Accessed 08 Aug 2020
7. Krol, R., Kawalec, W., Gladysiewicz, L.: An effective belt conveyor for underground ore transportation systems. IOP Conf. Ser. Earth Environ. Sci. **95**(4), 042047 (2017). https://doi.org/10.1088/1755-1315/95/4/042047

8. Bajda, M., Błażej, R., Jurdziak, L.: Analysis of changes in the length of belt sections and the number of splices in the belt loops on conveyors in an underground mine. Eng. Fail. Anal. **101**, 436–446 (2019). https://doi.org/10.1016/j.engfailanal.2019.04.003

9. Antoniak, J.: Energy-saving belt conveyors installed in Polish collieries. Transp. Prob. **5**, 5–14 (2010)

10. Pihnastyi, O.M., Khodusov, V.D.: Optimal control problem for a conveyor-type production line. Cybern. Syst. Anal. **54**(5), 744–753 (2018). https://doi.org/10.1007/s10559-018-0076-2

11. Bebic, M.Z., Ristic, L.B.: Speed controlled belt conveyors: drives and mechanical considerations. Adv. Electric. Comput. Eng. **18**(1), 51–60 (2018). https://doi.org/10.4316/AECE.2018.01007

12. Pihnastyi, O, Khodusov, V.: The optimal control problem for output material flow on conveyor belt with input accumulating bunker. Bull. South Ural State Univ. Ser. Math. Model. Program. Comput. Softw. **12**(2), 67–81 (2019). https://doi.org/10.14529/mmp190206

13. Wolstenholm, E.: Designing and assessing the benefits of control policies for conveyor belt systems in underground mines. Dynamica **6**(2), 25–35 (1980)

14. Alspaugh, M.: The evolution of intermediate driven belt conveyor technology over 20 years. Bulk Solids Hand. **23**(3), 168–172 (2003)

15. Masaki, M.S., Zhang, L., Xia, X.: A design approach for multiple drive belt conveyors minimizing life cycle costs. J. Cleaner Prod. **201**, 526–541 (2018). https://doi.org/10.1016/j.jclepro.2018.08.040

16. Marais, J.H.: Analysing DSM opportunities on mine conveyor systems. Dissertation, North-West University (2007)

17. Cousins, T.: Using time of use (TOU) tariffs in industrial, commercial and residential applications effectively. TLC Engineering Solutions, Johannesburg (2009)

18. Advanced Conveyor Technologies Inc.: Experience. https://www.actek.com/consulting/experience/, Accessed 20 Aug 2020

19. ConveyorBeltGuide (2020). https://conveyorbeltguide.com, Accessed 20 Aug 2020

20. Eskom: Tariffs & charges booklet 2020/2021. Eskom Holdings SOC Ltd. (2020)

21. Resource efficiency and cleaner production: A guide to understanding your industrial electricity bill. National Cleaner Production Centre, South Africa (2020)

22. Eskom: Megaflex gen# schedule of standard prices for non-local authority supplies – 1 April 2020 to 31 March 2021 (2020). https://www.eskom.co.za/CustomerCare/TariffsAndCharges/Documents/MegaflexGenschedule2020-21.pdf, Accessed 20 Aug 2020

23. Eskom: Strategic direction and tariff design principles for Eskom's tariffs (2017). https://www.eskom.co.za/CustomerCare/TariffsAndCharges/Strategicpricingdirection201725-07-2017.pdf, Accessed 20 Aug 2020

24. Iknet: Energy costs optimisation for Ukrainian households. https://iknet.com.ua/en/energy-costs-optimization-for-households/, Accessed 20 Aug 2020

25. Woo, C.K., Sreedharan, P., Hargreaves, J., et al.: A review of electricity product differentiation. Appl. Energy **114**, 262–272 (2014). https://doi.org/10.1016/j.apenergy.2013.09.070

26. Granell, R., Axon, C.J., Wallom, D.C.: Predicting winning and losing businesses when changing electricity tariffs. Appl. Energy **133**, 298–307 (2014). https://doi.org/10.1016/j.apenergy.2014.07.098

27. DIN 22101:2011–12. Continuous conveyors. Belt conveyors for loose bulk materials. Basis for calculation and dimensioning. Beuth-Vertrieb GmbH, Berlin (2011). https://doi.org/10.31030/1821227

28. Reutov, A.A.: Simulation of load traffic and steeped speed control of conveyor. IOP Conf. Ser. Earth Environ. Sci. **87**(8), 082041 (2017). https://doi.org/10.1088/1755-1315/87/8/082041

29. Lauhoff, H.: Geschwindigkeitsregelung bei Gurtförderern: spart das wirklich Energie? ZKG Int. **58**(12), 47–61 (2005)

30. Halepoto, I.A., Shaikh, M.Z., Chowdhry, B.S., Uqaili, M.A.: Design and implementation of intelligent energy efficient conveyor system model based on variable speed drive control and physical modeling. Int. J. Control Autom. **9**(6), 379–388 (2016). https://doi.org/10.14257/ijca.2016.9.6.36

Methods for Producing Nanostructures and Performance of Zirconium Alloys

Gennadiy Kostyuk[iD], Iryna Kantemyr[✉][iD], and Hanna Snitsar[iD]

National Aerospace University "Kharkiv Aviation Institute",
17 Chkalova Str., Kharkiv 61070, Ukraine
i.kantemyr@khai.edu

Abstract. According to the example of increasing the serviceability and reliability of the design of the parts of nuclear reactors from the zirconium alloy Zr1Nb due to the deposition of nanocoatings and the formation of nanostructures while they were bombarded with B^+, C^+, N^+, Si^+, Al^+, V^+, Cr^+, O^+, Fe^+, Ni^+, Co^+, Y^+, Zr^+, Mo^+, Hf^+, Ta^+, W^+, Pt^+ (with charge numbers 1, 2, 3) with energies of 200, 2000, 20000 eV the volume of grains and the minimum and maximum depth of their occurrence has been determined. This allows you to design the complex structure in depth, depending on the requirements for the reinforced layer. It is found that nanostructures are formed practically for all types of ions, except for boron and carbon, in the considered range of charge-number which ensures the reliability and performance of parts made of zirconium alloys.

Keywords: Reliability · Resource · Nanostructures · Nanocoatings · Zirconium alloys

1 Introduction

The most important safety conditions of nuclear power plants in emergency situations, especially in case of loss of coolant, which require the improved properties to the materials of fuel elements (FA) [1, 2]. Currently, the main materials used for them are zirconium alloys [3]. In event of a crash, and an inclusion of passive cooling the reactor core generates heat and water which is converted to steam oxidation of the zirconium alloy accelerates, and hence will produce explosive oxygen and compounds Z_2O_2, it is also the decomposition which will release heat that intensify care process, and therefore increases the probability of an explosion that occurred at the Japanese Fukushima nuclear power plant [4].

It can be assumed that new materials for their creation and testing will take at least 10 years.

Currently, the creation of nanostructures and nanocoatings on zirconium alloy can solve these issues quickly enough, it indicates the reality of use of nanostructures and nanocoatings.

M. Nechyporuk et al. (Eds.): ICTM 2020, LNNS 188, pp. 377–388, 2021.
https://doi.org/10.1007/978-3-030-66717-7_32

2   Status of Issue

Currently, there is a significant number of works under study of nanostructures and their preparation, reviews of these are in monographs [5–8]. But there are practically no works devoted to the study of obtaining nanostructures on zirconium alloys, because the attempts to obtain a number of coatings on these alloys have already been carried out [9, 10].

Nanostructures allow improving the physical and mechanical characteristics and working capacity of parts and cutting tools, as well as corrosion resistance [11–13]. The works [14–16] can also be referred to theoretical studies. Everything mentioned above indicates the prospects for obtaining nanocoatings on parts and cutting tools made of zirconium alloy. And also, considering the possibility of changing the chemical composition of the zirconium alloy, it is possible to solve the issue of a significantly increase in resource and reliability of parts made of zirconium alloys [17].

All of these indicates the relevance and timeliness of the consideration of the application of nanocoatings and the creation of nanostructures on the structural elements of nuclear reactors.

3   Model of Interaction of Ions with Structural Materials

The joint problem of thermal conductivity and thermoelasticity [5] has been solved, that allows to obtain the field of temperatures and temperature stresses for which the material zones have been located, where the criteria for the formation of nanostructures has been realized: the required temperature range (500–1500K), the achievement of a temperature growth rate greater than 10^7 K/s and the presence of temperature stresses in the range of 10^7–10^9 N/m^2, accelerating their formation, or 10^{10} N/m^2, allowing to form structures directly.

4   Results and Discussions

The action of ions was considered B^+, C^+, N^+, Si^+, Al^+, V^+, Cr^+, O^+, Fe^+, Ni^+, Co^+, Y^+, Zr^+, Mo^+, Hf^+, Ta^+, Pt^+ with charge numbers 1, 2, 3 and energies 200, 2000, 20000 eV, calculated temperature fields, their growth rates and temperature stresses, which were selected the areas of the material, which were implemented the criteria for the formation of nanostructures: temperatures and lie in the range of 500–1500 K, their growth rates exceed 10^7 K/s or temperature stresses exceed 10^{10} N/m^2 and calculated the volume of grain Vi, the minimum h_{min} and maximum h_{max} depth of its occurrence. The results are shown in Fig. 1, 2, 3, 4, 5, 6, 7, 8, 9, 10, 11, 12, 13 and 14 for zirconium alloy Zr1Nb.

So, for the case of action when boron ions have a grain volume range of $1.57 \cdot 10^{-27}$ to $10.8 \cdot 10^{-22}$ m^3 with a minimum depth of $1.35 \cdot 10^{-9}$ to $6.39 \cdot 10^{-8}$ and the maximum of $1.9 \cdot 10^{-9}$ to $8.47 \cdot 10^{-8}$ m, and the temperature voltage from $8.1 \cdot 10^{6}$ to $1.3 \cdot 10^{-6}$ m (Fig. 1).

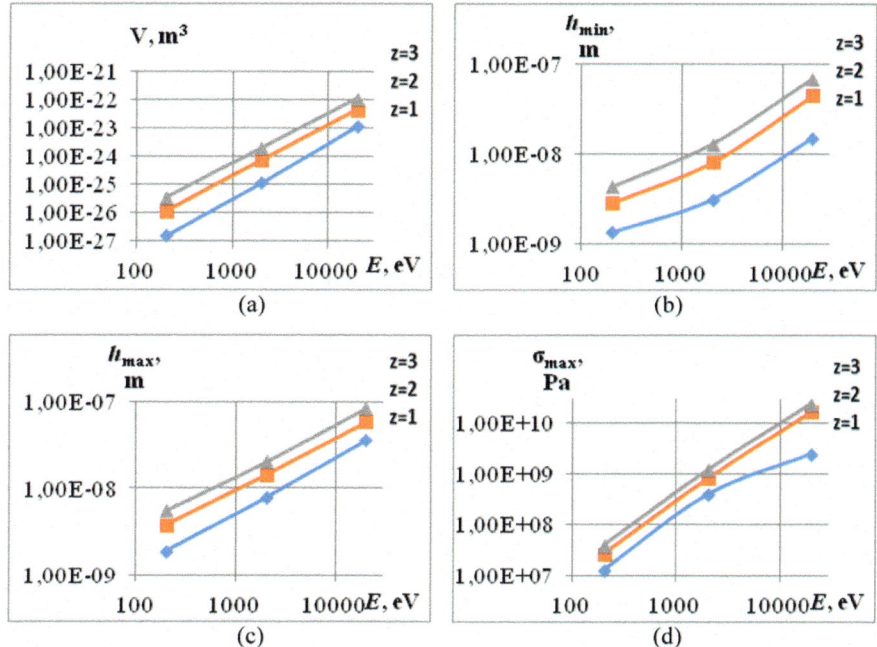

Fig. 1. Dependence of the nanocluster (NC) volume (a), minimum (b) and maximum (c) depth of occurrence of NC, maximum temperature stresses (d), under the action of boron ions (B^+) with different charge ($z = 1$, $z = 2$, $z = 3$) on zirconium alloy Zr1Nb

For carbon ions (C^+) in Fig. 2 we have a change in the volume of NS from $1.58 \cdot 10^{-27}$ to $7.9 \cdot 10^{-23}$ m³, and the depth of occurrence: minimum from $1.16 \cdot 10^{-9}$ to $5.94 \cdot 10^{-8}$ m and maximum from $1.81 \cdot 10^{-9}$ to $7.47 \cdot 10^{-8}$ m, and σ from $1.6 \cdot 10^7$ to $1.2 \cdot 10^9$ Pa.

For N^+ ions we have a range of grain volumes from $1.51 \cdot 10^{-27}$ to $6.24 \cdot 10^{-23}$ m³, and the voltage σ_{max} from $7.14 \cdot 10^7$ to $4.8 \cdot 10^8$ Pa. As well as for boron and carbon ions, with an increase in energy or a charge number, the NS volume increases with temperature stresses (Fig. 3).

It can be seen that with the growth of energy and charge and the value of temperature stresses, the volume significantly increases.

Transition to aluminum ions (Al^+) (Fig. 4) the values of the volumes of NS lie within the range of $1.4 \cdot 10^{-27}$ to $4 \cdot 10^{-23}$ m³, and the depth from $7.5 \cdot 10^{-10}$ to $4 \cdot 10^{-9}$ m minimum and from $1.7 \cdot 10^{-10}$ to $5.2 \cdot 10^{-8}$ m. The growth of NS and temperature stresses and depths of the ion energy and its charge remains.

For vanadium ion (V^+) (Fig. 5), values of the volumes are in the range of $1.06 \cdot 10^{-27}$ to $2.45 \cdot 10^{-23}$ m³ and the depth of it: minimum of $5.02 \cdot 10^{-10}$–$3.29 \cdot 10^{-8}$ m, maximum $1.69 \cdot 10^{-9}$–$4.72 \cdot 10^{-8}$ m, and $\sigma_{max} = 6.77 \cdot 10^6$–$7 \cdot 10^7$ Pa.

The growth of these values with increasing energy and ion charge is maintained. The growth of vanadium by ion mass is significantly reduced.

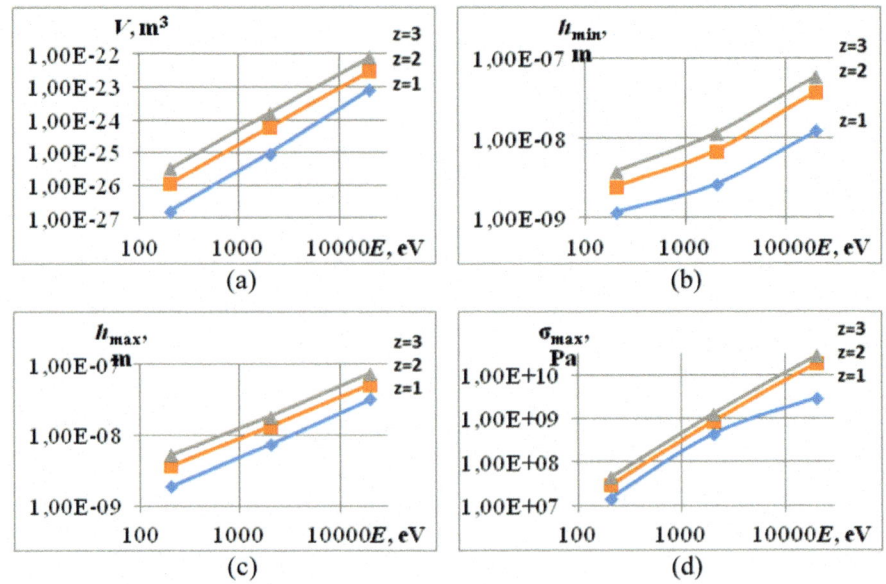

Fig. 2. Dependences of the NC volume (a), minimum (b) and maximum (c) depth of occurrence of NC, maximum temperature stresses (d), under the action of carbon ions (C^+) with different charge ($z = 1$, $z = 2$, $z = 3$) for zirconium alloy Zr1Nb.

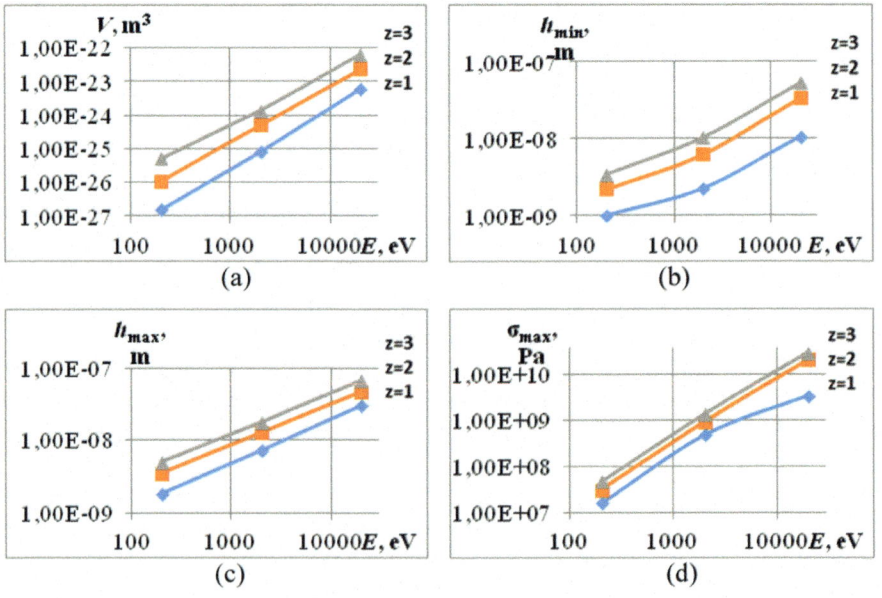

Fig. 3. Dependences of NC volume (a), minimum (b) and maximum (c) depth of occurrence of NC, maximum temperature stresses (d), under the action of nitrogen ions (N^+) with different charge ($z = 1$, $z = 2$, $z = 3$) for zirconium alloy Zr1Nb

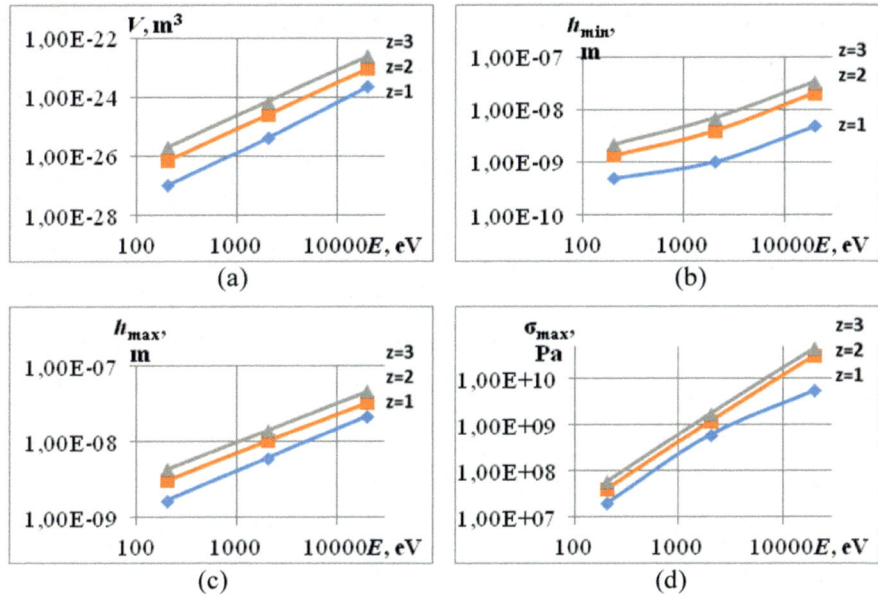

Fig. 4. Dependences of NC volume (a), minimum (b) and maximum (c) depth of occurrence of NC, maximum temperature stresses (d), under the action of aluminum ions (Al$^+$) with different charge (z = 1, z = 2, z = 3) for zirconium alloy Zr1Nb

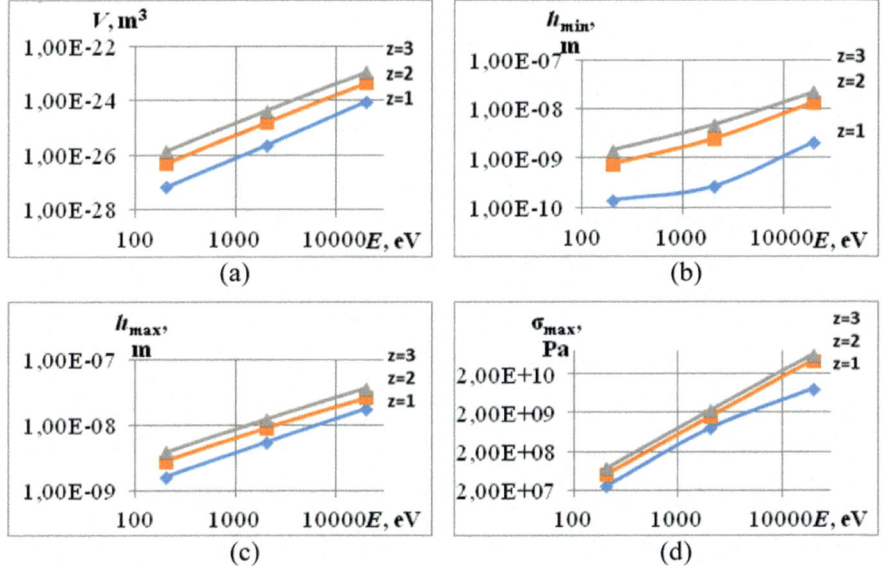

Fig. 5. Dependences of NC volume (a), minimum (b) and maximum (c) depth of occurrence of NC, maximum temperature stresses (d), under the action of vanadium ions (V$^+$) with different charge (z = 1, z = 2, z = 3) for zirconium alloy Zr1Nb

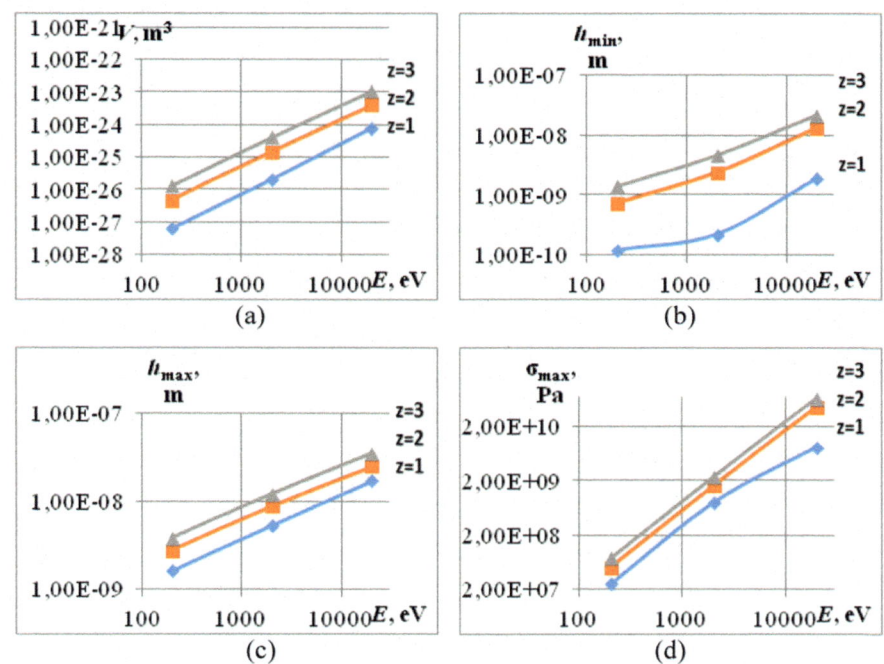

Fig. 6. Dependences of NC volume (a), minimum (b) and maximum (c) depth of occurrence of NC, maximum temperature stresses (d), under the action of chromium ions (Cr^+) with different charge ($z = 1$, $z = 2$, $z = 3$) for zirconium alloy Zr1Nb

Similar features for chromium ion (Cr^+) (Fig. 6) show that the volume of NS lies within $6.25 \cdot 10^{-28}$ to $1.14 \cdot 10^{-23}$ m^3, and the depth of $1.41 \cdot 10^{-10}$–$2.23 \cdot 10^{-8}$ m and $1.6 \cdot 10^{-9}$–$3.58 \cdot 10^{-8}$ m, respectively, minimum and maximum, $\sigma_{max} = 1.48 \cdot 10^7$–$7.69 \cdot 10^7$ Pa. Character of dependence was saved.

At transition to oxygen ions the range of volumes of NS lies within $6.25 \cdot 10^{-28}$ to $1.09 \cdot 10^{-23}$ m^3, and depths of occurrence within $1.3 \cdot 10^{-10}$–$2.18 \cdot 10^{-8}$ m and $1.61 \cdot 10^{-9}$–$3.52 \cdot 10^{-8}$ m, $\sigma_{max} = 4.7 \cdot 10^6$–$7.5 \cdot 10^7$ Pa accordingly (Fig. 7).

For iron ion (Fe^+) (Fig. 8) amounts of NS are in the range of $1.43 \cdot 10^{-27}$ to $5.13 \cdot 10^{-23}$ m^3 and depth $8.9 \cdot 10^{-10}$–$1.7 \cdot 10^{-8}$ m and $1.78 \cdot 10^{-9}$–$6.25 \cdot 10^{-8}$ m, and $\sigma_{max} = 1.47 \cdot 10^7$–$1.07 \cdot 10^8$ Pa, respectively. The values correlate with the ion mass: the greater the mass, the smaller the value.

The NS volume for the case of the action of nickel ions (Ni^+) (Fig. 9) is the value of $5.8 \cdot 10^{-28}$ m^3, and the depth of $8.09 \cdot 10^{-11}$–$2.98 \cdot 10^{-8}$ m^3 and $1.61 \cdot 10^{-9}$–$3.4 \cdot 10^{-8}$ m, a $\sigma_{max} = 4.64 \cdot 10^6$–$7.28 - 10^7$ Pa respectively.

For cobalt ion (Co^+) (Fig. 10) the volumes are in the range $5.42 \cdot 10^{-28}$–$9.28 \cdot 10^{-24}$ m^3 and depth $4.61 \cdot 10^{-10}$–$1.99 \cdot 10^{-8}$ m and $5.45 \cdot 10^{-9}$–$3.32 \cdot 10^{-8}$ m, $\sigma_{max} = 4.54 \cdot 10^6$–$7.2 \cdot 10^7$ Pa respectively.

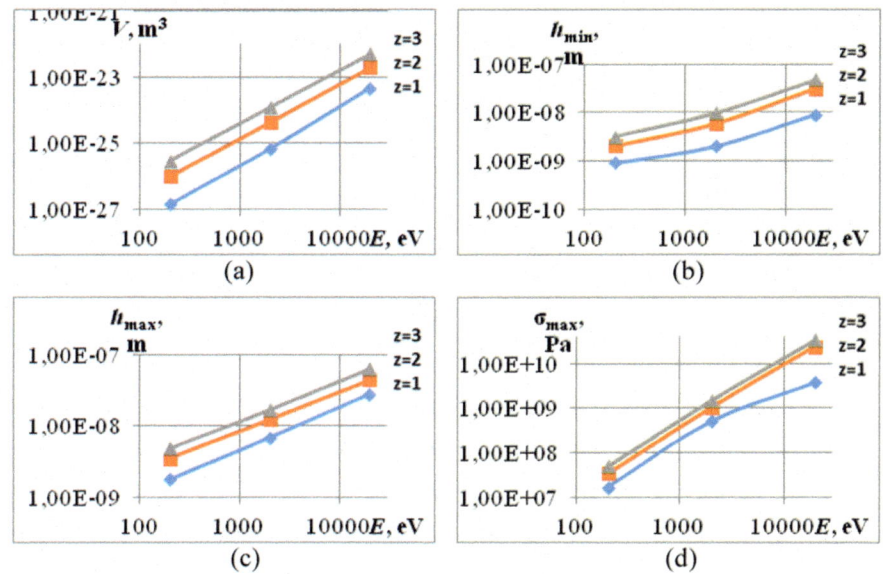

Fig. 7. Dependences of NC volume (a), minimum (b) and maximum (c) depth of occurrence of NC, maximum temperature stresses (d), under the action of oxygen ions (O^+) with different charge ($z = 1$, $z = 2$, $z = 3$) for zirconium alloy Zr1Nb

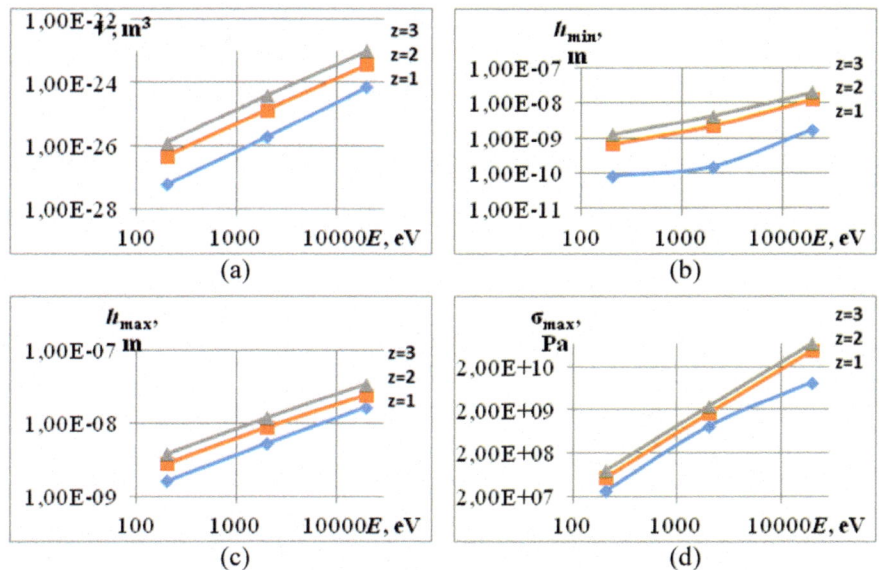

Fig. 8. Dependences of NC volume (a), minimum (b) and maximum (c) depth of occurrence of NC, maximum temperature stresses (d), under the action of iron ions (Fe^+) with different charge ($z = 1$, $z = 2$, $z = 3$) for zirconium alloy Zr1Nb

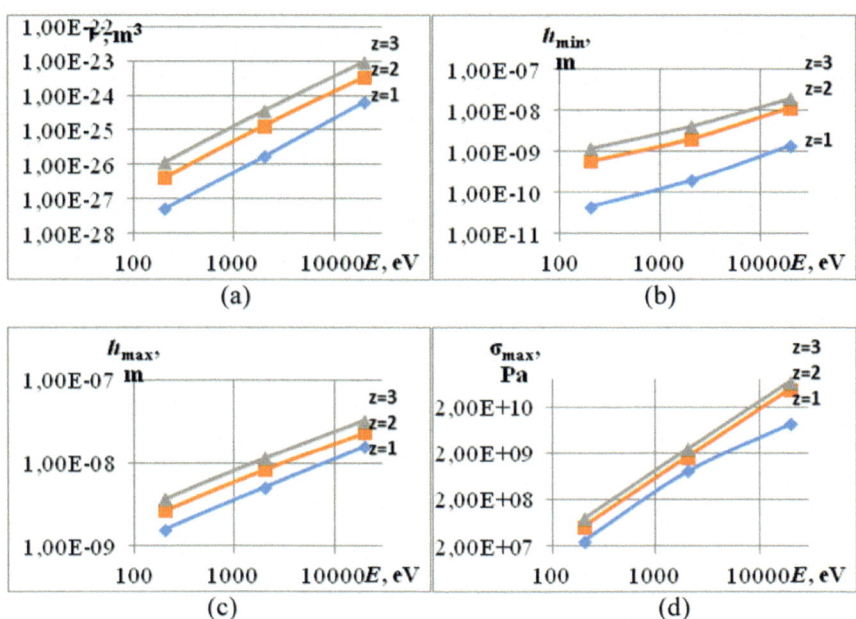

Fig. 9. Dependences of NC volume (a), minimum (b) and maximum (c) depth of occurrence of NC, maximum temperature stresses (d), under the action of nickel ions (Ni^+) with different charge (z = 1, z = 2, z = 3) for zirconium alloy Zr1Nb

Similar dependencies for yttrium ion (Y^+) (Fig. 11) give the values of volume NS $5.1 \cdot 10^{-28} – 9.59 \cdot 10^{-24}$ m^3 and depth $6.33 \cdot 10^{-10} – 2.04 \cdot 10^{-8}$ m and $4.16 \cdot 10^{-9} – 3.36 \cdot 10^{-8}$ m, $\sigma_{max} = 4.59 \cdot 10^6 – 7.41 \cdot 10^7$ Pa respectively.

The use of zirconium ions leads to change of the grain volume HS V = $4.15 \cdot 10^{-28} – 6.9 \cdot 10^{-24}$ m^3, $\sigma_{max} = 4.8 \cdot 10^6 – 6.69 \cdot 10^6$ Pa that is, the trend continues (Fig. 12).

For molybdenum ions have V = $4.48 \cdot 10^{-28} – 6.7 \cdot 10^{-24}$ m^3, $\sigma_{max} = 4.4 \cdot 10^6 – 6.88 \cdot 10^7$ Pa (Fig. 13).

Transition to hafnium ions gives volume HS V = $3.24 \cdot 10^{-28} – 5.22 \cdot 10^{-24}$ m^3, $\sigma_{max} = 3,15 \cdot 10^7 – 6,48 \cdot 10^7$ Pa (Fig. 14).

For tantalum, tungsten, and platinum ions we have a close value.

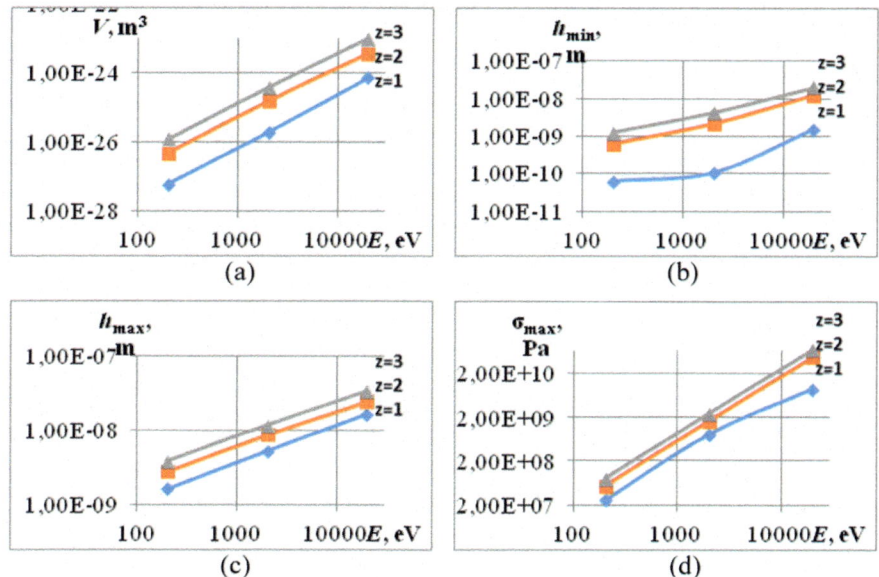

Fig. 10. Dependences of NC volume (a), minimum (b) and maximum (c) depth of occurrence of NC, maximum temperature stresses (d), under the action of cobalt ions (Co^+) with different charge ($z = 1$, $z = 2$, $z = 3$) for zirconium alloy Zr1Nb

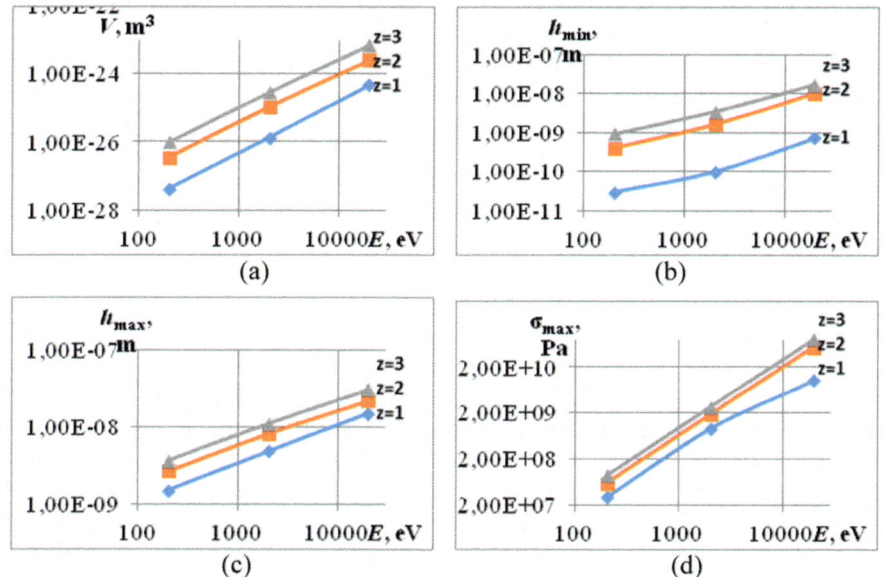

Fig. 11. Dependences of NC volume (a), minimum (b) and maximum (c) depth of occurrence of NC, maximum temperature stresses (d), under the action of yttrium ions (Y^+) with different charge ($z = 1$, $z = 2$, $z = 3$) for zirconium alloy Zr1Nb

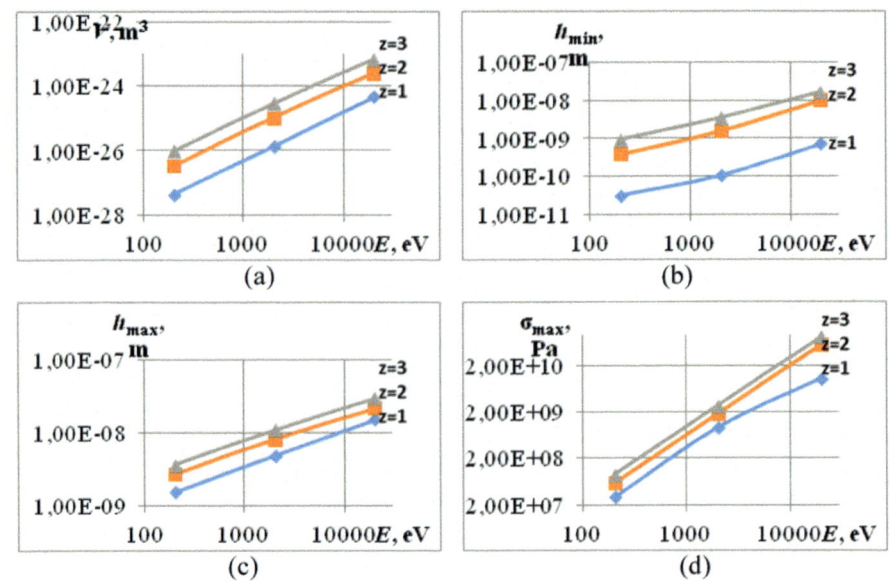

Fig. 12. Dependences of NC volume (a), minimum (b) and maximum (c) depth of occurrence of NK, maximum temperature stresses (d), under the action of zirconium ions (Zr^+) with different charge (z = 1, z = 2, z = 3) for zirconium alloy Zr1Nb

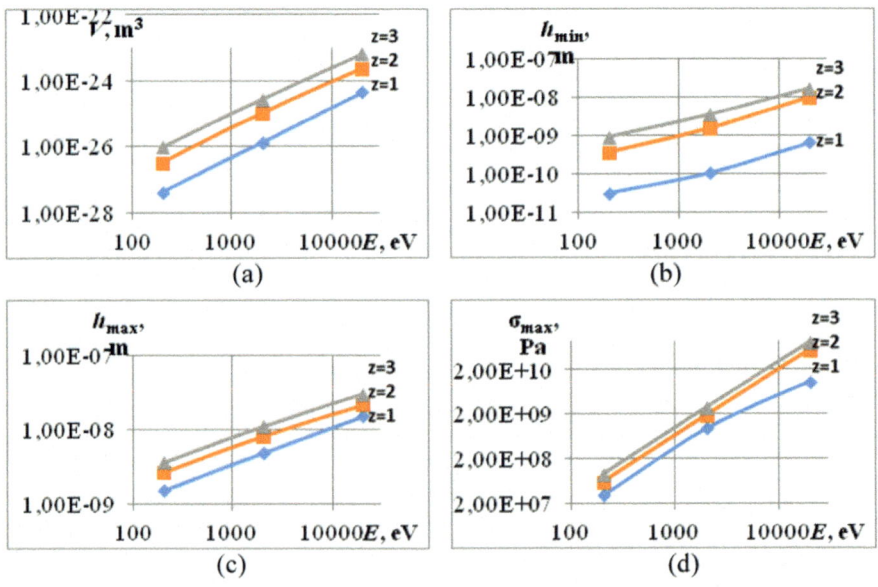

Fig. 13. Dependences of NC volume (a), minimum (b) and maximum (c) depth of occurrence of NC, maximum temperature stresses (d), under the action of molybdenum ions (Mo^+) with different charge (z = 1, z = 2, z = 3) for zirconium alloy Zr1Nb

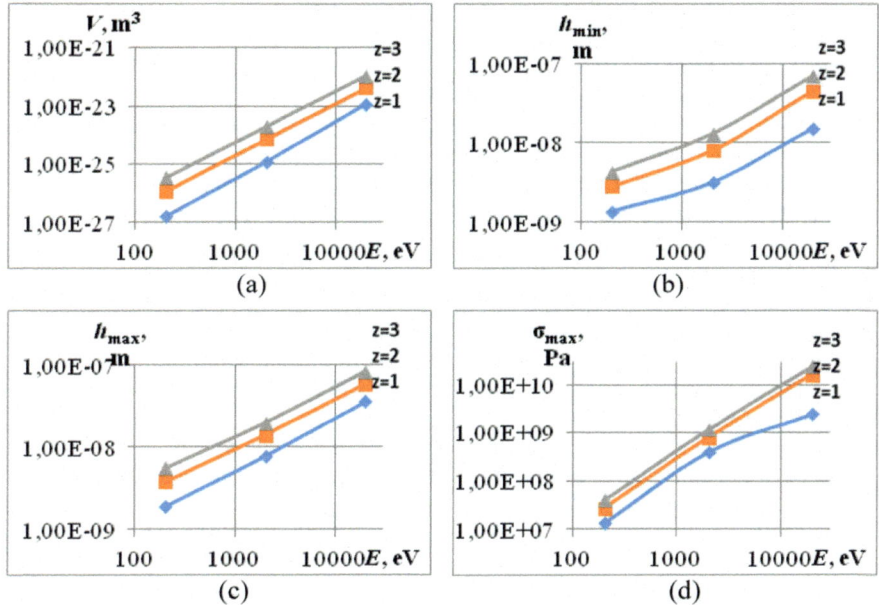

Fig. 14. Dependences of NC volume (a), minimum (b) and maximum (c) depth of occurrence of NC, maximum temperature stresses (d) of hafnium ions (Hf^+) with different charge ($z = 1$, $z = 2$, $z = 3$) for zirconium and its alloy Zr1Nb

5   Conclusions

According to the research, the following conclusions can be drawn:

- it is shown that it is possible to obtain layers of nanostructures of sufficient size up to 0.1 mm by varying the energy of ions, its varieties and charges;
- to reduce the probability of passing unnecessary reactions it can be used zirco-nium ions and niobium, it is presented in the zirconium alloy Zr1Hb;
- considering the neutrality of materials to water, oxygen and other reagents, you can choose the cheapest ions of heavy metals for the treatment of zirconium alloys.

References

1. Park, D.J., Kim, H.G., Jung, Y.I., et al.: Behavior of an improved Zr fuel cladding with oxidation resistant coating under loss-of-coolant accident conditions. J. Nucl. Mater. **482**, 75–82 (2016). https://doi.org/10.1016/j.jnucmat.2016.10.021
2. Krukovskyi, P.G., Diadiushko, Y.V., Garin, V.O., et al.: CFD model as a digital twin of the radiation state of the new safe confinement of the Chernobyl NPP. Prob. At. Sci. Technol. **128**(4), 54–62 (2020)

3. Nikulina, A.V.: Zirconium alloys in nuclear power engineering. Met. Sci. Heat Treat. **46**(11–12), 458–462 (2004). https://doi.org/10.1007/s11041-005-0002-x

4. Charit, I.: Accident tolerant nuclear fuels and cladding materials. JOM **70**(2), 173–175 (2018). https://doi.org/10.1007/s11837-017-2701-3

5. Kostyuk, G.I.: Nanotechnology: The Choice of Technological Parameters and Installations, Processing Performance, Physical and Mechanical Characteristics of Nanostructures. Engineering Academy of Ukraine Publ, Kharkiv (2014).[in Russian]

6. Sharma, K.R.: Nanocoatings. Momentum Press Engineering, New York (2017)

7. Goyal, R.K.: Nanomaterials and Nanocomposites: Synthesis, Properties, Characterization Techniques and Applications. CRC Press, Boca Raton (2018)

8. Syrkov, A.G., Levine, K.L.: New Materials: Preparation, Properties and Applications in the Aspect of Nanotechnology. Nova Science Publishers, New York (2020)

9. Yu, H., Yao, Z., Idrees, Y., et al.: Accumulation of dislocation loops in the α phase of Zr Excel alloy under heavy ion irradiation. J. Nucl. Mater. **491**, 232–241 (2017). https://doi.org/10.1016/j.jnucmat.2017.04.038

10. Rafique, M., Chae, S., Kim, Y.S.: Surface, structural and tensile properties of proton beam irradiated zirconium. Nucl. Instrum. Methods Phys. Res., Sect. B **368**, 120–128 (2016). https://doi.org/10.1016/j.nimb.2015.12.001

11. Popov, V., Kostyuk, G., Tymofyeyev, O., et al.: Design of new nanocoatings based on hard alloy. In: Ivanov, V. et al. (eds.) Advances in Design, Simulation and Manufacturing III. DSMIE 2020. LNME, pp 522–531. Springer, Cham (2020). https://doi.org/10.1007/978-3-030-50794-7_51

12. Plankovskyy, S., Shypul, O., Tsegelnyk, Y., et al.: Simulation of surface heating for arbitrary shape's moving bodies/sources by using R-functions. Acta Polytechnica **56**(6), 472–477 (2016). https://doi.org/10.14311/AP.2016.56.0472

13. Callisti, M., Polcar, T.: Combined size and texture-dependent deformation and strengthening mechanisms in Zr/Nb nano-multilayers. Acta Mater. **124**, 247–260 (2017). https://doi.org/10.1016/j.actamat.2016.11.007

14. Yan, P., Chen, K., Wang, Y., et al.: Design and performance of property gradient ternary nitride coating based on process control. Materials **11**(5), 758 (2018). https://doi.org/10.3390/ma11050758

15. Jiang, L., Xiu, P., Yan, Y., et al.: Effects of ion irradiation on chromium coatings of various thicknesses on a zirconium alloy. J. Nucl. Mater. **526**, 151740 (2019). https://doi.org/10.1016/j.jnucmat.2019.151740

16. Yang, H.L., Kano, S., McGrady, J., et al.: Microstructural evolution and hardening effect in low-dose self-ion irradiated Zr-Nb alloys. J. Nucl. Mater. **542**, 152523 (2020). https://doi.org/10.1016/j.jnucmat.2020.152523

17. Garner, A.J.: Investigating the effect of oxide texture on the corrosion performance of zirconium alloys. Dissertation, University of Manchester (2015)

Stress State of Two Glued Coaxial Tubes Under Nonuniform Axial Load

Sergey Kurennov⬥ ID, Kostiantyn Barakhov[(✉)] ID, Daria Dvoretska ID,
and Oleksandr Poliakov ID

National Aerospace University "Kharkiv Aviation Institute", 17 Chkalova Str.,
Kharkiv 61070, Ukraine
kurennov.ss@gmail.com, kpbarakhov@gmail.com

Abstract. The deflected mode problem for the structure composed of two glued coaxial cylindrical tubes under nonuniform longitudinal load is considered. The purpose of the paper is to obtain a mathematical model of a stress state that allows us to find an analytic solution to the problem. Pipes are considered as thick-walled rods, which are joined by a non-zero thickness adhesive layer. The tangent stresses in the glue are considered constant by the thickness of the adhesive layer. The tangent stresses in the adhesive layer are proportional to the difference of the longitudinal shifts of the tube sides, which are faced to the adhesive layer. The stresses in the pipes are considered to be constant in thickness, that is, in radial direction and dependent on the axial and circumferential coordinates. It is considered that the structural elements only move in the axial direction, i.e. the effects related to Poisson deformations are neglected. The deflected mode problem for the joint is reduced to a system of two partial differential equations relatively to longitudinal shifts of layers. The solution is obtaining using the classical variable separation method in the form of Fourier series by circumferential coordinate. The convergence of the solutions is proved. The model problem is solved, results are compared to calculations produced by the finite element method. The tangent stresses in the glue reach maximum values at the edges of the adhesive line. The mathematical model of the joint under certain constraints has a sufficient accuracy for engineering problems and can be used to solve structural design problems.

Keywords: Adhesive joint · Analytical solution · Two-dimensional model · Variable separation method

1 Introduction

Adhesive joints of coaxial cylindrical tubes are widespread structural elements. These joints have several advantages over other classical junctions, such as leak tightness, high aerodynamic efficiency, manufacturability, low weight, etc. However, the quality control of adhesive joints is difficult, and the experience of the composite structures usage shows that the loss of the bearing capacity of a structure often occurs due to the destruction of just the joining nodes. This is due to the fact that it is difficult to achieve uniform transfer of forces from one structural element to another in a lap joint. Even if

© The Author(s), under exclusive license to Springer Nature Switzerland AG 2021
M. Nechyporuk et al. (Eds.): ICTM 2020, LNNS 188, pp. 389–400, 2021.
https://doi.org/10.1007/978-3-030-66717-7_33

the load is applied uniformly across the joint width, the stress concentration occurs in the glue line at the joint edges [1–3]. This effect is exacerbated if the load is applied uniformly across the joint width [4, 5]. Therefore, the computation of the stress state of the joining elements is an important component of the structure strength calculation.

The analytical solution problem for the the deflected mode (DM) of the joint of cylindrical tubes in the general formulation has not yet been solved. Axisymmetric models are most often used to describe analytically the DM of the coaxial tubes joint. There are considered connections which transmit a torque [6, 7] or a longitudinal load [8–10]. As a rule, jointed tubes are considered as thin-walled cylindrical shells [11, 12] or as thick-walled rods, the bending of which can be neglected [13, 14]. In order to be able to construct an analytical solution in the latter case, it is assumed that the normal stresses in the jointed tubes in the axial direction are constant in thickness and equal to zero in the radial direction. There are known papers where nonaxisymmetric problems are considered [15], but the solution is constructed using numerical methods.

So far as it is assumed that the stress state in the joint varies along the length of the joint and depends on the angular coordinate, this model is two-dimensional. The well-known two-dimensional adhesive joint models of DM suggest either a numerical solution of the problem [16], or the introduction of additional hypotheses that simplify the mathematical formulation of the problem without significant model accuracy reduction. One of these approaches is the introduction of the small transversal deformation hypothesis, i.e. the infinitesimality of the Poisson's ratio. Such approach [4] allows us to eliminate transversal shifts and reduce the number of unknowns and obtain a solution in an analytical form. Of course, the finite element method (FEM) is also used to investigate the DM of the coaxial pipes adhesive joint, as, for example, in [10, 17–19] and experiments [20]. Finite element modeling allows us to investigate a wide kind of problems, however, it has difficulties in parametric investigation, modeling complex structures containing structural elements of various sizes, solving design and optimization problems, etc.

In this paper, the previous simplified two-dimensional DM model of the joint is developed for the two coaxial tubes joint. In the proposed model, the influence of longitudinal (axial) deformations on transverse (tangential) deformations is ignored. That is, it is assumed that all joint elements can only move in the longitudinal direction. The distribution of stresses across the thickness of the layers is considered uniform. However, unlike classical one-dimensional solutions, it is assumed that the axial load or shifts are distributed nonuniformly along the circumferential coordinate.

2   Formulation of the Problem

The joint diagram is shown in Fig. 1. The thicknesses of the outer and inner tubes are denoted as δ_1 and δ_2. The thickness of the adhesive layer is denoted by δ_0. Let us suppose that the elements of the base layers can only move along the joint axis, as the adhesive layer acts only in shear, and the stresses are uniformly distributed over the thickness of the layers.

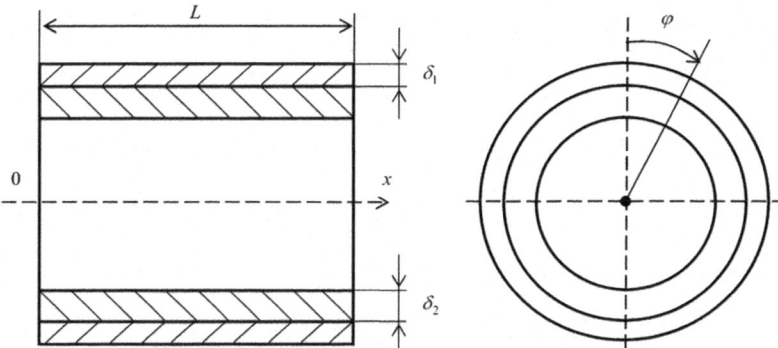

Fig. 1. Joint diagram.

Differential joint elements and acting forces are shown in Fig. 2. In this figure: N_m, q_m are normal (in longitudinal direction) and tangential forces in the base layer m, ($m = 1, 2$) which are products of a corresponding stresses by layer thickness, $N_m = \delta_m \sigma_x^{(m)}$, $q_m = \delta_m \tau^{(m)}$; τ_0 are tangential stresses in the adhesive layer in longitudinal direction. Moreover, x is the axial coordinate; circumferential curvilinear coordinates $s_1 = R_1 d\varphi$, $s_2 = R_2 d\varphi$, $s_0 = R_0 d\varphi$; φ is the angular coordinate, counted from some plane, R_m is the radius of the middle surface of the layer m, R_0 is the radius of the middle surface of the adhesive layer. That is, we think that the tangential stresses in the adhesive layer act in its middle surface by a radius R_0.

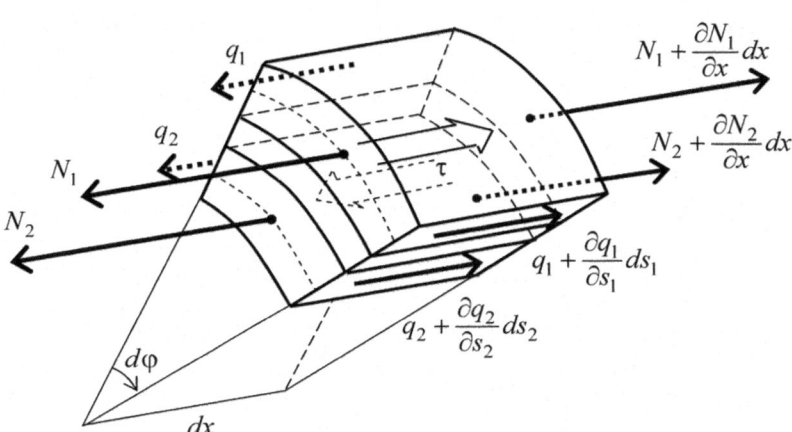

Fig. 2. Differential joint element.

The equilibrium equations for the differential elements of a joint have the form:

$$\tau_0 + \frac{R_1}{R_0}\frac{\partial N_1}{\partial x} + \frac{1}{R_0}\frac{\partial q_1}{\partial \varphi} = 0; \quad -\tau_0 + \frac{R_2}{R_0}\frac{\partial N_2}{\partial x} + \frac{1}{R_0}\frac{\partial q_2}{\partial \varphi} = 0. \tag{1}$$

From the Cauchy relations on the assumption of zero transverse displacements and the Hooke law as well, it follows:

$$N_m = \delta_m E_m \frac{\partial U_m}{\partial x}, \quad q_m = \delta_m G_m \frac{\partial U_m}{\partial s_m} = \frac{\delta_m G_m}{R_m}\frac{\partial U_m}{\partial \varphi}, \tag{2}$$

where U_m are longitudinal shifts of m-th layer.

Tangential stresses in the adhesive layer are assumed proportional to the difference of the layers shift:

$$\tau_0 = P \cdot (U_2 - U_1), \tag{3}$$

where P is a shift rigidity of the adhesive layer.

Substituting given relations in (1), we get the system of equations

$$\alpha_1 \left(\frac{\partial^2 U_1}{\partial x^2} + \mu_1^2 \frac{\partial^2 U_1}{\partial \varphi^2} \right) - (U_1 - U_2) = 0;$$
$$\alpha_2 \left(\frac{\partial^2 U_1}{\partial x^2} + \mu_2^2 \frac{\partial^2 U_1}{\partial \varphi^2} \right) + (U_1 - U_2) = 0, \tag{4}$$

where $\alpha_m = \delta_m \frac{E_m}{P}\frac{R_m}{R_0}$; $\mu_m = \frac{1}{R_m}\sqrt{\frac{G_m}{E_m}}$; $m = 1, 2$.

Boundary conditions:

$$N_2|_{x=0} = E_2 \delta_2 \frac{\partial U_2}{\partial x}\Big|_{x=0} = F^{(2)}(\varphi), \quad N_m|_{x=L} = E_m \delta_m \frac{\partial U_m}{\partial x}\Big|_{x=L} = H^{(m)}(\varphi),$$

$$U_1|_{x=0} = u^{(1)}(\varphi), \tag{5}$$

In addition, periodicity conditions are imposed on shifts:

$$U_m(x, \varphi) = U_m(x, \varphi + 2\pi n), \tag{6}$$

The adhesive layer rigidity, which was introduced in (3) can be given in several ways. The classic approach means, that the shift is concentrated only in the adhesive layer. In this case $P = G_0 \delta_0^{-1}$, where G_0 is a glue shift module, δ_0 is an adhesive layer thickness. However, this approach gives us a few overvalued stresses in the adhesive. For flat joints, a satisfactory approximation shows a model of the linear distribution of tangential stresses over the thickness of the parts to be joined [21, 22]. A linear distribution of stresses across the thickness of the joined layers is observed in the regular zone, far from the edges of the gluing. Whereas the stresses in the adhesive are maximum precisely at the edges of the joint. Therefore, the question of the effective dependencies choosing for the shift compliance calculation of the adhesive layer, especially in the case of relatively large thicknesses of the joined layers, cannot be considered completely solved.

The same considerations can be applied in this case. I.e. since the tubes are relatively thin, then the more accurate rigidity of the adhesive layer can be calculated in the same way as for flat joints: $P = \left(\frac{\delta_0}{G_0} + \frac{\delta_1}{2G_1} + \frac{\delta_2}{2G_2} \right)^{-1}$. The distribution of tangential stresses over the thickness of relatively thick tubes will differ from linear, so the formula can be transformed, for example, as follows

$$P = \left(\frac{\delta_0}{G_0} + \frac{\delta_1}{K_1 G_1} + \frac{\delta_2}{K_2 G_2} \right)^{-1},$$

where $1 < K_m < 2$. The choice of K coefficient values is open.

3 Constructing the Solution

From the first equation of (4) we get

$$U_2 = U_1 - \alpha_1 \left(\frac{\partial^2 U_1}{\partial x^2} + \mu_1^2 \frac{\partial^2 U_1}{\partial \varphi^2} \right), \tag{7}$$

Substituting into the second equation of (1.4) we obtain an equation.

$$\frac{\partial^4 U_1}{\partial x^4} + \beta_1^2 \frac{\partial^4 U_1}{\partial x^2 \partial \varphi^2} + \beta_2^2 \frac{\partial^4 U_1}{\partial \varphi^4} - \beta_3^2 \frac{\partial^2 U_1}{\partial x^2} - \beta_4^2 \frac{\partial^2 U_1}{\partial \varphi^2} = 0, \tag{8}$$

where $\beta_1^2 = \mu_1^2 + \mu_2^2$, $\beta_2^2 = \mu_1^2 \mu_2^2$, $\beta_3^2 = \frac{1}{\alpha_1} + \frac{1}{\alpha_2}$, $\beta_4^2 = \frac{\mu_1^2}{\alpha_2} + \frac{\mu_2^2}{\alpha_1}$.

A particular solution to Eq. (8) will be sought in the form of the product of two functions, one of which depends only on the angular, and the second only on the linear coordinate $U_1^* = \Phi(\varphi) X(x)$. Due to the physical meaning, the each particular solution U_1^* must be 2π periodical function. So, functions $\Phi(\varphi)$ are periodical, of period 2π. Any continuous periodic function of period 2π can be expanded in Fourier series. It can be assumed that the functions $\Phi(\varphi)$ in particular solutions (8) take values $\{1, \cos n\varphi, \sin n\varphi\}$. We find the corresponding functions $X_n(x)$.

If $\Phi(\varphi) = const$, then substituting $U_1^* = \Phi_0(\varphi) X_0(x)$ in (8), we obtain

$$\frac{d}{dx^2} \left[\frac{d^2 X_0(x)}{dx^2} - \beta_3^2 X_0(x) \right] = 0.$$

The solution to this equation can be written as:

$$X_0 = A_0 + A_1 x + A_3 ch(\beta_3 x) + A_4 sh(\beta_3 x).$$

If $\Phi_n(\varphi) = \sin n\varphi$ or $\Phi_n(\varphi) = \cos n\varphi$, then substituting partial solutions $U_n^* = \Phi_n(\varphi)X_n(x)$ in (8), we get

$$\frac{d^4X_n}{dx^4} - (\beta_1^2 n^2 + \beta_3^2)\frac{d^2X_n}{dx^2} + (\beta_2^2 n^4 + \beta_4^2 n^2)X_n = 0,$$

i.e. fourth-order ordinary differential equation with constant coefficients. The corresponding characteristic equation has the form

$$k^4 - (\beta_1^2 n^2 + \beta_3^2) \cdot k^2 + (\beta_2^2 n^4 + \beta_4^2 n^2) = 0.$$

The roots of this characteristic equation have the form $\pm k_{1,n}$ and $\pm k_{2,n}$, where

$$k_{m,n} = \frac{1}{\sqrt{2}}\sqrt{\beta_1^2 n^2 + \beta_3^2 + (-1)^m \cdot \sqrt{(\beta_1^2 n^2 + \beta_3^2)^2 - 4(\beta_2^2 n^4 + \beta_4^2 n^2)}}.$$

As $n \to \infty$ the dependence $k_{1,n}$ and $k_{2,n}$ of n tends to be linear. It can be proved, that $k_{1,n}$ and $k_{2,n}$ are real numbers for $n > 0$.

In light of the above, we represent shifts U_1 as a superposition of corresponding particular solutions U_1^*. Then, using (7) we find shifts U_2. After taking the appropriate action, we get:

$$U_m = A_0 + A_1 x + \gamma_0^{(m)}[A_3\mathrm{ch}(\beta_3 x) + A_4\mathrm{sh}(\beta_3 x)] +$$

$$+ \sum_{n=1}^{\infty} R_n(x)\cos n\varphi + \sum_{n=1}^{\infty} T_n(x)\sin n\varphi,$$

$$R_n(x) = \frac{\gamma_{1,n}^{(m)}}{k_{1,n}}\left(a_{1,n}\frac{\mathrm{ch}k_{1,n}x}{\mathrm{sh}k_{1,n}L} + a_{2,n}\frac{\mathrm{sh}k_{1,n}x}{\mathrm{ch}k_{1,n}L}\right) + \frac{\gamma_{2,n}^{(m)}}{k_{2,n}}\left(a_{3,n}\frac{\mathrm{ch}k_{2,n}x}{\mathrm{sh}k_{2,n}L} + a_{4,n}\frac{\mathrm{sh}k_{2,n}x}{\mathrm{ch}k_{2,n}L}\right),$$

$$T_n(x) = \frac{\gamma_{1,n}^{(m)}}{k_{1,n}}\left(b_{1,n}\frac{\mathrm{ch}k_{1,n}x}{\mathrm{sh}k_{1,n}L} + b_{2,n}\frac{\mathrm{sh}k_{1,n}x}{\mathrm{ch}k_{1,n}L}\right) + \frac{\gamma_{2,n}^{(m)}}{k_{2,n}}\left(b_{3,n}\frac{\mathrm{ch}k_{2,n}x}{\mathrm{sh}k_{2,n}L} + b_{4,n}\frac{\mathrm{sh}k_{2,n}x}{\mathrm{ch}k_{2,n}L}\right),$$

where $A_1, ..., A_4$, $a_{1,n}, ..., a_{4,n}$ and $b_{1,n}, ..., b_{4,n}$ are unknown coefficients, obtaining from boundary conditions; multipliers $k_{m,n}\mathrm{sh}k_{m,n}L$ and $k_{m,n}\mathrm{ch}k_{m,n}L$ in the denominator are used to normalize and to simplify the solution convergence analysis; coefficients $\gamma_0^{(1)} = \gamma_{1,n}^{(1)} = \gamma_{2,n}^{(1)} = 1$, $\gamma_0^{(2)} = 1 - \alpha_1\beta_3^2$, $\gamma_{1,n}^{(2)} = 1 - \alpha_1\left(k_{1,n}^2 - n^2\mu_1^2\right)$, $\gamma_{2,n}^{(2)} = 1 - \alpha_1\left(k_{2,n}^2 - n^2\mu_1^2\right)$.

Note that in the shift formulas for U_1 and U_2 the first terms outside the sum sign represent the classic one-dimensional Wolkersen solution [1] and describe shifts, caused by resulting uniform forces. And the terms under the sum signs, which describe the stresses caused by self-balanced forces, exponentially decrease with distance from the edge of the joint into the gluing area. The same can be said of forces (2). This property of the solution is fully consistent with the principle of Saint-Venant.

Suppose that a longitudinal load is applied to the outer tube along the right butt $x = L$, on the left butt of the inner tube $x = 0$ shifts are given (5).

Longitudinal forces in the base layers (2):

$$N_m = E_m \delta_m \left[A_1 + \gamma_0^{(m)} [A_3 \beta_3 \mathrm{sh}(\beta_3 x) + A_4 \beta_3 \mathrm{ch}(\beta_3 x)] + \right.$$

$$\left. + \sum_{n=1}^{\infty} r_n(x) \cos n\varphi + \sum_{n=1}^{\infty} t_n(x) \sin n\varphi \right],$$

$$r_n(x) = \gamma_{1,n}^{(m)} \left(a_{1,n} \frac{\mathrm{sh}\, k_{1,n} x}{\mathrm{sh}\, k_{1,n} L} + a_{2,n} \frac{\mathrm{ch}\, k_{1,n} x}{\mathrm{ch}\, k_{1,n} L} \right) + \gamma_{2,n}^{(m)} \left(a_{3,n} \frac{\mathrm{sh}\, k_{2,n} x}{\mathrm{sh}\, k_{2,n} L} + a_{4,n} \frac{\mathrm{ch}\, k_{2,n} x}{\mathrm{ch}\, k_{2,n} L} \right),$$

$$t_n(x) = \gamma_{1,n}^{(m)} \left(b_{1,n} \frac{\mathrm{sh}\, k_{1,n} x}{\mathrm{sh}\, k_{1,n} L} + b_{2,n} \frac{\mathrm{ch}\, k_{1,n} x}{\mathrm{ch}\, k_{1,n} L} \right) + \gamma_{2,n}^{(m)} \left(b_{3,n} \frac{\mathrm{sh}\, k_{2,n} x}{\mathrm{sh}\, k_{2,n} L} + b_{4,n} \frac{\mathrm{ch}\, k_{2,n} x}{\mathrm{ch}\, k_{2,n} L} \right).$$

The forces and shifts at the butts (5) are expanded in Fourier series:

$$u_1(\varphi) = \frac{c_0^{(1)}}{2} + \sum_{n=1}^{\infty} \left(c_n^{(1)} \cos n\varphi + s_n^{(1)} \sin n\varphi \right),$$

$$F^{(2)}(\varphi) = \frac{c_0^{(2)}}{2} + \sum_{n=1}^{\infty} \left(c_n^{(2)} \cos n\varphi + s_n^{(2)} \sin n\varphi \right),$$

$$H^{(m)}(\varphi) = \frac{C_0^{(m)}}{2} + \sum_{n=1}^{\infty} \left(C_n^{(m)} \cos n\varphi + S_n^{(m)} \sin n\varphi \right).$$

Satisfying the boundary conditions, we obtain systems of linear algebraic equations relatively to the coefficients $A_1, ..., A_4$, $a_{1,n}, ..., a_{4,n}$ and $b_{1,n}, ..., b_{4,n}$. It can be shown that the rate of decrease of the coefficients $a_{1,n}, ..., a_{4,n}$ and $b_{1,n}, ..., b_{4,n}$ exceeds the rate of decrease of coefficients $c_n^{(m)}$, $C_n^{(m)}$, $s_n^{(m)}$, $S_n^{(m)}$, which are proportional to $n^{-\theta}$, where $\theta \geq 1$.

4 Model Problem

Let's consider the adhesive joint of two tubes of length $L = 50$ mm, outer radii are $R_1 = 28$ mm, $R_2 = 33$ mm, thicknesses are $\delta_1 = 4$ mm, $\delta_2 = 2,9$ mm. The adhesive layer thickness is $\delta_0 = 0,1$ mm. Elastic characteristics of joint materials are $E_1 = E_2 = 70$ GPa, $G_1 = G_2 = 27$ GPa, $G_0 = 0,34$ GPa. We set the boundary conditions at the butts of the glued tubes:

$$U_1|_{x=0} = u^{(1)}(\varphi) = 0, \quad F^{(2)}(\varphi) = 0, \quad H^{(1)}(\varphi) = 0.$$

$$H^{(2)}(\varphi) = \begin{cases} F_0, & -\dfrac{\pi}{8} \leq \varphi < \dfrac{\pi}{8}; \quad \dfrac{3\pi}{8} \leq \varphi < \dfrac{5\pi}{8}; \quad \dfrac{7\pi}{8} \leq \varphi < \dfrac{9\pi}{8}; \quad \dfrac{11\pi}{8} \leq \varphi < \dfrac{13\pi}{8}; \\[2mm] 0, & \dfrac{\pi}{8} \leq \varphi < \dfrac{3\pi}{8}; \quad \dfrac{5\pi}{8} \leq \varphi < \dfrac{7\pi}{8}; \quad \dfrac{9\pi}{8} \leq \varphi < \dfrac{11\pi}{8}; \quad \dfrac{13\pi}{8} \leq \varphi < \dfrac{15\pi}{8}. \end{cases}$$

The load application diagram is shown more clearly in Fig. 3, which shows the butts of the glued structure under consideration.

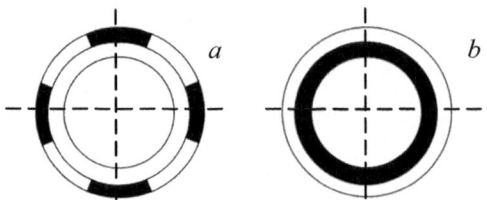

Fig. 3. Boundary conditions at $x = L$ (a), and $x = 0$ (b).

The loaded areas are highlighted in black. Sites without load are white. The forces applied to the outer tube in four sectors are transmitted through the adhesive layer to the inner tube, rigidly fixed around the butt. In Fig. 4 shows a graph of shift stresses $\tau_0(\varphi, x)$ (3) as a surface in coordinates (φ, x).

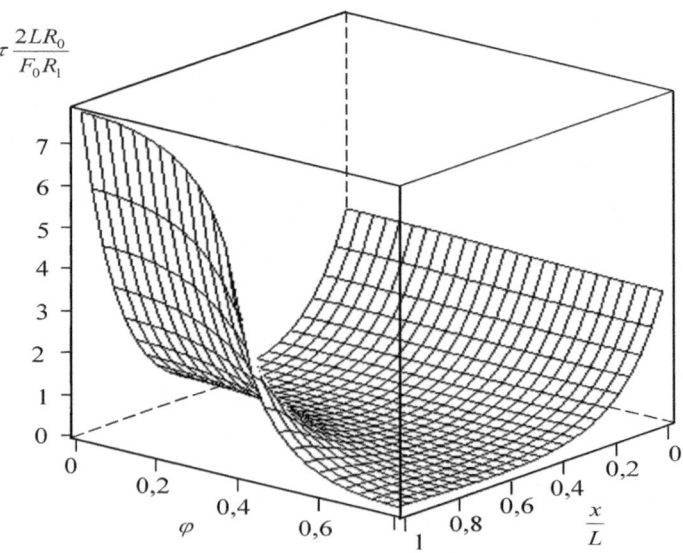

Fig. 4. Stresses in the adhesive layer.

Taking into account the symmetry of the problem, the figure shows the stress on one eighth of a circle, $\varphi \in [0; \pi/4]$. The stress graphs are given in dimensionless form, as a ratio of acting tangential stresses τ to hypothetical stresses $T_0 = \frac{F_0 R_1}{2LR_0}$, that would occur in the adhesive layer with a uniform distribution of the applied load along the length of the joint. The two in the denominator is due to the fact that the load F_0 is applied only on the half length of the outer tube circle, (see. Fig. 3a).

From the presented figure it is seen that the highest stress values reach in the middle of the loaded sections. And at the supported butt, the stress state is almost uniform along the circumferential coordinate. In the calculations, the rigidity of the adhesive layer was calculated as $P = \left(\delta_0 G_0^{-1} + 0,5\,\delta_1 G_1^{-1} + 0,5\,\delta_2 G_2^{-1}\right)^{-1}$.

To estimate the accuracy of the proposed model, the stress state of the structure under consideration was calculated using the finite element method (FEM) in three-dimensional formulation. Boundary conditions allow us to consider the sector $\pi/4$ with boundary conditions of the type roller on lateral sides. The forces at the butt are replaced by equivalent normal stresses. In Fig. 5 it is shown the distribution of tangential stresses in the middle surface of the adhesive layer (also in dimensionless form).

For a clear comparison of the results, we present on one graph the stresses calculated by the proposed method and by the FEM. Consider the stress distribution along the length of the joint. In Fig. 6 it is shown the stresses in the adhesive at $\varphi = 0$ and $\varphi = \pi/4$. Moreover, it is shown a graph of tangential stresses in the middle of the adhesive layer at $\varphi = 0$, obtained using the finite element method.

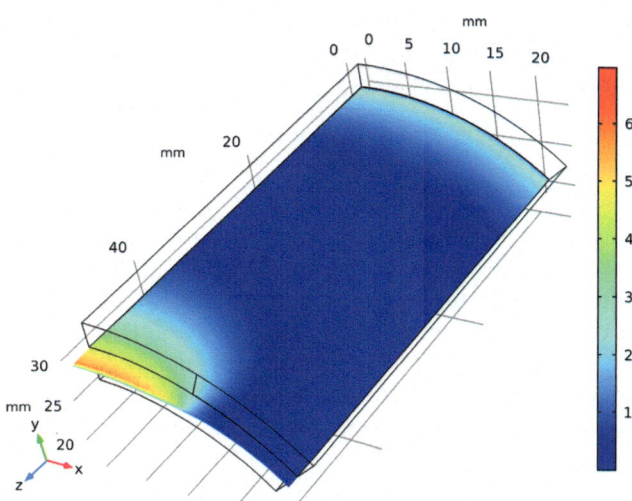

Fig. 5. Stresses in the adhesive layer calculated using FEM.

It is easy to see, that graphs "a" and "b" almost coincide, differ only at the edges of the gluing, where the adhesive is in a composite stress state. According to the model of the adhesive stress state (3), tangential stresses are maximal at the gluing edges, that

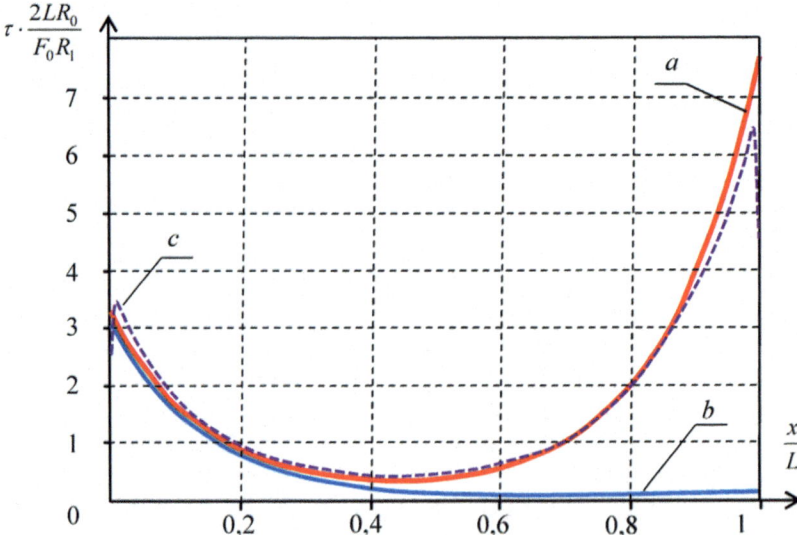

Fig. 6. Stresses in the adhesive layer, computed by proposed method at $\varphi = 0$- "a", $\varphi = 0.25 \pi$ - "b", and calculated by FEM – "c" ($\varphi = 0$).

contradicts the law of tangential stresses pairing, since the outer edge of the adhesive layer is free of load. This property of model (3) is well known, and to eliminate this contradiction, more accurate one-dimensional models of the joint stress state have been created [17, 19]. However, the usage of such adhesive layer stress state models in two-dimensional joint models is still unknown.

5   Conclusions

A mathematical model of the adhesive joint stress state for two coaxial cylindrical tubes is proposed. An analytical solution to the problem of determining the DM of this structure is developed. The main assumption, which allows us to simplify the problem, is the assumption of small transverse deformations and shifts of the jointed elements. The problem is reduced to a system of two partial differential equations relatively to the longitudinal shifts of the base layers. The solved model problem showed that the proposed model has an accuracy sufficient for design problems.

References

1. da Silva, L.F., das Neves, P.J., Adams, R.D., Spelt, J.K.: Analytical models of adhesively bonded joints – part I: literature survey. Int. J. Adhes. Adhes. **29**(3), 319–330 (2009). https://doi.org/10.1016/j.ijadhadh.2008.06.005
2. Ebnesajjad, S., Landrock, A.H.: Adhesives Technology Handbook. Elsevier/William Andrew, Amsterdam (2015). https://doi.org/10.1016/c2013-0-18392-4

3. Yousefsani, S.A., Tahani, M.: Relief of edge effects in bi-adhesive composite joints. Compos. B Eng. **108**, 153–163 (2017). https://doi.org/10.1016/j.compositesb.2016.09.099
4. Kurennov, S.S.: A simplified two-dimensional model of adhesive joints. Nonuniform Load. Mech. Compos. Mater. **51**(4), 479–488 (2015). https://doi.org/10.1007/s11029-015-9519-2
5. Rapp, P.: The numerical modeling of adhesive joints in reinforcement of wooden elements, subjected to bending and shearing. Drewno **60**(199), 21–36 (2017). https://doi.org/10.12841/wood.1644-3985.192.02
6. Adams, R.D., Peppiatt, N.A.: Stress analysis of adhesive bonded tubular lap joints. J. Adhes. **9**(1), 1–18 (1977). https://doi.org/10.1080/00218467708075095
7. Aimmanee, S., Hongpimolmas, P.: Stress analysis of adhesive-bonded tubular-coupler joints with optimum variable-stiffness composite adherend under torsion. Compos. Struct. **164**, 76–89 (2017). https://doi.org/10.1016/j.compstruct.2016.12.043
8. Aimmanee, S., Hongpimolmas, P., Ruangjirakit, K.: Simplified analytical model for adhesive-bonded tubular joints with isotropic and composite adherends subjected to tension. Int. J. Adhes. Adhes. **86**, 59–72 (2018). https://doi.org/10.1016/j.ijadhadh.2018.08.010
9. Bakulin, V.N., Vinogradov, Y.I., Men'kov, G.B.: The stressed state of a stiffened conical shell with thermal protective coating with temperature-dependent properties. Russ. Aeronaut. **61**(2), 156–164 (2018). https://doi.org/10.3103/S1068799818020022
10. Dragoni, E., Goglio, L.: Adhesive stresses in axially-loaded tubular bonded joints – part I: critical review and finite element assessment of published models. Int. J. Adhes. Adhes. **47**, 35–45 (2013). https://doi.org/10.1016/j.ijadhadh.2013.09.009
11. Lubkin, J.L., Reissner, E.: Stress distribution and design data for adhesive lap joints between circular tubes. Trans. ASME **78**, 1213–1221 (1956)
12. Kurennov, S.S., Barakhov, K.P., Poliakov, A.G.: Stressed state of the axisymmetric adhesive joint of two cylindrical shells under axial tension. Mater. Sci. Forum **968**, 519–527 (2019). https://doi.org/10.4028/www.scientific.net/MSF.968.519
13. Nemeş, O., Lachaud, F., Mojtabi, A.: Contribution to the study of cylindrical adhesive joining. Int. J. Adhes. Adhes. **26**(6), 474–480 (2006). https://doi.org/10.1016/j.ijadhadh.2005.07.009
14. Shishesaz, M., Tehrani, S.: Interfacial shear stress distribution in the adhesively bonded tubular joints under tension with a circumferential void or debond. J. Adhes. Sci. Technol. **34**(11), 1172–1205 (2020). https://doi.org/10.1080/01694243.2019.1701894
15. Yang, C., Huang, H., Guan, Z.: Stress model of composite pipe joints under bending. J. Compos. Mater. **36**(11), 1331–1348 (2002). https://doi.org/10.1177/0021998302036011167
16. Rapp, P.: Mechanics of adhesive joints as a plane problem of the theory of elasticity. Part II: displacement formulation for orthotropic adherends. Archives of Civil and Mechanical Engineering 15(2), 603–619 (2015). https://doi.org/10.1016/j.acme.2014.06.004
17. Wang, J., Zhang, C.: Three-parameter, elastic foundation model for analysis of adhesively bonded joints. Int. J. Adhes. Adhes. **29**(5), 495–502 (2009). https://doi.org/10.1016/j.ijadhadh.2008.10.002
18. Baishya, N., Das, R.R., Panigrahi, S.K.: Failure analysis of adhesively bonded tubular joints of laminated FRP composites subjected to combined internal pressure and torsional loading. J. Adhes. Sci. Technol. **31**(19–20), 2139–2163 (2017). https://doi.org/10.1080/01694243.2017.1307498
19. Amidi, S., Wang, J.: An analytical model for interfacial stresses in double-lap bonded joints. The Journal of Adhesion **95**(11), 1031–1055 (2019). https://doi.org/10.1080/00218464.2018.1464917

20. Jairaja, R., Naik, G.N.: Single and dual adhesive bond strength analysis of single lap joint between dissimilar adherends. Int. J. Adhes. Adhes. **92**, 142–153 (2019). https://doi.org/10.1016/j.ijadhadh.2019.04.016

21. Kurennov, S.S., Koshevoi, A.G., Polyakov, A.G.: Through-thickness stress distribution in the adhesive joint for the multilayer composite material. Russ. Aero. (Iz VUZ) **58**(2), 145–151 (2015). https://doi.org/10.3103/S1068799815020026

22. Fernández-Cañadas, L.M., Ivañez, I., Sanchez-Saez, S., Barbero, E.J.: Effect of adhesive thickness and overlap on the behavior of composite single-lap joints. Mech. Adv. Mater. Struct. (2019). https://doi.org/10.1080/15376494.2019.1639086

Information Technology in Design and Manufacturing of Engines

The Validation of the Bird-Impactor Model for Mathematical Modelling of Damage Processes in Turbofan Engine Parts

Dmitry Ivchenko[1] and Natalia Smetankina[2(✉)]

[1] SE "Ivchenko-Progress", Ivanova Street, Zaporizhzhia 69068, Ukraine
[2] A. Pidgorny Institute of Mechanical Engineering Problems of the National Academy of Sciences of Ukraine,
2/10 Pozharskogo Street, Kharkiv 61046, Ukraine
nsmetankina@ukr.net

Abstract. The paper presents a validation of the authors' model of a bird-impactor for bird masses M_B = 0.7–3.65 kg, bird impact velocities V_B = 100–200 m/s and impact angles α = 30–90°. Validation was performed by computations of bird impact with an obstacle by using a bird-impactor model and comparing computational data with available experimental data of bird impact field tests. The obstacles considered were steel plates: a weakly deformable (rigid) plate and a deformable plate. Computations were performed using the explicit LS-DYNA Solver. The bird-impactor model was validated for the load created on the rigid plate by using experimental research data of the Dayton Research Institute. The bird-impactor model was validated for plate strains by using data of experimental research at A. Podgorny Institute for Mechanical Engineering Problems of the National Academy of Sciences of Ukraine. Validation found that the bird-impactor model is capable of adequately reproducing impact loads acting like those as in the case of a strike by a real bird. This enables using the bird-impactor model for mathematical modelling of the processes of bird impact and of the damage to turbofan engine parts.

Keywords: Bird-impactor model · Validation · Collision modelling

1 Introduction

One of the problems in mathematical modelling of the processes of bird impact and damage to turbofan engine flow parts is using a proven bird-impactor model (BIM). One of the main stages of developing a BIM is its validation that determines whether a BIM can adequately reproduce impact loads like those as in the case of impact by a real bird.

In the papers published to date [1–5], BIM of different masses were used for mathematical modelling of bird impact processes and of damage to turbofan engine parts. As a rule, the results of their validation are presented for one impact velocity and one impact angle, making it impossible to assess their adequacy in full.

The purpose of this study is the validation of the authors' BIM for bird masses M_B = 0.7−3.65 kg (medium-size birds of small mass and large-size birds) according to

M. Nechyporuk et al. (Eds.): ICTM 2020, LNNS 188, pp. 403–414, 2021.
https://doi.org/10.1007/978-3-030-66717-7_34

the requirements of airworthiness standards for aviation engines CS-E, FAR-33, and AP-33 [6]. Bird impact velocities $V_B = 100-200$ m/s and impact angles $\alpha = 30°-90°$ were considered.

2   Formulation of the Problem

BIM was validated by computations of bird impact with an obstacle by using a BIM and comparing computational data with available experimental data. The obstacles considered were steel plates: a weakly deformable (rigid) plate and a deformable plate.

The study used numerical methods: smoothed particle hydrodynamics (the SPH method) for modelling birds, and the finite element method for modelling steel plates, and Simpson's rule and polynomial regression for analysing computational and experimental data.

Computations were performed using explicit LS-DYNA Solver of the ANSYS LS-DYNA software package, for which the numerical BIM was realised. Also, the pre- and postprocessor LS-PrePost program was used for numerical modelling of bird impact with a plate and generating computational results.

3   Numerical Modelling of Bird Impact with a Plate

3.1   Bird-Impactor Model

BIM was represented as a cylinder with hemispheric ends. This geometric form is quite close to the forms of bird bodies used for experimental research in bird impact, including testing a turbofan engine with casting of birds into the flow part.

Since a bird, with impact loading, behaves identical to a fluid [7], the fluid dynamic theory, NULL material [8] and the equation of state, which was specified using a linear polynomial [8] and [9], were used for bird modelling.

The meshless SPH method was used for the BIM. It provides for discretisation of the design volume with an assembly of particles, allowing to model big strains of the bird body during impact. All particles were assigned an initial velocity by using its components.

The BIM are shown in Fig. 1. BIM parameters are presented in Table 1.

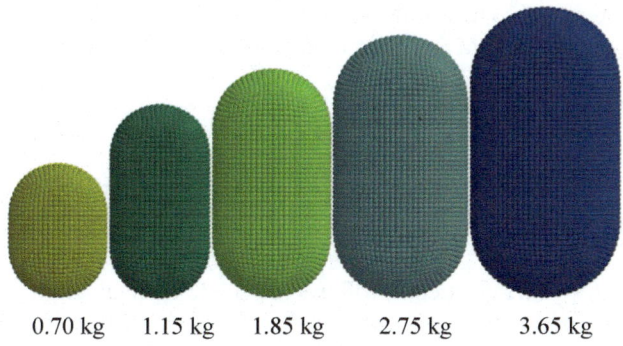

0.70 kg   1.15 kg   1.85 kg   2.75 kg   3.65 kg

Fig. 1. Bird-impactor models.

Table 1. BIM parameters.

Parameter	Medium bird		Large bird		
M_B, kg	0.70	1.15	1.85	2.75	3.65
Mass density, kg/m^3	969	955	942	931	924
Diameter of hemispherical ended cylinder, D_B, mm	95	97	114	131	145
Length of hemispherical ended cylinder, L_B, mm	133	194	228	262	290
Ratio L_B/D_B	1.4	2.0	2.0	2.0	2.0
Dynamic viscosity, Pa·s	996.7·10^{-6}				
Coefficients for a linear polynomial equation of state, MPa	$C_0 = 0$; $C_1 = 28$; $C_2 = -85$; $C_3 = 35000$ [10]				
Number of particles, thousand pieces	15.7	23.9	34.8	50.5	70.4
Particle mass, mg	45	48	53	54	52
Conditional particle diameter, mm	445	4.58	4.76	4.82	4.75

3.2   Plate Model

The following steel plate models were developed for BIM validation: the rigid plate model (RIGPM) with a sensor for determining the load during bird impact; the deformable plate model (DEFPM).

The plate and sensor material is steel. During bird impact, the plates and the sensor undergo elastic deformation. The DEFPM thickness was half of that of the RIGPM one. The volumes of the plates were divided into FE. In the impact zone, the FE sizes were approximately 5 mm.

A sensor with the dimensions 14 × 14 × 5 mm was installed in the point of intersection of the diagonals of the RIGPM front face and flush with it. The sensor is modelled with a single FE. It was assumed that, upon impact, the BIM acts on the sensor with a force equal to that of the sensor on the RIGPM. With account of this, the pressure acting on the RIGPM in the impact zone was found by the formula.

$$P = \frac{F_{norm}}{S},\tag{1}$$

where F_{norm} is the normal component of the force acting in the contact between the sensor and the RIGPM; S is the contact area, taken to be the area of the sensor's back face.

The plate models were secured as follows: on the horizontal axis – over one of the two side faces; on the vertical axis – on the bottom face; along the impact direction – on the back face edges.

Fig. 2 illustrates the RIGPM and DEFPM. Table 2 shows the RIGPM and DEFPM parameters.

RIGPM   DEFPM

Fig. 2. Plate models.

Table 2. RIGPM and DEFPM parameters.

Parameter	RIGPM	DEFPM
Dimensions, mm	700 × 700 × 100	700 × 700 × 50
Mass density, kg/m³	7850	
Young's modulus, GPa	210	
Poisson's ratio	0.3	
The number of finite elements, thousand pieces	233.3	112.4

4   The Bird-Impactor Model Validation for Loading Applied to a Plate

For validating the BIM by the load created on the plate, experimental research data of the University of Dayton Research Institute (research by Wilbeck et al.) [11] were used. This experimental research was conducted to determine the impact of the mass, velocity and angle of impact of the bird on the loads (pressures) that act on a rigid plate during bird impact. Birds of different mass were used. Pressures were measured using pressure sensors installed in the plate.

To compare with experimental research data, the BIM impact by the RIGPM with a sensor was computed using the following parameters: M_B = 0.7, 1.15, 1.85, 2.75, 3.65 kg; $\alpha = 45°$, $90°$; V_B = 100; 150; 200 m/s.

An RIGPM with a sensor was used for computing pressures created by birds upon impact.

The BIM impact deformation process by the RIGPM with the parameters $M_B = 1.85$; $\alpha = 45°$, $90°$; V_B = 200 m/s is shown in Fig. 3. For BIM with other masses, computations yielded similar BIM deformation processes.

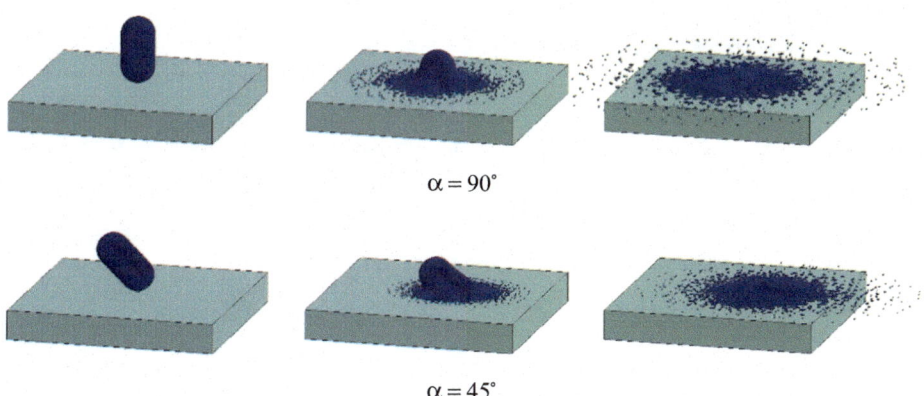

$\alpha = 90°$

$\alpha = 45°$

Fig. 3. BIM impact deformation process by the RIGPM.

Both computations and experimental research yielded pressure curves. Figure 4 shows pressure curves obtained for BIM impact by the RIGPM with the parameters $M_B = 1.85$ kg; $\alpha = 45°$, $90°$; V_B = 200 m/s. For BIM of other masses, computations yielded similar pressure curves.

Pressure curves were used to determine the Hugoniot pressure (impact pressure) P_H and the stagnation pressure P_S (Fig. 4). The Hugoniot pressure was determined as the maximum curve pressure. The stagnation pressure was computed for the curve section where a low-level time-constant pressure was observed (approximately at $T_{norm} = 0.28 - 0.70$) by the formula [12]:

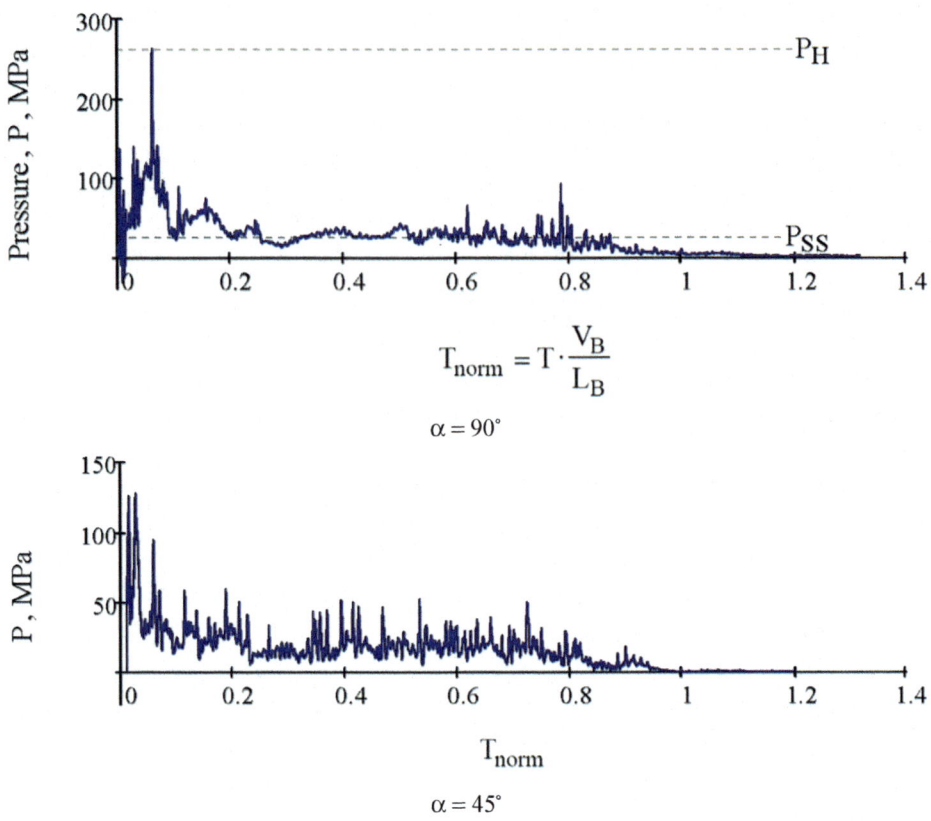

$$T_{norm} = T \cdot \frac{V_B}{L_B}$$

$$\alpha = 90°$$

$$\alpha = 45°$$

Fig. 4. Pressure curves.

$$P_S = \frac{I_P}{\Delta T_{norm}}, \qquad (2)$$

where I_P is specific impulse of stagnation pressure; ΔT_{norm} is time of action of stagnation pressure.

$$\Delta T_{norm} = T_{norm2} - T_{norm1}, \qquad (3)$$

T_{norm1} and T_{norm2} are normalised times of the start and end of action of stagnation pressure.

Stagnation pressure impulse is.

$$I_P = \int_{T_{norm1}}^{T_{norm2}} P(T_{norm})dT_{norm}. \qquad (4)$$

Simpson's rule was used to determine the stagnation pressure impulse.

BIM was validated by pressures P_H and P_S by comparing computational data when using BIM and the data of experimental research. In so doing, polynomial data regression was used. BIM validation by pressures P_H and P_S is illustrated in Figs. 5, 6 and 7 and in Tables 3 and 4.

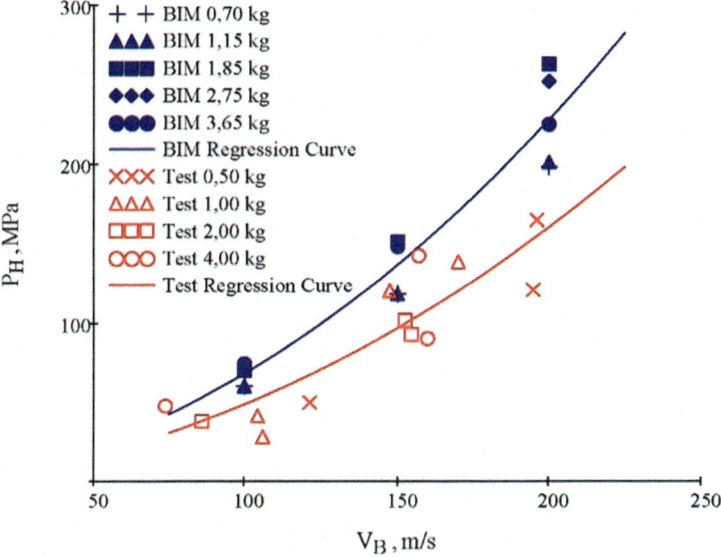

Fig. 5. BIM validation by pressure P_H ($\alpha = 90°$).

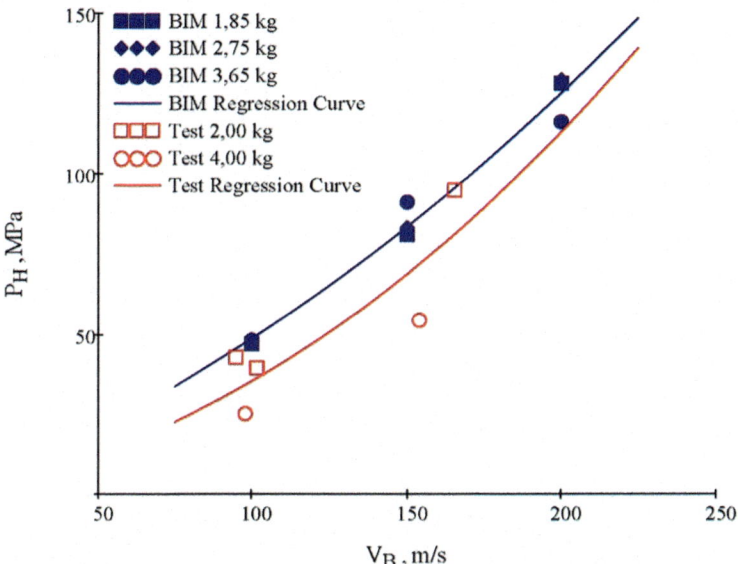

Fig. 6. BIM validation by pressure P_H ($\alpha = 45°$).

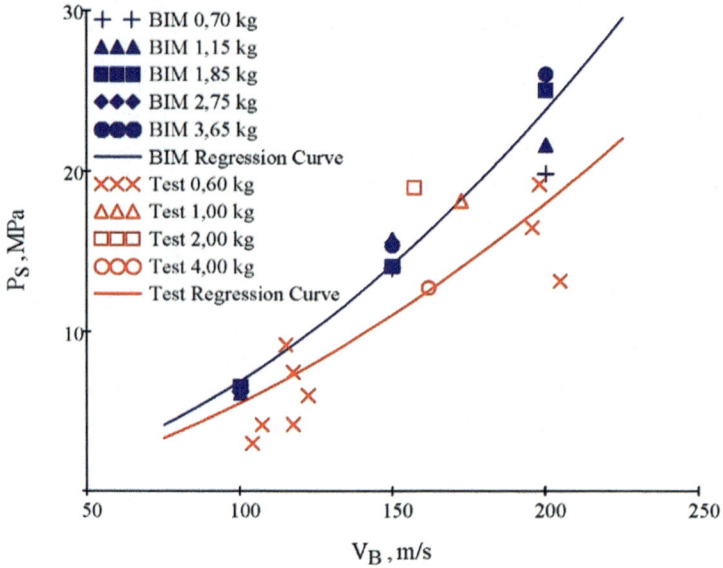

Fig. 7. BIM validation by pressure P_S ($\alpha = 90°$).

Table 3. BIM validation by pressure P_H.

M_B, kg	α, deg	V_B, m/s	P_H, MPa		δ, %
			BIM	Test	
1.85–3.65	45	100	49	35	40
		150	83	69	20
		200	125	113	11
0.70–3.65	90	100	68	48	42
		150	137	96	43
		200	228	160	43

Table 4. BIM validation by pressure

M_B, kg	α, deg	V_B, m/s	P_S, MPa		δ, %
			BIM	Test	
0.70–3.65	90	100	6.8	5.5	24
		150	14.1	11.0	28
		200	23.8	18.0	32

Analysing the results obtained, note that the dependencies of pressures P_H and P_S on V_B found by computation with BIM and from experimental research are qualitatively the same. The levels of pressures P_H and P_S obtained by computation with BIM are somewhat higher than those found by experimental research. The relative error of pressure computations with BIM was as follows: for P_H is $\delta = 11 - 43\%$; for P_S is $\delta = 24 - 32\%$.

These differences can be explained as follows. During computations with BIM, the interval between data recording was taken to be 0.001 ms (1 MHz frequency). The results were pressures P_H of higher levels and a bigger number of pressure peaks that increased P_S than those as in the case of experimental research conducted with a smaller frequency of pressure recording. Increasing the interval between data recording to 0.01 ms (100 kHz frequency) during computations with BIM decreases the levels of pressures P_H and P_S to those found by experimental research.

Hence, the results of BIM validation by the load acting on the plate demonstrates its adequacy.

5   The Bird-Impactor Model Validation for Plate Strains

BIM validation by plate strains used data of experimental research at A. Podgorny Institute for Mechanical Engineering Problems of the National Academy of Sciences of Ukraine [13–15]. This experimental research was conducted for determining the influence of velocity and angle of impact of the bird on the elastic strains of the steel plate. Large-size birds were used with the masses of 1.37–1.99 kg. The elastic strains were measured using strain gauge sensors that were glued under the impact point on the plate back face.

For comparison with experimental research data, computations of BIM impact by DEFPM were conducted with the following parameters: $M_B = 1.85$ kg; $\alpha = 30°$, 45°, 90°; $V_B = 100$, 150, 200 m/s.

Elastic DEFPM strains were determined in one of the finite elements located under the impact point on the DEFPM back face. Analysis of patterns of DEFPM strains shows that the zones of action of maximum strain values ε_x and ε_y are located in the centre and on the DEFPM side faces (in the points of fastening). DEFPM strains during BIM impact at point of time $T_{norm} = 0.5$ for $\alpha = 30°$, 45°, 90° are shown in Fig. 8.

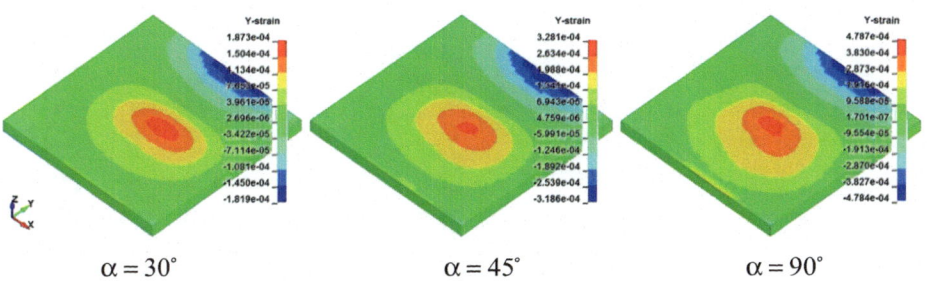

Fig. 8. DEFPM strains during BIM impact (view on back face).

BIM was validated by plate strains by comparing the data of computations of maximum values of elastic strain with the use of BIM and the data of experimental research. In so doing, for comparison, the values of strains were those found by using polynomial regression of experimental data.

The results of BIM validation by DEFPM strains are shown in Fig. 9 and Table 5.

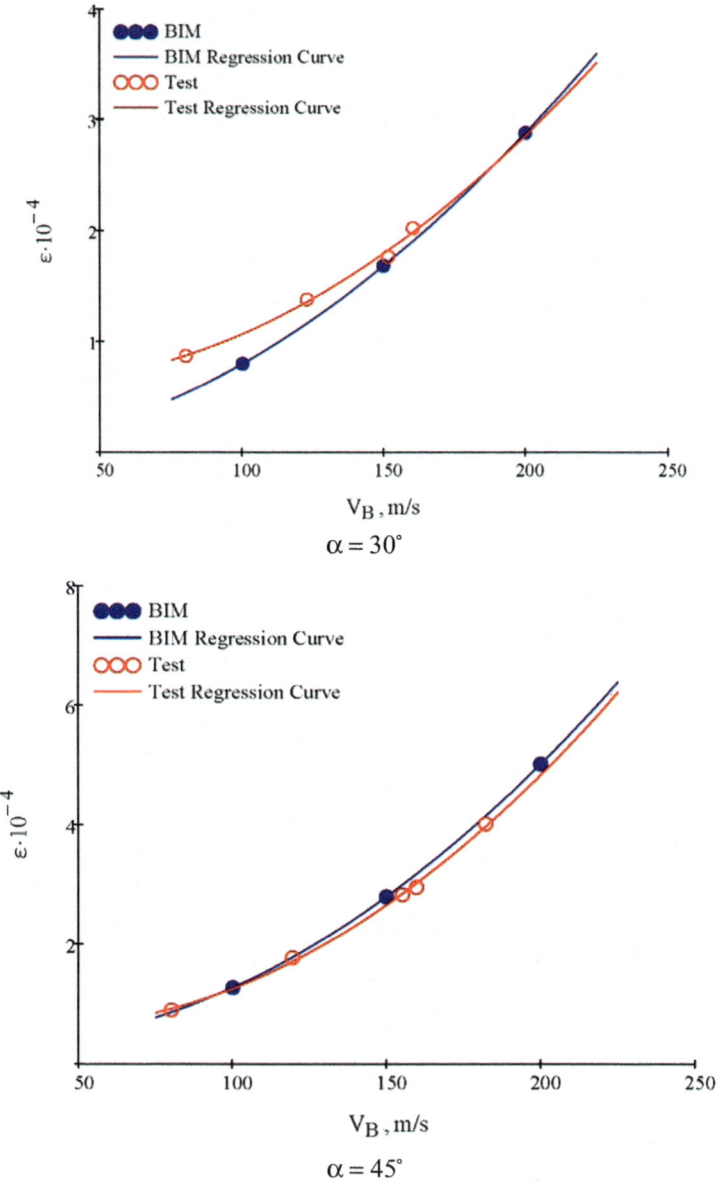

Fig. 9. BIM validation by DEFPM strains.

Table 5. BIM validation by DEFPM strains.

α, deg	V_B, m/s	$\varepsilon \cdot 10^{-4}$		δ, %
		BIM	Test	
30	100	0.80	1.06	25
	150	1.68	1.79	6
	200	2.88	2.86	1
45	100	1.27	1.25	2
	150	2.79	2.65	5
	200	5.01	4.82	4
90	150	4.41	4.09	8

The results show that there is a qualitative and quantitative match between computational data when using BIM and experimental research data. The maximum relative error of computations $\delta = 25$ % at $\alpha = 30°$ and $V_B = 100$ m/s. For all other impact parameters, the relative error $\delta = 1 - 8\%$. Thus, results of BIM validation by plate strains show its adequacy.

6 Conclusion

The BIM developed by the authors was validated for bird masses $M_B = 0.7 - 3.65$ kg, bird impact velocities $V_B = 100 - 200$ m/s and impact angles $\alpha = 30° - 90°$.

Validation was performed by computing bird impacts with an obstacle with the use of BIM and comparing computational data with available experimental data of bird impact field tests.

Validation demonstrated that the BIM is capable of adequately reproducing impact loads that act like those as in the case of impact by a real bird. This enables using the BIM for mathematical modelling of the processes of bird impact and damage of turbofan engine parts.

References

1. Niering, E.: Simulation of bird strikes on turbine engines. J. Eng. Gas Turbines Power **112**(4), 573–578 (1990). https://doi.org/10.1115/1.2906207
2. Moffat, T.J., Cleghorn, W.L.: Prediction of bird impact pressures and damage using MSC/DYTRAN. In: Turbo Expo: Power for Land, Sea, and Air, Paper No. 2001-GT-0280 (2001). https://doi.org/10.1115/2001-GT-0280
3. Chuan, Z., Xiang-hua, J., Xiang-hai, C., Tong-cheng, S.: TC4 hollow fan blade structural optimization based on bird-strike analysis. Procedia Eng. **99**, 1385–1394 (2015). https://doi.org/10.1016/j.proeng.2014.12.674
4. Kuzmin, M., Kirsanov, A.: Validation of aircraft gas-turbine engine inlet bird throw-in simulation. Civ. Aviation High Technol. **212**, 120–126 (2015). (in Russian)

5. Husainie, S.N.: Bird strike and novel design of fan blades. In: Proceedings of Science in the Age of Experience Conference, pp. 26–40. Dassault Systèmes, Chicago (2017)
6. Smetankina, N., Ugrimov, S., Kravchenko, I., Ivchenko, D.: Simulating the process of a bird striking a rigid target. In: Ivanov, V., et al. (eds.) Advances in Design, Simulation and Manufacturing II. DSMIE 2019. LNME, pp. 711–721. Springer, Cham (2020). https://doi.org/10.1007/978-3-030-22365-6_71
7. Hedayati, R., Sadighi, M.: Bird Strike: An Experimental, Theoretical and Numerical Investigation. Woodhead Publishing, Cambridge (2015). https://doi.org/10.1016/C2014-0-02336-2
8. LS-DYNA® Keyword User's Manual. LS DYNA R8.0. LSTC, vol. II. Livermore Software Technology Corporation, Livermore (2015)
9. Batchelor, G.: An Introduction to Fluid Dynamics. Cambridge University Press, Cambridge (2013)
10. Selezneva, M., Stone, P., Moffat, T., et al.: Modeling bird impact on a rotating fan: the influence of bird parameters. In: Proceeding of the 11th International LS-DYNA Users Conference, pp. 37–58 (2010)
11. Barber, J.P., Taylor, H.R., Wilbeck, J.S.: Bird impact forces and pressures on rigid and compliant targets. Technical report ADA061313. University of Dayton Research Institute, Dayton (1978)
12. Ugrcic, M.: Application of the hydrodynamic theory and the finite element method in the analysis of bird strike in a flat barrier. Sci. Tech. Rev. **62**(3–4), 28–37 (2012)
13. Rodichev, Y., Smetankina, N., Shupikov, O., Ugrimov, S.: Stress-strain assessment for laminated aircraft cockpit windows at static and dynamic load. Strength Mater. **50**(6), 868–873 (2018). https://doi.org/10.1007/s11223-019-00033-4
14. Smetankina, N., Kravchenko, I., Merkulov, V., Ivchenko, D.: Simulation of bird collision with aircraft laminated glazing. In: Ivanov, V., et al. (eds.) Advances in Design, Simulation and Manufacturing III. DSMIE 2020. LNME, pp. 179–188. Springer, Cham (2020). https://doi.org/10.1007/978-3-030-50491-5_18
15. Smetankina, N., Kravchenko, I., Merculov, V., et al.: Modelling of bird strike on an aircraft glazing. In: Nechyporuk, M., et al. (eds.) Integrated Computer Technologies in Mechanical Engineering. AISC, vol. 1113, pp. 289–297. Springer, Cham (2020). https://doi.org/10.1007/978-3-030-37618-5_25

Estimation of Strength of the Combustion Chamber of the ICE Piston with a TBC Layer

Andriy Marchenko⬡, Vyacheslav Pylyov$^{(\boxtimes)}$⬡, and Oleh Linkov⬡

National Technical University "Kharkiv Polytechnic Institute",
2 Kyrpychova Street, Kharkiv 61002, Ukraine
dvs@kpi.kharkov.ua

Abstract. An improvement of the performance of modern heat engines is ensured through the use of new materials, design and technological solutions, and a higher level of modeling the processes that inherent in IC engines. Theoretical study and a pilot research prove that use of components with heat insulation layers on the surfaces that form the combustion chamber is one of the promising directions of integrated improvement of IC engines. It is achieved by means of increasing the amplitude of the combustion chamber surface temperature high-frequency fluctuations, at the time that the strength of the surface layer exceeds the one of the basic material. Under such conditions, it becomes possible to determine the reserves for further power increase of IC engines. The approach based on the principles of guaranteeing the structural strength is proposed that uses the results of mathematical simulation with the maximum reduction of design routes. It presupposes integration of the 1-D simulation method of the high-frequent temperature field of the surface layer of the structure in characteristic zones, the 3-D simulation of stationary temperature field and thermal stressed state of the structure, as well as the 0-D simulation of total fatigue and creep damage in the critical piston zone for a given cyclic change of engine load. The comparative analysis of the results for the piston of the tractor diesel engine is presented.

Keywords: Piston TBC · Material strength loss · Thermal stresses · High-frequency temperature fluctuation · Heat conduction modeling

1 Introduction

With the improvement of internal combustion engines (ICE), the conditions imposed on their design become more and more stringent. These conditions are related to the global trend of increase of power per unit and specific output of engines, the limitations because of the regulations on emissions with exhaust gases, the improvement of combustion process organization, the search for and use of alternative types of fuels, etc. On the whole, the complex of these works is aimed at reducing every cost incurred by manufacturers and consumers to support the lifecycle of modern ICE designs. It is the levels of harmful emissions and thermal pollution that are considered to be the integrated performance criteria of such designs, first of all [1].

© The Author(s), under exclusive license to Springer Nature Switzerland AG 2021
M. Nechyporuk et al. (Eds.): ICTM 2020, LNNS 188, pp. 415–426, 2021.
https://doi.org/10.1007/978-3-030-66717-7_35

For almost half a century applying a layer of coating with low heat conductivity to the piston crown has been considered to be one of the integrated directions of ICE improving [2–4]. It is believed that the main difficulty of using such designs is due to the low reliability of the ceramic thermal barrier coatings (TBC). Therefore, researchers tend to consider piston designs having a main coating bonded to the piston crown with special material [5–8].

For aluminum alloy pistons, plasma electrolytic oxidizing of the surface may be a more promising method of creating a thermal barrier. As a result, it is not a traditional coating that obtained but a modified piston surface with a layer of aluminum oxide Al_2O_3 [9, 10]. This surface layer has a wide range of important properties, in particular, low heat conductivity, heat resistance, mechanical strength, wear resistance, catalytic activity, including burning of diesel soot, etc. [10–14]. Such positive effects drew researchers' attention as a possible means of further ICE power increase at a specified resource, as power is the main factor of increasing of labor productivity and a life cycle efficiency of the design integrally.

The piston of the ICE is one of the main parts, which determine the reliability of the engine. Therefore, great attention is paid to improving its quality and ensuring the resource. However, until today, for the pistons of engines with high specific output, especially diesel ones, there are cases of unpredicted loss of physical reliability that manifests itself in the cracking of the edges of the combustion chambers (CC) during the operation [15, 16]. Such facts indicate the necessity of application of new designs as well as the imperfection of the engineering technologies in use.

The analyses carried out in [15, 16] indicate that cracks in the heat-stressed zones of the pistons occur as a result of piston temperature drops caused by structural features and frequent changes in the load of the engine. And the data of [17, 18] show that the high-frequency fluctuations of temperature and thermal stresses of the surface layer of the piston crown have to be taken into account in the estimates of the loss of strength of the piston too. However, in the works [3–8] discussed above and others, it is the temperature and stress levels obtained as a result of solving the piston stationary thermal conduction problem that is taken to be the comparison criterion for the quality of designs. Thus, the other criterion for the quality of the design should be used, and the higher complexity level of the problems being solved is needed.

Solving such problems requires the use of nonlinear continuous medium models, in which the material of the part has not only elastic but also momentary plastic deformations with the possible appearing of creep deformations and stress relaxation with time [18, 19]. Modeling these processes is a considerably more complex task [20, 21]. At the same time, with the development of CAD systems, there was a clear indication that the cost, the time of execution of designing work, and the effectiveness of the system as a whole depend on efficiency of applied mathematical models.

Thus, the task of practical implementation of the approach based on the principle of guaranteeing the strength of the piston crown being designed that provides for mathematical simulation of the nonstationary influences, during maximum simplification of the models and design routes, should be considered of current importance.

2   Estimation of the Material Strength of the Combustion Chamber of the Piston at the Point of Designing

The facts of physical reliability loss of the edges of the piston CC in operation [15, 16] indicate that the piston endurance is different for instances of the same level of engine specific output. In critical cases it is less than the assigned resource. For this reason, the contradictory requirements should be complied with at the development engineering stage for promising ICE designs: taking into account non-stationary operational impacts on the design durability; maximal simplification of mathematical models and design routes reduction; ensuring the operation near the material breaking point. This approach conforms to the goal of increasing the efficiency of the lifecycle of ICE designs.

The issue of material strength loss in the case of non-stationary influences is believed to be related to conceptions of damage accumulation. The damage value $d(\tau)$ is assumed to represent the exhausted fraction of the resource of the part, so it may vary in the interval of 0 to 1. The value $d(\tau^*) = 1$ corresponds to the moment of total loss of material strength τ^*. Hence, for guaranteeing the design strength, the condition.

$$d(P) \leq 1, \tag{1}$$

where P is a specified resource of the engine, should not be violated under any accepted assumptions and simplifications.

In the conditions of cyclic loading being applied to the piston, the current fraction of exhaustion of its resource must be determined taking into account the fatigue damage $d_f(\tau)$ and the creep damage $d_s(\tau)$,

$$d_{fs}(\tau) = d_f(\tau) + d_s(\tau). \tag{2}$$

On the basis of (2), the calculation expression for the accumulated damage for the specified life P takes the form.

$$d_{fs}(P) = \sum_{k=1}^{N_P} \frac{1}{N_{fk}} + \sum_{k=1}^{N_P} \frac{1}{N_{sk}} \leq 1, \sum_{k=1}^{N_P} \tilde{P}_k \leq , P \tag{3}$$

where N_P is a whole of all piston loading cycles for the specified resource P; N_{fk} is the number of cycles until fatigue failure under a loading cycle of mode k only; N_{sk} is the number of cycles until failure over creep under a loading cycle of mode k only; $\tilde{P}_k$ is duration of the engine loading cycle of mode k.

It is important that due to the complex nature of hardening and loss of material strength over creep, each loading cycle of the edge of the piston CC is unique even for identical diesel engine operation cycles. So, there is a need to model every of k cycles over a specified resource P.

It is known that most accumulation of damage occurs in cycles from idle running to the mode of maximum load and back. On the basis of statistical data it is possible to determine the maximum possible time Π of operation of the engine in such cycles.

Then, for the goal of comparing alternative designs, expression (3) can be simplified to the form.

$$d_{fs}(\Pi) = \sum_{k=1}^{N_\Pi} \frac{1}{N_{fk}} + \sum_{k=1}^{N_\Pi} \frac{1}{N_{sk}} < 1, \quad \sum_{k=1}^{N_\Pi} \tilde{P}_k \ll P. \tag{4}$$

This makes significant shortening of the design time and route possible.

Then, the part of the problem that remains unsolved is to determine the temperature and thermal stresses dependencies for a particular cycle of engine operation:

$$t_k(\tau) = \bar{t}_k(\tau) + \tilde{t}_k, \tag{5}$$

$$\sigma_k(\tau) = \bar{\sigma}_k(\tau) + \tilde{\sigma}_k, \tag{6}$$

where the values $\bar{t}_k(\tau)$ and $\bar{\sigma}_k(\tau)$ are the mean values during the process of low-frequency change of temperature and thermal stress in the studied zone of the piston in the conditions of a single loading cycle of the mode k; $\tilde{t}_k$ and $\tilde{\sigma}_k$ are the instantaneous deflections of temperature and thermal stress from the mean value.

Formulation of the problem in the 0-D form (5), (6) with subsequent use of models (3) or (4) significantly increases the design time. Therefore, the simplification of the model (5), (6) is proposed, which does not contradict the principle of guaranteeing the piston crown strength at the point of designing. In the cases of surging and the operation of the engine in heavy loaded stationary mode it is proposed to use expressions.

$$t_k^{ab}(\tau) = \bar{t}_k(\tau) + \tilde{t}_k^{b\,max}, \tag{7}$$

$$\sigma_k^{ab}(\tau) = \bar{\sigma}_k(\tau) + \tilde{\sigma}_k^{b\,min}. \tag{8}$$

In the cases of load shedding and the operation of the engine in a less loaded stationary mode of the cycle the proposed expressions are.

$$t_k^{cd}(\tau) = \bar{t}_k(\tau) + \tilde{t}_k^{d\,max}, \tag{9}$$

$$\sigma_k^{cd}(\tau) = \bar{\sigma}_k(\tau) + \tilde{\sigma}_k^{d\,max}. \tag{10}$$

The graphical explanation of the used variables is presented in Fig. 1.

In the expressions the value $\tilde{t}_k^{b\,max}$ is the maximum deviation of temperature in the high-frequency change from the averaged value $\bar{t}_k^b(\tau)$ for the heavy loaded stationary operating mode; the value $\tilde{\sigma}_k^{b\,min}$ is the minimum deviation of thermal stress in the high-frequency change from the averaged value $\bar{\sigma}_k^b(\tau)$ for the heavy loaded stationary operating mode; the value $\tilde{t}_k^{d\,max}$ is the maximum deviation of temperature in the high-frequency change from the value $\bar{t}_k^b(\tau)$ for the less heavy loaded stationary operating mode; and the value $\tilde{\sigma}_k^{d\,max}$ is the maximum deviation of thermal stress in the high-frequency change from the value $\bar{\sigma}_k^b(\tau)$ for the less heavy loaded stationary operating mode.

The low-frequency change of parameters $\bar{t}_k(\tau)$, $\bar{\sigma}_k(\tau)$ can be found on the basis of the solution of a non-stationary low-frequency problem of the piston thermal conduction for the processes of the engine operating mode transition. For this purpose, the method [22] can be used. The values of $\tilde{t}_k^{b\,\max}$, $\tilde{t}_k^{d\,\max}$, $\tilde{\sigma}_k^{b\,\min}$, $\tilde{\sigma}_k^{d\,\max}$, in turn, can only be a result of solving the non-stationary high-frequency problem of the piston thermal conduction for the stationary modes of the engine operation, which takes an unacceptably long time to be used in engineering practice. Finding a simplified method of determining these parameters for an ICE piston is the paper main goal. In the general case, the problem has to be solved for a piston with TBC of its crown surface.

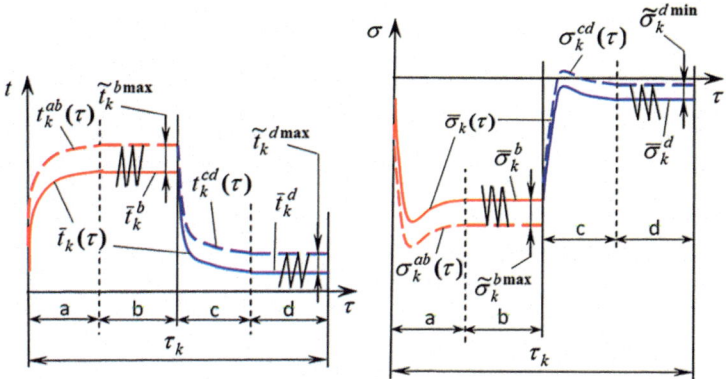

Fig. 1. Local temperature (left) and thermal stress (right) in a single loading cycle of the studied zone of the piston: a - surging; b - heavy loaded stationary mode; c - load shedding; d - less heavy loaded stationary mode.

3   Simplified Method for Modeling the High-Frequency Temperature Field of the Piston Crown

The proposed method involves dividing the 3-D body of the piston into two calculation areas denoted by Ω_1 and Ω_2, which have a perfect contact by the surface S. A schematic diagram of layout of the piston model, in accord to this, is shown in Fig. 2 (left). There, the area Ω_1 corresponds to a layer of material with thickness Δ_b, which extends along the surfaces of the CC and the piston crown.

Based on the assumption that heat fluxes parallel to the body surface in this zone are admissible to neglect, the temperature field is considered 1-D non-stationary there. It corresponds to the following form of the Fourier's heat conduction equation:

$$\rho(x)c(x,t(x,\tau))\frac{\partial t(x,\tau)}{\partial \tau} = \partial\left(\lambda(x,t(x,\tau))\frac{\partial t(x,\tau)}{\partial x}\right)/\partial x, \qquad (11)$$

where x is the spatial coordinate that represents an internal normal to the surface of the part, and τ is the time coordinate; λ, c, ρ are local heat conductivity, heat capacity, and density of the material depending on the current temperature and, formally, the spatial coordinate x as an indicator of belonging to the TBC surface layer or the base material of the part.

The value of the thickness Δ_b was suggested to be set as 1 mm. At this depth, if the TBC layer thickness is $\delta_c = 0.12$ mm, as in the case under test, the presence of high-frequency temperature fluctuations can already be neglected because the temperature wave propagating deeper into the piston body, induced by the cyclical nature of the processes in the cylinder, decays rapidly in the presence of a TBC layer [17].

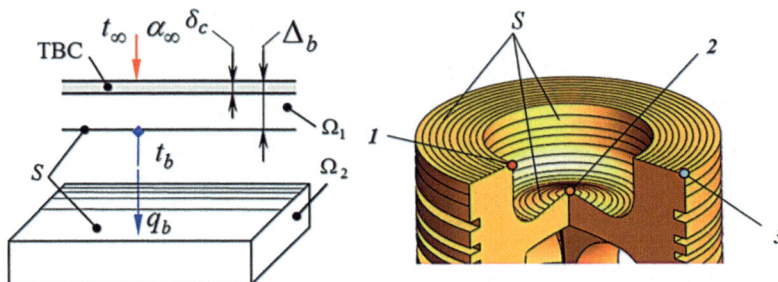

Fig. 2. Computational model of the piston with TBC layer on (left) and model of 3-D area Ω_2 of the piston with a layout of the surface S into 27 zones of contact with 1-D area Ω_1 (right).

The rest of the body of the part belongs to the area Ω_2. The formulation of the problem in that area is 3-D stationary:

$$\lambda(t(x,y,z))\left(\frac{\partial^2 t(x,y,z)}{\partial x^2} + \frac{\partial^2 t(x,y,z)}{\partial y^2} + \frac{\partial^2 t(x,y,z)}{\partial z^2}\right) = 0. \tag{12}$$

Simulation in the area Ω_1 is proposed to be performed using the finite difference method, whereas in the area Ω_2 – the finite element method. The area Ω_1 is further divided into a set of calculation subareas, which differ in heat transfer boundary conditions (BC) or values of local thicknesses of TBC. The simulations using Eq. (11) for each of these subareas are performed separately. The surface S of the area Ω_2 is also divided into zones with different BC corresponding to subareas of Ω_1. Geometric model of the piston area Ω_2 with these contact zones is shown in Fig. 2 (right).

According to the general problem being solved, the method involves the application of BC of the third type. On the surface of the part related to the area Ω_1 they are assigned in a modified form.

$$\rho(0)c(0,t(0,\tau))\frac{\partial t(0,\tau)}{\partial \tau} = \lim_{x \to +0} \frac{\lambda(x,t(x,\tau))\frac{\partial t(x,\tau)}{\partial x} + \alpha_\infty(\tau)(t_\infty(\tau) - t(0,\tau))}{x}, \tag{13}$$

where t_∞, α_∞ are the cylinder charge nonstationary temperature and heat emission coefficient.

That equation can be obtained by combining the thermal conduction equation and Newton-Richman's one. Due to the gained conservative qualities, using a mesh analogue of this equation during a computation leads to faster approaching of the temperatures in the mesh solution to the real ones. Advance to the result of the calculation is controlled by the repeatability of temperature waves; time length of that depends on the initial condition.

On the surface of the part related to the area Ω_2 the BC are assigned in the form of standard Newton-Richman's equations and depend only on the spatial coordinates.

Simulation of the temperature field in the areas Ω_1 and Ω_2 requires the specification of the BC on the surface of their contact S. The determination of these BC is achieved in the process of iterative approximation.

The initial calculation is rational to perform for the area Ω_2, specifying local parameters in the BC of third type at the surface S according to the formulas.

$$\alpha_{\infty b} = \alpha_{\infty} \Big/ \left(1 + \alpha_{\infty} \int_{\Delta_b} \frac{dx}{\lambda(x)}\right), \tag{14}$$

$$t_{\infty b} = t_{\infty}. \tag{15}$$

They are obtained by transformations performed over the system of equations that reflects the equality of heat fluxes through the surface of the CC, the piston material in the area Ω_1, as well as its boundary with the area Ω_2:

$$\alpha(\tau)(t_{\infty} - t_1) = \frac{t_1 - t_b}{\int_{\Delta_b} \frac{dx}{\lambda(x)}} = \alpha_{\infty b}(t_{\infty b} - t_b), \tag{16}$$

where t_1 is the local mean cycle temperature of the surface of the CC, which is subject to removal from the equations during their transformation.

In order to solve a complex of problems in the area Ω_1, the BC for each of them on the surface S is proposed to be assigned as of the first type, stationary:

$$t(\Delta_b) = t_b. \tag{17}$$

In turn, one of the results of solving problems in the area Ω_1 is the value of heat fluxes q_b across the surface S in each subarea. As they will not be equal to the similar fluxes intrinsic to temperature field in the area Ω_2, these heat fluxes are proposed to be used as BC of a second type in subsequent iterations for the solution in the area Ω_2:

$$q(x, y, z)|_S = q_b. \tag{18}$$

With this in mind, an iterative solution of the problem can be represented as a process of finding the intersection of two functions, $q_b = f_1(t_b)$ and $t_b = f_2(q_b)$ for each of the separate subarea of Ω_1. To find the values of these functions, simulation results in the areas Ω_1 and Ω_2 are used. The root that fulfills the BC of the fourth type at the intersection of these functions is only one; its achievement with certain accuracy can be considered a condition for termination of iterations.

It has to be noted that in each iteration, calculations should be performed for all subareas in the area Ω_1. The iteration results for three characteristic piston zones are shown in Fig. 3. There, the iteration 1 corresponds to the design of the piston without TBC.

The temperature field in the area Ω_1 obtained in the iteration 2, like in [3–8], takes into account only the thermal resistance of heat insulation. The resulting dynamic effect of temperature decrease because of TBC is achieved in computation after the matching of solutions in the areas Ω_1 and Ω_2 (iteration 6). It is over 1.65 times higher as compared to the decrease obtained with stationary heat conduction models.

An example of resulting temperature oscillation for a one zone of the CC in partial load operating mode can be seen in Fig. 4.

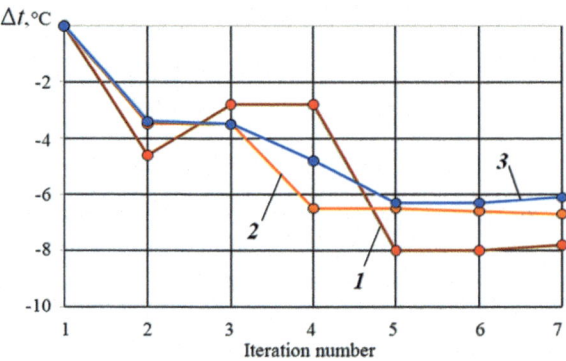

Fig. 3. Example of decrease of the temperature on the surface S with respect to the temperature of the piston without TBC in iterative calculations (1, 2, 3 correspond to characteristic piston zones shown in Fig. 2).

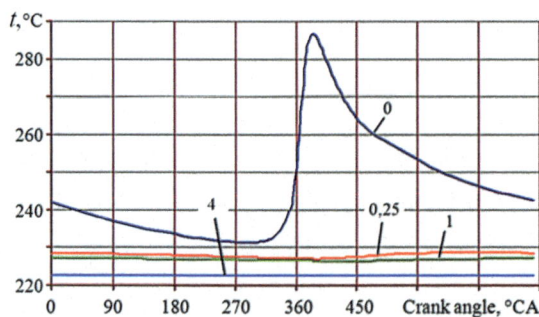

Fig. 4. Example of the high-frequency temperature change in the piston crown with TBC at the specific output of 16 kW/l (numbers near the curves are surface internal normal x values, mm).

On the basis of the presented method, to simplify calculation of the high-frequency change parameters of thermal stress in (8), (10) the following expressions [19] can be used:

$$\tilde{\sigma}_k^{b\,min} = \alpha_{tk}^b E_k^b \tilde{t}_k^{b\,max} / \left(1 - \nu_k^b\right), \tag{19}$$

$$\tilde{\sigma}_k^{d\,max} = \alpha_{tk}^d E_k^d \tilde{t}_k^{d\,max} / \left(1 - \nu_k^d\right), \tag{20}$$

where α_{tk}^b, α_{tk}^d are linear thermal expansion coefficient values depending on local mean cycle temperature of the piston material; E_k^b, E_k^b are elastic modulus values; ν_k^b, ν_k^d are Poisson's ratio values.

4 The Results of the Piston Crown TBC Using

Based on the methods proposed in the previous sections, the pilot passing of the design route has been performed by authors for a diesel engine of SMD model, the piston heat transfer BC of which are well known; and the damage accumulated by the edge of the CC has been evaluated. The piston of the conventional design and one with a corundum layer on the surface of the CC are examined. The basic material of the piston is the alloy AK12M2MgH (Al25). The thickness of the Al_2O_3 TBC layer is 0.25 mm. The calculations are performed for the diesel engine specific output of 25 kW/l. The rated engine speed is 2000 min^{-1}. For the piston with TBC the design analysis is performed using the parameters of the material under heat insulation layer. The operation model having been used is typical for engines of agricultural tractors. For instance, duration of all engine load cycles is assumed to be identical; according to the mean value for Ukrainian conditions, $\tau_k = 0.1$ h. The durations of operation in heavy and less heavy loaded modes within cycles are assumed to be equal as well (a + b = c + d, as shown in Fig. 1). The duration of simulation according to model (4) is taken $\Pi = 1040$ h based on the resource P = 10000 h [19]. The accumulated damage was calculated using the software tool RESURS [23]. Fatigue damage was determined by the Pospishil's equation, and creep damage was determined by the Sosnin's method [19].

The temperature field of the piston was modeled according to the proposed method. On this basis, the thermal stress of the basic material of the piston CC edge in the form (7)–(10) was determined. The main parameters of the loading cycle are presented in Table 1. The table shows that, for the case of piston without a TBC, increasing the specific output of the engine from 25 to 29 kW/l leads to the rise in the temperature of the CC edge by 10% and the thermal stress by 20%. The temperature level reaches 343 °C there, which is a critical value for this alloy. However, a comparison of the pistons without and with the TBC layer shows that the temperature of the CC edge under this layer decreases by 20 °C, while the thermal stress decreases more than 5 times. At that, in relation to the less loaded original design without a TBC layer, the temperature of the CC edge increases, and the thermal stress decreases. Such a change in these parameters indicates that the damage criterion d_{fs} requires further analysis.

Table 1. Parameters of thermal and mechanical loading of the piston CC edge in the operating modes of the engine load cycle.

	TBC is absent		TBC is present
Engine specific output, kW/l	25	29	29
Temperature in the most heavy loaded stationary operating mode (leg b), °C	314	343	322
Temperature in the least heavy loaded stationary operating mode (leg d), °C	191	193	188
Stress in the most heavy loaded stationary operating mode (leg b), MPa	− 36	−42,5	−7,6
Stress in the least heavy loaded stationary operating mode (leg d), MPa	−0,15	0	−2,9
Accumulation of damage	0.328	5.3	0.314

It can be seen in the table that $d_{fs}(\Pi) = 0.328$ for the test piston without TBC in the case of the engine specific output 25 kW/l. However, in the case of the specific output equal 29 kW/l, the specified resource is not guaranteed as the amount of damage exceeds 1, $d_{fs}(\Pi) = 5.3$. In the case of the specific output equal to 29 kW/l, for the design with the TBC, the value of the basic material damage is 0.312, i.e. not higher than for design without TBC with specific output equal to 25 kW/l. This is an evidence of the current importance of use of the pistons with a TBC on the CC surface as a means of reliability increase.

5   Conclusion and Future Scope

Using of technologies producing a reliable adhesion of a coating layer with low heat conductivity to a piston basic material is considered to be one of the directions for integrated improving the economic and environmental performance of ICE. During modernization of diesel engines and piston designs, the application of a TBC layer should also be considered as a means of resource increase for the basic material of the thermal stressed edges of the CC. When in attempt to achieve the last goal, conflicting problems emerge, specifically holding to the principle of guaranteeing the structural strength as early as at the initial stages of designing, maximal simplification of mathematical models, and the reduction of design routes.

The criterion for the design quality chosen in the paper is the total amount of fatigue and creep damage for the critical zone of the piston. It allows performing of the comparative analysis of designs as respects to their breaking point. The application of this criterion requires solving of non-stationary high-frequency problems of heat conduction and thermal elasticity of the surface layer of the piston crown in the conditions of cyclic change of the engine load modes in operation. The simplification of the problem is proposed, which does not contradict the principle of guaranteeing the strength of the piston crown at the point of designing.

The simplified method of simulation proposed in the paper takes into account the dynamic effect caused by TBC. It is based on the dividing the piston body into the set of areas of 1-D non-stationary heat conduction and the common area of 3-D stationary heat conduction. Finding a general solution is carried out by iterations.

A comparative analysis of the damages of the piston CC edge of the tractor diesel engine without and in the presence of a TBC layer on was performed. It has been found that the use of a TBC allows achieving the increase of an engine power by 16% without increasing of the accumulated damage level in the zone of the edge of the CC.

Further direction of the authors' research work is aimed to obtaining of refined non-stationary models of diesel engine load on the basis of statistics on the set of the engine transient processes in operation and the development of methods for formulating requirements for the properties of the basic piston material for the specified power of ICE. Performing of these works leads to the formation of a sufficient set of design routes for the effective carrying out of projects, while holding to the concepts of guaranteeing the specified resource and operating near the breaking point of material.

References

1. Reitz, R.D., Ogawa, H., Payri, R., et al.: IJER editorial: the future of the internal combustion engine. Int. J. Engine Res. **21**(1), 3 (2020). https://doi.org/10.1177/1468087419877990
2. Kostin, A.K., Larionov, V.V., Mikhailov, L.I.: Thermal Loading in Internal Combustion Engines. Mashinostroenie, Leningrad (1979). (in Russian)
3. Sinha, D., Sarkar, S., Mandal, S.C.: Thermo mechanical analysis of a piston with different thermal barrier coating configuration. Int. J. Eng. Trends Technol. **48**(6), 335–349 (2017). https://doi.org/10.14445/22315381/IJETT-V48P260
4. Bakthavathsalam, S., Gounder, R.I., Muniappan, K.: The influence of ceramic-coated piston crown, exhaust gas recirculation, compression ratio and engine load on the performance and emission behavior of kapok oil–diesel blend operated diesel engine in comparison with thermal analysis. Environ. Sci. Pollut. Res. **26**(24), 24772–24794 (2019). https://doi.org/10.1007/s11356-019-05678-x
5. Buyukkaya, E., Cerit, M.: Thermal analysis of a ceramic coating diesel engine piston using 3-D finite element method. Surf. Coat. Technol. **202**(2), 398–402 (2007). https://doi.org/10.1016/j.surfcoat.2007.06.006
6. Soltic, P., Bach, C.: Catalytic piston coating. Paper presented at the Gas Powered Vehicles Conference, Berlin, 30 September–1 October 2010 (2010)
7. Kumar, S.: Design and thermal analysis of MgZrO3 ceramic coated IC engine piston based on Finite Element Analysis (FEA). In: Kumar, A., et al. (eds.) Advanced Numerical Simulations in Mechanical Engineering, pp. 156–176. IGI Global, Hershey (2018). https://doi.org/10.4018/978-1-5225-3722-9.ch009
8. Ghazaly, N., Fouad, G., Abd-El-Tawwab, A., Abd El-Gwwad, K.A.: Evaluation of gasoline engine piston with various coating materials using Finite Element Method. Int. J. Mech. Mechatron. Eng. **13**(3), 189–193 (2019). https://doi.org/10.5281/zenodo.2643563
9. Shpakovsky, V., Shpakovsky, I., Beleske, A.: Method of producing corundum layer on metal parts. US Patent 0207884A1, 21 September 2006
10. Parsadanov, I.V., Sakhnenko, N.D., Ved', M.V., et al.: Increasing the efficiency of intra-cylinder catalysis in diesel engines. Issues Chem. Chem. Technol. **6**, 82–88 (2017)

11. Shpakovsky, V.V.: Results of studies of the wear of ring lands of pistons with a corundum surface layer of a diesel locomotive ChME-3. Intern. Combust. Engines **2**, 132–136 (2012). (in Russian)

12. Shpakovsky, V.V.: The introduction of pistons with a corundum layer in the repair of engines of diesel locomotives ChME-3. Intern. Combust. Engines **2**, 112–115 (2013). (in Russian)

13. Parsadanov, I.V., Polivyanchuk, A.P.: Evaluation of the effect of galvanoplasma coating of the piston on the emissions of solid particles with diesel exhaust gases. Intern. Combust. Engines **2**, 97–100 (2009). (in Russian)

14. Parsadanov, I.V., Sakhnenko, N.D., Khyzhniak, V.A., Karakurkchi, A.V.: Improving the environmental friendliness of diesels by in-cylinder neutralization of pollutants. Intern. Combust. Engines **2**, 63–67 (2016). [in Ukrainian]

15. MS Motor Service International GmbH: Piston damage – recognising and rectifying (2015). https://www.ms-motorservice.com/en/news/latest-news/article/news/piston-damage-recognising-and-rectifying/. Accessed 17 July 2020

16. DFC Diesel: Failure analysis. https://www.dfcdiesel.com/warranty-info/failure-analysis. Accessed 17 July 2020

17. Shpakovsky, V.V., Marchenko, A.P., Pylyov, V.V.: Results of mathematical modeling of the temperature field of the combustion chamber of a piston with a ceramic surface layer. Aerosp. Technic Technol. **3**, 63–67 (2009). (in Russian)

18. Pylyov, V.A., Belogub, A.V.: Features of thermomechanical loading and taking into account the durability of a thin-walled piston of a gasoline engine. Intern. Combust. Engines **2**, 74–81 (2010). (in Russian)

19. Pylyov, V.O.: Automated Designing of Pistons of High-Speed Diesels with a Given Level of Long-Term Strength. KhPI-Press, Kharkiv (2001). (in Ukrainian)

20. Yoon, J.W., Yang, D.Y., Chung, K.: Elasto-plastic finite element method based on incremental deformation theory and continuum based shell elements for planar anisotropic sheet materials. Comput. Methods Appl. Mech. Eng. **174**(1–2), 23–56 (1999). https://doi.org/10.1016/S0045-7825(98)00275-8

21. Raenko, M.I., Chainov, N.D.: Application of a finite element model of a nonlinear continuous medium for the analysis of the stress-strain state of structural elements. Proc. High. Educ. Inst. Mach. Build. **5**, 28–35 (2018). https://doi.org/10.18698/0536-1044-2018-5-28-35. (in Russian)

22. Mordvintseva, I.A., Klimenko, A.N., Aryan, R., et al.: Specifics of setting the boundary conditions for the non-stationary problem of heat conduction of a diesel piston. Intern. Combust. Engines **1**, 33–41 (2017). https://doi.org/10.20998/0419-8719.2017.1.07. (in Russian)

23. Pylyov, V.V., Prokopenko, M.V., Shekhovtsov, A.F.: Rerurs. UA Computer Software 5915, 16 July 2002

Numerical and Experimental Research of Radial-Axial Pump-Turbine Models with Spliters in Turbine Mode

Andrii Rusanov⦿, Oleg Khorev⦿, Yevgen Agibalov⦿,
Yurii Bykov⦿, and Pavlo Korotaiev⁽⊠⁾⦿

A. Pidgorny Institute of Mechanical Engineering Problems of the National
Academy of Sciences of Ukraine, 2/10 Pozharskogo Street,
Kharkiv 61046, Ukraine
{rusanov,agibalov,bykow}@ipmach.kharkov.ua,
oleg_xo@ukr.net, korotaiev@gmail.com

Abstract. The ways of efficiency increase of the Francis type pump-turbines and widening of their operation range are considered. It is shown that one of the perspective ways is the usage of runners with additional shortened blades – splitters. Experimental studies of influence of the splitter geometric parameters on the energy performance of medium specific speed pump-turbines in turbine mode were conducted at the hydrodynamic test stand of A. Pidgorny Institute of Mechanical Engineering Problems NAS of Ukraine. The splitter length influence on the value of maximum efficiency, characteristics and parameters of the optimum mode is established. The simulation of the incompressible fluid viscous flow in the flow part is performed using authentic software package *IPMFlow* on the basis of numerical integration of Reynolds-averaged Navier-Stokes equations with an additional term that contains artificial compressibility. A two-parameter differential Menter's *SST* turbulence model is applied to account the turbulent effects. The numerical integration of equations is performed using the implicit quasimonotone Godunov scheme with second order accuracy in space and time. Pressure distribution on the runner blades, comparison of the results of numerical and experimental studies are presented. The developed software package IPMFlow for the study of viscous compressible and incompressible fluid and gas flows, as well as the obtained results of the influence of geometrical parameters of the pump-turbine runner with splitters on energy characteristics can be used in the design and research of power machines for aviation and space technology.

Keywords: Pump-turbine · Runner · Splitter · Hydrodynamic test stand · Experimental studies · Numerical simulation

M. Nechyporuk et al. (Eds.): ICTM 2020, LNNS 188, pp. 427–439, 2021.
https://doi.org/10.1007/978-3-030-66717-7_36

1 Introduction

The operation of Pumped Storage Power Station (PSPS), which are used as a source of peak power in power systems, usually occurs for a limited period of time in the modes of maximum and rated generator capacity, with a possible reduction of capacity up to 70% of the rated.

In recent years, there's been a need to operate the PSPS pump-turbines in the generation mode with a wider capacity range with the possibility of reduction to 25–30% of the rated.

Analysis of worldwide experience shows that such a significant reduction in the capacity of existing pump-turbines leads to an increase in the level of unsteadiness in flow parts associated with the appearance of a vortex rope at the runner outlet and separation at the blade inlet. This increases vibrations of the turbine cover and generator and increases the hydraulic unit shaft beating, resulting in reduced operational reliability and equipment life.

Nowadays the following technical solutions have been applied in the worldwide practice to widen the pump-turbine operation range while increasing energy performance [1, 2]:

1. Development of new pump-turbines flow parts, whose geometrical parameters allow to widen the capacity regulation range due to the use of runner with increased number of blades and their special spatial profiling [3, 4].
2. Application of asynchronous motor-generator with adjustable rotation frequency within −30%/+30% of synchronous, for pump-turbine hydraulic unit [5, 6]. At this most experienced country is Japan, where variation of the pump-turbine runner rotation frequency is applied by more than 10 PSPSs. In Europe, this method is used at the PSPS Goldisthal (Germany) [1, 7], where 4 Francis pump-turbines with a maximum turbine capacity of $N = 265$ MW, two of which have asynchronous generators, are installed.
3. Development of hydromachines with runner, which have intermediate shortened blades – splitters.

For the first time traditional runner was replaced by the splitter runner at the Azumi PSPS (Japan) [8], with number of blades $z = 8$ (4 long blades and 4 short blades). The modernization was preceded by a considerable amount of computational work with the usage of numerical simulation for the comprehensive flow analysis and for obtaining a highly efficient flow part. In 2007, Voith developed a project for the Jin Ping II PSPS (China), which used a 610 MW Francis hydroturbine with runner diameter of 6,5 m and splitters [9]. TEPCO and Toshiba have installed 2 units of pump-turbines with splitters with capacity of 470 MW in generation mode at the Kannagawa PSPS. The splitter installation allowed to increase the unit capacity by 20 MW. The results of numerical research of the splitter length effect on the Francis hydroturbine flow structure and energy characteristics for ultra-high pressures are considered in paper [10]. The results of an experimental study of effect of the splitter installation in the Francis turbine 15-blade runner on energy and pulsation parameters at heads from 164 to 243 m are considered in [11].

However, the problems of the splitter length effect on the energy characteristics of pump-turbines with medium specific speed in a wide range of operating modes have not been fully explored and need further research. The paper presents the results of the first experimental and numerical steps of the runner pump-turbine with splitters studies at heads up to 200 m in turbine mode. Sample Heading (Third Level). Only two levels of headings should be numbered. Lower level headings remain unnumbered; they are formatted as run-in headings.

2   Description and Characteristics of Hydrodynamic Test Stand of A. Pidgorny Institute of Mechanical Engineering Problems National Academy of Science of Ukraine

Experimental studies of pump-turbine models with splitters were performed at the hydrodynamic test stands of A. Pidgorny Institute of Mechanical Engineering Problems National Academy of Science of Ukraine. These stands are intended for carrying out research works on the working process studying in various hydromachines (hydroturbines, pumps, pump-turbines, micro HPP), conducting research and obtaining experimental characteristics [12]. In terms of their parameters and equipment, they represent unique facility that have no analogues in Ukraine and received the status of "national heritage" in 2011.

Since 1975, more than 50 variants of pump-turbine runner models, as well as Francis, Kaplan and diagonal types of hydromachines have been developed and investigated in the laboratory.

The laboratory of hydromachines includes two closed hydrodynamic stands ECS-15 and ECS-30. They represent universal units that provide complex experimental studies for development of the highly efficient flow parts of hydromachines [13]. The hydraulic system of the two stands is made joint, it is located on three levels of the stand section. This allows to use both circulation pumps for each of the stands. Those pumps can be connected both sequentially or in parallel in each of the stands to provide the required test parameters. All power, hydromechanical and electrical equipment placed in the basement.

Table 1. Comparison of hydraulic stand model hydraulic machines measurement error with requirements of IEC 60193

Parameter	Measurement error	
	Requirements IEC 60193, [%]	Achieved value, [%]
Rotation frequency	±0,075	±0,03
Pressure	±0,100	±0,10
Torque	±0,100	±0,10
Flow rate	±0,200	±0,20
Model efficiency	±0,250	±0,25

Due to expansion of the requirements for carrying out acceptance tests of hydro-machine models, which are regulated by the International Standard IEC 60193 [14], in 2006–20013, jointly with KharkivTurboEngineering, reconstruction and modernization of main and auxiliary equipment of the stands was carried out. That included also reconstruction and modernization of power equipment, test modes stability control and maintenance systems, primary sensors that convert physical quantities (pressure, flow rate, hydromachine shaft torque, rotation frequency, etc.) into electrical signals for computer-aided measurement system [15, 16].

Table 1 shows the compliance of the stand measurement parameters with the requirements of IEC 60193.

Main parameters of the experimental stands research equipment of the A. Pidgorny Institute of Mechanical Engineering Problems NAS of Ukraine are given in Table 2.

Table 2. Parameters of IMEP NAS of Ukraine hydrodynamic stands

Stand name	Runner model diameter, [mm]	Pressure, [m]	Flow rate, [m³/s]	Power of DC circulation pumps drive motors, [kW]	Power of balancing motor-generator, [kW]
ECS-30	350–400	≤ 25 (30)	≤ 0,3 (≤ 0,5)	≤ 160	200
ECS-15	350–380	≤ 12 (15)	≤ 0,56 (≤ 0,7)	≤ 160	≤ 200

3 Model Unit for Study of the Francis Pump-Turbines Models with Splitters

The spiral case of studied pump-turbine flow part model was calculated according to the law Vu = const and has an angle of coverage of 360° and round sections. The stator has 20 columns, the stator rings are flat. The guide vane has 20 moving blades. The blade profile is asymmetric, positive curvature, the blade height is $b_0 = 0,14D_1$. The runner has a diameter $D_1 = 350$ mm. The blade height at the leading edge is the same as in the guide vanes. Throat diameter on suction side $D_0 = 0,685D_1$. The draft tube has a height $h = 3,14D_1$ and length $L = 4,5D_1$. The vertical section represents a conical diffuser, standard elbow, KU-3RO type. Exit section represents a straight-axis pipe.

4 Selection and Production of Runners for Research

The researched pump-turbine runners consist of blade set that have different lengths and, accordingly, different coverage angles. As an origin was taken the model of the pump-turbine runner with a diameter of 350 mm, developed for the conditions of the Dnister PSPS that has 7 bronze blades. The blades of other runners were made using 3D printing technology.

The second variant of the runner has 8 blades identical to the blades of the original version. In this runner, the length of splitters is conventionally assumed to be 100% of the main blade length. The third, fourth and fifth runner variants had 8 blades – 4 blades same as the original runner and 4 shortened. In the third variant, the shortened blades had a length equal to 80% relative to the length of the original version, in the fourth 65%, in the fifth 50%. The blades shortening was carried out by cutting its trailing edges. After the blade shortening, its outlet part thinning was held according to the same law for the thickness distribution as in the original blade.

All runner blades with splitters were manufactured at A. Pidgorny Institute of Mechanical Engineering Problems National Academy of Science of Ukraine on a PLA plastic [17] 3D printer.

Blades were printed layer by layer with a thickness of 0,1 mm, which ensured high quality of the blade surface (Fig. 1).

Figure 2 shows photos of Francis pump-turbine runner models with splitters, which have been tested on a hydrodynamic stand at IMEP NAS of Ukraine.

Fig. 1. The process of the blades printing on 3D printer.

Fig. 2. Runner model with splitter.

5   Results of Experimental Studies

Studies in turbine mode were carried out at a head of 6 m and 9 values of the guide vane opening in the range of the reduced rotation frequency $n_I^{'}$ from 50 min^{-1} to 110 min^{-1}. Shell diagram were built for all flowing parts variants. Table 3 shows the parameters of the optimum modes and the values of the relative efficiency, which were obtained from the tests of five runner variants in turbine mode. The relative efficiency in this case is equal to the ratio of the current efficiency value to the maximum efficiency value in the flow part with the original runner (version 1).

As can be seen from the Table 3, the position of shell diagram optimum in turbine mode shifts towards increasing of the mass flow by about 25% - from 0.315 m^3/s to 0.402 m^3/s, with a reduction of the splitter blades length from 100% to 50% (runner variants 2 and 5). It can be explained by the increase of the flow angle at the runner outlet. At the same time, the change of the reduced rotation frequency is negligible. The relative efficiency decreases by almost 4% as the splitter length decreases. The eight-blade runner

Table 3. Maximum efficiency and optimal turbine modes parameters in the studied runner

Runner No.	Runner parameters	Q_1', [m³/s]	n_I', [min⁻¹]	η_{max}^*, [%]
1	$z = 7$, 100% L	0,365	80,0	100,00
2	$z = 8$, 100% L	0,315	78,0	100,35
3	$z = 8$, 80% L	0,352	78,3	97,78
4	$z = 8$, 65% L	0,397	80,5	97,66
5	$z = 8$, 50% L	0,402	79,0	96,37

has a slightly higher efficiency – 0.35% – relative to the original seven-blade runner. At the same time, the reduced flow rate at the optimum decreased by 0.050 m³/s, and the reduced rotation frequency by 2 min⁻¹.

Figure 3 and 4 presents shell diagram sections $n_I' =$ const for the original variant and four other variants of eight-blade runners with splitters at different values of the reduced rotation frequency $n_I' = 75$; 80; 85; 90 min⁻¹.

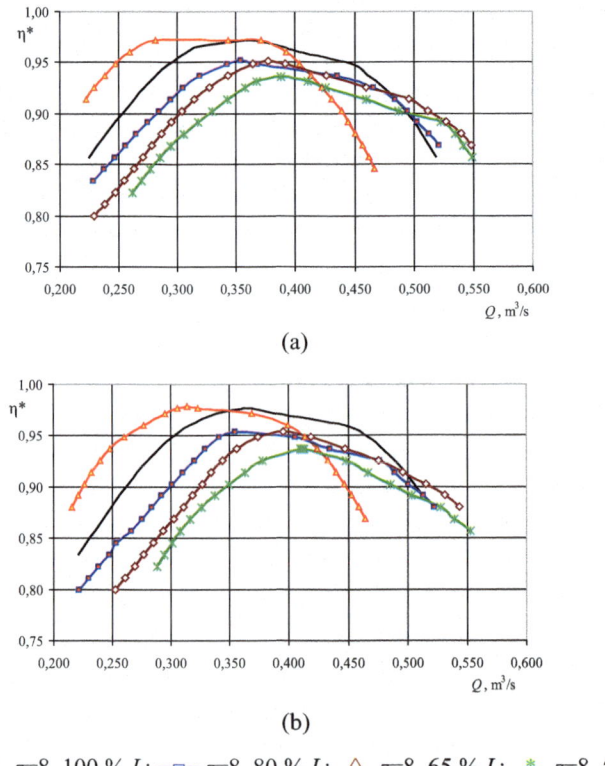

(a)

(b)

— z=7; –△– z=8, 100 % L; –□– z=8, 80 % L; –◇– z=8, 65 % L; –*– z=8, 50 % L

Fig. 3. Shell diagram sections $n_I' = 75$ min⁻¹ (a) and $n_I' = 80$ min⁻¹ (b) for original variant and four variants of the runners with splitters.

The increase of runner blades number from 7 to 8 at 100% of the splitter blades length (runner version 2) significantly reduces the flow rate of the flow part – from 0.365 m^3/s to 0.315 m^3/s in the optimum ($n'_1 = 80$ min^{-1}). With increase of the rotation frequency from 80 min^{-1} to 90 min^{-1}, the presence of an additional eighth blade is even more marked – the difference in the reduced flow reaches a value of 0.090 m^3/s. The efficiency of runner versions 1 and 2 is almost equal at the values of the reduced rotation frequency $n'_1 = 87.5\%$. Further increase of the rotation frequency leads to a noticeable decrease in the efficiency of flow part with an eight-blade runner relative to the original one.

The trimming of the four blades to 80% of the original length (version 3) is accompanied by an increase in flow rate up to 0.350 m^3/s, while the efficiency is reduced by about 2.5% compared to version 2. Further trimming of the 4 blades up to 65% of length (version 4) increases the flow rate at the optimum up to 0.375 m^3/s. The efficiency level in the range of reduced rotation frequency from 77.5 min^{-1} to 85.0 min^{-1} is the same as for the runner with splitters of 80% of the length, and with a further increase in the rotation frequency there is even a slight increase in efficiency by about 0.7%.

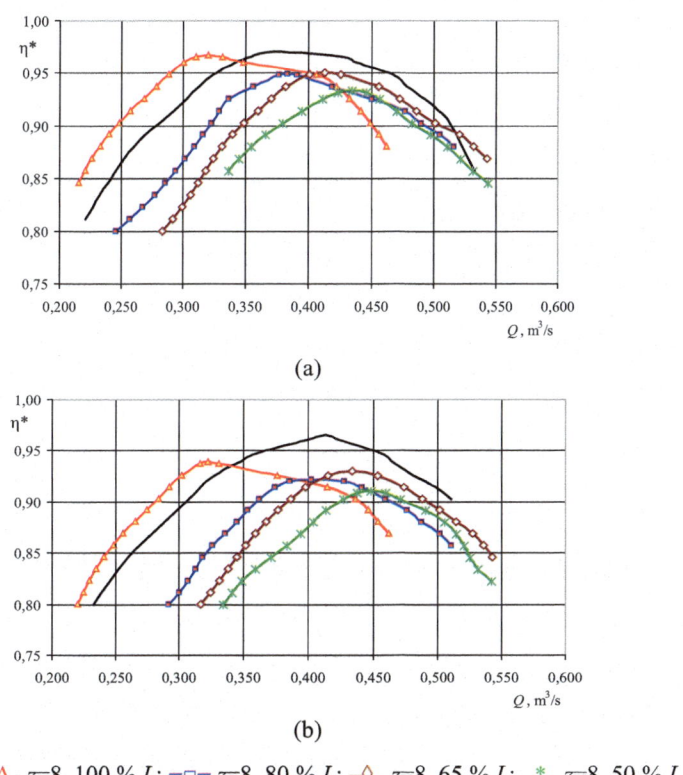

(a)

(b)

—— $z=7$; –△– $z=8$, 100 % L; –□– $z=8$, 80 % L; –◇– $z=8$, 65 % L; –*– $z=8$, 50 % L

Fig. 4. Shell diagram sections $n'_1 = 85$ min^{-1} (a) and (b) $n'_1 = 90$ min^{-1} for original variant and four variants of the runners with splitters.

The use of runner with 50% trimming of four blades (version 5) relative to runner with 65% trimming led to a slight increase in flow rate, which is more pronounced with increasing of the rotation frequency to $n'_I = 90$ min^{-1}, and also to the efficiency reduce to the value that is approximately 96.37% relative to the original.

The blades number increasing from 7 in the original runner to 8 in turbine mode led to a slight increase in the maximum efficiency at sections of the rotation frequency $n'_I = const$ in the range from 75 min^{-1} to 80 min^{-1} compared to the original version, but could not lead to the flow rate increase. With a flow rate less than 0.350 m^3/s, the flow part with such runner has a higher efficiency, and with the flow rate increase – lower.

6   Numerical Studies

The viscous incompressible fluid flow simulation in the flow part of the runners with splitters was performed by using the IPMFlow software. The software is based on the numerical integration of the Reynolds-averaged Navier-Stokes equations with an additional term that contains artificial compressibility. The Reynolds equations with the two-parameter Menter's SST turbulence model are used.

The equations numerical integration is carried out using the implicit quasimonotone Godunov scheme of second-order accuracy in space and time.

More detailed description of the mathematical model and numerical method is presented in papers [18, 19].

The calculation area contained one channel of guide vane and one channel of runner. Computational grid with six-side cells is structured. Number of cells per channel: in guide vane about 330 thousand, in runner – 520 thousand. The calculations are performed for the modes, which were taken from the experimental research protocols on the hydrodynamic stand and correspond to the optimal values of the guide vanes opening for each of the runner variants.

Initial data of calculation area are the next: at the guide vane inlet – velocity vector corresponding to the required mass flow rate:

- on the walls – the no-slip condition, i.e. the velocity is zero;
- at the runner outlet – static pressure of 100 000 Pa.

Figure 5, 6, 7 and 8 shows the pressure distribution on the pressure surfaces and the suction of the main runner blades and splitters of all studied flow part versions under the modes of maximum efficiency at the guide vane optimum opening.

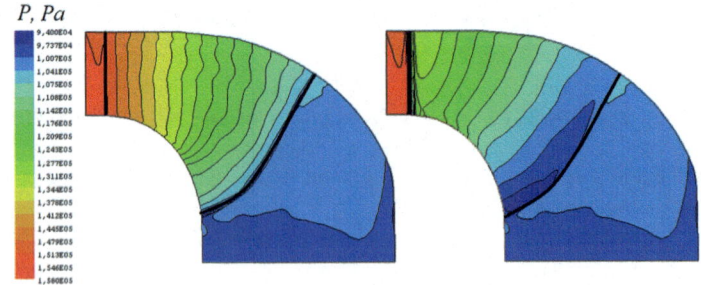

Fig. 5. Pressure distribution on runner blades pressure side and suction side $z = 8$, 100% L.

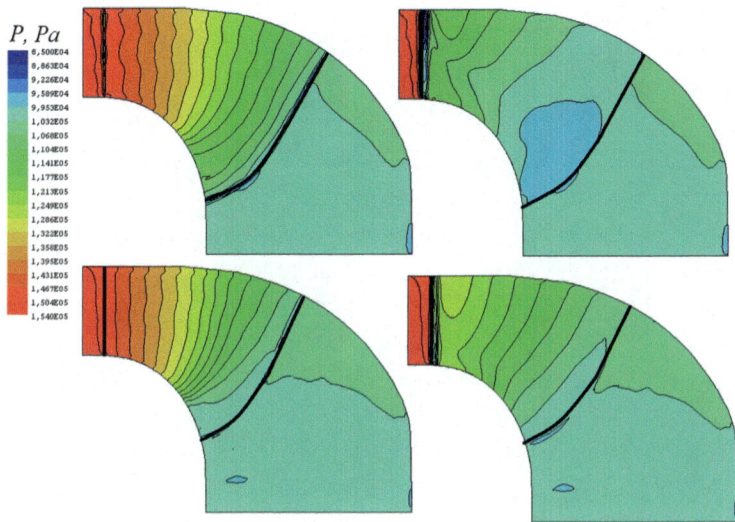

Fig. 6. Pressure distribution on runner blades pressure side and suction side $z = 8$, 80% L.

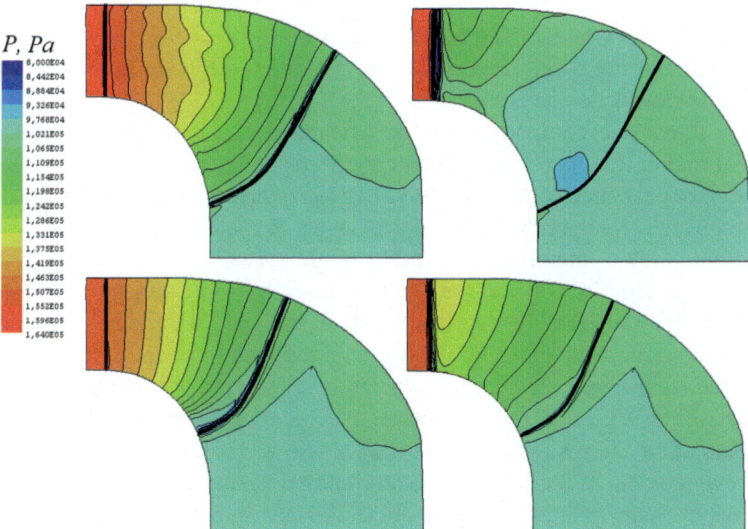

Fig. 7. Pressure distribution on runner blades pressure side and suction side $z = 8$, 65% L.

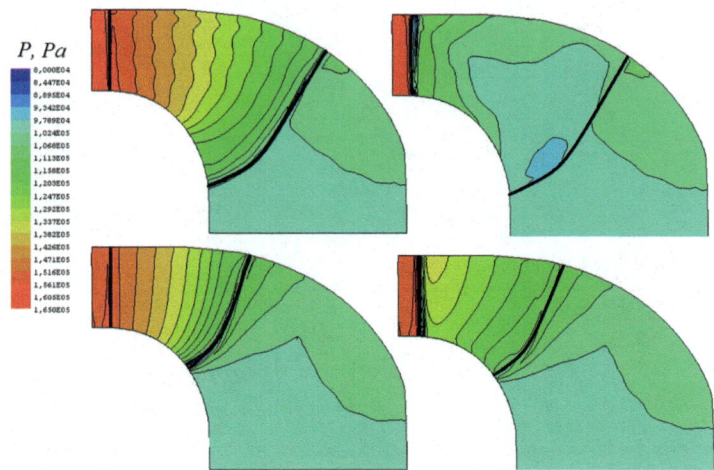

Fig. 8. Pressure distribution on runner blades pressure side and suction side $z = 8$, 50% L.

As can be seen from the figure, for the eight-blade runner on the pressure side (left) the pressure distribution along the blade from inlet to outlet is uniform on both the main blades and the splitters of all lengths. On the splitter blades suction side (right) the pressure distribution from the inlet to the outlet is uniform, and on the main ones – less advantageous. In runner with 8 identical blades (splitter 100%) the pressure distribution is uniform on both blade surfaces.

Figure 9 shows the comparison of the calculated and experimental energy characteristics of runner flow part with splitters at optimal values of the guide vane opening in the form of dependence of $\Delta\eta^{**}$ on the rotation frequency. The $\Delta\eta^{**}$ is stated as the difference between the maximum calculated value of the efficiency of each flow part and its current value – numerical or experimental.

The values of the *CFD* curves are obtained directly from studies of the calculation area with usage of the *IPMFlow* software, which included the guide vane and runner channels. To determine the efficiency values that take into account other losses in flow part (*CFD full losses* curves), hydraulic losses in spiral case with stator and draft tube [19, 20], as well as disc and volume losses [21] have been calculated.

The analysis of the results presented above showed a satisfactory qualitative and quantitative agreement of the obtained data of numerical and experimental studies of all runner versions with splitters. The smallest difference between the calculated and experimental curves of efficiency in all versions of flow part is observed at modes with maximum efficiency. As the rotation frequency increases, this difference increases slightly, as it decreases – it remains almost constant.

Thus, the calculated method of the research of the splitter length effect on the pump-turbine energy performance in turbine mode using *IPMFlow* software allows to obtain reliable results. This makes it possible to apply it reasonably in further works in which it is planned to research the runner blades with splitters number effect and their location in runner channels.

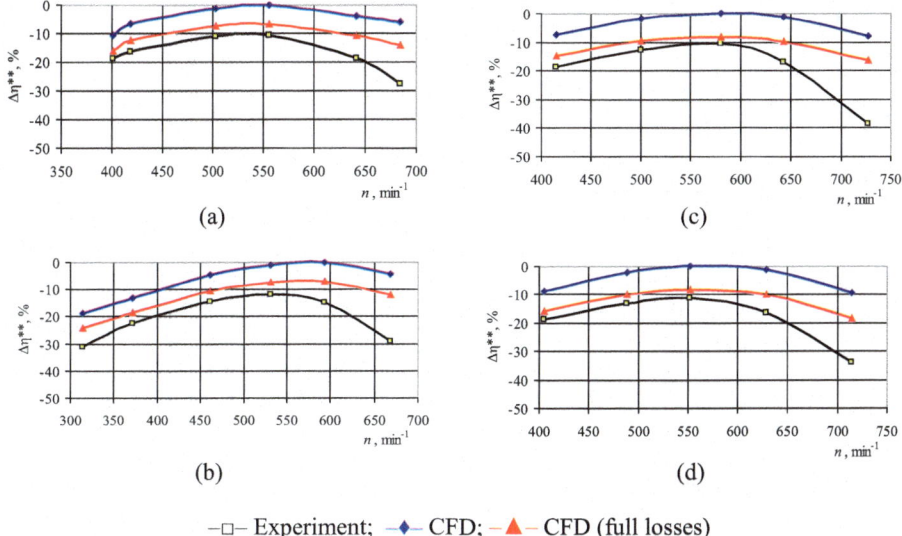

−□− Experiment;   −◆− CFD;   −▲− CFD (full losses)

Fig. 9. Comparison of the calculated and experimental characteristics of the runner flow part with splitters in turbine mode: (a) $z = 8$, 100% L, $a_0 = 18$ mm; (b) $z = 8$, 80% L, $a_0 = 18$ mm; (c) $z = 8$, 65% L, $a_0 = 22$ mm; (d) $z = 8$, 50% L, $a_0 = 22$ mm.

7   Conclusions

Experimental studies of runner flow part models with a splitter in turbine mode showed that the blade shortening in the eight-blade version of the runner leads to a decrease in the level of the flow part maximum efficiency and a shift of its position on the shell diagram to the greater flow rate by 0.085 m³/s at almost unchanged rotation frequency.

In turbine mode, the flow part with an eight-blade runner at the reduced flow rate $Q_I' < 0{,}350$ m³/s at all sections $n_I' = $ const of the shell diagram has a higher efficiency than seven-blade runner, and at a flow rate $Q_I' > 0{,}350$ m³/s – lower.

Comparison of hydrodynamic experimental research of models on hydrodynamic test stand with the data of numerical research in turbine mode shows their satisfactory qualitative and quantitative agreement. It allows the *IPMFlow* software package reasonable usage for further study of effect of splitter geometrical parameters on the operation characteristics of the Francis pump-turbines.

The developed software package *IPMFlow* for the study of viscous compressible and incompressible fluid and gas flows, as well as the obtained results of the influence of geometrical parameters of the pump-turbine runner with splitters on energy characteristics can be used in the design and research of power machines for aviation and space technology.

References

1. Vennemann, P., Thiel, L., Funke, H.C.: Pumped storage plants in the future power supply system. VGB PowerTech **1–2**, 44–49 (2010)
2. Ruppert, L., Schürhuber, R., List, B., et al.: An analysis of different pumped storage schemes from a technological and economic perspective. Energy **141**, 368–379 (2017). https://doi.org/10.1016/j.energy.2017.09.057
3. Nowicki, P., Sallaberger, M., Bachmann, P.: Modern design of pump-turbines. In: 2009 IEEE Electrical Power & Energy Conference (EPEC), Montreal, pp. 1–7. IEEE (2009). https://doi.org/10.1109/EPEC.2009.5420368
4. Brekke, H.: Design, performance and maintenance of Francis turbines. Glob. J. Res. Eng. **13**(5), 28–40 (2013)
5. Kuwabara, T., Shibuya, A., Furuta, H., et al.: Design and dynamic response characteristics of 400 MW adjustable speed pumped storage unit for Ohkawachi power station. IEEE Trans. Energy Convers. **11**(2), 376–384 (1996). https://doi.org/10.1109/60.507649
6. Artyukh, S.F., Galat, V.V., Kuz'min, V.V., et al.: Improving the energy efficiency of Pumped-storage power plants. Power Technol. Eng. **48**(5), 396–399 (2015). https://doi.org/10.1007/s10749-015-0542-1
7. Bessa, R., Moreira, C., Silva, B., et al.: Role of pump hydro in electric power systems. J. Phys. Conf. Ser. **813**, 012002 (2017). https://doi.org/10.1088/1742-6596/813/1/012002
8. Nishiwaki, Y.: A proposal to realize sustainable development of large hydropower project. Soc. Soc. Manag. Syst. Internet J. **5**(1), 1–10 (2009)
9. Fischer-Aupperle, B.: Current opportunities for hydro power in China. HyPower **16**, 24–27 (2007)
10. Hou, Y., Li, R., Zhang, J.: Research on the length ratio of splitter blades for ultra-high head Francis runners. Procedia Eng. **31**, 92–96 (2012). https://doi.org/10.1016/j.proeng.2012.01.996
11. Jia, Y., Wei, X., Wang, Q., et al.: Experimental study of the effect of splitter blades on the performance characteristics of Francis turbines. Energies **12**(9), 1676 (2019). https://doi.org/10.3390/en12091676
12. Veremeenko, I.S., Rusanov, A.V., Dedkov, V.N., et al.: Development of experimental base of hydroturbine construction in IPMash of NAS of Ukraine. Bull. Natl. Tech. Univ. "KhPI". Ser. Energy Thermal Eng. Process. Equip. **1**, 12–21 (2014). (in Russian)
13. Veremeenko, I.S., Dedkov, V.N., Agibalov, E.S.: Improvement of hydrodynamic stands of the laboratory of hydromachines of IPMash of NAS of Ukraine. J. Mech. Eng. **1**, 24–31 (2006). (in Russian)
14. IEC 60193:2019. Hydraulic turbines, storage pumps and pump-turbines – Model acceptance tests. International Electrotechnical Commission, Geneva (2019)
15. Veremeenko, I.S., Gladyishev, S.V., Dedkov, V.N.: Modernization of energy cavitation stands of the laboratory of hydromachines of IPMash of NAS of Ukraine. J. Mech. Eng. **13**, 3–12 (2010). (in Russian)
16. Veremeenko, I.S., Gladyishev, S.V., Dedkov, V.N.: Installation of UG-1 for calibration of flowmeters of energy cavitation stands. Metrol. Devices **2**, 42–47 (2010). (in Russian)
17. Zalohin, M.Y., Skliarov, V.V., Dovzhenko, J.S., Brega, D.A.: Experimental determination and comparative analysis of the PPH030GP, ABS and PLA polymer strength characteristics at different strain rates. Sci. Tech. **18**(3), 233–239 (2019). https://doi.org/10.21122/2227-1031-2019-18-3-233-239

18. Rusanov, A., Rusanov, R., Lampart, P.: Designing and updating the flow part of axial and radial-axial turbines through mathematical modeling. Open Eng. **5**, 399–410 (2015). https://doi.org/10.1515/eng-2015-0047

19. Rusanov, A., Shubenko, A., Senetskyi, O., et al.: Heating modes and design optimization of cogeneration steam turbines of powerful units of combined heat and power plant. Energetika **65**(1), 39–50 (2019). https://doi.org/10.6001/energetika.v65i1.3974

20. Anup, K.C., Thapa, B., Lee, Y.H.: Transient numerical analysis of rotor–stator interaction in a Francis turbine. Renew. Energy **65**, 227–235 (2014). https://doi.org/10.1016/j.renene.2013.09.013

21. Qian, J., Zeng, Y., Guo, Y., Zhang, L.: Reconstruction of the complete characteristics of the hydro turbine based on inner energy loss. Nonlinear Dyn. **86**(2), 963–974 (2016). https://doi.org/10.1007/s11071-016-2937-4

Increasing Accuracy of the Gas Temperatures Pattern Calculation for GTE Combustor Using CFD

Serhii Yevsieiev$^{(\boxtimes)}$, Dmytro Kozel , and Igor Kravchenko

SE "Ivchenko-Progress", 2 Ivanova Street, Zaporizhzhia 69068, Ukraine
EvseevSA@ivchenko-progress.com

Abstract. It was found that the accuracy of the radial and circumferential non-uniformity modeling for the gas temperature pattern at the exit of the combustion chamber is unsatisfactory when using the k–ε turbulence model with the initial settings for the Ansys Fluent program. To reduce the non-uniformity of the temperature pattern at the exit of the combustion chamber, the degree of turbulent diffusion of gas components was increased with respect to the initial version of calculation, performed using the k–ε model of turbulence with the initial settings, by reducing the turbulent Schmidt number Sc. A numerical experiment was performed for the values of the Schmidt number Sc = 0.85 (default), Sc = 0.6, Sc = 0.4, and Sc = 0.2. The results of a numerical experiment confirmed the reductions of radial and circumferential non-uniformities with decreasing Sc, but theirs levels are different. Therefore, to ensure high accuracy in calculating both the circumferential and radial non-uniformities of the gas temperature pattern, it was proposed to use a variable value of Sc, depending on the gas temperature. The functional dependence of the turbulent Schmidt number Sc on the gas temperature was implemented in the Ansys Fluent program using the user function (UDF). The results of the gas temperature pattern modeling using the proposed UDF function for the turbulent Schmidt number Sc are in satisfactory agreement with the experimental data for both radial and circumferential non-uniformities of the gas temperature pattern at the exit of the combustion chamber.

Keywords: Combustion chamber · Turbulent Schmidt number · Gas temperature pattern · Computer simulation · UDF · ANSYS Fluent

1 Introduction

The gas temperature non-uniformity at the exit of the combustor has a significant impact on the reliability and resource of turbine parts. Therefore, the task of the radial and circumferential non-uniformity modeling for the gas temperature pattern at the exit of the combustion chamber is important and relevant. Given the variety of constructions of the combustor, the many physical processes that occur in the combustor and the complexity of the mathematical modeling of these processes, it is difficult to calculate the gas temperature pattern at the exit from the combustor by analytical and empirical methods. Analytical calculation methods, as a rule, are based on a

M. Nechyporuk et al. (Eds.): ICTM 2020, LNNS 188, pp. 440–450, 2021.
https://doi.org/10.1007/978-3-030-66717-7_37

generalization of experimental data using empirical and theoretical mathematical models, and empirical models are built solely on the basis of studying the reactions of an object to changing external conditions. In this case, the theory of the object is not considered. Therefore, the task of calculating the temperature pattern non-uniformity can be solved by the methods of computational aerohydrodynamics (CFD). The CFD uses the physical model of a viscous gas and its mathematical model as a system of Reynolds-averaged Novier-Stokes equations (RANS), including the equation of conservation of mass (equation of continuity), the equation of conservation of momentum and energy. To close the system of Reynolds averaged equations, various turbulence models and their modifications were used. Discretization of partial differential equations is carried out using finite difference methods, finite (control) volumes, or finite elements.

However, the results of the calculation using the CFD methods depend on the selected parameters of the applied turbulence model. This paper presents the results of a study of the influence of the turbulent Schmidt number Sc applied by the k-e turbulence model on the accuracy of the gas temperature pattern at the combustor exit calculation. To ensure high accuracy of calculation, it is proposed to use the functional dependence of the turbulent Schmidt number Sc on the gas temperature instead of the constant value Sc = 0.85, which is proposed in the Ansys Fluent program by default.

The object of the study is the gas flow with the combustion of atomized fuel in the combustor (sector 1/18) of the gas turbine engine (see Fig. 1).

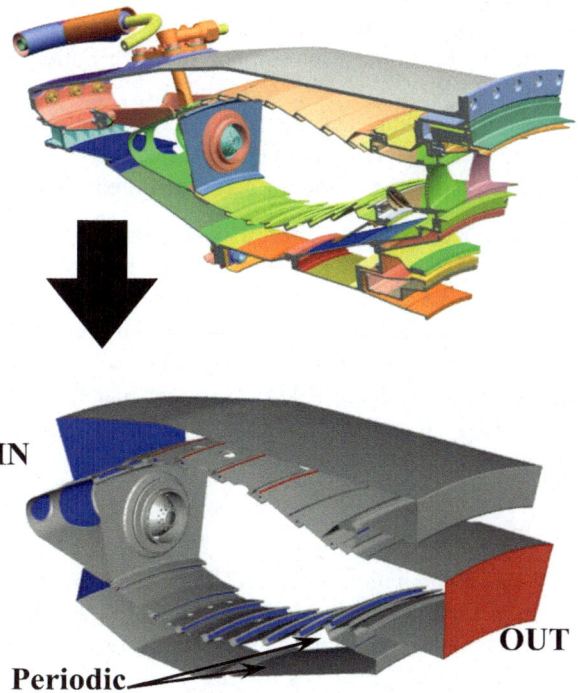

Fig. 1. The geometry of the computational domain.

2 Numerical Method and Boundary Conditions

The temperature of the elementary volume of gas at the exit of the combustion chamber depends on the entire history of its movement, starting from the exit of the compressor. The temperature and composition of the gas change under the influence of combustion, heat transfer and turbulent mixing. These processes depend on the distribution of air flows inside the combustion chamber. The geometry of the flame tube and the design of the front device are responsible for the formation of jets that determine the structure of the flame, the displacement pattern as a whole and, as a result, the distribution of the gas temperature pattern at the exit section of the combustion chamber [1].

2.1 Physical and Mathematical Statement of the Problem

To simulate the gas flow with kerosene burning in the studied area, we used the ANSYS Fluent 2020 R1 software package, which implements a numerical solution of the Reynolds averaged Navier-Stokes equations (RANS), including the mass conservation equation (continuity equation), momentum conservation equations, and energy. To close the system of Reynolds averaged equations, we used the Realizable k-e model turbulence of the Lunder-Spalding. Discretization of partial differential equations is carried out using finite (control) volume methods.

The boundary conditions on the solid wall for finding shear stresses, kinetic energy of turbulence (TKE) and dissipation rate of TKE were determined using the wall function (Standard Wall Functions).

The combustion model of the unmixed mixture was used (the Schwab Zeldovich function/probability density function of the probability distribution (PDF (PDF))/pre-calculated PDF tables).

When modeling the movement, heat transfer and evaporation of droplets of atomized fuel, the Discrete Phase Model was used.

In relation to the gas phase, the following basic assumptions were made:

- the gas phase is a multicomponent chemically reacting mixture of thermodynamically perfect gases that are part of the fuel, oxidizer and combustion products;
- the gas flow is three-dimensional, turbulent, incompressible, substantially subsonic;
- bulk viscosity, viscous heating and radiant heat transfer are neglected;
- the volume occupied by the droplets and the influence of the droplets on the turbulence characteristics are neglected. Please note that the first paragraph of a section or subsection is not indented. The first paragraphs that follows a table, figure, equation etc. does not have an indent, either.

2.2 Building a Computational Grid

The construction of the computational grid refers to the key points of numerical modeling by methods of computational aerohydrodynamics. A rational choice of the grid can achieve a reasonable compromise between the accuracy of the numerical model, computational costs and labor costs for constructing the grid.

Theoretically, the highest accuracy of calculations can be obtained on hexahedral grids of a special type (the so-called "cartographic"), the shape and location of cells in which are consistent with the direction of the flow and the gradients of independent variables. In practice, this condition can only be partially satisfied, since the structure of the flow is not known beforehand. With the complex form of the computational domain or its subdomains, the construction of cartographic grids using modern grid generators is often impossible. Therefore, the only way to generate a cartographic grid is to divide the computational domain into elementary subdomains whose geometric complexity does not exceed the capabilities of the grid generator.

For geometrically complex computational domains, it is more convenient to use tetrahedral computational grids: their construction is much less laborious. However, due to the fundamental inconsistency with the flow direction, tetrahedral meshes provide lower calculation accuracy (this deficiency can be partially compensated by increasing the order of accuracy of the approximation scheme for convective terms and calculating gradients from the values at the cell nodes).

In addition, with the same resolution, the tetrahedral mesh has significantly more elements than the hexahedral mesh. Therefore, the use of a tetrahedral mesh is not economical in terms of the use of computing resources.

The computational region of space, including the single-burner sector of the combustion chamber, is geometrically complex, in particular, multiconnected and multiscale. A compromise solution is to use a combined (hybrid) grid consisting of hexahedrons, tetrahedral, connecting their pyramids, as well as prisms. Hybrid meshes provide significant savings in the number of cells and increase the accuracy of the calculation compared to purely tetrahedral ones.

To build hybrid grids, it is necessary to decompose the flow parts of the combustion chamber in the subregion (blocks). In most blocks inside the flame tube and in the annular channel a hexahedral mesh should be used, where possible (in particular, in the flow core), a "cartographic" grid. The location of the calculated cells in the "cartographic" grid should be consistent with the prevailing direction of the gas flow as a whole (from the frontal device to the exit from the combustion chamber) and the local directions of the air jets originating from the main openings of the flame tube and its layered cooling openings. In the holes in the walls of the flame tube and the front device, it is worth using a quadrangular prismatic grid. The most complex sections of the computational domain, where the construction of hexahedral and prismatic grids causes difficulties (for example, near all openings of the flame tube and front-end device), should be covered with a tetrahedral grid.

In accordance with the principles of computational grids stated above, a grid was constructed for the gas temperature pattern calculation at the exit of the combustor.

The computational grid, taking into account adaptation in the area of mixing holes, contains ~ 36 million cells (see Fig. 2). It should be noted that the value $y^+ \sim 30$ is also most desirable due to the fact that it corresponds to the lower boundary of the completely turbulent zone of the boundary layer, where the logarithmic law of averaged velocity is valid for the first wall cell [2].

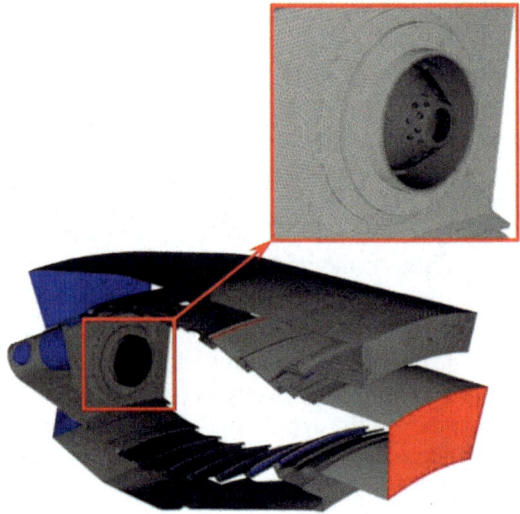

Fig. 2. Calculation grid.

2.3   Boundary Conditions

In Fig. 1 shows a conditional scheme for setting boundary conditions.

At the inlet boundary (IN), the mass air flow rate, the total temperature, the turbulence intensity, and the ratio of turbulent viscosity to dynamic viscosity were set.

At the outlet boundary (OUT), the overpressure, the full temperature of the return flow, the intensity of the turbulence of the return flow, and the ratio of the turbulent viscosity to the dynamic viscosity of the return flow were set. On the lateral faces (Periodic), the type of boundary conditions Rotational-periodic was set.

2.4   Numerical Solution Method

The numerical method for solving control equations is shown in Table 1.

The method of calculating gradients by values in nodes is more accurate than the method of calculating gradients by values in the centers of cells, especially on irregular (beveled) unstructured grids.

It is advisable to change the values of a number of lower relaxation coefficients compared with the default values to ensure convergence and improve the stability of the solution [3].

3   Research Results

The problem of numerical simulation of the gas flow with the combustion of atomized liquid fuel in the annular combustion chamber of a gas turbine engine was solved. As a result of the calculations, it was found that the accuracy of the radial and circumferential non-uniformity modeling for the gas temperature pattern at the exit of the

Table 1. Numerical solution method.

Aspect of the numerical procedure	Fluent program option
Solver	Segregated
Gradient option	Node-based
Pressure-velocity coupling	SIMPLE
Equations	Flow (4 equations)
	Turbulence (2 equations)
	Energy (1 equations)
	PDF (2 equations)
	Total: (9 equations)
Discretization	First order upwind or third order MUSCL
Under-relaxation	Pressure − 0,2;
	Momentum − 0,5;
	Energy − 0,9;
	Temperature − 0,9;
	Mean mixture fraction − 0,9;
	Mixture fraction variance − 0,8;
	Discrete phase sources − 0,2;
	Rest by default

combustion chamber is unsatisfactory when using the k-e turbulence model with the initial settings for the Ansys Fluent calculation complex. Moreover, the maximum value of the radial non-uniformity of the gas temperature pattern at the exit of the combustion chamber exceeded the value obtained in the experiment by 12.61%, and the maximum value of the circumferential non-uniformity by 12.69%. To increase the accuracy of the non-uniformity of the temperature pattern at the exit of the combustion chamber modeling, a numerical experiment was conducted to study the effect of the degree of turbulent diffusion of gas components on the value of the non-uniformity of the temperature pattern [4]. To reduce the non-uniformity of the temperature pattern at the exit of the combustion chamber, the degree of turbulent diffusion of gas components was increased with respect to the initial version of the calculation performed with the initial setting for the Ansys Fluent program Sc = 0.85 by reducing the turbulent Schmidt number Sc in the turbulence model. A decrease in Sc number does not correspond to the fundamental physics of the processes occurring in the combustion chamber, but it simulates the intensification of mixing due to large-scale turbulence during the injection of transverse streams of dilution air in a flame tube, which is not taken into account in the RANS simulation of turbulent flows. It should be noted that a decrease in Sc also overestimates the rate of mixing of fuel with an oxidizing agent upstream from the place of injection of diluting air jets in the flame tube.

A numerical experiment was performed in the Ansys Fluent program for turbulent Schmidt number Sc = 0.85 (default); Sc = 0.6; Sc = 0.4 and Sc = 0.2 (see Fig. 3 and 4) [5]. The figures show the influence of the turbulent Schmidt number Sc on the result of the gas temperature pattern at the exit of the combustion chamber calculating, with a decrease in Sc, the level of the circumferential and radial non-uniformities of the gas temperature pattern decreases. However, the degree of reduction of radial and circumferential non-uniformities with a decrease in Sc are different.

Therefore, in order to improve the accuracy of calculation for both circumferential and radial non-uniformities of the gas temperature pattern, it was proposed to use a variable value of Sc depending on the gas temperature instead of a constant value of the turbulent Schmidt number Sc. The functional dependence of the turbulent Schmidt number Sc on the gas temperature was implemented in the Ansys Fluent program using the user function (UDF – User Defined Functions).

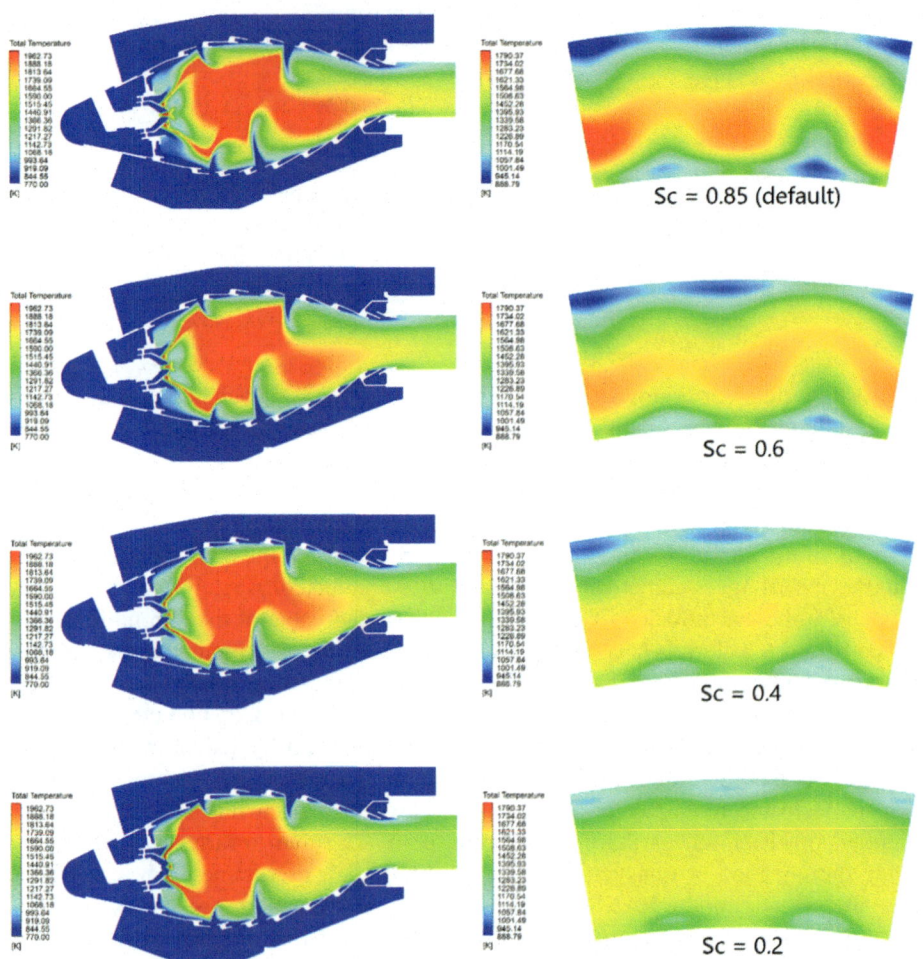

Fig. 3. The temperature pattern of the gas in the meridional and exit sections of the combustion chamber with various turbulent values of the Schmidt number Sc = 0.85 (default); 0.6; 0.4 and 0.2, respectively.

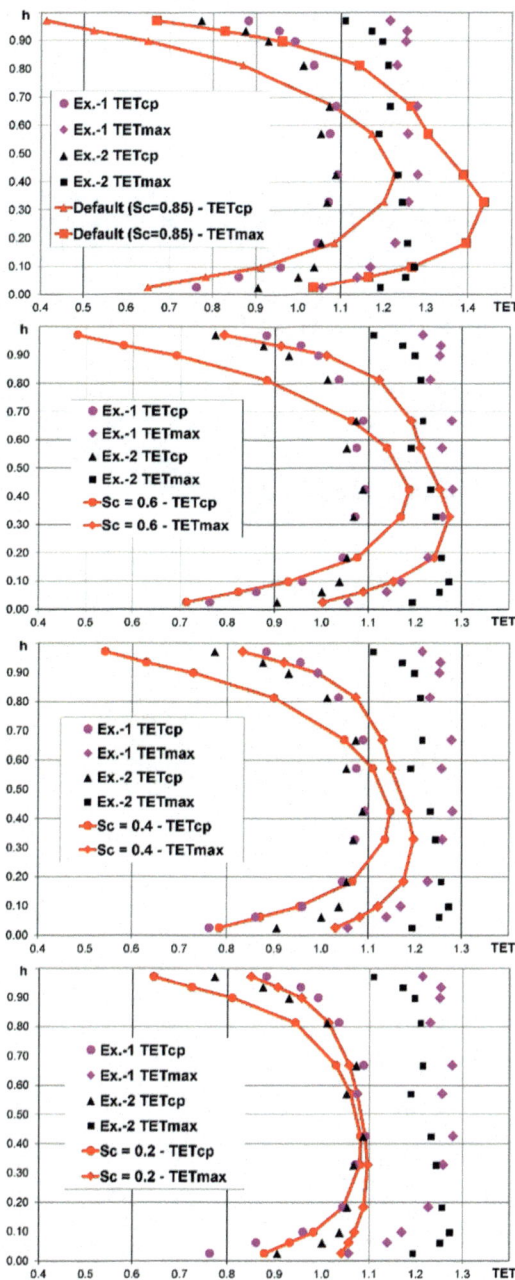

Fig. 4. Plots of the relative gas temperature at the exit of the combustion chamber with various turbulent values of the Schmidt number Sc = 0.85 (default), 0.6, 0.4 and 0.2, respectively.

The results of modeling the gas temperature pattern using the proposed UDF function for the turbulent Schmidt number Sc are in satisfactory agreement with the experimental data for both the radial and circumferential non-uniformities of the gas temperature pattern at the exit of the combustion chamber (see Fig. 5).

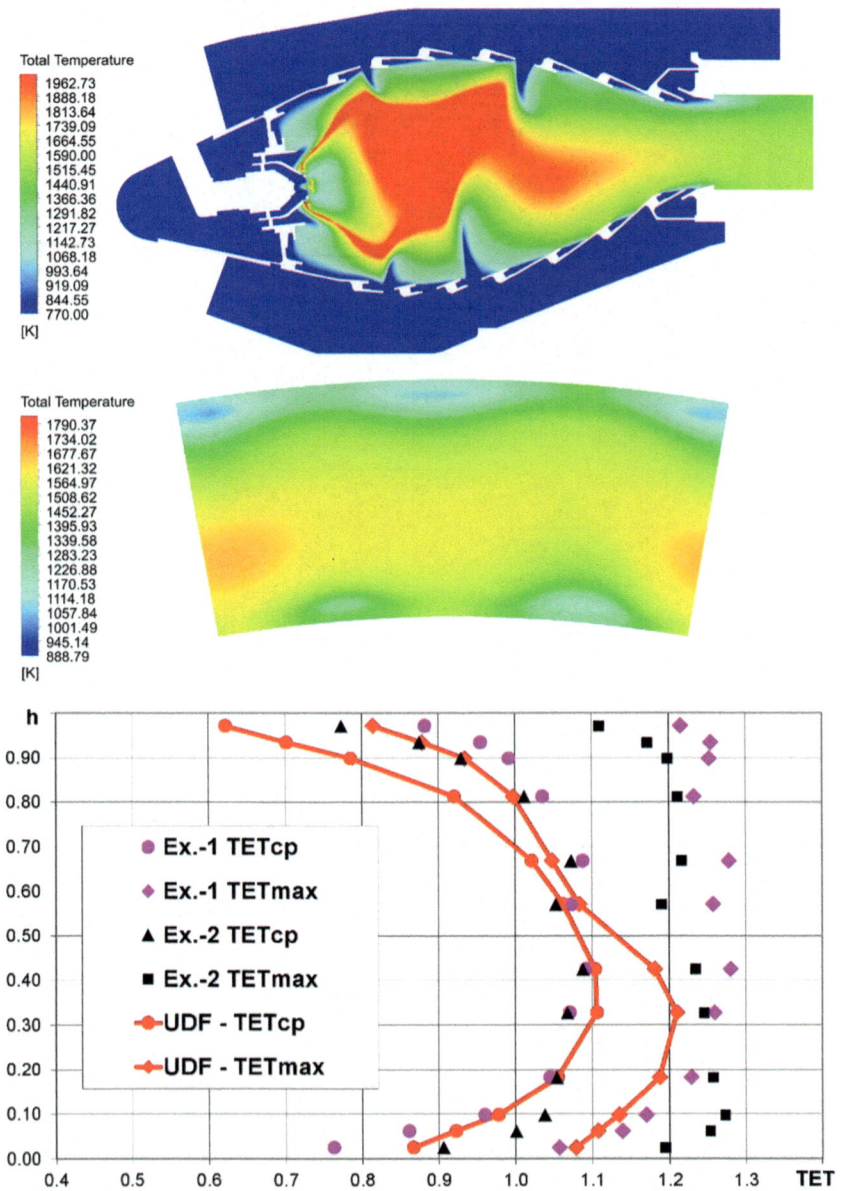

Fig. 5. The results of modeling the gas temperature pattern at the exit of combustor using the UDF.

Figure 6 shows the results of the radial and circumferential plots modelling with the default Schmidt number (Sc = 0.85) and the proposed UDF function.

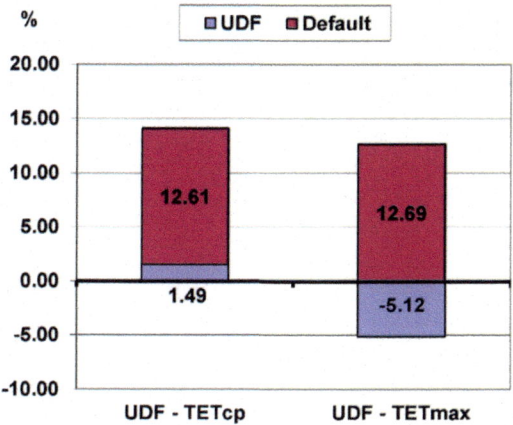

Fig. 6. The results of radial and circumferential non-uniformities modeling accuracy for the gas temperature pattern.

4   Conclusions

Based on the results of numerical studies, the following conclusions can be done:

– the effect of the turbulent Schmidt number Sc on the result of the gas temperature pattern calculation at the exit of the combustion chamber is confirmed. A decrease in the value of the turbulent Schmid number leads to a decrease in the level of the circumferential and radial non-uniformities of the gas temperature pattern at the exit of the combustion chamber:
– the results of modeling the gas temperature pattern using the functional dependence for the turbulent Schmidt number Sc on the gas temperature are in satisfactory agreement with the experimental data for both radial and circumferential non-uniformities of the gas temperature pattern at the exit of the combustion chamber.

References

1. Lefebvre, A.H.: Gas Turbine Combustion, 2nd edn. Taylor & Francis, Philadelphia (1999)
2. Averin, S.I., Minaev, A.N., Shvydkii, V.S., Yaroshenko, Y.G.: Fluid and Gas Mechanics. Metallurgy Publ., Moscow (1987). (in Russian)
3. Users Guide FLUENT. Fluent Co. (2005)

4. He, G., Guo, Y., Hsu, A.T., et al.: The effect of Schmidt number on turbulent scalar mixing in a jet-in-crossflow. In: Turbo Expo: Power for Land, Sea, and Air, vol. 78590, p. V002T02A029. ASME (1999). https://doi.org/10.1115/99-GT-137
5. King, P.T., Andrews, G.E., Pourkashanian, M.M., McIntosh, A.C.: CFD predictions of isothermal fuel-air mixing in a radial swirl low NO_x combustor using various RANS turbulence models. In: Turbo Expo: Power for Land, Sea, and Air, vol. 44687, pp. 973–983. ASME (2012). https://doi.org/10.1115/GT2012-69299

Identification of Computational Models of the Dynamics of Gas Turbine Unit Rotors with Magnetic Bearings by Incomplete Data for Design Automation

Gennadii Martynenko$^{(\boxtimes)}$ ⓘ and Volodymyr Martynenko ⓘ

National Technical University "Kharkiv Polytechnic Institute",
2 Kyrpychova Street, Kharkiv 61002, Ukraine
gmartynenko@ukr.net, martynenko.volodymyr@gmail.com

Abstract. The paper discusses simulation of the rotor dynamics phenomena in a power gas turbine unit (GTU) on the basis of numerical approaches and presents the results of a finite element analysis of the GTU rotor dynamics with identification of computational models based on incomplete data. They include parametric modeling and identification of the design finite element model of the rotor of a gas turbine unit in active magnetic bearings (AMBs) in terms of geometric and dynamic parameters. The parametric models of the gas turbine engine and generator rotors are created using beam finite elements. Depending on the section of the shaft line, the final element was assigned either with circular or annular cross-sections. Finite elements of an elastic connection were used for AMB modeling. The search for the geometric and physical parameters for the GTU rotor model was carried out in two stages – varying the geometric parameters and selecting the reduced properties of materials for individual sections. The criterion for the adequacy and suitability of the computational model for the further execution of various dynamic analyzes was the satisfaction of the known data (the values of natural frequencies and critical speeds, the amplitude-frequency characteristics of the entire GTU shaft line and isolated rotors). The results of the analysis of forced vibrations allowed verifying the models by the values of the resonance mode frequencies according to the data known from the open literature at the given parameters of imbalance and damping. Models are further suitable for computer-aided design using information technology.

Keywords: Gas turbine unit · Turbocompressor · Turbogenerator · Magnetic bearing · Identification of a model · Rotor dynamics · Computer-aided design

1 Introduction

Gas turbine engines (GTE) are internal combustion engines with the Brayton thermodynamic cycle and are widely used in modern technology from aircraft to power machines. For example, the gas turbine engine NK-16ST for the gas industry and power engineering is based on the aircraft engine NK-8-2U [1]. It is used in gas pumping units GPA-Ts-16 [2]. Another application of gas turbine engines in power

© The Author(s), under exclusive license to Springer Nature Switzerland AG 2021
M. Nechyporuk et al. (Eds.): ICTM 2020, LNNS 188, pp. 451–463, 2021.
https://doi.org/10.1007/978-3-030-66717-7_38

engineering is their use in gas turbine heat and power plants GT HPP. These are heat power plants for the combined production of electrical energy in a gas turbine unit (GTU) and heat energy in a waste heat boiler. One of the problems requiring study during designing such gas turbine units (as well as other systems with gas turbine engines) is the dynamic behavior of their rotors. These machines are classified as medium-sized and large-sized systems. Therefore, resonance phenomena associated with imbalance of their rotors can lead to failure and destruction.

Automation of the design process of such complex rotor machines as gas turbine units requires adequate mathematical models including rotors with bearing units. To build them, detailed drawings with the values of almost all parameters and material properties are needed. However, there are frequent situations when these data are not fully known in practice, but results characterizing the behavior of the system are available. Then the problem of identifying mathematical computational models arises. This paper is devoted to this issue from the point of view of modeling the dynamics of GTU rotors with one of the modern bearing units, namely magnetic bearings.

2   Literature Review

The problem of identifying mathematical models based on incomplete data consists in the use of special methods for constructing mathematical models of a dynamical system based on observational data [3]. In this case, a model is understood as a mathematical description of the behavior of a physical system in the frequency or time domain. Identification of dynamic objects and processes consists in determining their structure and parameters [4]. This is done according to the initial data (input action) and results (output values). The object is a "black box", and the task is to find a mathematical description of its structure and parameters by external influences and reactions to them, that is, to turn the "black box" into a "white box". The parameter identification problem can be either a subproblem of general identification, or it can be solved independently [5].

The following approaches to solving similar problems for various rotary machines are known as applied to the mathematical description of processes and phenomena of different physical nature. The paper [6] considers identifying the models for underwater turbine power plant. The same problem is solved for the turbine engine [7], permanent magnet synchronous machine [8], DC motor [9], steam turbine [10], fan blades [11], aero-derivative gas turbine for power generation [12], and other machines. Due to the increased practical interest, a very urgent issue is the identification of dynamic models of various nodes of wind generators [13]. This can involve different methods [14]. Identification methods can be based on both standard approaches of control theory [15, 16] and on modern data mining algorithms [17]. A separate task is to build models for identifying various faults [18], in particular cracks in rotor shafts with magnetic bearings (MBs) [19]. Solving problems of rotor dynamics for the construction of an adequate dynamic model is often connected with a problem of identifying the parameters of various types of supports [20, 21] including magnetic ones [22, 23]. When identifying models for describing the dynamics of rotors of a gas turbine with magnetic bearings, various approaches can also be used, including the finite element method (FEM) [24, 25]. The review of information sources indicates that there is no general approach to the construction of mathematical models of the dynamics of rotors taking into account MBs. The problem of identifying models is especially acute when, for various reasons, there is no detailed technical documentation, and an assessment of the state of structural variants of an object is necessary to ensure proper functionality as a result of design. These two facts determine the relevance of the topic of the work.

3   The Object of Research and Problem Formulation

The paper poses and solves the problem of identifying a mathematical model suitable for conducting in-depth studies of the dynamic behavior of a gas turbine rotor based on incomplete data. The problem is solved for a specific large-sized rotary machine, however, the research results in terms of the approach to the setting and the execution technique can be applied to a wide class of rotary machines.

3.1   Gas Turbine Unit Design

At the moment, the use of low-power gas-turbine power plants based on a gas-turbine installation for the generation of electricity and heat for domestic and industrial consumers located in areas with an impossibility or a difficulty of centralized supply from large cogeneration plants is topical problem.

The paper considers gas turbine heat and power plant (GT HPP), designed to generate electricity and heat [26]. The object of research is the dynamics of the rotor of a gas turbine unit (GTU) GTE-009M with a single electric capacity of 9 MW, based on which a GT HPP was made. Fig. 1 presents its appearance [27].

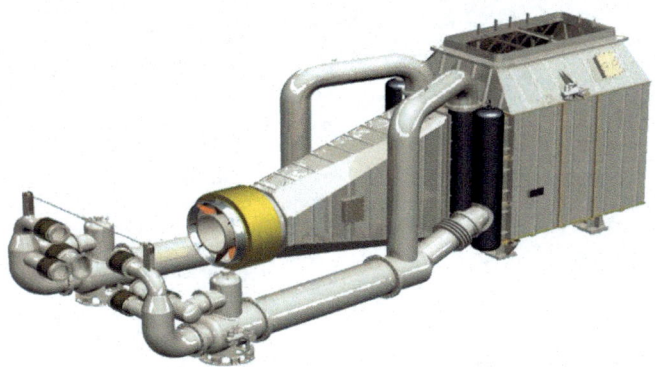

Fig. 1. Gas turbine heat and power plant [27].

This unit is driven by a gas turbine engine similar to aircraft engines. This is a GT-009M regenerative cycle gas turbine engine with a recuperative air heater and a system of air and gas ducts [27]. It also includes: turbogenerator TFE-10–2(3 × 2)/6000U3 (operational frequency is 101.6 Hz) with split stator winding, with built-in air cooling and static thyristor excitation system and automatic control and regulation system; an automatic control system for magnetic bearings, on which rotors of a gas turbine engine and a turbogenerator are installed, a thyristor frequency converter, a waste-heat boiler and other elements. Thyristor frequency converter is used to supply electricity to an external network with a frequency of 50 Hz. Two gas turbine installations, consisting of gas turbine units of the GTE-009M type, recuperative air heaters, heat recovery boilers and hot water boilers, are installed at the GT HPP [26]. The GTE-009M is single-shaft with a rotational speed of 6096 rpm, with MBs including sectional combustion chambers and with the axial gas output after the turbine.

The use of an electromagnetic suspension system, that is, active magnetic bearings, replaces oil bearings in the turbo unit, which increases the durability and environmental friendliness of the power plant. An appearance of this installation and of the support units (magnetic bearings) is shown in Fig. 2. The shaft line of the installation is also shown in Fig. 2. It consists of a rotor of a turbocompressor (a gas turbine engine) (1), a rotor of a generator (2) and an intermediate shaft (3) [26, 28]. The shaft line is installed in active magnetic bearings (Fig. 2) – one radial (4) and one radial-axial (5) AMBs of the turbocompressor rotor, two radial (6 and 7) AMBs of the generator rotor. Each radial AMB is additionally equipped with two safety bearings with ceramic rolling elements without lubrication (Fig. 2). They are necessary to ensure the run-out of the rotor in case of failure of the magnetic suspension system [26].

Fig. 2. An appearance of the gas turbine unit GTE-009M [26, 28].

3.2 Initial Data for the Problem of the Mathematical Model Identification of the GTU Rotor Dynamics

The initial data for solving the stated technical problem are the scheme of the entire GTU rotor (Fig. 3); schemes of rotors of the turbocompressor (gas turbine engine) and the generator; the mass of the turbocompressor rotor with a part of the intermediate shaft

(~ 6300 kg), the mass of the generator rotor with a part of the intermediate shaft (~ 4700 kg), the mass of the entire GTU shaft line (~ 11000 kg); transverse (J_e) and equatorial (J_p) moments of inertia and centres of gravity, set separately for each rotor and rotor assembly, ($J_e = 93130$ kg·m^2 and $J_p = 606$ kg·m^2) [26, 29].

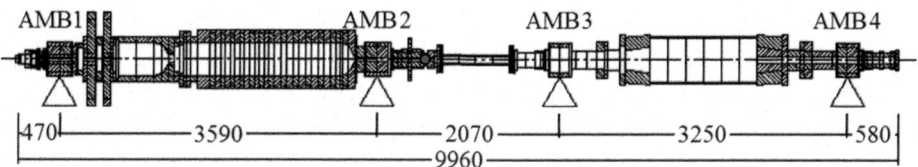

Fig. 3. Scheme of the GTU rotor (sectional lengths in mm) [26, 27, 29].

3.3 Research Objectives

Incomplete data from the open literature, which are the basis of the mathematical model identification of the GTU rotor dynamics, are the natural frequencies of the non-rotating rotor, which are in the range with the upper limit, that one and a half times exceeds the operating frequency (6096 rpm or 101.6 Hz), and the modes corresponding to them [26–29]. These data characterize the design at a test value of the stiffness of all AMBs equal to 1 MN/m. For the final verification of the model, the dependences of the natural frequencies on the rotor speed (Campbell diagram) and the critical rotor speeds are used. Also, the recovered data are the characteristics of forced vibrations which include the values of the resonance mode frequencies.

The purpose of the research is to develop a methodology for constructing an adequate mathematical model according to known geometric and physical parameters, as well as dynamic characteristics using the FEM. The structure of the model should be suitable for in-depth analysis of the rotor dynamics under harmonic loads of the rotor imbalance. The parameters of this model should ensure the coincidence of the results of the corresponding analyzes with the reference values. Testing of the model should be carried out according to the responses of the design to dynamic influences, which can be obtained during the analysis of forced vibrations of the GTU rotors.

4 Calculation Models Construction Based on Incomplete Data

Mathematical modeling of the dynamics of rotors in magnetic bearings can use various approaches, for example, analytical ones [30]. But their disadvantages include inflexibility to changes in geometric and physical parameters which need to be specially recalculated, as well as neglect of geometric design features. Numerical approaches, for example, based on the finite element method, allow creating computational mathematical models that automatically take into account all the nuances of designs. The method provides automatic calculation of such parameters as mass, moments of inertia, positions of centers of mass, and others. Therefore, such models are more universal and therefore they are used for the automation of the design of gas turbine system rotor elements. The process of identifying mathematical computational finite element models of GTU rotors with magnetic bearings in this case is also divided into the identification of the structure and parameters.

4.1   The Structure of the Finite Element Mathematical Model of GTU Rotors

An integrated software tool was developed that uses the finite element method to solve this problem. The detailed element-wise structure of the rotor is unknown, and only the scheme is known (Fig. 3). Therefore, the parametric rotor model was created using three-node beam finite elements (FE) with six degrees of freedom at each node. Depending on the section of the shaft line, the final element was assigned a circular or annular cross-section. For each section, an internal division into subdomains was also specified to approximate the sought quantities (displacements) over the section. Active magnetic bearings (AMB) as elastic-damping bearings were modeled using finite elements of a special type. For radial bearings, finite elements were used to simulate the radial arrangement of elastic-damping elements with a common node on the rotor axis. For these elements, the spring stiffness and the damping coefficients were set in the vertical and horizontal directions, as well as in the directions located at 45° to them. The placement of these elements was carried out according to the design scheme (Fig. 3) in the middle of the trunnions of the radial AMB. Their nodes, not lying on the rotor axis, were fixed along all six degrees of freedom, which are also inherent in this element. The axial bearing was modeled by a finite element of linear stiffness, one node of which was located on the rotor axis in the center of the AMB2 rotor part, and the other also laid on the axis of the bulk model, but did not belong to it. It was fixed in all six degrees of freedom. Finite element models of the GTU rotor, as well as individual structural elements – the rotors of the turbocompressor and generator – are presented in Fig. 4.

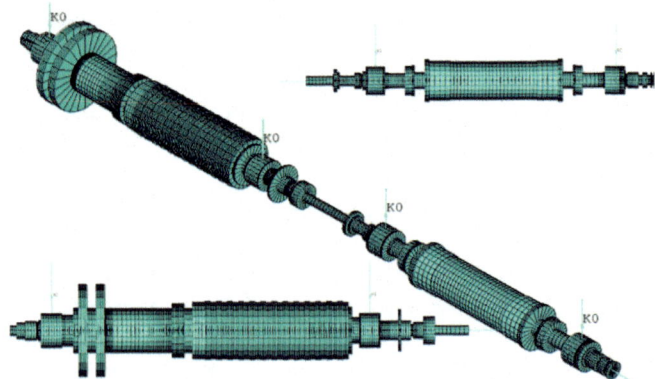

Fig. 4. Beam finite-element model of the GTU rotor and its elements.

It contains the visualization of beam finite element sections with different constants of sections, and elastic-damping FEs are designated as K0. This model includes about 125 beam FEs, 4 elastic-damping FEs that simulate radial AMBs, and one that simulates an axial support. The total number of degrees of freedom for this model is about 2000. The quality of this model was confirmed by the results of a series of static calculations under the action of constant gravity forces on the rotor. For further studies, a variant with the calculated error of less than 0.5% was adopted.

4.2   Automation Technique for Parametric Modeling

Automatic parametric construction of the finite element model uses the created program code in the parametric programming language, which allows performing both static and dynamic variant calculations. The following set of parameters was varied: types of cross-section and diameters of individual sections of each rotor, physical constants of the materials of the turbocompressor and generator rotors (elastic modulus and density), as well as stiffness and damping coefficients of the supports.

4.3   Identification of the Computational Mathematical Model of the GTU Rotor in AMBs by Geometric and Dynamic Parameters

At the first stage, the initial version of the model was found by varying the types of sections, their diameters and the material density values of the turbocompressor and generator rotors. The coincidence was ensured with known data affecting only the static behavior of the structure, that is, the results of the static analysis when time-independent loads are applied to the structure. These known data were the masses of the turbocompressor, the generator, the entire GTU shaft line, the transverse, equatorial moments of inertia and the centers of gravity of each rotor and the rotor assembly. The deviation from these values for the model version found after a series of calculations was no more than 0.5%. In this case, the total reduced values of the material density of compressor and generator are 7511 and 8134 kg/m³ respectively.

At the second stage of research, the equality of natural frequencies (NFs) to the reference values was satisfied at the same modes of natural vibrations (NMs) with the stiffness of all supports of 1 MN/m. Table 1 and Fig. 5 contains the results of these analyses (TCR – turbocompressor rotor, GR – generator rotor). This was done by

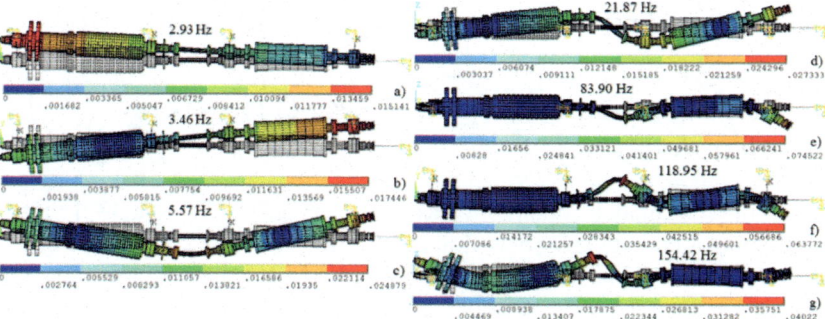

Fig. 5. Modes of natural vibrations of the shaft line with the stiffness of the supports 1 MN/m.

selecting the general reduced values of the material elastic moduli of the individual GTU rotor structural elements.

In the case of translational vibration modes, the rotor performs rigid body motions of cylindrical or conical precessions. The natural frequencies of these forms practically do not depend on the stiffness properties of the rotor, but are determined only by its mass characteristics and the support stiffness. The required values of these NFs were obtained automatically (Table 1, #1–4). For the deformational modes (Table 1, #5–7), coincidence of frequencies was achieved by selecting the values of the elastic moduli. The found quantities, which provide the required values of natural frequencies, are: for the turbocompressor – $1.87 \cdot 10^{11}$ Pa, and for the generator – $1.065 \cdot 10^{11}$ Pa. The analysis of the results shows that the discrepancies for all natural frequencies that are in the rotor operational speed range are acceptable (Table 1).

Table 1. Results of the analysis of the GTU rotor natural vibrations.

#	Natural frequencies (NF), Hz			Vibration character at NF	NF type (Fig. 5)
	Calculated value	Reference value [29]	Deviation, %		
1	2.93	2.9	0.2	TCR translational	a
2	3.46	3.4	0.2	GR translational	b
3	5.57	5.9	5.4	TCR tilting	c
4	21.87	23.1	5.3	GR tilting	d
5	83.90	79.3	5.8	GR bending	e
6	118.95	117.1	1.6	GR bending	f
7	154.42	151.7	1.8	TCR bending	g

The final identification of the GTU rotor computational model parameters was carried out according to the critical speed values of the rotating rotor with a change in the rotational speed in the range from 0 to maximum possible operation (6600 rpm) [29]. The Campbell frequency diagram enabled determining the critical speed values corresponding to the forward and reverse precessions. It is designed for a rotor on MBs with an average reduced stiffness of 1 MN/m and is shown in Fig. 6. Table 2 summarizes the values of the GTU rotor critical speeds for the comparative analysis.

Analysis of the calculation results of NFs and critical rotation speeds confirms the adequacy of the constructed mathematical finite element model of the GTU rotor.

5 Model Verification Based on the Results of Forced Vibration Analysis

Validation of any model should be based on experimental data [30]. However, numerical simulation data can be used to verify the mathematical model. Fig. 7 illustrates the results of the analysis of the GTU rotor dynamic behavior under the influence of harmonic forces in the form of amplitude-frequency characteristics and orbits of the

Table 2. The results of an analysis of the GTU rotor critical speeds.

#	NF, Hz ($\omega = 0$)	Critical speeds, rpm					
		Reverse precession			Direct precession		
		Calculation	Reference	Error, %	Calculation	Reference	Error, %
1	2.9	175.5	177.8	1.3	175.8	178.3	1.4
2	3.5	206.8	211.1	2.0	208.1	212.4	2.0
3	5.6	327.1	319.4	2.4	342.3	335.1	2.1
4	21.9	1284.9	1299.4	1.1	1341.4	1363.1	1.6
5	83.9	4883.0	4768.8	2.4	5195.6	5067.2	2.5

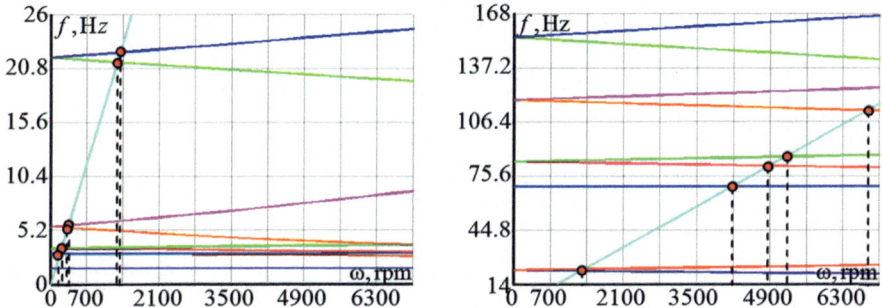

Fig. 6. Campbell diagram for the GTU rotor indicating the critical speeds (at the support stiffness of 1 MN/m).

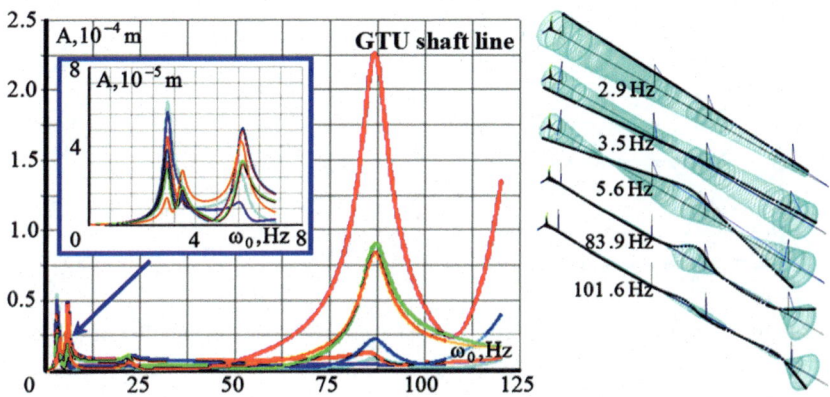

Fig. 7. Amplitude-frequency characteristics of the entire GTU shaft line and the orbit of the rotor axis points.

axis points. These forces are caused by the rotor imbalance ($6.3 \cdot 10^{-6}$ kg·m for the turbocompressor and $4.7 \cdot 10^{-6}$ kg·m for the generator), at relative damping of 4%.

The given orbits correspond to the critical and operating (101.6 Hz) frequencies (Table 2). Fig. 8 and Fig. 9 presents similar harmonic analysis results for isolated turbocompressor and generator rotors. These data fully correlate with the parameters of natural vibrations (Table 2) and confirm the correctness of the procedure for identifying the structure of the model and its parameters.

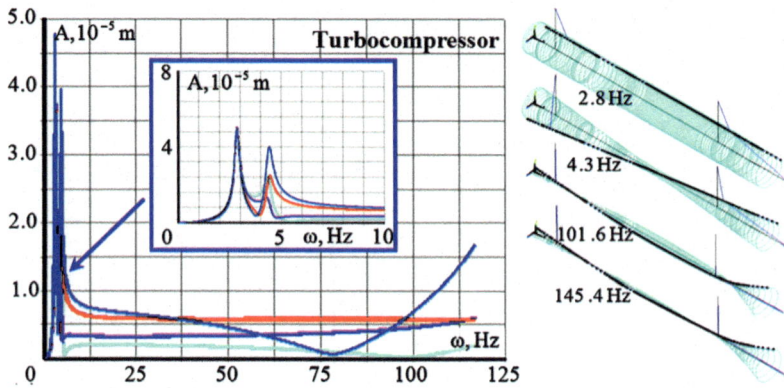

Fig. 8. Amplitude-frequency characteristics of the turbocompressor rotor and the orbit of the rotor axis points.

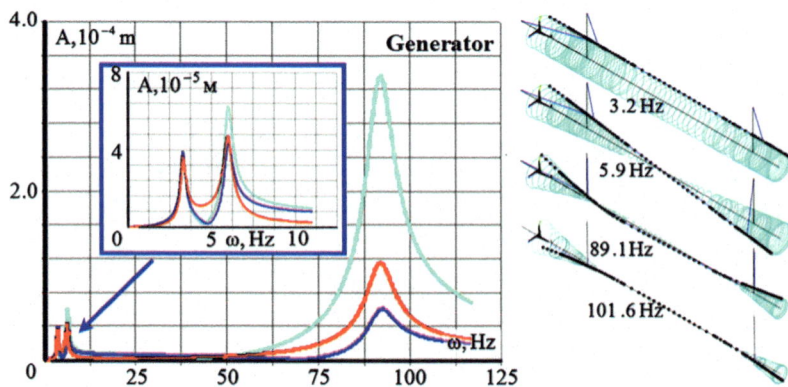

Fig. 9. Amplitude-frequency characteristics of the generator rotor and the orbit of the rotor axis points.

6 Conclusion and Discussion

The paper proposes a methodology and identifies the mathematical computational finite element model of gas turbine rotors. It consists of beam FEs (which simulate the rotors of a turbocompressor and generator) and elastic-damper FEs (which simulate AMBs). For this model, the identity of the known integral data of the structure (masses, moments of inertia, location of the centers of mass) and dynamic characteristics (frequencies and modes of natural vibrations, critical speeds) has been achieved. They are verified using data from open literature.

The constructed computational models of the rotors are suitable for the use in imitation computational studies, for example, in numerical experiments to study the features of the rotor dynamics in operating modes when using supports with variable stiffness and damping properties. The results obtained by means of simulations can also serve as a reference for verification of other mathematical models.

References

1. Devold, H.: Oil and Gas Production Handbook. An Introduction to Oil and Gas Production. ABB ATPA Oil and Gas, Oslo (2006)
2. Artemova, T.G.: Gas pumping unit GPA-Ts-16. USTU, Yekaterinburg (2002). (in Russian)
3. Ljung, L.: System Identification: Theory for the User. Prentice Hall, Upper Saddle River (2012)
4. Liu, W.: System identification: state and parameter estimation techniques. In: Introduction to Hybrid Vehicle System Modeling and Control, pp. 325–363. Wiley, Hoboken (2013)
5. Moeller, D.P.F.: Parameter identification of dynamic systems. In: Mathematical and Computational Modeling and Simulation, pp. 257–310. Springer, Heidelberg (2004). https://doi.org/10.1007/978-3-642-18709-4_5
6. Tzelepis, V., VanZwieten Jr, J.H., Xiros, N.I., Sultan, C.: Modeling, system identification and linearization of underwater turbine power plant dynamics. In: ASME International Mechanical Engineering Congress and Exposition, vol. 6B: Energy, paper No. 53455. ASME, Houston (2015). https://doi.org/10.1115/IMECE2015-53455
7. Li, H., Wu, L., Li, Y., Li, C.: Identification of turbine engine dynamics with the governor in the loop. In: 2016 IEEE International Conference on Systems, Man, and Cybernetics (SMC), Budapest, pp. 4200–4205. IEEE (2016). https://doi.org/10.1109/SMC.2016.7844891IEEE
8. Hall, S., Márquez-Fernández, F.J., Alaküla, M.: Dynamic magnetic model identification of permanent magnet synchronous machines. IEEE Trans. Energy Convers. **32**(4), 1367–1375 (2017). https://doi.org/10.1109/TEC.2017.2704114
9. Romero, D., Vélez, A., Gómez-Mendoza, J.: Non-linear grey-box models applied to DC motor identification. In: 2019 IEEE 4th Colombian Conference on Automatic Control (CCAC), Medellín, pp. 1–5. IEEE (2019). https://doi.org/10.1109/CCAC.2019.8921181
10. Rusanov, A., Martynenko, G., Avramov, K., Martynenko, V.: Detection of accident causes on turbine-generator sets by means of numerical simulations. In: 2018 IEEE 3rd International Conference on Intelligent Energy and Power Systems (IEPS), Kharkiv, pp. 51–54. IEEE (2018). https://doi.org/10.1109/IEPS.2018.8559546
11 Martynenko, V., Hrytsenko, M., Martynenko, G.: Technique for evaluating the strength of composite blades. J. Inst. Eng. (India) Ser. C **101**(3), 451–461 (2020). https://doi.org/10.1007/s40032-020-00572-9

12. Shen, L., Hu, Z., Zheng, Q.: Modeling and simulation of aero-derivative gas turbine for power generation. In: 2018 10th International Conference on Modelling, Identification and Control (ICMIC), Guiyang, pp. 1–6. IEEE (2018). https://doi.org/10.1109/ICMIC.2018.8529995

13. Murillo, J., Herrera, F.B., León, L.T.: Modelado e identificación en un sistema de generación eólica basado en Dfig. Revista Ibérica de Sistemas e Tecnologias de Informação 2(E18), 165–180 (2019)

14. Chu, J., Yuan, L., Hu, Y., et al.: Comparative analysis of identification methods for mechanical dynamics of large-scale wind turbine. Energies 12(18), 3429 (2019). https://doi.org/10.3390/en12183429

15. Coraça, E.M., Junior, M.D.: Model-based identification of rotor-bearing system parameters employing adaptive filtering. In: Cavalca, K., Weber, H. (eds.) Proceedings of the 10th International Conference on Rotor Dynamics – IFToMM. MMS, vol. 62, pp. 236–249. Springer, Cham (2019). https://doi.org/10.1007/978-3-319-99270-9_17

16. Xu, Y., Zhou, J., Jin, C., Guo, Q.: Identification of the dynamic parameters of active magnetic bearings based on the transfer matrix model updating method. J. Mech. Sci. Technol. 30(7), 2971–2979 (2016). https://doi.org/10.1007/s12206-016-0606-7

17. Pavlenko, I., Simonovskiy, V., Ivanov, V., et al.: Application of artificial neural network for identification of bearing stiffness characteristics in rotor dynamics analysis. In: Ivanov, V., et al. (eds.) Advances in Design, Simulation and Manufacturing. DSMIE 2018. LNME, pp. 325–335. Springer, Cham (2019). https://doi.org/10.1007/978-3-319-93587-4_34

18. Ovchinnikov, V., Nikolaev, M., Litvinov, V., Kapitanov, D.: Identification of the structural deviations impacting the dynamics of a flexible multispan rotor on full electromagnetic suspension. Appl. Comput. Electromagn. Soc. J. 34(4), 528–534 (2019)

19. Sarmah, N., Tiwari, R.: Dynamic analysis and identification of multiple fault parameters in a cracked rotor system equipped with active magnetic bearings: a physical model based approach. Inverse Probl. Sci. Eng. 28(8), 1103–1134 (2020). https://doi.org/10.1080/17415977.2019.1700982

20. Theisen, L., Niemann, H., Santos, I., et al.: Modelling and identification for control of gas bearings. Mech. Syst. Signal Process. 70–71, 1150–1170 (2016). https://doi.org/10.1016/j.ymssp.2015.09.016

21. Martynenko, G., Ulianov, Y.: Combined rotor suspension in passive and active magnetic bearings as a prototype of bearing systems of energy rotary turbomachines. In: 2019 IEEE International Conference on Modern Electrical and Energy Systems (MEES), Kremenchuk, pp. 90–93. IEEE (2019). https://doi.org/10.1109/MEES.2019.8896571

22. Vuojolainen, J., Nevaranta, N., Jastrzebski, R., Pyrhönen, O.: Comparison of excitation signals in active magnetic bearing system identification. Model. Identif. Control 38(3), 123–133 (2017). https://doi.org/10.4173/mic.2017.3.2

23. Jiang, K., Zhu, C.: Parameter identification for stiffness and damping of active magnetic bearing in flexible rotor system. J. Vib. Eng. 30(6), 883–892 (2017). https://doi.org/10.16385/j.cnki.issn.1004-4523.2017.06.001. (in Chinese)

24. Martynenko, G., Martynenko, V.: Numerical determination of active magnetic bearings force characteristics taking into account control laws based on parametric modeling. In: 2019 IEEE International Conference on Modern Electrical and Energy Systems (MEES), Kremenchuk, pp. 358–361. IEEE (2019). https://doi.org/10.1109/MEES.2019.8896501

25. Giagopoulos, D., Arailopoulos, A., Zacharakis, I., Pipili, E.: Finite element model developed and modal analysis of large scale steam turbine rotor: quantification of uncertainties and model updating. Eccomas Proceedia 5349, 32–44 (2017). https://doi.org/10.7712/120217.5349.16898

26. Anurov, Yu.M., Litvinov, E.V.: Development and operation of serial energy gas turbines with magnetic bearings. Eastern-European J. Enterp. Technol. **4**(4), 20–24 (2009). (in Russian)
27. Harcii, O.V.: Experience in the construction and operation of a GT-CHPP as part of the implementation of a strategic project in small energy by the Energomash group of companies. Paper presented at the Round table "Development of Small Distributed Energy in the Belgorod Region", Belgorod Institute of Alternative Energy, 13 March 2014. https://www.altenergo-nii.ru/docs/harcii.pdf. Accessed 09 July 2020. (in Russian)
28. SKF-S2M. Magnetic bearings for combined heat and power generation plant. Products leaflets PUB MT/S9 15571 EN, SKF Group, Marcel (2015)
29. Kashtanov, D.: SKF-S2M. Magnetic systems. Technology. General presentation S2M, SKF/S2M (2010)
30. Martynenko, G.: Accounting for an interconnection of electrical, magnetic and mechanical processes in modeling the dynamics of turbomachines rotors in passive and controlled active magnetic bearings. In: 2018 IEEE 3rd International Conference on Intelligent Energy and Power Systems (IEPS), Kharkiv, pp. 326–331. IEEE (2018). https://doi.org/10.1109/IEPS.2018.8559518

Numerical Estimation of the Residual Life-Time of the Elements of the Centrifugal Pump of the Energy Station Due to Corrosion Wear

Oleksiy Larin[1] , Kseniia Potopalska[1($\boxtimes$)] , Evgen Grinchenko[2] ,
and Andrii Kelin[3]

[1] National Technical University "Kharkiv Polytechnic Institute",
2 Kyrpychova Street, Kharkiv 61002, Ukraine
ks.potopalskaya@gmail.com
[2] Kharkiv National University of Internal Affairs, 27 L. Landau Avenue,
Kharkiv 61080, Ukraine
[3] Engineering and Technical Center "KORO", 1 Akademika Proskury Street,
Kharkiv 61070, Ukraine

Abstract. This paper considers the issue of numeral estimating the residual strength of a centrifugal pump operating in excess of the design life in the line of pumps of the power station. Using modern technologies of computer and mathematical modeling the calculations of pump body for define its stress-strain state taking into account the change of geometry of body details had been carried out. The three-dimensional CE models are made, which take into account the actual geometry of the pump parts and the predict of its possible change for the period of extended time work. Estimation of static strength was performed for the main operating mode of the pump (under normal operating conditions). Taking into account the predicted values of the percentage of thinning of the wall of the pump, the statistical deformed state of the structure had been investigated. On the basis of statistical data, the parameter of damage and the probability of failure-free operation of the structure had been determined.

Keywords: Life-time · Corrosion wear · Reliability · Power equipment · Numeral calculation

1 Introduction

At the heart of Ukraine's energy efficiency and energy security are important issues related to energy components transportation. The use of modern technologies of computer and mathematical modeling makes it possible to effectively assess the residual resource and predict the period of no-failure operation of the power supply elements. Important elements of the system that play a significant role in meeting the needs of the end consumer are the accompanying power machines, in particular centrifugal pumps [1–5]. The pumps that are used at stations in Ukraine have already

© The Author(s), under exclusive license to Springer Nature Switzerland AG 2021
M. Nechyporuk et al. (Eds.): ICTM 2020, LNNS 188, pp. 464–474, 2021.
https://doi.org/10.1007/978-3-030-66717-7_39

worked out their design resource, but as practice shows, they can still be exploited. But for their safe use, it is necessary to assess their life-time [1, 2], taking into account the thinning of the walls of the case as a result of long-term operation. In operation, these structures are subject to significant cyclic overloads, as well as experiencing the action of an aggressive environment, influence of aggressive temperature extremes [6], which for some time of operation leads to accumulation of damage [7–11], and as a result to system failures [8, 12, 14]. Untimely detection can cause accidents, environmental disasters, cause significant consumer damage, as well as be a threat to human life [12]. Preventing these events through timely maintenance and repair is an extremely important task.

This paper considers the issue of estimating the residual strength of a centrifugal pump operating in excess of the design life in the line of pumps of the Nuclear station. The results of theoretical researches of its stress-strain state taking into account the change of geometry of case details which was observed after the end of design term of operation are resulted. Estimation of static strength was performed for the main operating mode of the pump (under normal operating conditions).

The study was conducted in the framework of numerical computer modeling based on the finite element method using modern software. Estimated three-dimensional CE models have been developed, which take into account the actual geometry of the pump parts and the predict of its possible change for the period of the extended life-time. The change in the geometry of the structure is taken into account on the basis of extrapolation of the data of the thickness of the walls of the body of pump, obtained in the process of long service life.

2 Defining Parameters of Deformed State

The pump under consideration has exhausted its design resource. Expert assessment of the operating organization of its technical condition shows that there is a thinning of the walls of the body in comparison with the design values. When analyzing the rate of erosion and corrosion wear, it was found that during the operation of the wall thickness of the housing, cover and tubulures, they will thin linearly over time evenly throughout the housing by 1% per 100 h of operation.

Operating experience shows that the average operating time for the pump is 20 h/year. Based on expert estimates of the rate of thinning of the walls and the assessment of the average operating time of the pump per year, the thinning of the walls of the body parts is predicted.

Strength calculations were performed using three-dimensional CE modeling technology. For adequate deformed state assessment, fragments of 1 m long pipes with elastically suspended ends were added to each pipe.

Linear elements of hexagonal and tetrahedral shapes were used for CE sampling (Fig. 1). The pipelines have a less dense mesh than the body parts, because their stress state is not the object of study – they are modeled only to transfer adequate boundary conditions to the pump tubulures. In fact, these elements are needed to level the possible edge effects near the pump. To assess the quality of the constructed model, a series of calculations with mesh of different densities was performed. The wall thicknesses of the body parts are set in accordance with the predicted data on thinning.

The parts of the pump are divided so that the main structural elements have at least two elements in thickness, everywhere the ratio of the sides of the CE is maintained so that the mesh does not have degenerate elements. CE mesh, which is used for calculations, is presented in Fig. 1.

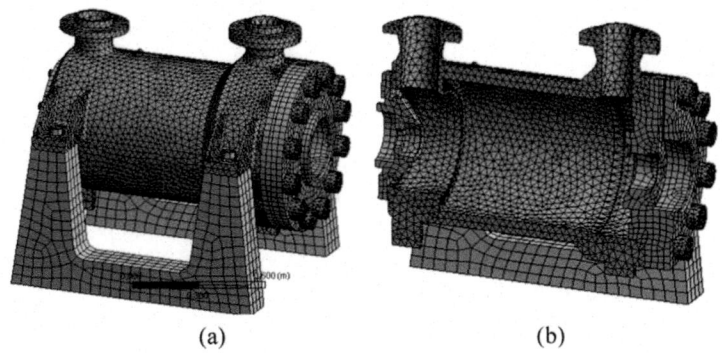

(a)   (b)

Fig. 1. CE mesh of the pump body (a) (horizontal sectional view (b)).

To select the size and number of elements for the CE model, test calculations were performed up to a permissible error of 5% for equivalent stresses.

Physical and mechanical characteristics of materials 12X18H10T, 40X, according to RNNE G 7-002-86 were used for calculations. The following boundary conditions were used:

- rigid sealing on the bearing surfaces of the pump legs;
- limiting the possibility of radial compression on the inner circles of the pressure and inlet covers;
- the base of the pressure and inlet flange in the axial direction had an elastic support, which simulates the effect of discarded pipes.
- volumetric force – the force of gravity;
- convective heat exchange water – steel was set on the inner surfaces (heat transfer coefficient, 27900 W/m^2 °C);
- on the outer surfaces was set convective heat transfer steel – wind (heat transfer coefficient, 5 W/m^2 °C);
- the ambient temperature was taken as 22 °C.

In calculations was taken into account the presence of pre-tightening of threaded connections (Fig. 2). The values of the axial forces for the studs of the inlet cover and the pressure head were 100 kN and 35 kN, respectively.

Thus, at the first stage the problem of determination of the prestressing condition caused by tightening of hairpins was solved. In Fig. 3 shows the distribution of the intensity of stresses (equivalent stresses according to the Mises test), which are formed in the pump under this load mode.

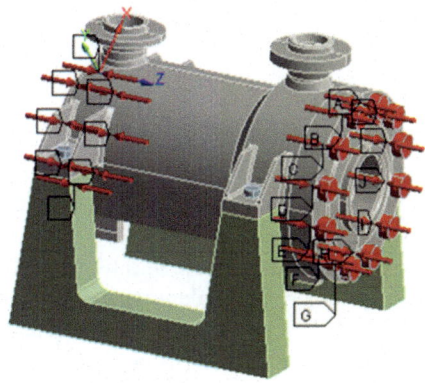

Fig. 2. Modeling of conditions of tightening of threaded connections.

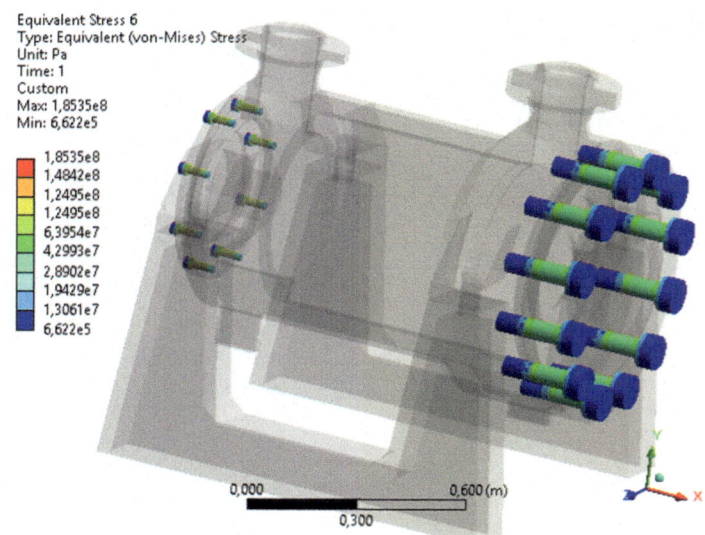

Fig. 3. Pre-stressed state of the studs caused by their mounting tightening.

According to RNNE G 7-002-86 (p. 3.4), the rated stress allowed for the elements of equipment and pipelines loaded with internal pressure, take the minimum of the following values:

$$\sigma = \min\{R_m/2.6; R_{0.2}/1.5\}, \tag{1}$$

where R_m is the tensile strength; $R_{0.2}$ is the yield point.

For bolted connections

$$[\sigma]_w = R_{0.2}/1.5. \tag{2}$$

Analyzing the previous stress state of the studs, tightening led to the presence of compressive stresses in the studs of a significant level (the level of maximum stresses is formed under the nut and at the entrance of the stud into the body and is 185 MPa, which is less than the allowable value of 295 MPa).

A high level of stresses in the places of concentration was formed on the covers of the pump (Fig. 3), and the obtained equivalent stress (72 MPa) are less than the nominal allowable stresses (131 MPa). To estimate the residual resource, the results of deformed state calculation at NCE pressure were used (Fig. 4).

Thus, the values of equivalent Mises stresses (88 MPa in the housing, 83 MPa in the pressure cap) are used as the minimum cycle stresses.

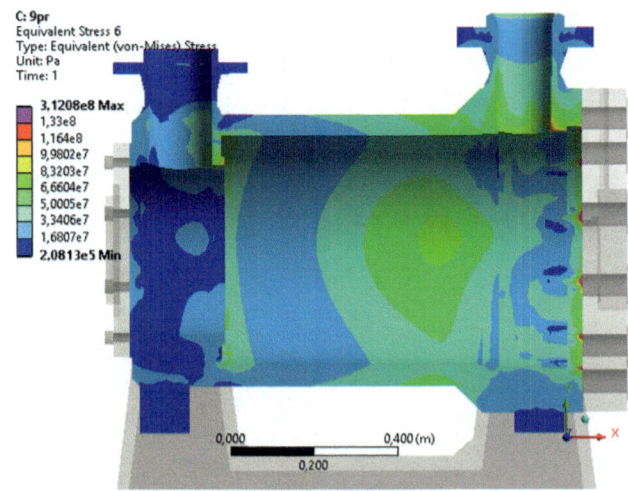

Fig. 4. Distribution of equivalent stresses according to Mises on the pump housing.

At the welds, the cyclic strength was checked separately, taking into account the decrease in their strength by increasing the equivalent stress amplitudes by dividing them by the coefficient of decrease in the strength of the weld. This coefficient for austenitic grade steel in manual welding followed by visual inspection corresponds to 0.8 (the most conservative estimate) in accordance with the recommendation in the RNNE. Thus, together with the weld on the paw, the minimum cycle stresses are 143/0.8 = 178 MPa, and the maximum – 155/0.8 = 193 MPa.

From the data on the operation of the centrifugal pump at the power plant it is known that after the design period the walls of the housing were thinned by 10% due to erosion and corrosion wear. In subsequent calculations, the predicted value of the percentage of thinning of the hull wall was considered to be a random variable that changes over time.

Thinning of the pump wall is proposed to be taken into account in time in the form of a power function, which connects the operating time and the percentage of thinning $h(t)$:

$$h(t) = \widehat{h} + k\left(t - \widehat{t}\right)^{\alpha}, \tag{3}$$

where t_0 is the design life is 30 years, α and k are indicators of thinning kinetics. The process of wall thinning due to corrosion-erosion wear is random, which can be taken into account if the indicators of the equation of its growth kinetics (3): α and k are considered random variables.

The parameter k is able to change significantly, even under more or less the same external factors. The parameter α in many studies is considered to be a constant deterministic value equal to 0.73. From the statistical data it is also known that the distribution of values with sufficient accuracy can be considered subject to the log-normal distribution law:

$$f(h, t) = \frac{1}{h(t)S(t)\sqrt{2\pi}} \exp\left[\frac{(-\ln h - \mu(t))^2}{S^2(t)}\right], \tag{4}$$

where $S(t)$ and $\mu(t)$ are the parameters of the law, which depend on the value of the percentage of thinning in the current operation time and are determined from the coefficient of variation and mathematical expectation of the percentage of thinning as follows:

$$\mu(t) = \ln\left(\frac{m(t)}{\sqrt{1 + \frac{v(t)}{m^2(t)}}}\right), \quad S^2(t) = \ln\left(1 + \frac{v(t)}{m^2(t)}\right), \tag{5}$$

$$m(t) = \left\langle \widehat{h} + k\left(t - \widehat{t}\right)^{\alpha} \right\rangle = \widehat{h} + \langle k \rangle \left(t - \widehat{t}\right)^{\alpha}, \tag{6}$$

$$Var_h(t) = \left\langle \left(\widehat{h} + k\left(t - \widehat{t}\right)^{\alpha}\right)^2 \right\rangle = Var(k)\left(t - \widehat{t}\right)^{2\alpha}, \tag{7}$$

$$v(t) = \frac{\sqrt{Var_h(t)}}{m(t)} = \frac{\sqrt{Var_k(t)}}{\langle k \rangle} = const, \tag{8}$$

where $\langle \ldots \rangle$ is the averaging operator, $m(t)$ is the mean value, $Var()$ is the variance, $v(t)$ is the coefficient of variation. Which depends on time. In Fig. 4 schematically shows the development over time of the mean of corrosion damage $m(t)$ and the possible scatter of its values over time (Table 1 and Fig. 5).

Table 1. Probabilistic characteristics of the thinning.

Parameter	Characteristics of the growth rate of corrosion thinning
Coefficient of variation	0.075
Average value, mm	0.853

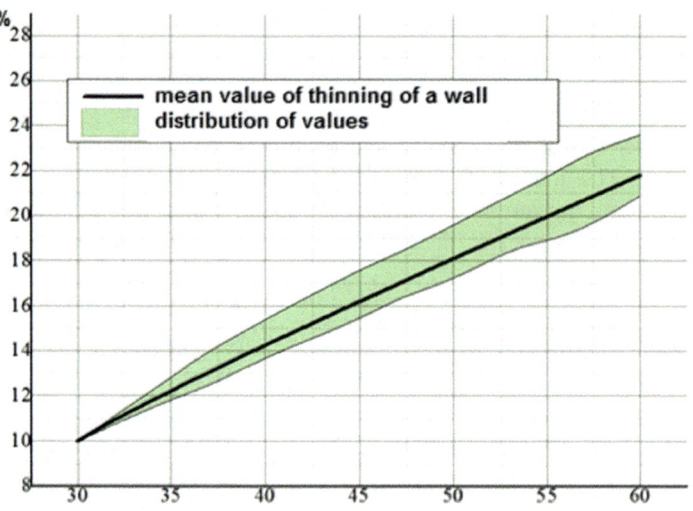

Fig. 5. Distribution of values of percent of thinning of a wall of the case of the pump depending on time of operation.

Displayed equations are centered and set on a separate line.

The change of plastic deformations depending on the service life and thinning of the pump housing is shown in Fig. 6

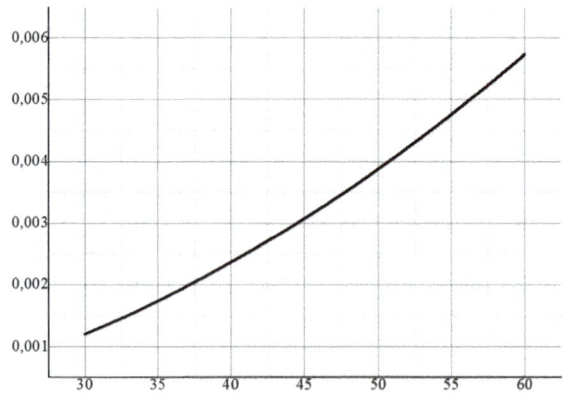

Fig. 6. Plastic deformations in the pump housing depending on design time of operation.

$$\varepsilon(h,t) = K_1 h(t)^2 + K_2 h(t) + K_3 \text{ for } 30 \le t \le 60. \tag{9}$$

The level of plastic deformation is shown that in structure could accumulated fatigue die to cyclic load and it for defending safe life-time of pump it should taken into account.

3 Estimation of Pump Body Life-Time

Accordingly, to the probable values of the percentage of thinning of the wall was investigated statistical deformed state of the structure, on the basis of which the parameter of damage was determined. For this purpose, the step law of kinetics of damage accumulation is used within the framework of the Rabotnov-Kachanov concept of effective stresses. Taking into account the processes of accumulation of high- or low-cycle fatigue, the damage parameter is determined for each of the predicted cases in accordance with expression 10.

$$D = \begin{cases} 1 - [1 - (n + N_D)\delta_D(2s+1)]^{1/2s+1}, dp > 0, \\ 1 - [1 - (n + N_D)\delta_M(m+1)]^{1/m+1}, dp = 0, \end{cases} \tag{10}$$

where D is the damage parameter, n is the current value of the cycle, N_D is the value of the cycle at which the accumulation of damage begins, δ_D is a variable that depends on the deformed state parameters of the structure when accumulating low-cycle fatigue, δ_M is a variable that depends on the deformed state parameters with the accumulation of multicycle fatigue, s, m are the parameters of the material, which are determined from the Wehler curve, dp is the value of plastic deformations [13]. From the data on the operation of centrifugal pumps at the power plant it is known that the number of cycles per year is 1000. Taking into account the developed mathematical model, the parameter of damage to the thinned wall was obtained when predicting further operation for 15 years (Fig. 7).

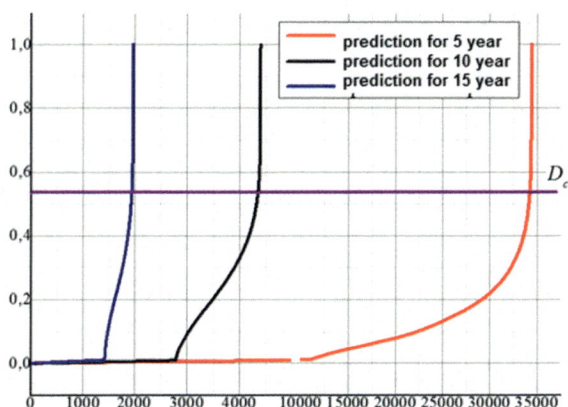

Fig. 7. Mean value of the damage parameter when predicting 5, 10 and 15 works over the design period.

From the analysis of the received data it is defined that at the predict at thinning of a wall in 5 years of operation the number of cycles to failure is equal 30,000 cycles that is equivalent to service life in 30 years, at the predict at thinning of a wall in 10 years of operation the number of cycles to failure is equal – 4200, which is equivalent to 4 years, and the predict for thinning of the wall after 15 years of operation, the number of cycles to failure is equal to 1000 cycles (1 year of operation). That is, operation of the pump after this level of corrosion and erosion wear is dangerous, and this object should be removed from operation.

Thus, the probabilistic characteristics of the accumulation of fatigue damage during operation were obtained. These results can be used to estimate the probabilistic characteristics of the time to failure as a random variable.

Using the obtained statistical data of the damage parameter, the probability of trouble-free operation in predicting the extended service life for 5, 10 and 15 years was determined. For that was used the next equation

$$P(t) = \Pr[D \in (0, D_c)], \tag{11}$$

where $P(t)$ is probability of non-failure operation.

Using the numerical procedure, the probability of failure-free operation was determined as the probability that the parameter D is equal to the value of D_c (0.5 for the pump material). The results of the calculations are shown in Fig. 8.

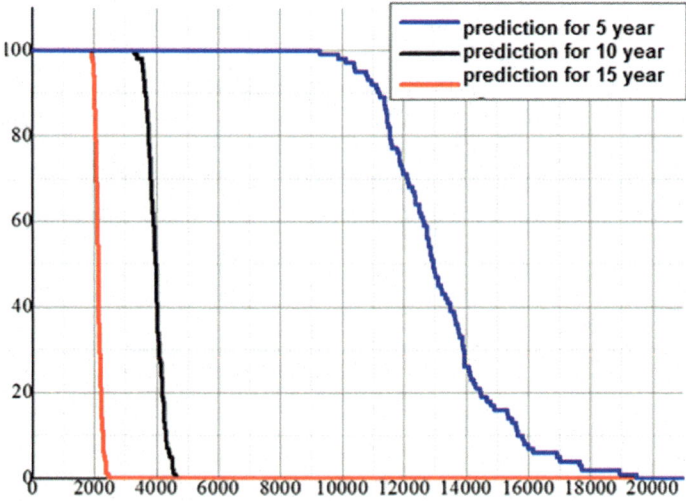

Fig. 8. Probability of non-failure operation of the pump body with a predict of 5, 10 and 15 works over the design period.

4   Conclusion

In this work the analysis of the residual life-time of the centrifugal pump which has fulfilled the design resource is carried out. Numerical calculations had been performed to determine the parameters of the deformed state, taking into account the thinning of the wall of the pump housing. Based on these studies, the parameter of damage during the accumulation of fatigue in the material of the pump had been obtained. The probabilistic characteristics for the design had been determined, which allow to predict the failure time at the corresponding thinning of the wall of the pump housing. Therefore, it can be concluded that with further operation of the centrifugal pump, after 15 years of operation, its efficiency will be reduced by 15 times, and its further use will be dangerous.

References

1. Kelin, A., Larin, O., Naryzhna, R., et al.: Estimation of residual life-time of pumping units of electric power stations. In: 2019 IEEE 14th International Scientific and Technical Conference on Computer Sciences and Information Technologies (CSIT), Lviv, vol. 1, pp. 153–159. IEEE (2019). https://doi.org/10.1109/STC-CSIT.2019.8929748
2. Kelin, A., Larin, O., Naryzhna, R., et al.: Mathematical modelling of residual lifetime of pumping units of electric power stations. In: Nechyporuk, M., et al. (eds.) Integrated Computer Technologies in Mechanical Engineering. AISC, vol. 1113, pp. 271–288. Springer, Cham (2020). https://doi.org/10.1007/978-3-030-37618-5_24
3. Larin, O., Kelin, A., Naryzhna, R., et al.: Analysis of the pump strength to extend its lifetime. Nucl. Radiat. Saf. 3, 30–35 (2018). https://doi.org/10.32918/NRS.2018.3(79).05
4. Nechuiviter, M.M., Shelepov, I.G.: Increasing the reliability of the operation of feed pumps for the deaerating plants of the steam-turbine blocks of electric power stations. Bull. Natl. Tech. Univ. "KhPI" Ser. Power Heat Eng. Process. Equip. 15, 151–155 (2015). (in Russian)
5. Leskin, S.T., Slobodchuk, V.I., Shelegov, A.S.: Analysis of VVER-1000 main circulation pump condition in operation. Nucl. Energy Technol. 3(1), 10–14 (2017). https://doi.org/10.1016/j.nucet.2017.03.002
6. Zaitsev, R.V., Kirichenko, M.V., Khrypunov, G.S., et al.: Operating temperature effect on the thin film solar cell efficiency. J. Nano- Electron. Phys. 11(4), 04029 (2019). https://doi.org/10.21272/jnep.11(4).04029
7. Noon, A.A., Kim, M.H.: Erosion wear on centrifugal pump casing due to slurry flow. Wear 364, 103–111 (2016). https://doi.org/10.1016/j.wear.2016.07.005
8. Xing, D., Hai-lu, Z., Xin-yong, W.: Finite element analysis of wear for centrifugal slurry pump. Procedia Earth Planet. Sci. 1(1), 1532–1538 (2009). https://doi.org/10.1016/j.proeps.2009.09.236
9. Sarvestani, H.Y., Hoa, S.V., Hojjati, M.: Three-dimensional stress analysis of orthotropic curved tubes – Part 1: single-layer solution. Eur. J. Mech. – A/Solids 60, 327–338 (2016). https://doi.org/10.1016/J.EUROMECHSOL.2016.06.005
10. Tarodiya, R., Gandhi, B.K.: Hydraulic performance and erosive wear of centrifugal slurry pumps – a review. Powder Technol. 305, 27–38 (2017). https://doi.org/10.1016/j.powtec.2016.09.048

11. Poberezhnyi, L., Maruschak, P., Prentkovskis, O., et al.: Fatigue and failure of steel of offshore gas pipeline after the laying operation. Arch. Civ. Mech. Eng. **16**, 524–536 (2016). https://doi.org/10.1016/j.acme.2016.03.003

12. Rämä, T., Toppila, T., Kelavirta, T., Martin, P.: CFD analysis of the temperature field in emergency pump room in Loviisa NPP. Nucl. Eng. Des. **279**, 104–108 (2014). https://doi.org/10.1016/j.nucengdes.2014.03.002

13 Potopalska, K.E., Larin, O.O.: Evaluation and forecasting of resource elements of the pipeline taking into account the processes of accumulation of fatigue and corrosion development. Herald of Khmelnytskyi national university. Ser. Tech. Sci. **1**, 46–52 (2019). https://doi.org/10.31891/2307-5732-2019-269-1-46-52. (in Ukrainian)

14. Larin, A.A., Vyazovichenko, Y.A., Barkanov, E., Itskov, M.: Experimental investigation of viscoelastic characteristics of rubber-cord composites considering the process of their self-heating. Strength Mater. **50**(6), 841–851 (2018). https://doi.org/10.1007/s11223-019-00030-7

The Design of Elements of Systems with Gas-Turbine Engines Based on Information Technology

Lyudmyla Rozova$^{(\boxtimes)}$ and Gennadii Martynenko

National Technical University "Kharkiv Polytechnic Institute",
2 Kyrpychova Street, Kharkiv 61002, Ukraine
luda.rozova@gmail.com, gmartynenko@ukr.net

Abstract. Computer modeling and solving of elements of gas-turbine engines allows taking into account the influence of various factors, and ensuring the reliable operation of the structure as a whole. The considered in this work dry gas seals are used as end seals in the gas-turbine engine-compressor system. This system is the main part of the gas-pumping unit, which maintains the required gas pressure in main gas pipelines. End seals are sufficiently important units, which ensure system tightness. The paper presents the developed specialized software package GasDin, which implements the proposed iterative algorithm for solving of interrelated gasdynamic, heat transfer and thermoelasticity problems for dry gas seals. The main part of this software package is the gasdynamic solution module, which makes it possible to obtain the gas pressure distribution in the gap between the rings taking into account the gas temperature changes. The developed software package GasDin also manages all stages of the iterative solution process. Testing of the developed software package took place in several stages for real designs of dry gas seals.

Keywords: Engine-compressor system · Dry gas seal · Engineering software · Integrated technologies

1 Introduction

Information technology has recently been widely used in the design of engineering objects. On the one hand, this is due to a sharp increase of computing technics, the presence of a large number of specialized software packages and environments for software products development. On the other hand, due to the expensive cost and shortage of resources of different nature, the use of information technology in the objects modeling allows minimizing experimental researches and finishing works, and, in some cases, to avoid them. A large number of physical processes can be simulated with sufficient accuracy during design and researches. At the same time, the obtained results can be presented in visual form, convenient for further analysis [1].

The use of information technology in the modeling, researching and production of aircraft engines is advanced direction. Aircraft gas-turbine engines are widely used not only in the aviation industry.

M. Nechyporuk et al. (Eds.): ICTM 2020, LNNS 188, pp. 475–486, 2021.
https://doi.org/10.1007/978-3-030-66717-7_40

One of the areas for the use of gas-turbine engines is their use in the natural gas transportation system through main gas pipelines. Gas-turbine engines are used as a part of gas-pumping unit in the engine-compressor system, which maintains the necessary gas pressure and its further transportation through the system. Designed on the base of aviation engine, gas-turbine engine provides high operation reliability and efficiency (see Fig. 1) [2–4].

Modern requirements for the reliability of such engine-compressor systems is the use of non-contact dry gas seals as end seals. This type of seals has been widely used in recent years due to the large number of advantages over oil seals [5–7]. The advantages of using of dry gas seals are: the possibility of operation at high rotation speeds with minimal gas leakages; no need to use an oil supply system; smaller dimensions and weight.

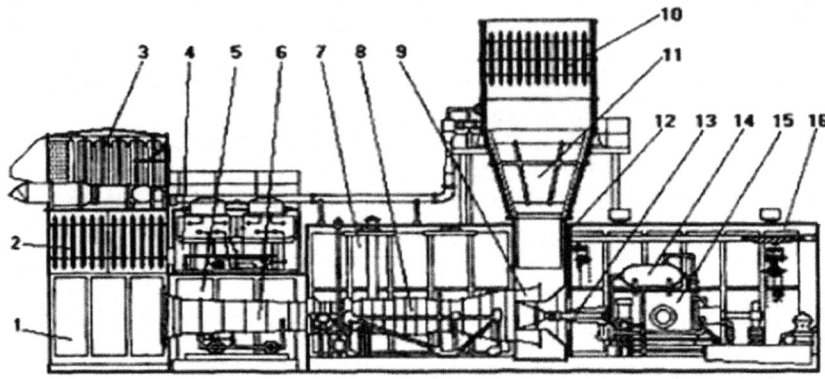

Fig. 1. Gas-pumping unit configuration: 1 – suction chamber; 2 – acoustic filter; 3 – air-cleaning facility; 4 – ventilation unit; 5 – intermediate unit; 6 – duct; 7 – engine section; 8 –gas-turbine engine NK-16ST; 9 – exhaust volute; 10 – exhaust muffler; 11 – diffuser; 12 – sealed bulkhead; 13 – subshaft; 14 – accumulator; 15 – compressor; 16 – compressor section.

Dry gas seal consists of two rings: rotating and axially movable (Fig. 2) [8]. Operational feature of this type of seal is existence of the necessary gas layer in the gap between the seal rings in the steady-state operating mode. The creation and maintenance of such gas layer with a thickness of $3–4 \cdot 10^{-6}$ m is provided by special grooves with a depth of $6–7 \cdot 10^{-6}$ m. The type of grooves is selected from the functional necessity: spiral, T-shaped, etc. [9, 10].

The use of information technology and computer modeling in design and servicing of this type of seal can ensure its reliable and long-term operation, and, consequently, the operation of mentioned above engine-compressor system.

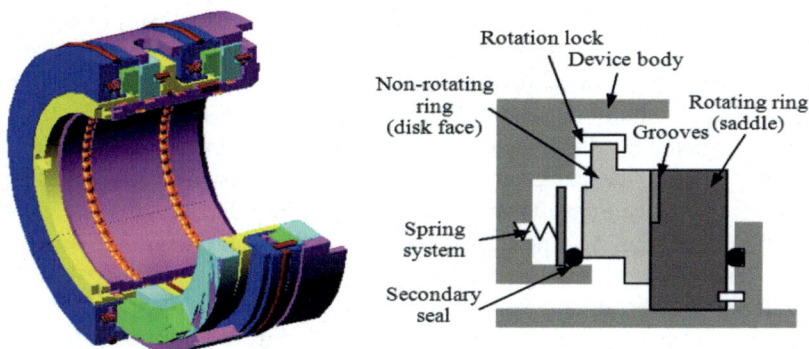

Fig. 2. Dry gas seals.

2   Literature Review

Today, a large number of researchers and scientific schools study the issues of modeling and solving of dry gas seals. Due to the scientific intensity of analysis problem in general formulation, existing works are divided into the study of influence of different physical processes in dry gas seals, separately. Of course, information technologies are widely used, especially, the existing engineering powerful software packages for modeling and solving. However, the main thing here is the understanding of physical processes that occur during dry gas seal operation and their rather accurate mathematical modeling.

Under steady-state seal operation, when the compressor shaft reaches operating speeds, gas pressure, uneven seal rings heating by heat generation in gas layer lead to the appearance of strains in working rings. The strains of working rings are commeasurable with the size of working gap, therefore, they can significantly change its configuration and size.

That is why, the improved methodology of dry gas seal analysis under steady-state operation should include simultaneous solution of interrelated gas dynamics, heat transfer and thermoelasticity problems, and cannot be solved separately [8].

Literature review shows, that most of the available works on this topic are devoted to the issues gasdynamic solution and getting the gas pressure distribution in the gap between seal rings [11–15]. In this case, the equations of gas lubrication theory are considered, which related to a special field of gas dynamics, but most of them do not take into account heat generation in gas layer.

At the same time, various types of grooves and their influence on the resulting gas-dynamic force, opening the rings, are studied [9, 13–15]. The results of solutions in works are in good agreement with each other. The specialized software packages are mainly used.

But only few works consider the heat generation in gas layer and the influence of temperature loads on seal rings [16, 17]. However, the gas temperature changes in the gap lead to the appearance of temperature strains of working rings and to a redistribution of gas pressure in the gap [8].

Also, in the available works, the issues of appearance of non-steady-state operating modes of seals are considered, which are relevant mainly for transient processes that occur at the time of starting-stopping the engine-compressor system as a whole. The issues of dynamic stability during seal operation are studied in works [18, 19].

It should be noted that the high accuracy of modeling the processes occurring in the dry gas seals provides the stable operation of system in transient regimes. Mainly, the existing general engineering software packages are used for this purpose, which, due to their general orientation, may not take into account the features of physical nature of considered processes.

Considering the above, this work is devoted to the use of information technologies for computer modeling of the processes that occur during the dry gas seals operation, and the creation of the specialized software package for it.

3   Formal Problem Statement

The aim of this paper is to present the developed software package GasDin, which implements created iterative algorithm of simultaneous solution of gasdynamic, heat transfer and thermoelasticity problems for dry gas seals used as end seals in the engine-compressor system.

4   Theoretical Basis

The gasdynamic solution for dry gas seals leads to the consideration of the gas lubrication equation, which describes the distribution of gas pressure in the gap between the seal rings. This equation, after some conversions taking into account the temperature changes in the gas layer, is given relative to the square of the pressure [8]:

$$\frac{\partial}{\partial x}\left[\frac{1}{\mu T_{av}}\left(h^3\frac{\partial p}{\partial x} - 12\,\mu\,h\,\omega\sqrt{p}\,z\right)\right] + \frac{\partial}{\partial z}\left[\frac{1}{\mu T_{av}}\left(h^3\frac{\partial p}{\partial z} + 12\mu\,h\,\omega\sqrt{p}x\right)\right] = 0 \quad (1)$$

where x and z are coordinates in the plane of the gap; $p = P^2$ is the square of the gas pressure; $T_{av} = T_{av}(x,z)$ is an average integral function of gas temperature changes over the gap thickness (along the y coordinate); h is the thickness of the gas layer outside the groove; μ is the dynamic coefficient of gas viscosity; ω is the angular velocity of rotating ring.

The gas temperature distribution in the working gap is described by the heat equation, taking into account the heat generation in the gas layer due to viscosity and convective heat transfer [8]:

$$\frac{\partial}{\partial x}\left[k_T\frac{\partial T}{\partial x}\right] + \frac{\partial}{\partial y}\left[k_T\frac{\partial T}{\partial y}\right] + \frac{\partial}{\partial z}\left[k_T\frac{\partial T}{\partial z}\right] + \mu\left[\left(\frac{\partial v_x}{\partial y}\right)^2 + \left(\frac{\partial v_z}{\partial y}\right)^2\right] - \frac{PC_v}{RT}\left(v_x\frac{\partial T}{\partial x} + v_z\frac{\partial T}{\partial z}\right) = 0$$

$$(2)$$

where $T = T(x,y,z)$ is the gas temperature; v_x and v_z are gas velocity components in the plane of the gap; C_v is the specific gas heat intensity ratio at a constant volume; k_T is the thermal conductivity coefficient of the gas film; R is the universal gas constant. The values of gas temperature and pressure at the inlet, on the outer radius, (T_2, P_2) and outlet, on the inner radius, (T_0, P_0) of the seal are the boundary conditions for Eqs. (1), (2).

The heat equation for working rings is:

$$\nabla(k_{T_{1,2}}\nabla T) = 0, \tag{3}$$

where k_{T1} and k_{T2} are thermal conductivity coefficients of axially movable and rotating seal rings respectively.

It should be noted, that the heat transfer tasks for gas layer and seal rings are solved simultaneously. Having solved Eqs. (1)–(3), we obtain the pressure and temperature loads acting on the seal rings and after that determine the strains of seal working rings.

However, the nonlinearity of the mentioned above tasks for dry gas seals and their tight coupling significantly complicate the solution. It is also necessary to take into account the fact that the solution of each of the above tasks separately has its own features. Due to the nonlinearity of considered tasks, an iterative algorithm for their joint solution has been developed.

Here, at each global iteration, at first stage for a plane-parallel gap, the gasdynamic and heat transfer tasks are solved by the method of simple iterations, and the field of gas pressures and temperatures in the gap and in the working rings is established. After that, the thermoelasticity and gasdynamic tasks are solved iteratively for each ring separately at a steady temperature field. In this case, the configuration of the working gap is established, at first, taking into account the strains of rotating ring, after that, taking into account the strains of axially movable ring.

Having obtained the final form of working gap in iterative way taking into account the strains of the seal rings, the gasdynamic problem is solved again and the convergence condition (1) is checked:

$$\left| F_{gd}^{(s)} - F_{gst} \right| \leq \varepsilon_1, \tag{4}$$

where $F_{gd}^{(s)}$ is the value of resulting gasdynamic force in the gap at the s-th iteration; F_{gst} is the value of resulting gas-static force; ε_1 is the force accuracy.

Then the heat transfer problem is solved and the convergence condition (2) is checked:

$$\left| T_k^{(s)} - T_k^{(s-1)} \right| \leq \varepsilon_2, \tag{5}$$

where $T_k^{(s)}$ and $T_k^{(s-1)}$ are the temperature values at the nodes of finite elements at the s-th and $(s-1)$-th iterations; ε_2 is the temperature accuracy.

If the convergence conditions are not verified, the iterative process continues further. The unique solution algorithm has also been developed to solve each of the tasks included in the iteration cycle.

The solution of gasdynamic task is carried out using a specially developed algorithm based on the application of the Bubnov-Galerkin method in combination with the finite element method [8].

Heat transfer and thermoelasticity tasks are solved by using finite element method in a variational formulation.

5   Program Implementation

It should be noted, that for realization of proposed iterative algorithm of gasdynamic, heat transfer and thermoelasticity solution for dry gas seals the software package GasD has been developed by using the C++ programming language (Fig. 3) [8].

The program implementation of iterative solution algorithm took place in several stages. The main part of the developed software package GasDin is gasdynamic solution module.

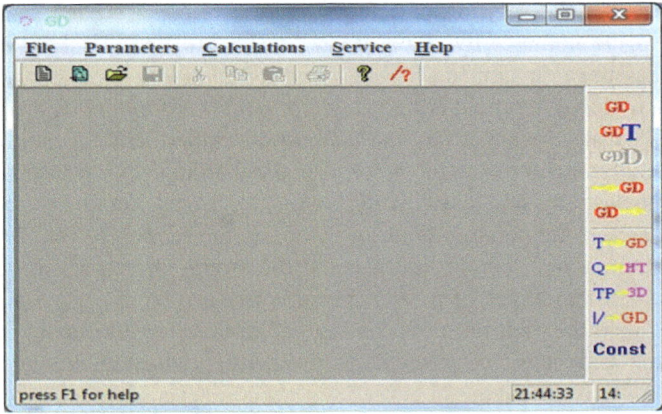

Fig. 3. Main window of program package GasDin.

The software package GasDin provides gasdynamic solutions at a constant temperature of the gas layer, and taking into account the temperature changes in the gas layer.

It should be noted that the heat transfer and thermoelasticity tasks are solved by using the universal finite element software package. Data exchange between the software packages during iterative process is carried out by specially designed programs. The automatic implementation of created iterative algorithm is controlled by specialized programs (Fig. 4), which also organize the exchange of data between software packages [8].

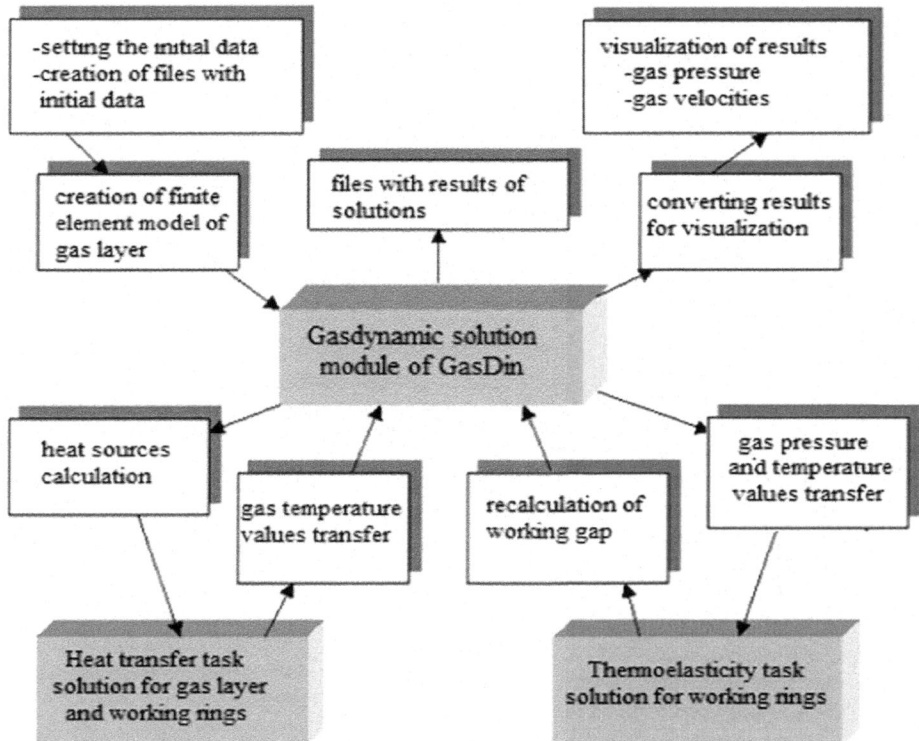

Fig. 4. Scheme of developed program package GasDin.

6 Testing and Discussions

Testing of the developed software package GasDin was carried out in several stages, in the process of increasing the complexity of the general solution algorithm. Testing of the gasdynamic solution module is given in the work [8].

Real designs of dry gas seals have been used for testing developed software package GasDin and created iterative algorithm. These types of seals are used as end seals in engine-compressor system of gas-pumping unit at the Joint Stock Company "Sumy Machine-Building Science-and-Production Association", Sumy, Ukraine.

Let us present the obtained results using the developed software package GasDin and an iterative solution algorithm for one of the working designs of seals with spiral grooves. It should be noted that solutions can be carried out for seals with any type of grooves.

Initial data for the considered seal with spiral grooves are: $r_0 = 101 \cdot 10^{-3}$ m; $r_1 = 112.1 \cdot 10^{-3}$ m; $r_2 = 125 \cdot 10^{-3}$ m; $P_0 = 1.3 \cdot 10^5$ Pa; $P_2 = 60.8 \cdot 10^5$ Pa; the thickness of the gas layer is $h_1 = 3 \cdot 10^{-6}$ m; in the groove $h_2 = 10 \cdot 10^{-6}$ m; $\omega = 555$ rad/s, $T_2 = 303$ K, number of grooves $\eta_k = 12$; $R = 509$ N·m/(kg·K), operating environment – natural gas; $F_{gst} = 88347.9$ N. Accuracy of finding the resulting force is $\varepsilon_1 = 1 \cdot 10^{-1}$; temperature finding accuracy is $\varepsilon_2 = 1 \cdot 10^{-1}$; pressure finding accuracy $\varepsilon_3 = 1 \cdot 10^{-5}$ in relative values.

The rotating seal ring is made of tungsten carbide, the axially movable one is made of carbographite. The physical properties of materials are given in Table 1.

Table 1. Physical properties of ring materials.

Type of material	Elasticity modulus E, Pa	Thermal-expansion coefficient α, 1/K	Thermal conductivity, k_T, W/(m·K)
Tungsten carbide	$7 \cdot 10^{11}$	$5.6 \cdot 10^{-6}$	50.2
Carbographite	$0.11 \cdot 10^{11}$	$4.5 \cdot 10^{-6}$	23

The gasdynamic problem is solved for the region of the gas layer shown on Fig. 5. On subiteration, the gasdynamic and thermoelasticity tasks are solved in iterative way, and then, on each global iteration, the gasdynamic and heat transfer tasks are solved in iterative way. The convergence of the iterative process for the considered dry gas seal is achieved in 4 global iterations and 24 subiterations.

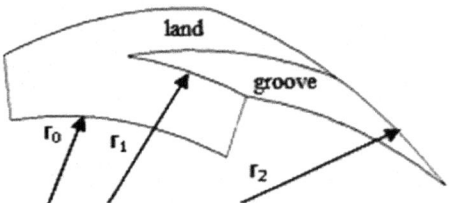

Fig. 5. Solution model of gas layer with cyclic periodicity conditions.

The form of working gap between the rings was obtained as a result of whole iterative process, taking into account strains of working rings from action of gasdynamic pressure in the gap, and gas-static pressure on outer surfaces of seal rings, the mechanical action of springs, heat release in the gas layer and uneven heating of seal working rings. A schematic representation of final view of working gap is shown in Fig. 6.

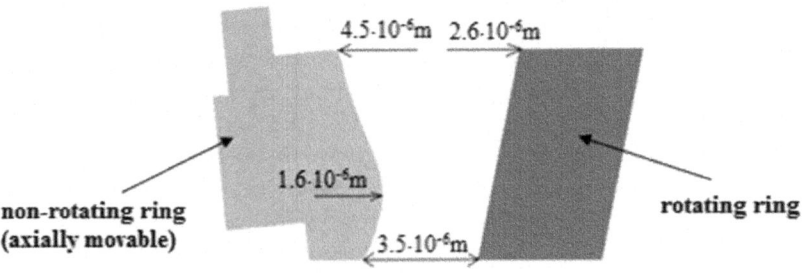

Fig. 6. Schematic representation of final view of working gap.

The flatness of working gap is significantly changed because of the strains working rings (see Fig. 6). The resulting axial displacements (by gap thickness) of working rings are shown on Fig. 7. It should be noted that three-dimensional models of working rings are used for the solution of thermoelasticiy task, taking into account the conditions of cyclic symmetry for spiral grooves [8]. Heat transfer problem for gas layer and rings is solved for axisymmetric model.

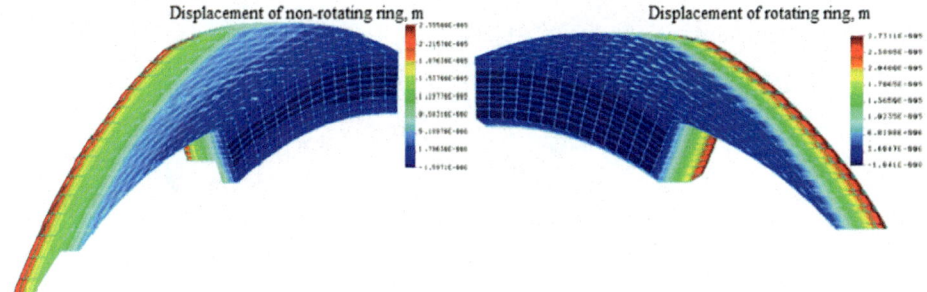

Fig. 7. Axial displacements of non-rotating and rotating rings.

The changes of gap size at the outlet of the seal (parallel component of working gap) during iterative process, while maintaining the equality of the resulting gas-dynamic and gas-static forces, are presented on Fig. 8.

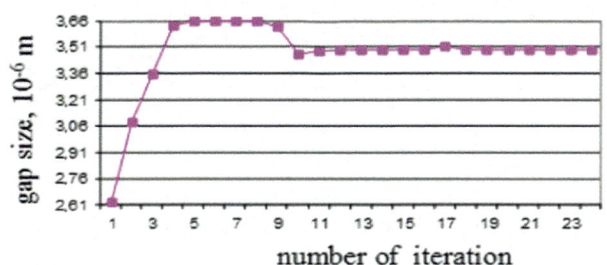

Fig. 8. Changes of gap size on each iteration.

The parallel component of working gap as a result of the iterative process was established at $3.5 \cdot 10^{-6}$ m, and the gas leakages through the seal is $1.03 \cdot 10^{-3}$ m kg / s. The calculated gas leakages differs from the experimental values, obtained on experimental stand on Joint Stock Company "Sumy Machine-Building Science-and-Production Association", for this seal design by 9%. It can be considered a good testing result. But given gas leakages trough the seal exceeds the norm, which indicates the need to improve the geometric parameters of this type of seal. The inlet and outlet gas pressures, outlet gas temperature, gas velocities also can be determined on experimental stand. The calculations for considered type of seal show the gas heating at the outlet of the seal up to 46 °C [8]. It also agrees with the experimental data.

The distribution of gas pressure at a plane-parallel gap and at the final iteration are shown on Fig. 9 and Fig. 10. Gas pressure in groove-land zone increases 1.05 times at a plane-parallel gap (see Fig. 9). However, with a steady gap as a result of strains of working rings, no excess gas pressure over inlet pressure is observed (see Fig. 10). The grooves on the rotating ring no longer have this effect.

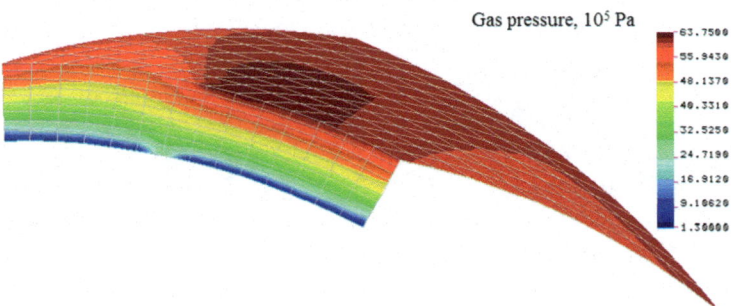

Fig. 9. The distribution of gas pressure at a plane-parallel gap.

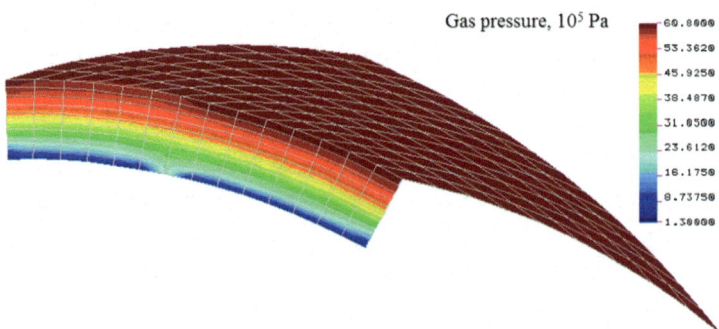

Fig. 10. The final distribution of gas pressure.

7   Conclusions

The use of information technology in the design and production of gas-turbine engines allows ensuring their non-failure operation and significantly extend their service life. This can be achieved through more accurate computer modeling of individual elements. Sealing units are critical units that ensure reliable operation of the system as a whole.

The considered in this work dry gas seals are used as end seals in the engine-compressor system. This system is the main part of the gas-pumping unit, which maintains the required gas pressure in main gas pipelines. The analysis of the processes occurring in dry gas seal showed the need to take into account many factors that affect the stable operation of the sealing units. It was found that a complete mathematical solution model of dry gas seals on steady state operation should include the solution of the interrelated of gasdynamic, heat transfer and thermoelasticity tasks, based on the features of their work.

The specialized software package presented in the work implements the developed iterative algorithm of combined solution of mentioned above problems. Improved computer modeling and solving of this type of seal will allow selecting the optimal geometric and physical parameters of this type of seal.

References

1. Martynenko, G.Y., Marusenko, O.M., Ulyanov, Y.M., Rozova, L.V.: The use of information technology for the design of a prototype engine with rotor in magnetic bearings. In: Nechyporuk, M., et al. (eds.) Integrated Computer Technologies in Mechanical Engineering. AISC, vol. 1113, pp. 301–309. Springer, Cham (2020). https://doi.org/10.1007/978-3-030-37618-5_26

2. Falaleev, S., Bondarchuk, P., Tisarev, A.: Development of advanced carbon face seals for aircraft engines. IOP Conf. Ser. Mater. Sci. Eng. **302**, 012004 (2018). https://doi.org/10.1088/1757-899X/302/1/012004

3. Flitney, R.K.: A description of the types of high speed rotary shaft seals in gas turbine engines and the implications for cabin air quality. J. Biol. Phys. Chem. **14**, 85–89 (2014). https://doi.org/10.4024/17FL14R.jbpc.14.04

4. Artemova, T.G.: Gas pumping unit GPA-Ts-16. USTU-UPI, Yekaterinburg (2002). (in Russian)

5. Forsthoffer, W.E.: Best Practice Handbook for Rotating Machinery. Elsevier, Amsterdam (2011). https://doi.org/10.1016/C2009-0-64191-X

6. Boyce, M.P.: Bearings and seals. In: Gas Turbine Engineering Handbook, pp. 557–604. Elsevier, Amsterdam (2012). https://doi.org/10.1016/B978-0-12-383842-1.00013-5

7. Cao, S., Chen, Y.: A review of modern rotor/seal dynamics. Eng. Mech. **26**, 68–79 (2009). (in Chinese)

8. Rozova, L., Martynenko, G.: Information technology in the modeling of dry gas seal for centrifugal compressors. CEUR Workshop Proc. **2608**, 536–546 (2020)

9. Sun, J., Liu, M., Xu, Z., Liao, T.: Research on operating parameters of T-groove cylindrical gas film seal based on computational fluid dynamics. Adv. Compos. Lett. **28**, 1–7 (2019). https://doi.org/10.1177/0963693519864373

10. Du, Q., Zhang, D.: Research on the performance of supercritical CO_2 dry gas seal with different deep spiral groove. J. Therm. Sci. **28**, 547–558 (2019). https://doi.org/10.1007/s11630-019-1139-z

11. Blasiaka, S., Zahorulko, A.V.: A parametric and dynamic analysis of non-contacting gas face seals with modified surfaces. Tribol. Int. **94**, 126–137 (2016). https://doi.org/10.1016/j.triboint.2015.08.014

12. Shahin, I.: Gas seal performance and start up condition enhancing with different seal groove geometries. J. Aeronaut. Aerosp. Eng. **5**(4), 1000177 (2016). https://doi.org/10.4172/2168-9792.1000177

13. Jing, X., Xudong, P., Shaoxian, B., Xiangkai, M.: CFD simulation of micro scale flow field in spiral groove dry gas seal. In: Proceedings of 2012 IEEE/ASME 8th IEEE/ASME International Conference on Mechatronic and Embedded Systems and Applications, Suzhou, pp. 211–217. IEEE (2012). https://doi.org/10.1109/MESA.2012.6275564

14. Hu, J., Tao, W., Zhao, Y.: Numerical analysis of general groove geometry for dry gas seals. Appl. Mech. Mater. **457–458**, 544–551 (2013). https://doi.org/10.4028/www.scientific.net/AMM.457-458.544

15 Gao, L.: Steady simulation of T-groove and spiral groove dry gas seals. Int. J. Heat Technol. **37**(3), 839–845 (2019). https://doi.org/10.18280/ijht.370321

16. Chen, Z., Jiang, L., Li, J., Wu, B.: Numerical analysis of temperature field on the sealing rings of a dry gas seal. J. Sichuan Univ. (Eng. Sci. Ed.) **46**(3), 175–181 (2014). (in Chinese)

17. Zhu, W., Li, N., Wang, H.: Finite element analysis on thermal deformation of T-shape groove dry gas seal. In: 2010 International Conference on Measuring Technology and Mechatronics Automation, Changsha City, pp. 284–288. IEEE (2010). https://doi.org/10.1109/ICMTMA.2010.43

18. Chen, Y., Jiang, J., Peng, X., et al.: Dynamic performance of dry gas seals and analysis of interactions among its influencing factors. Tribology **39**(3), 269–278 (2019). https://doi.org/10.16078/j.tribology.2018144. (in Chinese)

19. Badykov, R., Falaleev, S.: Advanced dynamic model development of dry gas seal. Procedia Eng. **176**, 344–354 (2017). https://doi.org/10.1016/j.proeng.2017.02.331

Analysis of Ship Main Engine Intake Air Cooling by Ejector Turbocompressor Chillers on Equatorial Voyages

Andrii Radchenko[1]([⊠]) , Andrii Andreev[2] ,
Dmytro Konovalov[2] , Zhang Qiang[3] , and Luo Zewei[3]

[1] Admiral Makarov National University of Shipbuilding,
9 Heroes of Ukraine Avenue, Mykolayiv 54025, Ukraine
nirad50@gmail.com
[2] Kherson Branch of Admiral Makarov National University of Shipbuilding,
44 Ushakova Avenue, Kherson 73000, Ukraine
[3] Jiangsu University of Science and Technology,
2 Mengxi Road, Zhenjiang 212003, China

Abstract. The efficiency of cooling the inlet air of the main ship diesel engine by a refrigeration ejector chiller (ECh) transforming the exhaust gas heat into the cold was analyzed for equatorial voyages. Refrigeration ECh is used as the most simple in design and reliable. However, the efficiency of converting the exhaust heat to cold by ECh is low: their coefficients of performance (COP) are 0.2–0.3, that requires a lot of available heat. The turbocompressor chillers (TCCh) are characterized by high COP of about 0.7 and comparatively small dimensions, that is of very importance for ship application. But because of high technology of their manufacturing nowadays they are quite expansive. The effect of cooling engine cyclic air was determined compared with engine without its cooling, taking into account the changing climatic conditions during the routes of the vessel. It is shown, that because of comparatively low COP the cooling capacity of ECh, produced by using the available exhaust gas heat, is not sufficient to cool the inlet air to a potentially possible minimum temperature of 15 °C when operating the ship engine in tropical climates. Meantime, due to high efficiency of TCCh they consume less exhaust heat than needed for cooling inlet air to 15 °C. An innovative combined cooling system is proposed that enables to use the excessive heat of exhaust gas, remained after TCCh, for subcooling air from 15 °C to 10 °C in ECh with gaining additional fuel saving.

Keywords: Ship engine · Air cooling · Combined chiller · Fuel consumption

1 Introduction

Low-speed diesel engines (LSDE) are the most widespread main engines on the sea ships [1, 2]. A fuel efficiency of diesel engines strictly depends on the temperatures of air at the inlet of engine turbocharger (TC), falling with its raising [3, 4]. A need of engine cyclic air cooling is extremely acute when engines running in tropical climatic conditions [5, 6]. So as the temperatures of exhaust gas are increased too, it is quite reasonable to apply the waste heat recovery technologies [7], firstly to convert the exhaust gas heat to refrigeration: absorption, refrigeration vapor compression [8, 9] or ejectors [10]. The lithium-bromide absorption chillers (ACh) are high efficient heat transformer with coefficient of performance $\zeta = 0.6$–0.7 [19, 20]. But ACh are characterized by increased dimensions and require a quite large space for ACh unite placement aboard ships, that is a real problem. The refrigeration ejector chillers (ECh) are the most simple in design, generally consist of heat exchangers, easy for distributed arrangements aboard ships. However, their efficiency is comparatively low: their coefficients of performance are approximately 0.2–0.3, that requires a lot of available heat [10]. The turbocompressor chillers (TCCh) [11, 12] are characterized by high coefficients of performance of about 0.7 and comparatively small dimensions, that is of very importance for ship application. But because of high technology of their manufacturing nowadays they are quite expansive. So, the combined application of TCCh and ECh might be considered as a compromised decision of ship main engine intake air cooling problem while running in tropical climatic conditions. Such innovative waste heat recovery technologies would provide efficient operation of ship engine at lowered cyclic air temperatures, fuel saving due to cooling engine intake air and are well implemented for ship application with limited space.

2   Literature Review

Nowadays, it is promising to use technologies that would ensure an increase in fuel and energy efficiency of combustion engines: gas turbines [13, 14], gas reciprocating engines [15], diesel engines [5, 6]. They would provide engine operation with high fuel efficiency through using alternative fuels and water-fuel emulsion [16], green power technologies, afterburning [17], thermochemical recuperation [18], catalytic reduction of NOx [19], waste heat utilization [20, 21], including the use of Organic Rankine Cycle [22, 23], Kalina and Chistyakov-Plotnikov cycles [24, 25], combined utilization [26], engine cyclic air cooling by applying refrigeration [27, 28], jet technologies [29]. Statistical treatment of data on influence of actual climatic conditions on performance characteristics to determine rational cooling loads [30] are used.

The aim of research is development of innovative system for cooling intake air of the ship main engine based on turbocompressor and ejector chillers using the heat of hot water, obtained by utilizing the heat of the exhaust gas, to provide efficient operation of ship engine at lowered cyclic air temperatures on equatorial voyages and fuel saving as result.

3   Research Methodology

The object of investigation is the system for cooling intake air at the inlet of low-speed two-stroke diesel engine MAN B&W 6G60ME-C9.2 with rated power output of 12800 kW and specific fuel consumption of 176 g/(kWh) as the main ship engine. There are two types of chillers are considered: ejector chiller as the most simple and cheap but with low coefficient of performance $\zeta = 0.2$, that requires a large value of

waste heat to be converted in refrigeration capacity, and a high efficient turbocompressor chiller with coefficient of performance $\zeta = 0.7$ using the waste heat of exhaust
gas but much more complex in design and expensive.

In order to evaluate the fuel reduction due to engine intake air cooling the CEAS
software package of MAN Diesel Turbo was used [4]. According to calculations carried
out by the CEAS software package, when cooling intake air for every 1 °C air temperature drop a reduction in specific fuel consumption is $\Delta b_e/\Delta t = 0.11\ldots0.12$ g/(kWh°C).

The calculation of the characteristics of the engine intake air cooling systems was
carried out on the operating mode during the voyage of the dry-cargo ship from
Shanghai to Karachi and return.

The variation in climatic conditions (ambient air temperature t_{amb+5} and relative
humidity φ_{amp}) during the vessel's voyage for climatic conditions on the voyage line
"Shanghai (PRC)-Karachi (Pakistan)-Shanghai", 01.07.2019–28.07.2019, is presented
in Fig. 1.

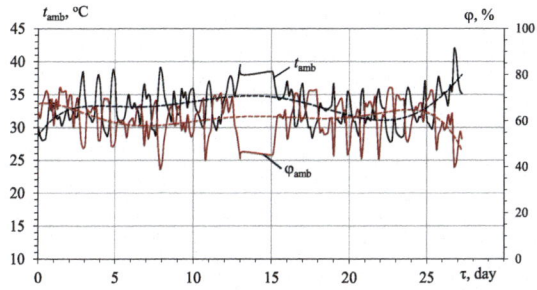

Fig. 1. The variation of temperature t_{amb} and relative humidity φ_{amb} of ambient along the vessel
trade routes "Shanghai-Karachi-Shanghai" (1.07.2019–28.07.2019).

The current temperatures of air at the inlet of engine turbocharger, taking into account
a heat influx from engine room surroundings, were calculated as $t_{amb+5} = t_{amb} + 5$ °C,
and decrease of air temperature due to cooling to target chilled air temperatures $t_{a2} = 15$
or 10 °C at the air cooler outlet – as $\Delta t_{15} = t_{amb+5} - 15$ °C or $\Delta t_{10} = t_{amb+5} - 10$ °C and
from $t_{a2} = 15$ °C to 10 °C – as $\Delta t_{15-10} = 5$ °C.

The vessel route fuel saving $\sum B$ is calculated by summarizing the current values B
of fuel reduction on hour by hour basis: $\sum B = \sum[(\Delta t \cdot \tau) (\Delta b_e/\Delta t) \cdot P_e]$.

A total refrigeration capacity Q_0, when cooling engine cyclic air with flow rate G_a:
$Q_0 = c_a \xi \cdot \Delta t_a \cdot G_a$, where $\Delta t_a = t_{amb+5} - t_{a2}$ is the decrease of cyclic air temperature; t_{a2}
is the target chilled air temperature at the air cooler outlet; ξ is the specific heat ratio as
ratio of the total heat (air enthalpy depression) extracted from the humid air during
cooling (including the latent heat of water steam condensation and sensible heat) to
sensible heat extracted; c_a is the specific heat of humid air [kJ/(kg·K)].

The heat required for cooling air at the inlet of engine to $t_2 = 15$ °C in ECh with
$\zeta = 0.2$ is calculated as $Q_{h.in15(0.2)} = Q_{0.in15}/\zeta$ and accordingly $Q_{0.in15(0.7)}$ for TCCh
with $\zeta = 0.7$.

The following characteristics of the ECh for cooling the ship engine intake air were chosen: refrigerant – R142b; refrigerant evaporation temperature in the evaporator-air cooler $t_0 = 5$ °C and in the generator $t_g = 80...85$ °C, refrigerant condensing temperature $t_c = 45$ °C. The values of coefficient of performance for ECh: $\zeta = 0.2$.

4   Results

The schemes of the systems for cooling the air at the inlet of the turbocharger of marine diesel engine by ECh and TCCh using the exhaust gas heat are shown in Fig. 2.

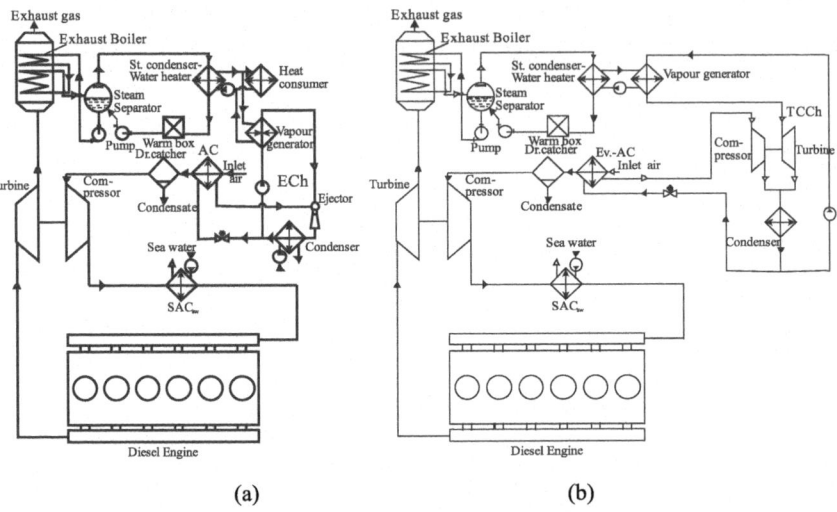

(a)   (b)

Fig. 2. Schemes of systems for cooling air at the inlet of ship diesel engines in ECh (a) and TCCh (b): SAC – scavenge air cooler.

Current values of refrigeration capacity $Q_{0.in15}$ required for cooling air at the inlet of engine from current temperatures in engine room t_{amb+5} to $t_2 = 15$ °C and available refrigeration capacities $Q_{0.exh(0.3)}$, that can be produced in ECh with coefficient of performance $\zeta = 0.2$ by using a heat of exhaust gas (and potentially possible reduction in the air temperature at the inlet of engine while its cooling down to 15 °C Δt_{in15}) along the vessel routes "Shanghai-Karachi-Shanghai" (1.07.2019–28.07.2019) are shown in Fig. 3.

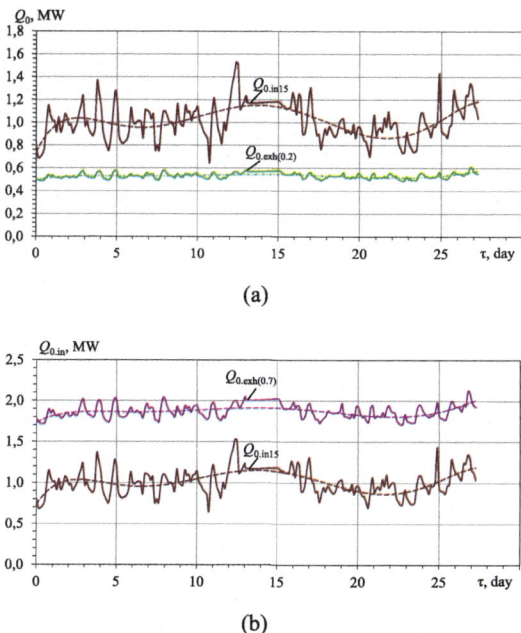

Fig. 3. Current values of refrigeration capacity $Q_{0.in15}$ required for cooling air at the inlet of engine from current inlet temperatures t_{amb+5} to $t_2 = 15\ °C$ and available refrigeration capacities $Q_{0.exh(0.2)}$ produced in ECh with coefficient of performance $\zeta = 0.2$ (a) and the values $Q_{0.exh(0.7)}$ produced in TCCh with $\zeta = 0.7$ (b) by using a heat of exhaust gas.

As it is seen, the refrigeration capacities $Q_{0.exh(0.2)}$ produced in ECh with $\zeta = 0.2$ are much less than the refrigeration capacities $Q_{0.in15}$ required for cooling air from current inlet temperatures t_{amb+5} to $t_2 = 15\ °C$, whereas the refrigeration capacities $Q_{0.exh(0.7)}$ produced in TCCh with $\zeta = 0.7$, vice versa, are much higher than $Q_{0.in15}$. This is caused low coefficient of performance $\zeta = 0.2$ of ECh compared with its value $\zeta = 0.7$ for TCCh, that results in a deficit of available heat of engine exhaust gas for ECh, $Q_{h.in15(0.2)}$ and, vice versa, its excess for TCCh $Q_{h.in15(0.7)} = Q_{0.in15}/\zeta$ (Fig. 4).

The improvement of the fuel efficiency of the ship LSE due to decreasing the inlet air temperature by $\Delta t_{in.exh(0.2)}$ in ECh with $\zeta = 0.2$, which uses the heat of exhaust gas $Q_{h.exh}$, as corresponding values of reduction in specific fuel consumption $\Delta b_{in.exh(0.2)}$, hourly fuel consumption $B_{in.exh(0.2)}$ and a total summarized reduction of fuel consumption $\Sigma B_{in.exh(0.2)}$ due to inlet air cooling in ECh and its value $\Sigma B_{in.exh(0.7)}$ due to inlet air cooling to $t_{in15} = 15\ °C$ in TCCh with $\zeta = 0.7$ for MAN B&W LSE 6G60ME-C9.2 (with a power output of 12800 kW, specific fuel consumption of 176 g/(kW·h)) during the voyage routes "Shanghai-Karachi-Shanghai" (01.07.2019–28.07.2019) are presented in Fig. 5.

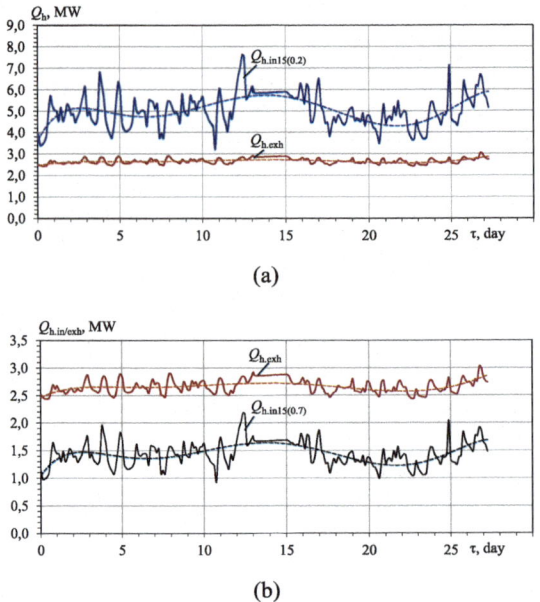

(a)

(b)

Fig. 4. Variations in heat $Q_{h.in15(0.2)}$ required for cooling ambient air at the inlet of engine to $t_2 = 15\ °C$ in ECh with $\zeta = 0.2$ (a), heat $Q_{0.exh(0.7)}$ required for TCCh with $\zeta = 0.7$ (b) and available heat of exhaust gas $Q_{h.exh}$: $Q_{h.in15(0.2)} = Q_{0.in15}/\zeta$ at $\zeta = 0.2$; $Q_{h.in15(0.7)} = Q_{0.in15}/\zeta$ at $\zeta = 0.7$.

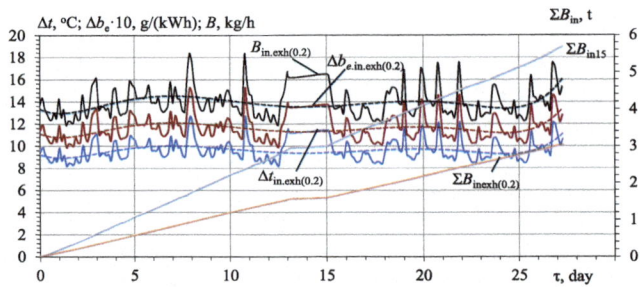

Fig. 5. Current values of reduction in air temperature at the inlet of engine $\Delta t_{in.exh(0.2)}$ due to its cooling by ECh with $\zeta = 0.2$, using exhaust gas heat $Q_{h.exh}$, corresponding values of reduction in specific fuel consumption $\Delta b_{in.exh(0.2)}$, hourly fuel consumption $B_{in.exh(0.2)}$ and total summarized reduction of fuel consumption $\sum B_{in.exh(0.2)}$ due to inlet air cooling in ECh and its value $\sum B_{in.exh(0.7)}$ due to inlet air cooling to $t_{in15} = 15\ °C$ in TCCh with $\zeta = 0.7$ for LSE MAN B&W 6G60ME-C9.2 of power output 12800 kW.

As Fig. 5 shows, due to reducing the air temperature at the inlet of LSE by $\Delta t_{in.exh(0.2)} = 9$–$12\ °C$ in ECh, which uses the heat of the exhaust gas $Q_{h.exh}$, the specific fuel consumption is reduced by $\Delta b_{in.exh(0.2)} = 1.1$–$1.4$ g/(kW·h), hourly fuel consumption – by $B_{in.exh(0.2)} = 13$–16 kg/h and the total summarized reduction of fuel consumption – by $\sum B_{in.exh(0.2)} \approx 3$ t, whereas due to inlet air cooling to $t_{in15} = 15°C$ in TCCh with $\zeta = 0.7$ the total fuel reduction is $\sum B_{in15} \approx 5.6$ t for MAN B&W 6G60ME-C9.2 along the voyage routes.

The results of calculations presented in Fig. 6 prove the existence of a considerable excessive exhaust gas heat $\Delta Q_{h.in15(0.7)ex}$ over the heat $Q_{h.in15(0.7)}$ required for cooling the air at the inlet of engine to $t_2 = 15$ °C in TCCh with $\zeta = 0.7$.

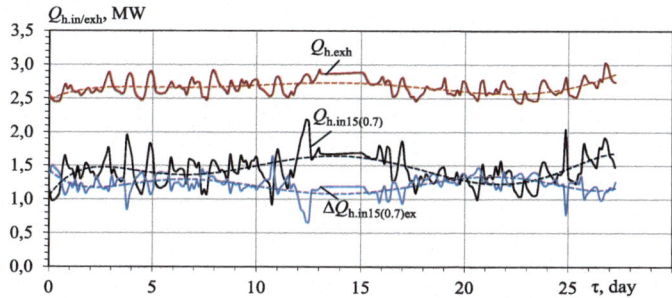

Fig. 6. Variations in available heat of exhaust gas $Q_{h.exh}$, heat $Q_{h.in15(0.7)}$, required for cooling ambient air at the inlet of engine to $t_2 = 15$ °C in TCCh with $\zeta = 0.7$ and excess of available exhaust heat $\Delta Q_{h.in15(0.7)ex}$ over its required values for TCCh: $\Delta Q_{h.in15(0.7)ex} = Q_{h.)exh} - \Delta Q_{h.in15(0.7)}$.

The excessive available exhaust gas heat $\Delta Q_{h.in15(0.7)ex} = Q_{h.exh} - \Delta Q_{h.in15(0.7)}$ over the value consumed by TCCh for cooling inlet air to $t_2 = 15$ °C might be used to produce addition refrigeration capacity $\Delta Q_{0.in10-15}$ needed for deeper cooling air, for instance, below $t_2 = 15$ °C down to $t_2 = 10$ °C (Fig. 7a).

As Fig. 7b shows, in order to provide subcooling air from 15 °C to $t_2 = 10$ °C with spending corresponding refrigeration capacity $\Delta Q_{0.in10-15}$ by using the excess of available exhaust heat $\Delta Q_{h.in15(0.7)ex}$ the low temperature chiller should have the coefficient of performance $\zeta_{in15-10(h.ex)} = \Delta Q_{0.in10-15}/\Delta Q_{h.in15(0.7)ex}$ not less than 0.3 or the addition waste heat is to be used to generate required refrigeration capacity $\Delta Q_{0.in10-15}$ with coefficient of performance $\zeta_{in15-10(h.ex)}$ lower than 0.3, for instance, a high potential heat of scavenge air at the outlet of turbocharge or of engine case high temperature cooling water.

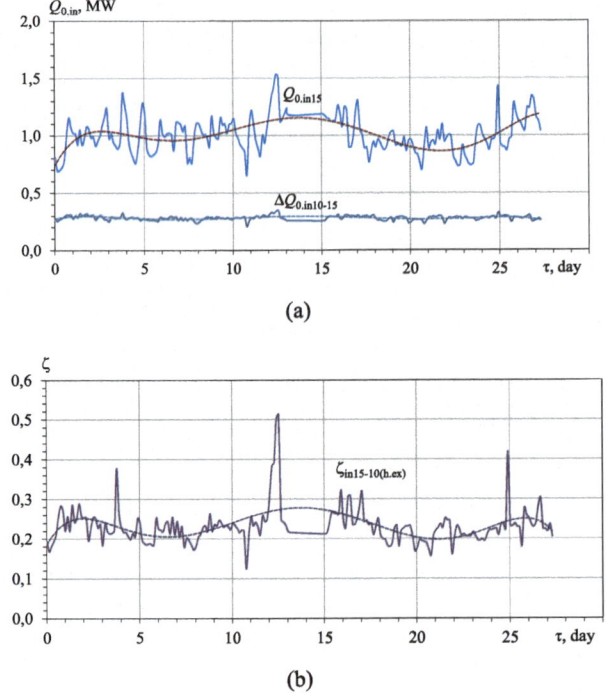

Fig. 7. The refrigeration capacities $\Delta Q_{0.in10\text{-}15}$ needed for cooling air from 15 °C to $t_2 = 10$ °C (a) and corresponding coefficients of performance $\zeta_{in15\text{-}10(h.ex)}$ for low temperature chiller (b).

5   Conclusions

The efficiency of using the heat of exhaust gas for cooling air at the intake of engine turbocharger was analyzed for the low speed engine MAN B&W 6G60ME-C9.2 of power output 12800 kW with the account of climatic conditions in July 2019 on the vessel route line Shanghai-Karachi-Shanghai. The effect of cooling engine intake air was estimated by fuel saving compared with its value without cooling.

There are two types of chillers are considered: ECh as the most simple and cheap but with low coefficient of performance $\zeta = 0.2$ when using hot water instead, that requires a large value of waste heat to be converted in refrigeration capacity, and a high efficient turbocompressor chiller with $\zeta = 0.7$, but much more complex in design.

It is shown that because of comparatively low coefficient of performance of about 0.2 the obtained cooling capacity in ECh is not sufficient to cool the engine inlet air to 15 °C when operating the ship engine in tropical climate.

Meanwhile the results of calculations reveal a considerable excessive heat of exhaust gas over the heat required for cooling air to $t_2 = 15$ °C in a high efficient TCCh with $\zeta = 0.7$. The excessive exhaust heat might be used in ECh to produce addition refrigeration capacity for deeper cooling air, for instance, below 15 °C down to 10 °C.

It was shown that in order to provide subcooling air from 15 °C to 10 °C due to excess of available exhaust heat, the low temperature ECh should have the coefficient of performance ζ not less than 0.3 or the addition waste heat is to be used with ζ lower than 0.3, for instance, a high potential heat of scavenge air at the outlet of turbocharger or of engine case high temperature cooling water.

References

1. MAN Diesel & Turbo: MAN B&W S50MC-C8.2-TII project guide: Camshaft controlled two-stroke engines. https://marine.man-es.com/applications/projectguides/2stroke/content/epub/S50MC-C8_2.pdf. Accessed 15 April 2020
2. Wärtsilä: Environmental Product Guide. Wärtsilä (2017)
3. MAN Diesel & Turbo: Thermal efficiency system for reduction of fuel consumption and CO2 emission. MAN Diesel, Augsburg (2009)
4. MAN Diesel & Turbo: CEAS engine calculations. https://marine.man-es.com/two-stroke/ceas. Accessed 15 Apr 2020
5. Pham, V.V.: Advanced technology solutions for treatment and control noxious emission of large marine diesel engines: a brief review. J. Mech. Eng. Res. Dev. **42**(5), 21–27 (2019)
6. Cao, T., Lee, H., Hwang, Y., et al.: Performance investigation of engine waste heat powered absorption cycle cooling system for shipboard applications. Appl. Therm. Eng. **90**, 820–830 (2015). https://doi.org/10.1016/j.applthermaleng.2015.07.070
7. Shu, G., Liang, Y., Wei, H., et al.: A review of waste heat recovery on two-stroke IC engine aboard ships. Renew. Sustain. Energy Rev. **19**, 385–401 (2013). https://doi.org/10.1016/j.rser.2012.11.034
8. Trushliakov, E., Radchenko, M., Bohdal, T., Radchenko, R., Kantor, S.: An innovative air conditioning system for changeable heat loads. In: Tonkonogyi, V., Ivanov, V., Trojanowska, J., Oborskyi, G., Edl, M., Kuric, I., Pavlenko, I., Dasic, P. (eds.) InterPartner 2019. LNME, pp. 616–625. Springer, Cham (2020). https://doi.org/10.1007/978-3-030-40724-7_63
9. Radchenko, N.I.: On reducing the size of liquid separators for injector circulation plate freezers. Int. J. Refrig **8**(5), 267–269 (1985). https://doi.org/10.1016/0140-7007(85)90004-0
10. Butrymowicz, D., Gagan, J., Śmierciew, K., et al.: Investigations of prototype ejection refrigeration system driven by low grade heat. In: E3S Web of Conferences, vol. 70, p. 03002 (2018). https://doi.org/10.1051/e3sconf/20187003002
11. Alshammari, F., Pesyridis, A., Karvountzis-Kontakiotis, A., et al.: Experimental study of a small scale organic Rankine cycle waste heat recovery system for a heavy duty diesel engine with focus on the radial inflow turbine expander performance. Appl. Energy **215**, 543–555 (2018). https://doi.org/10.1016/j.apenergy.2018.01.049
12. Hoang, A.T.: Waste heat recovery from diesel engines based on Organic Rankine Cycle. Appl. Energy **231**, 138–166 (2018). https://doi.org/10.1016/j.apenergy.2018.09.022
13. Mikielewicz, D., Kosowski, K., Tucki, K., et al.: Gas turbine cycle with external combustion chamber for prosumer and distributed energy systems. Energies **12**(18), 3501 (2019). https://doi.org/10.3390/en12183501
14. Mikielewicz, D., Kosowski, K., Tucki, K., et al.: Influence of different biofuels on the efficiency of gas turbine cycles for prosumer and distributed energy power plants. Energies **12**(16), 3173 (2019). https://doi.org/10.3390/en12163173

15. Radchenko, A., Mikielewicz, D., Forduy, S., Radchenko, M., Zubarev, A.: Monitoring the fuel efficiency of gas engine in integrated energy system. In: Nechyporuk, M., Pavlikov, V., Kritskiy, D. (eds.) Integrated Computer Technologies in Mechanical Engineering. AISC, vol. 1113, pp. 361–370. Springer, Cham (2020). https://doi.org/10.1007/978-3-030-37618-5_31

16. Radchenko, M., Radchenko, R., Kornienko, V., Pyrysunko, M.: Semi-empirical correlations of pollution processes on the condensation surfaces of exhaust gas boilers with water-fuel emulsion combustion. In: Ivanov, V., Trojanowska, J., Machado, J., Liaposhchenko, O., Zajac, J., Pavlenko, I., Edl, M., Perakovic, D. (eds.) DSMIE 2019. LNME, pp. 853–862. Springer, Cham (2020). https://doi.org/10.1007/978-3-030-22365-6_85

17. Kornienko, V., Radchenko, R., Stachel, A., Andreev, A., Pyrysunko, M.: Correlations for pollution on condensing surfaces of exhaust gas boilers with water-fuel emulsion combustion. In: Tonkonogyi, V., Ivanov, V., Trojanowska, J., Oborskyi, G., Edl, M., Kuric, I., Pavlenko, I., Dasic, P. (eds.) InterPartner 2019. LNME, pp. 530–539. Springer, Cham (2020). https://doi.org/10.1007/978-3-030-40724-7_54

18. Cherednichenko, O., Serbin, S., Dzida, M.: Application of thermo-chemical technologies for conversion of associated gas in diesel-gas turbine installations for oil and gas floating units. Pol. Maritime Res. 26(3), 181–187 (2019). https://doi.org/10.2478/pomr-2019-0059

19. Larsen, U., Pierobon, L., Baldi, F., et al.: Development of a model for the prediction of the fuel consumption and nitrogen oxides emission trade-off for large ships. Energy 80, 545–555 (2015). https://doi.org/10.1016/j.energy.2014.12.009

20. Guohui, J., Jianxin, F.: Technology development of gross energy utilization system for marine diesel engine. Diesel Engine 6, 7–10 (2010)

21. Radchenko, M., Radchenko, R., Ostapenko, O., et al.: Enhancing the utilization of gas engine module exhaust heat by two-stage chillers for combined electricity, heat and refrigeration. In: 2018 5th International Conference on Systems and Informatics (ICSAI), Nanjing, pp. 240–244. IEEE (2018). https://doi.org/10.1109/ICSAI.2018.8599492

22. Yang, M.H.: Payback period investigation of the organic Rankine cycle with mixed working fluids to recover waste heat from the exhaust gas of a large marine diesel engine. Energy Convers. Manag. 162, 189–202 (2018). https://doi.org/10.1016/j.enconman.2018.02.032

23. Chen, T., Zhuge, W., Zhang, Y., Zhang, L.: A novel cascade organic Rankine cycle (ORC) system for waste heat recovery of truck diesel engines. Energy Convers. Manag. 138, 210–223 (2017). https://doi.org/10.1016/j.enconman.2017.01.056

24. Liu, Z., Xie, N., Yang, S.: Thermodynamic and parametric analysis of a coupled $LiBr/H_2O$ absorption chiller/Kalina cycle for cascade utilization of low-grade waste heat. Energy Convers. Manag. 205, 112370 (2020). https://doi.org/10.1016/j.enconman.2019.112370

25. Chistyakov, F.M.: Refrigeration Turbo Aggregates. Mashinostroenie, Moscow (1967). (in Russian)

26. Forduy, S., Radchenko, A., Kuczynski, W., Zubarev, A., Konovalov, D.: Enhancing the gas engines fuel efficiency in integrated energy system by chilling cyclic air. In: Tonkonogyi, V., Ivanov, V., Trojanowska, J., Oborskyi, G., Edl, M., Kuric, I., Pavlenko, I., Dasic, P. (eds.) InterPartner 2019. LNME, pp. 500–509. Springer, Cham (2020). https://doi.org/10.1007/978-3-030-40724-7_51

27. Radchenko, N.: A concept of the design and operation of heat exchangers with change of phase. Arch. Thermodyn. 25(4), 3–18 (2004)

28. Radchenko, M., Radchenko, R., Tkachenko, V., Kantor, S., Smolyanoy, E.: Increasing the operation efficiency of railway air conditioning system on the base of its simulation along the route line. In: Nechyporuk, M., Pavlikov, V., Kritskiy, D. (eds.) Integrated Computer Technologies in Mechanical Engineering. AISC, vol. 1113, pp. 461–467. Springer, Cham (2020). https://doi.org/10.1007/978-3-030-37618-5_39

29. Konovalov, D., Trushliakov, E., Radchenko, M., Kobalava, H., Maksymov, V.: Research of the aerothermopressor cooling system of charge air of a marine internal combustion engine under variable climatic conditions of operation. In: Tonkonogyi, V., Ivanov, V., Trojanowska, J., Oborskyi, G., Edl, M., Kuric, I., Pavlenko, I., Dasic, P. (eds.) InterPartner 2019. LNME, pp. 520–529. Springer, Cham (2020). https://doi.org/10.1007/978-3-030-40724-7_53

30. Radchenko, A., Bohdal, L., Zongming, Y., Portnoi, B., Tkachenko, V.: Rational designing of gas turbine inlet air cooling system. In: Tonkonogyi, V., Ivanov, V., Trojanowska, J., Oborskyi, G., Edl, M., Kuric, I., Pavlenko, I., Dasic, P. (eds.) InterPartner 2019. LNME, pp. 591–599. Springer, Cham (2020). https://doi.org/10.1007/978-3-030-40724-7_60

Efficient Ship Engine Cyclic Air Cooling by Turboexpander Chiller for Tropical Climatic Conditions

Mykola Radchenko[1]([⊠]) [iD], Dariusz Mikielewicz[2] [iD],
Andrii Andreev[3] [iD], Serhiy Vanyeyev[4] [iD], and Oleg Savenkov[1] [iD]

[1] Admiral Makarov National University of Shipbuilding,
9 Heroes of Ukraine Avenue, Mykolayiv 54025, Ukraine
nirad50@gmail.com
[2] Gdansk University of Technology, 11/12 Gabriela Narutowicza Street,
80-233 Gdansk, Poland
[3] Kherson Branch of Admiral Makarov National University of Shipbuilding,
44 Ushakova Avenue, Kherson 73000, Ukraine
[4] Sumy State University, 2 Rymskogo-Korsakova Street, Sumy 40007, Ukraine

Abstract. The operation of the main ship diesel engines at high ambient temperatures are characterized by falling their fuel efficiency. In particular, the increased thermal loads on the engine cyclic air cooling systems are peculiar for tropical climatic conditions. This requires application of efficient waste heat recovery technologies. The cooling of the air at the inlet of engine by absorption lithium bromide chiller (ACh) is characterized by a high efficiency of transformation of waste heat into cold – by high coefficients of performance COP = 0.7–0.8. But the lowest temperature of air cooled by ACh of a simple cycle is limited by 15 °C, that is caused by a comparatively high temperature of its chilled water of about 7 °C. Meantime, the application of a refrigerant as a coolant enables deeper cooling air down to 10 °C and lower due to lower temperature of boiling refrigerant in intake air cooler. As alternative variant, the application of a refrigeration turboexpander chiller (TExpCh), characterized by a high COP of about 0.7 and comparatively small dimensions, that is very important for ship application, was investigated. The effect of cooling engine cyclic air for both ACh and TExpCh was estimated by fuel saving compared to the engine without intake air cooling and taking into account the changing climatic conditions during the vessel routes.

Keywords: Air cooling system · Absorption · Turboexpander · Chiller · Fuel

1 Introduction

The peculiarity of the main ship diesel engine operation is an increased fuel consumption at increased intake air temperature [1, 2]. This requires efficient transformation of the waste heat into the cold when waste heat recovery chillers are applied.

M. Nechyporuk et al. (Eds.): ICTM 2020, LNNS 188, pp. 498–507, 2021.
https://doi.org/10.1007/978-3-030-66717-7_42

The cooling of the air at the inlet of the engine by absorption lithium bromide chiller (ACh) is characterized by a high efficiency of transformation of waste heat into cold – high coefficients of performance (COP): $\zeta = 0.7$–0.8. However, the temperature of air cooled by ACh of a simple cycle is limited by 15 °C, that is caused by a comparatively high temperature of its chilled water of about 7 °C [3, 4]. Meanwhile, the application of a refrigerant as a coolant enables deeper cooling air down to 10 °C and lower due to lower temperature of boiling refrigerant in the intake air cooler [5, 6]. Therefore, as alternative variant, the application of a refrigerant turbo-expander chiller (TExpCh), characterized by a high COP of about $\zeta = 0.7$ and comparatively small dimensions, that is very important for ship application, was investigated [7, 8]. The effect of cooling engine cyclic air for both ACh and TExpCh was estimated by fuel saving [9, 10] compared to the engine without intake air cooling and taking into account the changing climatic conditions during the vessel routes.

2   Literature Review

Nowadays, it is promising to use technologies that would ensure an increase in fuel efficiency of gas engines [11, 12], gas turbines [13, 14], diesel engines [15, 16]. They would combine high fuel efficiency of engines with waste heat utilization [17, 18], using alternative fuels [19, 20] and water-fuel emulsion [21, 22]. Engine cyclic air cooling might be conducted by applying innovative refrigeration recirculation cycles for intensification of heat transfer [23, 24]. Modern methods are applied by using ANSIS [25] for optimizing the sizes and regimes of equipment operation [26, 27] and statistical methods for treating the data on influence of actual climatic conditions on performance characteristics of engines and cooling systems, including monitoring data on fuel efficiency [20] and cooling system rational designing [28, 29] with determining rational loads.

Such technologies provide engine cyclic air cooling by waste heat using chillers, for instance, absorption chiller (ACh) as a high efficient heat transformer with coefficient of performance (COP) $\zeta = 0.6$–0.7. So, it is quite reasonable to apply a compact and high efficient ($\zeta = 0.7$–0.8) refrigerant turbo-expander chiller (TExpCh) [6].

The *object* of the study is assessment of fuel efficiency of a ship main engine due to cooling the intake air by TExpCh using the waste heat of gas, as alternative to ACh, that enables deeper intake air cooling and additional fuel reduction as result with regard to variable climatic conditions during the vessel routes.

3   Research Methodology

The effectiveness of the application of proposed ship engine cyclic air cooling systems based on the TExpCh and ACh was analyzed for the ship low-speed two-stroke diesel engine MAN B&W 10L32/44CR of power output 5970 kW.

In order to analyze the parameters of the cooling system, as well as the characteristics of the main engine, the CEAS software package of the manufacturer MAN was used [10]. The calculation was made for the following initial data: the rated operation characteristics of the main engine MAN B&W 10L32/44CR under ISO conditions: power $P_e = 5970$ kW; speed $n_e = 96.5$ rpm; specific fuel consumption $b_e = 176$ g/(kWh).

There are two types of high efficient chillers with coefficient of performance of about $\zeta = 0.7$ are considered: ACh manufactured as mono block unit of quite large dimensions, that requires special refrigeration room, and TExpCh which is much more complex in design. Both chillers use the waste heat of engine exhaust gas.

The effectiveness of the application of proposed systems for cooling intake air at the inlet of the main ship low speed diesel engine (LSE) turbocharger by using the heat of exhaust gas was estimated by reduction of fuel consumption with regard to variable climatic conditions during the vessel route "Shanghai (PRC)-Karachi (Pakistan)-Shanghai" in July 2019.

According to calculations carried out by the CEAS software package [10], when cooling intake air for every 1 °C air temperature drop a reduction in specific fuel consumption is $\Delta b_e/\Delta t = 0.11...0.12$ g/(kWh).

The current temperature drops Δt_{15} during engine intake air cooling in ACh to a target temperature of cooled air $t_{a2} = 15$ °C along the vessel route were calculated. With this the current temperatures of air at the inlet of engine turbocharger, taking into account a heat influx from engine room surroundings, were accepted as $t_{amb+5} = t_{amb} + 5$ °C, and $\Delta t_{15} = t_{amb+5} - 15$ °C or $\Delta t_{10} = t_{amb+5} - 10$ °C and $\Delta t_{15-10} = 5$ °C.

The vessel route fuel saving $\sum B$ is calculated by summarizing current fuel reduction B on an hour by hour basis: $\sum B = \sum[(\Delta t \cdot \tau)(\Delta b_e/\Delta t) \cdot P_e]$. With this a climatic characteristic of engine as dependence of specific fuel consumption b_e on cyclic air temperature.

A total refrigeration capacity Q_0, when cooling engine cyclic air with flow rate G_a: $Q_0 = c_a \xi \cdot \Delta t_a \cdot G_a$, where $\Delta t_a = t_{amb+5} - t_{a2}$ – decrease of cyclic air temperature; t_{a2} – target chilled air temperature at the air cooler outlet; ξ – specific heat ratio as ratio of the total heat (air enthalpy depression) extracted from the humid air during cooling (including the latent heat of water steam condensation and sensible heat) to sensible heat extracted; c_a – specific heat of humid air [kJ/(kg · K)].

The heat required for cooling air at the inlet of engine to $t_2 = 15$ °C in ACh and TExpCh with $\zeta = 0.7$ is calculated as $Q_{h.in15(0.7)} = Q_{0.in15}/\zeta$. The parameters of the ACh are used according to operating characteristics of their manufactures: coefficient of performance $\zeta = 0.7$, temperatures chilled water (coolant) $t_c = 7$ °C, heating water $t_{hw} = 90$ °C.

4   Results

The schemes of the systems for cooling the air at the inlet of the turbocharger of marine low speed diesel engine (LSE) by ACh and TExpCh using the exhaust gas heat are shown in Fig. 1 and 2.

The characteristics of the engine intake air cooling systems were calculated for the operating a dry-cargo ship main engine in climatic conditions on the voyage line (vessel's voyage) "Shanghai-Singapore-Shanghai", 20.07.2019–9.08.2019 (Fig. 2).

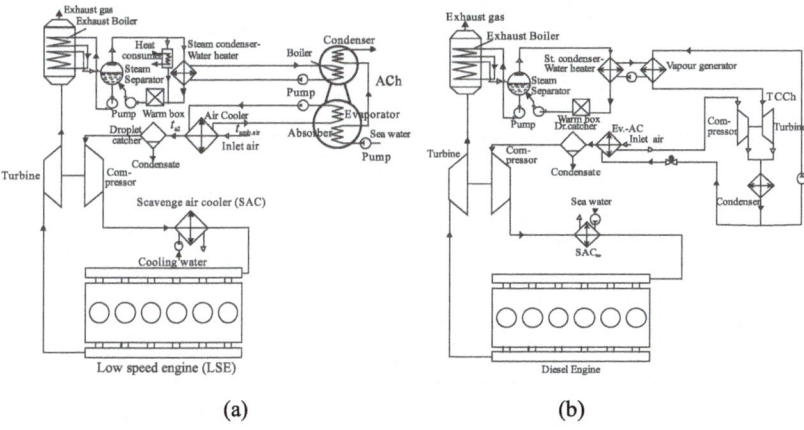

Fig. 1. Scheme of system for cooling air at the inlet of LSE in ACh: SAC – scavenge air cooler.

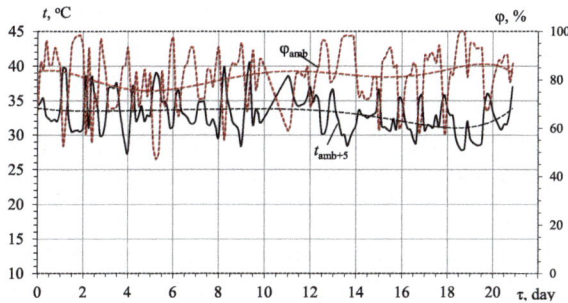

Fig. 2. The variation of intake air temperature t_{amb+5} and relative humidity φ_{amb} of ambient air during (along) the vessel trade routes "Shanghai-Singapore-Shanghai" (20.07.2019–9.08.2019).

The values of refrigeration capacity required to cool the LSE inlet air from the current temperatures at the inlet (without cooling) t_{amb+5} to $t_2 = 15$ °C and the available refrigeration capacities $Q_{0.exh(0.7)}$, that can be produced by ACh or TExpCh with coefficient of performance $\zeta = 0.7$ by using a heat of engine exhaust gas along the vessel routes Shanghai-Singapore-Shanghai (20.07.2019–9.08.2019) are shown Fig. 3.

The values of available heat potential of the exhaust gas $Q_{h.exh}$ (for LSE 10L32/44CR of power output 5970 kW) and the heat consumption $Q_{h.in15(0.7)}$ required to cool the LSE inlet air from the current temperatures at the inlet (without cooling) t_{amb+5} to $t_2 = 15$ °C in ACh or TExpCh with a thermal coefficient $\zeta = 0.7$ along the vessel routes Shanghai-Singapore-Shanghai (20.07.2019–9.08.2019) are shown Fig. 4.

The values of available heat of exhaust gas $Q_{h.exh}$ and its excess $\Delta Q_{h.in15(0.7)ex} = Q_{h.exh} - Q_{h.in15(0.7)}$ over the heat $Q_{h.in15(0.7)}$ (Fig. 4), required for cooling ambient air at the inlet of engine from current temperatures t_{amb} to $t_2 = 15$ °C in ACh with coefficient of performance $\zeta = 0.7$ along vessel route Shanghai-Singapore-Shanghai (20.07.2019–9.08.2019) can be estimated by the results in Fig. 5.

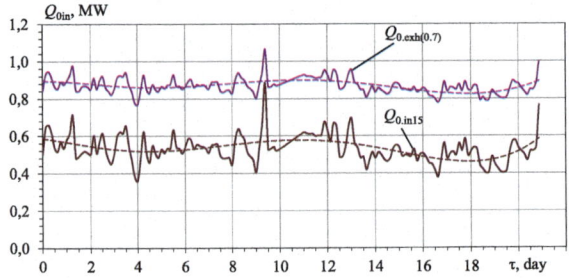

Fig. 3. Variations in refrigeration capacity $Q_{0.in15}$ required for cooling ambient air at the inlet of engine from the current temperatures t_{amb+5} to $t_2 = 15$ °C and available refrigeration capacities $Q_{0.exh(0.7)}$, that can be produced by ACh or TExpCh with coefficient of performance $\zeta = 0.7$.

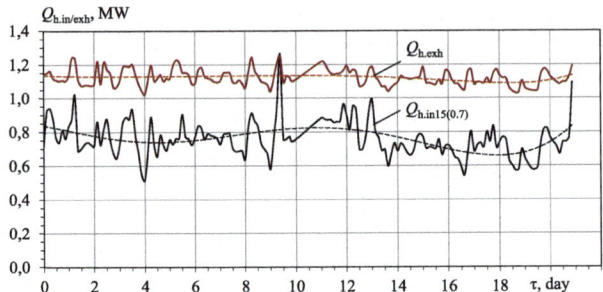

Fig. 4. Variations in heat $Q_{h.in15(0.7)}$, required for cooling ambient air at the inlet of engine from current temperatures t_{amb+5} to $t_2 = 15$ °C in ACh or TExpCh with coefficient of performance $\zeta = 0.7$ and available heat of exhaust gas $Q_{h.exh}$ for LSE 10L32/44CR of power output 5970 kW.

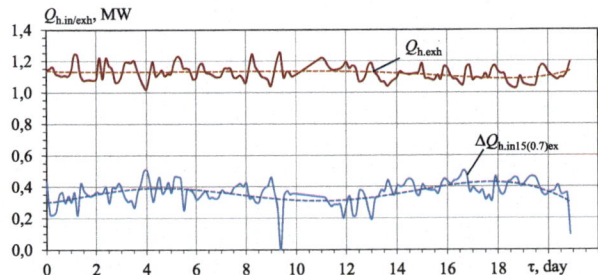

Fig. 5. The values of available heat of exhaust gas $Q_{h.exh}$ and its excess $\Delta Q_{h.in15(0.7)ex}$ over the heat $Q_{h.in15(0.7)}$, required for cooling ambient air at inlet of engine from current temperatures t_{amb} to $t_2 = 15$ °C in ACh with coefficient of performance $\zeta = 0.7$.

As can be seen from Fig. 5, due to the high efficiency of heat transformation in ACh and TExpCh ($\zeta = 0.7$) there is a significant amount of excess heat of exhaust gas $\Delta Q_{h.in15(0.7)ex}$ over the heat $Q_{h.in15(0.7)}$ required for cooling ambient air at the inlet of the

LSE to 15 °C (Fig. 4), which reaches 30–40% of the available exhaust gas heat $Q_{h.exh}$ along the vessel route Shanghai-Singapore-Shanghai.

This reveals the reserves for deep cooling the engine intake air in TExpCh down to 10 °C (Fig. 6), which is impossible in ACh because of comparably high temperature of its chilled water $t_{cw} = 7$ °C not allowing cooling air to the temperatures lower than $t_{a2} = 15$ °C with account of the temperature difference between cooled air and chilled water of about 8 °C at the outlet of intake air cooler.

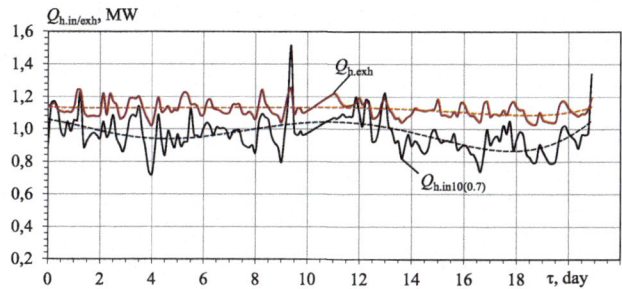

Fig. 6. Variations in heat $Q_{h.in10(0.7)}$ required for deep cooling ambient air at the inlet of engine from current temperatures t_{amb+5} to $t_2 = 10$ °C in TExpCh with coefficient of performance $\zeta = 0.7$ and available heat of exhaust gas $Q_{h.exh}$ for LSE 10L32/44CR of power output 5970 kW along the vessel routes Shanghai-Singapore-Shanghai (20.07.2019–9.08.2019).

Current values of reduction in the air temperature at the inlet of engine while its cooling down to 15 °C Δt_{in15} in ACh and to 10 °C Δt_{in10} in TExpCh, corresponding values of reduction in specific fuel consumption Δb_{in15} and Δb_{in10} and in the summarized fuel consumption $\sum B_{in15}$ and $\sum B_{in10}$ for LSE10L32/44CR of power output 5970 kW during "Shanghai-Singapore-Shanghai" (20.07.2019–9.08.2019) are shown in Fig. 7.

As can be seen, due to reducing the ambient air temperatures at the inlet of the LSE from t_{amb+5} to $t_2 = 15$ °C with their reductions $\Delta t_{in15} = 18$–23 °C in ACh which uses the exhaust gas heat $Q_{h.in15(0.7)}$ the specific fuel consumption is reduced by $\Delta b_{in(0.7)} = 2.0...2.7$ g/(kWh) and the total summarized reduction of fuel consumption – by $\sum B_{in15} \approx 2$ t for LSE 10L32/44CR of power output 5970 kW during the vessel routes Shanghai-Singapore-Shanghai (20.07.2019–9.08.2019).

Meantime due to deep cooling engine intake air from t_{amb+5} to $t_2 = 10$ °C with air temperature reductions $\Delta t_{in10} = 23$–28 °C in TExpCh which uses the exhaust gas heat $Q_{h.exh}$ the specific fuel consumption is reduced by $\Delta b_{in(0.7)} = 2.5...3.3$ g/(kWh) and the total summarized reduction of fuel consumption – by $\sum B_{in15} = 2.5$ t.

It should be noted that in spite of the fact that ACh is able to transfer the overall exhaust gas heat $Q_{h.exh}$ into refrigeration, but it is impossible to use all the refrigeration capacity produced in ACh for deep cooling engine intake air lower than $t_2 = 15$ °C because of comparably high temperature of its chilled water $t_{cw} = 7$ °C, whereas in TExpCh the temperature of boiling refrigerant can be about $t_0 = 2$–3 °C and lower that enables deep cooling intake air to $t_2 = 10$ °C and lower.

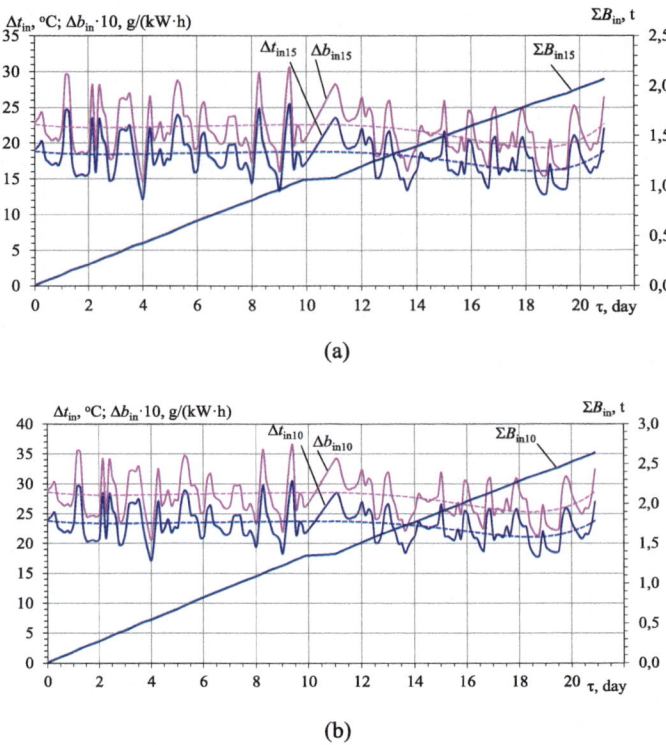

(a)

(b)

Fig. 7. Current values of reduction in the air temperature at the inlet of engine while its cooling down to 15 °C Δt_{in15} in ACh (a) and to 10 °C Δt_{in10} in TExpCh (b), corresponding values of reduction in specific fuel consumption Δb_{in15} and Δb_{in10} and in the summarized fuel consumption $\sum B_{in15}$ and $\sum B_{in10}$ for LSE10L32/44CR of power output 5970 kW along the vessel routes "Shanghai-Singapore-Shanghai" (20.07.2019–9.08.2019).

5 Conclusions

The efficiency of the main low speed engine cyclic air cooling of a transport vessel during operation in tropical climatic conditions on the Shanghai-Singapore-Shanghai routes was analyzed. The cooling of the air at the inlet of engine by absorption lithium bromide chiller of a simple cycle to 15 °C and by refrigeration turbo expander chiller (TExpCh) as a low-temperature chiller to 10 °C through using the heat of exhaust gas were investigated.

It was shown, that due to the high efficiency of heat transformation in ACh and TExpCh ($\zeta = 0.7$) there is a significant amount of excess heat of exhaust gas over the heat required for cooling ambient air at the inlet of the LSE to 15 °C, which reaches 30–40% of the available exhaust gas heat along the vessel route Shanghai-Singapore-Shanghai.

This reveals the reserves for deep cooling the engine intake air in TExpCh down to 10 °C, which is impossible in ACh because of comparatively high temperature of its chilled water $t_{cw} = 7$ °C not allowing cooling air to the temperatures lower than 15 °C.

It was shown that application of ACh allows to reduce specific fuel consumption by 2.0…2.7 g/(kWh) and the total summarized reduction of fuel consumption – by about 2 t for LSE 10L32/44CR of power output 5970 kW during the vessel routes Shanghai-Singapore-Shanghai (20.07.2019–9.08.2019).

Meantime due to deep cooling engine intake air to 10 °C in TExpCh the specific fuel consumption is reduced by 2.5…3.3 g/(kWh) and the total reduction of fuel consumption – by 2,5 t.

The scheme solutions of the air cooling system at the inlet of the ship main engine using the heat of exhaust gases by ACh and TExpCh are proposed.

References

1. Zhang, Q., Luo, Z., Zhao, Y., Cao, R.: Performance assessment and multi-objective optimization of a novel transcritical CO_2 trigeneration system for a low-grade heat resource. Energy Convers. Manag. **204**, 112281 (2020). https://doi.org/10.1016/j.enconman.2019.112281
2. Wärtsilä: Environmental Product Guide. Wärtsilä (2017)
3. Cao, T., Lee, H., Hwang, Y., et al.: Performance investigation of engine waste heat powered absorption cycle cooling system for shipboard applications. Appl. Therm. Eng. **90**, 820–830 (2015). https://doi.org/10.1016/j.applthermaleng.2015.07.070
4. Pham, V.V.: Advanced technology solutions for treatment and control noxious emission of large marine diesel engines: a brief review. J. Mech. Eng. Res. Dev. **42**(5), 21–27 (2019)
5. Liu, Z., Xie, N., Yang, S.: Thermodynamic and parametric analysis of a coupled LiBr/H_2O absorption chiller/Kalina cycle for cascade utilization of low-grade waste heat. Energy Convers. Manag. **205**, 112370 (2020). https://doi.org/10.1016/j.enconman.2019.112370
6. Chistyakov, F.M.: Refrigeration Turbo Aggregates. Mashinostroenie, Moscow (1967). (in Russian)
7. Bohdal, T., Kuczyński, W.: Boiling of R404A refrigeration medium under the conditions of periodically generated disturbances. Heat Transfer Eng. **32**(5), 359–368 (2011). https://doi.org/10.1080/01457632.2010.483851
8. Forduy, S., Radchenko, A., Kuczynski, W., Zubarev, A., Konovalov, D.: Enhancing the gas engines fuel efficiency in integrated energy system by chilling cyclic air. In: Tonkonogyi, V., Ivanov, V., Trojanowska, J., Oborskyi, G., Edl, M., Kuric, I., Pavlenko, I., Dasic, P. (eds.) InterPartner 2019. LNME, pp. 500–509. Springer, Cham (2020). https://doi.org/10.1007/978-3-030-40724-7_51
9. MAN Diesel & Turbo: MAN B&W S50MC-C8.2-TII project guide: Camshaft controlled two-stroke engines. https://marine.man-es.com/applications/projectguides/2stroke/content/epub/S50MC-C8_2.pdf. Accessed 15 Apr 2020
10. MAN Diesel & Turbo: CEAS engine calculations. https://marine.man-es.com/two-stroke/ceas. Accessed 15 Apr 2020
11. Payrhuber, K., Trapp, C.: GE's new jenbacher gas engines with 2-stage turbocharging. In: 7 Internationale Energiewirtschaftstagung an der TU Wien IEWT (2011)

12. Trapp, C., Laiminger, S., Chvatal, D., et al.: Die neue gasmotorengeneration von GE jenbacher – mit zweistufiger aufladung zu höchsten wirkungsgraden. In: International Vienna Motor Symposium, pp. 281–297 (2011)

13. Dizaji, H.S., Hu, E.J., Chen, L., Pourhedayat, S.: Using novel integrated Maisotsenko cooler and absorption chiller for cooling of gas turbine inlet air. Energy Convers. Manag. **195**, 1067–1078 (2019). https://doi.org/10.1016/j.enconman.2019.05.064

14. Chacartegui, R., Jiménez-Espadafor, F., Sanchez, D., Sanchez, T.: Analysis of combustion turbine inlet air cooling systems applied to an operating cogeneration power plant. Energy Convers. Manag. **49**(8), 2130–2141 (2008). https://doi.org/10.1016/j.enconman.2008.02.023

15. Jie, M.: Ways and measures of waste gas utilization of super-large ships. J. North China Electr. Power Univ. (2010)

16. Shu, G., Liang, Y., Wei, H., et al.: A review of waste heat recovery on two-stroke IC engine aboard ships. Renew. Sustain. Energy Rev. **19**, 385–401 (2013). https://doi.org/10.1016/j.rser.2012.11.034

17. Guohui, J., Jianxin, F.: Technology development of gross energy utilization system for marine diesel engine. Diesel Engine **6**, 7–10 (2010)

18. Radchenko, R., Kornienko, V., Pyrysunko, M., Bogdanov, M., Andreev, A.: Enhancing the efficiency of marine diesel engine by deep waste heat recovery on the base of its simulation along the route line. In: Nechyporuk, M., Pavlikov, V., Kritskiy, D. (eds.) Integrated Computer Technologies in Mechanical Engineering. AISC, vol. 1113, pp. 337–350. Springer, Cham (2020). https://doi.org/10.1007/978-3-030-37618-5_29

19. Cherednichenko, O., Serbin, S., Dzida, M.: Application of thermo-chemical technologies for conversion of associated gas in diesel-gas turbine installations for oil and gas floating units. Pol. Marit. Res. **26**(3), 181–187 (2019). https://doi.org/10.2478/pomr-2019-0059

20. Trushliakov, E., Radchenko, A., Forduy, S., Zubarev, A., Hrych, A.: Increasing the operation efficiency of air conditioning system for integrated power plant on the base of its monitoring. In: Nechyporuk, M., Pavlikov, V., Kritskiy, D. (eds.) Integrated Computer Technologies in Mechanical Engineering. AISC, vol. 1113, pp. 351–360. Springer, Cham (2020). https://doi.org/10.1007/978-3-030-37618-5_30

21. Kornienko, V., Radchenko, R., Konovalov, D., Andreev, A., Pyrysunko, M.: Characteristics of the rotary cup atomizer used as afterburning installation in exhaust gas boiler flue. In: Ivanov, V., Pavlenko, I., Liaposhchenko, O., Machado, J., Edl, M. (eds.) DSMIE 2020. LNME, pp. 302–311. Springer, Cham (2020). https://doi.org/10.1007/978-3-030-50491-5_29

22. Kornienko, V., Radchenko, R., Mikielewicz, D., et al.: Improvement of characteristics of water-fuel rotary cup atomizer in a boiler. In: Tonkonogyi, V., et al. (eds.) Advanced Manufacturing Processes II. InterPartner 2020. LNME. Springer, Cham (2021)

23. Trushliakov, E., Radchenko, M., Bohdal, T., Radchenko, R., Kantor, S.: An innovative air conditioning system for changeable heat loads. In: Tonkonogyi, V., Ivanov, V., Trojanowska, J., Oborskyi, G., Edl, M., Kuric, I., Pavlenko, I., Dasic, P. (eds.) InterPartner 2019. LNME, pp. 616–625. Springer, Cham (2020). https://doi.org/10.1007/978-3-030-40724-7_63

24. Bohdal, T., Sikora, M., Widomska, K., Radchenko, A.M.: Investigation of flow structures during HFE-7100 refrigerant condensation. Arch. Thermodyn. **36**(4), 25–34 (2015). https://doi.org/10.1515/aoter-2015-0030

25. Bohdal, Ł., Kukiełka, L., Legutko, S., et al.: Modeling and experimental analysis of shear-slitting of AA6111-T4 Aluminum alloy sheet. Materials **13**(14), 3175 (2020). https://doi.org/10.3390/ma13143175

26. Bohdal, L., Kukielka, L., Świłło, S., et al.: Modelling and experimental analysis of shear-slitting process of light metal alloys using FEM, SPH and vision-based methods. In: AIP Conference Proceedings, vol. 2078, no. 1, p. 020060 (2019). https://doi.org/10.1063/1.5092063
27. Bohdal, L., Kukielka, L., Radchenko, A.M., et al.: Modelling of guillotining process of grain oriented silicon steel using FEM. In: AIP Conference Proceedings, vol. 2078, no. 1, p. 020080 (2019). https://doi.org/10.1063/1.5092083
28. Radchenko, A., Bohdal, L., Zongming, Y., Portnoi, B., Tkachenko, V.: Rational designing of gas turbine inlet air cooling system. In: Tonkonogyi, V., Ivanov, V., Trojanowska, J., Oborskyi, G., Edl, M., Kuric, I., Pavlenko, I., Dasic, P. (eds.) InterPartner 2019. LNME, pp. 591–599. Springer, Cham (2020). https://doi.org/10.1007/978-3-030-40724-7_60
29. Radchenko, A., Stachel, A., Forduy, S., Portnoi, B., Rizun, O.: Analysis of the efficiency of engine inlet air chilling unit with cooling towers. In: Ivanov, V., Pavlenko, I., Liaposhchenko, O., Machado, J., Edl, M. (eds.) DSMIE 2020. LNME, pp. 322–331. Springer, Cham (2020). https://doi.org/10.1007/978-3-030-50491-5_31

Modelling of Condenser Circuit of the Geothermal Heat Pump

Svitlana Matus$^{(\boxtimes)}$ ⓘ, Bohdan Sydorchuk ⓘ,
and Oleksandr Naumchuk ⓘ

National University of Water and Environmental Engineering, 11 Soborna Street,
Rivne 33028, Ukraine
{s.k.matus, b.p.sydorchuk, o.m.naumchuk}@nuwm.edu.ua

Abstract. An approach to the study of heat exchange processes of the heat pump condenser circuit has been proposed. A mathematical model of the heat exchange process of the heat pump condenser circuit and the consumption line circuit has been constructed. The mathematical model is based on the equations of material and energy balances of the working fluid of the heat pump. Transfer function by the temperature of the freon in the condenser due to the regulating action of the temperature of the liquid of the consumption line has been constructed as well as the transfer function by the temperature in the consumption line due to the control action of the working fluid in the heat pump circuit. The transient responses for the freon temperature in the condenser and the temperature of the consumption line liquid have been obtained, respectively, and their research has been carried out with the help of simulation modeling. The transient responses of the studied circuits have been built. The results and graphs of transient processes and dynamic responses of the object have been given. It has been established that at lower consumption of liquid in the line of consumption the constant of time of transient process is of inertial character, and, therefore, in this case, the investigated conditions necessitate the increase of expenditures in the line of consumption, or in performance control of a compressor.

Keywords: Heat pump · Condenser · Transfer function

1 Introduction

At the present stage of technology development the use of renewable heat sources is especially topical. Geothermal heat pumps use ground heat, i.e. low-potential heat source and provide autonomous heating and high-temperature water supply.

The main elements of the heat supply system of geothermal heat pumps are heat exchangers which transfer low-potential heat sources from the ground to the consumers of heat. An important part of such a system is the heat exchanger, in which heat is transferred from the condenser of the heat pump to consumers.

In fundamental research in the calculations of heat pumps, usually all possible thermodynamic cycles, efficiency of installations and their constructive solutions are considered [1–3]. In [1] the general model for different technologies of heat compression is proposed. With the help of the developed model productivity of heat pumps

© The Author(s), under exclusive license to Springer Nature Switzerland AG 2021
M. Nechyporuk et al. (Eds.): ICTM 2020, LNNS 188, pp. 508–518, 2021.
https://doi.org/10.1007/978-3-030-66717-7_43

is analyzed, as well as the choice of cooling agent and its operational analysis. In [2] it is proposed to divide the heat exchangers of the evaporator and condenser circuits into sections for supplying coolant with increased pressure to obtain a higher heat transfer temperature. In [3] it is proposed to add an intermediate heat exchanger with a control valve for the possibility of pressure control in the case of cascading inclusion of compressors in the heat pump circuit. However, little attention is paid to the processes occurring, in particular, in the condenser circuit from the point of view of heat transfer.

Modeling of separate parts of heat pumps is investigated in [4, 5]. In [4], it was proposed to represent the dependence of the torque of a single-piston compressor on time as the sum of sine waves dependent on the angle of rotation of the electric shaft, and in [5] separate models of compressor, heat exchangers of evaporator and condenser were built and numerical calculations were performed.

In [6] a model for the study of thermodynamic properties of liquids is proposed, the results of research and their consistency with experimental data are given. In [7], a heat pump with several condensers operating at different consumption lines with the equal pressure and at constant loads is described. Also the focus is drawn to the research of high-temperature heat pumps. Thus, in [8] heat pumps with heat transfer temperature up to 160 °C have been investigated, where the main attention has been paid to the analysis of operating ranges of heat pumps.

A number of works are devoted to the study of heat transfer from the ground heat exchanger to the inner circuit of the heat pump. The work [9] analyzes the operating conditions and prediction of the conditions of heat transfer from the ground to the horizontal ground heat exchanger. The approach to the assessment of the main reasons of the loss and deterioration of the pump performance as well as for the change in operating parameters have been given. In [10], a three-dimensional numerical model was developed that allows us to study the heat transfer process from the ground heat exchanger horizontally – from the spiral heat exchanger to the heat pump. In [11], an algorithm for determining the approximate transfer function of the ground heat exchanger is constructed taking into account the convective heat transfer in the heat exchanger tube and the soil temperature by depth.

2 Description of Heat Pump Operation and Statement of the Research Problem

The principle of operation of all heat pumps is the same and is based on the well-known so-called «the Carnot cycle» [1]. Freon, a liquid with a low boiling point, circulates in a closed circuit. The main elements of the internal circuits of heat pumps (Fig. 1) include 1 – evaporator, 2 – compressor, 3 – condenser and 4 – throttling device. Liquid freon enters the evaporator through the throttling device, in which due to the low-potential heat from the ground circuit a transition from liquid to gas state occurs. Once in the condenser, the freon is compressed to the desired value and the temperature of the freon increases. After that, under high pressure in the condenser the heated freon gives the received heat away through the heat exchanger in a heating circuit.

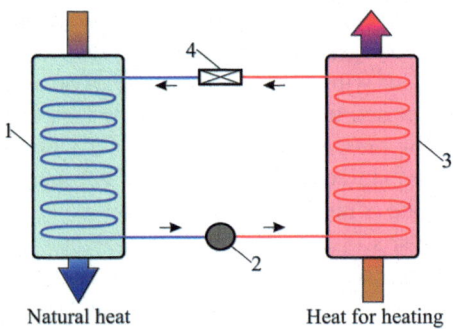

Fig. 1. Principal diagram of the heat pump operation.

In order to visualize the process of heat transfer from a low-potential power source to consumers, let us consider a logarithmic diagram of pressure/enthalpy of the physical state of a freon. In the section 1–2 (Fig. 2) refrigerant (freon) evaporates. The energy required for this (enthalpy of evaporation) will be released into the environment. Then in the section 2–3 the compressor increases the pressure with the help of mechanical energy and, accordingly, the temperature of the refrigerant. Thus, the enthalpy increases. In the section 3–4, the refrigerant condenses as a result of compression and gives off the absorbed energy during evaporation and the energy obtained from mechanical compression. In 4–1, the refrigerant expands with the throttle device and returns to its original state with certain temperature and pressure. The problem consists in building a mathematical model of the section 3–4 and studying with its help the processes of heat transfer from the condenser circuit of the heat pump to the circuit of the consumption line.

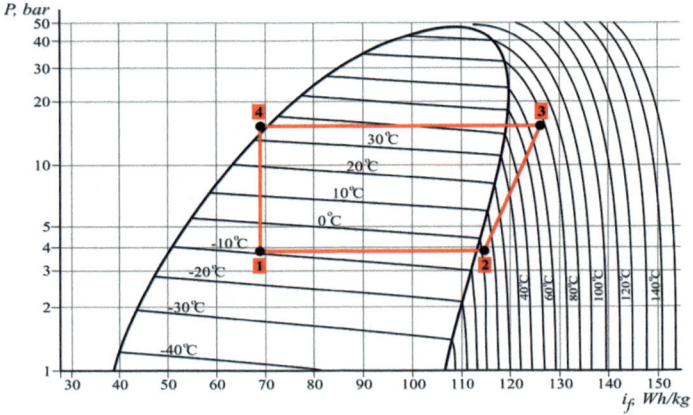

Fig. 2. Pressure/enthalpy diagram.

3   The Mathematical Model of Heat Transfer Between Condenser Circuit of Heat Pump and Circuit of Consumption Line

Let us consider a mathematical model that describes the process of heat transfer in the heat pump circuit, namely in the heat exchanger, where heat is transferred from the heat pump condenser to the consumers.

A refrigerant (freon), which flows through the condenser (Fig. 3), condenses as a result of compression and gives off the absorbed energy during evaporation and the energy obtained from mechanical compression. At the same time the temperature of the consumer circuit increases. We have adopted a condenser in the form of a heat exchanger of the type "pipe in a pipe" under the assumption that heat loss to the environment does not occur.

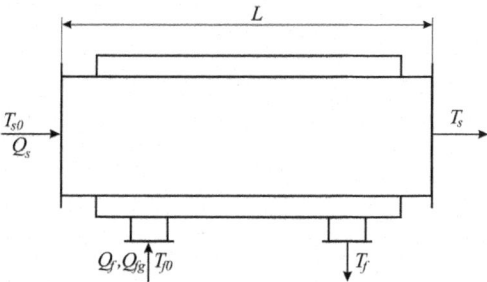

Fig. 3. Heat-exchange scheme from the condenser of the heat pump to consumers.

Mass balance equation for freon in the condenser has the form.

$$\frac{dm_f}{dt} = Q_f - Q_{fg}, \tag{1}$$

where m_f is the mass of freon in the condenser, Q_f and Q_{fg} are corresponding freon consumptions in the liquid and gaseous state.

Energy balance equation for a freon in the liquid state has the form.

$$c_{pf}\frac{d(m_f T_f)}{dt} = kF(T_s - T_f) + Q_f c_{pf} T_{f0} - Q_{fg} i_f, \tag{2}$$

where k is the heat transfer coefficient through the heating surface, c_{pf} is the specific heat of freon, T_f, T_{f0} are the temperatures of freon inside the condenser and at its entrance point, T_s is the temperature of the consumption fluid in the heat exchanger, F is the heat transfer surface between the consumption fluid and the freon, $Q_{fg} i_f$ is the amount of heat expended on the condensation of freon, i_f is the specific enthalpy of freon, which has turned into a liquid state, $i_f = \lambda + c_{pf} T_f$, λ is the specific heat of condensation of freon.

Differentiating the left hand side of the Eq. (1) as a function of two arguments and equating to (2), we obtain.

$$c_{pf}m_f\frac{dT_f}{dt} = kF(T_{sp} - T_f) - Q_fc_{pf}(T_f - T_{f0}) - Q_{fg}(i_f - c_{pf}T_f). \tag{3}$$

A similar equation is obtained for the consumption fluid that passes through the heat exchanger of the condenser.

$$c_{ps}m_s\frac{dT_s}{dt} = kF(T_f - T_s) - Q_sc_{ps}(T_s - T_{s0}), \tag{4}$$

where c_{ps}, m_s are the specific heat and the mass of the consumption fluid respectively, T_{s0} is the temperature of the consumption fluid at the entrance point to the heat exchanger, Q_s is the flow rate of the consumption fluid.

Therefore, the following system will be used to solve this problem.

$$\begin{cases} c_{pf}m_f\frac{dT_f}{dt} = kF(T_s - T_f) - Q_fc_{pf}(T_f - T_{f0}) - Q_{fg}(i_f - c_{pf}T_f), \\ c_{ps}m_s\frac{dT_s}{dt} = kF(T_f - T_s) - Q_sc_{ps}(T_s - T_{s0}). \end{cases} \tag{5}$$

Using the Laplace transform, we convert Eqs. (5) to the form.

$$\begin{cases} (c_{pf}m_fp + kF + Q_fc_{pf} - Q_{fg}c_{pf})T_f = kFT_s + f_1, \\ (c_{ps}m_sp + kF + Q_sc_{ps})T_s = kFT_f + f_2, \end{cases} \tag{6}$$

where $f_1 = Q_fc_{pf}T_{f0} - Q_{fg}i_f$, $f_2 = Q_sc_{ps}T_{s0}$.

Let us define the transfer functions for each equation of the system (6), and for the first equation let us set T_s as the control effect, and for the second equation – T_f.

$$\begin{cases} W_{Tf}(p) = \frac{kF}{c_{pf}m_fp + kF + Q_fc_{pf} - Q_{fg}c_{pf}}, \\ W_{Ts}(p) = \frac{kF}{c_{ps}m_sp + kF + Q_sc_{ps}}. \end{cases} \tag{7}$$

In the first transfer function of the system of Eqs. (7) the output is T_f, due to the control action of T_s under the condition of zero perturbation $f_1 = 0$. In the second transfer function the output is T_s, due to the control action of T_f under the condition of zero perturbation $f_2 = 0$.

From (7) we obtain the equations of aperiodic transition characteristics h_f (freon temperature in the condenser) and h_s (fluid temperature of the consumption line), in particular

$$h_f = \frac{kF}{kF + Q_fc_{pf} - Q_{fg}c_{pf}}\left(1 - e^{\frac{-(kF + Q_fc_{pf} - Q_{fg}c_{pf})t}{c_{pf}m_f}}\right), \tag{8}$$

$$h_s = \frac{kF}{kF + Q_s c_{ps}} \left(1 - e^{\frac{-(kF + Q_s c_{ps})t}{c_{ps} m_s}}\right). \tag{9}$$

4   Numerical Experiments of the Heat Exchange Process Between the Heat Pump Condenser Circuit and the Consumption Line Circuit

Let us study the obtained transfer functions and transient responses (Fig. 4). For the condenser the parameters of the coolant at the points on the line of saturation and condensation of freon brand $R407C$ were determined by a given value $T_{f0} = 60\,°C$. The flow rate of the consumption fluid through the heat exchanger of the condenser circuit $Q_s = 2.24\ m^3/h$.

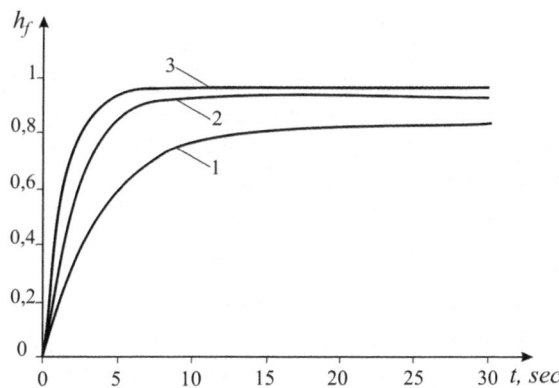

Fig. 4. Transient responses for various mass consumptions of freon through the condenser: 1 – 3.4 m^3/h; 2 – 5.1 m^3/h; 3 – 8.9 m^3/h.

As can be seen from Fig. 4, the increase of mass consumption of freon through the condenser leads to the improvement of the quality of the transient process and to the decrease of the inertia of such a circuit, namely to the increase of the gain from 0.83 to 0.95 and to the decrease of the time constant from 3.9 to 1.5.

Let us build the graphs of the amplitude-frequency (AFR) and phase-frequency responses (PFR) of the transfer function of the condenser circuit (Fig. 5).

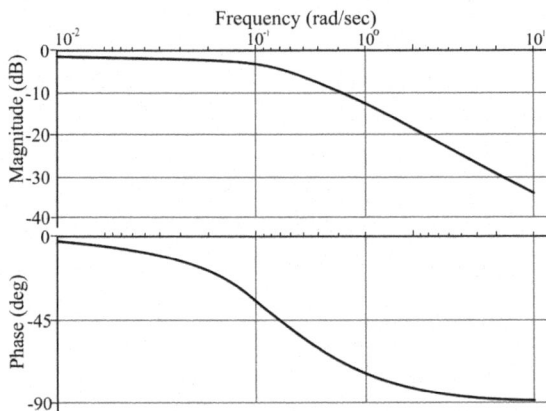

Fig. 5. The graphs of AFR and PFR of the transfer function of the condenser circuit for $Q_f = 3.4$ m³/h.

Analyzing the amplitude-frequency response, we see that this circuit has the property of a filter, i.e. it transmits signals of low frequencies well and of large frequencies badly; with the increase of frequency the amplitude of the output signal decreases. The analysis of the phase response shows that the output oscillations lag behind the input and this lag varies between 0 and 90°.

Similarly, let us study the transfer function of the consumption line under the condition of constant consumption of freon through the condenser (Fig. 6).

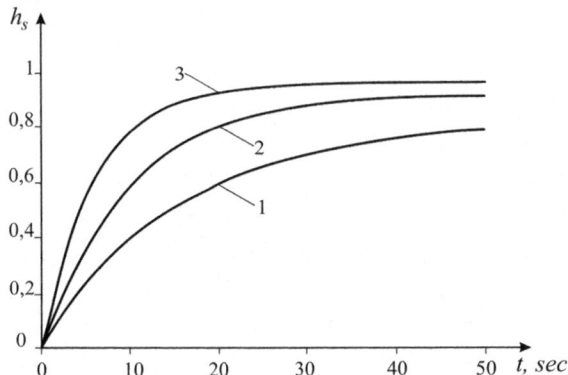

Fig. 6. Transient responses for various mass consumptions of a fluid in consumption line: 1 – 1.08 m³/h; 2 – 1.53 m³/h; 3 – 2.24 m³/h.

As can be seen from Fig. 6 at a lower consumption of fluid in the consumption line, the time constant of the transient process is inertial, and, therefore, for these conditions it is more optimal to increase the consumption in the consumption line to $Q_s = 2.24$ m³/h.

Let us plot the graphs of the amplitude-frequency (AFR) and phase-frequency responses (PFR) using the transfer function of the consumption circuit (Fig. 7).

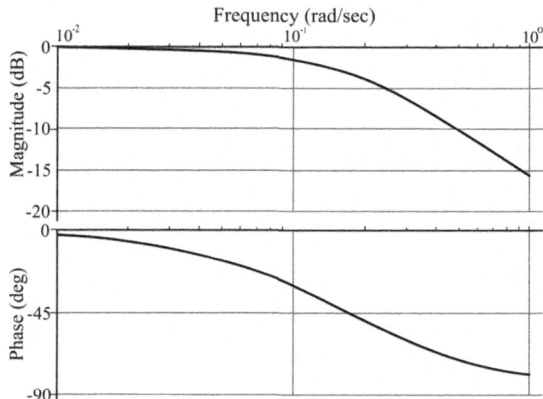

Fig. 7. The graphs of AFR and PFR of the transfer function of the consumption line circuit for $Q_s = 2.24$ m^3/h.

We observe that in the circuit of the consumption line as the frequency increases the amplitude of the output signal decreases. Therefore, a better transient process is obtained with small temperature fluctuations in the consumption line.

5   Simulation Modeling of the Heat Transfer Process from the Heat Pump Condenser to the Consumers

Let us study the heat exchange in the heat pump circuit, namely in the heat exchanger, where heat is transferred from the heat pump condenser to the consumers. In order to do this, we convert the system of Eqs. (5) to the form.

$$\begin{cases} \frac{dT_f}{dt} = \frac{1}{c_{pf}m_f}\left(\left(-kF - Q_f c_{pf} + Q_{fg}c_{pf}\right)T_f + kFT_s + Q_f c_{pf}T_{f0} - Q_{fg}i_f\right), \\ \frac{dT_s}{dt} = \frac{1}{c_{ps}m_s}\left(\left(-kF - Q_s c_{ps}\right)T_s + kFT_f + Q_s c_{ps}T_{s0}\right). \end{cases} \tag{10}$$

With the help of Simulink, the Matlab program we obtained a model (Fig. 8) that allows us to plot graphs of temperature changes in the heat exchanger: freon in the condenser circuit and fluid in the consumption line circuit (Fig. 9).

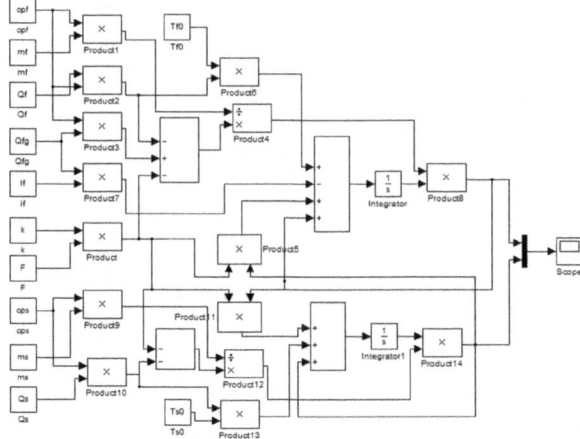

Fig. 8. Simulation model of the heat transfer process from the heat pump condenser to the consumers according to the equations system (10).

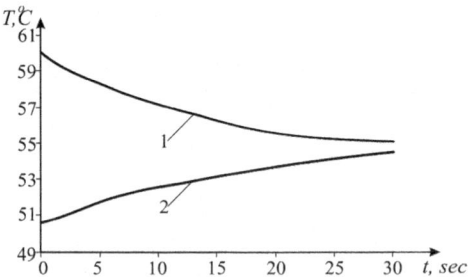

Fig. 9. Graphs of temperature changes: 1 – temperature T_f of freon; 2 – temperature T_s of the fluid in the consumption line for $Q_s = 2.24$ m^3/h, $Q_f = 3.4$ m^3/h.

From Fig. 9 we see, as expected, that as the temperature of the freon decreases, the temperature of the fluid in the consumption line increases.

The value T_{s0} depends on the need for heat consumption. In this case, there is a need to adjust the performance of the compressor. In particular, such an action is necessary both at seasonal fluctuations of ground temperature, and at the maximum need of heat consumption at the minimum ambient temperature. Such an action can be carried out in various technological ways, in particular, the most common in practice – changing the power of the compressor using inverter technology.

6   Conclusions

An approach to modeling the heat exchange process in the heat pump circuit, namely in the heat exchanger, where heat is transferred from the heat pump condenser to consumers, is proposed. The transfer functions and transient responses of the condenser circuit and the consumption line circuit are obtained. According to the results of numerical calculations, it is obtained that at a lower consumption of liquid in the consumption line the time constant of the transient process is inertial, and, therefore, for the studied conditions there is a need to increase consumption in the consumption line or control the condenser performance.

As the mass flow rate of freon through the condenser increases from 3.4 m^3/h to 8.9 m^3/h, the gain increases from 0.83 to 0.95 and the time constant decreases from 3.9 to 1.5, and as the frequency of freon temperature fluctuations increases, its amplitude decreases.

References

1. Jiang, S., Wang, S., Jin, X., Zhang, T.: A general model for two-stage vapor compression heat pump systems. Int. J. Refrig. **51**, 88–102 (2015). https://doi.org/10.1016/j.ijrefrig.2014.12.005
2. Márquez-Nolasco, A., Huicochea, A., Torres-Merino, J., et al.: Thermodynamic analysis into a heat exchanger for absorption at high temperatures. Appl. Therm. Eng. **103**, 1014–1021 (2016). https://doi.org/10.1016/j.applthermaleng.2016.04.035
3. Şit, M.L., Starikov, A.V., Zhuravleov, A.A., Timchenko, D.V.: Multi-temperature heat pump with cascade compressor connection. Probl. Reg. Energ. **2**(34), 91–98 (2017). https://doi.org/10.5281/zenodo.1189318. (in Russian)
4. Bukaros, A.Y., Bukaros, V.N., Onishchenko, O.A.: Simulation of the resistance moment of single-piston compressor of ship refrigeration unit. Technol. Audit Prod. Reserves **4**(1), 46–51 (2015). https://doi.org/10.15587/2312-8372.2015.47765. (in Russian)
5. Timofeev, D.V., Malyavina, E.G.: Computer model of heat pump with a constant rotation frequency of scroll compressor spiral. Vestnik MGSU **12**(4), 437–445 (2017). https://doi.org/10.22227/1997-0935.2017.4.437-445. (in Russian)
6. Coquelet, C., El Abbadi, J., Houriez, C.: Prediction of thermodynamic properties of refrigerant fluids with a new three-parameter cubic equation of state. Int. J. Refrig. **69**, 418–436 (2016). https://doi.org/10.1016/j.ijrefrig.2016.05.017
7. Sarkar, J.: Performance analyses of novel two-phase ejector multi-evaporator refrigeration systems. Appl. Therm. Eng. **110**, 1635–1642 (2017). https://doi.org/10.1016/j.applthermaleng.2016.08.163
8. Arpagaus, C., Bless, F., Uhlmann, M., et al.: High temperature heat pumps: market overview, state of the art, research status, refrigerants, and application potentials. Energy **152**, 985–1010 (2018). https://doi.org/10.1016/j.energy.2018.03.166
9. Verda, V., Cosentino, S., Russo, S.L., Sciacovelli, A.: Second law analysis of horizontal geothermal heat pump systems. Energy Build. **124**, 236–240 (2016). https://doi.org/10.1016/j.enbuild.2015.09.063

10. Go, G.H., Lee, S.R., Yoon, S., Kim, M.J.: Optimum design of horizontal ground-coupled heat pump systems using spiral-coil-loop heat exchangers. Appl. Energy **162**, 330–345 (2016). https://doi.org/10.1016/j.apenergy.2015.10.113

11. Sydorchuk, B., Naumchuk, O.: Modelling of vertical soil collectors of thermal pumps. Inform. Control Measur. Econ. Environ. Prot. **1**, 29–31 (2016). https://doi.org/10.5604/20830157.1194262

Numerical Simulation of an Aerothermopressor with Incomplete Evaporation for Intercooling of the Gas Turbine Engine

Halina Kobalava[1]([✉]) , Dmytro Konovalov[1] ,
Roman Radchenko[1] , Serhiy Forduy[2] , and Vitaliy Maksymov[1]

[1] Admiral Makarov National University of Shipbuilding,
9 Heroes of Ukraine Avenue, Mykolayiv 54025, Ukraine
g.lavamay@gmail.com
[2] PepsiCo, Inc., Kyiv, Ukraine

Abstract. Complex cycles with cyclic air intercooling are used to increase the energy efficiency of gas turbines. A modern and widespread way to improve the cooling process is to humidify the working fluid (cyclic air). The efficiency of wet compression primarily depends on the intensity of evaporation and heat exchange of droplets with the air flow, which begins to increase sharply when the effective diameter of droplet spraying decreases to 20 μm. It is proposed to use a contact heat exchanger to obtain a finely dispersed flow of water in the flow path of a gas turbine. The operation of such contact heat exchanger called aerothermopressor was investigated in this paper. CFD simulation of the water droplet evaporation process in the aerothermopressor airflow was carried out. Calculations were carried out for three variants of evaporation of water injected into the air flow: complete evaporation of water droplets in the evaporation chamber, additional evaporation of water droplets in the diffuser and incomplete evaporation, with obtaining smaller droplets at the outlet of the aerothermopressor diffuser. Efficiency of the aerothermopressor application in the gas turbine circuit for contact cooling of cyclic air is analyzed. It has been revealed that the aerothermopressor allows increasing the cyclic air pressure between the compressor stages by 2–10%, which will lead to a decrease in the compression work in the compressor stages and makes it possible to increase the gas turbine engine efficiency by 1–2%.

Keywords: Droplet diameter · Two-phase flow · Thermogasdynamic compression

1 Introduction

One of the most perspective ways to increase the power of modern power plants based on gas turbine engines (GTE) is to increase the flow rate of the working fluid (air) and the useful work of expanding combustion products (exhaust gases) simultaneously with a decrease in costs for compression in the compressor stage. This is achieved by cooling the working fluid (air) before or during the compression process, which reduces the specific fuel consumption. One of the most common ways to implement this principle is to use the contact air cooling with water injection. At the same time, there is a significant length of the evaporation section. This is not always constructively possible to implement and, if the workflow is ineffective, it can lead to droplets getting into the GTE compressors.

A promising development of the contact cooling principle in the gas turbine engine is to use the energy potential of the working fluid to accelerate the air flow to transonic speeds and for instant evaporation of water injected into the air flow, that is, the implementation of the thermogasdynamic compression effect.

2 Literature Review

A modern and widespread way to increase the efficiency of gas turbine plants is to humidify the working fluid (cyclic air). For this, the following technologies with water (steam) injection are used:

- air cooling at the compressor inlet [1–3];
- wet compression (contact air cooling during compression in a compressor stage) [2, 4];
- additional cooling at the compressor outlet with subsequent recuperation [5];
- utilization of the exhaust gases heat of the gas turbine plant (GTP) in the steam-turbine heat recovery circuit (gas-steam turbine technology "Vodoley") [6];
- steam injection (technology Steam Injected Gas Turbine (STIG), Cheng cycle) [2, 7];
- water injection into the air flow in the humidifying tower (Humidified Air Turbine (HAT) and Evaporative Gas Turbines (EvGT) technologies) [8, 9];
- superheated water injection [6].

To increase the energy efficiency of gas turbines, complex cycles with cyclic air intercooling in the units are used [2]. Intercooling can be carried out in the following way – contact cooling of the cyclic air with injection of the finely dispersed water flow, which was proposed in the 60s by Jones and Hawkins [10].

As a rule, water injection at the compressor inlet or between compressor stages is used to increase the power and efficiency of the gas turbine plant at high outside temperatures. Injection of steam generated in the heat recovery circuits increases the power and efficiency of the gas turbine plant by increasing the amount of the working fluid in the cycle [11, 12].

As a result of cyclic air intercooling during the compression process, the compressor efficiency is increased by isothermal air compression process (approaching the final compression temperature to the initial one) [2, 6, 12, 13]. Hence, without loss of the gas turbine performance, the gas temperature at the turbine inlet decreases. It favorably affects the gas turbine resource [14–16].

The efficiency of wet compression primarily depends on the intensity of evaporation and heat exchange of droplets with the air flow, which begins to increase sharply when the effective diameter of droplet spraying decreases to 20 µm. This occurs as a result of a significant increase in the total surface area of the droplets, which, in turn, is inversely proportional to its average diameter. At the same time, the separation of droplets of such a small size on the surface of the blades is almost absent. Accordingly, the associated mechanical losses are sharply reduced [3]. However, even the most modern mechanical and pneumatic nozzles do not allow water atomization with an effective droplet diameter less than 30 μm [2, 6].

For example, to supply water into the cyclic air flow, nozzles of a special design were developed (the outlet diameter is 0.1–0.4 mm), in which spraying is realized due to the impact action: water under high pressure (7–14 MPa) is fed to the nozzle head, after which it breaks down due to the "pin" into small droplets, the diameter of which does not exceed 20 μm [3, 17]. According to Mee Industries Inc, an increase in the efficiency of gas turbine plants using a contact cooling system in the summer season reaches 2–4%. Despite the relatively widespread use, such systems require rather complex pumping equipment for water supply and create difficulties in operating the system under high pressure [17].

A promising method for realizing more efficient atomization is to use a liquid superheated relative to the saturation temperature. In the process of the outflow of such a liquid through the nozzle of the spraying device, its explosive boiling occurs, as a result of which the liquid is crushed into small droplets [18]. In addition, during experimental studies it was found that if the diameter of a water droplet in the flow path of the compressor does not exceed 20 μm, then the effect of centrifugal forces on the droplet becomes almost imperceptible and droplets are directed along the air flow [19]. From this point of view, ideally nozzles installed at the compressor inlet should spray a liquid with a droplet diameter of no more than 20 μm.

A promising method for humidifying the working fluid of a gas turbine plant is to use a two-phase jet apparatus – an aerothermopressor [20, 21]. The operation of this apparatus is based on the process of thermogasdynamic compression. A feature of this process is the air pressure increase as a result of instantaneous evaporation of the water injected into the air flow accelerated to the speed of sound. In this case, the air pressure and air cooling are increased by removing heat from the air flow [22]. It should be noted that the aerothermopressor can have one more function – providing effective fine dispersion of liquid due to incomplete evaporation [22].

There are experimental studies of high- and low-flow aerothermopressors in which the authors have shown the effectiveness of the aerothermopressor using as an effective high-speed compact contact heat exchanger [20, 23]. For example, experimental studies [23] were carried out with the following data (a low-flow aerothermopressors): maximum gas flow rate of 1.63 kg/s; inlet temperatures 355–365 °C; the length of the evaporation chamber was regulated in the range $(l/D) = 2.15$–11.30; the inlet Mach number varied within the range $M = 0.32$–0.89; the relative water flow rate was 0–15%. The paper concludes that it is advisable to narrow the evaporation chamber. A conclusion is given on the possibility of reducing the back pressure in the gas turbine plant to 5 kPa at gas flow rates by 20 kg/s and $M = 0.7$–0.8. In addition, the possibility of increasing the total flow pressure in the aerothermopressor flow path to 20% was shown due to the appearance of the thermogasdynamic compression effect [20].

The following advantages of the aerothermopressor using are:

- increasing the pressure and cooling the working fluid will reduce the work on compression in the compressor;
- it also will provide effective atomization and humidification of water between compressor stages;
- it will reduce the additional compressor work when water droplets evaporate in the flow path during compression, and increase the amount of working fluid in the cycle.

The rational organization of thermophysical processes in the flow path of the aerothermopressor in the presence of the thermogasdynamic compression effect and incomplete evaporation is possible when the optimal geometric parameters are selected. These parameters include: the evaporation chamber diameter, the relative length of the evaporation chamber, confuser convergent angle, diffuser divergent angle, distance between the water injection point and the evaporation chamber inlet. The correct choice will ensure the evaporation of the amount of water (80–85%) in the evaporation chamber and the diffuser, and the additional evaporation of remaining water (15–20%) in the flow path of the high-pressure compressor. In this case, the water droplets diameter entering the compressor will not exceed 20 μm.

The choice of such optimal geometric parameters of the aerothermopressor, as well as the determination of the characteristics and injection mode (flow velocity; average, maximum and minimum droplet diameters; inlet air temperature; relative water flow rate, air pressure and air flow rate) should be carried out according to the results of an experimental study of working processes and in numerical modeling.

Based on the above analysis, the following research hypothesis can be formulated: to use an aerothermopressor for cooling the working fluid (air) of a gas turbine engine will provide effective fine dispersion of liquid due to intense heat and mass exchange processes. These processes are associated with incomplete evaporation in the flow path of the apparatus, hence a more effective isothermal compression process in the compressor will be provided. This, in turn, will make it possible to compensate for hydraulic pressure losses along the air path with a corresponding decrease in the compression work due to an increase in total pressure by 5–10%. As a result, it will increase the efficiency of the GTE with a corresponding decrease in specific fuel consumption by 1.0–1.5%.

3   Research Methodology

An aerothermopressor was designed for cyclic air intercooling in the cycle of a WR-21 Rolls Royce gas turbine engine ($N_e = 25{,}250$ kW). For the numerical simulation of the aerothermopressor operation an experimental model of the apparatus [22, 23] was used, which was designed and manufactured for carrying out experimental studies. The model has the following geometrical characteristics (Fig. 1): the aerothermopressor length $L_{atp} = 1324$ mm; the evaporation chamber diameter $D_{ch} = 68$ mm; air flow rate $G_{air} = 1.56$ kg/s. The confuser convergent angle ($\alpha_c = 35°$) and the diffuser divergent angle ($\beta_d = 5°$) were chosen taking into account the obtaining of minimum losses when overcoming the friction forces and local resistances in the narrowing-expansion sections of the apparatus [24, 25].

The finite volume method was chosen as a numerical method. This method is implemented using computer CFD modeling in the ANSYS Fluent software package [26]. It is a reliable, feature-rich and easy to use tool for designing, modeling and analyzing fluid flows of varying complexity. The computational domain is divided by means of a grid into a set of finite volumes. The nodes are in the center of these volumes for which the calculation is carried out. For each control volume, the laws of conservation of mass, momentum and energy must be satisfied [24].

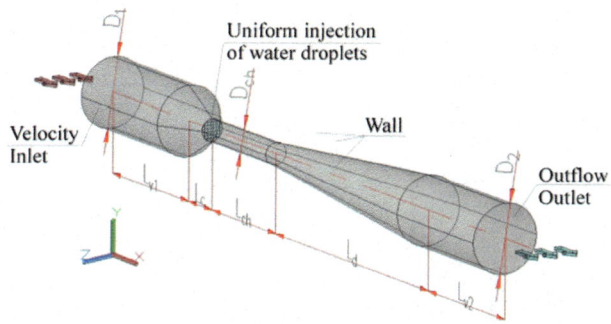

Fig. 1. Geometry model of the aerothermopressor.

The Eulerian-Lagrangian approach was used to simulate the interaction of injectable droplets and the air flow. Among existing turbulence models RANS, a two-parameter k-ε Realizable turbulence model was chosen [27]. Discrete Phase Model (DPM) has been used to model the motion of water droplets [28, 29]. A step-by-step grid solution with a simple algorithm for the relationship between velocity and pressure is carried out. The calculation was carried out taking into account the convergence of the results. The processing and visualization of the output data in the postprocessor was carried out, in the form of graphs, fields and streamlines for the main parameters of the working processes in the flow path of the aerothermopressor.

Air with the initial parameters of pressure, temperature and relative humidity, which correspond to the parameters of the GTP cyclic air after the first stage of the compressor, is considered as a working fluid: inlet air pressure $P_1 = 301325$ Pa; inlet air temperature $T_{air1} = 473$ K; inlet air velocity $w_{air1} = 55$ m/s; air mass flow $G_{air} = 1.56$ kg/s; Mach number at the evaporation chamber inlet $M = 0.55$.

The output parameters for numerical computer modeling are the main parameters of the two-phase flow (air-water) at the aerothermopressor outlet (at the inlet to the next stage of the gas turbine compressor): total air pressure P_{atp}, air velocity w_{air}, water velocity w_w, temperature T_{atp}, water droplet diameter in the air flow δ_p, air density ρ_{atp}, mass concentration of water m_{H2O}.

When analyzing the gas turbine scheme (Fig. 2), the well-known methods of calculating the cycles of gas turbines [6], as well as calculating the process of thermogasdynamic compression [20] were used. On the basis of these methods the corresponding software package was developed. The calculation of the GTP cycles was carried out for the degrees of pressure increase p_c = 12–42 and for the air parameters corresponding to ISO: t_0 = 15 °C, φ_0 = 30%. At the same time, instead of an air cooler (surface or contact with nozzle injection), it is proposed to use the aerothermopressor in the circuits.

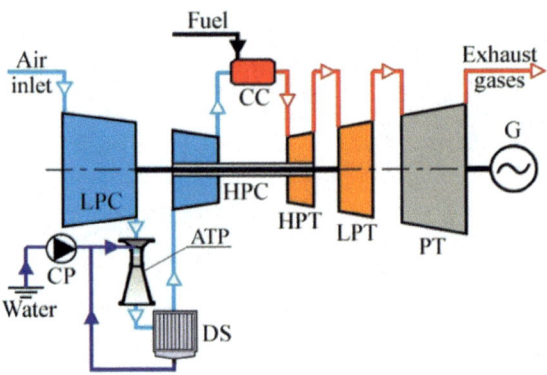

Fig. 2. A complex cycle with intercooling of air by using an aerothermopressor: LPC – low pressure compressor; HPC – high pressure compressor; CC – combustor; HPT – high pressure turbine; LPT – low pressure turbine; PT – power turbine; ATP – aerothermopressor; CP – circulation pump; DS – droplet separator; G – generator.

4   Results

The simulation of a "dry" aerothermopressor (without injection of water droplets into the evaporation chamber) was carried out to determine the pressure loss through the aerodynamic resistance in the flow path of the apparatus. It was found that the decrease in the air flow pressure (Fig. 3) due to friction losses is $\Delta P_{atp.dry}$ = 17 kPa (6%).

A discrete phase model was used when simulating the injection of water droplets into the flow path of the aerothermopressor (at the inlet to the evaporation chamber). Calculations were carried out for three variants of evaporation of water injected into the air flow: complete evaporation of water droplets in the evaporation chamber (δ_p = 3–30 μm, G_w = 0.031 kg/s); additional evaporation of water droplets in the diffuser (δ_p = 3–30 μm, G_w = 0.062 kg/s) and incomplete evaporation, with obtaining smaller droplets at the diffuser outlet of the aerothermopressor (δ_p = 3–30 μm, G_w = 0.156 kg/s).

A comparative analysis of the results obtained when simulating the aerothermopressor operation with and without water injection is carried out. Comparison of the pressure change P_{atp} along the length of the flow path with and without injection (Fig. 3a) shows that the pressure in the evaporation chamber due to losses decreases from 220 kPa to 215 kPa, that is, the pressure loss is up to 5 kPa (1.7%). However, the presence of thermogasdynamic compression increases the pressure and compensates for these losses. In this case, the total pressure in the evaporation chamber of the aerothermopressor increases by ΔP_{atp} = 30 kPa (10%).

The magnitude of the cyclic air cooling in the aerothermopressor is: with complete evaporation of water droplets ΔT_{atp} = 106 K; with additional evaporation of water droplets in the diffuser ΔT_{atp} = 118 K; with incomplete evaporation ΔT_{atp} = 133 K (Fig. 3b).

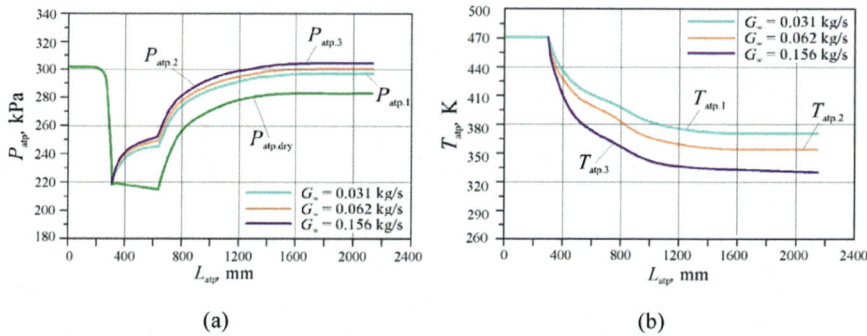

(a)   (b)

Fig. 3. Dependences of the total pressure P_{atp}, (a) and two-phase flow temperature T_{atp} (b) on the length L_{atp} of the aerothermopressor: $P_{atp.dry}$ – without injection of water droplets into the evaporation chamber.

With complete evaporation of water droplets in the evaporation chamber, the increase in total pressure (Fig. 3a) as a result of thermogasdynamic compression is ΔP_{atp} = –2.3 kPa (–0.76%) relative to the inlet pressure. With complete evaporation of water droplets in the diffuser, the increase in total pressure is already a positive value and reaches ΔP_{atp} = 0.25 kPa (0.08%). The greatest effect from thermogasdynamic compression is observed with incomplete evaporation, with obtaining smaller droplets at the outlet from the diffuser part of the apparatus, while the increase in total pressure (Fig. 4a) is ΔP_{atp} = 1.4 kPa (0.5%).

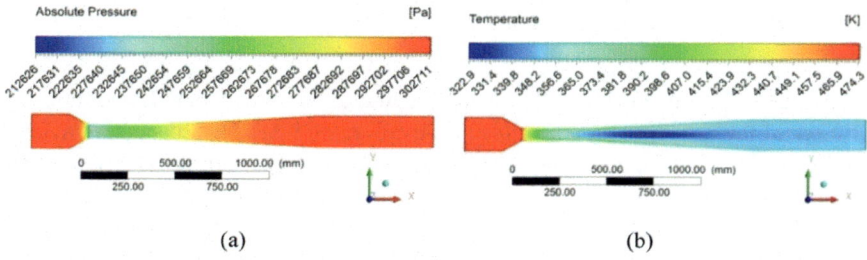

(a)   (b)

Fig. 4. The fields distribution of total pressure P_{atp} (a) and flow temperature T_{atp} (b) on the length L_{atp} of the aerothermopressor at incomplete evaporation, with obtaining smaller droplets at the outlet of the diffuser part of the apparatus (δ_p = 3–30 μm, G_w = 0.156 kg/s).

Thus, the inlet temperature T_{atp1} = 473 K (200 °C) decreases to the outlet temperature (Fig. 4b) by T_{atp2} = 340–367 K (67–94 °C).

The dispersion of water droplets at the evaporation chamber inlet is δ_p = 3–30 μm. The distribution of sprayed water droplets in the flowing part of the aerothermopressor is given: for complete evaporation of water droplets (G_w = 0.031 kg/s) (Fig. 5a); additional evaporation of water droplets in the diffuser (G_w = 0.062 kg/s) (Fig. 5b); and incomplete evaporation, with obtaining smaller droplets at the outlet of the diffuser part of the aerothermopressor (G_w = 0.156 kg/s) (Fig. 5c).

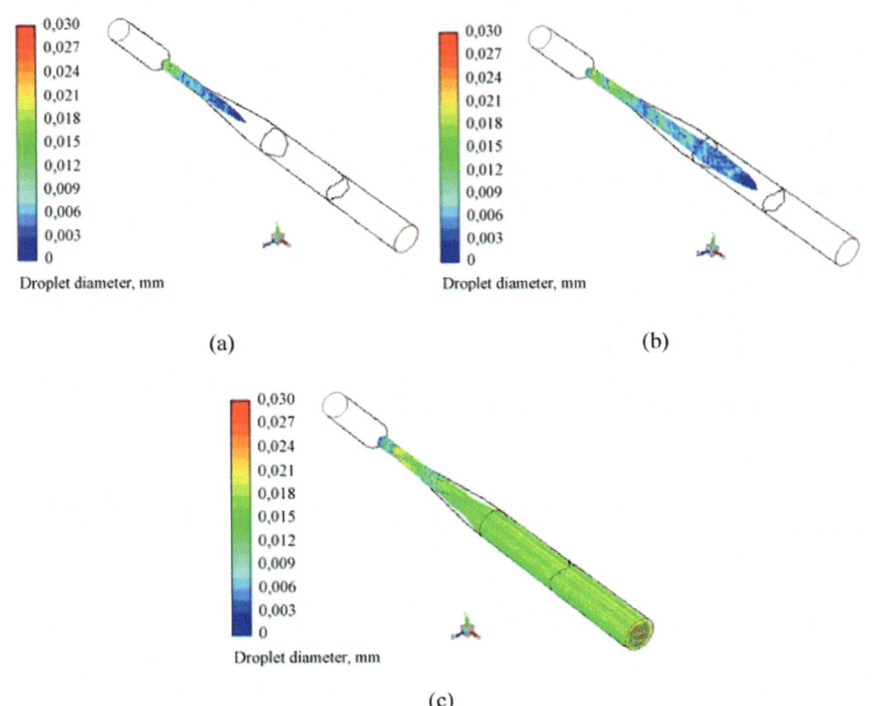

(a)   (b)

(c)

Fig. 5. Distribution of dispersion δ_p of sprayed water in the flow path of the aerothermopressor: a – complete evaporation of water droplets in the evaporation chamber (g_w = 2%); b – additional evaporation of water droplets in the diffuser (g_w = 4%); c – incomplete evaporation, with obtaining smaller droplets at the outlet of the diffuser part of the apparatus (g_w = 10%).

With incomplete evaporation, the droplet diameter decreased, and at the outlet from the diffuser of the aerothermopressor for non-evaporated water, the dispersion averaged δ_{p2} = 18 μm (Fig. 5c). In this case, the concentration of water droplets (dispersed two-phase flow) at the outlet is evenly distributed over the cross section. The obtained diameter of water droplets at the outlet of the aerothermopressor (18 μm) meets the requirements for the dispersion of the two-phase flow before the high-pressure compressor stage. Excess water will evaporate already in the flow path of the compressor. This amount is up to 4%.

Comparing the efficiency of the aerothermopressor operating regime with different relative flow rates of injected water (Fig. 6), it can be seen that an increase in the liquid amount to $g_w = 10\%$, leads to an increase in the relative increase in the total pressure by ΔP_{atp} to 1% and a decrease in the outlet temperature by $\Delta T_{atp} = 130$ K. This is due to the water injection in excess of the amount required for evaporation reduces the friction pressure loss and, accordingly, increases the total outlet pressure to practically the initial one (there is compensation to hydraulic pressure losses in the aerothermopressor). With an appropriate rational organization of the workflow associated with water injection, it is possible to obtain an increase in pressure relative to the initial one (more than 2–5%).

The average diameter of a dispersed flow droplet at the outlet is $\delta_{p2} = 18$ μm, with an excess amount of water injected to 10%. In this case, an increase in water flow rate leads to an increase in the average droplet diameter. With an increase in water flow rate g_w of more than 10%, it can be expected that the pressure loss due to the drag of the droplet in the flow will exceed all other losses. The positive effect of the pressure increase as a result of the thermogasdynamic compression effect will be significantly reduced. Therefore, the water injection into the aerothermopressor nozzle should be determined by the maximum pressure increase ΔP_{atp} and a decrease in the outlet temperature $T_2 < 343$ K.

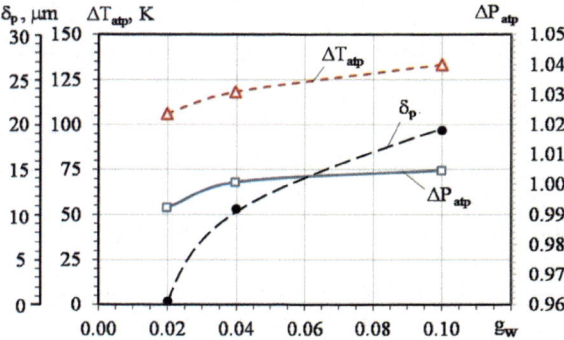

Fig. 6. Dependences of the temperature decrease ΔT_{atp}, the relative pressure increase ΔP_{atp}, the average droplet diameter at the diffuser outlet δ_p on the relative amount of water injected g_w into the aerothermopressor.

The aerothermopressor using in the GTP cycle for air intercooling allows reducing the cycle air temperature between compressor stages by 60–130 °C. Such a temperature decrease under thermogasdynamic compression conditions allows increasing the pressure by $P_2 - P_1 = 5$–35 kPa, that is, by $\Delta p_{atp} = 2$–10% (Fig. 7a).

The simulation of gas turbine plant operation by using the developed software package to calculate of the GTP cycles showed the following efficiency from the aerothermopressor using to provide intercooling of cyclic air. Injected water after evaporation is an additional working fluid, the increase of which, in turn, makes it possible to increase the GTP specific power. A decrease in the compressor operation and a simultaneous increase in the amount of the working fluid in the cycle makes it possible to increase the efficiency GTP by $\Delta\eta_e = 0.01$–0.02 (1–2%) in comparison with intercooling of air by using a surface air cooler. In this case, the specific fuel consumption will decrease by $\Delta g_e = 5$–10 g/(kW·h) (Fig. 7b). At the same time, the GTP specific power is increased by $\Delta N_s = 5$–30 kW/(kg/s), which is 1.5–7.0% (Fig. 7b). The simulation of the GTP operation was carried out for the range of degrees of pressure increase in compressor stages of the gas turbine plant $p_c = 12$–42, which are typical for the operation mode according to the classical cycle (Fig. 2).

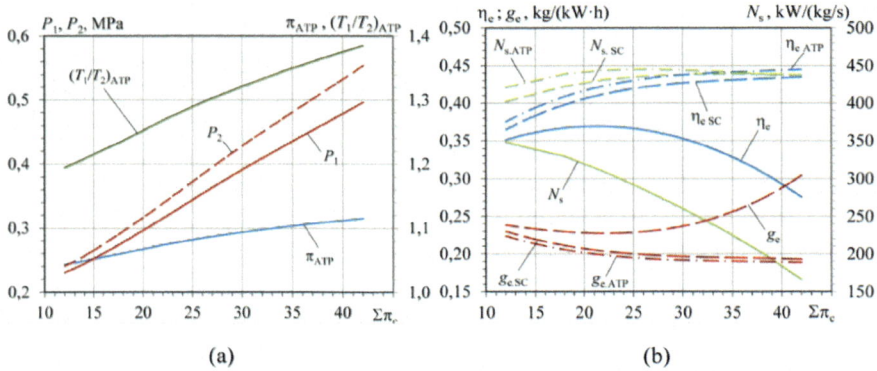

Fig. 7. Dependences of inlet pressure (P_1) and outlet pressure (P_2) of the aerothermopressor, relative air temperature (T_1/T_2)_atp, specific power output (N_s), specific fuel consumption (g_e) and efficiency (η_e) on the total compression ratio ($\Sigma\pi_c$) in compressors for the simple cycle, the complex cycle with a surface air cooler (SC) and the complex cycle with an aerothermopressor (ATP).

5   Conclusions

CFD simulation of the water droplet evaporation process in the airflow in the aerothermopressor was carried out. It has been established that the aerothermopressor provides effective fine atomization of water (the droplets diameter at the outlet from the apparatus does not exceed 18 µm), and, therefore, a more effective isotherm of the compression process in the high-pressure compressor occurs. In this case, the additional amount of water injected can be up to 10% relative to the flow rate of the cycle air.

The efficiency of the aerothermopressor application in the gas turbine cycle for contact cooling of cyclic air is analyzed. It has been revealed that the aerothermopressor allows increasing the pressure of the cyclic air between the compressor stages by $\Delta p_{atp} = 2$–10%, which will lead to a decrease in the compression work in the compressor stages. A decrease in the compressor operation and a simultaneous increase in the amount of the working fluid in the cycle makes it possible to increase the GTP efficiency by $\Delta\eta_e = 0.01$–0.02 (in comparison with intercooling of air by using a surface air cooler). In this case, the specific fuel consumption decreases by $\Delta g_e = 5$–10 g/(kW·h). At the same time, the gas turbine engine specific power is increased by $\Delta N_s = 5$–30 kW/(kg/s) (1.5–7.0%).

References

1. Bhargava, R.K., Meher-Homji, C.B., Chaker, M.A., et al.: Gas turbine fogging technology: a state-of-the-art review – Part I: Inlet evaporative fogging – analytical and experimental aspects. J. Eng. Gas Turbines Power **129**(2), 443–453 (2007). https://doi.org/10.1115/1.2364003
2. Jonsson, M., Yan, J.: Humidified gas turbines – a review of proposed and implemented cycles. Energy **30**(7), 1013–1078 (2005). https://doi.org/10.1016/j.energy.2004.08.005
3. Chaker, M.A.: Key parameters for the performance of impaction-pin nozzles used in inlet fogging of gas turbine engines. J. Eng. Gas Turbines Power **129**(2), 473–477 (2007). https://doi.org/10.1115/1.2364006
4. Sexton, W.R., Sexton, M.R.: The effects of wet compression on gas turbine engine operating performance. In: Turbo Expo: Power for Land, Sea, and Air, Atlanta, pp. 673–679. ASME (2003). https://doi.org/10.1115/GT2003-38045
5. Isaiah, T.G., Dabbashi, S., Bosak, D., et al.: Life cycle evaluation of an intercooled gas turbine plant used in conjunction with renewable energy. Propul. Power Res. **5**(3), 184–193 (2016). https://doi.org/10.1016/j.jppr.2016.07.005
6. Dykyi, M., Solomakha, A.: Features of the contact air cooling process in gas turbine plants and method of its implementation. Energy Econ. Technol. Ecol. **1**, 22–26 (2012). (in Ukrainian)
7. Kayadelen, H.K., Ust, Y.: Thermoenvironomic evaluation of simple, intercooled, STIG, and ISTIG cycles. Int. J. Energy Res. **42**(12), 3780–3802 (2018). https://doi.org/10.1002/er.4101
8. Orts-Gonzalez, P.L., Zachos, P.K., Brighenti, G.D.: Techno-economic analysis of a reheated humid air turbine. Appl. Therm. Eng. **137**, 545–554 (2018). https://doi.org/10.1016/j.applthermaleng.2018.03.094
9. Brighenti, G.D., Orts-Gonzalez, P.L., Sanchez-de-Leon, L., Zachos, P.K.: Design point performance and optimization of humid air turbine power plants. Appl. Sci. **7**(4), 413 (2017). https://doi.org/10.3390/app7040413
10. Jones, J.B., Hawkins, G.A.: Engineering Thermodynamics. Wiley, New York (1990)
11. Kornienko, V., Radchenko, R., Konovalov, D., Andreev, A., Pyrysunko, M.: Characteristics of the rotary cup atomizer used as afterburning installation in exhaust gas boiler flue. In: Ivanov, V., Pavlenko, I., Liaposhchenko, O., Machado, J., Edl, M. (eds.) DSMIE 2020. LNME, pp. 302–311. Springer, Cham (2020). https://doi.org/10.1007/978-3-030-50491-5_29
12. Trushliakov, E., Radchenko, M., Bohdal, T., Radchenko, R., Kantor, S.: An innovative air conditioning system for changeable heat loads. In: Tonkonogyi, V., Ivanov, V., Trojanowska, J., Oborskyi, G., Edl, M., Kuric, I., Pavlenko, I., Dasic, P. (eds.) InterPartner 2019. LNME, pp. 616–625. Springer, Cham (2020). https://doi.org/10.1007/978-3-030-40724-7_63
13. von Deschwanden, I., Benra, F.K., Brillert, D., Dohmen, H.J.: Droplet evaporation in the context of interstage injection. In: 16th International Symposium on Transport Phenomena and Dynamics of Rotating Machinery (ISROMAC 2016), pp. 1–11 (2016)
14. Trushliakov, E., Radchenko, A., Radchenko, M., Kantor, S., Zielikov, O.: The efficiency of refrigeration capacity regulation in the ambient air conditioning systems. In: Ivanov, V., Pavlenko, I., Liaposhchenko, O., Machado, J., Edl, M. (eds.) DSMIE 2020. LNME, pp. 343–353. Springer, Cham (2020). https://doi.org/10.1007/978-3-030-50491-5_33
15. Radchenko, A., Bohdal, L., Zongming, Y., Portnoi, B., Tkachenko, V.: Rational designing of gas turbine inlet air cooling system. In: Tonkonogyi, V., Ivanov, V., Trojanowska, J., Oborskyi, G., Edl, M., Kuric, I., Pavlenko, I., Dasic, P. (eds.) InterPartner 2019. LNME, pp. 591–599. Springer, Cham (2020). https://doi.org/10.1007/978-3-030-40724-7_60

16. Sun, J., Hou, H., Zuo, Z., et al.: Numerical study on wet compression in a supercritical air centrifugal compressor. Proc. Inst. Mech. Eng. Part A J. Power Energy **234**(3), 384–397 (2020). https://doi.org/10.1177/0957650919861490

17. Mee Industries Inc. http://www.meefog.com/fog-evaporative-cooling/gas-turbine-cooling. Accessed 14 June 2020

18. Solomakha, A.S.: Experimental study of the spraying of superheated water. Eastern Eur. J. Enterp. Technol. **1**(8), 20–25 (2013). (in Ukrainian)

19. Chaker, M., Meher-Homji, C.B., Mee III, T.: Inlet fogging of gas turbine engines – Part A: fog droplet thermodynamics, heat transfer and practical considerations. J. Eng. Gas Turbines Power **126**(3), 545–558 (2004). https://doi.org/10.1115/1.1712981

20. Fowle, A.: An experimental investigation of an aerothermopressor having a gas flow capacity of 25 pounds per second. Dissertation, Massachusetts Institute of Technology (1972)

21. Konovalov, D., Trushliakov, E., Radchenko, M., Kobalava, H., Maksymov, V.: Research of the aerothermopressor cooling system of charge air of a marine internal combustion engine under variable climatic conditions of operation. In: Tonkonogyi, V., Ivanov, V., Trojanowska, J., Oborskyi, G., Edl, M., Kuric, I., Pavlenko, I., Dasic, P. (eds.) InterPartner 2019. LNME, pp. 520–529. Springer, Cham (2020). https://doi.org/10.1007/978-3-030-40724-7_53

22. Konovalov, D., Kobalava, H., Maksymov, V., Radchenko, R., Avdeev, M.: Experimental research of the excessive water injection effect on resistances in the flow part of a low-flow aerothermopressor. In: Ivanov, V., Pavlenko, I., Liaposhchenko, O., Machado, J., Edl, M. (eds.) DSMIE 2020. LNME, pp. 292–301. Springer, Cham (2020). https://doi.org/10.1007/978-3-030-50491-5_28

23. Konovalov, D., Kobalava, H., Radchenko, M., et al.: Determination of hydraulic resistance of the aerothermopressor for gas turbine cyclic air cooling. In: E3S Web of Conferences, vol. 180, p. 01012 (2020). https://doi.org/10.1051/e3sconf/202018001012

24. Sirignano, W.A.: Fluid Dynamics and Transport of Droplets and Sprays, 2nd edn. Cambridge University Press, New York (2010)

25. Bergman, T.L., Incropera, F.P.: Fundamentals of Heat and Mass Transfer, 7th edn. Wiley, Hoboken (2011)

26. ANSYS Fluent Tutorial Theory Guide Release 17.0. ANSYS Inc. (2016)

27. Korkodinov, Y.A.: An overview of the k-ε family of models for modeling turbulence. Mech. Eng. Mater. Sci. **5**(2), 5–16 (2013). (in Russian)

28. Nijdam, J.J., Guo, B., Fletcher, D.F., Langrish, T.A.: Lagrangian and Eulerian models for simulating turbulent dispersion and coalescence of droplets within a spray. Appl. Math. Model. **30**(11), 1196–1211 (2006). https://doi.org/10.1016/j.apm.2006.02.001

29. Chen, Z., Xie, Q., Chen, G., et al.: Numerical simulation of single-nozzle large scale spray cooling on drum wall. Therm. Sci. **22**(1A), 359–370 (2018). https://doi.org/10.2298/TSCI170920243C

Improving the Ecological and Energy Efficiency of Internal Combustion Engines by Ejector Chiller Using Recirculation Gas Heat

Roman Radchenko[1](✉) , Maxim Pyrysunko[2] ,
Victoria Kornienko[2] , Ionut-Cristian Scurtu[3] ,
and Radosław Patyk[4]

[1] Admiral Makarov National University of Shipbuilding,
9 Heroes of Ukraine Avenue, Mykolayiv 54025, Ukraine
nirad50@gmail.com
[2] Kherson Branch of Admiral Makarov National University of Shipbuilding,
44 Ushakova Avenue, Kherson 73000, Ukraine
[3] "Mircea cel Batran" Naval Academy, 1 Fulgerului Street,
900218 Constanta, Romania
[4] Koszalin University of Technology, 2 Śniadeckich Street,
75-453 Koszalin, Poland

Abstract. The use of such techniques conflicts with the engine's energy efficiency and leads to increasing fuel consumption. It is promising to use technologies that would increase fuel and energy efficiency of ICE with EGR systems and combine high environmental efficiency with engine fuel efficiency. The technology of precooling intake air at the suction of turbocharger by waste heat using chiller (WHUCh) was developed for ICE with EGR system. The scheme-design solution of the exhaust gas recirculation system with using the heat of recirculation gas by an ejector chiller for cooling the air at the intake of main ship engine is proposed. The scheme-design solution of the EGR system with using the heat of recir-culation gas by an ejector chiller (ECh) for cooling the air at the intake of main ship engine is proposed. The effect of using the heat of recirculation gas for cooling engine intake air is analyzed taking into account the changing climatic conditions on a vessel's route line. It is shown that using the heat of recirculation gas for cooling engine intake air by ejector refrigeration machine reduces the air temperature at the entrance of the main engine by 5–15 °C, which decreases the specific fuel consumption by 0.5–1.5 g/kW · h. This reduces emissions of harmful substances (NOx by 26–39%; SOx by 9–14%) when the engine is running with recirculation of gas.

Keywords: Harmful emissions · Exhaust gas recirculation · Ejector chiller · Ecology

1 Introduction

The most sensitive environmental impact comes from ship power plants, in which the main source of energy (thermal, mechanical, electrical) is internal combustion engines (ICE) [1]. During the operation of internal combustion engines the greatest harm is caused by toxic substances contained in the exhaust gas [2]. The formation of harmful gases, such as carbon dioxide CO_2, nitrogen oxides NOx, carbon monoxide CO, sulfur oxides SOx, etc., depends on the organization of work processes in the internal combustion engine.

A very effective way of greening marine internal combustion engines is the artificial neutralization of harmful substances in exhaust gases, for example, by exhaust gas recirculation (EGR technology) [1]. However, the use of such technology is in conflict with the energy efficiency of the internal combustion engine and causes increasing fuel consumption compared with conventional engine operation without exhaust gas recirculation [3].

So it is reasonable to use the technologies that would ensure an increase in fuel and energy efficiency of internal combustion engines with EGR systems, that is, would combine high environmental efficiency with engine fuel efficiency [4]. These technologies would provide fuel saving due to cooling air at the suction [5] and scavenge air at the outlet [6, 7] of the turbocharger of combustion engines by waste heat using chillers (WHUCh) as an example [8, 9].

2 Literature Review

The International Maritime Organization (IMO) has released a series of protocols that limit emissions allowed by diesel engines. The protocols are called Tier I, II and III and relate to emissions of NOx, SOx and greenhouse gases. The Tier I and II protocols are global and entered into force in 2000 and 2011, respectively. The Tier III protocol has been applied to NOx in emission control zones since 2016 [10].

Currently, methods have been developed to reduce toxic emissions from exhaust gases, divided into primary and secondary measures. Primary measures are associated with the organization of the processes of mixture formation and combustion, in sufficient turbocharging and fuel injection systems, as well as the use of alternative fuels, such as natural gas. Secondary measures include: exhaust gas recirculation, selective catalytic reduction system and others. The effectiveness of various ways to reduce NOx emissions is shown in Fig. 1: 1 – organization of work processes; 2 – the use of natural gas; 3 – exhaust gas recirculation; 4 – water-fuel emulsion; 5 – humidification of charge air; 6 – selective catalytic reduction system [11].

Exhaust gas recirculation is a method that significantly reduces the formation of NOx in marine diesel engines [12]. Using this method fully meet Tier III requirements for NOx. In the EGR system, after the cooling and cleaning process, a part of the exhaust gas is recirculated to the scavenge air receiver. Thus, part of the oxygen in the air, that is used in the combustion process, is replaced by CO_2 oxide. This, in turn, reduces the oxygen content of O_2 and the burning rate, thereby reducing the maximum burning temperature, and thus reducing the intensity of NOx formation [13]. NOx emission reduction is almost linear to exhaust gas recirculation.

The increased amount of recirculation gases causes under-burning of the fuel and an increase in emissions of CO, CO_2 and soot. Therefore, the quantity (part in the total volume of exhaust gases) of the recirculation gases must be limited, that is, a compromise must be found between the effect of NO recirculation and combustion efficiency (in the form of low-grade hydrocarbons). This imposes restrictions on the upper limit of the recycle rate [14].

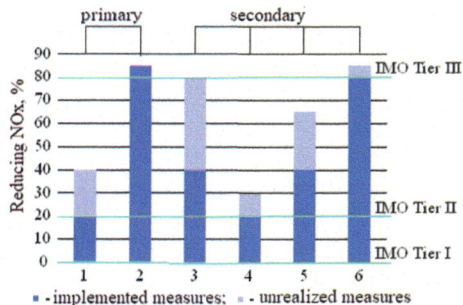

Fig. 1. Efficiency of NOx reduction methods.

When using EGR, there is an increased fuel consumption caused by the slowed rate of heat release. In the upper engine load range, fuel consumption gradually increases with recycle rate. At recirculation ratio Kr = 20% the fuel consumption increases by 10–20 g/(kWh), in the average engine load range, fuel consumption increases linearly by the Kr, and in the lower load range it slightly affects the fuel consumption.

Using recirculation with a Kr = 10% can reduce NOx by about 30% without a significant increase in fuel consumption – although the exhaust gases smoking slightly increases. When Kr = 20% the reduction in emissions of nitrogen oxides can reach 60%. However, already at Kr > 10–15%, the fuel economy deteriorates by 4–7%. One advantage of the EGR is low cost compared with other methods of reducing NOx. In the case of recycling there is no need to use complex and expensive devices, the manufacture of which for large-sized marine diesel engines causes great technical difficulties.

Since the CO_2 and water molecules have a higher heat capacity, the combustion temperature somewhat decreases. According to the data of [15], an increased mass flow rate gives about 93% of the effect of reducing the flue gas temperature, while an increase in the specific heat gives about 7%. The cooling of recirculation gases leads to a decrease in NOx and particulate emissions with comparable recirculation ratio Kr. This effect is more significant with large recirculation ratio [16].

The work [20] shows the results of using exhaust gas recirculation for the engine of the company MAN 12K90ME9, which is applied as the main engine on the container vessel. The recycling rate at nominal and partial engine loads was Kr = 30–40%, while NOx emissions decreased to 80%, SOx emissions decreased by 25%, particulate matter

(PM) by 50%, but this led to an increase in specific fuel oil consumption by 5–6 g/(kWh). The increase in specific fuel oil consumption of low-speed engines up to 5 g/(kWh) is also confirmed in [17]. It is also indicated that it is a technology for EGR at a concentration of fuel sulfur level S < 3.5%.

A typical EGR system for a ship's main low-speed engine includes a scrubber, a cooler, a water separator, a fan, and a support system for NaOH solution with a pump and tank. It should be noted that the components of the system are quite dimensional, the pumps of the gas cleaning system, cooling system and fan (or electric compressor) require a small amount of electrical energy [18].

It is promising to use technologies that would ensure an increase in fuel and energy efficiency of combustion engines with EGR systems, i.e. would combine high environmental efficiency with engine fuel efficiency due to deep waste heat utilization [19] and engine cyclic air cooling by applying refrigeration with boiling refrigerant for deep cooling [20], refrigerant recirculation for intensification of heat transfer in heat exchangers, injector, ejector [21] and other jet technologies and modern modelling methods and statistical treatment of monitoring data on influence of actual climatic conditions [22] and air cooling on performance characteristics of combustion engines [23] for rational designing of air cooling system and choosing optimum design heat loads. Such technologies provide engine cyclic air cooling by waste heat using chillers (WHUCh): ejector chillers (ECh) as the most simple in design [24] and absorption chillers (ACh) as high efficient heat transformer with coefficient of performance $\zeta = 0.6$–0.7 [25]. There were proposed some new concepts of combined exhaust heat conversion for cooling air in absorption-ejector chillers and stage air cooling [26], innovative methods for designing of cooling (conditioning) air systems to match changeable heat loads according to actual climatic conditions to provide maximum annual effect, in particular by dividing an overall heat load on cooling system in the range of air precooling with considerable heat load fluctuations and the range of air subcooling with practically stable heat loads.

The advantage of developed solution is the possibility of using the waste heat of recirculation gases for air cooling and hence to reduce the heat load on the scrubber recycling system and fuel consumption due to engine cyclic air cooling.

The aim of the study is assessment of the efficiency of ship's main engine air precooling by WHUCh using the heat of exhaust gases of the EGR system.

3   Research Methodology

The effectiveness of the application of proposed technical solution was analyzed on the basis of EGR system typical for MAN low-speed two-stroke diesel engines in accordance with the Tier III environmental conditions. Recirculation is provided by bypassing part of the exhaust gases purified from harmful gases in the scrubber after cooling in the heat exchanger-gas cooler. The EGR system comprises scrubber, chilling unit, condensate trap, the fan and the system is cleaned from the solution of NaOH.

Circuit solution with the use of the heat-using circuit of the ECh is considered for the ship's low-speed two-stroke diesel engine MAN B&W 6G50ME-C9.6. To analyze the parameters of the recirculation system, as well as the characteristics of the main engine, the CEAS software package of the leading manufacturer MAN was used [27]. The calculation was made for the following initial data: the performance characteristics of the main engine (under ISO conditions) – engine load – N_{MCR} = 90%; power – N_e = 9288 kW; speed – n_e = 96.5 rpm; specific fuel oil consumption (SFOC) – g_e = 166.0 g/(kWh); exhaust gas recirculation system (EGR) – as bypass with scrubber and gas cooler, responsible for Tier III environmental conditions.

The calculation of the characteristics of the engine was carried out on the operating mode during the voyage of the dry-cargo ship from Odessa to Shanghai. The variation of climatic conditions (ambient air temperature t_a, absolute humidity d_a and relative humidity φ_a, temperature of sea water t_w) during the vessel's voyage is presented in Fig. 2.

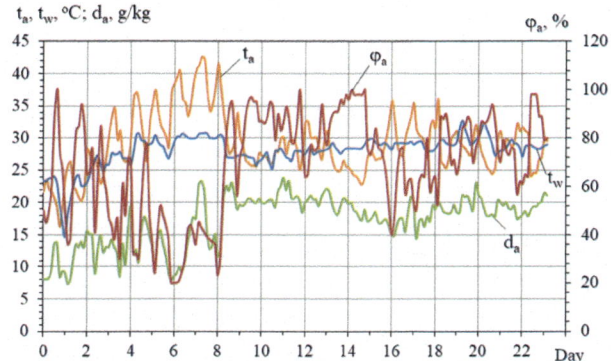

Fig. 2. The variation of temperature t_a, absolute humidity da and relative humidity φ_a of ambient air and temperature of sea water t_w during the vessel trade route Odessa-Shanghai.

The operating parameters of the heat-recovery contour based on the ECh were calculated using the well known equations, applied in the software complex developed at the Department of Conditioning and Refrigeration.

The following characteristics of the ECh for cooling the ship engine intake air were chosen: refrigerant – R142b; refrigerant evaporation temperature in the evaporator-air cooler t_0 = 5 °C and in the generator t_g = 80–120 °C, refrigerant condensing temperature in the condenser t_c = 25–45 °C.

The values of coefficient of performance for WHUCh: ζ = 0.30; 0.35.

4 Results

The solution with using the ECh was developed and analyzed (Fig. 3). The by-passing recirculation system runs as follows: exhaust gases from 10 to 40% in quantity are fed to the scrubber, where they are partially cooled and cleaned by spraying water with special nozzles. Then the exhaust gases are cooled in the heat exchanger – gas cooler (heater of water for refrigerant generator of ECh), condensed vapor from exhaust gases is drained through condensate trap and cooled gases are fed by the fan to the scavenge air receiver, where gases are mixed with the scavenge air coming from the turbocharger.

It is proposed to use the heat of the recirculating gases for high pressure liquid refrigerant evaporation in the generator of ECh with generation of high pressure refrigerant vapor as motive fluid for ejector to suck a low pressure refrigerant vapor from refrigerant evaporator – air cooler (AC-RE) at the intake of turbo-charger. Thus, cooling capacity of ECh is used for cooling air at the intake of the engine turbocharger.

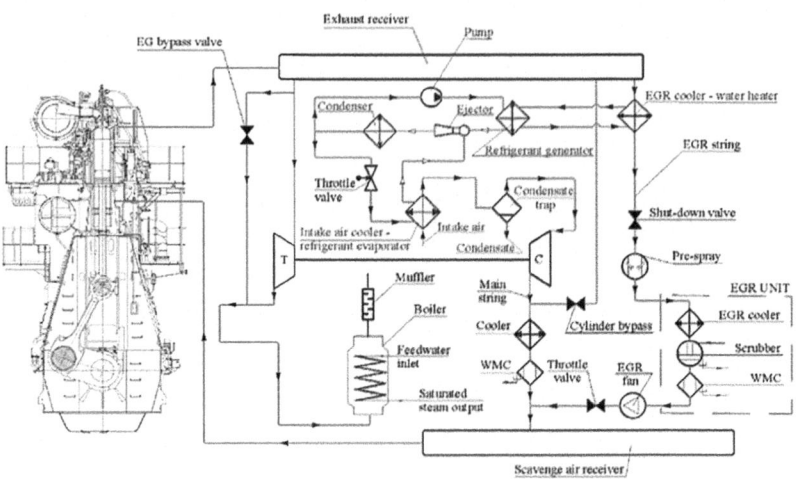

Fig. 3. Scheme of EGR-technology with bypass for the MAN marine diesel engine with ejector chiller: WMC – water mist catcher.

As the results of experimental research have shown, the time for stabilization of corrosion process is 2…3 h, especially when WFE is burnt. This allows to assess the level of corrosion process and to predict process for a long time from the results of kinetic studies for 2…12 h. In first 1…3 h, processes are accelerated, and then proceed with deceleration. After 4…5 h, dynamic balance is almost attained and the corrosion process proceeds at an almost constant rate. This allows, under the same working conditions to determine (predict) the value of ΔG_c by calculation during any period of operation, but while ensuring stable operating modes.

For MAN B & W 6G50ME-C9.6 engine considered, the specific fuel consumption decreases by about 0.5–1.6 g/(kWh) due to engine inlet air cooling during a voyage (Fig. 4). With this the number of recirculation during a voyage is K_r = 13–20%, the flow of recirculating flue gases is $G_{g.r}$ = 2.5–4 kg/s and a total exhaust gas flow G_g = 14–16 kg/s. The flow of "fresh" air to the engine turbocharger is $G_{a.egr}$ = 13–16 kg/s with exhaust gas recirculation and G_a = 17–19 kg/s – without recirculation of exhaust gases.

For the 6G50ME-C9.6 engine, according to the data of the MAN company (according to the calculations using the CEAS software package), when cooling intake air for every 10 °C a reduction in specific fuel consumption is 1.09 g/(kWh) or 0.109 g/(kWh·K) for every 1 °C air temperature drop.

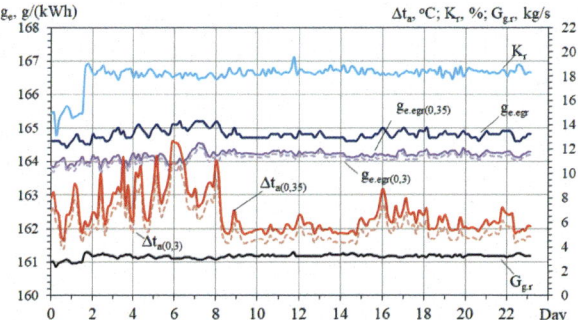

Fig. 4. Variation of specific fuel consumption g_e, number of recirculation K_r, decrease in air temperature at the engine intake Δt_a and recirculation gas flow $G_{g.r}$ during the vessel route Odessa-Yokohama at different coefficients of performance of chiller $\zeta = 0.30; 0.35$.

The results of analyzing the operation efficiency of recirculation gas heat-recovery chiller with different coefficients of performance $\zeta = 0.30; 0.35$ show the following cooling capacities received in chiller (Fig. 5): $Q_{0(0,3)} = 160{-}260$ kW ($\zeta = 0.30$) and $Q_{0(0,35)} = 180{-}270$ kW ($\zeta = 0,35$). The heat load on the ECh generator (heat consumption of chiller) is $Q_g = 515{-}805$ kW (Fig. 4) with appropriate cooling of the recirculation gas in the gas cooler (before the scrubber) from the temperature $t_{g2} = 370{-}420$ °C to the temperature $t_{g1} = 180$ °C (limited taken into account the danger of corrosion) provides decrease in air temperature at the inlet of the engine turbocharger, respectively (Fig. 4): $\Delta t_{a(0.3)} = 3.9{-}10.7$ °C ($\zeta = 0.30$); $\Delta t_{a(0.35)} = 4.8{-}12.6$ °C ($\zeta = 0.35$).

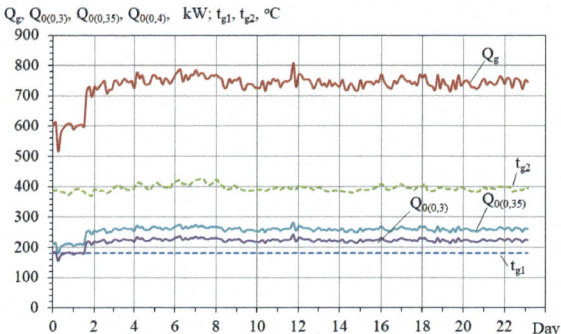

Fig. 5. The temperature of recirculation gases cooled by removing the heat for ECh generator (at the exit from heat exchanger- gas cooler) t_{g1}, temperature of the gases at the inlet of heat exchanger-gas cooler t_{g2}, thermal load on the generator of the chiller Q_g, cooling capacity of the chiller at different coefficients of performance $Q_{0(0,3)}$ and $Q_{0(0,35)}$ during the route of the vessel.

A decrease in the engine intake air temperature ensures a reduction in the specific fuel consumption (Fig. 4) in accordance with: $\Delta g_{e(0,3)} = 0.4–1.4$ g/(kWh) ($\zeta = 0.30$); $\Delta g_{e(0.35)} = 0.5–1.6$ g/(kWh) ($\zeta = 0.35$). The maximum efficiency of engine intake air cooling through recirculation exhaust gas heat-recovery corresponds to the coefficient of performance of the chiller $\zeta = 0.35$ and is $\Delta g_e = 0.9–1.1\%$, while the total specific fuel consumption will be $g_{e(0.35)} = 163.7–164.4$ g/(kWh), and without engine intake air cooling – $g_e = 163.9–165$ g/(kWh).

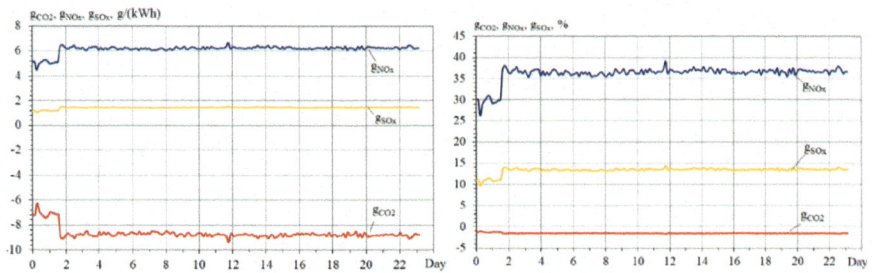

Fig. 6. The values of NOx, SOx and CO_2 emissions in absolute values (a) and relative (b) during the route of the vessel.

Reduction of emissions due to lowering the engine intake air temperature when using the heat of recirculation gases is insignificant and amounts to no more than 0.2–0.3% for NOx and SOx, but for the system with gas recirculation and $\zeta = 0.35$ (Fig. 6) is: $\Delta g_{NOx(0.6)} = 26.3–39.1\%$ (4.7–6.7 g/(kWh)); $\Delta g_{SOx(0.6)} = 9.6–14.3\%$ (1.1–1.5 g/(kWh)).

However, it should be noted that this enhance CO_2 emissions by $\Delta g_{CO2(0.6)} = 1.5–1.7\%$ (6.3–9.4 g/(kWh)).

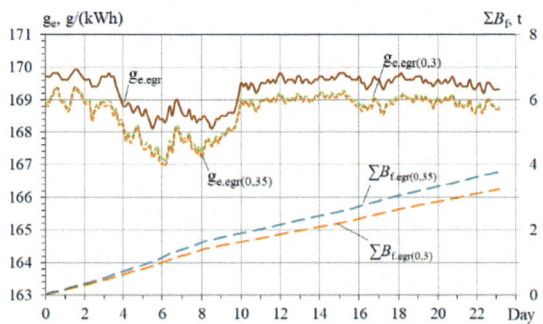

Fig. 7. Variation of specific fuel consumption $g_{e.egr}$ and total fuel economy $\sum B_{f.egr}$ during the vessel route Odessa-Shanghai at different coefficients of performance of chiller $\zeta = 0.30$; 0.35.

The total fuel economy $\sum B_f$ due to decrease in the temperature of intake air at the suction of turbocharger by ejector chiller using the heat of recirculation gas during the vessel route Odessa-Shanghai at different coefficients of performance of chiller is the following (Fig. 7): $\Sigma B_{f(0.3)}$ = 3.3 t (for ζ = 0.3); $\Sigma B_{f(0.35)}$ = 3.8 t (for ζ = 0.35).

For the price of heavy fuel oil 295 \$/t [28] the savings in consumed fuel cost during the vessel route Odessa-Shanghai are 973.5 \$ (ζ = 0.3) and 1121 \$ (ζ = 0.35). A vessel on this route can make up to 12 voyages per year. Then the annual savings in fuel cost will be 11682 \$ (ζ = 0.3) and 13452 \$ (ζ = 0.35). The ONDA heat exchanger MPE 1350 can be applied in ejector chiller [29]. Its cost is 27000 \$ taking into account the cost of its mounting. Proceeding from this a payback time of waste heat recovery installation will be about 2 years.

5 Conclusions

The technology of precooling intake air at the suction of turbocharger by waste heat using ejector chiller (ECh) is implemented in internal combustion engines with EGR system. The scheme solution of the ecological exhaust gas recirculation system with the use of its heat by an ejector refrigeration chiller for cooling the air at the intake of the ship's main engine is developed.

The effect of using the heat of recirculating gases for cooling air at the turbocharger intake was analyzed for the engine MAN 6G50ME-C9.6 with the account of changing climatic conditions on the vessel route line Odesa-Shanghai.

It is shown that the use of ecological recirculation gas heat in ejector refrigeration machine allows to reduce the air temperature at the intake of the ship's main engine by 5–15 °C, that provides a reduction of the specific fuel consumption by 0.5–1.5 g/kW · h. With this the emissions of harmful substances are reduced due to recirculation of gases: of NOx by 26–39%; SOx by 9–14%.

References

1. Kumar, J.T.S., Sharma, T.K., Murthy, K.M., Rao, G.A.P.: Effect of reformed EGR on the performance and emissions of a diesel engine: a numerical study. Alexandria Eng. J. **57**(2), 517–525 (2018). https://doi.org/10.1016/j.aej.2017.01.008
2. Liu, J., Sun, S., Xu, C.: Effects of external cooled EGR on particle number emissions under cold and warm spark-ignition direct-injection engine conditions. Therm. Sci. **22**(3), 1363–1371 (2018). https://doi.org/10.2298/TSCI180207119L
3. Reddy, E.R., Keerthana, B.V.S., Raju, V.D., et al.: Effect of EGR on diverse characteristics of diesel engine operated with corn seed biodiesel blend. Int. J. Ambient Energy (2020). https://doi.org/10.1080/01430750.2020.1745275
4. Forduy, S., Radchenko, A., Kuczynski, W., Zubarev, A., Konovalov, D.: Enhancing the gas engines fuel efficiency in integrated energy system by chilling cyclic air. In: Tonkonogyi, V., Ivanov, V., Trojanowska, J., Oborskyi, G., Edl, M., Kuric, I., Pavlenko, I., Dasic, P. (eds.) InterPartner 2019. LNME, pp. 500–509. Springer, Cham (2020). https://doi.org/10.1007/978-3-030-40724-7_51

5. Trushliakov, E., Radchenko, A., Forduy, S., Zubarev, A., Hrych, A.: Increasing the operation efficiency of air conditioning system for integrated power plant on the base of its monitoring. In: Nechyporuk, M., Pavlikov, V., Kritskiy, D. (eds.) Integrated Computer Technologies in Mechanical Engineering. AISC, vol. 1113, pp. 351–360. Springer, Cham (2020). https://doi.org/10.1007/978-3-030-37618-5_30

6. Grünenwald, B., Knödler, W., Saumweber, C.: A methodology for evaluation of charge air coolers for low pressure EGR systems with respect to corrosion. In: Vehicle Thermal Management Systems Conference and Exhibition (VTMS), pp. 57–67. Woodhead Publishing, Cambridge (2011). https://doi.org/10.1533/9780857095053.2.57

7. Kornienko, V., Radchenko, M., Radchenko, R., et al.: Improving the efficiency of heat recovery circuits of cogeneration plants with combustion of water-fuel emulsions. Therm. Sci. (2020). https://doi.org/10.2298/TSCI200116154K

8. Thangaraja, J., Kannan, C.: Effect of exhaust gas recirculation on advanced diesel combustion and alternate fuels – a review. Appl. Energy **180**, 169–184 (2016). https://doi.org/10.1016/j.apenergy.2016.07.096

9. MAN Diesel & Turbo: Tier III two-stroke technology. https://marine.mandieselturbo.com/docs/librariesprovider6/technical-papers/tier-iii-two-stroke-technology.pdf. Accessed 10 Apr 2020

10. Llamas, X.: Modeling and control of EGR on marine two-stroke diesel engines. Dissertation, Linköping University (2018)

11. De Serio, D., De Oliveira, A., Sodré, J.R.: Application of an EGR system in a direct injection diesel engine to reduce NO_X emissions. J. Phys: Conf. Ser. **745**(3), 032001 (2016). https://doi.org/10.1088/1742-6596/745/3/032001

12. Radchenko, R., Pyrysunko, M., Radchenko, A., et al.: Ship engine intake air cooling by ejector chiller using recirculation gas heat. In: Tonkonogyi, V., et al. (eds.) Advanced Manufacturing Processes II. InterPartner 2020. LNME. Springer, Cham (2021)

13. Nielsen, B.Ø.: 8500 TEU container ship. GSF container ship concept study. Odense Steel Shipyard Ltd. (2009)

14. Hadjab, R., Kadja, M.: Computational fluid dynamics simulation of heat transfer performance of exhaust gas re-circulation coolers for heavy-duty diesel engines. Therm. Sci. **22**(6B), 2733–2745 (2018). https://doi.org/10.2298/TSCI160830317H

15. Pham, V.V.: Advanced technology solutions for treatment and control noxious emission of large marine diesel engines: a brief review. J. Mech. Eng. Res. Dev. **42**(5), 21–27 (2019)

16. Nielsen, K.V.: Exhaust recirculation control for reduction of NOx from large two-stroke diesel engines. Dissertation, Technical University of Denmark (2016)

17. Larsen, U., Pierobon, L., Baldi, F., et al.: Development of a model for the prediction of the fuel consumption and nitrogen oxides emission trade-off for large ships. Energy **80**, 545–555 (2015). https://doi.org/10.1016/j.energy.2014.12.009

18. Radchenko, A., Radchenko, M., Konovalov, A., Zubarev, A.: Increasing electrical power output and fuel efficiency of gas engines in integrated energy system by absorption chiller scavenge air cooling on the base of monitoring data treatment. In: E3S Web of Conferences, vol. 70, p. 03011 (2018). https://doi.org/10.1051/e3sconf/20187003011

19. Radchenko, A., Mikielewicz, D., Forduy, S., Radchenko, M., Zubarev, A.: Monitoring the fuel efficiency of gas engine in integrated energy system. In: Nechyporuk, M., Pavlikov, V., Kritskiy, D. (eds.) Integrated Computer Technologies in Mechanical Engineering. AISC, vol. 1113, pp. 361–370. Springer, Cham (2020). https://doi.org/10.1007/978-3-030-37618-5_31

20. Bohdal, T., Sikora, M., Widomska, K., Radchenko, A.M.: Investigation of flow structures during HFE-7100 refrigerant condensation. Arch. Thermodyn. **36**(4), 25–34 (2015). https://doi.org/10.1515/aoter-2015-0030

21. Radchenko, A., Bohdal, L., Zongming, Y., Portnoi, B., Tkachenko, V.: Rational designing of gas turbine inlet air cooling system. In: Tonkonogyi, V., Ivanov, V., Trojanowska, J., Oborskyi, G., Edl, M., Kuric, I., Pavlenko, I., Dasic, P. (eds.) InterPartner 2019. LNME, pp. 591–599. Springer, Cham (2020). https://doi.org/10.1007/978-3-030-40724-7_60

22. Konovalov, D., Trushliakov, E., Radchenko, M., Kobalava, H., Maksymov, V.: Research of the aerothermopressor cooling system of charge air of a marine internal combustion engine under variable climatic conditions of operation. In: Tonkonogyi, V., Ivanov, V., Trojanowska, J., Oborskyi, G., Edl, M., Kuric, I., Pavlenko, I., Dasic, P. (eds.) InterPartner 2019. LNME, pp. 520–529. Springer, Cham (2020). https://doi.org/10.1007/978-3-030-40724-7_53

23. Radchenko, A., Radchenko, M., Trushliakov, E., et al.: Statistical method to define rational heat loads on railway air conditioning system for changeable climatic conditions. In: 2018 5th International Conference on Systems and Informatics (ICSAI), Nanjing, pp. 1294–1298. IEEE (2018). https://doi.org/10.1109/ICSAI.2018.8599355

24. Trushliakov, E., Radchenko, M., Bohdal, T., Radchenko, R., Kantor, S.: An innovative air conditioning system for changeable heat loads. In: Tonkonogyi, V., Ivanov, V., Trojanowska, J., Oborskyi, G., Edl, M., Kuric, I., Pavlenko, I., Dasic, P. (eds.) InterPartner 2019. LNME, pp. 616–625. Springer, Cham (2020). https://doi.org/10.1007/978-3-030-40724-7_63

25. Kornienko, V., Radchenko, R., Konovalov, D., Andreev, A., Pyrysunko, M.: Characteristics of the rotary cup atomizer used as afterburning installation in exhaust gas boiler flue. In: Ivanov, V., Pavlenko, I., Liaposhchenko, O., Machado, J., Edl, M. (eds.) DSMIE 2020. LNME, pp. 302–311. Springer, Cham (2020). https://doi.org/10.1007/978-3-030-50491-5_29

26. Kornienko, V., Radchenko, R., Mikielewicz, D., et al.: Improvement of characteristics of water-fuel rotary cup atomizer in a boiler. In: Tonkonogyi, V., et al. (eds.) Advanced Manufacturing Processes II. InterPartner 2020. LNME. Springer, Cham (2021)

27. MAN Diesel & Turbo: CEAS engine calculations. https://marine.man-es.com/two-stroke/ceas. Accessed 22 June 2019

28. Ship & Bunker: World bunker prices. https://shipandbunker.com/prices#VLSFO. Accessed 30 July 2020

29. Onda S.p.A.: Dry-expansion exchangers MPE. https://www.onda-it.com/eng/products/shell-and-tube-heat-exchangers/dry-expansions-evaporators-mpe. Accessed 30 July 2020

Integrated Computer Technologies in Aerospace Engineering

Analysis of Operation of Ambient Air Conditioning Systems with Refrigeration Machines of Different Types

Eugeniy Trushliakov[ID], Mykola Radchenko[(⊠)][ID],
Bohdan Portnoi[ID], Veniamin Tkachenko[ID], and Artem Hrych[ID]

Admiral Makarov National University of Shipbuilding,
9 Heroes of Ukraine Avenue, Mykolayiv 54025, Ukraine
nirad50@gmail.com

Abstract. The processing of ambient air by vapor compression and exhaust heat conversion refrigeration machines with heat removal from them to atmosphere by the circulating cooling systems are studied. Ambient air cooling is conducted by using different types of refrigeration machines: absorption lithium-bromide refrigeration machine (ARM), utilizing the exhaust gas heat of gas engine, combined absorption-ejector refrigeration machine (AERM) and electrically driven vapor compression refrigeration machine (VCRM). The data on current thermal loads on refrigeration machines of different types and their circulating cooling systems with dry cooling towers (air condensers in the case of VCRM) for different their project thermal loads were obtained through modelling their current loading in accordance with actual climatic conditions. The efficiency of performance of refrigeration machines of different types (ARM, combined AERM and VCRM) to provide the target temperatures of cooled air is estimated by annual production of refrigeration according to ambient air conditioning duties. It was shown, that the most energy-efficient refrigeration machine is the ARM, while the most energy-consuming was VCRM.

Keywords: Refrigeration machine · Air conditioning system · Dry cooling tower · Air condenser · Fuel consumption

1 Introduction

Energy saving technologies are applied to increase combustion engines efficiency, such as cogenerative internal combustion engines [1], afterburning installation in engine exhaust gas boiler [2, 3] and diesel engines [4]. In particular, the use of various kinds of fuel [5, 6] and utilization of exhaust heat [7, 8], including a production of refrigeration by exhaust heat conversion refrigeration machines (EHCRM) [9] are widespread. Waste heat recovery techniques [10, 11] is also widely used in the maritime energy [12, 13]. In particular, it is quite reasonable to generate refrigeration for feeding ambient air conditioning systems (ACS).

Energy consumption for processing ambient air in ACS depends on the ambient air temperature t_a and relative humidity φ_a, which changes significantly during the day [14, 15]. It is obvious, that the production of refrigeration by refrigeration machine

(RM) in response to air conditioning duties over a certain period of time, for example a year (annual refrigeration), can be calculated as $\sum(Q_0 \cdot \tau)$, kW·h, where Q_0 – the current refrigeration capacity of RM as thermal load on the ACS according to ambient air parameters, kW; τ – a duration of the ACS operation in hours at each current refrigeration capacity, characterizes the efficiency of yearly using the installed refrigeration capacities of the RM [16].

Since the heat removed from the EHCRM by dry cooling towers (DC) of the circulating cooling system or by air condensers (ACn) cooled by ambient air in the case of vapor compression refrigeration machine (VCRM), and thermal loads on DC and ACn, in turn, depend on the actual climatic conditions and coefficient of performance (COP) of RM, we have to determine design thermal loads on RM and DC or ACn in the whole and their distribution between RM of different types.

The aim of the research is to analyse the efficiency of ambient ACS operation with refrigeration machines of different types with regard to removing their rejected heat into surrounding to determine rational values of design thermal loads of cooling towers.

2 Literature Review

Many publications are devoted to improving air procession in ACS by intensification of heat transfer processes in air coolers (AC) and condensers [17], evaporators [18, 19], application of various refrigerant circulation contours [20, 21], waste heat recovery technics [22], including combined cooling, heating and power generation, modelling, experimental, monitoring and statistical methods to match current cooling demands [23]. Some of principal technical innovations and methodological approaches in waste heat recovery refrigeration might be successfully applied for vapor compression refrigeration technologies in air conditioning [24, 25]. A promising direction is the use of jet technologies due to their simplicity in design and low cost [26]. Modern methods with using ANSIS [27, 28] for optimizing the sizes and regimes of equipment operation as well as statistical treatment of experimental data [17, 29], the issues of deterioration of heat transfer in waste heat boilers have not been sufficiently studied, which leads to difficulty in assessing the efficiency of heat recovery.

A combination of heating, ventilation and air conditioning (HVAC) system used as the outdoor air processing (OAP) system in the variable refrigerant flow (VRF) and control strategies to enhance its energy performance and thermal comfort were proposed [14, 15]. The VRF system with energy recovery ventilation (ERV) and a dedicated outdoor air system (DOAS) was introduced [14]. The evaluation of indoor thermal environments and energy consumption of the VRF system [15] was conducted.

The HVAC system that processed outdoor air loads by supplying refrigerant from the outdoor unit performed simultaneously as an outdoor unit in the VRF system in contrast with the OAP, which had compressors. In the VRF-OAP system the multiple indoor units and the OAP were simultaneously connected to an outdoor unit.

A higher energy reduction compared with conventional operation without refrigerant flow regulation, revealed when the outdoor air temperature was closer to the indoor temperature set point, was quite evident due to superlative applying the variable speed compressor in part load modes.

3   Research Methodology

The analysis involves a simulation of the ambient ACS operation with EHCRM utilizing the exhaust heat of gas engine (GE) and with a vapor compression refrigeration machine (VCRM) driven by the electricity generated in the GE. The EHCRM act as a consumer of heat. In the first case, it is the absorption lithium-bromide refrigeration machine (ARM) witch cooling air from ambient temperatures down to 20 °C or to 15 °C. In the second case, the two-stage absorption-ejector refrigeration machine (AERM) is considered: the first stage AC receives cold water for cooling air from the ARM, and the second is refrigerant from the ejector refrigeration machine (ERM). Air cooling by the traditional VCRM as well as using the ARM is investigated down to 20 °C and 15 °C.

The schemes of ACS by the ARM and the two-stage AERM with utilization of the GE waste heat and the traditional vapor compression cycle are presented in Fig. 1.

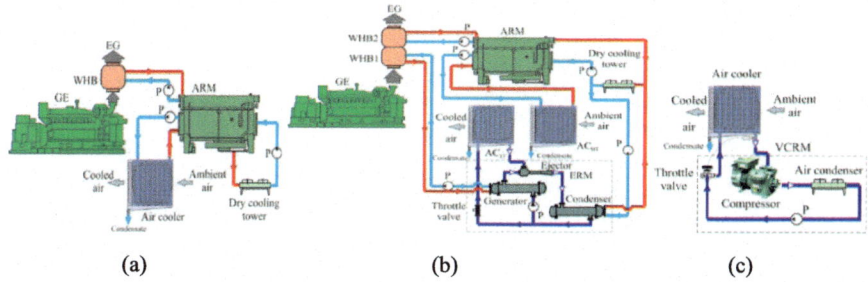

(a)   (b)   (c)

Fig. 1. The scheme of ACS: with the gas engine (GE) exhaust heat utilization (a, b) from ambient air temperatures down to t_{a2} °C by the ARM (a) and from ambient air temperatures down to 20 °C in the AC_{HT} by the ARM and further air cooling to 15 °C in AC_{LT} by the ERM (b) and the traditional ACS by the VCRM (c) with heat removal from the RM in the air cooled dry cooling tower (a, b) and air condenser (c): WHB – waste heat boiler; P – pump; EG – exhaust gases; AC – air cooler; ARM – absorption lithium-bromide refrigeration machine; ERM – ejector refrigeration machine; VCRM – vapor compression refrigeration machine.

A feature of the second scheme (Fig. 1b) is that a structurally simpler ERM, but with a lower coefficient of performance (COP) is $\zeta_{ERM} = 0.2$–0.25 [26] versus $\zeta_{ARM} = 0.7$–0.8 for the ARM [22], is used to cooling the air to 15 °C, but when the air is cooled to a relatively high temperature of 15 °C, the ERM COP increases significantly to $\zeta_{ERM} = 0.3$–0.4 (with evaporation temperature $t_0 = 8$–10 °C), which makes it possible to reduce the power of a more structurally complex ARM and, potentially with sufficient exhaust gases heat, provide deeper cooling of the ambient air. However, it is necessary to investigate the effect of changes in heat load due to a change in the RM type on the its condensers (and therefore dry coolers if using the exhaust heat conversion refrigeration machines) and to determine the possibility of using the two-stage AERM.

The efficiency of ACS and their RM performance depends on their current loading Q_0·and an hour duration τ of yearly operation. Issuing from this the efficiency of using the refrigeration capacity for covering the ambient air processing needs is estimated by annual refrigeration energy production $\sum(Q_0 \cdot \tau)$.

The proposed method to determine the rational design refrigeration capacity is based on the yearly loading characteristic cumulative curve of annual refrigeration production dependence on the design refrigeration capacity of the RM. The current refrigeration, generated by the RM at any time period for ambient air cooling down to the target temperature t_{a2} = 15 or 20 °C, has been summarized over the year. The rational design refrigeration capacity is selected according to yearly loading characteristic cumulative curve to provide closed to maximum annual refrigeration production.

In order to generalize the results and simplify calculations for any total refrigeration capacities Q_0, it is convenient to present the refrigeration capacity of the RM ACS not in absolute Q_0, but in relative (specific) values per unit air flow rate through the air cooler (G_a = 1 kg/s) – in the form of specific refrigeration capacity, $q_0 = Q_0/G_a$, kJ/kg, where Q_0 is the total refrigeration capacity when cooling the air with the flow rate G_a: $Q_0 = (c_a \xi \cdot \Delta t_a)G_a$, where $\Delta t_a = t_a - t_{a2}-$ decrease in air temperature.

The specific annual production of refrigeration:

$$\Sigma(q_0 \cdot \tau) = \Sigma(\xi \cdot c_{ma} \cdot (t_a - t_{a2}) \cdot \tau \cdot 10^{-3}), \tag{1}$$

where $q_0 \cdot$is the specific refrigeration capacity [kJ/kg]; $\sum(q_0 \cdot \tau)$ is the specific annual production of refrigeration [MJ·h/kg]; ξ is the moisture coefficient; t_a is the ambient air temperature [°C]; t_{a2} is the air temperature at the air cooler outlet [°C]; c_{ma} is the specific heat of moist air [kJ/(kg·K)]; τ is the time interval [h].

The specific refrigeration capacity is calculated as:

$$q_0 = \xi \cdot c_{ma} \cdot (t_a - t_{a2}). \tag{2}$$

A rational specific refrigeration capacity q_0 is determined to exclude unproductive expenses of refrigeration capacity q_0 caused by oversizing RM without obtaining a noticeable effect in increasing the annual refrigeration production $\sum(q_0 \cdot \tau)$.

The thermal load on the EHCRM dry cooling tower (DC) was calculated as:

$$q_{DC} = (q_{0.HT}/\zeta_{HT} + q_{0.HT}) + (q_{0.LT}/\zeta_{LT} + q_{0.HT}), \tag{3}$$

where $q_{0.HT}$ is the specific thermal load on AC_{HT} [kJ/kg]; $q_{0.LT}$ is the specific thermal load on AC_{LT} [kJ/kg]; ζ_{HT} is the refrigeration machine of AC_{HT} thermal coefficient; ζ_{LT} is the refrigeration machine of AC_{LT} thermal coefficient.

The thermal load on the VCRM air condensers (ACn) was calculated as:

$$q_{ACn} = q_0 \cdot (1 + \varepsilon)/\varepsilon, \tag{4}$$

where q_0 is the specific thermal load on the AC [kJ/kg]; ε is the VCRM COP.

4   Results

To determine the rational design heat load on the AC (the RM specific refrigeration capacity) that provides closed to maximum average annual values of specific production of refrigeration $\sum(q_0 \cdot \tau)_{av}$ for 2014–2018 years versus an AC design specific refrigeration capacity q_0 (G_a= 1 kg/s) at different temperatures of cooled air for various operation climatic conditions of Kiev region, Ukraine are presented in Fig. 2.

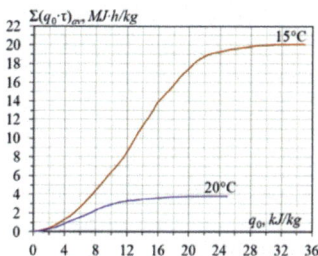

Fig. 2. The average annual values of specific production of refrigeration $\sum(q_0 \cdot \tau)_{av}$ for 2014–2018 years versus an air cooler design specific refrigeration capacity q_0 (air flow $G_a = 1$ kg/s) at different temperatures of cooled air t_{a2}: 15 °C and 20 °C for Kiev region, Ukraine.

From calculation results (Fig. 2), for the considered climatic conditions when the ambient air is cooled to the temperature $t_{a2} = 15$ °C a design specific refrigeration capacity of the RM $q_{0.15} = 23$ kJ/kg provides average annual refrigeration production $\sum(q_0 \cdot \tau)_{av} \approx 19$ MJ·h/kg, while maintaining its increment with a noticeable high rate.

A design specific refrigeration capacity $q_{0.15} = 23$ kJ/kg for cooling the ambient air to $t_{a2} = 15$ °C is assumed as rational one, this value is 32% less than maximum specific refrigeration capacity $q_{0.15max} = 34$ kJ/kg. Similarly, for cooling air to $t_{a2} = 20$ °C the rational value of specific refrigeration capacity is $q_{0.20} = 16$ kJ/kg is lower than its maximum value $q_{0.20max} = 24$ kJ/kg by 33% with corresponding reduction in the RM sizes.

To determine the rational design thermal load on DC q_{DC} and ACn q_{ACn} we will use the results of calculating the average annual values of specific production of refrigeration $\sum(q_0 \cdot \tau)_{av}$ versus an DC q_{DC} or ACn q_{ACn} design specific thermal load for various operation climatic conditions of Kiev region, Ukraine for 2014–2018 years (Fig. 3).

The DC design specific thermal load $q_{DC15A} = 56$ kJ/kg for providing cooling the ambient air to 15 °C is assumed as rational one and $q_{DC20A} = 38$ kJ/kg – to 20 °C for air cooling by the ARM, similar to determining the q_0, whereas for air cooling by the AERM – $q_{DC15AE} = 70$ kJ/kg and $q_{DC20AE} = 38$ kJ/kg, accordingly, since the ARM also acts as the first stage of air cooling. Whereas for VCRM ACn design specific thermal load – $q_{ACn15VC} = 30$ kJ/kg and $q_{ACn20VC} = 22$ kJ/kg. It is explained by the smaller amount of heat for the removal in ACn from the VCRM compressor, compared to the EHCRM generators heat removal in DC.

As seen, the necessary DC design specific thermal load q_{DC} for air cooling with the AERM is higher than for air cooling by the ARM (with DC q_{DC}) and the VCRM (with ACn q_{ACn}) on about 20 and 70…130%, respectively.

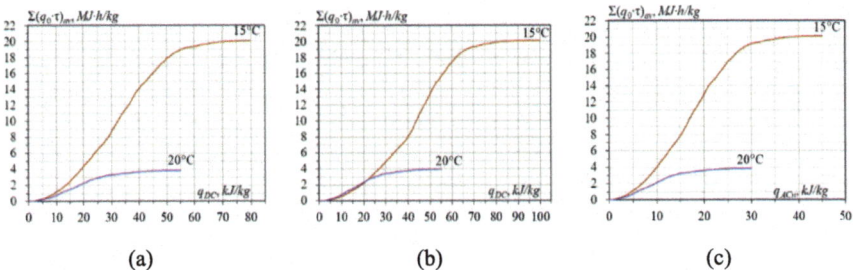

Fig. 3. The average annual values of specific production of refrigeration $\sum(q_0·τ)_{av}$ for 2014–2018 years versus dry cooling tower (a, b) and air condenser (c) design specific thermal load q_{DC} (a, b) and q_{ACn} (c) (air flow through the AC G_a = 1 kg/s) at different temperatures of cooled air t_{a2}: 15 °C and 20 °C for air cooling by the ARM (a), the AERM (b) and the VCRM (c) for Kiev region, Ukraine.

The current values of t_{amb} and $φ_{amb}$, specific heat loads on the AC q_0, actual specific heat consumed by the EHCRM generators q_g, actual specific heat load equivalent to the VCRM compressor electricity consumption q_{comp} removed by the ACn, specific heat load on the DC $q_{DC.RM}$ and ACn $q_{ACn.RM}$, project specific thermal load on DC $q_{DC.pr}$ and ACn $q_{ACn.pr}$, actual heat loads on the DC q_{DC} and ACn q_{ACn} at different temperatures of cooled air t_{a2}: 15 °C and 20 °C for Kiev region, Ukraine are presented in Fig. 4.

As seen from the Fig. 4, for all versions of the ACS, design specific thermal load provides the necessary heat removal from the RM. However, it is necessary to assess the electricity consumption, since the amount of heat required to divert from the generator q_g has significantly increasing for air cooling with using the AERM ($q_{g.20A}$ plus $q_{g.20-15E}$) (Fig. 4b) in comparison with $q_{g.15A}$ (Fig. 4a) for cooling the ambient air to 15 °C. As for VCRM, the main part of its electricity consumption should be from the drive of the compressor, while the heat amount for removal $q_{ACn.RM}$ is much lower than when using the EHCRM ($q_{DC.RM}$), and therefore the ACn fans electricity consumption is much less than the DC fans, since the amount of heat required to remove from the compressor q_{comp} is significantly less than the heat supplied to the EHCRM generator q_g.

The current values of ambient air temperatures t_{amb} and relative humidity $φ_{amb}$, values of (for air flow through the AC G_a = 1 kg/s) electricity consumption of the RM dry cooling towers $N_{eF.DC}$ and air condensers $N_{eF.ACn}$ fans and the VCRM $N_{e.comp}$ compressor at different temperatures of cooled air t_{a2}: 15 °C and 20 °C for Kiev region, Ukraine are presented in Fig. 5.

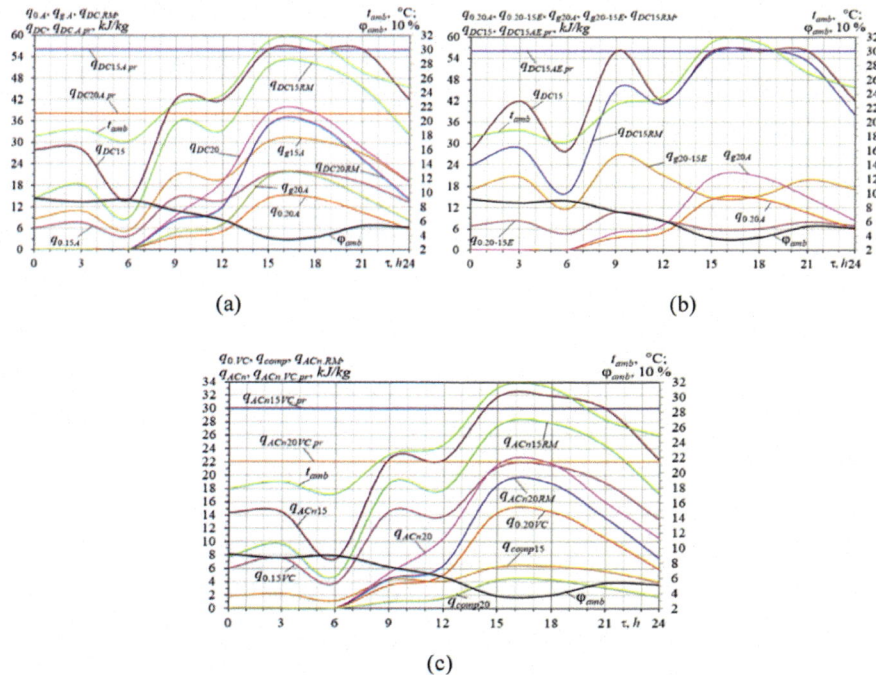

(a)

(b)

(c)

Fig. 4. Current values of t_{amb} and φ_{amb}, specific heat loads on the air cooler q_0, actual specific heat consumed by the EHCRM generators q_g to cover adequate needs q_0, actual specific heat load equivalent to the VCRM compressor electricity consumption q_{comp} removed by the ACn, specific heat load on DC $q_{DC.RM}$ (a, b) and ACn $q_{ACn.RM}$ (c) from the RM, project specific thermal load on DC $q_{DC.pr}$ (a, b) and ACn $q_{ACn.pr}$ (c), actual heat loads on DC q_{DC} (a, b) and ACn q_{ACn} (c) during July 21, 2017 for ambient air cooling by the ARM (a), by the AERM (b), by the VCRM (c) at different temperatures of cooled air t_{a2}: 15 °C and 20 °C for Kiev region, Ukraine.

As seen from Fig. 5, the daily total specific electricity consumption of DC $\Sigma N_{eF.DC15AE}$ fans for air cooling by the AERM is about 70 kJ·h/kg (Fig. 5b) (referred to 1 kg/s of AC mass air flow rate), which is 12% more than the daily total specific electrical power consumption of DC $\Sigma N_{eF.DC15A}$ when air cooling by the ARM, which is 62.5 kJ·h/kg (Fig. 5a) for cooling the ambient air to t_{a2} = 15 °C. While for the VCRM, the specific electrical power consumption by ACn fans $\Sigma N_{eF.ACn15VC}$ is 6.8 kJ·h/kg, however, it is necessary to take into account the specific electrical power consumption by the VCRM compressor motors $\Sigma N_{e.comp15}$, which is 144 kJ·h/kg for cooling the ambient air to t_{a2} = 15 °C. And the VCRM total specific electrical power consumption, for cooling the ambient air to t_{a2} = 15 °C, is about 150.8 kJ·h/kg, which is 141% more than for the ARM $\Sigma N_{eF.DC15A}$ and 115% more – the AERM $\Sigma N_{eF.DC15AE}$. However, further it is necessary to consider how much this change in power affects the fuel consumption by GE.

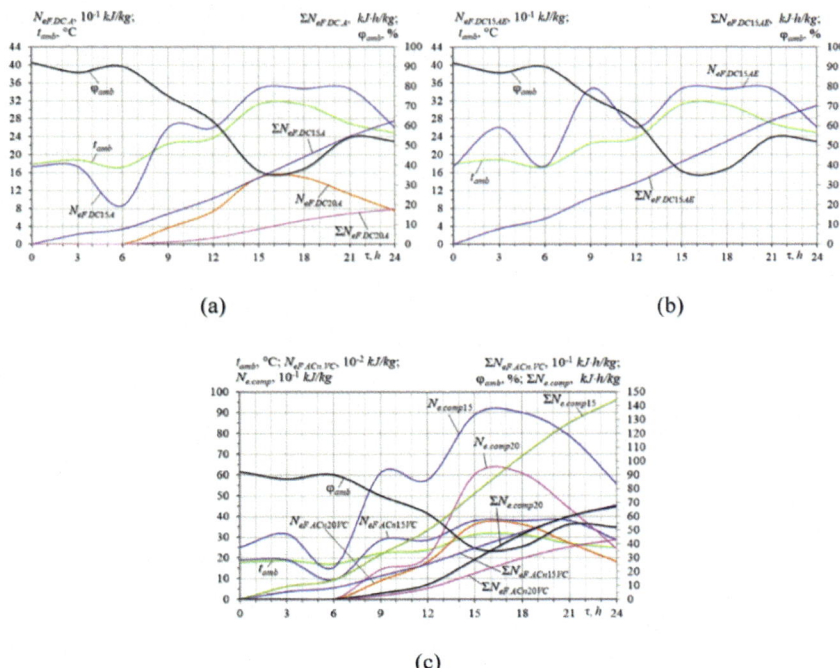

(a)   (b)

(c)

Fig. 5. Current values of t_{amb} and φ_{amb}, specific (for unit air flow through the AC $G_a = 1$ kg/s) electricity consumption of the RM dry cooling towers (DC) $N_{eF.DC}$ (a, b) and air condensers (ACn) $N_{eF.ACn}$ (c), fans and the VCRM $N_{e.comp}$ compressor (c), summarized specific electrical power consumption of fans for DC $\Sigma N_{eF.DC}$ (a, b) and ACn $\Sigma N_{eF.ACn}$ (c) and of the VCRM $\Sigma N_{e.comp}$ compressor (c) during July 21, 2017 for air cooling by the ARM (a), AERM (b) and VCRM (c) at different temperatures of cooled air: 15 °C and 20 °C for Kiev region, Ukraine.

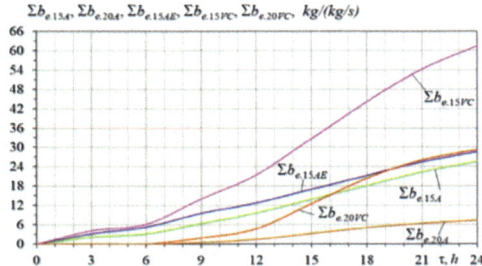

Fig. 6. The values of specific summarized gas engine fuel consumption Σb_e during July 21, 2017 for air cooling by the ARM $\Sigma b_{e.A}$, by the AERM $\Sigma b_{e.AE}$ and by the VCRM $\Sigma b_{e.VC}$ at different temperatures of cooled air t_{a2}: 15 °C and 20 °C for Kiev region, Ukraine.

The values of specific summarized gas engine (GE) fuel consumption Σb_e to drive dry cooling tower and air condenser electric fans and compressor motor at different temperatures of cooled air t_{a2}: 15 °C and 20 °C for Kiev region, Ukraine are presented in Fig. 6. The calculations are carried out for JMS 420 GS-N.LC GE produced by Jenbacher (electrical output 1415 kW, thermal output 1559 kW, electrical efficiency 41.5%) [1], for which the specific fuel consumption Δb_e is 0,183 kg/(kW·h).

As seen from Fig. 6, the daily specific summarized fuel consumption $\Sigma b_{e.A}$ is 24.5 kg/(kg/s) for air cooling by ARM (Fig. 6a) (referred to 1 kg/s of AC mass air flow rate), which is 14% less than when using the AERM ($\Sigma b_{e.AE}$= 28.5 kg/(kg/s)) (Fig. 6b) and 59% less when using the VCRM ($\Sigma b_{e.VC}$= 61 kg/(kg/s)) (Fig. 6c), at target temperature of cooled air 15 °C.

5   Conclusions

The data on actual current heat loads while cooling ambient air to the temperature of about 20 °C and to 15 °C in an absorption lithium-bromide refrigeration machine (ARM) and in a two-stage absorption-ejector refrigeration machine (AERM): the first stage air cooler is fed by chilled water from ARM for cooling air from ambient temperatures to 20 °C, and the second stage uses a boiling refrigerant from the ejector refrigeration machine (ERM) to cool the air further to 15 °C, as well as cooling ambient air to 20 °C and 15 °C in a vapor compression refrigeration machine (VCRM) for climatic conditions of Kiev region, Ukraine, is considered.

The rational design thermal loads on air coolers (AC), dry coolers and refrigerant condensers to provide closed to maximum annual refrigeration production, taking into account the current variable thermal loads according to actual site climatic conditions, are calculated by the methods developed.

It was shown that the application of VCRM leads to extremely high fuel consumption compared to any EHCRM. The use of AERM when cooling ambient air to 15 °C increases the load on the dry cooling towers, which increases the fuel consumption of the GE by about 14%, compared with the use of ARM. However, cooling system with AERM provides deeper air cooling that results in more fuel reduction.

References

1. Radchenko, A., Mikielewicz, D., Forduy, S., et al.: Monitoring the fuel efficiency of gas engine in integrated energy system. In: Nechyporuk, M., et al. (eds.) Integrated Computer Technologies in Mechanical Engineering. AISC, vol. 1113, pp. 361–370. Springer, Cham (2020). https://doi.org/10.1007/978-3-030-37618-5_31
2. Kornienko, V., Radchenko, R., Konovalov, D., et al.: Characteristics of the rotary cup atomizer used as afterburning installation in exhaust gas boiler flue. In: Ivanov, V., et al. (eds.) Advances in Design, Simulation and Manufacturing III. DSMIE 2020. LNME, pp. 302–311. Springer, Cham (2020). https://doi.org/10.1007/978-3-030-50491-5_29
3. Kornienko, V., Radchenko, R., Mikielewicz, D., et al.: Improvement of characteristics of water-fuel rotary cup atomizer in a boiler. In: Tonkonogyi, V., et al. (eds.) Advanced Manufacturing Processes II. InterPartner 2020. LNME. Springer, Cham (2021)
4. Cherednichenko, O., Serbin, S., Dzida, M.: Application of thermo-chemical technologies for conversion of associated gas in diesel-gas turbine installations for oil and gas floating units. Pol. Marit. Res. 26(3), 181–187 (2019). https://doi.org/10.2478/pomr-2019-0059

5. Mikielewicz, D., Kosowski, K., Tucki, K., et al.: Gas turbine cycle with external combustion chamber for prosumer and distributed energy systems. Energies **12**(18), 3501 (2019). https://doi.org/10.3390/en12183501

6. Mikielewicz, D., Kosowski, K., Tucki, K., et al.: Influence of different biofuels on the efficiency of gas turbine cycles for prosumer and distributed energy power plants. Energies **12**(16), 3173 (2019). https://doi.org/10.3390/en12163173

7. Kornienko, V., Radchenko, M., Radchenko, R., et al.: Improving the efficiency of heat recovery circuits of cogeneration plants with combustion of water-fuel emulsions. Therm. Sci. (2020). https://doi.org/10.2298/TSCI200116154K

8. Kornienko, V., Radchenko, R., Stachel, A., et al.: Correlations for pollution on condensing surfaces of exhaust gas boilers with water-fuel emulsion combustion. In: Tonkonogyi, V., et al. (eds.) Advanced Manufacturing Processes. InterPartner 2019. LNME, pp. 530–539. Springer, Cham (2020). https://doi.org/10.1007/978-3-030-40724-7_54

9. Gluesenkamp, K., Hwang, Y., Radermacher, R.: High efficiency micro trigeneration systems. Appl. Therm. Eng. **50**(2), 1480–1486 (2013). https://doi.org/10.1016/j.applthermaleng.2011.11.062

10. Radchenko, R., Kornienko, V., Pyrysunko, M., et al.: Enhancing the efficiency of marine diesel engine by deep waste heat recovery on the base of its simulation along the route line. In: Nechyporuk, M., et al. (eds.) Integrated Computer Technologies in Mechanical Engineering. AISC, vol. 1113, pp. 337–350. Springer, Cham (2020). https://doi.org/10.1007/978-3-030-37618-5_29

11. Cherednichenko, O., Serbin, S.: Analysis of efficiency of the ship propulsion system with thermochemical recuperation of waste heat. J. Mar. Sci. Appl. **17**(1), 122–130 (2018). https://doi.org/10.1007/s11804-018-0012-x

12. Cherednichenko, O., Tkach, M., Timoshevskiy, B., et al.: Improving the efficiency of a gas-fueled ship power plant using a Waste Heat Recovery metal hydride system. Zeszyty Naukowe Akademii Morskiej w Szczecinie **59**, 9–15 (2019). https://doi.org/10.17402/346

13. Konovalov, D., Trushliakov, E., Radchenko, M., et al.: Research of the aerothermopresor cooling system of charge air of a marine internal combustion engine under variable climatic conditions of operation. In: Tonkonogyi, V., et al. (eds.) Advanced Manufacturing Processes. InterPartner 2019. LNME, pp. 520–529. Springer, Cham (2020). https://doi.org/10.1007/978-3-030-40724-7_53

14. Im, P., Malhotra, M., Munk, J.D., Lee, J.: Cooling season full and part load performance evaluation of variable refrigerant flow (VRF) system using an occupancy simulated research building. In: Paper Presented at the 16th International Refrigeration and Air Conditioning Conference, Purdue University, West Lafayette, 11–14 July 2016

15. Lee, J.H., Yoon, H.J., Im, P., Song, Y.H.: Verification of energy reduction effect through control optimization of supply air temperature in VRF-OAP system. Energies **11**(1), 49 (2018). https://doi.org/10.3390/en11010049

16. Trushliakov, E., Radchenko, M., Bohdal, T., et al.: An innovative air conditioning system for changeable heat loads. In: Tonkonogyi, V., et al. (eds.) Advanced Manufacturing Processes. InterPartner 2019. LNME, pp. 616–625. Springer, Cham (2020). https://doi.org/10.1007/978-3-030-40724-7_63

17. Bohdal, T., Sikora, M., Widomska, K., Radchenko, A.M.: Investigation of flow structures during HFE-7100 refrigerant condensation. Arch. Thermodyn. **36**(4), 25–34 (2015). https://doi.org/10.1515/aoter-2015-0030

18. Dąbrowski, P., Klugmann, M., Mikielewicz, D.: Channel blockage and flow maldistribution during unsteady flow in a model microchannel plate heat exchanger. J. Appl. Fluid Mech. **12**, 1023–1035 (2019). https://doi.org/10.29252/jafm.12.04.29316

19. Dąbrowski, P., Klugmann, M., Mikielewicz, D.: Selected studies of flow maldistribution in a minichannel plate heat exchanger. Arch. Thermodyn. **38**(3), 135–148 (2017). https://doi.org/10.1515/aoter-2017-0020

20. Chua, K.J., Chou, S.K., Yang, W.M., Yan, J.: Achieving better energy-efficient air conditioning – a review of technologies and strategies. Appl. Energy **104**, 87–104 (2013). https://doi.org/10.1016/j.apenergy.2012.10.037

21. Śmierciew, K., Gagan, J., Butrymowicz, D., Karwacki, J.: Experimental investigations of solar driven ejector air-conditioning system. Energy Build. **80**, 260–267 (2014). https://doi.org/10.1016/j.enbuild.2014.05.033

22. Gomri, R.: Investigation of the potential of application of single effect and multiple effect absorption cooling systems. Energy Convers. Manag. **51**(8), 1629–1636 (2010). https://doi.org/10.1016/j.enconman.2009.12.039

23. Forduy, S., Radchenko, A., Kuczynski, W., et al.: Enhancing the fuel efficiency of gas engines in integrated energy system by chilling cyclic air. In: Tonkonogyi, V., et al. (eds.) Advanced Manufacturing Processes. InterPartner 2019. LNME, pp. 500–509. Springer, Cham (2020). https://doi.org/10.1007/978-3-030-40724-7_51

24. Chua, K.J., Yang, W.M., Wong, T.Z., Ho, C.A.: Integrating renewable energy technologies to support building trigeneration – a multi-criteria analysis. Renew. Energy **41**, 358–367 (2012). https://doi.org/10.1016/j.renene.2011.11.017

25. Freschi, F., Giaccone, L., Lazzeroni, P., Repetto, M.: Economic and environmental analysis of a trigeneration system for food-industry: a case study. Appl. Energy **107**, 157–172 (2013). https://doi.org/10.1016/j.apenergy.2013.02.037

26. Radchenko, R., Pyrysunko, M., Radchenko, A., et al.: Ship engine intake air cooling by ejector chiller using recirculation gas heat. In: Tonkonogyi, V., et al. (eds.) Advanced Manufacturing Processes II. InterPartner 2020. LNME. Springer, Cham (2021)

27. Bohdal, L., Kukielka, L., Świłło, S., et al.: Modelling and experimental analysis of shear-slitting process of light metal alloys using FEM, SPH and vision-based methods. AIP Conf. Proc. **2078**(1), 020060 (2019). https://doi.org/10.1063/1.5092063

28. Bohdal, L., Kukielka, L., Radchenko, A.M., et al.: Modelling of guillotining process of grain oriented silicon steel using FEM. AIP Conf. Proc. **2078**(1), 020080 (2019). https://doi.org/10.1063/1.5092083

29. Bohdal, Ł., Kukiełka, L., Legutko, S., et al.: Modeling and experimental analysis of shear-slitting of AA6111-T4 Aluminum alloy sheet. Materials **13**(14), 3175 (2020). https://doi.org/10.3390/ma13143175

Computer Simulation of Abnormal Glow Discharge in an Inverse Magnetron Sputtering System with Axial Plasma Flows

Denis Sliusar$^{(\boxtimes)}$ ⓘ, Oleksii Isakov ⓘ, Volodymyr Kolesnyk ⓘ,
Oleg Chugai ⓘ, Leonid Litovchenko ⓘ, and Mikola Stepanushkin ⓘ

National Aerospace University "Kharkiv Aviation Institute", 17 Chkalova Street,
Kharkiv 61070, Ukraine
d.sliusar@khai.edu

Abstract. Developed in NAU "Kharkiv Aviation Institute" inverse magnetron sputtering system with sectioned cathode units and axial plasma flows allows to obtain multicomponent, multilayer and functional-gradient coatings. The advantage is shown over currently used technological ion-plasma generators. It is concluded that it is necessary to develop a mathematical model that allows one to calculate the operating parameters of the sputtering system. These parameters are to be known to obtain coatings of a given composition and thickness. It is shown that to solve this problem it is necessary to develop a numerical model for calculating the distribution of local plasma parameters in the discharge gap of an inverse magnetron sputtering system with sectioned cathode units and axial plasma flows. To solve this problem, a fluid plasma model was used. The distributions of the plasma potential, electron temperature and density in the discharge gap of the investigated sputtering system were obtained. Based on comparison of these calculated parameters with experimental data, the conclusion was made about the correctness of the developed mathematical model. Noted that in order to complete the development of a mathematical model for calculation the deposition rate and composition of the coating for an inverse magnetron sputtering system with sectioned cathode units and axial plasma flows, it is necessary to develop a mathematical model for calculating the distribution of the ion current density over the surface of the target cathodes and the energy of the bombarding cathode ions.

Keywords: Plasma fluid model · Electron density · Electron temperature · Magnetic field

1 Introduction

One of the major challenges of the modern mechanical engineering is to improve the operational characteristics of mechanisms' and machines' parts. This problem is solved in two ways: by creating new materials (alloys, ceramics, etc.) or by modifying surface layer of the part. Nowadays, the part surface is supposed to have a number of properties such as: high mechanical strength and hardness, resistance to alternating loads, high corrosion and wear resistance, heat resistance, resistance to ionic and dust erosion wear.

M. Nechyporuk et al. (Eds.): ICTM 2020, LNNS 188, pp. 556–564, 2021.
https://doi.org/10.1007/978-3-030-66717-7_47

It is also important to ensure a low-cost coating formation on a part surface. In this case, the properties of the coating should differ significantly from the properties of the base material.

Modern functional coatings consist of numerous metallic and non-metallic components. Moreover, coatings can be not only single-layer, but also multi-layer. In order to obtain such coatings by electron beam evaporation methods, Ukrainian scientists [1] proposed to use composite ingots. Thus, gradient coatings can be formed from metallic and non-metallic components. To create multicomponent layers on a part surface one use cathodes with same, as layers, chemical composition [2, 3]. Creation of complex alloys with required component composition is quite labor-intensive and expensive process, and sometimes, due to significant difficulties, even impossible.

A few decades ago, native and foreign scientists proposed and implemented in practice the application of coatings by generating particle streams of pure metals or partial alloys, last ones included only part of the coating components [4–6]. It should be noted that the existing equipment for ion-plasma coating, due to its design features, has a limited number of traditional coating material flow generators, which can be simultaneously placed in the working area of the vacuum chamber.

At the National Aerospace University "Kharkiv Aviation Institute" new technological generators [7] were developed, which are a sort of inverse magnetron systems [8]. On their basis pilot industrial technological installations have been designed and new technologies were developed. These technologies do not require cathodes with complex composition any more. Due to sputtering of a large number of cathodes made either from separate coating components or partial alloys, these technologies allow to obtain multicomponent and gradient coatings of almost any component composition. Using various reactive gases in the working area of the technological chamber in addition to the plasma-forming gas (usually argon), enhances controlling opportunities of coating composition. Moreover, changing electrical parameters of the sputtering system makes it possible to form coatings with a given shift in the component composition both over the coating thickness and over its surface [9].

2 Research Objective Determination

For successful implementation of the coating formation process using the proposed sputtering systems, it is necessary to identify the parameters that affect the technological process, determine the nature of this influence and control these parameters.

It can be achieved by using a model of cathode material transfer to the substrates surface. For the inverse magnetron sputtering system with a gas anode, sectioned cathode units and axial plasma flows, the model presented in [9] can be used. Based on this model and taking into account sectioned cathodes arrangement according to Fig. 1, a formula was obtained that makes it possible to determine the deposition rate of individual components at any point on the substrate surface. With regard to discharge parameters, type of sputtered materials, and ion current density distribution on the cathodes depending on distance from the system axis the formula takes next form:

$$V_{dp}(l_n, r) = \frac{K_{pf}(E_i)m_{ap}H^2}{\pi e k} \int\limits_{R_{f1}}^{R_{f2}} \int\limits_{0}^{2\pi} \frac{j_{pf}r}{(r^2 + l_n^2 - 2l_n r \cos\Theta + H^2)^2} d\Theta dr, \qquad (1)$$

where $P = 1...k$, and $f = 1...n$; j_{pf} is ion current density at the cathodes, made of coating element No. p, grouped No. f, A/cm^2; $K_{pf}(E_i)$ is a cathode material sputter yield coefficient made of element No. p, grouped No. f, atom/ion; R_{f1} is a minimum radius of cathodes group No. f, m; R_{f2} is a maximum radius of cathodes group No. f, m; H is a distance from the cathode surface to the substrate, m; l_n is the distance from the axis of the sputtering system to the point on the substrate at which it is necessary to determine the coating deposition rate, m; e is an elementary charge; r is a distance from the system's axis, m; m_{ap} is an atom mass of a cathode made from a coating element No. p, kg.

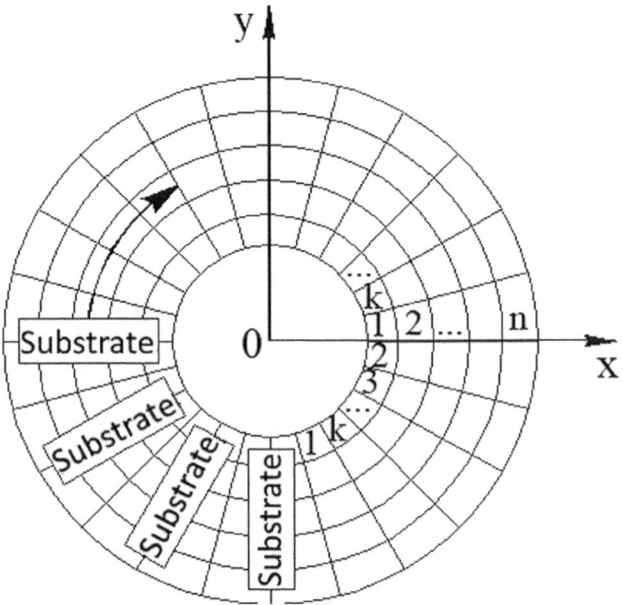

Fig. 1. Layout of cathode sections.

The ion current density can be considered as a constant over its entire area in the case of small size cathode sections:

$$j_{pf} = I_{pf}/S_{pf}, \qquad (2)$$

where I_{pf} is a cathode current, made of coating element No. p, combined into group No. f, A; S_{pf} is a cathodes area, made of coating element No. p, combined into group No. f, m^2.

Thus, the deposition rate and the composition of the coatings are determined by the mutual arrangement of the cathode sections and substrates, the current to each of the cathode sections, and sputter yield coefficient of the corresponding cathodes, which depends on the energy of the ions with which they are sputtered. It should be noted that the current of non-self-sustaining discharges can be regulated depending on the problem being solved within wide limits. At the same time, all these non-self-sustaining discharges are supported by the main self-sustained anomalous glow discharge in crossed electric and magnetic fields. Changing the discharge parameters at each separate section of the cathodes affects the discharge parameters at other sections. Thus, in order to achieve the required operating mode of the investigated sputtering systems, it is necessary either to carry out a sufficiently large number of experiments in order to select the required set of parameters, or to create a mathematical model that would make it possible to select promptly the required operating mode.

The solution of this problem is the development and verification of a mathematical model that would make it possible to calculate the distribution of the ion current density over the surface of the cathodes and the energy of the ions bombarding the cathodes, which can be obtained from the main parameters of the plasma in the discharge gap of the investigated sputtering system.

3   Results of Modeling

In this work, the calculation of the local plasma parameters is performed for the discharge gap of an inverse magnetron sputtering system with sectioned cathode units and axial plasma flows. For these purposes a fluid plasma model [7, 10, 11] was used, to calculate the operating modes of an inverse magnetron sputtering system with sectioned cathode units and radial plasma flows.

The plasma parameters were determined at pressure 0.05 Pa, discharge voltage of 900 V. Discharge current was up to 1.8 A. In the numerical model, the sectioned cathode was replaced by a solid disk with the same surface area.

The results of solving differential equations of fluid model are presented in Fig. 2, 3, 4.

Numerical model allows to describe cathodic potential drop near the electrodes with zero potential (see Fig. 2), which is necessary for ion energy determination.

From electron density distribution one can localize the main ionization area, which will be needed in the future to determine the energy of ions (see Fig. 3).

Temperature and density of charged particles near the electrodes allow to determine distributions of heat flows in discharge region.

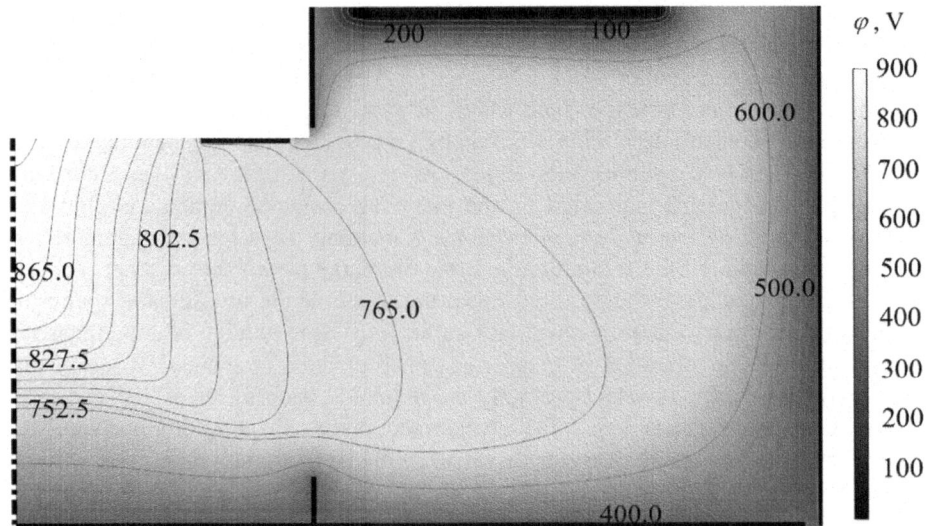

Fig. 2. Plasma potential distribution in discharge gap.

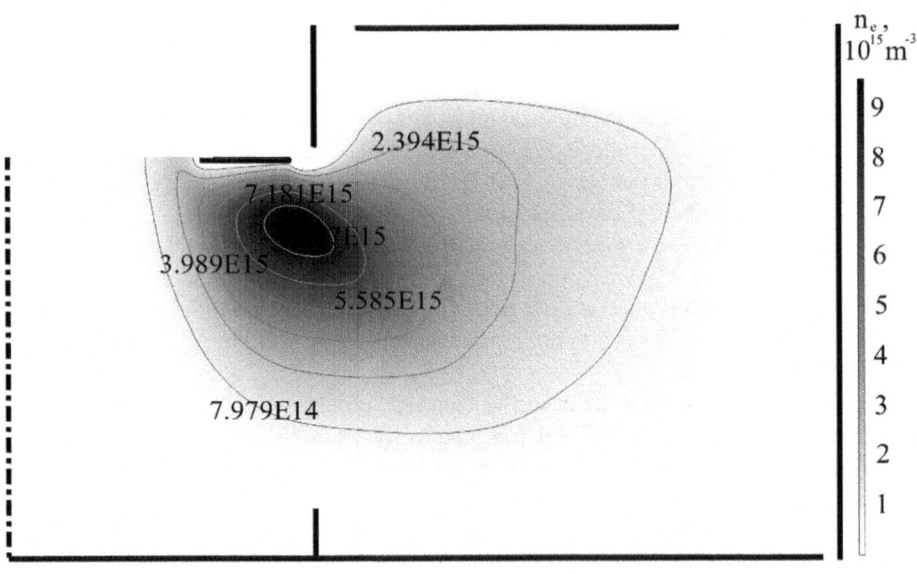

Fig. 3. Electron density distribution in the discharge gap.

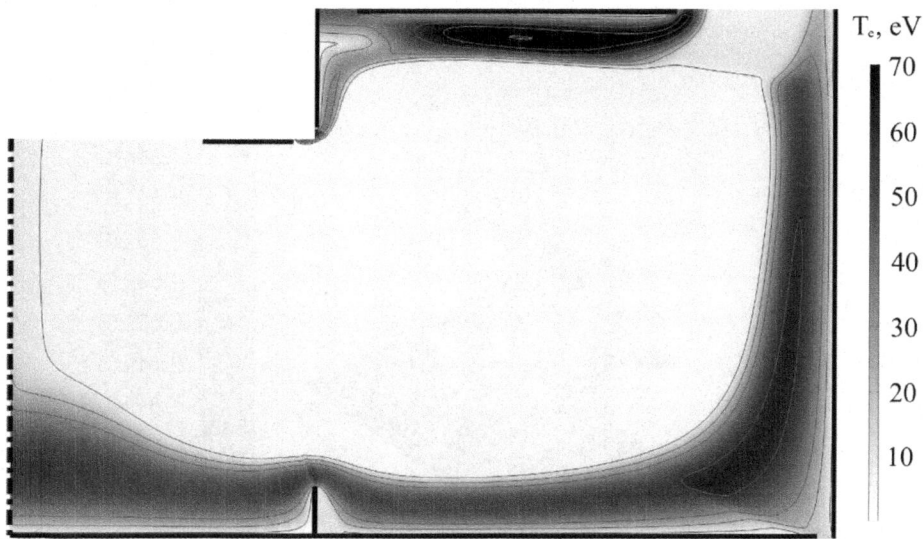

Fig. 4. Electron temperature distribution in the discharge gap.

4   Mathematical Model Verification

The experimental determination of the local plasma parameters was carried out for the discharge gap of the inverse magnetron sputtering system with sectioned cathode units and axial plasma flows at pressure of the plasma-forming argon gas $p = 0.05$ Pa, the discharge voltage between the anode and the sectioned cathode $U_d = 900$ V (voltage across all sections of the sectioned cathode was maintained at the same level), the

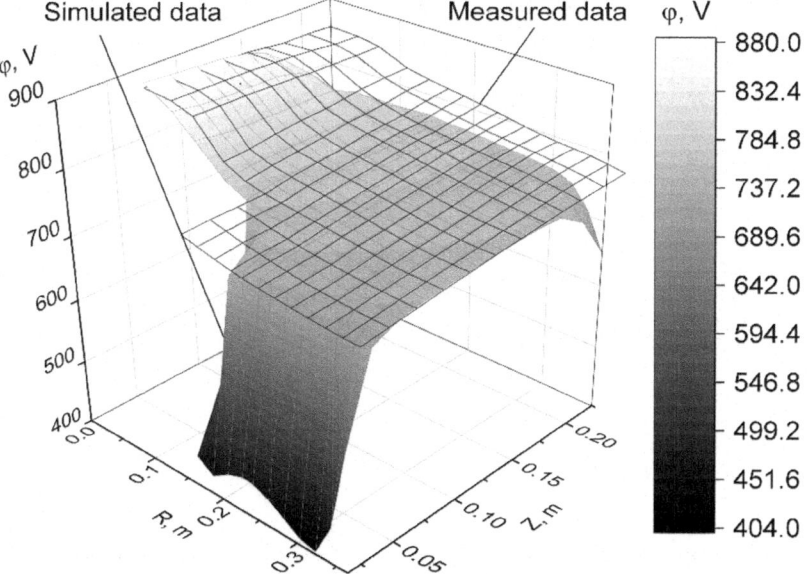

Fig. 5. Measured and calculated distributions of the plasma potential in the discharge gap.

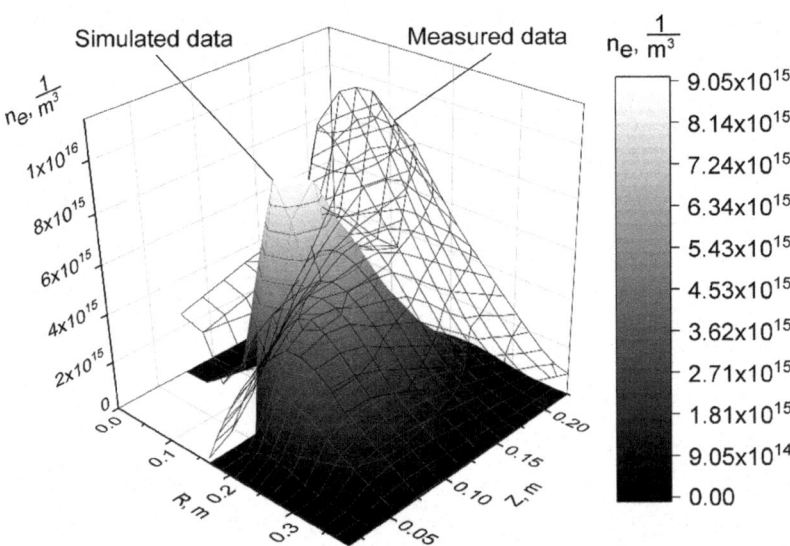

Fig. 6. Measured and calculated distributions of the electron density in the discharge gap.

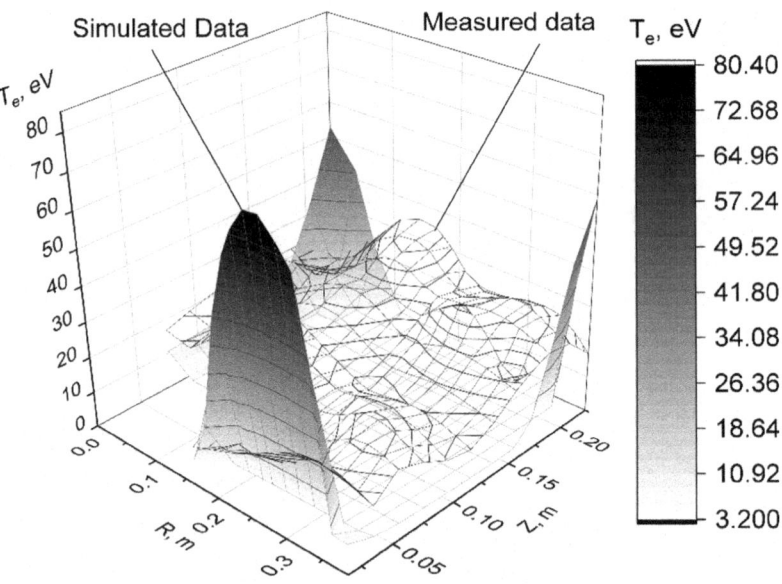

Fig. 7. Measured and calculated distributions of the electron temperature in the discharge gap.

discharge current was 1.8 A. Langmuir probe was used to measure the plasma potential, density and temperature of electrons. The measurements were carried out in twelve planes at distance of 0.015 m from the cathode surface. The results of probe measurements of the plasma potential, electron density and temperature are shown in Fig. 5, 6, 7, respectively.

The nature and magnitude of the potential distribution practically coincides in the discharge gap, only near the cathode there is a slight divergence. The divergence can be caused by a high error of probe measurements due to a number of reasons such as difference between collecting probe surface and the sum of the probe and Debye radii, because the probe surface area is greater than sum of radii. This is due to the penetration of the probe electric fields into the plasma; charged particles gain additional energy in the electric field of the probe, which is not taken into account during data processing [12–14].

The divergence between the main plasma parameters in the edge zones, near the electrodes, can be associated with the main limitation of the probe measurement technique, the Debye radius should be much smaller than the probe size. According to the simulation results, the Debye radius is larger than the size of the probe near the electrodes, which is caused by the low density of electrons and their high energy. The nature of the electron density distribution is almost identical. An insignificant divergence of the electron density peak along the R coordinate is 2.5% of the discharge gap size, and along Z – 12%, which does not exceed the accuracy of the probe measurements. The experimentally determined distribution of the electron temperature in the discharge gap also differs from the calculated one within the accuracy of the probe measurements.

5 Conclusions

This paper presents the results of numerical calculations using the fluid model for the inverse magnetron sputtering system with sectioned cathode units and axial plasma flows. The obtained results are compared with the distributions of local plasma parameters for the investigated sputtering system, obtained by the probe method. It is concluded that they differ from each other within the accuracy of the probe measurements. At the same time, the nature of these distributions is very similar. Thus, the first step has been taken towards solving the problem of numerical simulation of the process of multicomponent coatings formation using inverse magnetron sputtering systems with sectioned cathode units and axial plasma flows. In the future, it is necessary to verify the developed numerical model for other modes of operation of the sputtering system, whereupon it is necessary to calculate the distributions of the ion current density over the cathode surface and the energies of bombarding ions, depending on the operating modes of this sputtering system.

References

1. Movchan, B., Yakovchuk, K.: High-temperature protective coatings produced by EB-PVD. J. Coat. Sci. Technol. **1**(2), 96–110 (2014)
2. Chang, K.S., Chen, K.T., Hsu, C.Y., Hong, P.D.: Growth (AlCrNbSiTiV)N thin films on the interrupted turning and properties using DCMS and HIPIMS system. Appl. Surf. Sci. **440**, 1–7 (2018). https://doi.org/10.1016/j.apsusc.2018.01.110
3. Muboyadzhyan, S.A., Egorova, L.P., Gorlov, D.S., Bulavintseva, E.E.: Corrosion-resistant coating for GTE compressor parts made of steels with low tempering temperatures. Russ. Metall. (Met.) **2017**(1), 1–9 (2017). https://doi.org/10.1134/S0036029517010086
4. Li, W., Liu, P., Liaw, P.K.: Microstructures and properties of high-entropy alloy films and coatings: a review. Mater. Res. Lett. **6**(4), 199–229 (2018). https://doi.org/10.1080/21663831.2018.1434248
5. Yan, X.H., Li, J.S., Zhang, W.R., Zhang, Y.: A brief review of high-entropy films. Mater. Chem. Phys. **210**, 12–19 (2018). https://doi.org/10.1016/j.matchemphys.2017.07.078
6. Dolique, V., Thomann, A.L., Brault, P.: High-entropy alloys deposited by magnetron sputtering. IEEE Trans. Plasma Sci. **39**(11), 2478–2479 (2011). https://doi.org/10.1109/TPS.2011.2157942
7. Isakov, A., Kolesnik, V., Okhrimovskyy, A., et al.: Computer simulation of abnormal glow discharge processes in crossed electric and magnetic fields. Probl. At. Sci. Technol. **6**, 171–174 (2014)
8. Thornton, J.A., Penfold, A.S.: Cylindrical Magnetron Sputtering: Thin Film Processes. Academic Press, New York (1978)
9. Sliusar, D.: Method of formation of two-dimensional functionally graded coatings. Dissertation, National Aerospace University "Kharkiv Aviation Institute" (2014). [in Russian]
10. Markosyan, A., Teunissen, J., Dujko, S., Ebert, U.: Comparing plasma fluid models of different order for 1D streamer ionization fronts. Plasma Sources Sci. Technol. **24**(6), 065002 (2015). https://doi.org/10.1088/0963-0252/24/6/065002
11. Gudmundsson, J., Lundin, D., Brenning, N., et al.: An ionization region model of the reactive Ar/O$_2$ high power impulse magnetron sputtering discharge. Plasma Sources Sci. Technol. **25**(6), 065004 (2016). https://doi.org/10.1088/0963-0252/25/6/065004
12. Kudrna, P., Passoth, E.: Langmuir probe diagnostics of a low temperature non-isothermal plasma in a weak magnetic field. Contrib. Plasma Phys. **37**, 417–429 (1997). https://doi.org/10.1002/ctpp.2150370504
13. Hutchinson, I.H.: Principles of Plasma Diagnostics, 2nd edn. Cambridge University Press, Cambridge (2005)
14. Ryan, P.J., Bradley, J.W., Bowden, M.D.: Comparison of Langmuir probe and laser Thomson scattering for electron property measurements in magnetron discharges. Phys. Plasmas **26**(7), 073515 (2019). https://doi.org/10.1063/1.5109621

A Data-Driven Approach to the Prediction of Spheroidal Graphite Cast Iron Yield Surface Probability Characteristics

Mariya Shapovalova$^{(\boxtimes)}$ and Oleksii Vodka

National Technical University "Kharkiv Polytechnic Institute",
2 Kyrpychova Street, Kharkiv 61002, Ukraine
MiShapovalova@gmail.com, oleksii.vodka@gmail.com

Abstract. This work is aimed at studying material with a heterogeneous microstructure. The probabilistic characteristics of the yield surface are investigated. Statistically equivalent internal material structures are generated using computer simulations. The design takes into account the different amounts of spheroidal graphite inclusions concentration in the ferrite material. The stress state is calculated by the finite element method based on plane models. A series of experiments is calculated for each variant of the concentration of inclusions. The yield surfaces are determined. Based on the collected data, a study of the probabilistic characteristics of a random function is carried out. The radius function acts as a random variable. The number of intersections of the line with the yield surfaces is analyzed. The radii are constructed from the origin for each rotation angle along the closed circle. The proposed scheme takes into account the different behavior of composite materials under tensile and compressive loads. The probabilistic characteristics of the investigated quantity give a vision of the material operation modes at various loads. Going beyond the plasticity surface indicates the possibility of a product transition into a plastic state.

Keywords: Microstructure · Finite element method · Material properties · Probability · Yield surface

1 Introduction

An analysis of the collected data is one of the modern approaches to increasing the competitiveness of production. Nowadays, a data-driven approach helps to reduce the cost of production, allows predicting and avoiding equipment failures, contributes to improving the quality of products and processes. Big data analysis allows to track defects, identify the causes of inconsistencies, and eliminate them. A data-driven approach found application in artificial intelligence, engineering [1, 2, 5–7, 13, 14], strategy, marketing [3], policy, medicine [4], etc. It helps scientifically make decisions. In this article, such an approach applies to the prediction of yield surface probability characteristics.

Investigation of the probabilistic characteristics of the anisotropic materials yield surface [8–12], makes it possible to describe the behavior of a structure under a complex stress state. Identified weak points in the structure helps to avoid breakdowns.

M. Nechyporuk et al. (Eds.): ICTM 2020, LNNS 188, pp. 565–576, 2021.
https://doi.org/10.1007/978-3-030-66717-7_48

Carrying out numerous experiments is an expensive and laborious process. Computer modeling for predicting material behavior is an alternative to such an approach.

2 Objectives

As the initial data in this work, the results of previous studies are taken [15–17]. It is assumed that the microstructure is simulated synthetically based on real images of the material. The elastic properties and stress state of the sample are modeled by the finite element method.

The main objective of this study is to predict of spheroidal graphite cast iron yield surface probability characteristics. This objective requires the completion of such tasks:

- to create an experimental set of yield surfaces to ensure the absence of plastic deformations;
- to evaluate the influence of the inclusions concentration on the yield stress state during the tension and compression loading;
- to calculate the probability of plastic strain occurrence.

3 Generation of the Statistically Equivalent Artificial Microstructure

Image processing and artificial microstructure generation of statistically equivalent material have been implemented in previous works [15–17]. According to the results obtained in the articles, the creation of an equivalent structure is possible by establishing the dependence between the size and concentration of inclusions. The information about the quantity and size of inclusions located on a plane is collecting by using computer vision technology. The mathematical expectation data $M[R]$ and the variance $D[R]$ of the radii inclusions dependence on the concentration have been obtained by (1):

$$M[R] = 18.308 \cdot (\psi - 0.048)^{0.123} \; ; \; \sqrt{D[R]} = 9.683 \cdot (\psi - 0.045)^{0.314}. \quad (1)$$

The location is followed to a uniform distribution and the size of inclusions is followed to a normal distribution function of concentration. Concentration (ψ) is defined as the ratio of the area of the inclusion to the area of the sample, which varies in the range of [0.055...0.3].

The finite element model construction is based on the artificial generated geometric model of the spheroidal graphite cast iron microstructure. To create the mesh grid, a two-dimensional 8-node finite element with two degrees of freedom in each node is used [21]. For calculation, is assumed that the main matrix of the investigate sample is isotropic ferrite and the inclusions are an orthotropic graphite material. The corresponding materials properties and elastic constants are given in Table 1 and Table 2. Various material properties and their resistance to tension and compression are taken into account.

Table 1. Properties of ferrite material (matrix).

E, GPa	v	Yield strength, MPa
180	0.35	196

Table 2. Properties of graphite material (inclusions).

c_{11}, c_{22}	c_{12}	c_{13}, c_{23}	c_{33}	c_{44}, c_{55}	c_{66}	Yield strength	
						Compress	Tensile
GPa						MPa	
1060	290	109	46.6	2.3	435	31.9	3.9

To construct the yield surface, the model is considered under different loadings. Some of the typical load cases are presented in Fig. 1. The corresponding stress state of the model is shown in Fig. 2.

The model is represented by a square plate with a side – l. The deformation is set equal to $\varepsilon_\rho = \Delta l/l = 10^{-5}$, then the displacement is calculated by (2):

$$U_x = \varepsilon_\rho \cdot l \cdot \cos \Theta,$$
$$U_y = \varepsilon_\rho \cdot l \cdot \sin \Theta, \tag{2}$$

where U_x, U_y is the displacement along the corresponding axis, $\Theta = (0...360)°$ the angle changes in a range, with a step in $3.6°$.

Computer simulation methods are used to calculate the yield surface in a multi-dimensional stress state. This approach uses the hypothesis of yield strength under difficult loading conditions [8–12, 18]. Finding the yield surface is based on the hypothesis of the maximum distortion energy theory (the Huber - Mises - Hencky hypothesis) [19–21]. According to it, plastic strains of a sample in a complex stress state occurs when the specific formation energy becomes equal to or exceeds the specific formation energy of the material under the action of a uniaxial stress state.

For the microstructure which is consists of two types of materials (ferrite and graphite), the maximum stresses for each phase are found. For graphite, the tensile and compressive strengths differ significantly, therefore, separately for each type of stress state, the ratios maximum stresses to the corresponding allowable tensile strength are found. The yield surface is determined by the ratio of the principal stresses to the safety factor. The graphite material has a different tensile and compressive strength, therefore the dependence of the yield strength is calculated according to (3). Using the Heaviside step function of the first invariant of the stress tensor σ_0 with a coefficient equal to $k = 1\ \mu Pa^{-1}$ (4) and substituting it into (3), the final expression type for finding the yield stress of graphite is obtained by (5).

$$\sigma_{yield}^{graphite}(\sigma_0) \cong H(\sigma_0) \cdot \left(\sigma_{yield}^{tens} - \sigma_{yield}^{comp}\right) + \sigma_{yield}^{comp}, \tag{3}$$

$$H(\sigma_0) = \frac{1}{2} \cdot (1 + th(k \cdot \sigma_0)), \tag{4}$$

$$\sigma_{yield}^{graphite}(\sigma_0) \cong \frac{1}{2} \cdot (1 + th(k \cdot \sigma_0)) \cdot \left(\sigma_{yield}^{tens} - \sigma_{yield}^{comp}\right) + \sigma_{yield}^{comp}, \tag{5}$$

The dependence of the yield strength from the principal stresses for ferrite and graphite materials are shown in Fig. 3.

The calculation results of 250 random typical implementations of the yield surface are presented graphically in Fig. 4. This figure also contains results for $\psi = (0.055, 0.100, 0.300)$.

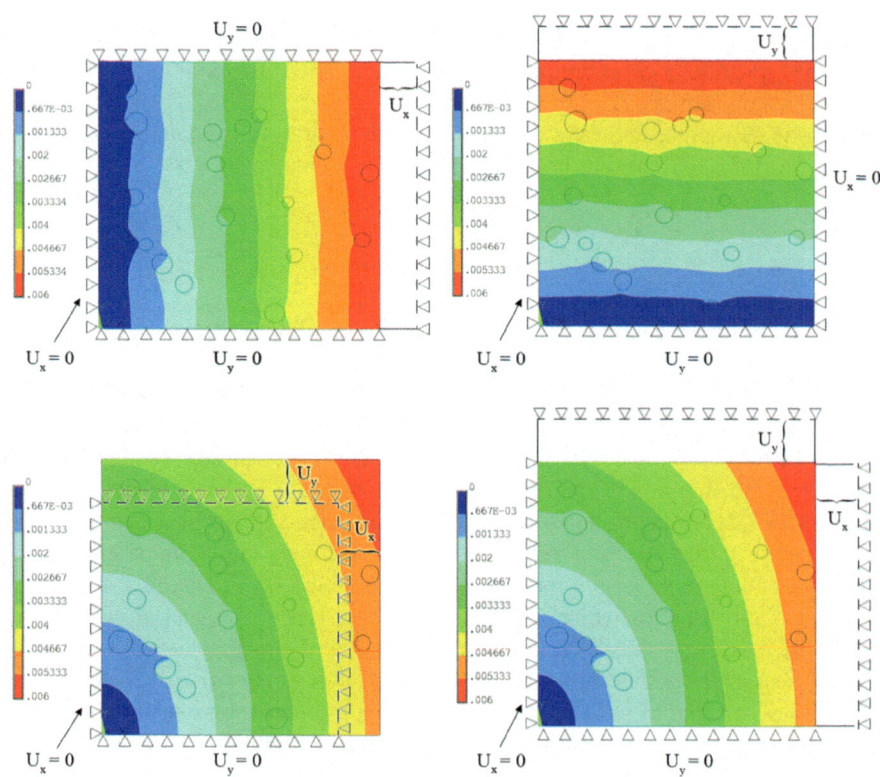

Fig. 1. Model displacement under different types of loads ($\psi = 0.055$).

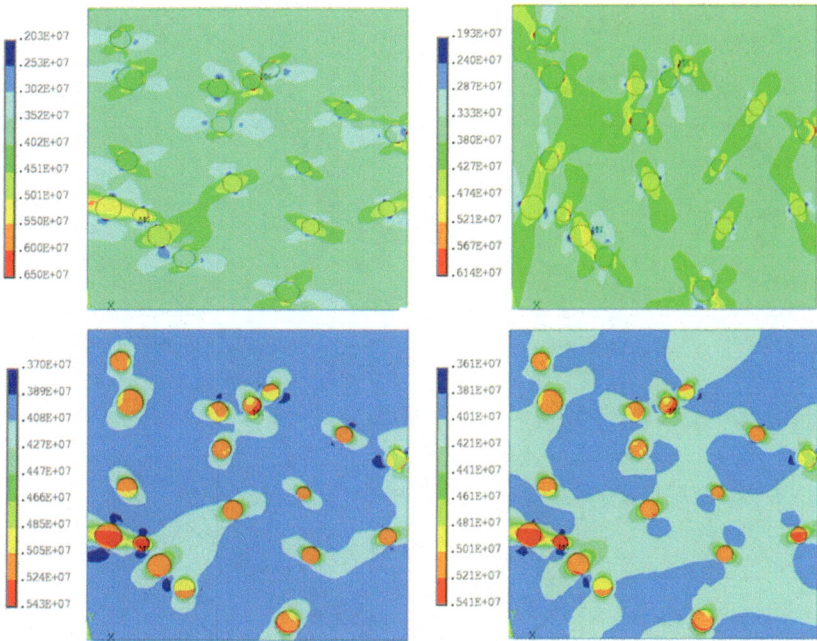

Fig. 2. Von Mises equivalent stresses under different types of loads ($\psi = 0.055$).

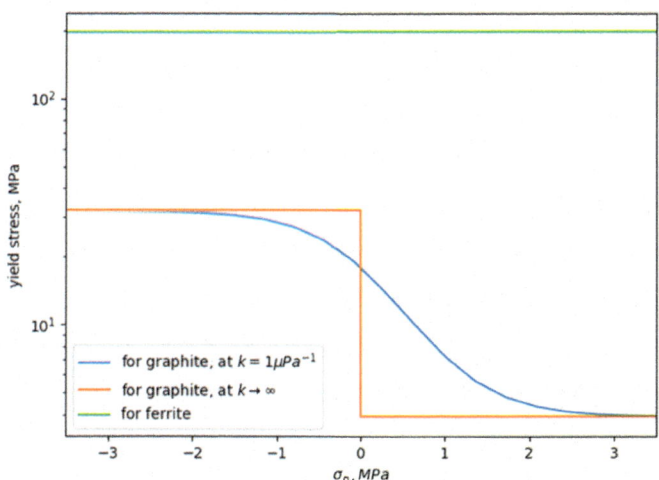

Fig. 3. The dependence of the yield strength from the principal stresses.

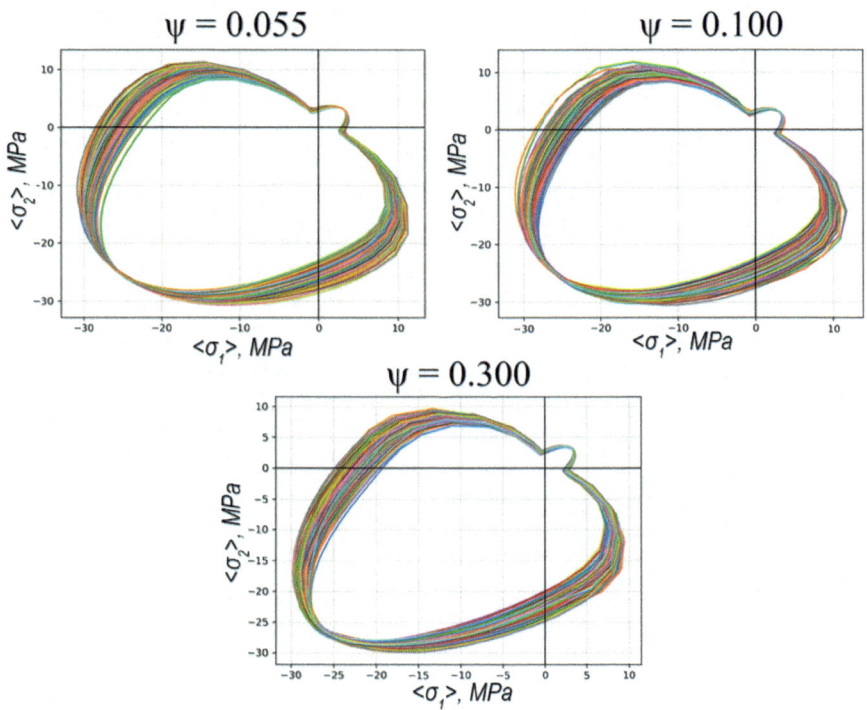

Fig. 4. Typical implementations of the yield surface arrangement.

4   Probability Estimation

The collected statistical data on the maximum allowable stresses for all components of the material under different types of loading leads to an assessment of the yield surface probabilistic characteristics.

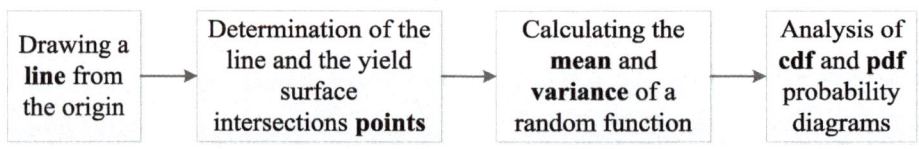

Fig. 5. Probability estimation schema.

A sequential diagram of the random variable estimation shown in Fig. 5. The accumulated statistical information on possible yield surface variants helps to determine the area of stress impact. In this work is using the construction of a line passing through the origin of the coordinates. Knowing the safe area of the stress, the measurement interval is set [1…40] MPa. The line intersects the graph of the yield surface is plotted point by point. The start point lies in the safe area, the endpoint is deliberately chosen to lie outside the safe area. Along the line are selected 10000 control points. Some cases of the stress state in which the maximum allowable principal stresses for ferrite and graphite materials are located along the corresponding straight lines presented in Fig. 6.

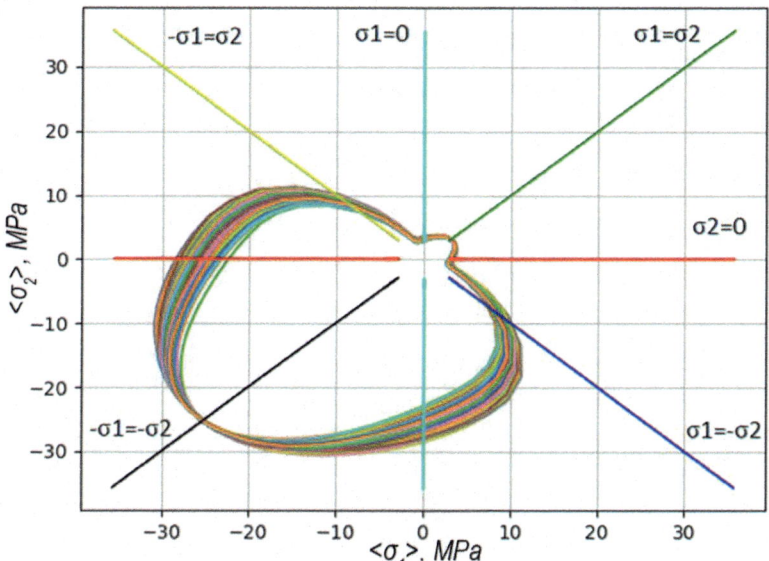

Fig. 6. Principal stresses location.

Information about the number of the yield surfaces that have fallen into the control points along the line are obtained. This method allowing to define the inverse cumulative distribution function $(1 - F(r))$, which in turn determining the parameters of descriptive statistics according to (6) and (7).

$$M[r] = \int_0^\infty (1 - F(r))dr, \tag{6}$$

$$\text{var}[r] = \int_0^\infty 2r(1 - F(r))dr - M[r]^2, \tag{7}$$

where $M[r]$ is the mathematical expectation (mean), and $var[r]$ is the variance of the random radius function.

The corresponding distributions have been approximated by the normal law with the defined parameters. The dependence graph of the mean, standard deviation, and variation coefficient of the random radius function for the following concentrations of inclusions $\psi = (0.055, 0.100, 0.300)$, presented in Fig. 7. It can be seen that an increase in the concentration of graphite inclusions leads to a decrease in the standard deviation, which, in turn, leads to an increase in the tensile strength of the material.

The coefficient of variation shows the extent of variability concerning the mean of the dataset. It's calculated as the ratio of the variance to the mathematical expectation. The obtained form of the coefficient of variation function corresponds to the different sample components material properties. The variance has the same tendency as the mean and depends proportionally.

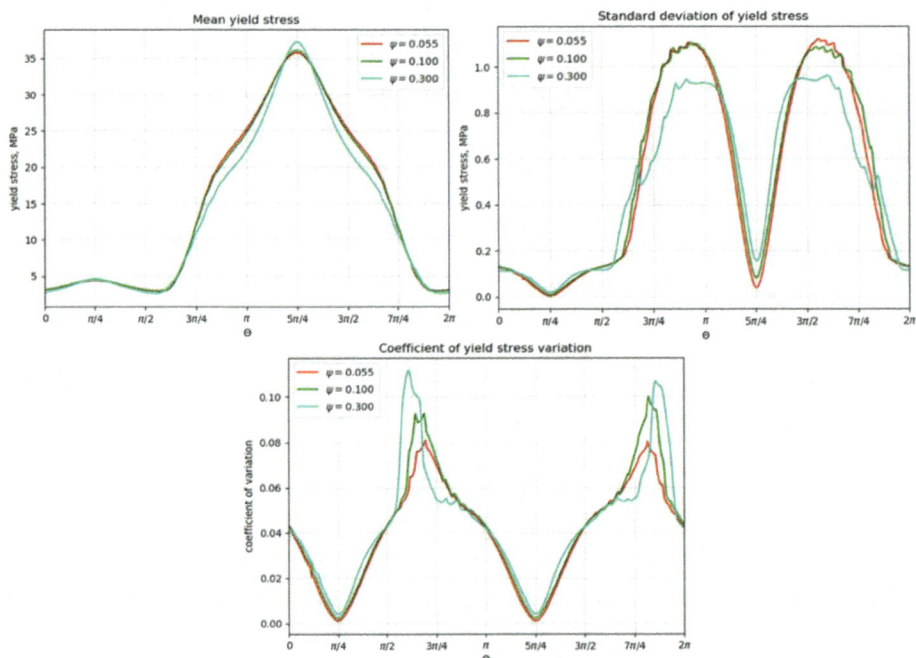

Fig. 7. Mean, standard deviation, and variation coefficient of the random radius function.

The reverse cumulative distribution functions of the yield surface intersection with a random function of the radii are shown in Fig. 8, and the probability distribution function is shown in Fig. 9. The analysis is carried out for several theta angles that correspond to the loading trajectory. The angles are calculated according to (8), and selected equal to $\Theta = (0, 45, 135, 300)°$.

$$tg\Theta = \frac{\sigma_2}{\sigma_1}, \tag{8}$$

where σ_1 and σ_2 are the principal stresses.

Lines of different colors correspond to different angles, and the solid and dashed line styles correspond to calculated and fitted distribution function by **scipy.stats. norm.cdf** (Fig. 8). Results similar in values with insignificant deviation are obtained by graphic comparison of both methods.

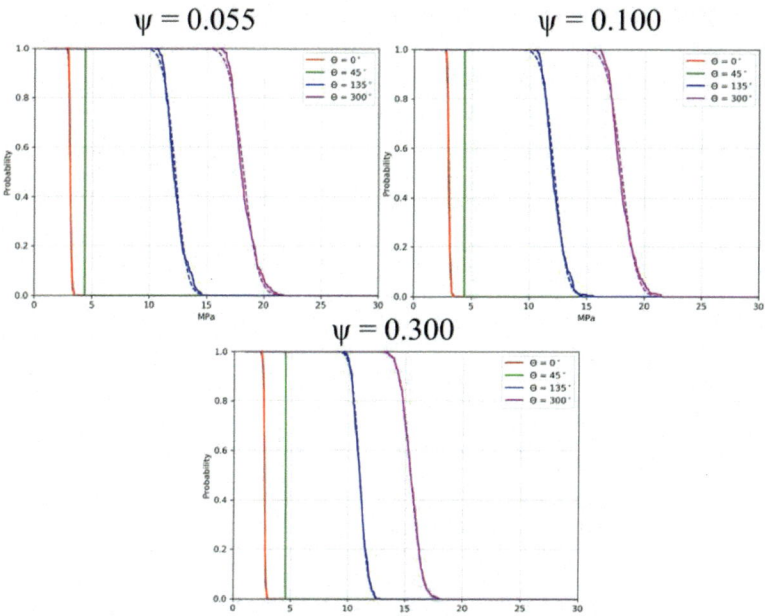

Fig. 8. Cumulative distribution function (cdf).

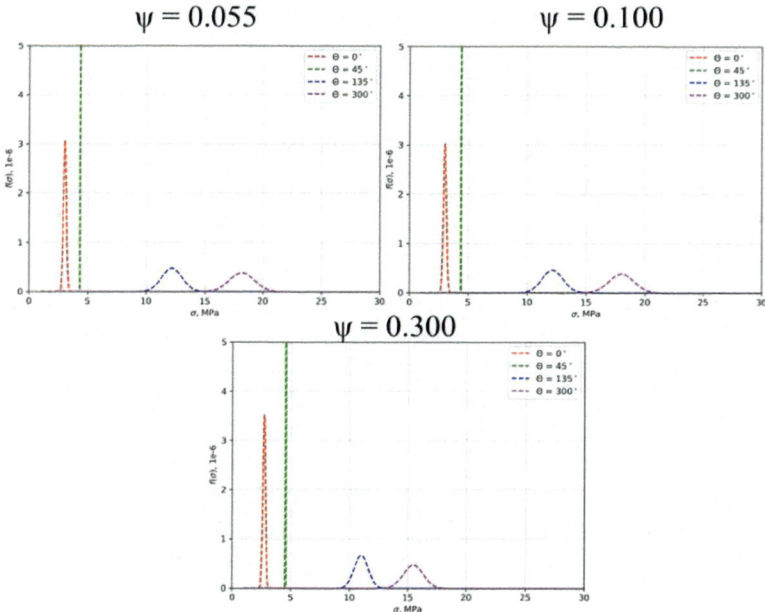

Fig. 9. Probability distribution function (pdf).

The dependence of the tensile and compression stress on the concentration of inclusions is shown in Fig. 10. The corresponding probabilistic characteristics at various concentrations of inclusions are given in Table 3.

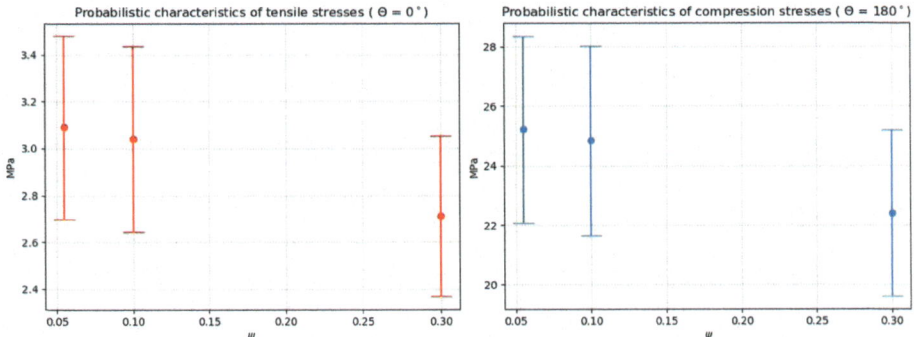

Fig. 10. Dependence of structural stress on the concentration of inclusions.

Table 3. Probabilistic characteristics of stresses at different concentration of inclusions.

Concentration, ψ	0.055	0.100	0.300
Tensile			
M, MPa	3.0879	3.0390	2.7103
$\sqrt{Var}$, MPa	0.1301	0.1320	0.1137
CV	0.0421	0.0434	0.0442
Compression			
M, MPa	25.224	24.840	22.403
$\sqrt{Var}$, MPa	1.0451	1.0598	0.9286
CV	0.0414	0.0427	0.0415

Analysis of the yield surface probabilistic characteristics shows the influence of the concentration of inclusions on the material strength characteristics. An increase in the concentration of graphite inclusions leads to the standard deviation decreases, which in turn of an increase in the material tensile strength. According to the results, the maximum spread of random typical implementations of the yield surface corresponds to the angle between stresses equal to $\varTheta = 300°$, minimum – equal to $\varTheta = 0°$. The spread in possible realizations of the yield surface is close to the minimum at equal values of the principal stresses ($\varTheta = 45°$). The obtained standard deviation functions are not smooth (Fig. 7), which indicates a lack of statistical data and the need to expand the experimental base.

5   Conclusions

The article discusses an algorithm for studying of the yield surface probabilistic characteristics. The internal structure model is generated synthetically based on real material images. An experimental set of yield surfaces has been created. A method for determining the location of stresses and the yield surface is proposed. The probabilistic characteristics are calculated and the graphs of the distribution density are constructed. Going beyond the surface indicates the appearance of plastic strains in the model. The influence of the spheroidal graphite inclusions concentration in the ferrite composition is determined. With an increase in the concentration of inclusions in the model, a decrease in stresses and a decrease in their spread were noted both under compression and under tension.

According to 3-sigma rule, a spread for stress under tension σ_{yield} = 3.0879 ± 0.3904 MPa for the smallest considered concentration (ψ = 0.055), which is decreases to σ_{yield} = 2.7103 ± 0.3411 MPa for the greatest one (ψ = 0.300), the spread of the random variable is decreased too. For the stress under compression, a spread for stress take a form σ_{yield} = 25.224 ± 3.1352 MPa for the smallest considered concentration (ψ = 0.055), which decreases to σ_{yield} = 22.403 ± 2.7858 MPa for the greatest one (ψ = 0.300). The spread of the random variable is also decreased. It is also noted that the difference between the stress spread for the case of tension and compression differ by two orders of magnitude. Despite such a difference, the coefficient of variation for different concentrations and different types of loading is approximately the same (is equal to 0.0426 ± 0.0030).

Acknowledgment. This work has been supported by the Ministry of Education and Science of Ukraine in the framework of the realization of the research project «Development of methods for mathematical modeling of the behavior of new and composite materials aims to structural elements lifetime estimation and prediction of engineering designs reliability» (State Reg. Num. 0117U004969).

References

1. Lvov, G., Kostromytska, O.: A data-driven approach to the prediction of plasticity in composites. In: Nechyporuk, M., et al. (eds) Integrated Computer Technologies in Mechanical Engineering. AISC, vol. 1113, pp. 3–10. Springer, Cham (2020). https://doi.org/10.1007/978-3-030-37618-5_1
2. Amir Siddiq, M.: Data-driven finite element method: theory and applications. In: Proceedings of the Institution of Mechanical Engineers, Part C: Journal of Mechanical Engineering Science (2020). https://doi.org/10.1177/0954406220938805
3. Shah, D., Murthi, B.P.S.: Marketing in a data-driven digital world: implications for the role and scope of marketing. J. Bus. Res. (2020). https://doi.org/10.1016/j.jbusres.2020.06.062
4. Barmparis, G.D., Tsironis, G.P.: Estimating the infection horizon of COVID-19 in eight countries with a data-driven approach. Chaos Solitons Fractals **135**, 109842 (2020). https://doi.org/10.1016/j.chaos.2020.109842

5. Bessa, M., Bostanabad, R., Liu, Z., et al.: A framework for data-driven analysis of materials under uncertainty: countering the curse of dimensionality. Comput. Methods Appl. Mech. Eng. **320**, 633–667 (2017). https://doi.org/10.1016/j.cma.2017.03.037

6. Latypov, M.I., Kalidindi, S.R.: Data-driven reduced order models for effective yield strength and partitioning of strain in multiphase materials. J. Comput. Phys. **346**, 242–261 (2017). https://doi.org/10.1016/j.jcp.2017.06.013

7. Chan, K.S.: Effects of plastic anisotropy and yield surface shape on sheet metal stretchability. Metall. Trans. A **16**(4), 629 (1985). https://doi.org/10.1007/BF02814237

8. Wu, B., Wang, H., Taylor, T., Yanagimoto, J.: A non-associated constitutive model considering anisotropic hardening for orthotropic anisotropic materials in sheet metal forming. Int. J. Mech. Sci. **169**, 105320 (2020). https://doi.org/10.1016/j.ijmecsci.2019.105320

9. Lu, D., Zhang, K., Hu, G., et al.: Investigation of yield surfaces evolution for polycrystalline aluminum after pre-cyclic loading by experiment and crystal plasticity simulation. Materials **13**(14), 3069 (2020). https://doi.org/10.3390/ma13143069

10. Tang, S., Li, Y., Qiu, H., et al.: MAP123-EP: a mechanistic-based data-driven approach for numerical elastoplastic analysis. Comput. Methods Appl. Mech. Eng. **364**, 112955 (2020). https://doi.org/10.1016/j.cma.2020.112955

11. Zhang, H., Diehl, M., Roters, F., Raabe, D.: A virtual laboratory using high resolution crystal plasticity simulations to determine the initial yield surface for sheet metal forming operations. Int. J. Plast. **80**, 111–138 (2016). https://doi.org/10.1016/j.ijplas.2016.01.002

12. Banabic, D., Barlat, F., Cazacu, O., Kuwabara, T.: Advances in anisotropy and formability. Int. J. Mater. Form. **3**(3), 165–189 (2010). https://doi.org/10.1007/s12289-010-0992-9

13. Larin, A.A., Vyazovichenko, Y.A., Barkanov, E., Itskov, M.: Experimental investigation of viscoelastic characteristics of rubber-cord composites considering the process of their self-heating. Strength Mater. **50**(6), 841–851 (2018). https://doi.org/10.1007/s11223-019-00030-7

14. Larin, O., Potopalska, K., Mygushchenko, R.: Statistical estimation of residual strength and reliability of corroded pipeline elbow part based on a direct FE-simulations. J. Serb. Soc. Comput. Mech. **12**(1), 80–95 (2018). https://doi.org/10.24874/jsscm.2018.12.01.06

15. Shapovalova, M., Vodka, O.: Image microstructure estimation algorithm of heterogeneous materials for identification their chemical composition. In: 2019 IEEE 2nd Ukraine Conference on Electrical and Computer Engineering (UKRCON), pp. 975–979. IEEE, Lviv (2019). https://doi.org/10.1109/UKRCON.2019.8879861

16. Shapovalova, M., Vodka, O.: Computer methods for constructing parametric statistically equivalent models of high-strength cast iron microstructure to analyze its elastic characteristics. Notes of the V.I. Vernadsky Tavrida National University. Ser. Tech. Sci. **30**(6), 179–187 (2019). https://doi.org/10.32838/2663-5941/2019.6-1/33 [in Ukrainian]

17. Shapovalova, M., Vodka, O.: Computer methods for modeling the synthetic structure of cast iron for statistical evaluation of its mechanical properties and strength characteristics. Theoret. Appl. Mech. **35**, 257–264 (2020). [in Russian]

18. Annin, B., Ostrosablin, N.: Anisotropy of the elastic properties of materials. Appl. Mech. Tech. Phys. **49**(6), 131–151 (2008). [in Russian]

19. Beliaev, N.: Strength of Materials. Nauka, Moscow (1965). [in Russian]

20. Ambatsumian, S.: Theory of Anisotropic Plates. Nauka, Moscow (1967). [in Russian]

21. Zienkiewicz, O.: The Finite Element Method in Engineering Science. McGraw-Hill, London (1971)

Amplification of Heat Transfer by Shock Waves for Thermal Energy Method

Sergiy Plankovskyy[3] , Olga Shypul[1] , Yevgen Tsegelnyk[1(✉)] ,
Alexander Pankratov[2] , and Tatiana Romanova[2]

[1] National Aerospace University "Kharkiv Aviation Institute",
17 Chkalova Street, Kharkiv 61070, Ukraine
y.tsegelnyk@khai.edu

[2] A. Pidgorny Institute of Mechanical Engineering Problems of the National
Academy of Sciences of Ukraine, 2/10 Pozharskogo Street,
Kharkiv 61046, Ukraine

[3] O.M. Beketov National University of Urban Economy in Kharkiv,
17 Marshala Bazhanova Street, Kharkiv 61002, Ukraine

Abstract. Despite the successful implementation of the Thermal Energy Method ('TEM'), the issues of simultaneous cleaning of internal cavities and deburring remain unresolved. Moreover, the deposition of oxides on the surfaces of parts is inevitable. In contrast, Impulse Thermal Energy Method with Shock Waves ('ITEMSW') is devoid of these disadvantages. This is due to its capabilities such as generating shock waves with controlled frequency and intensity. In addition, it is possible to control the time of release of combustion products from the chamber. In the current work, the essential parameters of the efficiency of this processing method were investigated. The performed numerical studies present the influence of the intensity and duration of shock waves action on the magnitude of the heat flux acting in the chamber. It was estimated that in case of the shock waves generation, the magnitude of the heat flux turned out to be 20–50 times greater. It is practically impossible to achieve such an increase in the heat flux by increasing the initial pressure of the fuel mixture. Thus, the simulation results confirm that the controlled generation of shock waves is an effective way to increase the intensity of heat transfer during TEM treatment. Because of the fact that the distribution of heat fluxes over the surfaces of the workpieces with processing by ITEMSW significantly depends on the frequency of exposure to shock waves the problem of optimal parts layout was considered. It was estimated that the best uniformity of heat flux distribution under ITEMSW processing is provided by the sparest layout with balancing parts and zones of shock wave generation to the centroid of the inner surface of the working chamber.

Keywords: Numerical simulation · Heat transfer · Shock waves · Balanced layout · Thermal deburring · Thermal Energy Method

M. Nechyporuk et al. (Eds.): ICTM 2020, LNNS 188, pp. 577–587, 2021.
https://doi.org/10.1007/978-3-030-66717-7_49

1   Introduction

A stable trend in the development of precision mechanisms is their miniaturization [1]. This leads to a decrease in the gaps in the friction and engagement pairs and requires an improvement in accuracy of the parts forming them. In this regard, edge finishing and surface cleaning technologies are of particular importance to ensure the reliability of precision mechanisms. Until now, abrasive technologies are widely used for these purposes, e.g. various flow abrasive processes [2, 3], and washing, increasingly ultrasonic [4]. However, the use of abrasives can lead to secondary contamination of surfaces. This is especially dangerous for the internal channels of parts of hydraulic and pneumatic systems, for which the remnants of micro-abrasive particles can lead to increased wear or even jamming of friction pairs. Therefore, more and more interest is attributed to deformation-free methods of edge finishing [5]. Such, for example, as electrochemical [6] electro erosive [7], laser [8]. These processes are difficult to automate. They require both high control accuracy and special methods for calculating processing modes [9]. In addition, the listed methods cannot combine edge finishing and surface cleaning. So they require additional washing to remove micro particles from the surfaces of the processed parts.

From this point of view, the Thermal Energy Method ('TEM') has unique advantages. This method is widely used for the edge finishing in mechanical engineering [10–12]. It is based on the use of the energy of lean gas mixtures combustion. The parts to be processed are placed in a sealed chamber filled with a fuel mixture. After the combustion the burrs on the edges, formed during machining, heat up and burn out when reacted with excessed oxygen. At the same time, the body of the part, due to the volume significantly larger than of the burrs, heats up very lightly (usually less than a hundred degrees). The duration of the processing cycle in TEM equipment does not exceed two minutes. Thanks to the ability to machine several parts at the same time, it is far superior to the other edge finishing methods in terms of productivity [13]. Since combustion products are in contact with all surfaces of the parts simultaneously, any mechanisms not are needed to control the deburring process. This provides opportunities for automation of processing based on fairly simple CNC systems. Such equipment (Fig. 1) is industrially produced by a number of firms [14, 15].

But as noted in [16], the TEM basic version does not allow combining edge finishing and surface cleaning. Generally micro particles of abrasive as oxides, nitrides or carbides of the metal cannot be removed at the moment and remains on surfaces. Moreover, due to the sufficiently long contact of the surfaces with the hot oxygen-containing gas, they are covered with an oxide of part material. For removing of metal oxides it is necessary to utilize washing [17] or by pickling [18]. Impulse Thermal Energy Method with Shock Waves ('ITEMSW') lacks these drawbacks and has possibility to combine edges finishing and surfaces cleaning in one processing cycle. It is provided by generating shock waves with controlled frequency and intensity and operating the processing time by controlled exhaust of combustion products from the chamber.

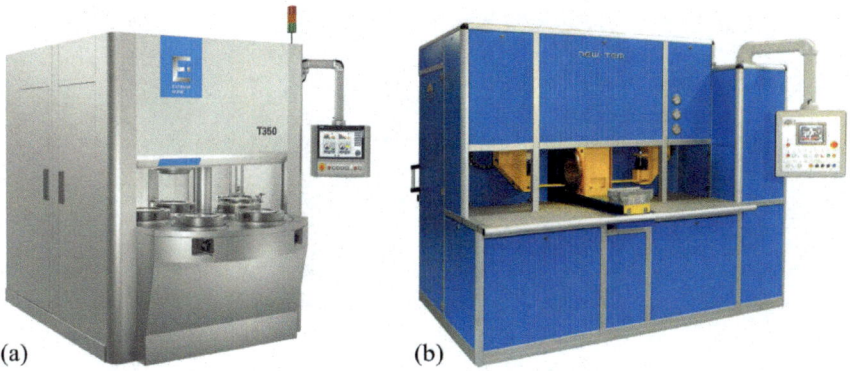

Fig. 1. (a) Thermal deburring machine T350 with vertical chamber by Extrude Hone® [14]; (b) Thermal deburring machine New-Tem XPF 450 with horizontal chamber by SGM® [15].

In [16] a method for assigning ITEM and ITEMSW processing modes is described. It is based on the appointment of a preliminary mode using the results of numerical simulation of heat transfer problem due to combustion as well as tasks of a plate burr fusion and micro particle removal. In addition to those presented, an important process parameter in ITEMSW is the intensity and duration of the shock waves action, which affect the magnitude of the heat flux. In turn, the impacts efficiency of heat flow depends on the uniformity of its distribution on the surface of the workpiece, so the localization of pieces in the working chamber is an important factor. Since these parameters have not previously been thoroughly studied, the purpose of this study is to investigate them.

2   Increasing Energy Efficiency of TEM by Shock Wave Generation

According to TEM principle where burrs are removed by combustion action, it is difficult to process parts from nickel, chromium, cobalt et al. alloys, which react poorly with oxygen, so TEM has limitations. As mentioned above due to a long time contact with hot gas the oxides are formed on the parts surfaces (Fig. 2) [19]. Their removal requires additional operations, which can be difficult for internal cavities with complex shapes. Some researchers have noted that during TEM processing of steel parts with a hardness of more than 40 HRC, surface cracks can be formed, and when processing parts made of stainless alloys, the corrosion resistance of the material can deteriorate [13, 20, 21].

Some of these problems can be eliminated by reducing the contact time of the combustion products with parts. For this, a time controlled exhaust of hot gas from the working chamber must be used. This idea underlies the Impulse Thermal Energy Method ('ITEM') [16]. By using water protected quick breathing valves on ITEM equipment, oxide layer formation and surface cracking are avoided.

Fig. 2. Typical stages of the TEM: (A) untreated workpiece; (B) deburring with Oxygen surplus; (C) oxide reduction with stoichiometric gas mixture; (D) washed workpiece [19].

But the controlled exhaust of combustion products cannot completely solve the problem of expanding the range of processed materials due to physical limitations. To ensure quality while reducing the processing time, it is required to increase of the heat flux on the parts surfaces. As shown in [22], during TEM treatment, the heat flux can be increased by raising the initial pressure of the fuel mixture. But the possibilities of such raising are limited: with an increase in the initial pressure of the fuel mixture by 25 times, the heat flux will increase by only 5 times. However, at such a high initial pressure of the mixture, the pressure of the combustion products can reach hundreds of bars. This dramatically increases the load on equipment and processed parts and significantly complicates the problem of hot combustion products deflation. The possibilities of increasing the heat flux by changing the composition of the fuel mixture are also limited. At the same time, it is known that the intensity of heat exchange between combustion products and solids depends very strongly on the gas velocity and reaches maximum values when shock waves appear [23]. This effect is at the basis of the Impulse Thermal Energy Method with Shock Wave generation ('ITEMSW') [16].

With spark ignition, shock waves in a closed volume can be generated by auto ignition of a part of the fuel mixture. Therefore, one of the ways to generate a series of shock waves can be the use of a working chamber of a special shape with specially provided cavities in which the conditions of a volumetric explosion will be realized. However, this method ensures the formation of shock waves only before the end of mixture combustion. The duration of the action of shock waves can be extended using additional devices. Some of them can generate shock waves with frequency up to 100 Hz [24]. One of the tasks of this work is to assess the possibilities of heat transfer control with such a method. For this, numerical simulation of TEM processing is fulfilled in two versions: for a regular cycle and with additional generation of shock waves.

Heat transfer between the combustion products of a stoichiometric methane-oxygen mixture with a part installed in the chamber is simulated. The chamber contained a channel for supplying shock waves. Gas flow parameters are determined by solving a system of equations presented in [22]:

$$\frac{\partial \rho}{\partial t} + \nabla \cdot (\rho \cdot \mathbf{u}) = 0, \quad \frac{\partial (\rho \cdot \mathbf{u})}{\partial t} + \nabla \cdot (\rho \cdot \mathbf{u} \cdot \mathbf{u}) = -\nabla P + \nabla \cdot \tau_{eff},$$

$$\frac{\partial (\rho h)}{\partial t} + \nabla \cdot (\mathbf{u} \rho h) = \frac{\partial \mathbf{P}}{\partial t} + \mathbf{u} \cdot \nabla P + \nabla \left(\lambda_{eff} \nabla T - \sum_i h_i \mathbf{J}_i + \tau_{eff} \cdot \mathbf{u} \right) + \sum_{i=1}^{N} Q^i,$$

$$\frac{\partial (\rho Y_i)}{\partial t} + \nabla \cdot (\rho \mathbf{u} Y_i) = \nabla \cdot \left(\left(\rho D_i + \frac{\mu_t}{Sc_t} \right) \nabla Y_i \right) + R_i,$$

where ρ is the density; $\mathbf{u}$ is the velocity vector; t is the time; P is the pressure; $\tau_{eff} = (\mu + \mu_t)[\nabla \mathbf{u} + (\nabla \mathbf{u})^T - 2/3 \, \mathbf{I} \cdot (\nabla \mathbf{u})]$ is the effective stress tensor; μ is the viscosity; μ_t is the turbulent viscosity; $\mathbf{I}$ is the unit tensor; h is the enthalpy of the mixture; $\lambda_{eff} = \lambda + \lambda_t$ is the effective thermal conductivity; λ is the laminar heat conductivity; $\lambda_t = C_p \mu_t \, \mathrm{Pr}_t^{-1}$ is the turbulent heat conductivity; C_p is the specific heat of the gas; Pr_t is turbulent Prandtl number; T is the temperature; $h_i = \int_{T_{ref}}^{T} C_{p,i} dT$, where T_{ref} is 298.15 K, is the enthalpy of species i; $\mathbf{J}_i$ is the diffusion rate vector of species i; $Q^i = h_i^0 R_i / M_i$ is the heat of chemical reactions that the species i participate in; Y_i is the mass fraction of species i; D_i is the diffusion coefficient of species i in the mixture; Sc_t is the turbulent Schmidt number; R_i is the rate of production of species i by chemical reaction.

For calculation of μ_t the SAS SST turbulence model is used. The heat flux on the part surface was calculated according to the standard method for SST based models [25].

An area with a value of temperature and reaction progress is created in the chamber to initiate combustion. The part was considered a wall with a constant temperature of 25 °C. The rest of the chamber boundary was set as an adiabatic wall. In both simulated cases, the heat exchange between the combustion products and the surface of the part is analyzed for same period time. But for the case with the generation of shock waves at the moment of time which is equal to 0.035 s a gas with a pressure 3 times greater than the pressure in the chamber is supplied to the channel entrance during 0.01 s. This ensures the formation of a shock wave in the channel. The static temperature and supplied gas composition corresponded to the combustion products.

A commercial code ANSYS CFX is used for numerical simulations. In the calculations a finite element mesh with the value $y^+ \approx 1$ near part surface is used. Time step is assigned based on the condition for Courant number $C \leq 10$. Figure 3 shows the results of calculating the value of the heat flux averaged over the surface of the part for both analyzed cases. During of the shock waves generation, the magnitude of the heat flux turned out to be 20–50 times greater. It is practically impossible to achieve such an increase in the heat flux by increasing the initial pressure of the fuel mixture. Thus, the simulation results confirm that the controlled generation of shock waves is an effective way to increase the intensity of heat transfer during TEM treatment.

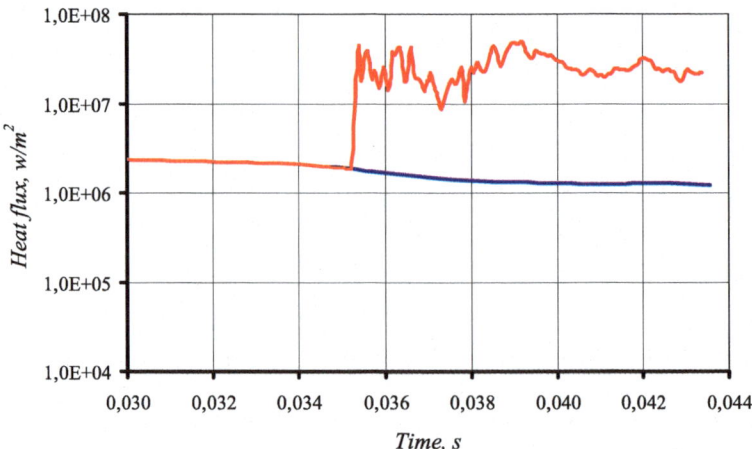

Fig. 3. Averaged heat fluxes as a function of time duration for regular TEM treatment (blue) and for shock wave generation (red)

3   The Problem of Optimal Parts Layout for ITEMSW

The distribution of heat fluxes over the surfaces of the workpieces with processing by ITEMSW significantly depends on the frequency of exposure to shock waves. When batch machining it requires correcting the placement of parts for consistent quality of finishing. A close frequency of the impact of shock waves on the parts can be provided when they are located at an equal distance from each other and from the walls of the chamber (Fig. 4) [26]. This leads to the problem of the most sparse packing, first formulated in [27] and developed in [28] for the 2D case and in [29] for the placement of three-dimensional objects.

In [27] it was shown that the condition of the greatest distance of the object from the boundaries of the placement area leads to the requirement that the centroids of thin shells coincide with their surfaces. Taking into account the balancing requirement the sparsest packing problem can be formulated as the following nonlinear optimization problem:

$$\max \rho \text{ s.t. } (u, \tau) \in W,$$

$$W = \{(u, \tau, \rho) : \Phi'_{ij}(u_i, u_j, \tau_{ij}, \rho) \geq 0, \ i < j \in I_n, \ \Phi'_i(u_i, \rho) \geq 0, \ i \in I_n, \ \Upsilon(v) \geq 0, \ \rho > 0\},$$

where $\tau = (\tau_{11}, \tau_{12}, \ldots, \tau_{ij}, \ldots, \tau_{(n-1),n})$, τ_{ij} is a vector of the auxiliary variables for the adjusted quasi phi-function $\Phi'_{ij}(u_i, u_j, \tau_{ij}, \rho)$, $i < j \in I_n$; $\Phi'_{ij}(u_i, u_j, \tau_{ij}, \rho) \geq 0$ represents the distance between objects $T_i(u_i)$ and $T_j(u_j)$, $i < j \in I_n$; $\Phi'_i(u_i, \rho) \geq 0$ determine the distance between $T_i(u_i)$, $i \in I_n$ and Ω^*; $\Upsilon(v) \geq 0$ states the balancing conditions. Here $\Upsilon(v) = \min\{f_1(v), f_2(v)\} \geq 0$, $f_1(v) = \min\{x_c + \delta, -x_c + \delta\}$, $f_2(v) = \min\{y_c + \delta, -y_c + \delta\}$.

Fig. 4. An example of the arrangement of parts during group TEM processing [26].

The feasible region W is defined by a system of non-smooth inequalities, which, due to the application of the apparatus of phi functions, can be reduced to a system of inequalities with differentiable functions. Formulated model is a non-convex and continuous nonlinear programming problem.

The articles [28, 29] described an algorithm for solving the problem of the sparse layout, examples of solving a number of problems for bodies of simple geometric shapes and composite bodies are given. In general, the method developed in [28, 29] makes it possible to solve the problem of optimal placement of parts during ITEMSW processing based on the condition of equidistance from each other and the chamber walls. But the questions of localization zones of shock waves generation were not considered in these works.

For the two-dimensional case, this problem was first solved by the authors and presented in [30]. The results of simulating heat transfer with attenuation of shock waves for a balanced and unbalanced layouts of circles with different diameters in a circular area are presented in the graphs of Fig. 5 and Fig. 6.

For the formation of shock waves in a circle centered in the centroid at the time $t = 0$, a high-pressure zone is specified, and the computational domain itself is assumed to be filled with hot gas. The balanced layout, taking into account the dimensions of the high pressure zone, showed the best results in terms of the uniformity of the averaged heat flux. For this case, the magnitude of heat fluxes on the biggest circle in averaging time is lower than the values for small circles and by 1.21 times, while with an unbalanced layout by 1.45 times. In the calculations for the balanced layout, it is checked whether the adopted location of the objects and high pressure zone provides optimal conditions for uniformity in the distribution of averaged heat fluxes. For this, additional calculations was carried out for the cases when the initial high-pressure zone is displaced relative to the centroid in relation to which the objects are balanced. Even with small (up to 2 mm) displacements of the high pressure zone relative to the centroid, the difference in the magnitude of averaged heat fluxes on circles increased by 1–2%.

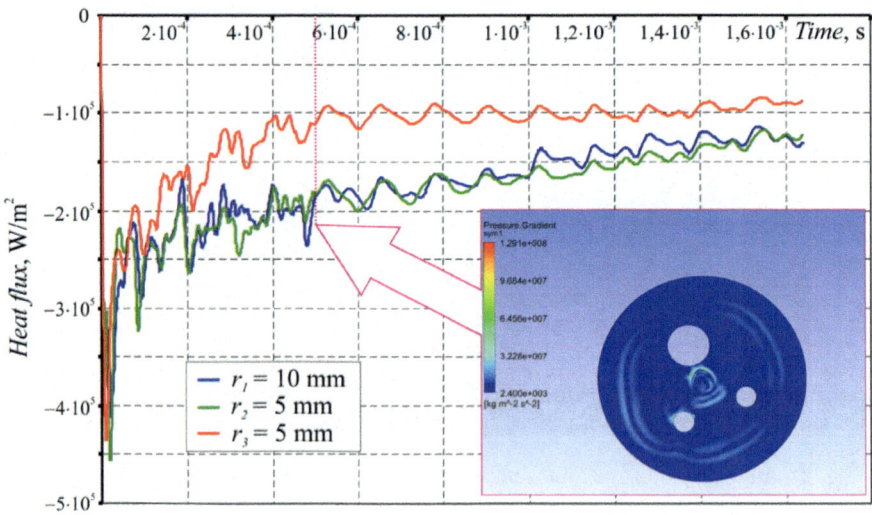

Fig. 5. Averaged heat fluxes for layout without balanced conditions as a function of time and the pressure gradient distribution at $t = 5 \cdot 10^{-4}$ s [30].

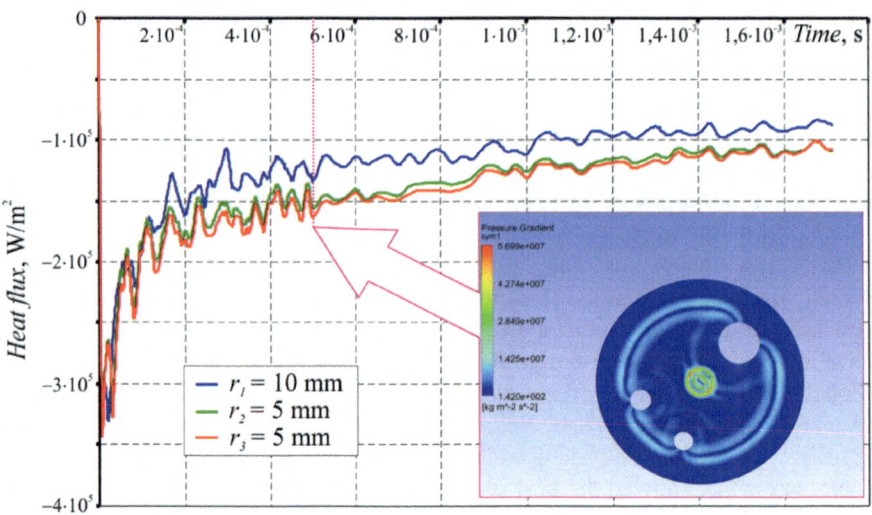

Fig. 6. Averaged heat fluxes for layout with balanced conditions as a function of time and the pressure gradient distribution at $t = 5 \cdot 10^{-4}$ s [30].

Thus, to ensure the same ITEMSW processing conditions, the parts in the chamber should be positioned so that the centroid of the thin shells coinciding with their outer surfaces coincides with the centroid of the shell, which coincides with the interior surface of the chamber. In addition, the alignment of the shock wave generation zones centroid with this point must be ensured. This placement is very difficult to implement for real parts, which have some limitations for fixation in space. However, the formulated principle is a good basis for the design of technological devices for ITEMSW.

4 Conclusions

Judging by the results of numerical simulations generating shock waves with special devices is a much more effective approach of heat transfer amplification for TEM treatment in comparison with enlarging an initial pressure of fuel mixture. It proves that ITEMS treatment with this principle is beneficial as an energy efficient edge finishing process.

It is also shown that the best uniformity of heat flux distribution during ITEMSW processing is provided by the sparest layout with balancing parts and zones of shock wave generation to the centroid of the inner surface of the working chamber.

The development of an approach for designing localization technological devices is the scope for future work.

References

1. Schellekens, P., Rosielle, N., Vermeulen, H., et al.: Design for precision: current status and trends. CIRP Ann. **47**(2), 557–586 (1998). https://doi.org/10.1016/S0007-8506(07)63243-0
2. Petare, A.C., Jain, N.K.: A critical review of past research and advances in abrasive flow finishing process. Int. J. Adv. Manufact. Technol. **97**(1–4), 741–782 (2018). https://doi.org/10.1007/s00170-018-1928-7
3. Kohli, R.: Microabrasive technology for precision cleaning and processing applications. In: Kohli, R., Mittal, K.L. (eds.) Developments in Surface Contamination and Cleaning: Applications of Cleaning Techniques, vol. 11, pp. 509–548. Elsevier, Amsterdam (2019). https://doi.org/10.1016/B978-0-12-815577-6.00013-X
4. Mason, T.J.: Ultrasonic cleaning: an historical perspective. Ultrason. Sonochem. **29**, 519–523 (2016). https://doi.org/10.1016/j.ultsonch.2015.05.004
5. Bhattacharyya, B., Doloi, B.: Advanced finishing processes. In: Bhattacharyya, B., Doloi, B. (eds.) Modern Machining Technology, pp. 675–743. Academic Press, London (2020). https://doi.org/10.1016/B978-0-12-812894-7.00008-6
6. Ruszaj, A., Gawlik, J., Skoczypiec, S.: Electrochemical machining – special equipment and applications in aircraft industry. Manage. Prod. Eng. Rev. **7**(2), 34–41 (2016). https://doi.org/10.1515/mper-2016-0015
7. Jeong, Y.H., HanYoo, B., Lee, H.U., et al.: Deburring microfeatures using micro-EDM. J. Mater. Process. Technol. **209**(14), 5399–5406 (2009). https://doi.org/10.1016/j.jmatprotec.2009.04.021
8. Lee, S.H., Dornfeld, D.A.: Precision laser deburring. J. Manuf. Sci. Eng. **123**(4), 601–608 (2001). https://doi.org/10.1115/1.1381007
9. Plankovskyy, S., Shypul, O., Tsegelnyk, Y., et al.: Simulation of surface heating for arbitrary shape's moving bodies/sources by using R-functions. Acta Polytechnica **56**(6), 472–477 (2016). https://doi.org/10.14311/AP.2016.56.0472

586 S. Plankovskyy et al.

10. George, W.E., McGeough, J.A.: Study of a thermal energy method of removal of surface irregularities. In: Davies, B.J. (ed.) Proceedings of the Twenty-third International Machine Tool Design and Research Conference, pp. 299–305. Palgrave, London (1983). https://doi.org/10.1007/978-1-349-06546-2_36
11. Lamikiz, A., Ukar, E., Tabernero, I., Martinez, S.: Thermal advanced machining processes. In: Davim, J.P. (ed.) Modern Machining Technology, pp. 335–372. Woodhead Publishing, Oxford (2011). https://doi.org/10.1533/9780857094940.335
12. Benedict, G.F.: Thermal energy method: deburring (TEM). In: Benedict, G.F. (ed.) Nontraditional Manufacturing Processes, pp. 349–361. CRC Press, Boca Raton (2017). https://doi.org/10.1201/9780203745410-22
13. Gillespie, L.K.: Deburring and Edge Finishing Handbook. Society of Manufacturing Engineers, Dearborn (1999)
14. Extrude Hone GmbH: T-series thermal deburring machines. https://extrudehone.com/t-series-thermal-deburring-machines. Accessed 21 July 2020
15. SGM s.r.l.: Adjust the shot. https://www.new-tem.com/en/. Accessed 21 July 2020
16. Plankovskyy, S., Popov, V., Shypul, O., et al.: Advanced thermal energy method for finishing precision parts. In: Pramanik, A., Gupta, K. (eds.) Advanced Machining and Finishing. Elsevier, Amsterdam (2021). https://doi.org/10.1016/B978-0-12-817452-4.00014-2
17. Lutz, T.: Oxide nach thermischem Entgraten effektiv entfernen. JOT J. für Oberflächentechnik 59(9), 102–104 (2019). https://doi.org/10.1007/s35144-019-0287-y
18. Treptow, F., Wulfestieg, K.P.: Gentle removal of oxide layers. IST Int. Surf. Technol. 10(3), 50–53 (2017). https://doi.org/10.1007/s35724-017-0076-1
19. moser-entgratungs ag: Verfärbung von Stahlteilen. https://www.moser-entgratung.ch/de/qualitaet-entgraten. Accessed 21 July 2020
20. Kelley, D.G., Schwarz, K.H.: Thermal energy deburring. Technical Paper MR91-136. Society of Manufacturing Engineers, Dearborn (1991)
21. Sonego, R.: Advances in TEM technology. Technical Paper MR93-322. Society of Manufacturing Engineers, Dearborn (1993)
22. Plankovskyy, S., Teodorczyk, A., Shypul, O., et al.: Determination of detonable gas mixture heat fluxes at thermal deburring. Acta Polytechnica 59(2), 162–169 (2019). https://doi.org/10.14311/AP.2019.59.0162
23. Quintens, H., Michalski, Q., Moussou, J., et al.: Experimental wall heat transfer measurements for various combustion regimes: deflagration, autoignition and detonation. In: AIAA Propulsion and Energy 2019 Forum, p. 4381. AIAA, Indianapolis (2019). https://doi.org/10.2514/6.2019-4381
24. Nikolaev, Y.A., Vasil'ev, A.A., Ul'yanitskii, B.Y.: Gas detonation and its application in engineering and technologies (review). Combust. Explos. Shock Waves 39(4), 382–410 (2003). https://doi.org/10.1023/A:1024726619703
25. Huang, H., Sun, T., Zhang, G., et al.: Modeling and computation of turbulent slot jet impingement heat transfer using RANS method with special emphasis on the developed SST turbulence model. Int. J. Heat Mass Transf. 126, 589–602 (2018). https://doi.org/10.1016/j.ijheatmasstransfer.2018.05.121
26. Benseler: Thermal energy machining. https://www.benseler.de/en/verfahren/entgratung/tem.php. Accessed 21 July 2020
27. Plankovskyy, S., Nikolaev, A., Shypul, O., et al.: Balance layout problem with the optimized distances between objects. In: Vasant, P., et al. (eds.) Data Analysis and Optimization for Engineering and Computing Problems. EAISICC, pp. 85–93. Springer, Cham (2020). https://doi.org/10.1007/978-3-030-48149-0_7

28. Romanova, T., Pankratov, A., Litvinchev, I., et al.: Sparsest packing of two-dimensional objects. Int. J. Prod. Res. (2020). https://doi.org/10.1080/00207543.2020.1755471
29. Romanova, T., Stoyan, Y., Pankratov, A., et al.: Sparsest balanced packing of irregular 3D objects in a cylindrical container. Eur. J. Oper. Res. (2020). https://doi.org/10.1016/j.ejor.2020.09.021
30. Plankovskyy, S., Shypul, O., Tsegelnyk, Y., et al.: Circular layout in thermal deburring. In: Shkarlet, S., et al. (eds.) Mathematical Modeling and Simulation of Systems (MODS 2020). AISC, vol. 1265, pp. 111–120. Springer, Cham (2021). https://doi.org/10.1007/978-3-030-58124-4_11

Transport Aircrafts Rear Cargo Door Ramp with Sealed Floor Main Parameters and Components Description and Determination

Dmytro Konyshev[1] , Andriy Humennyi[2] ,
Oleksandr Grebenikov[2] , Anton Chumak[2(✉)] , and Liliya Buival[2]

[1] Antonov Company, 1 Akademika Tupoleva Street, Kyiv 03062, Ukraine
d.s.konyshev@gmail.com
[2] National Aerospace University "Kharkiv Aviation Institute",
17 Chkalova Street, Kharkiv 61070, Ukraine
a.chumak@khai.edu

Abstract. Ramp and its units as one of the most important aggregates in the tail cargo door is reviewed. Such geometric parameters and units of the ramp as its maximum length Lp, the cutout in the ramp area, the theoretical line of the cargo door beam, the theoretical line of the ramp flooring, the ramp hinge units, its longitudinal and transverse beams, the ramp power locks, the ramp control cylinders and ramp rods, as well as the ramp sealing are presented. The main algorithms for determination are described using theoretical and practical data of most popular and used transport aircrafts such as An-124, An-26, Il-76, C-141, C-5 M. Most of presented information is collected considering time-tested best technical solutions as well as modern requirements of regulatory documents for transport category aircrafts (AP-25, CS-25, ICAO SRP A6, AC-25, OST).

Keywords: Cargo cabin · Inclined floor · Ramp · Cargo door beam · Cargo door doorway · Ramp power locks · Ramp sealing · Rods · Cylinders

1 Introduction

Aircraft configuration development [1–3], its geometric parameters coordination and description are among of the most important tasks of preliminary design stage [4–7]. Cargo door is one of the major transport airplane's units. Its size and technical characteristics are interconnected with airplane Technical Requirements Specification (TRS).

Accordingly to the methodology of integrated design and modeling of airplanes [4], the fuselage tail master geometry, cargo door type with its main units are interconnected and the result of one global design. But when the methods of master geometry design are described [8], the methods of tail cargo doors development didn't.

The main aim of the work is to describe construction of the cargo door ramp and its main components as one of the most important units. The construction of the ramp and its development are very important because ramp of the modern transport aircraft uses

M. Nechyporuk et al. (Eds.): ICTM 2020, LNNS 188, pp. 588–597, 2021.
https://doi.org/10.1007/978-3-030-66717-7_50

as hermetic fuselage door [9] as well as force floor during airdrop or loading/unloading cargoes and vehicles [10].

2 Cargo Door Ramp

One of the most important aggregates in the tail cargo door is ramp. During determining the geometric parameters of the ramp, it is necessary to choose its maximum length, which is limited by the aerodynamics of the aircraft when the maximum load is dropped from the end of the ramp [11]. The longer the ramp length Lp, the better the initial data for designing the cargo door (Fig. 1). Other parameters being equal, it is possible to increase the length of the ramp by reducing the length of the horizontal floor of the aircraft cargo compartment.

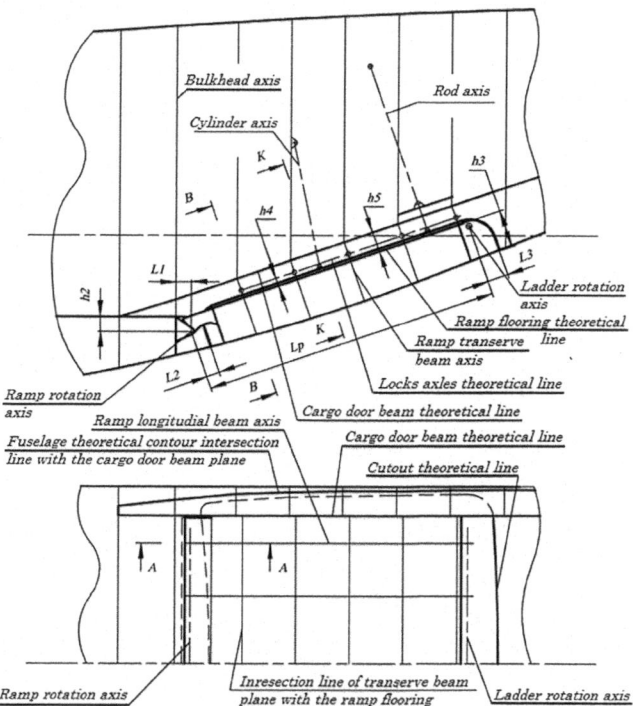

Fig. 1. The geometric parameters of the ramp with sealed floor in the side and plan views.

The ramp hitch, its geometric shape and design features depend on the way the cutout in the ramp area is formed.

The cut in the fuselage in the longitudinal direction is made parallel to the theoretical line of the ramp floor [12]. In the transverse direction, the cut is made perpendicular to the longitudinal line. Longitudinal and transverse cut lines are connected by radii $R1$ and $R2$.

The cut-out, made behind the axis of rotation of the ramp at a distance $L2$ (see Fig. 1 and Fig. 3), makes it possible to perform a continuous stream of sealing on the fuselage [13].

Dimensions $h2$ and $L1$ are selected from the condition of the ramp mounting, the $L3$ dimension is determined depending on the construction design and mounting the ladder. By setting the dimension $h3$, it is possible to define the fuselage cutout in the ramp area and the ramp configuration.

The theoretical line of the cargo door beam in the side view is drawn parallel to the theoretical line of the ramp flooring [7, 8] at a distance $h5$. The $h5$ value is determined depending on the construction of the ramp side locks.

Having coordinated the distance between the frames in the ramp area with its length, it is necessary to determine the breakdown of the load-bearing cross beams of the ramp. The planes of which are installed perpendicular to the theoretical line of the cargo door beam at the point of its intersection with the fuselage bulkhead.

After developing the ramp configuration, it became possible to start determining the length of the tilted floor Lop and its angle of inclination to the ground αop. The calculated value hn is the height of the cargo compartment sill for an empty equipped aircraft with a navigational fuel reserve. The Ψ-angle determines the calculated ground level (see Fig. 2).

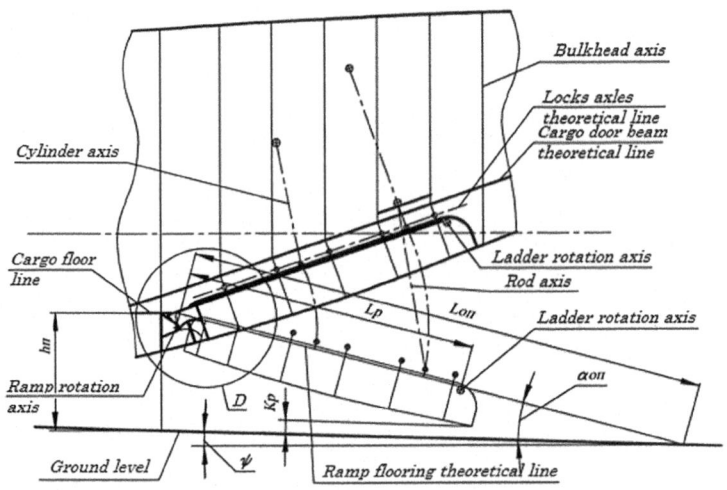

Fig. 2. Determination the geometric parameters of the ramp in open position.

The distance from the ground to the rear end of the ramp Kp determines the type of its fixation in the deflected position. In this version, the ramp is supported by the ramp rods. With a larger value of Lp, the ramp can be equipped with fixed supports made at its rear end.

The plan view (Fig. 1) shows the location of the ramp longitudinal beams, the axes of which are selected depending on [14] the track gauge of the loaded vehicles, and the design possibilities of the ramp hinge nodes.

Thus, the geometric parameters of the ramp and its structural-power scheme [15] allows to start developing the design of the ramp hinge units, its longitudinal and transverse beams, the ramp power locks, the ramp control cylinders and ramp rods, as well as the ramp sealing.

In Fig. 3 shown the view D of Fig. 2. To ensure smooth coupling of cargo compartment floor with the ramp flooring when it is deflected in any intermediate position, the front part of the ramp is radial, with $R = h2$. The ramp hinge brackets 1 are joined on the brackets 16 and fixed on the lower part [12, 13] of the threshold bulkhead.

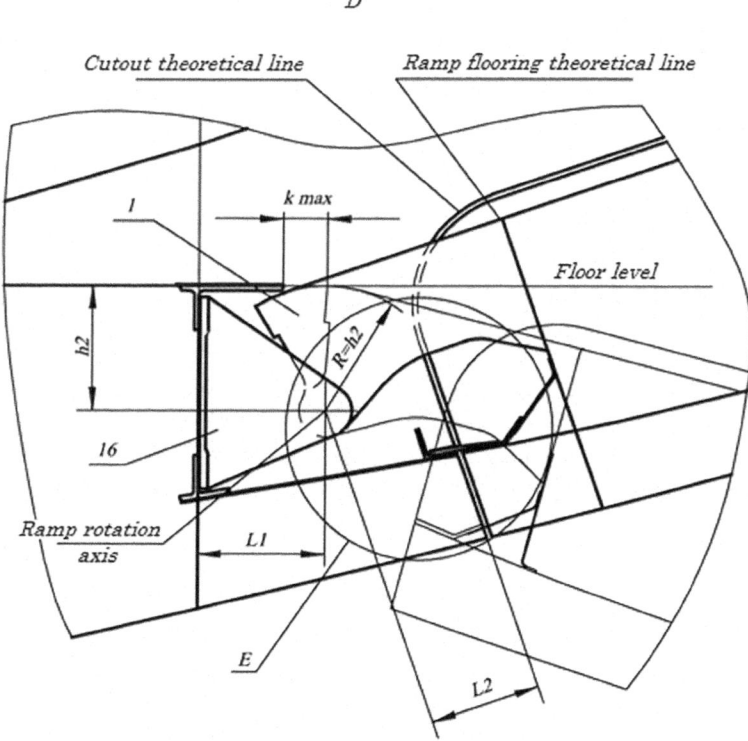

Fig. 3. View D of Fig. 2 – the main parameters of the ramp sill brackets.

Dimension *kmax* sets the maximum allowable clearance between the cargo compartment floor and the ramp deck.

In Fig. 4 shown the view E of Fig. 3. As can be seen from Fig. 3 and Fig. 4, the dimensions $h2$, $L1$ and $L2$ are selected from the condition of ensuring the necessary shape of the brackets 1, giving them a bearing capacity, ensuring a minimum clearance from the fuselage structure *kmin* during opening the ramp, as well as a clearance $k3$ from the hinge bracket 1 to the fuselage structure (corner profile 32, on which the sealing profile 33 is installed). It is also necessary to ensure a sufficient overlap of the sealing profile 33 with the corner profile 34 installed at the front end of the ramp.

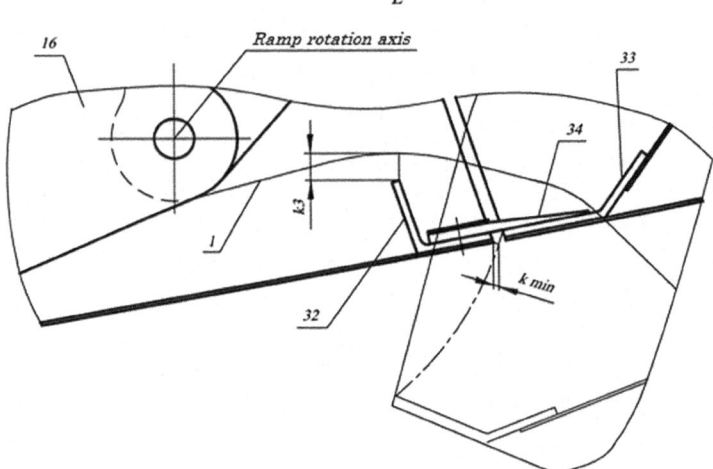

Fig. 4. View *E* of Fig. 3 – the sealing of the ramp.

Part of the pressurization load is redistributed through the longitudinal beams and, by means of the ramp hinge brackets, is transferred to the sill of the aircraft cargo compartment.

On Fig. 5 shown view *B-B* of Fig. 1. Racks of the ramp longitudinal beam are not shown conventionally. On the crossbeam of the ramp *7* there is a bracket with the axis of the side ramp lock. The beam *7* is made in the upper part with an undercut to ensure the passage of the upper rack *5* of the ramp longitudinal beam and installation on the upper chord of the ramp floor crossbeam stringers *12*.

3   Side Ramp Locks

In Fig. 6 shown the view *D* of Fig. 5, which shows the side lock of the ramp. The design of the lock determines the dimension *h5* from the theoretical ramp flooring to the theoretical cargo door beam line. The axis of the lock *21* is installed in the bracket *22* at a height *h4* from the theoretical line of the ramp floor. The bracket *22* is fixed on the crossbeam *7*. The value of the *h4* dimension and the distance of the bracket *22* from the fuselage symmetry axis are chosen to ensure the installation of the sealing profile *35* and the required clearance *k2* between the fuselage structure and the movable parts of the ramp lock. From Fig. 5 it can be seen that the section along the front lock is decisive for the layout and determination of the installation dimensions [16].

$$B - B$$

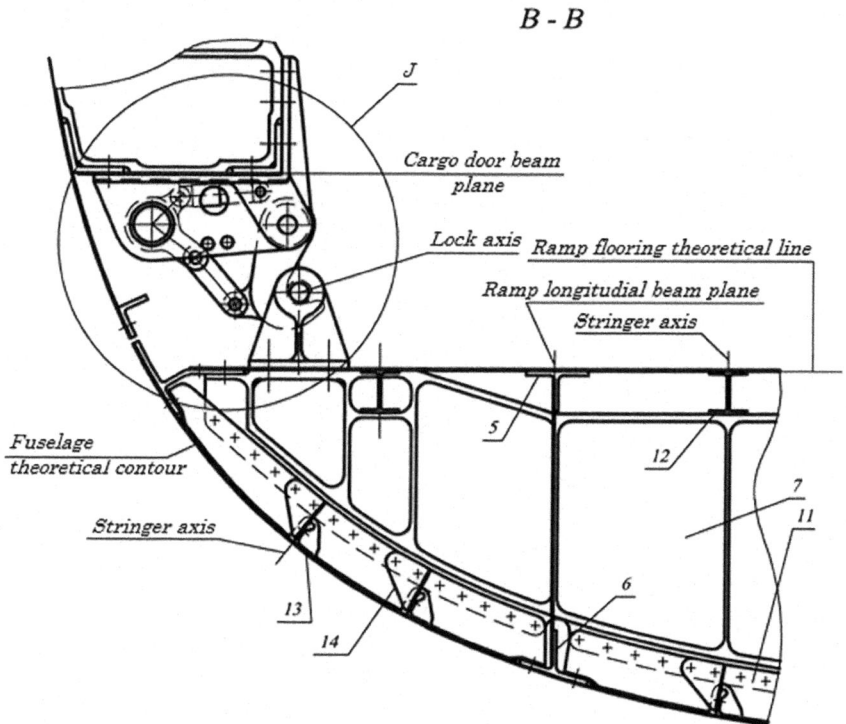

Fig. 5. View *B-B* of Fig. 1 – ramp lock side view.

In the closed position, the axle *21* of the lock interacts under load with the hook *23*, which is hinged on the axle *24* in the housing *25*. The housing *25* is fixed to the cargo door beam and the fitting which connects the beam to the power frame of the fuselage. A shaft *26* is hinged in the housing *25*, which is connected to the lock control cylinder. On the shaft *26*, a lever *27* is fixed, a link *28* is hinged to the hook *23*. The lever *27* is limited by a stop *30*.

On Fig. 6 shown that the hinged lever *27*, link *28* and hook *23* form a movable three-link [17]. When the lever *27* rotates on the shaft *26*, the hook *23* through the rocker 28 is disengaged from the axis 21 until it is mounted on the stop *31*. In the closed and open position, the lever *27* is spring-loaded by the shock absorber *29*. The shock absorber is made in the form of two cylindrical cavities, one of which is pivotally connected to the lever *27*, and the other is hinged in the housing *25*. A compression spring is installed inside the cylindrical cavities.

In the closed position, to prevent spontaneous opening of the lock, it is necessary to ensure the rotation of the hook *27* to the size *a1* from the line connecting the hinge axes of the lever *27* and the rocker joint *28* on the hook *23*. For free movement to the calculated position there is a groove in the lever *27*.

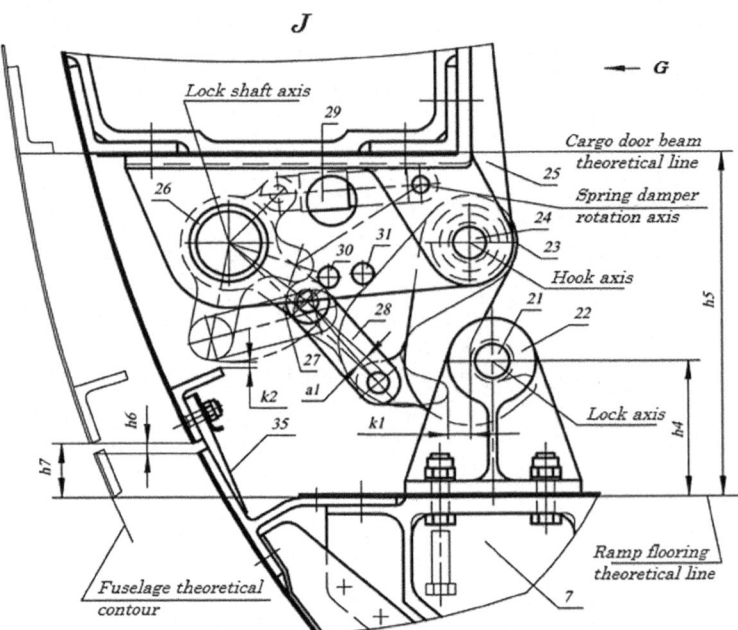

Fig. 6. View *J* of Fig. 5 – ramp lock kinematics.

The force taken by the side ramp lock is transferred to the aircraft fuselage. The kinematic lock formed as a result of setting the lever *27* to stop *30* prevents spontaneous opening of the lock under load. The opening of the lock is possible [9, 18] only when the shaft *26* rotates. In this case, the hook *23* by means of the shock absorber *29* is fixed on the stop *31* and in open position ramp moves with a gap *k1* from the axis *21*.

4   Ramp Control Rods and Cylinders

The main purpose of the rods is to keep the ramp open at the floor level of the cargo compartment of the aircraft. The rods can also be used to secure the ramp when it is deflected to the extreme position when it is not lean on the ground. The maximum load acting on the rods and their attachment points to the ramp and fuselage depends on the weight of the cargo dropped during airborne landing and the maximum weight of the monocargo loaded into the aircraft cargo compartment.

Under load, the ramp rods works in tension, so it can be made of a strip, hinged on the ramp, with the possibility of its movement in a slot installed on the fuselage board [19]. The rod is fixed with an easily removable head hinged to the rod. In this case, a preliminary transposition of the heads manually in the closed position, if an automated installation of the rods for fixing the ramp is not provided. It is depend on the working configuration of the cargo door.

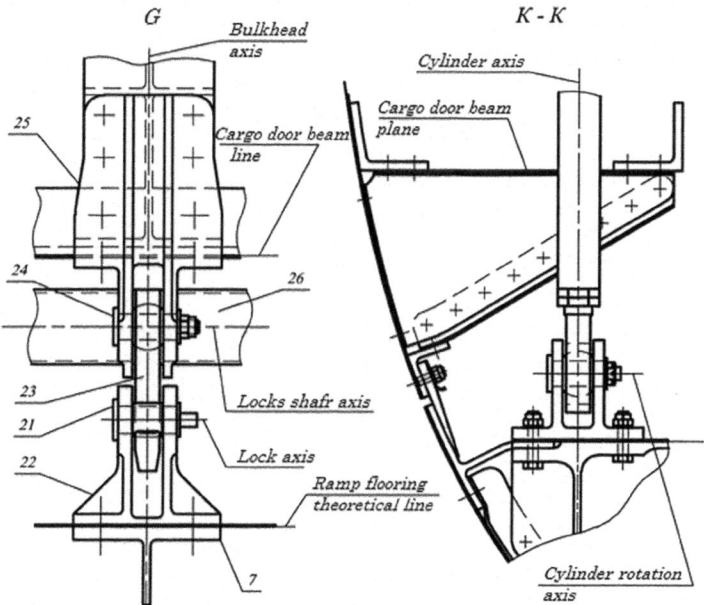

Fig. 7. Views *G* and *K-K* of Fig. 1 – ramp rod and cylinder mounting.

The ramp control cylinders during its opening and closing are installed on the fuselage board. In Fig. 7 in view *K-K* of Fig. 1 shown the mounting of the cylinder to the cargo door ramp.

When installing rods and cylinders in the ramp frame, additional elements of the structural power set are provided. The forces of the cylinders are determined depending on the mass of the ramp and the mass of the load lifted on the ramp. The optimal ratio of the cylinders power and its stroke is set by selecting the cylinder arm relative to the axis of rotation of the ramp when it is deflected to the maximum angle [20].

5   Ramp Sealing

Sealing of the cargo door in the ramp area is provided by installing special sealing profiles. Sealing profiles are made of rubber. As shown above, the laying of sealing profiles and the formation of their mating surface depend on the way of the cutout in the front of the cargo door and the hinge of the ramp.

Making a cutout behind the axis of the ramp rotation, as shown in Fig. 4 and Fig. 5 the sealing profile can be fixed with a continuous strand on the fuselage. This method of sealing is the most effective and reliable. In closed position the ramp fits the sealing profile in the same way over the entire mating surface. But this method can't be used always. In another ramp type, the ramp enters into the fuselage and at this point the sealing profile on the fuselage cannot be established.

On Fig. 6 shown the fastening of the sealing profile *35* to the fuselage sidewall. Dimension *h7* is selected from the requirement of providing the necessary overlap between the sealing profile and the structural longitudinal element of the ramp.

The configuration of the sealing profile and the method of its attachment to the fuselage, as well as the formation of the mating surface on the ramp, are determines depending on the design features of the ramp.

6   Conclusion

Transport category airplane rear cargo door ramp with its main components such as side ramp locks, ramp control rods and cylinders, sealing are presented. The description and methods of determination its main parameters provides such requirements to the developed aircraft as:

- Accommodation, load/unload and airdrop of payload and vehicles prescribed by the airplane technical specifications;
- Efficient structural integration of aircraft units providing minimum mass, high reliability and maintainability of cargo door and tail airframe part;
- Data for iterative formation fuselage tail part master geometry with high quality of external surface providing low aerodynamic drag;
- Compliance with the requirements of regulatory documents (AP-25, CS-25, ICAO SRP A6, AC-25, OST;
- A single source of information about the fuselage tail cargo door parameters, refined surface configuration for further aircraft design, preparation for its production and operation.

The method was tested using data from the An-1X8 transport category aircraft family, as well as in the process of local airliner frontend design.

The proposed theory allows to describe and develop the ramp with its unit as one of the main parts of cargo door. Which make it useful in further parametric studies of the aerodynamic, strength and mass characteristics of the fuselage with aid of computer engineering analysis systems.

References

1. Haidachuk, A.V., Wang, B., Bychkov, S.A., Andreev, A.V.: Development of an integrated criterion for the rational choice of polymeric composite materials. Mater. Sci. **55**, 899–907 (2020). https://doi.org/10.1007/s11003-020-00385-2
2. Plankovskyy, S., Tsegelnyk, Y., Shypul, O., et al.: Cutting irregular objects from the rectangular metal sheet. In: Nechyporuk, M., et al. (eds.) Integrated Computer Technologies in Mechanical Engineering, AISC, vol. 1113, pp. 150–157. Springer, Cham (2020). https://doi.org/10.1007/978-3-030-37618-5_14
3. Plankovskyy, S., Myntiuk, V., Tsegelnyk, Y., et al.: Analytical methods for determining the static and dynamic behavior of thin-walled structures during machining. In: Shkarlet, S., et al. (eds.) Mathematical Modeling and Simulation of Systems (MODS'2020). AISC, vol. 1265, pp. 82–91. Springer, Cham (2021). https://doi.org/10.1007/978-3-030-58124-4_8

4. Grebenikov, A.G.: Methodology for integrated design and modeling of assembly aircraft structures. National Aerospace University "Kharkiv Aviation Institute", Kharkiv (2006). [in Russian]
5. Kundu, A.K.: Aircraft Design. Cambridge University Press, New York (2014)
6. Ciliberti, D., Della Vecchia, P., Nicolosi, F., De Marco, A.: Aircraft directional stability and vertical tail design: a review of semi-empirical methods. Prog. Aerosp. Sci. **95**, 140–172 (2017). https://doi.org/10.1016/j.paerosci.2017.11.001
7. Grebenikov, A.G., Gumenniy, A.M., Buival, L.Y., et al.: Light civil turboprop airplane take-off weight preliminary design estimation method. In: Nechyporuk, M., et al. (eds.) Integrated Computer Technologies in Mechanical Engineering. AISC, vol. 1113, pp. 60–74. Springer, Cham (2020). https://doi.org/10.1007/978-3-030-37618-5_6
8. Chumak, A., Buival, L., Humennyi, A., et al.: Transport category airplane fuselage master geometry parametrical modeling method. In: Shkarlet, S., et al. (eds.) Mathematical Modeling and Simulation of Systems (MODS'2020). AISC, vol. 1265, pp. 101–110. Springer, Cham (2021). https://doi.org/10.1007/978-3-030-58124-4_10
9. FAR. Part 25. Airworthiness standards transport category aircraft. IAC (2015)
10. Dveirin, A.Z., Kostyuk, V.A., Rabichev, A.I., et al.: Systematization and classification of cargo door types of transport aircraft by main design features. Open Inform. Comput. Integr. Technol. **70**, 33–53 (2015). [in Russian]
11. C-141B Command aircraft systems training (CAST). Scott AFB (1994)
12. Aircraft An-124-100. Maintaince manual. Antonov Design Bureau (1993). [in Russian]
13. Aircraft Il-76-TD. Maintaince manual. Ilyushin Design Bureau (1978). [in Russian]
14. Shabana, A.A., Wang, G.: Durability analysis and implementation of the floating frame of reference formulation. Proc. Inst. Mech. Eng. Part K J. Multi-body Dyn. **232**(3), 295–313 (2018). https://doi.org/10.1177/1464419317731707
15. Fielding, J.P.: Introduction to Aircraft Design. Cambridge University Press, New York (2017)
16. Shabana, A.A.: Integration of computer-aided design and analysis: application to multibody vehicle systems. Int. J. Veh. Perform. **5**(3), 300–327 (2019). https://doi.org/10.1504/IJVP.2019.100707
17. Ji, A.M., Zhu, K., Huang, J.C., Dong, Y.P.: CAD/CAE integration system of mechanical parts. Adv. Mater. Res. **338**, 272–276 (2011). https://doi.org/10.4028/www.scientific.net/AMR.338.272
18. Young, T.M.: Performance of the Jet Transport Airplane: Analysis Methods, Flight Operations, and Regulations. John Wiley & Sons, Hoboken (2018)
19. Krivtsov, V.S., Voronko, V.V., Zaytsev, V.Y.: Advanced prospect for the development of aircraft assembly technology. Sci. Innov. **11**(3), 11–18 (2015). https://doi.org/10.15407/scine11.03.011
20. Patel, M.D., Pappalardo, C.M., Wang, G., Shabana, A.A.: Integration of geometry and small and large deformation analysis for vehicle modelling: chassis, and airless and pneumatic tyre flexibility. Int. J. Veh. Perform. **5**(1), 90–127 (2019). https://doi.org/10.1504/IJVP.2019.097100

Investigation into the Forming Process of Wing Panel Oblique Bending by Means of Rib Rolling

Valeriy Sikulskiy[1] , Stanislav Sikulskyi[2] ,
and Vadym Garin[1(✉)]

[1] National Aerospace University "Kharkiv Aviation Institute",
17 Chkalova Street, Kharkiv 61070, Ukraine
vsikulskij@gmail.com, v.garin@khai.edu
[2] Embry Riddle Aerospace University,
1 Aerospace Blvd, Daytona Beach, FL 32114, USA

Abstract. The method of forming, finishing and shape refinement of ribbed panels by local deformation of the panel ribs using the rolling process is considered. The purpose of the article is experimental check of panel forming technology and shape changing accuracy under nominal and critical technical conditions of rib rolling of the monolithic panel as well as drawing up technical recommendations for using this method. In order to shape the profile and panel bending it is proposed to use the method that utilizes two rollers to deform areas of the rib. It provides the significant reduce in the angle of springing and expands the technological capabilities of the equipment. The process was simulated using FEM in the ANSYS software package for the case of oblique bending of ribbed panels. The problem of analyzing local stress-strain state during panel shaping was considered as quasi-static. The values of stresses and deflections of the panel sample at the point of load action and after removing the load were determined. It is shown on the example of aircraft wing panel oblique bending that the practical use of panel bending method with compression of ribs by rollers is expedient in case of sequential deforming the ribs with a counter and a passing sequence. The surface geometry along the axes of ribs that were bent by rollers shows better accuracy compared to the geometry of samples obtained by free bending in a punch. The results were compared to the data obtained by forming an aluminum alloy panel sample in unwinding device. It is shown that the method local forming by rolling allows achieving high accuracy of panel shape by means of correcting one-time impacts.

Keywords: Monolithic panels · Shape refinement · Bending moment · Modeling · Deflection · Residual deflection · Rolling

1 Introduction

The design of modern aircrafts of the transport category employs the method of producing the wing torsion box without production breaks. This is achieved by assembling the wing box of monolithic panels with length of the wing and shape of the panel with

© The Author(s), under exclusive license to Springer Nature Switzerland AG 2021
M. Nechyporuk et al. (Eds.): ICTM 2020, LNNS 188, pp. 598–608, 2021.
https://doi.org/10.1007/978-3-030-66717-7_51

bends at an angle to the panel ribs due to the wing sweep. This option is the most economically advantageous in terms of reducing the structure mass, machining process and increasing the utilization rate of the material as well as reducing the cost of pressed panels. The process of panel bending at the angle to the rib axis belongs to the category of complex processes due to the system tendency to a minimum deformation energy, which leads to the deviation of the bending axis after unloading.

Traditional methods of forming and correction of ribbed parts do not always allow achieving needed shape accuracy mainly due to large overall dimensions and complex shape of the parts. The engineering capabilities of forming process can be significantly expanded by utilizing the methods of local plastic deforming, the method of rolling using rollers being one of them.

2   Analysis of Basic Research and Literature

The authors, when studying the process of shaping by rolling the parts of double curvature, developed a mathematical model for calculating biaxial bending and twisting under the action of tensile forces applied to the cross-sectional areas [1]. Such a scheme is suitable primarily for obtaining parts from a sheet, therefore, for bending parts with ribs, processes are used for joint deformation of the skin and ribs (thickenings) and provide a high degree of movements and deformations uniformity [2–4].

The part correction is understood as a technological operation of processing the metals by pressure in order to achieve the required shape accuracy of a part (work-piece). However, most of the processes used to achieve the required accuracy of parts claim the use of finishing processes. Final straightening (fine-tuning) of the general or local parts deflection from forgings, stampings and profiles is carried out by free bending on hydraulic presses, and profiles and sheets – by the method of planting or spreading parts of the details section in special dies. The final adjustment of the general or local deflection of parts made of monolithic panels is carried out by the method of free bending on presses and blasting. Especially there is often a need to straighten parts made of stampings or forgings with subsequent machining [5].

In connection with the technological difficulties arising during the bending of large-sized parts with rigid ribbing on pressing or roll equipment, as well as shot-percussion equipment, methods associated with local plastic deformation have become widespread [6].

Methods of plastic deformation of profile parts by local planting and spreading of the part material with special devices or rolling of edges significantly expands the possibilities of shaping and straightening. Eckold AG produces a range of special stationary and portable equipment for local planting and routing of sheet and profile parts [7]. This equipment fitted with various tools for different cold metal forming operations is used for forming parts and for correcting or refining the part shapes after forming by some other basic method.

The Airbus company is known to use the roll correction process. Due to the lack of industry-specific regulatory documentation for the rolling process using rollers, it is advisable to use the restrictions set in the industry-specific regulatory documentation for local shrinking and stretching processes in punches, whose loading patterns are similar to rolling loading pattern, for example, in [8]. According to them, the depth of traces from the action of working elements of the tool on the part after shrinking should not exceed the tolerance for the nominal size of the wall thickness of the part. The loading pattern of the part under correction by rolling is similar to the process of rolling on a cold-rolling mill with cylindrical, conical, and parabolic rollers [3]. As the result, the rolled area of a part section stretches out plastically, which leads to the change in the part shape and the formation of complexly distributed residual stresses.

Besides that, scientists do not pay sufficient attention to the accuracy of shape changing process that depends on conditions of minimal fitting works during the assembling process of the wing torsion box with separate panels [9]. The accuracy of the bending axis location considerably affects the scope of fitting works at the point of interconnecting the panels [10]. The process of producing the wing panels is usually divided into two stages, formation of transverse curvature along the wing contour and formation of longitudinal bend at the angle to the ribs while maintaining the transverse curvature [11, 12]. In light of this, it is crucial that the transverse curvature is maintained within the given tolerances on the second stage. The special literature provides no information on bending technology of large-sized panels and on the values of residual stresses after bending of complexly shaped profiles that are most often found in real structures.

3   The Purpose and Tasks of Investigation

The purpose of this article is an experimental test of the technology of panel forming and the accuracy of panel shape changing at nominal and critical technological modes of a monolithic panel rib rolling and the development of technological recommendations for the method application.

To achieve this goal, the following tasks were solved: the selection of an effective rolling pattern for shaping ribbed panels and creating special equipment; experimental studies of the peculiarities of local deformation process by rolling on samples of full-scale panels.

4   Research Materials and Results

Analyzing the deformation pattern of ribs during correction, the following basic patterns were considered: stretching; compressing the rib; rolling the part of the rib section; rolling the rib along its entire height. The latter pattern was recognized the most effective, although it requires a high accuracy in a rib section dimensions. Such pattern was implemented in device creating for local bending of panels by rolling, and the pattern of rolling the upper part of the section was used for correction and refining the panels. The study of the deformation mode was carried out according to the rolling pattern shown in Fig. 1.

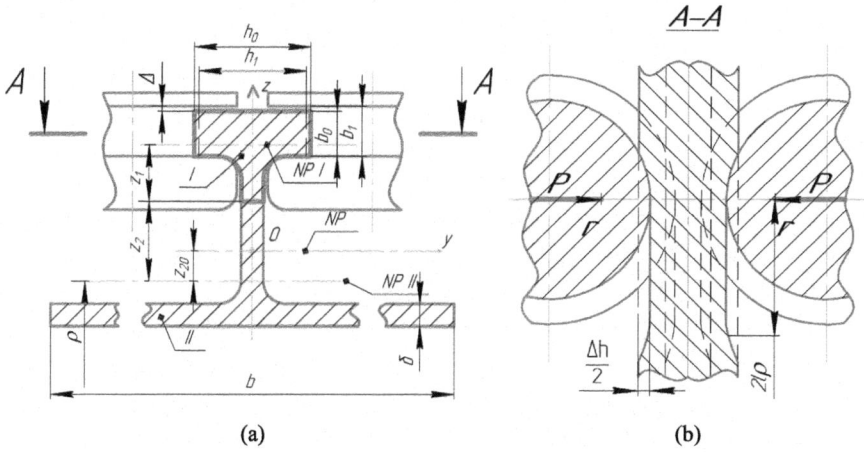

Fig. 1. The pattern of shaping the ribbed panels by rib rolling and location of the rollers: (a) – initial; (b) – during compression.

The calculation of technological parameters was carried out according to the dependences previously proposed by the authors. The value of rib compression Δh was calculated by the selection method using the following equation obtained from the rolling theory:

$$b_0\left(\frac{\Delta h}{h_0} - \frac{z_1 + z_2}{\rho}\right) = C_b C_\sigma a\left(\sqrt{r\Delta h} - \frac{\Delta h}{2\mu}\right)\ln\frac{h_0}{h_1},$$

where b_0 is the initial rib shelf height; h_0, h_1 are the initial rib shelf thickness and the rib shelf thickness after compression; z_1, z_2 are the coordinates of neutral surfaces of compressed and non-compressed parts of the section relative to their boundary; ρ is the curvature radius of non-compressed section neutral surface before unloading; C_b is the ratio of bandwidth effect on broadening; C_σ is the ratio of stretching effect under compression; r is the roller radius; μ is the Poisson's ratio; a is the ratio that depends on relative rib shelf compression ε

$$a = \frac{2(1 - \varepsilon)}{2 - 1.5\varepsilon}.$$

The curvature radius of non-compressed section neutral surface before unloading

$$\rho = R\left(1 - \frac{I_1 + I_2}{I}\right),$$

where R is a residual radius of rib bending; I_1, I_2 are the inertia moments of compressed and non-compressed areas of sections relative to their neutral surfaces; I is the inertia moment of the whole section relative to the neutral surface.

In order to check the main theoretical assumptions and to practice the process conditions of panel bending the experimental panel rib rolling device was created that operates on the principle of rib compression by rolling it between profiled rollers (Fig. 2).

Fig. 2. The experimental panel rib rolling device.

The following roller compression patterns were considered: with fixed rollers and constant compression rate; with constant roller pressing force. The second pattern is preferred because it provides stable compression under various dimensional drifts of panel ribs after machining.

Roller compression according to the second pattern is carried out by a high-pressure hydraulic cylinder, which is supported by a hydropneumatic accumulator, a reducer, and a controller. The experimental panel rib rolling device operates in a following manner. The rib to be bent is entered into the gap between the rollers when they are in open position. Then the rollers are set in working position until they touch the rib. The pump sets the required pressure in the roller compression cylinder, while the pressure value is controlled by a pressure gauge, and the rib compression value is controlled by a micrometric indicator built into the experimental device.

In the experiments, the oblique bend forming process was carried out by sequential deformation of individual ribs. The panel was placed on a hard bearer, namely a set of templates installed in a coordinate-controlled workbench (CCW) (Fig. 3). The rolling work tool made with an individual drive for rollers' rotation is installed at the bending point of the rib due to its low weight. The rolling device moves driven by the rollers' rolling over the rib during its compression. The compression routes are drawn beforehand based on calculations. The operator conducts the shape check and makes corrections according to the gaps between the front face of the panel leaf and template planes, as in the case of shot blasting.

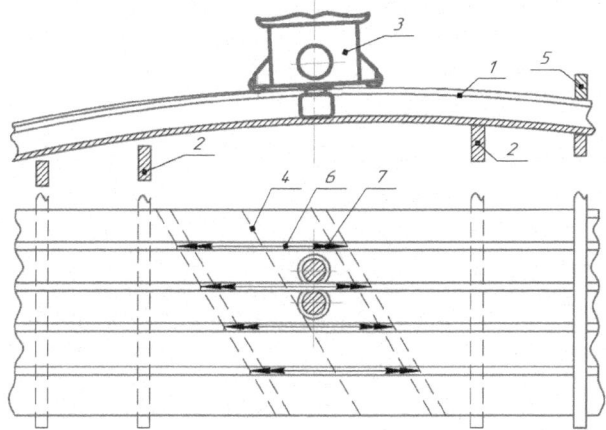

Fig. 3. The panel bending pattern on a CCW: 1 – panel: 2 – templates; 3 – rolling tip; 4 – bend line; 5 – clamp; 6 – routes of calculated compressions; 7 – routes of corrective compressions.

In the tests, samples of a monolithic panel with a natural rib cross-section with width of 800 mm, length of 1000 mm, leaf thickness of 4 mm, and a T-shaped ribbing height of 60 mm, made of aluminum alloys ENAW-2024 and ENAW-7075, were used. The profile of the rollers provided compression of the rib shelf and the part of the rib wall to ensure uniform deformation and minimal residual stresses. The roll mark and the boundary between non-compressed sector (left) and compressed sector (right) is shown on Fig. 4.

Fig. 4. The panel rib after compression in rolling device.

The portable coordinate measuring machine of the type ROMER Absolute Arm provided the control over geometric parameters of encircling surface of the studied sample in the conventional coordinate system by the layout grid with a pitch of 100 mm. The physical configuration of the sample after oblique bend forming is shown on Fig. 5.

Fig. 5. The physical configuration of the panel sample after oblique bend forming.

The discrepancy graphs of actual coordinates of encircling surface points of the monolithic panel due to oblique bend forming by rib rolling are presented on Fig. 6. The geometry of the panel surface along the rib axes is highly accurate compared to the traditional methods of oblique bend forming. The bend axis is located in a correct position relative to all of the ribs, which is confirmed by the coinciding minimums of all ribs at the graph on Fig. 6.

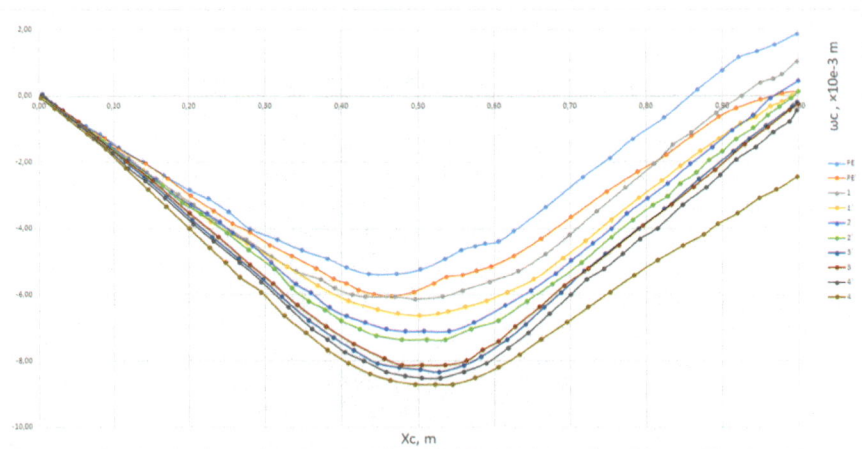

Fig. 6. The discrepancy of actual coordinates of encircling surface points of the monolithic panel along the panel ribs and free edges (FE).

The discrepancy graphs of actual coordinates of encircling surface points of the monolithic panel due to oblique bend forming by rib rolling in a section about the bending line are displayed on Fig. 7. The data are presented for the bending of considered panel sample with the rolled sector length of $2l_p = 0.3$ m; the oblique angle value of $\varphi = 30°$; the bending radius of $R = 10$ m; the transverse curvature radius of $R_t = 12.5$ m.

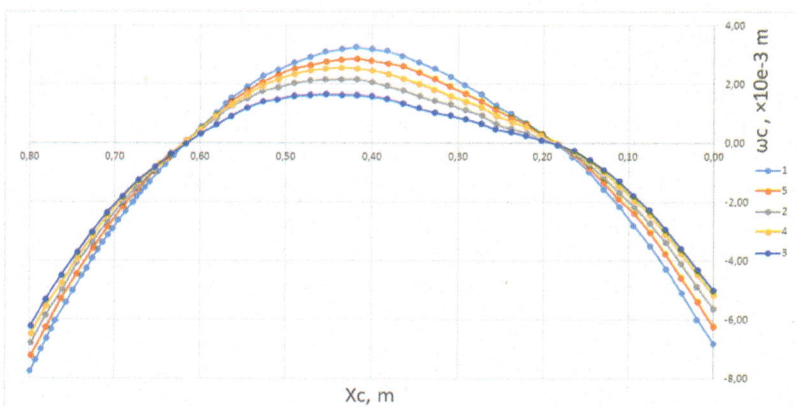

Fig. 7. The discrepancy of actual coordinates of encircling surface points of the monolithic panel due in a section about the bending line: 1 – initial contour; 2 – with the bending angle $\alpha = 1°43'$; 3 – with $\alpha = 2°30'$; 4 – the same with an increased height of compressed part of the peripheral ribs by 5 mm; 5 – the same with the compression of free edges of the leaf.

The discrepancy graphs of actual coordinates of encircling surface points of the monolithic panel due to oblique bend forming by rib rolling in an adjacent section to the bending line are shown on Fig. 8.

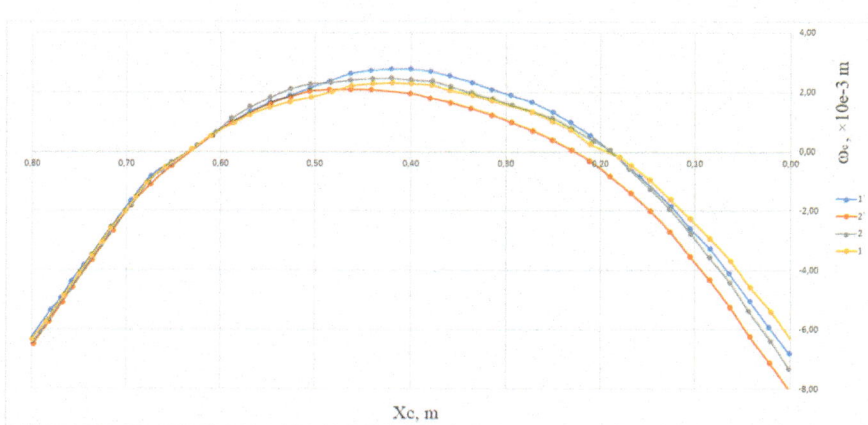

Fig. 8. The discrepancy of actual coordinates of encircling surface points of the monolithic panel with the bending angle $\alpha = 1°43'$: 1, 2 – initial contour; 1', 2' – after shape change; 1, 1' – in a section at 200 mm distance from the bend line; 2, 2' – in an adjacent section to the bend line.

As an accuracy appraisal criterion of fulfilling these conditions, the relative difference of deflections of the outer ribs being in opposite deformation conditions was taken. The specified relative difference of deflections in case of fulfilling the deformation conditions does not exceed 0.03.

The problem of determining the fields of deflections, deformations and stresses was solved using FEM in ANSYS software. The calculation used the effect of an experimental device on a panel sample in the form of bending moment application. The local stress-strain state problem was considered as quasi-static.

In order to create the finite element model of a panel the 8-node three-dimensional deformable solid finite elements SOLID185, presented in the ANSYS library, were used. The stresses and strains emerging in the studied model were defined as derivatives from obtained structure displacements.

The results were applied to the local method of a wing panel oblique bend forming. In this case local moments were applied to the intersection points of ribs and a line drawn at 60° angle to the rib axis.

The calculations reveal that the residual deformations reached the value of 0.25% only in a small area of panel ribs. The rest of the material of the panel leaf and adjacent ribs shows insignificant level of residual deformations after unloading. The graph obtained as a result of the calculations that shows residual deflections of the panel due to moment application after unloading, is shown at the Fig. 9. The values of the parameters in the image are deliberately increased many times for clarity.

The residual deformations occur only in panel ribs; the deformations of the rest of the elements are low and remain within the elastic zone.

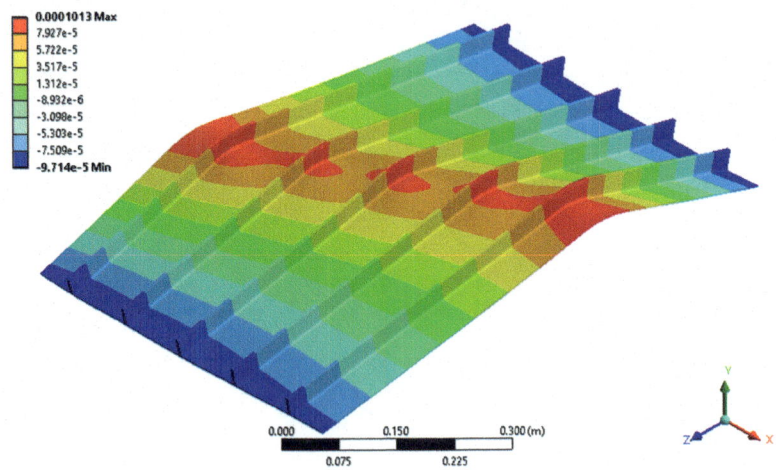

Fig. 9. The value of residual deflections in a panel due to applying local moments to the ribs.

The relative position of local effect areas on panel ribs allows obtaining the needed shape of a panel, namely, direct and oblique bending at different angles to the rib axes. The required bending angle is 1′ accurate because of one correcting procedure with an intermediate measurement. In case when the maximum length of local effect exceeds the length needed to obtain the specified angle, the compression of all ribs should be repeated.

The obtained accuracy of panel bending is several times higher than the accuracy of free forward bending. An exception is the shape of the free edge, which does not coincide with the shape of the ribs and takes an independent shape, which is clearly seen in Fig. 9. This disadvantage is easily overcome during the assembling procedure of the panels due to elastic deformation.

The experiments validated the calculation technique of basic technological parameters of ribbed panel forming by rib rolling. According to the technique, the residual curvature of the panel rib sector with the full-width leaf (width is equal to the distance between two adjacent ribs) attached on both sides is defined depending on compression Δh of the rolled rib section.

In case of sequential panel rib deformation, the condition must be fulfilled that the panel shape does not depend on the sequence of rib compression [13]. To ensure this, the following two conditions should be met. First, during the transition from rib to rib, the leaf should deform within elastic zone. Second, the effect of adjacent ribs and neighboring leaf areas on the process in the deformation zone must be insufficient.

5   Conclusion

The experimental investigation into the process of local shape changing on panel samples with the use of rolling devices has shown the efficiency of the panel shape forming, finishing and straightening processes. The practical use of the panel bending method by rib compression with rollers is advisable for the sequential deformation of the ribs with a counter and a passing sequence.

In the forming process of oblique bend using the correction method, the accuracy of the bending angle of ribs was about $1'$, and the deviation of the bending angle did not exceed $1°$.

The surface geometry along the rib axes bend by local methods shows better accuracy compared to the geometry of samples obtained by free bending in a punch. The relative ripple markings of a panel surface along the rib axis is $0.001...0.002$, which is $3...4$ times less than in case of free bending.

It is experimentally shown that at the bending angle under $2°30'$ and transverse curvature radii over 10 m the curvature change in the panel cross-section is insignificant.

References

1. Yoon, J.S., Kim, J., Kim, H.H., Kang, B.S.: Feasibility study on flexibly reconfigurable roll forming process for sheet metal and its implementation. Adv. Mech. Eng. **6**, 958925 (2014). https://doi.org/10.1155/2014/958925
2. Tan, J., Zhan, M., Liu, S.: Guideline for forming stiffened panels by using the electromagnetic forces. Metals **6**(11), 267 (2016). https://doi.org/10.3390/met6110267

3. Younes, W., Giraud, E., Fredj, M., et al.: Creep forming of an Al-Mg-Li alloy for aeronautic application. In: AIP Conference Proceedings, vol. 1769(1), p. 070016 (2016). https://doi.org/10.1063/1.4963469

4. Luo, H., Li, W., Li, C., Wan, M.: Investigation of creep-age forming of aluminum lithium alloy stiffened panel with complex structures and variable curvature. Int. J. Adv. Manuf. Technol. **91**(9–12), 3265–3271 (2017). https://doi.org/10.1007/s00170-017-0004-z

5. Plankovskyy, S., Myntiuk, V., Tsegelnyk, Y., et al.: Analytical methods for determining the static and dynamic behavior of thin-walled structures during machining. In: Shkarlet, S., et al. (eds.) Mathematical Modeling and Simulation of Systems (MODS' 2020). AISC, vol. 1265, pp. 82–91. Springer, Cham (2021). https://doi.org/10.1007/978-3-030-58124-4_8

6. Benedetti, M., Fontanari, V., Monelli, B., Tassan, M.: Single-point incremental forming of sheet metals: experimental study and numerical simulation. Proc. Institution Mech. Eng. Part B: J. Eng. Manuf. **231**(2), 301–312 (2017). https://doi.org/10.1177/0954405415612351

7. Eckold Blechumformung: Products for chipless coldforming of sheet metal, pipes and profiles. http://www.pressar.pt/pdf/eckold_prospekt_blechumformung.pdf. Accessed 30 Aug 2020

8. Sikulskiy, V., Kashcheyeva, V., Romanenkov, Y., Shapoval, A.: Study of the process of shape-formation of ribbed double-curvature panels by local deforming. East.-Eur. J. Enterp. Technol. **4**(1), 43–49 (2017). https://doi.org/10.15587/1729-4061.2017.108190

9. Belykh, S., Krivenok, A., Bormotin, K., et al.: Numerical and experimental study of multi-point forming of thick double-curvature plates from aluminum alloy 7075. KnE Mater. Sci. **1**(1), 17–23 (2016). https://doi.org/10.18502/kms.v1i1.556

10. Tan, J., Zhan, M., Gao, P., Li, H.: Electromagnetic forming rules of a stiffened panel with grid ribs. Metals **7**(12), 559 (2017). https://doi.org/10.3390/met7120559

11. Yan, Y., Wang, H.B., Wan, M.: FEM modelling for press bend forming of doubly curved integrally stiffened aircraft panel. Trans. Nonferrous Metals Soc. China **22**, s39–s47 (2012). https://doi.org/10.1016/S1003-6326(12)61681-1

12. Zhang, M., Tian, X., Li, W., Shi, X.: An equivalent calculation method for press-braking bending analysis of integral panels. Metals **8**(5), 364 (2018). https://doi.org/10.3390/met8050364

13. Sikulskiy, V.T., Sikulskyi, S.V., Kashcheeva, V.Y.: Modeling of the monolithic panels shaping process in ANSYS. Bull. National Tech. Univ. "KhPI". Ser.: Innov. Technol. Equipment Handling Mater. Mech. Eng. Metall. **23**, 67–73 (2018). (in Russian)

Numerical Analysis of Stress-Strain State of Fuel Tanks of Launch Vehicles in 3D Formulation

Pavlo Gontarovskyi⬤, Natalia Smetankina⬤,
Nataliia Garmash^(⊠)⬤, and Iryna Melezhyk⬤

A. Pidgorny Institute of Mechanical Engineering Problems of the National
Academy of Sciences of Ukraine, 2/10 Pozharskogo Street,
Kharkiv 61046, Ukraine
garm.nataly@gmail.com

Abstract. To model the kinetics of the thermal stress state in 3D structural elements with complex material properties, with account taken of plastic deformations, design features, and real loading conditions based on the finite-element method, a special calculation technique and software package have been developed. In this paper, the main relations of the developed calculation technique are presented and the main features of this technique are described. To study the stress-strain state of fuel tanks of launch vehicles, the software has been improved and upgraded. During the construction of different, in the degree of complexity and level of discretization, computational fine-element schemes, both the design and loading features of fuel tanks were taken into account. Calculation studies were performed based on data on the physical and mechanical properties of aluminum alloys and the corresponding strain diagrams. Calculations were performed with account taken of the internal and water pressures, the latter varying with the water level in the tank. The solution of elastic-plastic problems was carried out by the method of time steps with a given number of iterations in each of them. For different calculation models, values of plastic strains were obtained at fixed points of the tank. It was established that the results obtained by different calculation schemes are in good agreement with each other.

Keywords: Fuel tank · Finite-element method · Stress-strain state

1 Introduction

Modern competitive rocket and space vehicles require constant improvement of their structural elements and an increase in their reliability and operational safety [1]. An important task is also to reduce the cost of manufacturing new rocket and space vehicles and reduce their implementation time. One of the ways to implement these tasks is to replace the tests (especially destructive ones) of the vehicles designed with their computational studies [2, 3]. Therefore, the development of computational methods for determining the stress-strain state of elements of rocket and space vehicles is an urgent and important task [4–7]. Fuel tanks of launch vehicles (LV) are one of the

M. Nechyporuk et al. (Eds.): ICTM 2020, LNNS 188, pp. 609–619, 2021.
https://doi.org/10.1007/978-3-030-66717-7_52

most critical and expensive components of rocket and space hardware [2, 8], for which an important factor is a reduction in weight. Therefore, they are made of light aluminum alloys and are thin-walled structures that are supported by a large number of stringers and frames. Places of sharp difference in wall thickness are concentrators of stresses and strains, so it is possible to correctly estimate their stress-strain state by using the finite-element method only in 3D formulation [2].

The cylindrical body of a waffle-structure fuel tank is made of waffle-grid shells with increased wall thicknesses (10 mm) in welds. The diameter of the body is 3,900 mm, its length is 16,700 mm. On the inside, the tank is reinforced with 89 stringers, and along the axis, each of the 11 shells has 11 frames spaced at a distance of 137 mm from each other. The thickness of walls for different shells vary from 3 to 5.6 mm, and the thickness of frames and stringers, from 6 to 6.4 mm. The thickness of shells in the radial direction together with the height of frames and stringers is 25.5 mm. The general view of the tank is presented in Fig. 1.

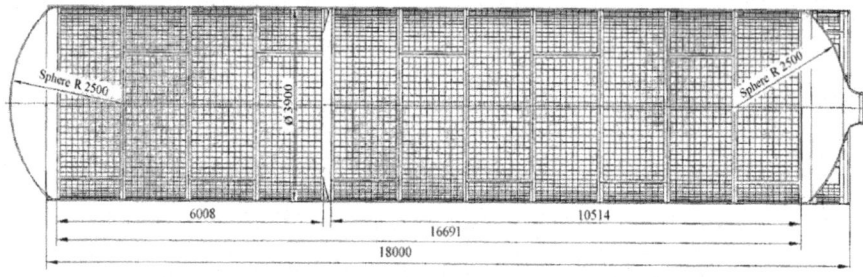

Fig. 1. General view of an LV waffle-structure fuel tank.

A finite-element calculation scheme of the cylindrical part of the tank with the discretization into elements in the form of arbitrary hexagons, which takes into account both design features and concentrators, has tens of millions of nodes. Therefore, in order to calculate the many variants of the stress-strain state of the structure, which requires that personal computers have large amounts of memory and sufficient speed, it is necessary to significantly simplify the calculation schemes. A separate problem is a detailed description of the geometry of the structure.

2 Methodical Ware and Software for the Investigation of the Thermostressed State of Structures in 3D Formulation

To study the stress-strain state of LV fuel tanks, researchers of IPMash of the NAS of Ukraine used the developed finite-element software complex "CUB" [9], which allows one to estimate the thermal stress of various structures [10, 11] in 3D formulation with account taken of plastic strains.

The main attention in the development of the software complex was paid to reducing the complexity of setting the input information and to the simplicity of analyzing the obtained results of calculations. In this regard, a special system of input data was developed for a wide class of structures. This system is based on a topologically regular decomposition of a body into macroelements in the form of arbitrary hexagons. The geometry of and the load on these hexagons can be set in different coordinate systems (Cartesian, cylindrical, spherical, toroidal and toroeliptic), arbitrarily oriented relative to the global Cartesian one. Hexagons in these coordinate systems can either be parallelepipeds, whose geometry requires only 6 numbers to be described, or have arbitrary geometry with rectilinear or parabolic edges, requiring, respectively, 24 and 60 members to be described. Macroelements are also allowed, for which the geometry of the sub-region is not given, but is determined in the process of constructing a finite-element computational model based on two faces of already given hexagons by connecting their vertices with rectilinear edges.

The discretization of the macroelements into finite elements is carried out automatically by the software complex according to the given number of the discretization points on edges in directions 1, 2, and 3 with account taken of the given non-uniformity of their discretization.

In addition, especially for the case of complex body structures with variable-thickness walls [9], when one constructs a computational model, cylindrical, conical, spherical, and toroidal surfaces are used, between which are macroelements with a geometry determined by rectangular-plan opposite faces of the surfaces given. The geometry of such macronutrients is given by eight numbers on two surfaces, between which a macroelement is located.

After the automatic calculation of the coordinates of finite-element nodes in a local coordinate system, their recalculation is performed in the global Cartesian system, where the stress-strain state of the structure is determined. The stress-strain state calculations by the finite-element method are possible only in a Cartesian coordinate system, because in cylindrical coordinates, the polynomial functions of the shape of an element as a solid bodies do not satisfy the conditions of motion without the occurrence of strains. As the calculations of the stress-strain state in cylindrical coordinates have shown, even relatively insignificant displacements of elements as solids lead to significant errors in the results.

In the case where the cylindrical body of a tank is described in a cylindrical coordinate system by macroelements in the form of parallelepipeds, the topologic array of macroelements can be significantly reduced by specifying the array of different layers of macroelements and the array of numbers of these layers. This allows one to significantly reduce the input information.

In a topologic array, the type of macroelement is determined by a number, which indicates the material number of the element, the type of its rheological properties, the method of taking into account changes in its geometry during deformation, geometric nonlinearity and other features specified in a special array of different types of macroelements.

On the faces of macronutrients, there can be set different types of boundary conditions in the form of distributed components either in global or local coordinate systems, including one-way interaction. In special arrays, the type of distribution of these boundary conditions on the faces of a macroelement (constant, linear, or quadratic) is specified. Taking into account the geometric nonlinearity, different types of loads are considered: normal to a surface (for example, pressure), dependent on displacements (centrifugal forces) and constant loads (gravity).

Structural loading is carried out by given components of stresses or displacements distributed along axes 1, 2 or 3 in the form of constant, linear or quadratic functions. The values of these components are specified in the corresponding arrays and can change over time. For this to be possible, special information is set on the values of these functions at the appropriate time points and the method of interpolation between them.

The non-stationary initial-boundary value problem of thermomechanics is solved by the method of time steps. At each of these steps, iterations are possible to determine the parameters of linearized problems. The system of equations of the finite-element method is solved by the square root method with account taken of the variable width of the tape, coefficients of the system of equations being determined using two-node Gaussian quadratures.

To calculate the thermophysical properties of material, a temperature forecast is allowed to be made for the end of each time step, which saves one iteration in solving a nonlinear non-stationary thermal conductivity problem by using an implicit Crank-Nicholson scheme [9].

To solve the problem of mechanics, a variant of the incremental theory of the modified Lagrange approach is applied using modified Kirchhoff's stress tensors and Green's strain tensors [12].

The linearized equation of the physical law

$$\Delta^* \sigma_{ij} = C_{ijkl} \Delta^* \varepsilon_{ij} + \Delta^* \sigma_{ij}^0,$$

represents the strain growth tensor in the form of the sum of elastic, temperature, and plastic components:

$$\Delta^* \varepsilon_{ij} = \Delta \varepsilon_{ij}^e + \Delta \varepsilon_{ij}^T + \Delta \varepsilon_{ij}^p.$$

The increase in elastic strains is determined by Hooke's law

$$\Delta \varepsilon_{ij}^e = A_{ijkm}(T^{(n+1)}) \Delta^* \sigma_{km} + \Delta A_{ijkm} {}^* \sigma_{km},$$

where A_{ijkm} is the tensor of the elastic pliability of material at the temperature at the end of $n + 1$ time step

$$\Delta A_{ijkm} = A_{ijkm}\left(T^{(n+1)}\right) - A_{ijkm}\left(T^{(n)}\right).$$

The increase in the components of high-temperature strains is

$$\Delta\varepsilon_{ii}^{T} = \alpha_i\left(T^{(n+1)}\right) \cdot T^{(n+1)} - \alpha_i\left(T^{(n)}\right) \cdot T^{(n)}; \quad \Delta\varepsilon_{ij}^{T} = 0, \ i \neq j.$$

The increase in the components of plastic strain depends on the theory used. For isotropic material with isotropic hardening under active load, the components of the increase in plastic strain are calculated by the relationship [13]:

$$\Delta\varepsilon_{ij}^{p} = \left(\frac{3}{2\sigma_i}\right)^2 \left(\frac{1}{E_k} - \frac{1}{E}\right) \cdot S_{ij}\left(S_{km}\Delta\sigma_{km} + \frac{2(\sigma_i - \sigma_T)}{3}\right),$$

where σ_i and σ_T are the stress intensity and yield strength at the beginning of the step; E, E_k are Young's tangent moduli, which are determined by the strain diagram of the material at the end of the time step; S_{ij} is the stress deviator. The use of modified strain and stress tensors allows using traditional theories of plasticity, as the conditions of non-compression of the material will be approximately fulfilled.

The CUB software package is implemented in the Visual C++ shell in the form of two modules, the first of which enters the input data, calculates finite-element coordinates, determines the structure of the finite-element method system matrix and other auxiliary values needed to solve the problem. Control of the input data is provided to detect errors and verify its authenticity. The second module solves the initial-boundary value problem.

For the graphical analysis of results, the CUT program is used, which allows one to control the geometry of the structure after entering the input information, and, after solving the problem, to obtain either the strained state of the structure as a whole or individual selected cross-sections at arbitrary scale as well as the isolines of temperature fields, stress components for the corresponding cross-sections of the structure in the selected layers and directions.

3 Investigation of the Stress-Strain State of an LV Fuel Tank According to Different Calculation Schemes

The simplest three-dimensional calculation scheme (scheme 1) of an LV waffle-structure fuel tank, the general view of which is presented in Fig. 1, is an ideal cyclically periodic in the circumferential direction model, which includes half of the sector 1/178 of a circle (about $2°$) between stringers with boundary conditions of symmetry in the meridional planes. This scheme was used in [2]. In the places where the shells are welded together in the circumferential direction, there is an increased wall thickness (up to 10 mm) with insignificant deformations, which is why the shells influence each other insignificantly and their stress-strain state without essential errors can be considered separately. The shells are loaded with both internal and water pressures, the latter varying with the water level, as well as tensile axial stresses resulting from the action of pressure on the upper bottom.

The finite-element discretization of the computational model of one shell, with the concentrators taken into account, sufficient to obtain the required accuracy, is about

26,000 finite elements and 34,000 nodes. With the plastic strains taken into account for all nine load steps, the problem time for one shell was 5.5 h. The waffle-grid shells of the tank are made of the AMg6NPP aluminum alloy, the material properties of which are the following: the tensile strength $\sigma_B = 380$ MPa, the yield strength $\sigma_{02} = 280$ MPa, the modulus of elasticity $E = 7 \cdot 10^4$ MPa, Poisson's ratio $v = 0.3$. The strain diagram used in the calculations is shown in Fig. 2.

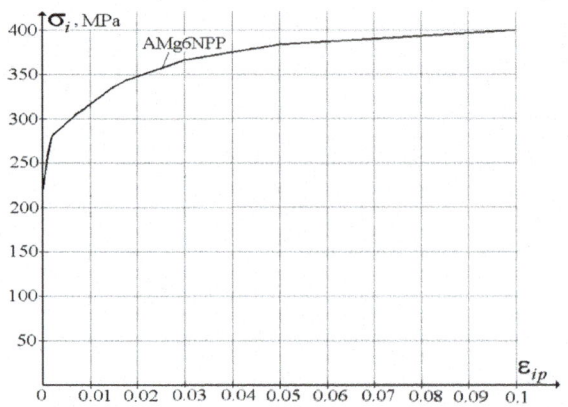

Fig. 2. The strain/diagram of the AMg6NPP aluminum alloy.

The maximum level of plastic strain intensity in the middle part of the fifth shell, starting from the top of the tank, is about 4.5% at an internal pressure of 0.8 MPa. The ultimate 6.5% plastic strain of the shell material will occur at a pressure of 0.82 MPa.

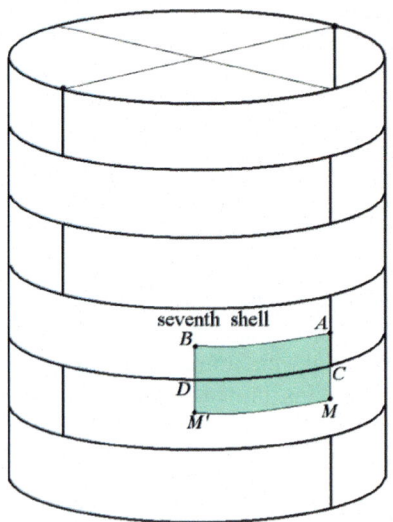

Fig. 3. A fragment of the tank in the area of the most heavily loaded shells.

To estimate the error that occurs when one uses the calculation scheme 1, a fragment of the tank was selected, which consists of two parts of the most loaded seventh and eighth shells, shown in Fig. 3.

It should be noted that each of the shells is welded of two parts by means of axial welds, where the wall thickness at the points of welding is also 10 mm. To assess the effect of increased stiffness in the area of axial welds (not taken into account in Fig. 1), it is necessary to consider the tank sector angle of 45° in the circumferential direction under the boundary conditions of symmetry in the extreme meridional planes (calculation scheme 2). The scheme of calculation model 2 is shown in Fig. 4. In the $Z = Z_S$ plane of the cylindrical tank wall with a thickness of 3 mm, applied are the stresses $\sigma_z = 3.2425 \cdot P$ MPa, where P is the pressure at the top of the tank, and in the $Z = Z_E$ plane, the displacement of the cylindrical wall $u_z = 0$. In the meridional planes $\theta = 0°$ and $\theta = 45°$, given are the conditions of symmetry $u_\theta = 0$ i $\tau_{r\theta} = \tau_{z\theta} = /0$.

During discretization, which corresponds to a cyclically symmetric calculation scheme, the amount of calculations in solving the system of equations of the finite-element method will increase by more than 10,000 times. To solve the problem for a reasonable period of time, the finite-element discretization was reduced. Figure 5 shows a fragment of calculation scheme 2 with the finite- element discretization for the cross-sections $r = 190$ cm (Fig. 5a) and $r = 195$ cm (Fig. 5b). The number of iterations was reduced twofold, with one iteration lasting about 50 min, and the entire elastic-plastic problem for calculation scheme 2 having been solved in 24 h.

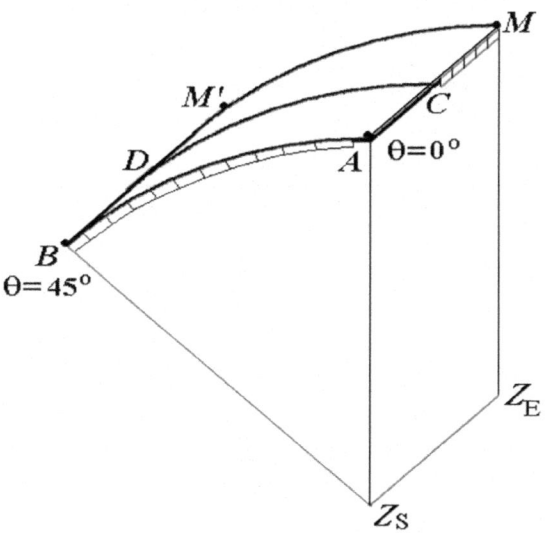

Fig. 4. A scheme of calculation model 2.

The inner surface of the fragment is loaded with the pressure P and the water pressure, the latter varying with the water level from 0.11 MPa to 0.125 MPa. The pressure P over time gradually increased from 0.4 MPa to 0.8 MPa, taking the values

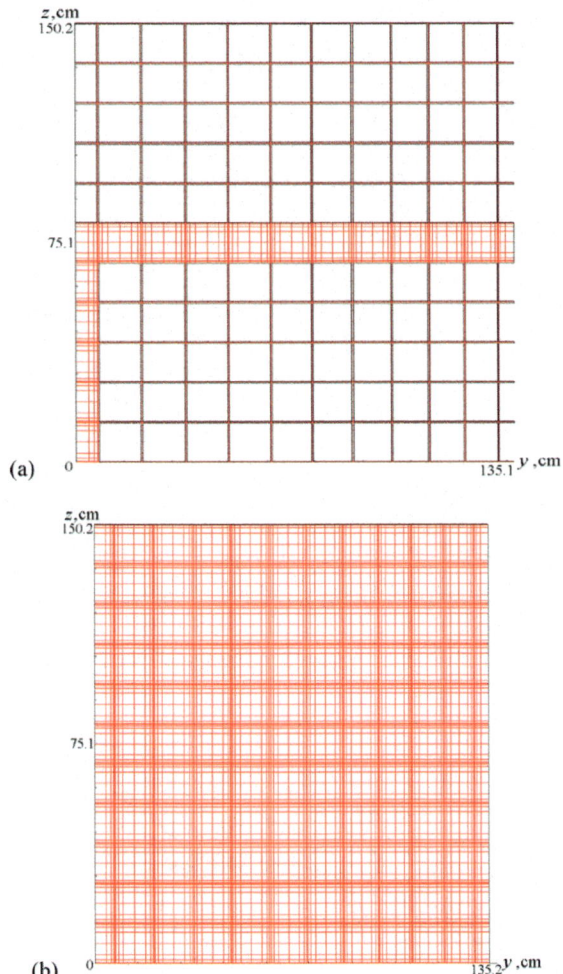

Fig. 5. A fragment of the calculation scheme with discretization into finite elements.

of 0.4; 0.5; 0.6; 0.7; 0.72; 0.74; 0.76; 0.78 and 0.8 MPa. When the pressure is 0.4 MPa, plastic strains are absent, at a pressure of 0.8 MPa, equivalent total strains on the outer surface of the tank at points M and M' are 6% and 6.7%, respectively, while at point B, it is only 5.6%. Excluding the axial weld, using a cyclic-symmetric model, the strains at points B and M' are 5.76% and 6.98%, respectively.

The influence of the thickening of the cylindrical wall under the weld along the tank axis on the stress-strain state can be estimated using a simplified calculation model 3 (Fig. 6), by replacing the stringers and frames with the layer with the orthotropic material properties $E_z = E_\theta = E \cdot 6.4/137 = E \cdot 0.0467$, where 6.4 is the thickness of stringer, and 137 is the distance between the stringers; the ordinates of the material strain diagram were reduced by the same value of 0.0467; $v = 0$.

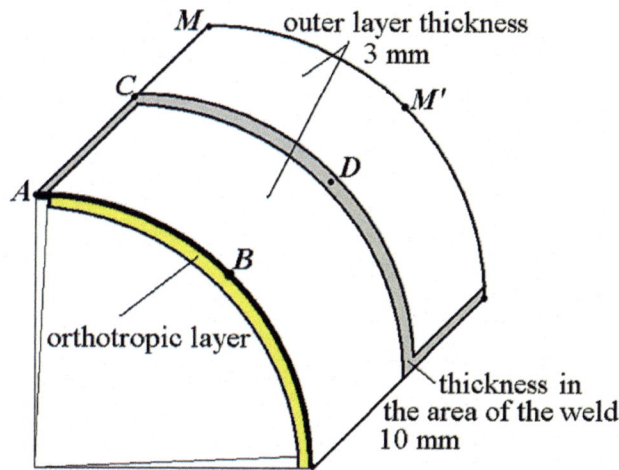

Fig. 6. A two-layer calculation model.

The use of the two-layer computational model 3 greatly simplifies the geometry, reduces the amount of input information and does not require the detailed finite-element discretization. The calculation time is significantly reduced, being 30 min.

The disadvantage of the two-layer calculation scheme is the impossibility of obtaining the concentration of the stress-strain state in the area of the connection of stringers and frames with the cylindrical wall of the shell.

In Table 1 are compared the displacements u_r (cm) of points A, B, C, D, M and M' obtained by calculation schemes 1–3.

Table 1. Radial displacements of points obtained by the calculation schemes 1–3.

		Pressure P					
		P = 0.5 MPa			P = 0.8 MPa		
Calculation scheme		1	2	3	1	2	3
Points	A	–	0.51	0.72	–	6.85	6.83
	B	0.877	1.013	1.0	9.06	9.14	9.35
	C	–	0.249	0.349	–	1.28	1.24
	D	0.464	0.614	0.56	1.71	1.88	1.98
	M	–	0.829	0.891	–	8.88	8.39
	M'	0.992	1.148	1.085	9.44	9.35	9.06

The displacements in the area of the longitudinal weld at point A are smaller than the maximum ones at a pressure of 0.5 MPa by about twofold, and at a pressure of

0.8 MPa, by 1.4 times. The strain state of the cross-section $\theta = 0$ for calculation scheme 2 at pressures of 0.5 MPa and 0.8 MPa is shown in Fig. 7 (the displacement is doubled).

Table 2 shows the intensity of strains $\varepsilon_i \cdot 10^5$ in the area of the outer surface of the tank for the same points A, B, C, D, M and M'.

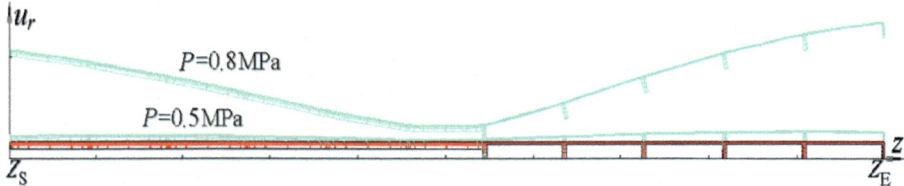

Fig. 7. The strain state of the cross-section $\theta = 0$.

Table 2. The intensity of strains at the above points, obtained by using different calculation schemes.

		Pressure P					
		P = 0.5 MPa			P = 0.8 MPa		
Calculation scheme		1	2	3	1	2	3
Points	A	–	196	181	–	387	317
	B	457	458	461	5,75	5,56	5,34
	C	–	186	197	–	622	610
	D	208	209	210	828	812	743
	M	–	388	485	–	595	612
	M'	470	472	483	6,97	6,70	5,83

4   Conclusions

This paper describes the main features of the calculation method developed on the basis of the finite-element method of studying the thermostressed state of structures in 3D formulation. A study of the stress-strain state of an LV fuel tank, using different, in the degree of complexity and level of discretization, computational fine-element schemes, was carried out. These schemes take into account the design features and conditions of loading the tank with internal pressure. The second calculation scheme, in contrast to the first, takes into account the thickening of the walls in the region of longitudinal welds, but it requires a significant increase in the amount of calculations. The third calculation scheme takes into account the thickening for longitudinal welds with a significant reduction in the volume of computational costs, but it does not take into account the concentrators in the region of the junction of the ribs with the cylindrical wall of the tank. The results obtained according to different schemes are consistent with each other.

It should be noted that the ultimate strains obtained according to calculation schemes 1–3 differ little from one another, while in the area of welds they are much smaller. The maximum stress points B and M′ are far from the longitudinal welds, which is why the stress-strain state assessment for the fuel tank can be done by using calculation scheme 1.

The developed calculation technique and the results of the research can be used to study the thermostressed state in the 3D formulation of structural elements of space and rocket hardware.

References

1. Zaytsev, B., Asayenok, A., Protasova, T., et al.: Dynamic processes during the through-plastic-damper shock interaction of rocket fairing separation system components. J. Mech. Eng. **21**(3), 19–30 (2018)
2. Gontarovskiy, P.P., Smetankina, N.V., Harmash, N.G., et al.: Researches of the stress-strained state of wafer-type fuel tanks of the launch vehicle. Prob. Comput. Mech. Strength Struct. **29**, 91–102 (2019). (in Russian)
3. Onhirsky, G.G., Shupikov, A.N., Ugrimov, S.V.: Influence of kinematic factors on the response of a deformable target in a bird collision. Issues Dev. Manuf. Aircr. Struct. **5**(56), 54–62 (2008). (in Russian)
4. Kubit, A., Trzepiecinski, T., Swiech, L., et al.: Experimental and numerical investigations of thin-walled stringer-stiffened panels welded with RFSSW technology under uniaxial compression. Materials **12**(11), 1785 (2019). https://doi.org/10.3390/ma12111785
5. Sim, C.H., Kim, G.S., Kim, D.G., et al.: Experimental and computational modal analyses for launch vehicle models considering liquid propellant and flange joints. Int. J. Aerosp. Eng. **2018**, 4865010 (2018). https://doi.org/10.1155/2018/4865010
6. Ugrimov, S.V.: Design of three-layer plates with composite skin. Issues Dev. Manuf. Airc. Struct. **3**(79), 47–56 (2014). (in Russian)
7. Zaytsev, B.F., Protasova, T.V., Larionov, I.F., et al.: Approximate model of elastic-plastic analysis of dynamics of the rocket fairing pyrotechnic separation system components. Prob. Comput. Mech. Strength Struct. **29**, 145–156 (2019). (in Russian)
8. Ţălu, M., Ţălu, Ş.: Numerical analysis of cylindrical fuel tanks thermal behaviour. Ann. Fac. Eng. Hunedoara **16**(3), 153–160 (2018)
9. Shulzhenko, N.G., Gontarowsky, P.P., Matyukhin, Y.I., Garmash, N.G.: Modeling of the kinetics of three-dimensional thermomechanical fields in the elements of turbomachines. Vib. Eng. Technol. **6**(38), 26–30 (2004). (in Russian)
10. Shul'zhenko, N.G., Gontarovskii, P.P., Matyukhin, Y.I., Garmash, N.G.: Numerical analysis of the long-term strength of the rotor disks of steam turbines. Strength Mater. **42**, 418–425 (2010). https://doi.org/10.1007/s11223-010-9232-2
11. Gontarovskiy, P., Shulzhenko, N., Garmash, N., Glyadya, A.: Creepage of the inner shell of the high-pressure cylinder of the steam turbine K-325-23.5. Bull. National Tech. Univ. "KhPI". Ser.: Power Heat Eng. Process. Equip. **2**, 19–23 (2019). (in Russian)
12. Washizu, K.: Variational Methods in Elasticity & Plasticity, 3rd edn. Pergamon Press, Oxford-New York (1982)
13. Birger, I.A., Schorr, B.F., Demyanushko, I.V., et al.: Thermal strength of machine parts. Mashinostroenie, Moscow (1975). (in Russian)

Active-Passive Radar Systems Using Radiation of HF Band Broadcasting Stations for Airborne Objects Detection

Vladyslav Lutsenko[1] , Irina Lutsenko[1]([✉]) , Ihor Popov[1],
Aleksandr Soboliak[1], Nguyen Xuan Anh[2], and Mikhail Babakov[3]

[1] O.Ya. Usikov Institute for Radiophysics and Electronics of the National
Academy of Sciences of Ukraine,
12 Academika Proskury Str., Kharkiv 61085, Ukraine
irene-lutsenko@ukr.net
[2] Institute of Geophysics Vietnam Academy of Science and Technology,
Hoang Quoc Viet, Cau Giay, Hà Noi A8-18, Vietnam
[3] National Aerospace University "Kharkiv Aviation Institute",
17 Chkalova Str., Kharkiv 61070, Ukraine

Abstract. The paper considers the possibility of using short-wave (SW) band broadcasting stations to detect airborne objects using the Doppler method. It has been shown that detection is possible at Doppler frequencies are spaced from the carrier frequency up to 60 Hz. Air objects moving at speeds up to 2000 km/h have such Doppler shifts in the decameter wavelength band. The relation is given and the detection range of air objects is estimated, the requirements for the degree of suppression of the direct illuminating signal are determined. The spectral characteristics of the signals of HF broadcasting stations have been studied experimentally. Theoretical estimates of the RCS of air objects of various types at different polarizations are given, including airplanes are made using the Stealth technology. It has been experimentally confirmed that their RCS in the resonant region can reach thousands of square meters and exceed the values in the microwave band by 4...5 orders of magnitude. Noise levels in the Doppler band of detection have been determined and the detection ranges of airborne objects have been estimated. It is shown that if the illumination signal is suppressed by 30...40 dB, the detection ranges can exceed 200 km.

Keywords: Monitoring of environment · Active-passive radar systems · Illumination by broadcasting stations in the HF band · RCS of aircraft in the resonant region · Stealth technology

1 Introduction

For effective airspace monitoring in the context of the rapid development of high-precision jamming facilities of radiation sources, new non-standard approaches are required. Existing electromagnetic radiation from television and broadcasting stations, artificial satellites can be used not only to solve problems of television, radio communication and navigation, but also to illuminate the airspace situation [1–3, 6–9].

M. Nechyporuk et al. (Eds.): ICTM 2020, LNNS 188, pp. 620–632, 2021.
https://doi.org/10.1007/978-3-030-66717-7_53

As known, in the most widely used for the detection and tracking of air targets in the active location mode, the radar station (radar) irradiates them, and then receives the necessary information from the scattered signals. The main disadvantage is that the radar can be detected quite easily by the enemy's technical means of reconnaissance, and then disabled. In the multi-position mode, the radar does not work for radiation, but only for reception (that is, it is used as a passive one), "extracting" information about the target from the signals scattered from it when it is irradiated by another station located in a different area of space. Radiation from radio and television facilities operating almost around the clock will provide overlap of near-earth space at significant distances [3, 8]. The receiving station will detect, recognize and track such potential targets as, for example, planes, helicopters, cruise missiles, by separating and processing the signals reflected from them. The advantages of this method of detecting air targets for areas with a highly developed infrastructure of television and radio broadcasting are obvious. First, the widely ramified network of television transmitters and repeaters, in contrast to the radars in service, practically makes it senseless to use anti-radar missiles and other weapons against them. Secondly, since the receiving stations do not work on radiation and are mobile, their detection and destruction are a difficult task. The choice of the optimal operating frequency range of the radar is based on the dependence of the RCS of the aircraft on the frequency of the irradiating signal. For example, the RCS of traditional fighters with a decrease in the frequency of the sounding signal grows according to a law close to linear. For low-observable aircraft, a similar dependence is even more pronounced – the RCS is proportional to the square of the wavelength of the probe signal. Calculations show that the detection range in free space of a low-observable aircraft in the 1...2 GHz range is 1.75 times greater than in the 2...4 GHz range, and 2.2 times greater than in the 4...8 GHz range. In this regard, foreign experts note an increased interest in VHF and UHF radars [3, 4, 8]. For several decades, one of the leading trends in radar has been the development of ever higher frequency bands, which was due to the possibility of obtaining higher resolution. The emergence of low-observable aircraft again attracted the attention of specialists to the meter, deci- and decameter bands. Conventional microwave radars provide detection of targets in the line-of-sight range, and if an aircraft is flying low above the ground, hiding behind mountains, behind the horizon or out of the detection zone, it remains "invisible" to control services. HF signals have the ability to be scattered from the ionosphere layer, after which they propagate further beyond the horizon line. As a result, the detection area of such radars significantly exceeds the detection area of a conventional radar [3, 8].

The use of broadband noise-like signals in bistatic radars, like information signals of broadcasting stations which are providing increased secrecy and survivability in a complex electromagnetic and target environment and the presence of a large number of sources of noises. Thus, the use of HF broadcast stations for illumination of the sky wave may allow detection of air objects for over-the-horizon. The scattering of these radio waves from aircraft occurs in the resonant region, which significantly increases their observed RCS. And the use of already existing radiation increases the secrecy and survivability of such active-passive systems. The purpose of this work is to consider the possibility and features of detecting air-borne objects, their RCS when used for airspace illumination the signals of broadcasting stations in the HF band for which the dimensions of the detected objects lie in the resonance region.

2 Features of the Construction of Bistatic Radars. Doppler Frequency Shift of the Target Signal in Active-Passive Systems

The system for detecting airborne objects using the radiation from short-wave communication stations can be classified as semi-active radar stations. Detection is based on the reception of electromagnetic fields emitted by broadcasting stations and scattered from aerial targets. The famous Soviet radio physicist I.S. Turgenev was one of the first to put forward the idea of constructing a bistatic decameter radar using illumination by a sky wave. To select the useful signal of the object against the background of passive interference created by reflections from the terrain, it is possible to use coherent processing with the extraction of the Doppler frequency signal caused by the change in the distance traveled by the electromagnetic wave along the path: the source of backlight – ionosphere – object – receiver. The simplest multi-position radar, consisting of one transmitter and one receiver, is called bistatic. A schematic representation of a bistatic system is shown in Fig. 1.

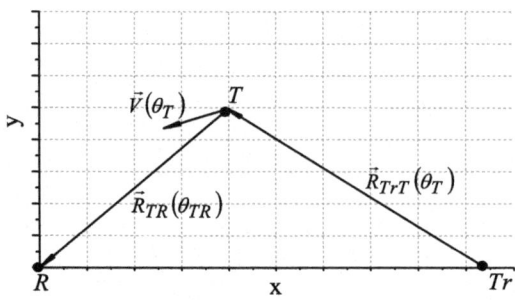

Fig. 1. The layout of the backlight station (*Tr*), target (*T*) and receiving point (*R*).

Where $\vec{V}(\theta_T)$ is the target speed vector; $\vec{R}_{TR}(\theta_{TR})$ is the distance between airborne object and receiver; $\vec{R}_{TrT}(\theta_{TrT})$ is the distance between source of backlight and air object.

Doppler frequency shift ω_D, is determined by the rate of change of the sum of the distances from the transmitter to the target and from the target to the receiver:

$$\omega_D = 2\pi f_D = \frac{2\pi}{\lambda_0}\frac{\partial(R_{TrT} + R_{TR})}{\partial t} = \frac{2\pi}{\lambda_0}V_D \le \frac{2\pi}{\lambda_0}2V_0, \tag{1}$$

where $\vec{R}_{TrT}(\theta_{TrT})$ is the distance between backlight station transmitter and object; $\vec{R}_{TR}(\theta_{TR})$ is the distance between object and receiving system; V_D is the doppler speed; $\lambda_0 = 2\pi c/\omega_0$ is the carrier wavelength; V_0 is the flight speed.

The rate of change in the distance is determined by the projection of the velocity vector onto the direction of propagation of the electromagnetic wave (the direction between the transmitter of the backlight station – $\vec{R}_{TrT}(\theta_{TrT})$, as well as directions between the object and the receiving system – $\vec{R}_{TR}(\theta_{TR})$). Wherein $\theta_{TrT}, \theta_{TR}$ are the corresponding angles characterizing the direction to the object from the backlight station and from the object to the receiver, and θ_T is the heading angle of the object.

An equation describing the behavior of the Doppler velocity, and hence the Doppler frequency shift:

$$v_D = \frac{V_D}{V_0} = \cos(\theta_{TrT} - \theta_T) + \cos(\theta_{TR} - \theta_T). \qquad (2)$$

The experimental study was carried out using professional radio receivers "Kanal-R" and "Katran", as well as an all-wave mobile receiver DEGEN, the technical characteristics of which are presented in Table 1.

Table 1. Main technical characteristics of the Kanal-R and Katran receivers. DEGEN, model: DE-1127.

Characteristic	Value		
	Kanal-R	Katran	DEGEN
HF band, MHz	2–32	1–32	2.30...23.0
MF band, KHz	–		522...1710
VHF band, MHz	–		64...108
Scale	Digital		
Step of tuning, Hz	5	0,1	9000
Modes			
AM/CW/SSB (USB/LSB)			AM/FM
Frequency instability: (after 4 h of warming up with "Hiacint-M" as reference oscillator)	<2.5 × 10^{-8}		–
Sensitivity:	<2.0 μV (AM)		<5 mV/m (MW)
SW	<0.6 μV (CW/SSB)		<2.5 μV
FM	–		<5 μV,
Selectivity (by mirror, and side channels of reception):	>80 dB		>40 dB
Dynamic range:	>70 dB		–
Efficiency of the manual amplification tuning on the IF: (relative to the initial level at the input of the receiver 1 μV)	>80 dB		–
Dynamic range (for two signals that differ from the main one by 5... 10 kHz)	>60 dB		–

Kanal-R is a two-channel tracking HF direction-finding radio receiver for auditory reception of signals from stations operating in telegraph (CW) and telephone (AM) modes and their recording on a tape recorder both from the LF amplifier output and from the 2nd IF output using a magnetic recording. The circuit of the Kanal-R radio receiver consists of two separate receiving paths with a common tunable smooth heterodyne.

3 Frequencies and Effective Scattering Surfaces of Air Objects in the Resonant Region

If the wavelength of the probing signal is commensurate with the size of the target, then the reflection will be resonant in nature, due to the interaction of the direct reflected wave and waves that envelop the target. This phenomenon contributes to the formation of strong echoes. The resonant frequency of the detection object is determined by the linear size depending on the type of polarization of the backlight signal. With horizontal polarization, the wavelength is approximately twice the horizontal dimensions (fuselage length and wingspan) $\lambda_0 \approx 2l_{x(y)}$. With vertical polarization, the resonant wavelength is approximately four times the vertical dimension (height) $\lambda_0 \approx 4l_z$. This makes it possible to estimate the resonant frequency.

It was experimentally found on models [5] that the RCS of an aircraft with horizontal polarization in the resonant region is determined by its length and wing span as $\sigma_H \approx 0,86\lambda_0^2 = 3,44l^2$. On vertical polarization, resonance occurs when the height of the aircraft is about a quarter of the wavelength of the irradiating field. At the same time, its EPR is insignificantly (by about 4...5 dB) lower than the EPR of a half-wave resonant dipole. Taking these considerations into account, the formulas for estimating the RCS of aircraft at horizontal and vertical polarizations in the resonance region will have the form:

$$\sigma_H \approx 3,44l^2, \quad \sigma_V \approx 4,44h^2, \tag{3}$$

where σ_H, σ_V are the RCS at horizontal and vertical polarizations; l is the aircraft length or span; h is the aircraft height.

Estimates of RCS and resonance frequencies for a number of aircraft using relation (3) show that on horizontal polarization RCS $(1...8)\cdot10^3$ m^2 is significantly larger than on vertical $(2...7)\cdot10^2$ m^2 by about 8...12 dB. The resonant frequency range for horizontal polarization is approximately 3...8 MHz and 6...12 MHz for vertical polarization. RCS for large aircraft is thousands of square meters for horizontal polarizations and hundreds of square meters for vertical polarizations.

Currently, special attention is paid to the creation of stealth aircraft and UAVs using stealth technologies. They suggest a set of methods for reducing the signature of combat vehicles in radar, infrared and other areas of the detection spectrum by means of specially developed geometric shapes and the use of radio-absorbing materials and coatings, which significantly reduces their observed RCS and, accordingly, the detection range. These approaches make it possible to significantly reduce the RCS by several orders of magnitude. So, at present, for fighter aircraft of the 5th generation, the achieved RCS values are 0.1...0.3 m^2, i.e. about 10...15 dB lower than that of 4th generation aircraft. Declared decrease in the observed RCS to 0.01...0.0001 m^2. These indicators are planned to be obtained in the microwave range. At the same time, the approaches proposed by this technology may turn out to be ineffective in the resonant region, when the entire structure of the aircraft reflects, as a system of crossed resonant dipoles [5].

The results of the RCS estimates of the 5-th generation stealth aircraft and their resonant frequencies are shown in Table 2.

Table 2. Typical dimensions of the frequency and RCS of aircraft in the resonant region.

Aircraft type	Appointment	Length aircraft, m	Swing wing, m	Height aircraft, m	Horizontal polarization		Vertical polarization	
					RCS/1000 m^2	Frequencies MHz	RCS/1000 m^2	Frequencies MHz
C-17A	T	53	51	16.79	9.7–9	2.8–2.9	1.25	4.5
«Mirage»	F	15.3	9.3	4.5	0.77–0.2	10–17.9	0.09	16.7
«Corsair2»	A	14.1	11.8	4.9	0.7–0.5	10.7–18	0.1	15.3
Be-200	T	32	32.7	8.9	3.7–3.5	4.6–4.7	0.35	8.4
An-24	P	23.5	29.2	8.3	1.9–2.9	6.4–5.1	0.3–0.33	9–8.7
An-26	T	23.8		8.6				
An-22	T	57.3	64.4	13.5	12.–13.5	2.6–2.3	0.81	5.6
An-74	T	28	31.9	8.7	2.7–3.5	5.4–4.7	0.34	8.6
Il-76	T	46.6	50.5	14.8	7.5–8.8	3.2–3	0.97	5.1
Tu160	SB	54.1	35.–55.7	13.1	4.4–10.7	4.2–2.7	0.76	5.7
Tu22M3	SB	41.46	23.–34.3	11	1.9–5.9	6.4–3.6	0.54	6.8
B-52	SB	49.05	56.39	12.4	8.3–10.9	3.1–2.7	0.68	6
Mig-29	F	17.32	11.36	4.73	0.4–1	13.2–8.7	0.099	15.9
Su-57	F	19.7	14/10.8	4.8	0.67–1.3	2.6–2.3	0.102	5.6
B2	SB	20.9	52.12	5.1	1.72–0.87	3.8–5.4	0.10	15.6
F-22	F	18.9	13.56	5.09	1.94–12.1	3.6–1.4	0.12	14.7
F-35	F	15.57	10.67	4.38	1.59–0.82	4.0–5.5	0.12	14.7
J-31	F	16.9	11.5	4.8	1.08–0.51	4.8–7.0	0.09	17.1
J-21	F	10.5	11.6	4.0	0.48–0.6	7.2–6.5	0.07	18.9

Notes: T – transport; SB – strategic bomber; A – attack aircraft; F – fighter

As can be seen, the resonant frequencies lie in the HF band, the frequencies of which are proposed to be used for radar tasks. In this case, the RCS values for horizontal polarization are one order of magnitude higher than for vertical polarization. This is due to the difference in the respective linear dimensions of the aircraft. In this case, the resonance values of the RCS on horizontal polarization, when the aircraft body resonates and its wing can reach in such large aircraft, like An-22 and B-52, the values from 10 to 15 thousand square meters, and on vertical polarization up to 1000 m^2. For aircraft of the 5th generation, made using stealth technologies, the RCS at horizontal polarization can exceed 1000 m^2, and for vertical polarization 100 m^2, i.e. be 4...5 orders of magnitude greater than the values of their RCS in the microwave range, for which the observed value of RCS was minimized, by applying the approaches of Stealth technology. This means that in the resonance region the use of various kinds of coatings, smoothing of corner shapes does not give an effect, since the thickness of the coatings to create the effect must be commensurate with the wavelength of the irradiating field, which is practically impossible for decameter waves (10…100 m).

The results of RCS estimates for some types of aircraft type UAVs, their resonant frequencies are shown in Table 3. From the analysis of the data presented, it can be seen that the RCS of the US strategic reconnaissance UAV RQ-4 Global Hawk can reach several thousand square meters and exceed its RCS in the microwave range by several orders of magnitude.

4   Determination of Detection Range

When used as a backlight signal, the signals of communication stations of the short-wave band, the signal scattered from the object has a structure similar to the structure of the backlight signal. The expression for estimating the detection range in bistatic radars using the signal of short-wave communication stations as backlight has the form [3, 4]:

$$R_{TR} = \sqrt{\frac{\mu_{SN}}{\mu} \frac{\sigma_T}{(4\pi)} \frac{F^2(\theta_T)}{F^2(\theta_{Tr})}}, \tag{4}$$

where σ_T is the target RCS; $F^2(\theta_T), F^2(\theta_{Tr})$ are the value of the radiation pattern of the receiving antenna in terms of power when directed to the target and to the transmitter.

When assessing the detection range of the radar, the approximate depth of the dip of the antenna directional pattern in the direction to the illumination station was chosen 30…40 dB, and the SNR required to detection is $\mu = 13$ dB. A dip in the direction to the backlight station can be provided by implementing two antenna interferometers.

For the selected backlight stations, the ratio of the carrier spectral line level to the noise spectral density in the Doppler frequency (velocity) range where the target is detected was estimated.

The velocities of the objects of observation are in the range of 150…300 mps. Estimated Doppler Frequency Offset Range:

$$f_{D_{max}} = \pm(1 \div 2) \cdot f_0', \tag{5}$$

where f_0 is the carrier frequency in MHz.

Table 3. Typical dimensions of the frequency and RCS of the UAV in the resonant region.

UAV aircraft type	Purpose	UAV length, m	UAV height, m	Wing-span, m	Horizontal polarization				Vertical polarization	
					RCS Wing, dB/m^2	RCS Hull, dB/m^2	Resonant frequency. MHz		RCS Hull, dB/m^2	Resonant frequency of Hull, MHz
							Hull	Wing, MHz		
1	2	3	4	5	6	7	8	9	10	11
Russia										
Dozor-85	R	2.6	1.2	4.6	18.6	13.7	57.69	32.6	8.1	62.5
Stingray	RC	10.3	2.7	11.50	26.6	25.6	14.63	13.04	15.1	27.8
Israel										
Hermes 180	MP R	4.43	1.8	6	20.9	18.29	33.86	25	10.47	41.67
IAI Heron	R	8.5	2.3	16.6	29.8	23.95	17.65	9.04	12.60	32.61
Skylark		1.83	–	2.74	14.1	10.61	81.97	54.74	–	–
USA										
Predator	MP	8.23	2.21	14.84	28.8	23.67	18.23	10.11	12.25	33.94
Global Hawk	SR	13.5	4.63	35.42	36.35	28.0	11.08	4.23	19.79	16.2
X-47 Pegasus	C	5.95	1.86	5.94	20.84	20.86	25.21	25.25	11.86	40.32
Reaper	RC	11		20	31.39	26.19	13.64	7.5		

SR – strategic reconnaissance, C – combat, MP – multipurpose, R – reconnaissance, RC – reconnaissance combat.

The results of an experimental study of the noise level of some backlight stations with detuning from the carrier by ±10 Hz are given in Table 4.

Signal to noise ratio estimation μ_{SN} of backlight stations were carried out experimentally as follows. For the selected broadcasting stations, which were subsequently used as backlight stations, the signals were recorded in the telephone (AM) and telegraph (CW) modes – Fig. 2.

Table 4. Carrier width and signal-to-noise ratio at different detuning relative to it.

Frequency, MHz	Polarization	Width of carrier, Hz	Signal-to-noise, dB when detuned	
			−10 Hz	+10 Hz
15130	VP	0.56	61.08	57.76
15130	HP	0.5	54.16	52.94
15130	VP	1.07	45.4	44.04
15130	HP	0.81	47.06	49.5
9490	HP	0.34	60.36	60.36
9490	VP	0.28	57.85	58.15
13904	VP	0.77	50.07	48.16
13904	VP	1.17	44.31	44.3
5909	VP	0.34	51.96	52.89
5909	HP	0.42	47.89	49.5

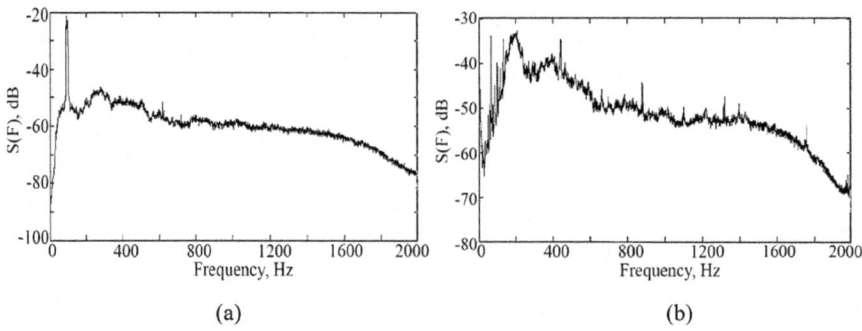

Fig. 2. Signal spectra of broadcasting stations: (a) – in telegraph mode (CW), (b) – in telephone mode (AM).

To determine the signal-to-noise ratio μ_{SN} first, the difference between the carrier level and the information signal level is determined μ_{CW}. When analyzing signals received in the telephone mode, the difference between the level at the Doppler frequencies and the level of the information signal is determined μ_{AM}. The final signal-to-noise ratio is defined as $\mu_{CW} - \mu_{AM}$.

$$\mu_{SN} = \mu_{CW} - \mu_{AM}. \tag{6}$$

Using the experimentally obtained data on the spectral densities of the signal in the telephone and telegraph modes, the actual signal-to-noise ratios for various backlight stations were estimated – Table 5. It should be noted that given in Table 5 the values of the signal-to-noise ratios were obtained when the backlight signal was received at a Hertz dipole with a gain of about 1. If you use a multi-element antenna with a gain G_a then the signal-to-noise ratio will be greater:

$$\hat{\mu}_{SN} = G_a \mu_{SN}. \tag{7}$$

Table 5. Signal to noise ratio for various backlight stations.

Backlight station	Frequency, kHz	μ_{CW}, dB	μ_{AM} dB	μ_{SN}, dB
Shijiazhuang	17735	35	−21	56
Kashi	17490	28	−16	44
Issoudun	17735	26	−17	43
Emirler	13635	26	−11	37
Sirjan	15150	32	−22	54
Islamabad	11645	26	−20	46

The expression for assessing the range of active-passive radars using signals from broadcasting stations for illumination, taking into account (8), will take the form:

$$R_{TR} = \sqrt{\frac{\mu_{SN}}{\mu} G_a \frac{\sigma_T}{(4\pi)} \frac{F^2(\theta_T)}{F^2(\theta_{Tr})}} \tag{8}$$

The results of estimating the detection range (5) for antennas with different gain and depth of suppression of the backlight signal on the target RCS are shown in Fig. 3. The signal-to-noise ratio required for detection was taken $\mu = 13$ dB. Carrier to noise ratio values in the detection band μ_{SN} for various backlight stations are taken from Table 5.

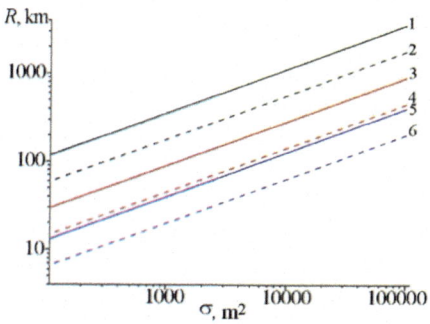

Fig. 3. Detection range on RCS of air targets and antenna gain G and signal-to-noise ratio μ_{SN}: 1, 3, 5 – $G = 9$ dB; 2, 4, 6 – $G = 3$ dB; 1, 2 – $\mu_{SN} = 56$ dB; 3, 4 – $\mu_{SN} = 44$ dB; 5, 6 – $\mu_{SN} = 37$ dB

Using the RCS data of the 5th generation aircraft in the resonance region and the results of estimates of the detection range of active-passive systems in the decameter range, it can be seen that in the resonant region, the detection ranges of aircraft manufactured using the stealth technology ("invisible") can reach hundreds of kilometers when illuminated with a sky wave of HF broadcasting radio stations.

5   Experiments on Aircraft Detection

Experiments have been carried out to assess the possibility of detecting airborne objects using the radiation of HF broadcast stations as an backlight signal. The objects of detection were aircraft of the An-74 and An-140 types.

A short-wave communication station broadcasting at a frequency of 15130 kHz was used as a backlight signal. The reception center was located at the O. Ya. Usikov Institute for Radiophysics and Electronics. The flight line of the aircraft, after takeoff from the airfield, crossed the base of the measuring complex at an angle of $\sim 40°$ and passed near the receiving point. The distance from the take-off/ landing point to the base line was ~ 1.5 km. In Fig. 4. shows the spectra of the signals of the carrier and the lines of the AN-74 aircraft body shifted to the Doppler frequency relative to it, obtained by us using the "Liberty" radio station (azimuth about 75°) as a backlight signal [6]. The polarization of the reception was changed by rotating the receiving antenna through an angle of 90°. A dipole with an effective length of about 40 cm was used as a receiving antenna. During the experiment, the signal scattered from the aircraft approaching the receiver was recorded.

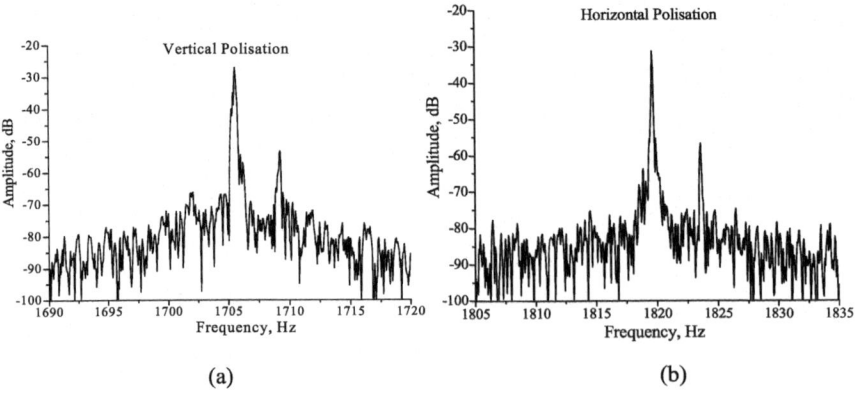

Fig. 4. Spectra of signals of reflections from aircraft at horizontal (a) and vertical (b) reception polarizations when used for illumination of a broadcasting station "Radio Liberty".

It can be seen that the spectral line of the carrier signal reflected at the Doppler frequency by the aircraft is more than 20 dB higher than the noise level in the Doppler frequency band relative to the carrier. And these ratios are sufficient for aircraft detection. At the same time, the carrier level, as can be seen, is also more than 20 dB higher than the level of scatterings from the aircraft. This means that in order to increase the detection range of aircraft when using broadcasting stations for back-lighting, it is necessary to reduce the level of the backlight signal at the receiver input by forming a zero of the receiving antenna directional pattern towards it.

Table 6 the numerical values of the characteristic for several experiments are given: the signal-to-noise ratio at various Doppler frequency offsets from the carrier, as well as the carrier levels in relation to the spectral line of the signal scattered from the aircraft and its width at various reception polarizations. Aircraft detection ranges in experiments were up to several kilometers. Using relation (5) and the experimentally obtained relationships between the carrier level and noise in the Doppler frequency band, as well as the realized detection ranges, it is possible to estimate the value of the observed aircraft RCS in experiments at about several thousand square meters – Table 6. This is consistent with the results of theoretical estimates. For the An-74 aircraft detected in the experiments, the calculated values of the RCS in the resonant region have values for the horizontal polarization of 2700–3500 m^2 for the vertical polarization, 340 m^2.

Table 6. Characteristics of scatterings from aircraft hull lines.

Frequency, MHz	Polarization	Hull line		Signal to noise ratio μ_{SN}, dB carrier offset at		Signal-to-noise ratio, dB	RCS, m^2
		Frequency, Hz	Width, Hz	10 Hz	3 Hz		
15,130	VP	4.22	1.12	38.8	24.8	−22.3	2153
15,130	HP	4.03	0.68	32.2	20.4	−22.0	2010
15,130	VP	3.58	1.05	34.7	23.6	−26.0	5048
15,130	HP	3.58	0.9	32.3	21.7	−26.1	5165

Moreover, the experiments did not re1veal a significant difference in the detection of a signal at the vertical and horizontal polarizations. This agrees with the results of model experiments on studying the scattering matrices of bodies of complex shapes.

6   Conclusion

The spectrum of the information signal of broadcasting stations is concentrated relative to the carrier at frequencies separated by more than 60 Hz. In the frequency range up to ±60 Hz with respect to the carrier, the noise spectral density level is approximately 30…40 dB below the carrier level. For aircraft with a cruising speed of 600–2000 km/h. in the decameter wavelength range of 60…20 m (frequencies 5…15 MHz), the range of Doppler frequency shifts is 3…60 Hz. It lies in the region between the carrier and the information signal of the broadcast station. This means that the frequency interval between the carrier and the information frequencies of the transmitted message spectrum can be used to detect moving air objects.

For many types of aircraft, their dimensions are resonant in the frequency range of the decameter range, which means that scattering from them will occur in the resonant region with a significant increase in the observed RCS. Even for the latest 5th generation aircraft, made using stealth technology, in contrast to the microwave band, where their RCS can be hundredths of a square meter, in this band their RCS will be hundreds of square meters, which will facilitate the conditions for their detection. The use of various types of coatings will be ineffective for reducing the RCS of aircraft and UAVs in the resonant frequency range. The most promising for these purposes is the widespread use in the construction of composite materials and carbon nanotubes.

For backlighting, it is necessary to choose broadcasting stations for which the ratio of the carrier level to the spectral noise density is maximum in the frequency range relative to it up to 50 Hz.

The scattered signal will have spectral components shifted to the Doppler frequency, not only associated with the carrier frequency, but also with the transmitted information signal. This can be used to determine the coordinates of an object, for which it is necessary to use the correlation principles of signal processing.

Acknowledgment. This work was supported by the Ministry of Industry and Trade of Vietnam.

References

1. Sedyshev, Y.N., Sedyshev, P.Y., Tyutyunnik, V.A.: Bistatic noise radars with coherent spatio-temporal processing of echoes and active interference. Appl. Radioelectron. **1**(2), 189–194 (2002). (in Russian)
2. Lobochko, S.E.: Construction of a detection system using VHF radiation and TV transmitters. In: Proceedings of the International Scientific Conference "Radiation and Scattering of Electromagnetic Waves", pp. 287–290. Taganrog (2003). (in Russian)
3. Lutsenko, V.I., Lutsenko, I.V., Popov, I.V.: Illumination of the air environment using radiation of HF broadcast stations. Radiophys. Quantum Electron. **58**(1), 9–18 (2015). https://doi.org/10.1007/s11141-015-9576-3
4. Bokov, A.: The problem of detecting stealth aircraft. Foreign Mil. Rev. **7**, 37–42 (1989). (in Russian)
5. Lutsenko, V.I., Tolstel, S.Y.: Frequency dependences of scattering matrices in the resonance domain. Telecommun. Radio Eng. **55**(4), (2001). https://doi.org/10.1615/TelecomRadEng.v55.i4.30
6. Sedyshev, Y.N., Dudush, A.S.: Evaluation of the impact of the time synchronization accuracy of multistatic radar positions on errors in determining the spatial coordinates of aerial objects. Radioelectron. Commun. Syst. **56**(4), 178–185 (2013). https://doi.org/10.3103/S07352 7271304002X
7. Syedyshev, Y.N., Tyutyunnik, V.A.: Information technologies of creating modems of multiposition active and passive radar systems. Appl. Radio Electron. **14**(1), 105–110 (2015). (in Russian)
8. Lutsenko, V.I., Lutsenko, I.V., Sobolyak, A.V., et al.: Interference to active-passive radar systems created by emissions from HF and VHF broadcasting stations. Telecommun. Radio Eng. **79**(10), 829–845 (2020). https://doi.org/10.1615/TelecomRadEng.v79.i10.10
9. Hu, P., Bao, Q., Chen, Z.: Weak target detection method of passive bistatic radar based on probability histogram. Math. Probl. Eng. **2018**, 8243686 (2018). https://doi.org/10.1155/2018/8243686

3D-Modeling of the Dynamics of Real Processes of Different Nature

Valeriy Mygal$^{(\boxtimes)}$ ⬥, Ihor Klymenko ⬥, and Galyna Mygal ⬥

National Aerospace University "Kharkiv Aviation Institute",
17 Chkalova Street, Kharkiv 61070, Ukraine
{valeriy.mygal, igor.klymenko, g.mygal}@khai.edu

Abstract. It is difficult to make decisions based on the large number of temporal and spatial signals received from physical, biological, chemical and other sensors. These signals differ in nature, dimension, complexity of form, etc., and the decision-making criteria are very diverse. The influence on the sensors reveals hidden features of functional characteristics. They create in information flows spatio-temporal inhomogeneities, the hidden nature of which is technologically or biologically inherited. Artifacts are observed in the sensor response-signals under extreme conditions, which indicate the fractal nature of the signals. Technological or biological heredity is inherent in all fractal objects and increases the risk of making erroneous decisions. The random nature of individuality finding required the development of a convergent approach based on 3D-modeling. For this, the means of processing, visualization and analysis of the dynamics of information flows are unified. They reveal the dynamic, energetic and informational characteristics of processes of different nature. This article describes the development of tools for modeling the dynamics of real processes of various natures that occur at different scale levels. This is relevant for solving the problems of the viability of the functioning of complex dynamic systems in abnormal conditions.

Keywords: Functioning models · Sygnayures · Patterns · Space-time signatures · Interdisciplinary communications · Space of dynamic events

1 Introduction

The Problem of the Complexity of Control and Modeling. Control of complex dynamic systems requires decision making based on a large number of temporal and spatial signals received from physical, biological, chemical and other sensors, controllers. These signals differ in physical content, dimension, complexity of shape, scale, etc., and the criteria for making decisions based on these signals are often very diverse [1]. When various stress factors act on sensors as sources of information, their responses begin to show individual features of functional characteristics, the nature of which is associated with technological or biological heredity. They indicate the presence of spatio-temporal inhomogeneities in information flows. Their occurrence is difficult to predict and such inhomogeneities cannot be modeled using traditional methods. In addition, it is under extreme conditions in the signals-responses of sensors

M. Nechyporuk et al. (Eds.): ICTM 2020, LNNS 188, pp. 633–643, 2021.
https://doi.org/10.1007/978-3-030-66717-7_54

that artifacts are often observed that are poorly reproduced and determine the fractal nature of the signals. As a result, a number of problems arise with the temporal stability of the characteristics of sensors in extreme conditions, which complicates control and increases the risk of making erroneous decisions [2].

Traditionally, in complex systems where control errors can have fatal consequences, attempts are made to solve these problems by selecting stress-resistant sensors and duplicating them. This in turn is accompanied by the complication of information processing algorithms in the system control functionality. However, such duplication increases the information processing time, which reduces its relevance at the time of making a decision. Therefore, difficulties arise in modeling and predictive analytics. After all, sensors of different types are traditionally described by different models that do not agree well with each other (different types of models with different ideologies, various differential equations, which are often nonlinear and do not have unambiguous solutions, etc.). As a result, the duplication of information sources without changing approaches to its processing, analysis and without unifying the applied models complicates the control function so much that the uncertainty of the system's behavior under extreme conditions only grows. On the other hand, heredity (technological or biological) is inherent in all self-organized objects and the random nature of the manifestation of individuality related to heredity requires the development of new universal models based on a single methodology. Within its framework it is necessary to unify the means of processing, visualization and analysis of the dynamics of information flows of different nature in the structure of which the features of physical, biological, technological and information processes are hidden.

Viability of Dynamic Systems in Difficult Conditions. The inevitable increase in the number of sources of relevant information in complex dynamic systems leads to the complexity of the perception of the spatio-temporal inconsistency of information flows. These cognitive problems are indirectly related to the law of G. Miller: the average person can keep seven plus or minus two different objects in the operative memory [3]. The min/max ratio is $(7-2)/(7+2) = 5/9$ which in our opinion reflects the probability of human error when exposed to stress factors of the environment and activity. Consequently the latent individuality of the perception of information sources and cognitive distortions must be identified at the stage of selecting information sources and operators.

A variety of modeling methods reduces the reliability of information analysis and, consequently, the efficiency of human-machine interaction. In the emergency conditions of functioning of complex dynamic systems the individuality of all sources of information (sensors, bio-sensors, etc.) is discovered. However there are difficulties in identifying and assessing the degree of individuality of different sources of information [2]. It is also difficult to classify semiconductor sensors which are resistant to force actions, radiations, fields and other actions. But the effectiveness of this selection depends on the safety, stability and reliability of the functioning of a complex dynamic system. Human-machine interaction problems are also associated with simplified mathematical models, which are given by differential equations and are effective only in natural sciences.

The Problem of Increasing Complexity of Human-Machine Interaction. The use of well-known computer graphics methods for the analysis of a large volume of poorly formalized data does not facilitate effective human-machine interaction. The key problem is obtaining graphic images that do not evoke direct subject associations in the decision-maker [4]. Therefore for holistic perception it is necessary to unify the means of cognitive graphics, which should exclude ambiguity in the interpretation of the received information and stimulate critical thinking.

Considering different signals of sensors and biosensors as natural information flows one can analyze their dynamics using the same means. Then the latent structure of different signals can be analyzed in one space. After all the processes of organization (structure) of control in objects of animate and inanimate nature are similar [5]. In addition most stressors are informational attacks. In this regard the aim of the work is to develop a convergent approach to modeling the dynamics of real processes of different nature occurring at different scale levels.

2   Results and Discussion

2.1   Visualization of Spatio-Temporal Relationships of a Dynamic Process

An Interdisciplinary Approach to Modeling by Geometrization. The individuality of signals from sensors, detectors, spectrometers and other sources of information are found in the complexity of the form, as well as in locally concentrated features. These are distortions (fluctuations, inhomogeneities, instabilities, etc.), which are internal micro-sources of information. Their diversity gives rise to spatially heterogeneous information flows. Hidden relationships of these flows affect the efficiency of human-machine interaction. The individuality of the space-time structure of signals of different nature is most found in parametric geometrization [6–12].

The visualization of interaction is based on the extreme principles of dynamics that are used in various sciences (continuum mechanics, thermodynamics, electrodynamics, field theory, synergetics, etc.). So, from the complementarity of the principles of least action (Hamilton, Lagrange, Jacobi, etc.) follows the differential principle of least curvature or distance of Hertz [13]. And the Gauss principle of least compulsion is associated with the symmetry of physical phenomena and fundamental conservation laws (theorem of E. Noether) [14]. The consequence of the LeChatelier - Brown thermodynamic principle is the dynamic similarity of the reaction processes that occur during the self-organization of objects of different nature. This can explain the wide-spread use of extreme principles of dynamics in synergetics, nonequilibrium thermo-dynamics, etc.

3D-Trajectory of Dynamic Events as a Topological 3D-Model. Extreme principles allowed us to move from a deterministic description of the dynamics of information flows to probabilistic methods. Geometrization of the Jacobi principle of least action in the form of a geodesic curve made it possible to reconstruct a topological 3D-model based on the measured signal. The consistent application of extreme principles to small changes in the dynamic state $X(t)$, as well as to the velocity of its change $dX(t)/dt$ and acceleration $d^2X(t)/dt^2$ transforms the scalar time series $X(t)$ (signal, response, sensor or biosensor characteristic) into the trajectory of causally related dynamic events [6–9]. Therefore, in space (state - velocity - acceleration) $X(t)$ is a 3D-trajectory, which can serve as a topological model of the dynamics of the time series (information flow) [10–12].

Orthogonal projections of a 3D-model are individual graphic images - signal signatures of the 1st and 2nd orders [6–10]. The configurations of these signatures are formed by geometrically ordered sections that differ in length, steepness, or curvature. In addition statistically these areas differ in the linear density of dynamic events and distribution in space. This is a manifestation of the multidimensionality of time, which determines the dynamic, energetic and informational features of the signal. The areas of configurations of signatures of the 1st and 2nd orders of different sources of information have dimensions of action, energy and power, which can be characterized by the power of a subset of microstates [6]. The similarity of geometric constituents of signature configurations simplifies the analysis of the functioning of different sources of information.

Advantages of a Topological 3D Signal Model. The relationship between the 3D-model and the configurations of the three signatures makes it possible to analyze the information sources of the elements of a complex dynamic system from complementary points of view. In particular this is evidenced by a comparative analysis of the signatures of the time and spectral photoresponse of semiconductor sensors (photodetectors, detectors, spectrometers) [11, 12].

In [15–18], using the example of signals from semiconductor and biological sensors, it is shown that the developed approach is effective precisely because it allows one to identify dynamic and statistical patterns of signals and analyze the relationship between them. So, parametric visualization of the dynamics of a signal not only makes it possible to reveal its fine structure, but also to use in analytics various types of entropy (Boltzmann, Kolmogorov, Shannon, etc.) [19]. As well as integrative indicators of the dynamic ordering of the components and their energy balance [6]. However, despite the similarity of 3D-models of various objects and their individual components, the problem of identifying patterns of processes of various nature and scale remains unresolved. Therefore, ***the practical task*** of this work was 3D-modeling of oscillatory and relaxation processes and the search for their patterns in 3D-models of processes in real objects.

2.2   The Dynamics Modeling of Real Processes of Different Nature, Occurring at Different Scale Levels

Processes at Different Scale Levels. The functioning of many natural and artificial systems is associated with periodic processes. Often there are several such processes and they proceed at different rates (relaxation processes in semiconductors, data transmission in radio communication systems, cardiocycles and other biological cycles, etc.) [20]. At the most fundamental level, modern physics presents us with the world of structures [21]. A large number of interrelated information sources are easier to analyze if their structures are similar (spectral analysis) [22]. Understanding this point of view assumes that information about the features of the interaction of fields of all inhomogeneities of the environment is hidden in small distortions of the structure of the information flow. Thus in the electromagnetic continuum mechanics, stationary and dynamic fields of different nature are interconnected [23]. At the same time, the components of complex processes often manifest themselves in different ways in the resulting signal: some of them may be "hidden" against the background of the dominant ones (see Fig. 1). In this regard, the identification in complex processes of components proceeding at different velocities has a practical interest. The figure shows a 3D-model of the superposition of two harmonic processes, the periods of which differ by an order of magnitude. It consists of a sequence of loop-like sections, each of which corresponds to one period of the high-frequency process. As can be seen from the figure, the scale of the model is determined by the fastest component. The slow component manifests itself as a factor causing a phase shift between successive oscillations of a higher frequency.

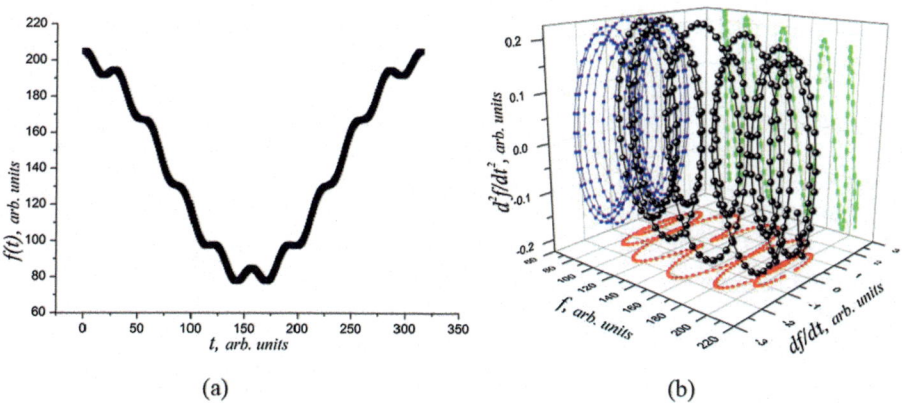

(a)   (b)

Fig. 1. Superposition of two harmonic signals (a) and their 3D model (b).

With an increase in the number of components in the resulting signal, its 3D-model and its projections spatially separate the processes proceeding with different velocities (Fig. 2b). The patterns of these processes are similar (they look like loop-like areas), but differ in the areas they cover in the projections, as well as in the curvature and steepness of the corresponding areas. In fact, the model illustrates the multidimensionality of time, which manifests itself in different ways in the dynamic, energetic and informational features of the structure of signals of different nature. It is these patterns that are observed in the previously obtained 3D-models of the transient photoresponse of ZnSe crystals and 3D-models of the photoresponse spectra of these crystals at opposite directions of the bias electric field [10]. It is important to note that these processes are almost imperceptible in the traditional form of signal presentation: they are either masked by processes of a different scale level, or look like weak fluctuation noise. Parametric 3D-modeling of dynamics allows "fast" processes to be identified, and also to highlight against the background of slower ones [24]. It is they that

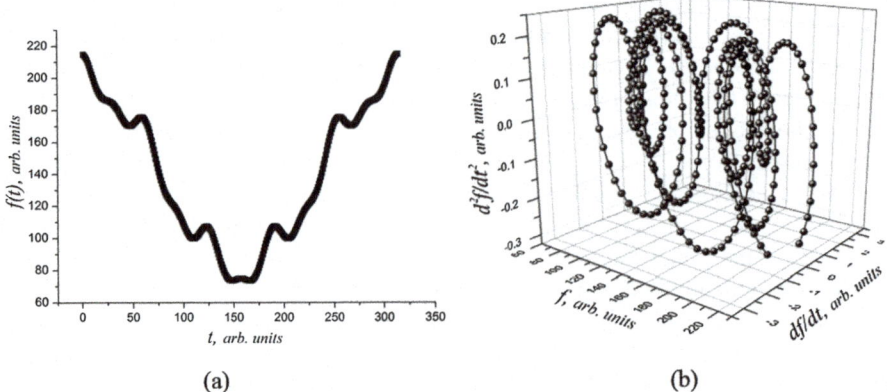

(a)   (b)

Fig. 2. Superposition of three harmonic signals (*a*) and their 3D-model (b).

determine the "fine" structure of the signal, which is determined by the individuality of its source and contains information about its inherited properties.

Modeling the System's Response to Stimuli. A separate problem of complex systems control in extreme conditions is the identification and analysis of the system's response to unsystematic rapid stimuli localized in time. In Fig. 3 shows a relaxation signal "distorted" by two local fast perturbations and its 3D-model. In it, the perturbation patterns are spatially distinguished against the background of the "slow" relaxation component of the signal. They determine the scale of the model and the individuality of the 3D-topology of the response dynamics. At the same time, their parameters (deviation range and covered area) depend on the curvature of the relaxation signal.

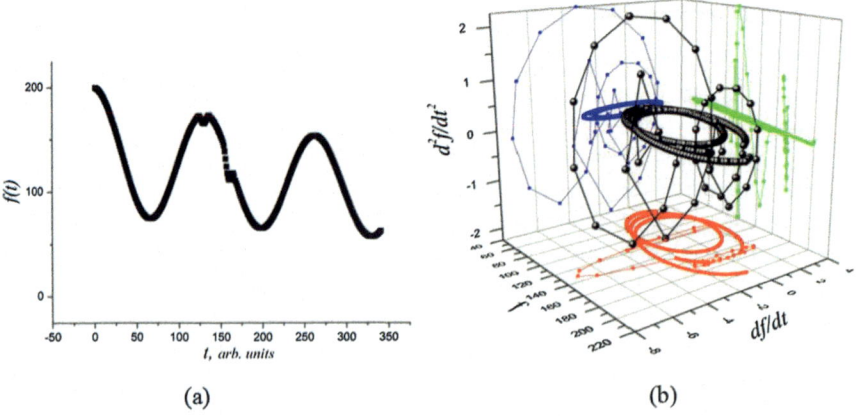

(a)   (b)

Fig. 3. Relaxation signal with local harmonic perturbations (a) and its 3D model (b).

Solar Activity Modeling. Similar patterns can be observed in the 3D-model of the solar cycle of the smoothed radio flux ($\lambda = 10.7$ cm) (Fig. 4), obtained on the basis of astronomical observations obtained from open sources [24, 25]. Interest in it is primarily due to the increasing influence of climatic factors on the functioning of critical cyberphysical systems and the difficulty of predicting this influence associated with the multifractality of solar activity [26]. It is characteristic that the patterns of fast "bursts" of the smoothed radio flux in the obtained 3D-model of the solar cycle have the form of loop-like areas. At the same time, not only patterns of signal components appear in the 3D-model, which are visible in the traditional form of its presentation, but also patterns of components masked by a slower cyclic process. This allows us to consider solar activity as a cyclical process, which includes a superposition of processes of different nature. They proceed at different velocities, and the configuration of the 3D-model in the space of dynamic events is determined primarily by the fastest processes.

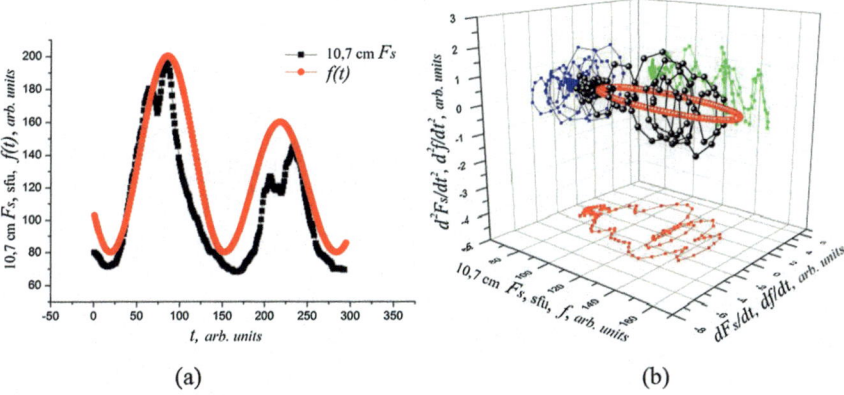

(a)   (b)

Fig. 4. Solar cycle of the smoothed radio flux ($\lambda = 10.7$ cm) (a) and its 3D model (b).

Multifractality of the Solar Cycle. Parametric 3D-modeling allows to take a fresh look at the dynamics of the development of many natural processes, which are characterized by a certain cyclicality in time or space. In particular, it is possible to compare the dynamics of the early stages of cyclical processes as well as interconnected processes and find hidden "trends" in them. E.g., a comparison of the dynamics of the observed and smoothed solar radio flux ($\lambda = 10.7$ cm) and the monthly mean Sunspot number and the monthly smoothed Sunspot number reveals similar patterns (Fig. 5). In this case, multifractality is found both in the time dependences of the observed values and in the series of averaged parameters.

Natural Visualization of Dynamics. The geometrization of processes has always played the role of a "generator of revolutionary ideas" in physics, biology and chemistry. The proposed approach is based on:

- spatio-temporal ordering of the signal structure [5], which underlies cybernetics and synergetics;

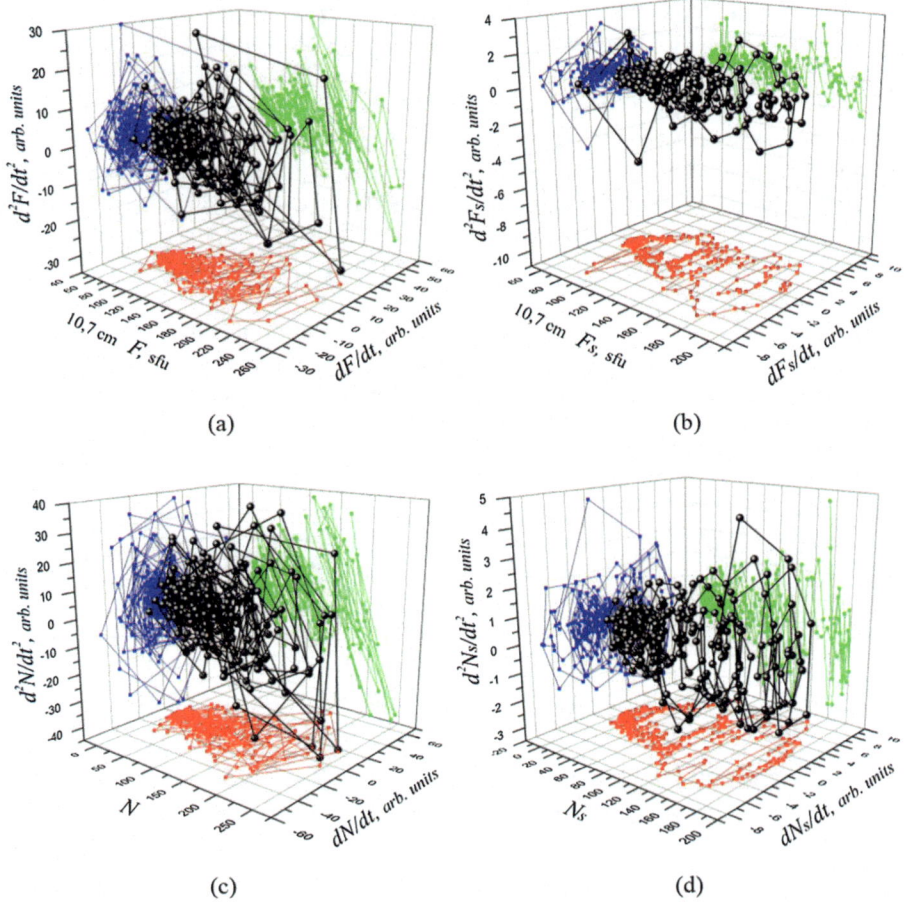

(a)   (b)

(c)   (d)

Fig. 5. 3D-models of solar cycles of the observed radio flux ($\lambda = 10.7$ cm) (a), the smoothed radio flux ($\lambda = 10.7$ cm) (b), the monthly mean Sunspot number (c) and the monthly smoothed Sunspot number (d).

- model reconstruction from a one-dimensional measured signal (Packard, 1980) [27], formalized by Takens [28];
- principles of system dynamics of complex systems (computer modeling of connections, feedback loops and response delays) [29, 30].

In conclusion, it should be noted that interdisciplinary approach we are developing allows visualizing the structure of various sources of information is in one universal space. This visualization provides a holistic perception of the spatio-temporal relationships of opposite phases of functioning and allows quantifying them. Revealing the connection between hidden dynamic and statistical patterns in real time is very important for predictive analytics. Other advantages of 3D-modeling of dynamics of fractal signals:

- 3D-model provides space-time separation of processes occurring at different velocities;
- the scale of the model is determined by the fastest processes, and in the traditional signal they are masked by the slowest processes;
- the individuality of the system is determined to the maximum extent by its "quick" reactions to external perturbations. The superposition of fast processes and their phase shifts determine the stochastic nature of the dynamics of the system.

In general, the application of the approach will reduce the problems of increasing complexity and uncertainty.

3 Conclusions

In non-equilibrium conditions of functioning of complex dynamic systems, the main problems are: a) the manifestation of the complexity and individuality of the functional characteristics of information sources (sensors, biosensors, etc.); b) ineffectiveness of methods for selecting sources of information; c) the manifestation of cognitive distortions in the system analysis of a large amount of poorly formalized data. To solve these and other problems, we are developing an interdisciplinary approach to modeling information sources of a different nature, by means of geometrizing their dynamics.

The approach is based on the transition from the complex dynamics of information sources and the variety of their graphic images (models) to source signatures and their cognitive patterns of perception. Signature configuration packages reveal the nature of the restructuring under the influence of external and internal stress factors. They contain information about the features of the functioning of the elements of complex dynamic systems and are cognitive patterns. As you can see, an interdisciplinary approach to cognitive visualization of information sources provides new opportunities for analysis, synthesis and comparison. These operations are the main components of analytical-synthetic thinking.

The creation of a knowledge base of signatures and typical patterns of self-organized information sources and their classification will allow identifying of the functional state of elements of complex dynamic systems in real time. Geometrization of antiphase processes in the space of dynamic events makes it possible to effectively use flexible logic of antonyms in the analysis. Cognitive visualization of the dynamic structure of information flows of different nature expands the possibilities of cognition and will help to increase the efficiency of human-machine interaction.

References

1. Fainzilberg, L.S.: Information Technologies for Processing Signals of Complex Form: Theory and Practice. Naukova Dumka, Kyiv (2008). [in Russian]
2. Mygal, G.V., Mygal, V.P.: Cognitive and ergonomic aspects human interactions with a computer. Radioelectron. Comput. Syst. 1, 90–102 (2020). https://doi.org/10.32620/reks. 2020.1.09, [in Ukrainian]

3. Miller, G.A.: The Psychology of Communication. Basic Books, New York (1967). https://doi.org/10.1002/hrm.3930060308
4. Mygal, V.P., Mygal, G.V., Balabanova, L.M.: Visualization of signal structure showing element functioning in complex dynamic systems – cognitive aspects. J. Nano- Electron. Phys. **11**(2), 02013 (2019). https://doi.org/10.21272/jnep.11(2).02013
5. Wiener, N.: Cybernetics, or Control and Communication in the Animal and Machine. Nauka, Moscow (1983). [in Russian]
6. Mygal, V.P., But, A.V., Mygal, G.V., Klimenko, I.A.: An interdisciplinary approach to study individuality in biological and physical systems functioning. Sci. Rep. **6**, 387–391 (2016). https://doi.org/10.1038/srep29512
7. Mygal, V., Mygal, G.: Problems of digitized information flow analysis: cognitive aspects. Inf. Secur. Int. J. **43**(2), 134–144 (2019). https://doi.org/10.11610/isij.4312
8. Mygal, V., Mygal, G.: Interdisciplinary approach to the human factor problem. Municipal Utilities **3**, 149–157 (2020). https://doi.org/10.33042/2522-1809-2020-3-156-149-157, [in Ukrainian]
9. Migal, V.P., Klymenko, I.A., Migal, G.V.: Individuality of photoresponse dynamics of semiconductor sensors. FuncT. Mater. **24**(2), 212–218 (2017). https://doi.org/10.15407/fm24.02.212
10. Mygal, V.P., Silevytch, V.Y., Mygal, G.V.: Integrated diagnostics of the functional state of the elements of transport systems based on visualization of the noise signal. Sci. Notes V. I. Vernadsky Taurida Natl. Univ. Ser. Tech. Sci. **31**(3), 102–108 (2020). https://doi.org/10.32838/TNU-2663-5941/2020.3-2/18, [in Ukrainian]
11. Mygal, V.P., But, A.V., Phomin, A.S., Klimenko, I.A.: Geometrization of the dynamic structure of the transient photoresponse from Zinc chalcogenides. Semiconductors **49**(5), 634–637 (2015). https://doi.org/10.1134/S1063782615050152
12. Mygal, V.P., Klymenko, I.A., Mygal, G.V.: Influence of radiation heat transfer dynamics on crystal growth. Funct. Mater. **25**(3), 574–580 (2018). https://doi.org/10.15407/fm25.03.574
13. Hertz, G.: Principles of Mechanics Set Out in a New Connection. USSR Academy of Sciences, Moscow (1959).[in Russian]
14. Noether, E.: Invariant variational problems. In: Variational Principles of Mechanics. Fizmatlit, Moscow (1959). [in Russian]
15. Mygal, G., Mygal, V.: The viability of dynamic systems in difficult conditions: cognitive aspects. In: 2020 IEEE 11th International Conference on Dependable Systems, Services and Technologies (DESSERT), pp. 224–229. IEEE, Kyiv (2020). https://doi.org/10.1109/DESSERT50317.2020.9125063
16. But, A.V., Migal, V.P., Fomin, A.S.: Structure of a time-variable photoresponse from semiconductor sensors. Tech. Phys. **57**, 575–577 (2012). https://doi.org/10.1134/S1063784212040044
17. Migal, V.P., But, A.V., Migal, G.V., Klymenko, I.A.: Hereditary functional individuality of semiconductor sensors. Funct. Mater. **22**(3), 387–391 (2015). https://doi.org/10.15407/fm22.03.387
18. Mygal, V., Mygal, G., Chukhray, A., Havrylenko, O.: Application of space-time patterns in tutoring. In: 16th International Conference on ICT in Research, Education and Industrial Applications (ICTERI). KHNU, Kharkiv (2020)
19. Vstovskij, G.V.: Elements of Information Physics. MGIU, Moscow (2002).[in Russian]
20. Gilbert, K.J.: Visualization in Science Education. Springer, Dordrecht (2006)
21. French, S.: The Structure of the World: Metaphysics and Representation. Oxford University Press, Oxford (2016)
22. Haken, G.: How the brain works. A macroscopic approach to brain activity, behavior and cognitive performance. PerSe, Moscow (2001). [in Russian]

23. Maugin, G.A., Achenbach, J.D.: Continuum Mechanics of Electromagnetic Solids. Elsevier Science, Amsterdam (1988). https://doi.org/10.1002/zamm.19890691106
24. LISIRD: LASP penticton solar radio flux at 10.7 cm: Time series. https://lasp.colorado.edu/lisird/data/penticton_radio_flux/, Accessed 30 Aug 2020
25. SpaceWeatherLive: Solar cycle progression. https://www.spaceweatherlive.com/en/solar-activity/solar-cycle, Accessed 30 Aug 2020
26. Afanasov, L.S., Maslovskaya, A.G.: Application of methods of time-frequency and multifractal analysis to study the dynamic characteristics of solar activity. Bull. AmSU **87**, 29–33 (2019). [in Russian]
27. Packard, N.H., Crutchfield, J.P., Farmer, J.D., Shaw, R.S.: Geometry from a time series. Phys. Rev. Lett. **45**(9), 712. https://doi.org/10.1103/PhysRevLett.45.712
28. Takens, F.: Distinguishing deterministic and random systems. In: Barenblatt, G.I. (ed.) Nonlinear Dynamics and Turbulence, pp. 314–333. Pitman, New York (1983)
29. Forrester, J.W.: System dynamics – a personal view of the first fifty years. Syst. Dyn. Rev. J. Syst. Dyn. Soc. **23**(2–3), 345–358 (2007). https://doi.org/10.1002/sdr.382
30. Schwaninger, M., Ríos, J.P.: System dynamics and cybernetics: a synergetic pair. Syst. Dyn. Rev. J. Syst. Dyn. Soc. **24**(2), 145–174 (2008). https://doi.org/10.1002/sdr.400

Stochastic Optimization Algorithms for Data Processing in Experimental Self-heating Process

Yuliia Viazovychenko$^{(\boxtimes)}$ ⓘ and Oleksiy Larin ⓘ

National Technical University "Kharkiv Polytechnic Institute",
2 Kyrpychova Street, Kharkiv 61002, Ukraine
viazovychenko.julia@gmail.com

Abstract. The search for parameters is carried out using stochastic optimization algorithms, namely, by the double annealing method. Self-heating simulation is based on finite element model of direct simulation of stationary heating of the specimen. The criterion for convergence of the optimization process is the data on the heating temperature obtained from field experiments. The verification of the reliability of the results is carried out by comparing the data on the area of the hysteresis loop obtained by numerical calculation and from a full-scale experiment. The paper presents the results of identification of the parameters of calculations of the thermal state, as well as the process of convergence of the search algorithm.

Keywords: Composites · Experiment · Self-heating · Loss modulus · Annealing

1 Introduction

Nowadays, composite and smart materials are widely used in various fields of engineering due to they can withstand the wide range of loads. Among of different types of composites, rubber-like composites reinforced with glass, metals, textiles, required on special attention. Such composites are found in aviation, automotive and power structural engineering. Often, they are used in the design of assemblies and parts operating in increased vibration loads conditions due to good damping properties.

Cyclic loading of rubber-cord composites is associated with dissipative processes arising inside the material, which leads to the self-heating phenomenon. A significant increase in the temperature of the material inevitably affects its properties, in particular, its ability to resist various types of loading, significantly influences to fatigue processes and stress-strain state formation and distribution [1].

All these factors require solving of the thermal state problem such composite materials in order to identify the pattern of it influence on material performance and its lifetime. Life-time of composite elements can significantly affect general lifetime of important mechanical engineering elements [2, 3].

The solving of this problem is based on the simulation of the thermal heating processes in composite materials under various types of loading and working conditions. Such modeling is inevitably based on the experimental data. Experiments are carried out on specimens for cyclic deformation model cases, which allow to determine the dependences of thermal behavior. Experimental data processing and implementation its results into the model is related with certain difficulties, because of the multi-parameter model complexity and experimental conditions inaccuracies. This is due to natural bias of material properties, the uncertainty in environment properties, as well as non-linearity of heat processes. In addition, these factors are interconnected and able to influence each other.

Thus, a peculiar difficulty is the veracious identification of each of the parameters separately.

Therefore, the processing of data obtained during in experiments in order to use it in the construction of models requires applying of optimization procedures.

2   Rationale Optimization Method Choosing

The using of regression optimization methods is not appropriate due to model non-linearity. Gradient methods are also inefficient due to the possible presence of local minimums and jumps (leaps) in the response surface. Moreover, the response surface is generating an implicit computational modeling. These leads to the need to use stochastic or evolutionary optimization methods.

So, to solve of this problem, one of the stochastic optimization methods was chosen, namely, the simulated annealing method (SA). It belongs to the metaheuristic global methods and is one of the most used, in particular, it is implemented in software packages for major programming languages. Since, this is an intermediate method between strong (gradient-based) and weak (random search) optimization methods. Besides, this method has sufficient convergence to the global optimum [4].

The only parts that must be provided by the user are an evaluation function, a procedure to generate a new trial state and so-called cooling scheme. In our case, the values of the objective function (solving of self-heating problem) are calculated outside the optimization process (using a third-party program), respectively, they are discrete quantities. This fact also testifies in favor of choosing this method.

3   Experiment Setting

Experiments were carried out on rubber-cord unidirectional composite specimen cyclic loaded along the textile cord reinforcement. The geometry and dimensions of the specimen (Fig. 1a) comply with the ISO 527-2-1B standard for mechanical testing of these materials. Tensile tests of the specimens were carried out under prolonged cyclic loading on the INSTRON ElectroPuls E3000 Test System experimental setup (Fig. 1b). These tests were held within 7th framework program INNOPIPES in laboratory of Riga Technical University [5].

The studies were carried out at various frequencies (5, 7.5, 10, 12.5, 15 Hz) and strain amplitudes (2, 4, 6%), with preliminary stretching up to 10%. Also, in order to minimize the Mullins effect, all specimens previously passed the 50 cycles loading stage at a frequency of 0.5 Hz and a strain value of 10%.

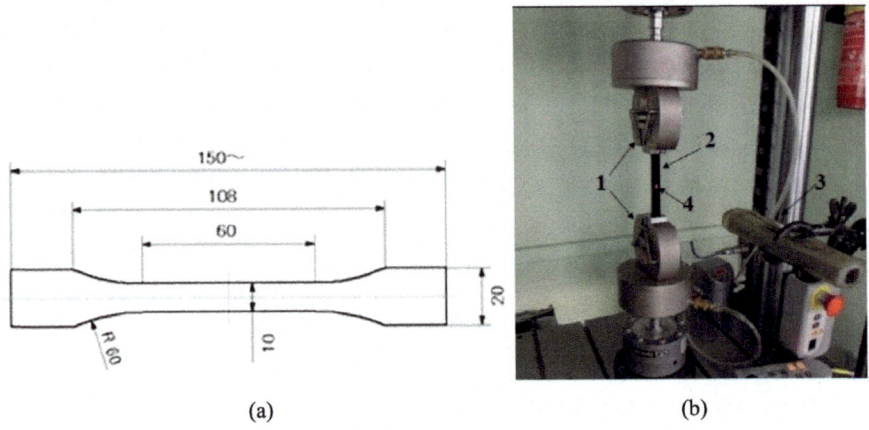

(a)   (b)

Fig. 1. General view of the specimen (a) and experimental installation INSTRON ElectroPuls E3000 Test System (b): 1 – clamps; 2 – specimen; 3 – laser thermometer, fixed on tripod; 4 – temperature measuring point.

The test results are obtained in the form of a deformation curve for each loading cycle (automatically recorded every 0.5 s). The temperature of specimen working surface was measured every 0.5 min using a SNR brand laser thermometer (Fig. 2).

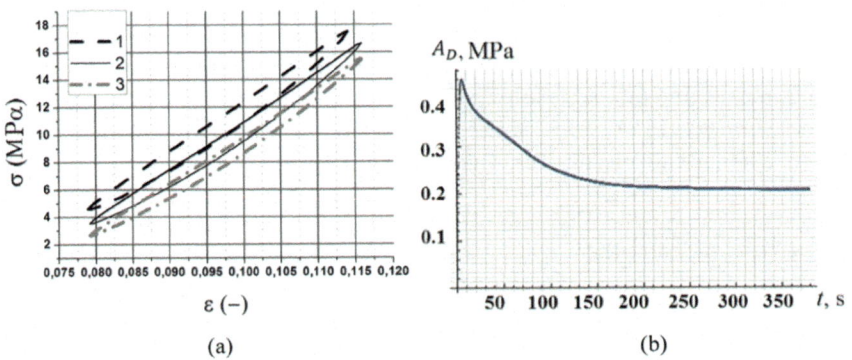

(a)   (b)

Fig. 2. Changing of the shape, slope (a) and area (b) of the hysteresis loop of the rubber-cord composite in time during cyclic deformation with $\hat{\varepsilon} = 2\%$ and $\omega = 7.5$ Hz: 1 – hysteresis loop on the 60th deformation cycle; 2 – on the 300th cycle; 3 – on the 3500 cycle.

According to the results of the tests, the deformation curves of the specimen under the conditions of cyclic loading formed hysteresis loops. Also, a changing of area and slope of the hysteresis loop was observed over time.

The observed effect occurs due to an increase in the length of the specimen (residual Malins effect) and its self-heating (experimentally confirmed). Identified features confirm the non-stationarity of the passing processes and the mutual influence of the parameters.

The time dependences of the heating of specimens at various loading frequencies ω and strain amplitudes ε̂ are shown in Fig. 3. We observe the result until the stabilization of temperature and the extinction of heating process.

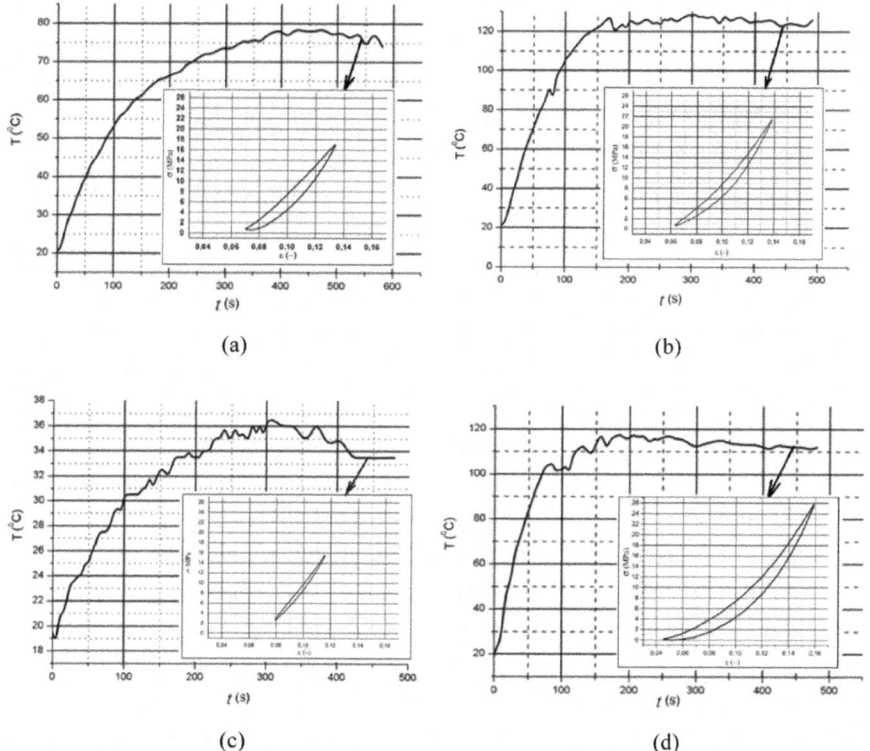

(a)   (b)

(c)   (d)

Fig. 3. Temporal dependences of specimen heating at ε̂ = 4% with ω = 5 Hz (a) and 12.5 (b) Hz; ω = 7.5 Hz with ε̂ = 2% (c) and 6% (d).

These graphs show that increasing of value of strain amplitude from 2% to 6% leads to increasing of temperature nearly by 4 times. In this case, an increase of the loading frequency from 5 to 12.5 Hz also leads to more intense heating (temperature values increase by more than 1.5 times). Moreover, in absolute terms, the heating reaches the value of temperature (>100 °C) which can significantly influence the properties and behavior of rubber-like materials [6–8]. So, the taking into account all of these features of heating during process of computational modeling of experiment is a separate rather difficult problem.

4 Processing of Experimental Data

4.1 Generalized Scheme

For optimization by the simulated annealing method, we need to generate a new state of the variable parameter and calculate the objective function for it at each iteration. In this case, the variable parameter is the values of the energy released during the loading cycle. So, data processing contained two main blocks of execution Dual annealing module which include ANSYS calculation as sub-module.

First module generates new state of variable parameters), makes a decision on the acceptance of the variable parameter's current value using the specified function, and also controls the convergence process using the cooling scheme. Second module calculates the objective function by solving thermal transient problem for each value of variable parameter and save it to be given to first module. See Fig. 4.

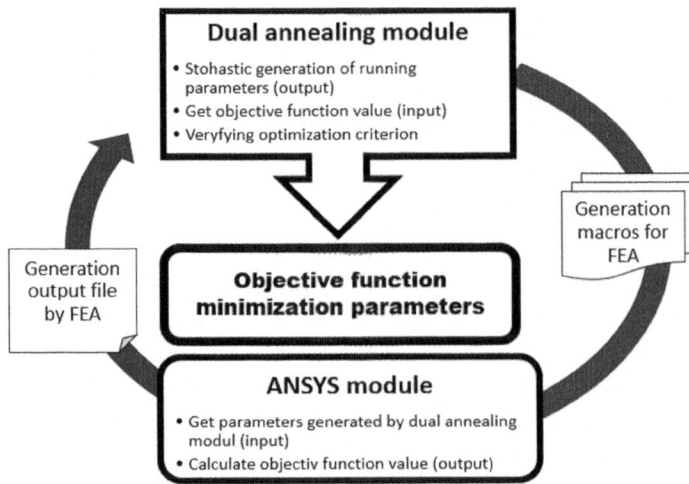

Fig. 4. Generalized scheme of experimental data processing.

4.2 Dual Annealing Optimization Method

For solving optimization problem was used scipy library including function implements the Dual Annealing optimization. This stochastic approach derived from [9] combines the generalization of CSA (Classical Simulated Annealing) and FSA (Fast Simulated Annealing) [10, 11] coupled to a strategy for applying a local search on accepted locations [12]. An alternative implementation of this same algorithm is described in [13] and benchmarks are presented in [14]. This approach introduces an advanced method to refine the solution found by the generalized annealing process.

This algorithm uses a distorted Cauchy-Lorentz visiting distribution, with its shape controlled by the parameter q_v

$$g_{q_v}(\Delta x(\tau)) \propto \frac{\left[\Theta_{q_v}(\tau)\right]^{\frac{-D}{3-q_v}}}{\left[1+(q_v-1)\frac{(\Delta x(\tau))^2}{\left[\Theta_{q_v}(\tau)\right]^{\frac{2}{3-q_v}}}\right]^{\frac{1}{q_v-1}+\frac{D-1}{2}}}, \tag{1}$$

where t is the artificial time. This visiting distribution is used to generate a trial jump distance $\Delta x(\tau)$ of variable $x(\tau)$ (changing in ranges presented in Table 1) under artificial temperature $T_{q_v}(\tau)$.

Table 1. Parameters ranges.

Parameters	Range
Material constant parameter	
Convection coefficient $h_c, \frac{W}{m^{20}C}$	10–30
Heat generation parameter	
Loss modulus G, MPa	0.1–100

From the starting point, after calling the visiting distribution function, the acceptance probability is computed as follows:

$$p_{q_v} = \min\left\{1, [1-(1-q_a)\beta\Delta E]^{\frac{1}{1-q_a}}\right\}, \tag{2}$$

where q_a is the acceptance parameter. For $q_a < 1$, zero acceptance probability is assigned to the cases where

$$1-(1-q_a)\beta\Delta E < 0 \tag{3}$$

The artificial temperature $\Theta_{q_v}(\tau)$ is decreased according to

$$\Theta_{q_v}(\tau) = \Theta_{q_v}(1)\frac{2^{q_v-1}-1}{(1+\tau)^{q_v-1}-1}, \tag{4}$$

where q_v is the visiting parameter.

4.3   Computational Simulation of Experiment (ANSYS Module)

The numerical simulation of the experiment was carried out based on the finite element method. The model of the test specimen is shown in the Fig. 5.

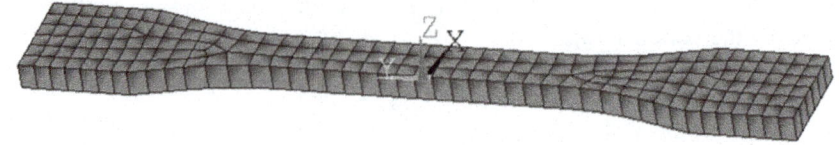

Fig. 5. FE-model of specimen.

Data about hysteretic loss was obtained from experiments and has used as boundary conditions for specimen in numerical simulations

$$[K]\{T\} = \{Q^a\}, \tag{5}$$

where $[K]$ is the conductivity matrix, $\{T\}$ the vector of nodal temperatures and $\{Q^a\}$ the applied heat flows.

The applied heat flows vector consists of three parts, in general:

$$\{Q^a\} = \{Q^c\} + \{Q^g\}, \tag{6}$$

where $\{Q^c\}$ is the element convection surface heat flow vector, $\{Q^g\}$ is the element heat generation load.

Where the components of heat generation vector defined as

$$\{Q^g\} = 2G_2\varepsilon^2\pi\omega. \tag{7}$$

On boundary B1: the convection surface heat flow ($\{Q^c\}$) and heat generation load ($\{Q^g\}$) was applied; B2: the convection surface heat flow ($\{Q^c\}$). Application of boundary conditions is shown on Fig. 6. The result of numerical simulation of experiment is represented on Fig. 7 as temperature distribution on specimen.

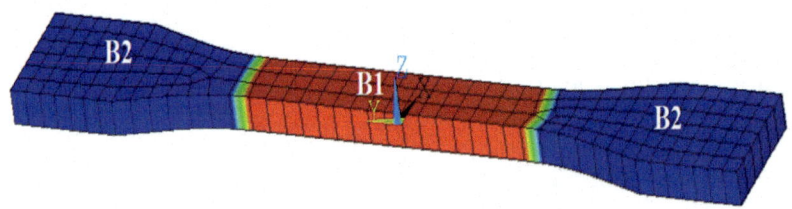

Fig. 6. Boundary conditions applied to the specimen: conditions of convective heat transfer and heat generation rate.

For connecting with python's dual annealing procedure the heat build-up calculation of specimen was realized by creating APDL-macros. It is consisted of several parts. The first one calls geometry model of specimen which was previously created. The next part builds the FE-model with mesh settings. The third one applies boundary

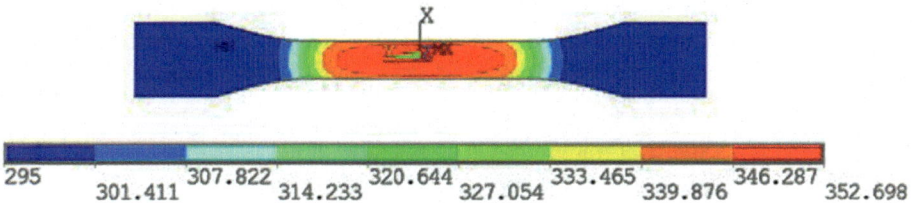

295   301.411   307.822   314.233   320.644   327.054   333.465   339.876   346.287   352.698

Fig. 7. Temperature distribution on specimen.

conditions, sets analysis settings and launch the solving. The last one saves the value of objective function. Also, we add material and environmental properties and heat generation function constants, which are generated by optimization procedure as variable parameters. Then all parts are joined to main macros. The generalized scheme is shown in Fig. 8.

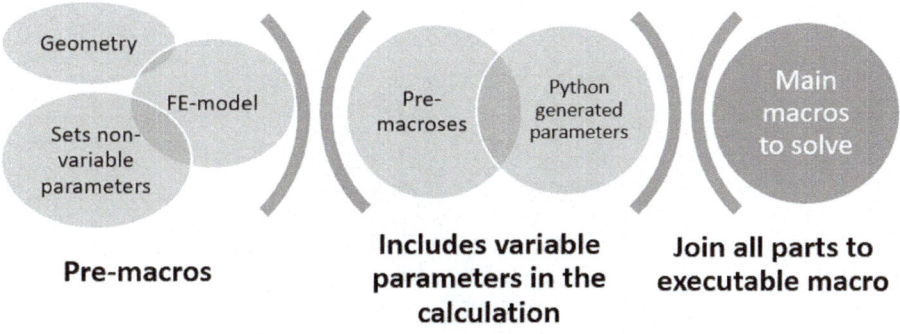

Pre-macros   **Includes variable parameters in the calculation**   **Join all parts to executable macro**

Fig. 8. Scheme of getting objective function value.

For the selected parameters, a sensitivity analysis was carried out in order to determine their degree of influence on the final heating temperature. It was found that all parameters from the selected set have a significant impact on the result. The first step was to determine the parameters by means of the above optimization procedure, independent of each other. The second stage involved taking into account the mutual influence of the parameters.

The 2-parameter annealing method was used to solve this problem. The parameters values have identified by SA optimization method are shown in Table 2.

Table 2. Result of optimization procedure.

Parameters	Value of parameters
Material constant parameter	
Convection coefficient $h_c, \frac{W}{m^2 {}^\circ C}$	25.7
Heat generation parameter	
Loss modulus G, MPa	16283644

The verification of results was carried out by comparing experimental data and numerical simulation about hysteretic loops areas. The results of verification procedure are represented in Table 3.

Table 3. Error estimation of optimization procedure.

Parameters	Strain amplitude	Error value %
Loss modulus G, MPa	2%	9
	4%	11

The converging of dual annealing optimization process for searching of loss modulus is shown on Fig. 9.

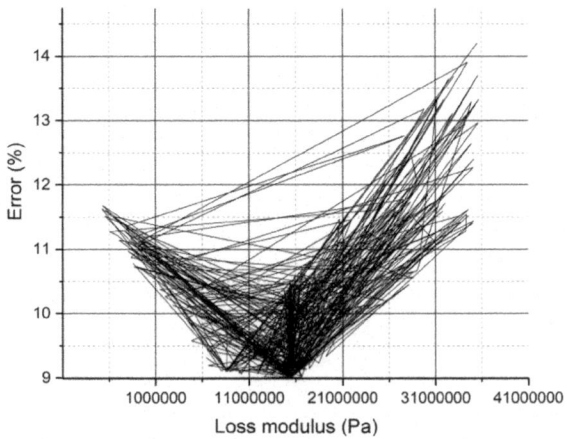

Fig. 9. Process of searching convergence.

5   Conclusions

An algorithm for obtaining data on the parameters of numerical modeling of an experiment to study the processes of heat generation in rubber cord composites under cyclic loading is proposed in the course of the work. The parameters were determined by applying the optimization procedure of double annealing simulation using the SciPy library in Python. The choice of the optimization method is substantiated taking into account the peculiarities of the optimum search in this case, namely, obtaining the value of the objective function implicitly (with the involvement of third-party software – ANSYS). The values of the objective function were obtained by direct numerical simulation of stationary heating of the sample with different amplitudes of deformations and frequencies.

As a criterion of convergence was the stabilized temperature of the cyclically loaded sample obtained during the experiment. The calculation error was estimated by comparing the experimental data with the results of numerical calculations for hysteresis losses for one sample loading cycle.

References

1. Zaitsev, R.V., Kirichenko, M.V., Khrypunov, G.S., et al.: Operating temperature effect on the thin film solar cell efficiency. J. Nano- Electron. Phys. **11**(4), 04029 (2019). https://doi.org/10.21272/jnep.11(4).04029

2. Larin, O., Kelin, A., Naryzhna, R., et al.: Analysis of the pump strength to extend its lifetime. Nuclear Radiat. Saf. **3**, 30–35 (2018). https://doi.org/10.32918/nrs.2018.3(79).05, [in Russian]

3. Kelin, A., Larin, O., Naryzhna, R., et al.: Mathematical modelling of residual lifetime of pumping units of electric power stations. In: Nechyporuk, M., et al. (eds.) Integrated Computer Technologies in Mechanical Engineering. AISC, vol. 1113, pp. 271–288. Springer, Cham (2020). https://doi.org/10.1007/978-3-030-37618-5_24

4. Wehrens, R., Buydens, L.M.:. Classical and nonclassical optimization methods. In: Meyers, R.A., Brown, S.D. (eds.) Encyclopedia of Analytical Chemistry. John Wiley & Sons, Chichester (2006). https://doi.org/10.1002/9780470027318.a5203

5. Larin, A.A., Vyazovichenko, Y.A., Barkanov, E., Itskov, M.: Experimental investigation of viscoelastic characteristics of rubber-cord composites considering the process of their self-heating. Strength Mater. **50**(6), 841–851 (2018). https://doi.org/10.1007/s11223-019-00030-7

6. Plagge, J., Klüppel, M.: A physically based model of stress softening and hysteresis of filled rubber including rate-and temperature dependency. Int. J. Plast. **89**, 173–196 (2017). https://doi.org/10.1016/j.ijplas.2016.11.010

7. Choi, J.H., Kang, H.J., Jeong, H.Y., et al.: Heat aging effects on the material property and the fatigue life of vulcanized natural rubber, and fatigue life prediction equations. J. Mech. Sci. Technol. **19**(6), 1229–1242 (2005). https://doi.org/10.1007/BF02984044

8. Seichter, S., Archodoulaki, V.M., Koch, T., et al.: Investigation of different influences on the fatigue behaviour of industrial rubbers. Polym. Test. **59**, 99–106 (2017). https://doi.org/10.1016/j.polymertesting.2017.01.018

9. Xiang, Y., Sun, D.Y., Fan, W., Gong, X.G.: Generalized simulated annealing algorithm and its application to the Thomson model. Phys. Lett. A **233**(3), 216–220 (1997). https://doi.org/10.1016/S0375-9601(97)00474-X

10. Tsallis, C.: Possible generalization of Boltzmann-Gibbs statistics. J. Stat. Phys. **52**(1–2), 479–487 (1988). https://doi.org/10.1007/BF01016429

11. Tsallis, C., Stariolo, D.A.: Generalized simulated annealing. Phys. A Stat. Mech. Appl. **233** (1–2), 395–406 (1996). https://doi.org/10.1016/S0378-4371(96)00271-3

12. Xiang, Y., Gong, X.G.: Efficiency of generalized simulated annealing. Phys. Rev. E **62**(3), 4473–4476 (2000). https://doi.org/10.1103/PhysRevE.62.4473

13. Xiang, Y., Gubian, S., Suomela, B., Hoeng, J.: Generalized simulated annealing for global optimization: the GenSA package. R J. **5**(1), 13–28 (2013). https://doi.org/10.32614/rj-2013-002

14. Mullen, K.M.: Continuous global optimization in R. J. Stat. Softw. **60**(6), 1–45 (2014). https://doi.org/10.18637/jss.v060.i06

Project Management and Business Informatics

Comparison of Metoheuristic Search Methods for the Task of Choosing a Rational Set of Measures to Risks' Respond

Olha Pohudina[1] , Anastasiia Morikova[1(✉)] ,
Bohdan Haidabrus[2] , Sergey Kiyko[3] , and Evgeniy Druzhinin[1]

[1] National Aerospace University "Kharkiv Aviation Institute",
17 Chkalova Street, Kharkiv 61070, Ukraine
olga.pogudina@khai.edu
[2] RISEBA University of Applied Sciences, 3 Meža Street, Riga 1048, Latvia
[3] PrJSC "Electrometallurgical Works "Dniprospetsstal" Named
After A. M. Kuzmin", 81 Pivdenne Highway, Zaporizhzhia 69008, Ukraine

Abstract. The relevance of project risk assessment has been analyzed. The processes of project risk management including process of planning a response to risk, which is substantial one, are described. It is proposed to carry out the selection of an optimal measures set for responding to risks using metoheuristic search methods. This search is a way to solve especially complex problems and is used in situations of significant lack of information. Two methods of metoheuristic search such as genetic algorithm and method of simulating annealing are considered. A feature of the genetic algorithm is the ability to simultaneously manipulate many parameters and generate new solutions through the utilization of crossing over and mutation operators. A feature of the simulated annealing method is the use of an ordered random search based on an analogy with the formation of a crystal structure in a substance, with a minimum energy upon cooling. It is chosen the project, which provides the minimum average duration in case of manifestation of risks, taking into account the assessment of their probability and consequences in the case of using a set of measures to respond to risks. The result is a vector whose length is equal to the number of risks planned in the analyzed project. Vector values correspond to one of the risk response options. The simulation is implemented using a computer program developed in the MATLAB system.

Keywords: Risk · Risk management · Project management · Simulated annealing · Genetic algorithm

1 Introduction

1.1 Relevance of Risk Assessment in Projects

Any project is associated with certain risks. They are formed as a result of the impact of a set of conditions, which, in turn, have a direct or indirect impact on the outcome of the project and its management processes in various areas (finance, quality, staff, communications, contracts and purchases, time, risks and etc.). On the other hand, decision making is associated with uncertainty, which entails the possibility of risky events.

In project work risks mean the possibility of an event occurring that has a bad effect on a project at all possible phases of its life cycle due to the uncertainty of the characteristics of the internal and external environment of the project. The risk management procedure presupposes to develop strategies for behavior in risk situations on a base of random factors' analysis. A preliminary risk assessment is needed already in the project initiation phase. The decision to launch the project largely depends on its results. A detailed discussion of risks begins in the planning phase and it accompanies the project until its completion. Among all other projects, R&D and innovation projects are the most at risk.

Project risks are characterized by three main parameters [1]:

1. Risk event – a random event that disrupts the project execution process and causes adverse consequences for the project as a whole.
2. Risk event probability – reflects the possibility of a risk event.
3. Damage from the risk event's occurrence – possible losses arising from the occurrence of a risk event.

So the reason that project risks are generated by the environment in which it is implemented, it is customary to divide them into internal and external. Internal risks are generated by internal project work and have every chance to manifest themselves in various active areas of the project – finance, technology, management, etc. In accordance with this, external risks are considered the result of the influence of all possible environmental factors – political, financial, social, natural, cultural, etc. Apart from the fact that risks are systematized according to a number of other principles: by the nature of threats, by spheres of manifestation, by the ability to foresee, by sources of occurrence, by the amount of probable damage, by financial consequences, by the nature of the manifestation in time, by the ability of insurance, by the frequency of implementation, etc. [2].

Risk management is aimed at minimizing losses from the occurrence of an undesirable event. The basis for making decisions on risk management is risk assessment – a characteristic linking the amount of damage with the probability of an event leading to this damage. The main sub-processes of the project risk management process are:

- risk management planning – defining approaches and planning actions to manage project risks;
- risks identification – identifying risks that could affect the project and documenting their characteristics;
- qualitative risk analysis – conducting a qualitative analysis of risks and conditions to determine their priorities in terms of impact on project goals;
- quantitative risk analysis – assessment of the possibility and consequences of risks and their impact on the goals of the project;
- risk response planning – developing procedures and methodologies to mitigate threats to project objectives;

- monitoring and risk management –remaining risks' monitoring, identifying new risks, executing risk mitigation plans and assessing its effectiveness throughout the project life cycle [3].

Risk planning needs to consider:

- risk description (number, type, name). For example: 1. Additional work, unforeseen calculations for the characteristics of engine design options;
- a risk management model with basic parameters (probability, consequences, management method, time, cost) For example: 0.5, 0.2, bias, 1 week, 40 thousand UAH;
- measures and changes in the project plan: For example: clarification of the technical assignment with the customer to include all potential alternatives (revision of 1 project work) or access to previous projects that considered similar options (revision of works 2, 10).

1.2   Overview of Metoheuristic Search Methods

Currently, there are a large number of methods for assessing project risks: analytical (risk-free equivalent; adjustments to the discount rate; expert assessments); statistical (measurement of variance, variation and correlation; capital asset pricing model (CAPM) and the like); stress testing (sensitivity analysis; scenario analysis; decision tree); imitation; non-traditional (fuzzy sets; simulated annealing; genetic algorithm). These methods are used independently and in common.

The complex of measures to risks' respond should reduce the uncertainty of the project or reduce the duration (cost) of the project when risks are manifested [4]. In this article, it is proposed to use the generalized indicator of risk management effectiveness (IRME):

$$IRME = \frac{\int_0^{T_r} U_r(t)dt - \int_0^{T_{rm}} U_{rm}(t)dt}{\int_0^{T_r} U_r(t)dt - \int_0^{T_{wr}} U_{wr}(t)dt}, 0 \le IRME \le 1,$$

where U_{wr} is an indicator of project uncertainty with the influence of risks, but without the variable uncertainty; U_r is an indicator of project uncertainty with the influence of all risks; U_{rm} is an indicator of project uncertainty with risk management; T_{wr}, T_r, T_{rm} are the project's duration for these cases.

To select a rational set of measures to the risks respond of a project methods it is required that it is able to take into account the specifics of the project, makes it possible to quickly respond to changes in the prerequisites for using the method, and reduces the negative impact of the human factor [5–9]. Among unconventional methods there are popular such as the simulated annealing (SA) method and the genetic algorithm-based method (GA). Compare the result of their work on the example of risk management of a project to create complex equipment.

2 Simulated Annealing Method

The application of the simulated annealing method to select a rational set of measures to respond to risks was proposed in article [10]. In the work, the implementation of each risk management measure is determined by four main parameters: the amount of change in the risk probability (ΔP) and risk consequences (ΔI), the time (T) and cost (C) required for the implementation of the proposed measure. The objective function of an optimization problem can be an indicator of risk management effectiveness (*IRME*), an indicator of project uncertainty, duration or cost of a project.

The method's designation is derived from the process of annealing metals in metallurgical science, in which controlled cooling of the metal occurs in such a way that at the end of the process the molecules are in the most regular mode of their crystalline network (this state has the minimum energy). In the SA method, this thermodynamic process is associated with the problem of finding the global minimum in the discrete optimization problem.

A set of points is found by imitating this process and the minimum value of the objective function is reached for these points $F(\overline{m})$, where $\overline{m} = (m_1, m_2, ..., m_n) \in M$; M is the the set of all possible cases (all possible options for project risk management); n is the size; $\overline{m}$ is the total number of project risks.

The optimal solution will be obtained by sequential calculations for the points $\overline{m_0}, \overline{m_1}, ...$ of the set M; starting from a point $\overline{m_1}$, each subsequent point approaches the optimal interchange better than the previous points. The algorithm treats the point $\overline{m_0}$ as output. At each step of the algorithm, the temperature value is decreased and calculations are performed for a new point.

Based on the current point $\overline{m_i}$, the next point $\overline{m_{i+1}}$ is obtained as follows. The operator A is applied to a point $\overline{m_i}$ and randomly changes that point and generates a new point $\overline{m^*}$. A point $\overline{m^*}$ becomes a point with a probability $P(\overline{m^*}, ... \overline{m_{i+1}})$ that is calculated according to the Gibbs distribution.

The initial temperature and its reduction schedule depend on the characteristics of the problem and the choice of the programmer and can be determined in various ways. One of the most famous and widely used ratios for determining the temperature at each step is exponential decramentation, in which we have:

$$T_{i+1} = \alpha T_i, i = \overline{1, k},$$

$$T_k = \alpha^k T_0, 0 < \alpha < 1,$$

where α is the reduction ratio, the best range for this ratio is in the range [0.7,.., 0.95] (mandatory threshold value 0.9); T_0, T_k are the initial and final temperatures respectively (selected value)$T_0 = 1000$).

The simulated annealing approach provides means that are less likely to fall into local minima, allowing you to move up the hill. It should be noted that simulated annealing, like other metaheuristic methods, does not guarantee finding the global minimum of the objective function. The correct policy of generating random points and choosing the appropriate solution parameters improves the quality of the algorithm results.

Boundary point generation: whereas the annealing optimization method is based on the search for points, one of the main stages of the implementation of this method is the creation of a random point in the vicinity (neighborhood) of the current point [7]. Neighboring points are essentially new conditions that arise when the current condition is carefully changed. Neighboring point generation is done by defining and applying certain operators at the current point. The type of operators and their functioning in each task depend on the characteristics and features of the optimization problem and may differ from others [8]. A good random point generator should be able to cover an acceptable range of an adjacent area (from close to relatively distant points), and also be able to adjust (tune) depending on the nature of the task.

Towards this end of view, we define an operator that converts the current point m_i^- to a point $\overline{m^*}$ using a set of random transitions. These points are members of the set M containing all possible combinations of project risk management:

$$A : \overline{m_i} \rightarrow \overline{m^*}, \overline{m_i}, \overline{m^*} \in M.$$

Both points $\overline{m_i}$ i $\overline{m^*}$ are vector size n, where n equal to the number of project risks.

Applying the operator A to a point $\overline{m_i}$, r with n terms of the vector $\overline{m_i}$ will change, while the rest of the terms (n-r terms) remain unchanged. To determine the number r, three suboperators A_1, A_2, A_3 are defined that are also operators for generating near, middle and distant random points, respectively. For each suboperator, the probability of its choice is determined, namely p_1, p_2, p_3. Therefore, we have: $A = <A_1, p_1, A_2, p_2, A_3, p_3>$ where $p_1 + p_2 + p_3 = 1$.

Each time the operator is used, a random number is generated and in a result of its comparison with the options for choosing suboperators (p_1, p_2, p_3), one of the three suboperators is selected. The value is obtained from the interval that corresponds to the selected suboperator.

Thus, for the calculation we have:

$$p_1 = P(r = \alpha_1), p_2 = P(r = \alpha_2), p_3 = P(r = \alpha_3),$$

where $\alpha_1, \alpha_2, \alpha_3$ are random numbers that are generated in the vicinity of near, middle, and far points $\overline{m_i}$ from a uniform probability distribution. We define the boundaries of these neighborhoods using three parameters l_1, l_2, l_3

$$\alpha_1 \sim U(1, \lceil n.l_1 \rceil), \alpha_2 \sim U(\lceil n.l_1 \rceil, \lceil n.l_2 \rceil), \alpha_3 \sim U(\lceil n.l_2 \rceil, \lceil n.l_3 \rceil).$$

Hence we have $0 < l_1 < l_2 < l_3 \leq 1$.

There is possible to find out that the generation of a new random point depends on six variables that the user of the optimization program selects, depending on the characteristics of the problem, in order to get the proper result from the algorithm. The suggested values for these parameters are as follows:

$$p_1 = 0.5, p_2 = 0.3, p_3 = 0.2;$$

$$l_1 = 0.2, l_2 = 0.5, l_3 = 0.9.$$

For example, to make the algorithm more sensitive and create points closer to the current point $\bar{m}_i$, it's necessary either increase p_1 or decrease the value, l_1, l_2 or l_3.

It should be noted that, in addition to the parameters of the operator for generating a neighboring point, the principle of temperature decrease and the choice of the setting-out datum point for solving the problem also affect the efficiency of obtaining the results of the method, and it is necessary to adjust these variables.

In this article, it is proposed to select the characteristics of the project taking into account all risks (without control) as the starting point for solving the problem $\bar{m}_0$.

As a result of the annealing method's implementation, an effective set of risk management measures is obtained, which shows the optimal values of the four main parameters of measures ($\Delta p, \Delta I, C, t$) for all risks.

Thus, having found the variables of the optimal set of measures, the project manager, with the help of the project team and experts, must form and introduce into the project plan specific changes that will ensure the achievement of the found parameters of the optimal solution.

Consider the main steps of the proposed SA method.

Step 1. Determine possible measures to manage each project risk and estimate the costs of their implementation (parameters ΔP, ΔI, T, C for all measures).

Step 2. Initialize parameters (by using the discrete-event modeling method for the project taking into account all risks without managing them).

Step 3. Generate a random neighbor point (a new combination of risk management measures) using random selection and applying random point generation suboperators. The operator A is applied to a point $\bar{m}_i$, randomly changes that point and generates a new point $\bar{m}^*$. Here $\bar{m}$ is a vector with size n (the number of project risks), in which for each risk the number of one of the possible measures to manage it is given.

Step 4. Calculate the objective function $F(\bar{m}^*)$ for the generated point (using discrete event project simulation).

Step 5. Refuse or take a new point based on the Metropolis criterion. Point $\bar{m}^*$ becomes $\bar{m}_{i+1}$ with probability P

$$P(\bar{m}^* \rightarrow \bar{m}_{i+1}|\bar{m}_i) = \begin{cases} e^{-\frac{F(\bar{m}^*)-F(\bar{m}_i)}{T_i}}, & \text{if } F(\bar{m}^*) - F(\bar{m}_i) \geq 0; \\ 1, & \text{if } F(\bar{m}^*) - F(\bar{m}_i) < 0, \end{cases}$$

where T_i is the added a number, slowly decreases and tends to zero at the end of the solution and is an analogue of the temperature parameter during annealing.

Step 6. Reduce the temperature and repeat steps 3–5 with the required number of iterations.

Step 7. Save the results and parameters of the found set of activities.

3   Genetic Algorithm

Taking into account the criteria considered in the simulated annealing method, we will consider a method based on a genetic algorithm for choosing a rational set of measures to respond to risks.

Assignment of input simulation parameters:

- the number of iterations for modeling projects by the Monte Carlo method with the manifestation of risks and the impact of reactive measures (max_iteration = 100 was used);
- the number of iterations of the genetic algorithm (it will be MaxIt = 50 iterations);
- population size in each iteration (nPop = 10). In this article a population unit is understood as a vector containing measures to respond to project risks;
- the number of individuals that will be generated in the new population by the crossover operation (pc = 0.4);
- the number of individuals that will be generated in the new population by the mutation operation (pc = 0.2).

The algorithm starts by creating a matrix containing numbers from the table of measures for responding to risks of size n, where n is the number of project risks. Further the first population is formed, consisting of randomly generated individuals. The project is being modeled. For this, the following data from work [10] are used: matrix of duration of work, matrix of relationship of work, matrix of probability and consequences of risks.

Simulation is carried out for each individual using the max_iteration parameter by the Monte Carlo method. As a result of modeling, the value of the objective function is formed. In this article, we will consider the value of the most likely duration of the project. After that, for all individuals of the population, the minimum value of the objective function is found. However, this will not be the most optimal solution for responding to risks in order to reduce the project execution time, but only the best of the first, randomly generated. For further search, a cycle is started with the specified number of MaxIt iterations. In this cycle, a new population will be generated, based on the previous one, then sorting will be performed again, and so on. To generate a new population, two individuals are crossed in a cycle, taken from the population formed at the previous iteration, then the next ones, and so on until the number of new individuals remains less than the percentage of new individuals obtained by the crossover operation specified in the input parameters. When two individuals are crossed, two individuals are also obtained. All new individuals are recorded at the end of the new population array.

The algorithm by which individuals are crossed is called "partially displayed crossing" [6]. The essence of this algorithm is as follows: let two parental chromosomes be given $ch_{1,1} = \{1, 2, 3, 4, 5, 6, 7, 8, 9, 10, 1\}$ and $ch_{1,2} = \{1, 4, 6, 2, 9, 8, 7, 3, 10, 5, 1\}$. Randomly find two break points: $p_1 = 4$ and $p_2 = 8$ $(1 \leq p_1 \leq p_2 \leq n)$. Believe that the gap occurred as follows: $ch_{1,1} = \{1, 2, 3, 4|5, 6, 7, 8|9, 10, 1\}$ and $ch_{1,2} = \{1, 4, 6, 2|9, 8, 7, 3|10, 5, 1\}$. Recall that positions 1 and 11 are not involved in the operation, as if the real chromosome is located between these positions. When forming descendants $ch_{2,1} = \{1, x, x, x | x, x, x, x | x, x, 1\}$ and $ch_{2,2} = \{1, x, x, x | x, x, x, x | x, x, 1\}$, first we exchange the parts that are between the break points: $ch_{2,1} = \{1, x, x, x | 9, 8, 7, 3 | x, x, 1\}$ and $ch_{2,2} = \{1, x, x, x | 5, 6, 7, 8 | x, x, 1\}$ (i.e. for the child $ch_{2,1}$ we take the middle part from the parent $ch_{1,2}$, and for the child $ch_{2,2}$ we take the middle part part from parent $ch_{1,1}$). Further arrange the remaining positions from the corresponding chromosomes (in $ch_{2,1}$ from $ch_{1,1}$, for $ch_{2,2}$ from $ch_{1,2}$) from left to right. We receive: $ch_{2,1} = \{1, 2, 6, 4 | 9, 8, 7, 3 | 5, 10, 1\}$ and $ch_{2,2} = \{1, 4, 3, 2 | 5, 6, 7, 8 | 10, 9, 1\}$. However, if two identical individuals are crossed, it turns out that their descendants will be identical to their parents. As a result, the population will already have four identical individuals and since the next population is formed on the basis of the previous one, sooner or later the entire new population will consist of identical individuals. To avoid this, after creating two new individuals, a check is made for their identity to each other. If it turns out that they are identical, then each individual is mutated. A mutation is a permutation of two randomly selected positions in each individual.

The general algorithm consists of the following steps.

Step 1. Determine possible measures to manage each risk of the project and estimate the costs of their implementation (parameters for all measures).

Step 2. Create an initial set (in our case 10) of projects using random combinations of risk management measures.

Step 3. Create a set using Cross-over population.

Step 4. Create a set using mutation operation.

Step 5. Find the best combination according to the criterion of the selected optimal value and repeat steps 3–5 with the required number of iterations.

Step 6. Save the results and parameters of the found set of activities.

4   Comparison of Simulation Results

Both methods were automated and applied to a project described in [4]. The computer program was developed in the MATLAB system.

Compare the results. The parameters' values of the found point corresponding to the optimal set of risk management measures using the simulated annealing method are shown in Fig. 1. The result of the best variant of the average project duration using the risk response complex is 254.96 weeks. In this case, the optimal *IRME* value is 0.57, which means that 57% of the uncertainty associated with project risks can be eliminated by implementing this found set of risk response methods. Figure 1 shows this set of activities in the form of 41 element vector "Optimum set". 41 elements are equal to the number of project risks, and each element is the number of the best response method for the corresponding risk.

The parameters' values of the optimal set of measures for responding to risks, obtained using the genetic algorithm, are shown in Fig. 2. The result of the best variant of the average duration of the project using the risk response complex is 242.899 weeks. Figure 2 also shows this set of activities in the form of 41 element vector "Optimum set". For both cases, the number of search iterations, the optimal variant is taken 50.

```
Best variant of risk management which is found after 50 iterations:
Objective function was maximising IRME
Optimum set = [1 1 1 1 1 1 1 1 2 1 2 1 1 1 1 1 1 2 1 2 1 1 1 1 1 3 2 1 1 1 2 2 1 1 1 1 1 1 3 1]
Project cost (mean value) for this variant = 152352 thousand UAH
Project duration (mean value) for this variant = 254.96 weeks
Index of risk management effectiveness (IRME) =   0.57
```

Fig. 1. Characteristics of the optimal set of risk management measures for the simulated annealing algorithm.

```
Best Solution up to Iteration = 50
M = [1 1 1 1 1 1 1 1 2 1 1 1 1 1 1 1 1 1 1 1 1 1 2 1 1 1 3 1 2 1 1 2 1 1 1 1 3 2 2 1 1], Cost function= 242.899

Best variant of risk management which is found after 50 iterations:
Optimum set = [1 1 1 1 1 1 1 1 2 1 1 1 1 1 1 1 1 1 1 1 1 1 2 1 1 1 3 1 2 1 1 2 1 1 1 1 3 2 2 1 1]
```

Fig. 2. Characteristics of an optimal set of risk management measures for a genetic algorithm.

The research results help managers make targeted decisions during planning and implementing projects based on the analysis of the impact of risks and changes in uncertainty by planning risk management activities. The selection criterion can be not only the implementation time, but also other recommended parameters [11]. Simulation models of rational options selection for attracting additional resources and optimal risk management measures, which allow implementing energy-saving projects are descri-bed [12–14].

References

1. Mills, A.: A systematic approach to risk management for construction. Struct. Surv. **19**(5), 245–252 (2001). https://doi.org/10.1108/02630800110412615
2. Ahlemann, F., El Arbi, F., Kaiser, M.G., Heck, A.: A process framework for theoretically grounded prescriptive research in the project management field. Int. J. Project Manag. **31**(1), 43–56 (2013). https://doi.org/10.1016/j.ijproman.2012.03.008
3. Klimova, J.: Risk-based project management audit. In: Antipova, T. (ed.) Integrated Science in Digital Age. ICIS 2019. LNNS, vol 78, pp. 39–49. Springer, Cham (2020). https://doi.org/10.1007/978-3-030-22493-6_5
4. Martens, A., Vanhoucke, M.: The impact of applying effort to reduce activity variability on the project time and cost performance. Eur. J. Oper. Res. **277**(2), 442–453 (2019). https://doi.org/10.1016/j.ejor.2019.03.020
5. Yehorchenkova, N.: Models of project management in electronic project management. Bull. Natl. Tech. Univ. "KhPI". Ser. New Solutions Mod. Technol. **9**, 112–117 (2018). https://doi.org/10.20998/2413-4295.2018.09.16, [in Ukrainian]
6. Junghans, L., Darde, N.: Hybrid single objective genetic algorithm coupled with the simulated annealing optimization method for building optimization. Energy Build. **86**, 651–662 (2015). https://doi.org/10.1016/j.enbuild.2014.10.039
7. Hu, X., Demeulemeester, E., Cui, N., et al.: Improved critical chain buffer management framework considering resource costs and schedule stability. Flex. Serv. Manuf. J. **29**(2), 159–183 (2017). https://doi.org/10.1007/s10696-016-9241-y

8. Peng, W., Huang, M.: A critical chain project scheduling method based on a differential evolution algorithm. Int. J. Prod. Res. **52**(13), 3940–3949 (2014). https://doi.org/10.1080/00207543.2013.865091

9. Zhang, J., Song, X., Chen, H., Shi, R.: Optimisation of critical chain sequencing based on activities' information flow interactions. Int. J. Prod. Res. **53**(20), 6231–6241 (2015). https://doi.org/10.1080/00207543.2015.1043157

10. Chenarani, A., Druzhinin, E.A., Kritskiy, D.N.: Simulating the impact of activity uncertainties and risk combinations in R&D projects. J. Eng. Sci. Technol. Rev. **10**(4), 1–9 (2017). https://doi.org/10.25103/jestr.104.01

11. Schwalbe, K.: Project management techniques. In: Bidgoli, H. (ed.) The Internet Encyclopedia (2004). https://doi.org/10.1002/047148296X.tie145

12. Kiyko, S., Druzhinin, E., Prokhorov, O., et al.: Logistics control of the resources flow in energy-saving projects: case study for metallurgical industry. Acta Logistica **7**(1), 49–60 (2020). https://doi.org/10.22306/al.v7i1.159

13. Vasiljeva, T., Novinkina, J.: Is robotics a solution for banking business process reengineering and automation?. In: Soliman, K.S. (ed.) Proceedings of the 33rd International Business Information Management Association Conference IBIMA 2019, pp. 5613–5620. IBIMA, Granada (2019)

14. Ingham, B., Chirijevskis, A., Carmichael, F.: Implications of an increasing old-age dependency ratio: The UK and Latvian experiences compared. Pensions Int. J. **14**(4), 221–230 (2009). https://doi.org/10.1057/pm.2009.16

Project Management in Universities Under the Global Pandemic: A Focus on Finance

Olena Zhykhor[1] (ID), Valeriy Ryeznikov[1] (ID), Olena Iafinovych[2] (ID),
Nataliia Pohribna[2] (ID), and Nataliia Miedviedkova[2(✉)] (ID)

[1] V. N. Karazin, Kharkiv National University,
4 Svobody Sq., Kharkiv 61022, Ukraine
{olena.zhykhor, valeriy.reznikov}@karazin.ua
[2] Taras Shevchenko National University of Kyiv,
60 Volodymyrska Str., Kyiv 01033, Ukraine
{olenayafinovych, nataliapohribna, nsmedvedkova}@knu.ua

Abstract. This paper is intended to provide an insight into the importance of project management success in implementing strategies and new financial packages within the context of university education. The discussion will call upon the universities to review and improve their strategies and detailed measures taking into account the international experience and the current situation. This paper builds up on the academic capitalism approach to show how universities can adapt to the new conditions simultaneously being under the continuous onward change processes in the context of Ukraine. The paper is structured as follows: first, it reveals the framework of the project management approach application in the university under the changing conditions with focus on the funding and financial aspect. Second, it investigates the domestic and international experience of responding to the challenges of the COVID-19 pandemic by the universities. It's worth noting that this paper provides an important source to analyze different types of support for solving problems of higher education institutions and students during challenging times.

Keywords: Project management in universities · Higher education institutions · Global pandemic · Online and in-person learning · Funding · Strategy · Support

1 Introduction

Project activity is a strategic development tool designed to create unique results (products, services) that transforms strategies into actions and goals into reality; accordingly, project implementation should affect key areas of the organization's development, defined by its long-term strategy. The general principles of project management are true both for private business and for the public administration sector, however, the introduction of the project approach in a university under the implications of COVID-19 pandemic has its own characteristics, on which we would like to focus. The university is a complex system where current global changes require careful preparation, detailed planning and willingness to invest large resources (material,

M. Nechyporuk et al. (Eds.): ICTM 2020, LNNS 188, pp. 667–679, 2021.
https://doi.org/10.1007/978-3-030-66717-7_57

human, financial) for a long time to obtain fast results going on with its routine activities and delayed results in a strategic perspective.

Over the years, a limited set of modern management tools has been used at the university, however, a number of serious challenges today have led to the need for such a reform. Rivalry between universities as the public sector organizations is also intensified by the influence of consumers of services. The increasing mobility of the population (especially academic), the exponential development of information and communication technologies and other factors have been already gradually erasing traditional barriers.

The intensive involvement of science and education in a market domain confirms the researchers' opinion on the transformation of a modern university into a business unit and testifies to new forms of development that have various expected and unexpected consequences. Following Slaughter and Leslie [1], market relations in the field of higher education are commonly referred to as an academic capitalism. Under the academic capitalism, universities, with the influence of external factors, become similar to economic corporations associated with the production and dissemination of knowledge, and all parts of the university structure – faculties, laboratories, research centers and even individual teachers – are determined by signs of competitiveness and profitability, and their contribution to the university brand in the market of educational services. The new economic nature of education is expressed not only in a change in the educational concept and ideology, it also manifests itself in the management of universities, in the development of specific curricula and ways of organizing the educational process and controlling its quality.

Given this, the **objective of this paper** is twofold. First, it will look into the framework of the project management approach application in the university under the changing conditions with focus on the funding and financial aspect. Second, it will investigate the domestic and international experience of responding to the challenges of the COVID-19 pandemic by the universities.

The academic environment is driven by other set of forces in adopting the project management in comparison with business environment, due to the understandable lack of marketplace forces.

Analysis of recent research and publications. There is a growing evidence on the merits produced by the use of project management approaches within the academic environment. Bickers [2] analyses the possibilities of improving the admissions process using project management methods. Murphy [3] argues the enhanced elaboration of teaching materials via traditional project management activities. Conway [4] investigates the successful evidence of fundraising activities via project management approach. Henry [5] looks into the management of the construction projects expenditures. Yerk-Zwickl [6] studied development of the information system at the educational institution with project management approach.

In the Ukrainian economic system, the development of academic capitalism is both promoted by the reform in educational system and complicated by the bureaucratic processes. In a short time, the budget system of higher education institutions has gone from a hundred percent state funding, support and control to a significant reduction in the managerial role of the state, which has necessitated self-financing and self-governance. At the same time, state universities received much more freedom in making their own decisions than before. These new realities have created a need and enabled universities to overcome budgetary difficulties, attracting additional financial resources in the educational market. The increasing decentralization of the budget process leads to serious changes in the system of budgeting universities and the conditions of their financial management. Fundamental changes in the nature of the university's relations with external entities led to the evolutionary transformations of the university micro- and macro-environment.

At present, resistance to the restrictions produces by the COVID-19 pandemic and increasing the global competitiveness of universities are the priority areas for the development of education. Notably, educational projects are a set of events with a specific goal, limited in time and resources and taking into account the specifics of educational institutions. Modern management of an educational project is a special type of management that, one way or another, can be applied to the management of any objects, and not just objects that have explicit project characteristics.

Methodology. To fulfill the aim of describing and analyzing the project management practices in universities, the article has used the stages of the formula application for the state budget expenditures distribution on higher education between higher education institutions, possible revenue reductions and unexpected costs in Ukrainian universities. Analysis of legal, professional and regulatory texts has been conducted. These documents provided an important source to analyze different types of support for solving financial problems of higher education institutions and students faced with financial difficulties. Also, the analysis of project management practices in the world was done based upon mass media and information available on the websites of universities, with analytical reports on the impact of COVID-19 on Higher Education around the world and strategies developed by universities.

2   Research Results

2.1   The Essence of Project Management in Universities

The widespread use of project management was a certain response to the challenges of the time, namely the accelerating socio-economic processes in the world, where universities, as other hierarchical organizations, do not have time to adapt quickly enough to the changes. However, under the unprecedented conditions produced by the COVID-19 pandemic in the world, there is no other way or second choices. Hence, the solution of many problems requires the involvement of professionals from various fields, the creation of global teams, and work in a distributed form.

The definition of the project says that it is a set of measures that must be implemented with time and resource constraints to obtain any unique result (changes in the system). So, the development and implementation of new educational programs meets the criteria of the project, while their implementation is an operational activity. At the same time, we may encounter projects that are initially little distinguishable from current operations. For example, the adaptation and application of typical educational programs in different universities is each time a project, because its performers are faced with a new external environment, stakeholders, etc., which makes the final product unique in its own way.

However, it is obvious that not all activities in a particular organization can and should be performed in a project form. In the classical project management, two types of the human activity organization are distinguished: operational and project ones. When the external environment is well studied and stable, and the functions of the performers are defined, repeatedly tested in practice and constant, they talk about operational activities. In this regard, universities can restrict itself to classical management methods for solving problems of ensuring the normal functioning of an organization. Regularly renewed activities are effective while reproducing the result already repeatedly obtained under similar conditions.

Infrastructure development is an essential condition for effective introduction of the project management approach at the university. So, a separate structural unit is to be created - the project office performing the functions of the project and program management, which develops the methodological basis for the planned changes. A standard project charter is to be approved normally containing a description of the goals and objectives of the project, an analysis of stakeholders and risks, as well as a network schedule for the implementation of activities. Later on, a corporate standard for project management can be established, which defines the relationship of employees involved in the implementation of projects, and also prescribes how, in what sequence and in what time frame it is necessary to perform certain actions when managing a project within the university.

Improving the skills of project management among the university management with time allows creating a pool of competent employees of various fields for performing projects in various subject areas. In addition, it provides for the university project office to abandon the role of the sole coordinator of strategic initiatives, moving to supervisory and consulting functions after the stabilization of the COVID-19 pandemic. It is worth noting that the emergence of a significant layer of employees with skills in project management allows university to formally set strategic goals and select initiatives aimed at achieving them.

Research universities are a special category of universities, where the most important product of their activities are new knowledge, technologies and competencies, which form the basis of economic and social development. Under the current implications of the COVID-19 pandemic the ability of society to generate, select, adapt, apply knowledge is crucial for sustainable economic growth and improving living standards.

Analyzing the features of project management in the university, exploring the specifics of project management under the COVID-19 pandemic, we can highlight certain features against the traditional management system. Firstly, project management is more focused on meeting the needs of citizens, society and the labor market in quality education, flexibly responds to changing needs and the emergence of new requests under the quarantine requirements. Secondly, the project approach stimulates the improvement of the educational services quality by improving educational technologies and financial mechanisms in the field of education. Then, the personal activity of the teaching staff is stimulated by involvement in project activities, personal responsibility for the result of work is established, the motivating mechanisms of the teachers' and management activities are being improved. Also, there is an effective promotion of innovation in the educational environment of the university, the rapid inclusion of the innovation results in the functional work. And last but not least, at the level of the educational process, methods and means of educational activity are updated, students are involved in project activities, joint work is aimed at obtaining practical results.

The feature of the project management in the university covers the transparency principle of managerial procedures for obtaining a synergistic effect in the implementation of interdisciplinary projects. The transparency of management procedures is based, in particular, on the principle of complementarity, within the framework of project management considered as an opportunity to comprehensively solve the problems of one project by including additional results of projects that are an independent object of management of research activities. This mechanism also contributes to the growth of collaborations of the university scientific teams, an increase in the circle of interactions with scientific and educational institutions, large research centers and business structures.

Financial planning is the key element of financial management of the university, where the development and execution of estimates of units is carried out by their managers, who perform the functions of economists and suppliers.

The importance of financial planning as one of the main elements of the strategic management of the university has increased due to the conditions of the COVID-19 pandemic. The procedure for determining and distributing the total financial resources of a university is presented as a sequence of management decisions (budget planning). However, the main direction of development is the improvement of the university's financial function, as well as the digitalization of routine processes. All expenses for the maintenance of buildings, dormitories, maintenance of the entire classroom fund (major and ongoing repairs), communication costs, payment Internet access is provided by the university centrally. The central budget of the university bears the main burden on financing the remuneration of employees of all categories working at the university.

The stable financial position of the university overall involves the following strategic task - creation of financial reserves and accumulated funds (endowment), allowing to survive a one-year or two-year decline in solvent demand, which may arise in education systems, as well as in research organizations.

2.2 National Practices to Finance Higher Education System Under the Changing Environment

Not only the development of the system itself, but also the efficiency and effectiveness of the country's socio-economic development directly depend on the financing of the higher education system. Therefore, this issue is the subject of attention of many Ukrainian and international researchers, and becomes especially relevant under the limited financial resources against the background of the COVID-19 crisis (IMF forecasts a 4.9% decline in the world economy in 2020, while real GDP of developed countries will decrease by 8%, and countries with emerging markets – by 3%).

Global research and practice define a multi-source funding structure for higher education, which consists of public/local, private and other funding. As a rule, public funding is one of the largest parts in the structure of financial resources of universities. At the same time, the financial crisis of 2008–2009 significantly disrupted the sustainability and generosity of public sources of funding for higher education, forcing countries to modernize and further develop forms of funding. The issue of the effectiveness of university funding is becoming increasingly important, so countries are beginning to be more demanding of the higher education institutions' performance and their search for new sources of income. If the older models of financing the higher education institutions were largely based on indicators of the level of activity (for example, the number of students), then in the transformed models attention is paid to the performance [7].

Ukraine, like many other countries, features the traditional dependence of public higher education institutions on budget funding. Under the chronic shortage of financial resources, expenditures for the maintenance of Ukrainian universities are constantly reduced, which leaves little room to develop and provide high quality education. While the funding formulas were introduced (in the 1990s) and reformed (early 2000s) in European higher education systems, the funding mechanism for domestic institutions remained unchanged in Ukraine and was based on the number of entrants per full-time position of research and teaching staff using the methods of calculating the approximate average cost to educate one student.

In recent years, scientists, representatives of the Ministry of Education and Science of Ukraine, the Parliament of Ukraine have been actively discussing to reform the financing model of the domestic higher education system, which would be consistent with European practices and based on a strategy of higher education. Finally, in 2020, Ukraine made the transition to "formula" funding. Now the total amount of funding provided to the institution of higher education at the expense of expenditures of the general fund of the state budget in the year, is the sum of three components, namely: the amount of funding for stable activities of the institution of higher education; the amount of funding provided depending on the performance and the reserve.

Each of the components has its own purpose. Thus, the financing of stable activities of a higher education institution allows supporting the fixed assets of state institutions of higher education, to carry out systematic work on their logistics. Without this, it is impossible to ensure the proper level of development of universities. Part of the funding provided depending on the performance of institutions stimulates them to intensify certain areas of their activities, in particular: to develop science, partnership with business, improve the quality of educational services, educate professionals not only in competitive but also in highly demanded by economy fields, etc. According to the newly created formula, the distribution of state budget expenditures on higher education between higher education institutions will take into account the specific performance indicators of each institution – its comprehensive performance.

From now on, the actual number of students financed by the state budget is only one of the six indicators (scale of the university, contingent, regional coefficient, indicator of international recognition, indicator of scientific activity, employment rate of graduates) that affect the university budget from the state source of funding. Over the next three years, the weight of this indicator will decrease. At the same time, the state encourages universities to maintain their profile, in particular, for technical universities - to support technical specialties by applying an increased index of specialty for them.

However, privately owned higher education institutions also have the right to receive funding to educate the students via budget funding in accordance with this formula.

The new formula for calculating the amount of budget funding for the universities of Ukraine will be introduced in stages from 2020 to 2022, depending on the category of institution (Fig. 1). In addition, the stages are defined with the application of some indicators of the university. Thus, currently the employment rate of graduates is at 1 for all higher education institutions until the system of monitoring the employment of graduates is introduced. It is planned that from 2021 the Ministry of Education and Science of Ukraine will monitor employment through an online system. In addition, in 2020, for a gradual transition, restrictions were introduced for the minimum and maximum change in the budget of each university - 95% and 120% from 2019, respectively.

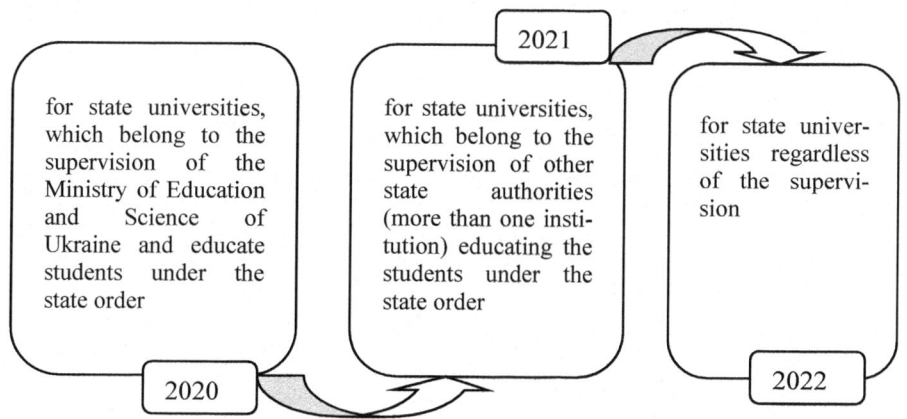

Fig. 1. Stages of the formula application for the state budget expenditures distribution on higher education between higher education institutions in Ukraine [8].

As to the plan for 2020, 136 higher education institutions of Ukraine and 12 of their branches, which have their own estimates, should receive funding according to the formula. At the same time, 94 institutions will receive 100–120% of the 2019 budget, and 54 institutions – 95–99% of the 2019 budget. Due to the calculation of funding according to the formula in 2020, the budgets of 17 universities of Ukraine received an increase in funding from the state budget by more than 15 million UAH [9].

Under the influence of the COVID-19 pandemic, states are forced to revise their budgets for 2020 and beyond in view of the problems of execution of state and local budget revenues in the current year and the need to redirect funds. In such circumstances, there is a likelihood of reducing government spending on higher education. For example, in Ukraine in April of 2020, changes were made to the state budget, which in some way affected the higher education system. In particular, they have effectively stopped funding the Higher Education Development Fund, reduced programs to support priority research at universities, will not fund a number of renovations of higher education institutions planned for 2020, and "withdraw" funds from the National

Research Fund. However, the allocations for salaries of university teachers, utilities and scholarships were not reduced [10].

Thus, the current and future financial situation of higher education institutions is a cause for concern in terms of both possible revenue reductions and unexpected costs (Fig. 2).

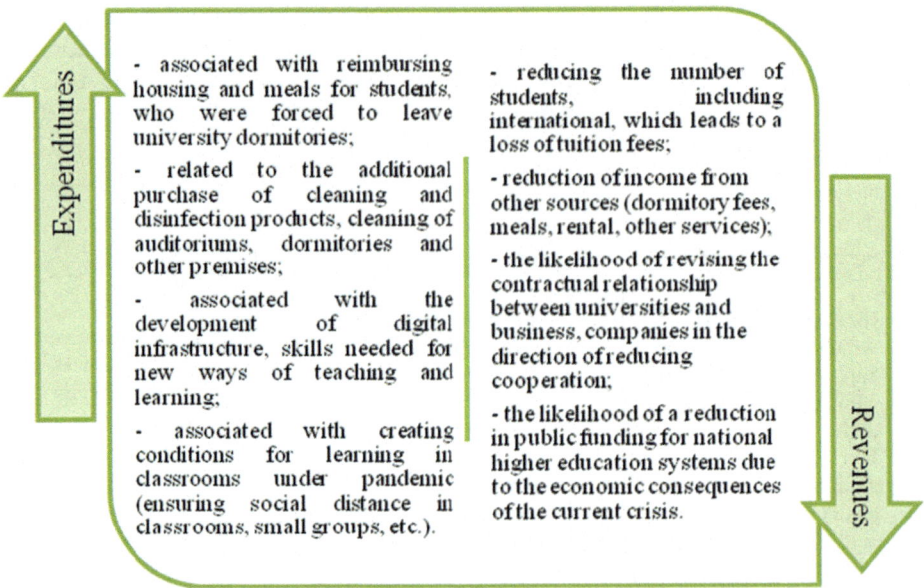

Fig. 2. Causes of financial problems of higher education institutions.

The financial situation of a significant number of students has been shaken by the global recession, which prevents them from continuing or starting their studies. Thus, the American Board of Education estimates that the number of entrants in the 2020 academic year will decrease by 15%, including a decrease in the number of international students by 25% [11]. In addition, a significant proportion of international students will be asked about their place of study due to travel restrictions, declining family incomes and ongoing health risks. Places to study will be chosen as close as possible to home. For example, the Beijing Research Service found out that more than a third of Chinese students change their plans to study abroad [12].

2.3   International Experience in Implementing Strategies and Specific Measures Developed by Higher Education Institutions

Universities around the world are facing a major funding crisis due to the sudden loss of revenues caused by the COVID-19 pandemic. The virus is causing immediate financial pain for institutions everywhere – with the potential for broader upheaval to follow.

The impact of COVID-19 on higher education around the world is as follows:

– universities in Australia hit by COVID-19 sooner than other universities of the world;
– US is most exposed to loss of foreign students, but big campuses with Silicon Valley partnerships may dominate;
– northern states of EU are in a better situation than south states;
– economy of Poland has suffered from coronavirus less than most countries, but the drop in foreign demand is predicted;
– UK – crisis has exposed limitations of the high-tuition fee model [13].

To overcome the negative situation caused by COVID-19, most countries of the world have introduced immediate response packages, which included the developed strategies and recovery plans with specific measures. But the common features for all countries are as follows: the attention paid to foreign students and optional/flexible study conditions (online and in-person studying; the opportunity to start the semester later).

The experience of the 2008–2009 crisis showed that public funding of European higher education systems was gradually reduced. Most likely, the consequences of COVID-19 will lead to the repetition of the same scenario to different degrees for each national system. According to a report by the International University Association, almost half (48%) of higher education institutions worldwide said their government/ministry of education would support their institution in disrupting the educational process caused by the pandemic, 24% said there would be no state support and the rest is unaware. Among those that receive support from the government or the relevant ministry, the most common type of support is end-of-school assistance (67%), with other types of support having a much lower share. Thus, only 13% will receive financial support from the state due to unexpected losses in the revenue of university budgets. If we look at this issue at the regional level, respondents showed that higher education in Europe has the highest rate of support by the government or the Ministry of Education (over 50%), and Africa is the region with the lowest rate of state support (39%). Asia, the Pacific and America are close to Europe in this indicator. At the same time, in both the United States and Africa, the largest number of surveyed universities report that their government/Ministry of Education will not support their institutions (29% in the United States and 31% in Africa) [14].

Currently, countries are studying, evaluating and implementing various solutions to financial problems for higher education institutions, namely:

1. Development of new financial packages to support students (subsidizing educational loans, unemployment benefits and other types of student assistance):

– *delays in payment and interest-free loan are allowed* (USA). For example, in March 2020, the US Secretary of Education announced that all borrowers with federal education loans would automatically receive 0%interest rates for at least 60 days. In addition, each of these borrowers will be able to suspend their payments for at least two months. In New York State, the accrual of interest on government student debt has been suspended [15];

- *support for students who have lost their jobs as a result of COVID-19* (Ireland): students who have lost their job as a result of the pandemic might be eligible for financial help in Ireland's "COVID-19 Pandemic Unemployment Payment" scheme [16];
- *material support for students* (Poland). Taking into account that young people faced with difficult psychological situation and require support at all levels, the university intends to increase the fund of material support for students, given the difficult financial situation of many families [17];
- *student support services* (UK). Some 78 universities of UK will offer in-person social opportunities to students, including outside events and sporting activities, along with government and public health guidance. Student support services including mental health support, careers advice, and study skills will be offered in a combination of online and in-person options, while five institutions said they would be available online [18]; American University also emphasizes, that student services including advising, academic support, and counselling, and the health center will be operational [19].

2. Establishment of higher education funds by higher education institutions, usually through private donations, to support students in the costs they incur in connection with the closure of campuses.
3. Creation of state/local funds to help higher education institutions covering their income losses and incurring unexpected expenses caused by the COVID-19 pandemic. For example, in the United States, the Coronavirus Aid, Relief, and Economic Security Act (CARES) provides for a package of regulations to provide emergency assistance to higher education institutions during a coronavirus pandemic worth $ 14.25 billion. In Massachusetts, for example, an emergency fund has been initiated to support public higher education institutions. The bill provides for a budget of $ 125 million [20].
4. Support for DAAD grant programs (Germany). The current situation everyone faces in times of the coronavirus raises many questions, especially for international students in Germany.

The German Academic Exchange Service (DAAD) has informed that they will be increasing the use of digital options for the winter semester and start with grants online. This is done to support the process of studying in Germany for international students during current difficult times. The global implemented restrictions have made studying abroad a challenging process.

Using the online scholarship option offered by DAAD, international students and researchers, who are unable to enter Germany because of coronavirus, have the opportunity to apply for the scholarship online. These students can start their scholarship online and afterwards, change to their desired institution after travel restrictions are cancelled. Another option is postponing the start of their scholarship for up to 6 months [21].

3 Concluding Remarks

Subject to these conditions and the full inclusion of all university services in the logic of project activities, project management becomes a highly effective mechanism to stimulate the activities of the university under the COVID 19 pandemic conditions, being an element of effective decision-making and a conductor of innovative developments in the university practices.

The key features of project and program technologies helping the universities to overcome the current challenges of education are the following: systematic vision of the problem via elaboration of a single program of development and functioning; embedding the connected projects in solving strategic problems; interdependence of all indicators of project implementation within the framework of one program; "overview vision" of the desired state of the system (programmed (predicted) values) based on and taking into account the initial state and the current requests.

Reopening universities during the COVID-19 pandemic poses a special challenge worldwide. A set of developed policy measures to the current crisis should help mitigate the negative impact on the university sector.

Therefore, realizing that the impact of the current crisis will be significant and long-lasting, higher education institutions must prepare for financial difficulties by developing a strategy for their activities quickly and decisively adapting to modern conditions. Thus, in the process of developing the strategies and detailed measures, universities should pay attention to the following points:

- at least two strategy options should be developed (including optimistic and pessimistic forecasts). According to the each strategy, appropriate measures should be prepared;
- the developed strategy options, as well as the corresponding stages with approximate dates, should be openly accessible;
- universities should offer online and in-person student support services in line with government and public health guidance. First of all it is important for first-year students who need the support much more than others;
- the detailed and perspective governance, as well as the income diversification should become the priority areas. Crisis management and Risk assessment should be a part of institutional operational planning;
- universities should take into account the social dimension and provide different kinds of support for students whose families suffered from the coronavirus and faced with financial difficulties;
- opportunities offered by digital transformation to find the balance between online and in-person learning should be investigated;
- universities should review funding and make decisions which activities are most important for investing (R&D, developing digitally enhanced learning, support for infrastructure). Due to changes in modes of learning and teaching, universities need additional funds for implementing digitally enhanced learning. Thus, investment in both skills development and digital infrastructure with the aim to develop new modes of learning and teaching is actual and requires further research.

References

1. Slaughter, S., Leslie, L.: Academic Capitalism: Politics, Policies, and the Entrepreneurial University. Johns Hopkins University Press, Baltimore (1999)
2. Bickers, D.: The Application of project management techniques to college and university admissions activities. College Univ. **68**(2), 86–92 (1993)
3. Murphy, C.: Utilizing project management techniques in the design of instructional materials. Perf. Instr. **33**(3), 9–11 (1994)
4. Conway, J.E.: Evolution of SOLAR, Harvard's client/server-based fundraising management system. Cause/Effect **18**(1), 46–52 (1995)
5. Henry, R.: Under control: careful planning and monitoring can help to contain costs in your next school construction project. American School & University (2001). https://www.asumag.com/mag/article/20842961/under-control, Accessed 30 Aug 2020
6. Yerk-Zwickl, S.: Project implementation using a team approach. Campus-Wide Inf. Syst. **12** (2), 27–30 (1995). https://doi.org/10.1108/10650749510798756
7. Claeys-Kulik, A.-L., Estermann, T.: DEFINE Thematic Report: Performance-Based Funding of Universities in Europe. European University Association, Brussels (2015)
8. Cabinet of Ministers of Ukraine: Resolution No. 1146 from 24.12.2019. On the distribution of state budget expenditures between higher education institutions on the basis of indicators of their educational, scientific and international activities (2019). https://zakon.rada.gov.ua/laws/show/1146-2019-%D0%BF#Text, Accessed 30 Aug 2020. [in Ukrainian]
9. Ministry of Education and Science of Ukraine: More money is for the strong. The Ministry of Education and Science published the distribution of state funding for universities in 2020 (2020).   https://mon.gov.ua/ua/news/bilshe-groshej-silnishim-mon-opublikuvalo-rozpodil-derzhfinansuvannya-universitetiv-u-2020-roci, Accessed 30 Aug 2020. [in Ukrainian]
10. UkrInform: Coronavirus Government budget-2020 cuts: it could be worse (2020). https://www.ukrinform.ua/rubric-polytics/3004922-koronavirusne-skorocenna-derzbudzetu2020-moglo-buti-girse.html, Accessed 30 Aug 2020. [in Ukrainian]
11. Amtrican Council on Education: Higher Education Fourth Supplemental. Letter President Amtrican Council on Education (2020). https://www.acenet.edu/Documents/Letter-Senate-Higher-Ed-Supplemental-Request-040920.pdf, Accessed 30 Aug 2020
12. China's Study Abroad Industry: Measured impacts caused by the outbreak of novel coronavirus COVID-19. Beijing Overseas Study Service Association (2020)
13. Kelly, E.: Universities prepare for the new post-pandemic world. Science|Business (2020). https://sciencebusiness.net/COVID-19/news/universities-prepare-new-post-pandemic-world, Accessed 30 Aug 2020
14. Marinoni, G., van't Land, H., Jensen, T.:The impact of Covid-19 on higher education around the world. IAU Global Survey Report (2020)
15. Abroadz: Admission to universities during the coronavirus pandemic (2020). https://abroadz.com/articles/postuplenie_v_vuzi_v_period_pandemii_koronavirusa, Accessed 30 Aug 2020. [in Russian]
16. ICOS: Information for international students on COVID-19 (2020). https://www.internationalstudents.ie/information-international-students-COVID-19, Accessed 30 Aug 2020
17. Inter Slavonic: Jagiellonian University will begin the new academic year with remote lectures (2020). https://isttravel.ru/index.pl?act=NEWSSHOW&id=2020070101, Accessed 30 Aug 2020. [in Russian]

18. Stacey, V.: Most UK unis preparing for in-person teaching in 2020, survey finds. The Pie News (2020). https://thepienews.com/news/uk-universities-prepare-for-in-person-teaching-in-2020/, Accessed 30 Aug 2020

19. American University: COVID-19 (Coronavirus): Summary information (2020). https://www.american.edu/coronavirus/, Accessed 30 Aug 2020

20. Smalley, A.: Higher education responses to coronavirus (COVID-19). National Conference of State Legistatures (2020). https://www.ncsl.org/research/education/higher-education-responses-to-coronavirus-COVID-19.aspx, Accessed 30 Aug 2020

21. Studying in Germany: The DAAD is offering an online scholarship option for international students (2020). https://www.studying-in-germany.org/the-daad-is-offering-an-online-scholarship-option-for-international-students/, Accessed 30 Aug 2020

Human Resource Management Tools in a Multiproject Environment

Nataliia Dotsenko[1] , Dmytro Chumachenko[2(✉)] ,
Igor Chumachenko[1] , Yuliia Husieva[1] , Dmytro Lysenko[1] ,
Iryna Kadykova[1] , and Nataliia Kosenko[1]

[1] O. M. Beketov National University of Urban Economy in Kharkiv,
17 Marshala Bazhanova Street, Kharkiv 61002, Ukraine
nvdotsenko@gmail.com, ivchumachenko@gmail.com,
yulia.y.guseva@gmail.com, lysenko.d@gmail.com,
irina.kadykova@gmail.com, kosnatalja@gmail.com
[2] National Aerospace University "Kharkiv Aviation Institute",
17 Chkalova Street, Kharkiv 61070, Ukraine
dichumachenko@gmail.com

Abstract. An urgent scientific problem is the development of methodological foundations of human resource management in a multiproject environment. The aim of this research is to create tools for managing human resources in a multproject environment. The following tasks are solved in the article: to develop a method for analyzing the coherence of human resource management strategies in a multiproject environment; analyze the causes of changes and their impact on human resource management processes; build a model of the process of redistribution of resources; develop tools for team building and reallocation of resources in a multiproject environment. In the study, methods of system analysis, the apparatus of the theory of optimization, Boolean algebra and set theory were used to solve the problems of team formation and redistribution of resources. An analysis of the literature on resource management. A method for analyzing the consistency of human resource management strategies in a multiproject environment has been developed. The analysis of the causes of changes in projects and their impact on the processes of managing human resources is carried out. The application of the donor-acceptor approach to resource support in a multiproject environment is considered. A model of the process of reallocation of resources is proposed. Tools were developed for team building and redistribution of resources in a multi-project environment.

Keywords: Project management · Multiproject environment · Project portfolio · Project team · Project-oriented management · Stakeholders · Reservation

M. Nechyporuk et al. (Eds.): ICTM 2020, LNNS 188, pp. 680–691, 2021.
https://doi.org/10.1007/978-3-030-66717-7_58

1   Problem Statement

The implementation of the principles of project management without adaptation to a specific organization, without taking into account the corporate culture, the specifics of the projects being implemented, the industry does not lead to increased management efficiency, and in some cases is a demotivating factor.

2   Analysis of Recent Researches and Publications

When managing a project portfolio, Capacity and capability management is an important process. Capacity and capability management is a comprehensive structure consisting of a set of tools and methods for identifying, distributing and optimizing resources to maximize the use of resources and minimize resource conflicts when executing a portfolio. Capacity planning allows you to determine resource requirements by measuring portfolio components in comparison with the available capacity of the organization's resources [1].

A change in the strategic direction taking place in the organization in connection with a change in the market, the focus of the organization, competition and other environmental factors has an impact on the management of human resources [2].

The approaches to changes affecting the management of human resources in projects, projects, portfolio and programs are different [2–4].

According to the PRiSM methodology, the formation of a project team and analysis of the impact of P5 (People, Planet, Profit, Process, Product) are an integral part of the "project initiation" phase [5].

In the framework of corporate knowledge management, at the end of the program, knowledge is transferred in accordance with the approved procedure, which significantly reduces the risk of loss of knowledge. Effective and proper release of program resources is an important event of program closure [4].

Promising areas for the development of the project team are [6–8]:

- improving focus;
- closely aligning talent strategy with strategy formulation, implementation, and execution.
- developing and managing key leaders through strategic partnerships.
- deepening C-suite involvement in talent strategy.

The implementation of HR Management Information Systems (HRIS) leads to increased productivity of Human Resource Management (HRM) [9–12]. When implementing flexible methodologies, it is necessary to invest in tools that facilitate and support rapid iterative development.

Innovations in the human resource management of the project are [13]:

- selection software (candidate search (sourcing), applicant tracking system tracks);
- applicant's pre-assessment;
- candidate management systems;
- online interviewing software;

- the use of artificial intelligence in recruitment;
- unlock the capabilities of passive candidates.

The use of self-organizing teams, flexible management methods have led to a number of changes in the traditional roles and practices of project management [14]. Hoda et al. identifies the main problems for each level of management: delay/change of requirements and attraction of senior management to financing at the project level; achievement of cross-functionality and effective assessments at the team level; affirmation of independence and independence on an individual level; and the absence of acceptance criteria and dependencies at the task level [14].

When considering the project portfolio as a multi-agent system, we use the An auction-based negotiation approach [15].

The importance of a holistic approach to planning and project management is noted, which will ensure the achievement of the strategic goals of the company, the preservation of critical knowledge in the company [14, 16, 17]. Critical competency management tools are [17]:

- formation of the register of critical competencies,
- multiproject teams formation,
- teams adaptation,
- monitoring of critical competencies,
- development of multi-project teams,
- PPA.

The main reasons for the formation of a personnel reserve in Ukrainian companies:

- focus on the internal resources of the company;
- staff shortage;
- aging staff;
- migration;
- cost savings;
- staff motivation;
- the ability to maintain knowledge and experience in the company;
- staff diagnostics.

The transition to Agile project management methodologies revealed that some companies were not ready to implement flexible project resource management in a multi-project environment.

Refusal from long-term planning, decentralized management, lack of an agreed human resources management strategy, expansion of project managers' powers along with a lack of coordination at the level of project management offices reduces the resource feasibility of projects and can lead to the loss of critical knowledge of project-oriented companies.

Changing requirements for portfolio projects (both exogenous and endogenous) lead to the need for reallocation of resources in a multi-project environment. Issues of providing project resources in a multiproject environment remains an urgent task [18–22].

3   The Allocation of Previously Unresolved Parts of a Common Problem. Aim of Research

An urgent scientific problem is the development of methodological foundations of human resource management in a multiproject environment. The aim of this article is to create tools for team building and reallocation of resources in a multiproject environment.

The following tasks are solved in the article:

- develop a method for analyzing the consistency of human resource management strategies in a multiproject environment;
- analyze the causes of changes and their impact on human resource management processes;
- build a model of the process of reallocation of resources;
- develop tools for team building and reallocation of resources in a multiproject environment.

In the study, methods of system analysis, the apparatus of the theory of optimization, Boolean algebra and set theory were used to solve the problems of team formation and redistribution of resources.

4   Main Research Material

4.1   Methodological Foundations of Human Resource Management in a Multi-project Environment

Despite the fact that projects carried out in a multi-project environment of companies are implemented within the framework of a single human resources management strategy, the human resources management strategies of individual projects can vary significantly.

In the case of projects in a company whose main activity is operational activities, there is a need to introduce a bimodal approach to human resource management.

The influence of stakeholders on human resource management processes arises due to the availability of expectations from the human resource management process [19]:

- definition of a human resources management strategy;
- fixed appointment – the appointment of a specific member of the project team to perform certain functions;
- determining the composition of the project teams;
- definition of resource constraints;
- determination of the amount of resources that can be allocated to a project in a multiproject environment or with bimodal management;
- prohibition of combinations;
- level of resource involvement in the project;
- lobbying for the interests of stakeholders (appointment of a protege).

According to the PMI PMBoK standard, multiple resource management processes for the i-th project can be represented as follows:

$$P_i = \{P_{i,1}, P_{i,2}, P_{i,3}, P_{i,4}, P_{i,5}, P_{i,6}\}, \tag{1}$$

where $P_{i,1}$ is the resource management planning; $P_{i,2}$ is the resource assessment operations; $P_{i,3}$ is the resource acquisition; $P_{i,4}$ is the team development; $P_{i,5}$ is the team management; $P_{i,6}$ is the resource control.

Each human resource management process is characterized by indicators:

- $LS^p_{i,j}$ is the element of an $(n \times 6)$ matrix of strategy assessments corresponding to the planned value of indicators of resource management processes in a multiproject environment;
- $LS^f_{i,j}$ is the element of an $(n \times 6)$ matrix of strategy assessments corresponding to the actual value of indicators of resource management processes in a multiproject environment;
- $LS^d_{i,j}$ is the an element of an $(n \times 6)$ matrix of permissible mismatches of indicators for evaluating strategies of resource management processes in a multiproject environment;
- $K^{cr}_{i,j}$ is the process criticality indicator.

The main stages of the method for analyzing the consistency of human resource management strategies are:

Stage 1. Determining the set S of human resource management strategies for projects implemented in a multi-project environment.
Stage 2. Checking strategies for consistency [23].
Stage 3. Determination of planned indicators of the resource management processes in the project for each project, taking into account the coordination with the selected strategy $LS^p{i,j}$ and process criticality indicator $K^{cr}{i,j}$.
Step 4. Determination of the main acceptable values of the mismatch between the planned and actual indicators.
Stage 5. Diagnostics of human resource management processes – determination of current actual values of indicators $LS^f_{i,j}$.
Step 6. Determining the mismatch of indicators for given projects D:

$$D_{i,j} = LS^f_{i,j} - LS^p_{i,j}.$$

Step 7. Evaluation of the mismatch (comparison with valid mismatch values, $LS^d_{i,j}$).
Stage 8. Development of recommendations for human resource management.

Based on the developed recommendations, a vector of changes in the human resources management processes of projects implemented in a multiproject environment is formed.

The human resources management unit must also consider the organizational and resource implications of major changes associated with portfolio components.

In the process of implementing projects in a multi-project environment, changes related to human resource management are initiated. The reasons for the changes and their impact on human resource management processes are given in Table 1.

Table 1. The causes of changes and their impact on human resource management processes.

Change	Process	Influence
Project Product Change	$P_{i,1}$, $P_{i,2}$, $P_{i,3}$, $P_{i,4}$, $P_{i,5}$, $P_{i,6}$	Changing requirements for the project team, changing the composition of the project team, reallocating resources
Change in project constraints (cost, time)	$P_{i,1}$, $P_{i,2}$, $P_{i,3}$, $P_{i,4}$, $P_{i,5}$, $P_{i,6}$	Calendar and resource rescheduling, reallocation of resources, change in project cost, change in project team financing
Changing a company's resource pool	$P_{i,1}$, $P_{i,3}$, $P_{i,4}$, $P_{i,5}$, $P_{i,6}$	Changing the resource availability function, changing the profile of competencies, the need for reallocation of resources
Change in the portfolio of projects	$P_{i,1}$, $P_{i,2}$, $P_{i,3}$, $P_{i,4}$, $P_{i,5}$, $P_{i,6}$	Redistribution and optimization of resources of a multi-project organization. Change competency profile
Change Project Priorities	$P_{i,1}$, $P_{i,2}$, $P_{i,3}$, $P_{i,4}$, $P_{i,5}$, $P_{i,6}$	Change in funding, reallocation of resources, change of priorities when accessing a common resource pool
Changing Human Resource Management Strategies	$P_{i,1}$, $P_{i,2}$, $P_{i,3}$, $P_{i,4}$, $P_{i,5}$, $P_{i,6}$	Changing requirements for the project team, changing the composition of the project team, methods for forming the project team, reallocating resources, changing staff development strategies
Conflicts between stakeholders	$P_{i,1}$, $P_{i,3}$, $P_{i,5}$	Changing requirements for project teams (prohibition of attracting human resources, combining work, participation in several projects, etc.)
Changing the legislative framework	$P_{i,1}$, $P_{i,2}$, $P_{i,3}$, $P_{i,4}$, $P_{i,5}$, $P_{i,6}$	Changing Human Resource Management Processes
Force Majeure	$P_{i,1}$, $P_{i,2}$, $P_{i,3}$, $P_{i,5}$	Revision of project teams and resource pool, reallocation of resources

The task of redistributing resources in the project portfolio is to ensure conditions for the feasibility of the project portfolio while minimizing the cost of changes.

As a result of the implementation of donor-acceptor interaction in a multi-project environment, resources are redistributed between projects in the project portfolio [24].

Redistribution of resources – a process that has a certain duration and cost.

The release of an employee and transfer to another project requires time and financial costs (transferring work to another employee in a donor project, time for transferring an employee, introduction to an acceptor project (informing, "immersion in a project", training at the workplace), reducing team effectiveness project related to the introduction of a new employee).

The cost of reallocation of resources is significantly affected by the procedure for reallocation (identifying the initiator project, choosing the order of reallocation).

The criterion for optimizing the process of reallocation of resources will be the total cost of reallocation of resources between projects.

When determining the total cost of the reallocation of resources between projects, the components are taken into account:

- the cost of calculating the redistribution of resources;
- costs necessary to withdraw an employee from a donor project;
- costs necessary to enter an employee into an acceptor project;
- costs caused by a decrease in the effectiveness of the teams associated with the introduction of new employees;
- losses associated with changes in the cost of the project.

Since matrix models of donor – acceptor interactions are often sparse matrices, it is proposed to use serial sequences for storing information on donor – acceptor interactions.

To select the optimal reallocation of resources, it is proposed to use the method of transforming combinatorial plans described in [25] and the developed software:

- Agile resource planning in a multiproject environment program;
- a program for solving the problem of providing donor-acceptor resource interaction in a multiproject environment;
- Agile project team program.

In order to formalize the process of redistribution of resources, the notation presented in the Table 2 is introduced.

Table 2. Elements of the model of the resource reallocation process.

Group	Designation	Title
Procedures	F1	Formation of a resource allocation matrix in a project portfolio (AS-IS model)
	F2	Formation of a matrix of redistribution of resources in the project portfolio (TO-BE model);
	F3	Formation of a matrix of changes
	F4	Character sequence definition
	F5	Formation of options for redistributing resources based on symbolic sequences
	F6	Definition of resource reallocation sequence
	F7	Reallocation of resources

(*continued*)

Table 2. (*continued*)

Group	Designation	Title
Inputs	I1	Resources information
	I2	Resource availability information
	I3	Change cost matrix
Control inputs	M1	Corporate standards
	M2	Reallocation tools
	M3	Resource requirements (updated)
Input process resources	R1	HR PMO
Outputs	O1	Resource allocation matrix in the project portfolio (AS-IS model);
	O2	Reallocation matrix in the project portfolio (TO-BE model);
	O3	Change matrix
	O4	Character sequences
	O5	Reallocation options
	O6	Reallocation sequence
	O7	Redistributed resources
	O8	Donor project registry (update)
	O9	Register of acceptor projects (update)

A model of the process of redistribution of resources during donor-acceptor interaction in a multiproject environment is shown in Fig. 1.

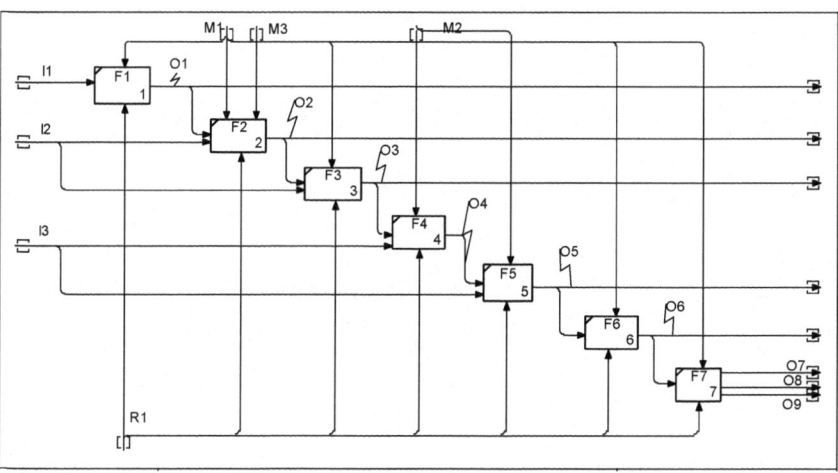

Fig. 1. Resource redistribution process model.

In the presented research the donor-acceptor approach to resource support in the multiproject environment allows the redistribution of resources between the projects included in the multiproject.

Redistribution of resources is carried out taking into account the cost and time criteria, which allows minimizing the cost of the multiproject.

4.2   Development of Human Resources Management Tools in a Multi-project Environment

Depending on the chosen human resource management strategy, various methods of forming project teams are used:

- formation of a functionally reserved team;
- formation of adaptive teams;
- team building taking into account the degree of involvement in projects;
- formation of teams with a fixed assignment of performers (prescriptive assignment of resources);
- formation of teams with a ban on combining (ban on combining functions, participating in several projects, participating in projects with different steak holders).

Since the task of redistributing resources in a multiproject environment is an NP-difficult task, there is a need to develop a software package that will automate this process. The developed complex received certificates of state registration of the author's rights to a work on a computer program.

Analysis of existing test problems and the results of experimental studies show that most of the test problems are designed to solve the classical problem of covering sets and their application to the problem of functional reservation and the task of reallocating resources in a multi-project environment is difficult.

To build only correct matrices, which will avoid the need for verification, a method for generating test problems (a method for generating a matrix with random values with a given number of units in columns) has been developed. For problems of large dimension, test cases were formed depending on the type of array of reservation coefficients.

Agile Resource Planning in a Multiproject Environment

The program is designed to solve the problem of flexible resource planning in a multiproject environment. The basis of the program is the formation of options for the distribution of the resource pool between projects, taking into account the established requirements, which will provide the ability to adapt existing teams to change.

As a result of the program's work, the generation and optimization of options for project teams implemented in a multiproject environment is carried out.

In order to ensure flexible resource planning at the generation stage, feasibility matrices of each function are generated and options for project teams and multiprojects are generated. The availability of resource allocation options will allow for the reallocation of resources between projects.

The Program for Solving the Problem of Providing Donor-Acceptor Resource Interaction in a Multiproject Environment

The program is designed to solve the problem of providing donor-acceptor resource interaction in a multiproject environment by generating options for reallocating project resources.

Based on the requirements for project teams, the availability of competencies of a certain level for applicants, resources are redistributed between donor projects and acceptor projects with cost optimization.

The redistribution of resources between projects in a multi-project environment is carried out taking into account the functional limitations and criticality of competencies, the specifics of the multi-project environment.

At the optimization stage, a multi-project team is determined that meets the requirements. The optimization criterion is the cost of performing work by members of the multi-project team.

Agile Project Team

The program is designed to form flexible teams for extreme control tasks in a multi-project environment, capable of carrying out projects under given restrictions due to the ability to adapt the team and reallocate resources between projects in a multi-project resource pool.

The program is based on the analysis of options for the distribution of functions between members of the project team, an assessment of their characteristics, the selection of the best option, and the determination of the functions implemented by each team member.

5 Conclusions and Prospects for Further Development

The article discusses and analyzes the features of human resource management in a multi-project environment.

In the study, methods of system analysis, the apparatus of the theory of optimization, Boolean algebra and set theory were used to solve the problems of team formation and redistribution of resources.

An analysis of the literature on resource management. A method for analyzing the consistency of human resource management strategies in a multiproject environment has been developed. The analysis of the causes of changes in projects and their impact on the processes of human resource management is carried out. The application of the donor-acceptor approach to resource support in a multiproject environment is considered. A model of the process of reallocation of resources is proposed. Tools were developed for team building and reallocation of resources in a multi-project environment.

A promising area is the development of a human resource management system for projects in a multiproject environment, which will reduce the impact of the subjective factor on human resource management processes.

References

1. PMI: The Standard for Portfolio Management, 4th edn. Project Management Institute, Newtown Square (2018)
2. PMI: The Standard for Program Management, 3rd edn. Project Management Institute, Newtown Square (2013)
3. PMI: A Guide to the Project Management Body of Knowledge (PMBOK Guide). Project Management Institute, Newtown Square (2017)
4. Pavlov, A.: Projects portfolio management based on PMI The Standard for Portfolio Management. Laboratoriya znaniy, Moscow (2015). (in Russian)
5. GPM Global P5 Standard, 1st edn. GPM Global (2016)
6. PMI: Talent Management: Powering Strategic Initiatives in the PMO. Project Management Institute, Newtown Square (2014)
7. PMI: Spotlight on Success: Developing Talent for Strategic Impact. Project Management Institute, Newtown Square (2014)
8. PMI: Rally the Talent to Win: Transforming Strategy into Reality. Project Management Institute, Newtown Square (2014)
9. Shahreki, J.: The use and effect of human resource information systems on human resource management productivity. J. Soft Comput. Decis. Support Syst. **6**(5), 1–8 (2019)
10. Bratton, J., Gold, J.: Human Resource Management: Theory and Practice. Palgrave, Basingstoke (2017)
11. Kavanagh, M.J., Johnson, R.D.: Human Resource Information Systems: Basics, Applications, and Future Directions. Sage Publications, Thousand Oaks (2017)
12. Dotsenko, N., Chumachenko, D., Chumachenko, I.: Modeling of the processes of stakeholder involvement in command management in a multi-project environment. In: 2018 IEEE 13th International Scientific and Technical Conference on Computer Sciences and Information Technologies (CSIT), vol. 1, pp. 29–32. IEEE, Lviv (2018). https://doi.org/10.1109/STC-CSIT.2018.8526613
13. Lazarova, T.: Innovations in human resources management. Econ. Bus. J. **13**(1), 215–223 (2019)
14. Hoda, R., Murugesan, L.K.: Multi-level agile project management challenges: a self-organizing team perspective. J. Syst. Softw. **117**, 245–257 (2016). https://doi.org/10.1016/j.jss.2016.02.049
15. Adhau, S., Mittal, M.L., Mittal, A.: A multi-agent system for distributed multi-project scheduling: an auction-based negotiation approach. Eng. Appl. Artif. Intell. **25**(8), 1738–1751 (2012). https://doi.org/10.1016/j.engappai.2011.12.003
16. Dotsenko, N., Chumachenko, D., Chumachenko, I.: Management of critical competencies in a multi-project environment. In: CEUR Workshop Proceedings, vol. 2387, pp. 495–500 (2019)
17. Dotsenko, N., Chumachenko, D., Chumachenko, I.: Modeling of the process of critical competencies management in the multi-project environment. In: 2019 IEEE 14th International Conference on Computer Sciences and Information Technologies (CSIT), vol. 3, pp. 89–93. IEEE, Lviv (2019). https://doi.org/10.1109/STC-CSIT.2019.8929765
18. Beşikçi, U., Bilge, Ü., Ulusoy, G.: Different resource management policies in multi-mode resource constrained multi-project scheduling. In: 2011 World Congress on Engineering and Technology, vol. 2, pp. 64–67. IEEE Press, Beijing (2011)
19. Erina, I., Ozolina-Ozola, I., Gaile-Sarkane, E.: The importance of stakeholders in human resource training projects. Procedia-Soc. Behav. Sci. **213**, 794–800 (2015). https://doi.org/10.1016/j.sbspro.2015.11.477

20. Mashtalir, V.P., Yakovlev, S.V.: Point-set methods of clusterization of standard information. Cybern. Syst. Anal. **37**(3), 295–307 (2001). https://doi.org/10.1023/A:1011985908177
21. Chumachenko, D., Chumachenko, K., Yakovlev, S.: Intelligent simulation of network worm propagation using the code red as an example. Telecommun. Radio Eng. **78**(5), 443–464 (2019). https://doi.org/10.1615/TelecomRadEng.v78.i5.60
22. Mashtalir, V.P., Shlyakhov, V.V., Yakovlev, S.V.: Group structures on quotient sets in classification problems. Cybern. Syst. Anal. **50**(4), 507–518 (2014). https://doi.org/10.1007/s10559-014-9639-z
23. Dotsenko, N., Chumachenko, D., Chumachenko, I.: Project-oriented management of adaptive commands formation resources in multi-project environment. In: CEUR Workshop Proceedings, vol. 2353, pp. 911–923 (2019)
24. Babayev, I.: Priorities management in the portfolio of projects in complex and dynamically variable environments. In: 2017 12th International Scientific and Technical Conference on Computer Sciences and Information Technologies (CSIT), vol. 2, pp. 187–191. IEEE, Lviv (2017). https://doi.org/10.1109/STC-CSIT.2017.8099446
25. Koshevoy, N.D., Kostenko, E.M., Pavlik, A.V., et al.: The methodology of cost-effective and time-consuming experiment design. Poltava State Agrarian Academy, Poltava (2017). (in Russian)

An Approach for Creating Learning Content from Knowledge Management System

Volodymyr Sokol📖, Mariia Bilova$^{(\boxtimes)}$📖, and Artem Kharin📖

National Technical University "Kharkiv Polytechnic Institute",
2 Kyrpychova Street, Kharkiv 61002, Ukraine
vlad.sokol@gmail.com, missalchem@gmail.com,
artem.kharin96@gmail.com

Abstract. Knowledge management plays significant role in company innovation processes since idea generation and solution development depend on knowledge and experience not available up to date. Therefore, main focus of this research is to improve effectiveness of employee knowledge management in IT company by creating appropriate learning content for personnel training. In order to accelerate that process and decrease effort necessary, we propose an approach for automatic content import from Knowledge Management System (KMS) into Learning Management System (LMS). The article includes a survey of recent approaches to knowledge management in the context of organizing e-learning process in a company. Moreover, for task of content import it was necessary to overview modern e-learning standards with their pros and cons taking into account our research goals. As a result, we have selected SCORM standard, which serves as a basis for software component and consequently decreases time for learning content creation and import.

Keywords: Knowledge management system · Learning management system · E-learning · Knowledge management · Content importing · Software component · IT company

1 Introduction: Problem Actuality and Research Goal

Domain knowledge and expertise play significant role in business processes of any company, including IT ones. Every company employee possesses relevant knowledge, which belongs to his area of responsibility and additional knowledge in different subjects. Therefore, knowledge management [1] becomes crucial for company management processes. This area of management has been developing very intensively in recent times, mainly due to comprehensive informatization of production processes, creation of data and knowledge bases, application of Internet technologies in enterprise management [2].

The introduction of knowledge management in the company allows you to systematize skills and technologies that an employee must have in order to meet the goals of the enterprise. The use of e-learning systems [3] also helps to improve the production processes. Moreover, the integration of knowledge management and distance learning approaches allows to flexibly manage knowledge of individual employees and

M. Nechyporuk et al. (Eds.): ICTM 2020, LNNS 188, pp. 692–703, 2021.
https://doi.org/10.1007/978-3-030-66717-7_59

helps to retrain and improve their skills. As a result of such integration company increases market efficiency and product competitiveness, which is especially important in the field of IT, where time to market and quality of software is the key to economic growth of the enterprise.

Thus, it is critically important to define approaches to the integration of KMS with LMS, which allows to combine their functionality for more efficient use of enterprise resources in order to create integrated knowledge-based methodological framework for staff training in IT companies [3, 4].

The system enables more effective management of employees' knowledge in the learning process at the enterprise, which is especially important in IT companies. The large amount of information that developers have to work with makes their learning process of new material rather slow [5]. Consequently, that increases time of project development, and risk of bugs and errors in the final version of software. Therefore, a manager in IT company needs a tool to systematize the knowledge required by developers to fulfil their daily tasks, as well as to solve the problem of knowledge consistency in the company and further training of new employees.

The main goal of this research is to optimize the creation of training courses through the reuse of knowledge from the KMS of the enterprise.

2   Related Works

Knowledge management processes reduce the lack of knowledge, giving employees freedom to communicate through many different channels, thus providing access to knowledge to different groups: senior executives, lower-level executives, project managers, research and development professionals.

The main goal of knowledge management in any organization is to turn data into information, and information into knowledge. A knowledge management strategy can allow IT company to utilize its knowledge resources more effectively. At the same time, the idea of e-learning lies in transferring explicit knowledge through the creation of learning and training modules, discussion boards, polls and other methods of learner participation [3]. Thus, both disciplines deal with knowledge capture, sharing, application, and potentially knowledge generation [6].

Knowledge management and e-learning define a certain kind of cycle, which consists in increasing the knowledge of employees through training, which in its turn contributes to the accumulation of knowledge in the company. Thus, manager may easily create courses that are tailored to meet organizational needs for increasing capability of each individual within the company [7]. From the standpoint of knowledge management, the costs required for the development and further implementation of training materials can be reduced through effective staff training [8, 9].

According to studies, e-learning is more effective than knowledge transfer through other channels such as e-mail and teleconference [6]. At the same time, there are currently no tools that will allow the integration of the Knowledge Management System (KMS) and the Learning Management System (LMS) to ensure the formation of training courses in accordance with the knowledge of the organization's employees. It is necessary to develop software that will eliminate this problem by transferring content from KMS to LMS and further formation of educational material in the form of appropriate courses.

3   Analysis of E-Learning Standards

To ensure the effectiveness of interaction with existing knowledge and the ability to use the knowledge of former employees for rapid and effective training of new staff, it is necessary to be able to import content from KMS to LMS. Therefore, we need to use existing LMS course standards to ensure that process. The following e-learning standards can be used for that purpose.

AICC. The Aviation Industry CBT Committee (AICC) standard is based on the use of HTTP AICC communication protocol (HACP) to facilitate two-way communication between the existing LMS and the developed course content. The AICC was developed to meet the standardization of computer training in the aviation industry, and in 1998 was adapted for e-learning [10]. The HACP technique is quite simple and involves the use of HTML-forms and simple text strings to transfer information to and from the LMS [11].

In AICC, course elements are divided into three groups: Assignable Units (AUs), Blocks and Objectives. AUs are the smallest educational elements that can be presented to a student, like an HTML page. AUs represent the lessons of a course. Blocks are used for nesting. Objectives are used to define simple or complex course requirements and represent goals that must be achieved in the course [12].

One of the most important advantages of AICC standard is data security through the use of HTTPS for data transfer between content and LMS. Disadvantages include the lack of course tracking capabilities, limited reporting capabilities, compatibility issues, and a general decline in support among content providers for relevant systems [11, 12].

SCORM. SCORM (Shareable Content Object Reference Model) represents a collection of standards and specifications for web-based learning and defines standards for interoperability between LMS and learning content, so that Sharable Content Objects (SCOs) which contain the learning content, can be exchanged among LMSs using XML files and manifests for describing the course content [11]. It helps to create the connection between general metadata specifications and specific content models [12].

In general, SCORM is a model containing set of technical specifications and procedures for building e-Learning software. When applied to e-Learning course content, produces small, reusable e-Learning objects. The online learning content and LMS software communicates with each other to bring the specific courses to the learner; the SCORM facilitates to do that.

SCORM consists of several modules such as Content Aggregation Model (CAM), Run-time Environment (RTE), and Sequencing and Navigation (SN). These modules are the core of the SCORM [13]. SCORM elements include Raw media (like Components, Fragments), Content (like Lessons, Modules, Units) Course. SCORM also specifies the packaging of SCOs into a ZIP file with the Package Interchange Format [11].

We have selected this standard as a basis for the research, despite a number of disadvantages, including the lack of support for offline learning (it is impossible to work with the course in the absence of the Internet), limited number of parameters to track, use of Flash to present course content (transition to HTML 5 is accompanied by reduced digital content quality). The choice has been done is due to simplicity of implementation and availability of a wide support.

TinCan (xApi). The TinCan standard, later revised into the Experience API (xAPI), was developed to address common SCORM shortcomings. It is very similar to its predecessor, however extends its functionality to reporting multiple scores instead of single scores, improved security, platform transitions, and tracking of numerous types of learning objects and resources. In addition, no web browser and no LMS are required, and the interaction from the learner with the learning object can be recorded as a grammar with noun, verb and object [11].

This standard is based on the people abilities to learn constantly, whether they are communicating with other people on the web, reading news blocks, or searching for some information The data is recorded in the Learning Record Store (LRS) and can be shared with other LRS systems [14].

Despite the provision of offline and mobile learning, as well as the ability to work with complex learning scenarios (simulations, games, blended learning), this standard has not yet found stable support, and requires to use additional LRS-tools, which complicates its further implementation.

CMI5. One of the modern e-learning standard, which uses the definition of the xAPI standard with interoperability determined between different systems [15]. CMI5 was created to replace both the AICC and SCORM specifications with a more feature-rich and robust solution. However, both specifications had technical issues and constraints as well as significant overlap [16].

CMI5, as an add-on to xAPI, narrows the format, making it more suitable for use in the familiar way of organizing distance learning online: course + LMS. At the same time, more advanced users get a wide range of opportunities provided by xAPI: mobile learning, receiving data from various sources and aggregating them in LRS.

At the same time, CMI5 is not yet widespread enough, so it is proposed to focus on the use of SCORM as a standard for content transfer.

4 Software Component Development for Creating and Importing Learning Content into LMS

4.1 Architecture Design

The component is an application for the integration of LMS and KMS, the main task of which is to enable the reuse of existing knowledge and reduce the time for the formation of educational content.

The stakeholders of the system for importing educational content into LMS in IT companies are the IT companies themselves, which have goals such as increasing the company's profits through more qualified staff; reduction of staff training costs through reuse of knowledge presented in KMS; reducing the use of resources to perform a unit of work.

The main users of the system are the company's employees and their technical managers or teachers, who act as an expert group to assess the level of knowledge of each employee. Managers should also choose the final competence and priorities in the choice of training content for each group or individual worker. Employees, in turn, play the role of a student who uses created training.

To create the correct SCORM package, it is necessary to generate the correct manifest file, which describes the content of the course in XML format.

The main structural elements of a simple course are resources – a list of "parts" that make up the course. There are two types of resources, Shareable Content Objects (SCO) and assets.

An asset is a collection of one or more files that make up a logical unit. Assets can be either separate learning units ("parts" of a course) or logical sets of files that are reused in other parts of the course.

SCOs are training units that also consist of one or more files. SCOs are almost always the teaching parts of the course.

The key difference between an SCO and an Asset is that an SCO can communicate with an LMS, while an Asset is simply static content that is delivered to a user. Any resource that a student can launch contains a pointer to the page to which the LMS must redirect the student to launch the resource. Resources must also contain an exhaustive list of all files required for that resource to function properly, so that they can be transferred to new environments and continue to function.

The process of creating a SCORM course is shown in Fig. 1.

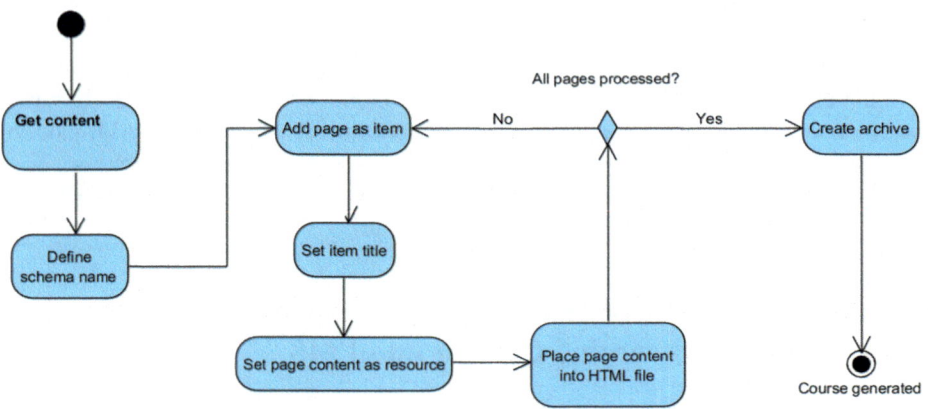

Fig. 1. SCORM course creation.

Figure 2 shows a sequence diagram of the creation of Learning Objects (LO).

The system assumes only one type of users - a teacher or a manager (expert) (Fig. 3). The main function of the application is to import information from KMS,

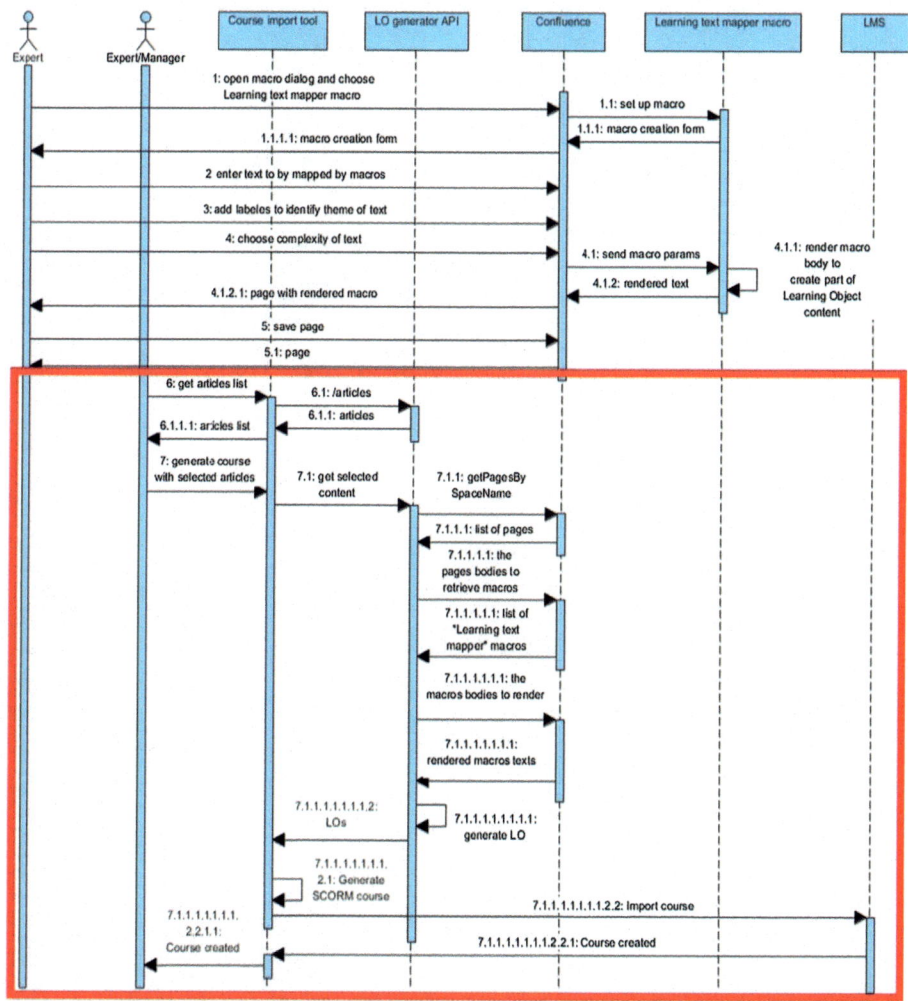

Fig. 2. Sequence diagram for Learning Object generation.

process it, generate SCORM or HTML package for the course and create a corresponding course in LMS.

The architecture of the software component defines several layers of the software package: the level of data access, the level of business logic, and the level of data representation. The level of data access will be both LMS and KMS, as the main owners and users of data and databases for additional information.

The main task of the system is to work with data stored in the database and on remote servers of LMS and KMS. The software client will visually interpret the information received from the server in a user-friendly form.

Conceptual data model is given in the form of UML class diagram (Fig. 4).

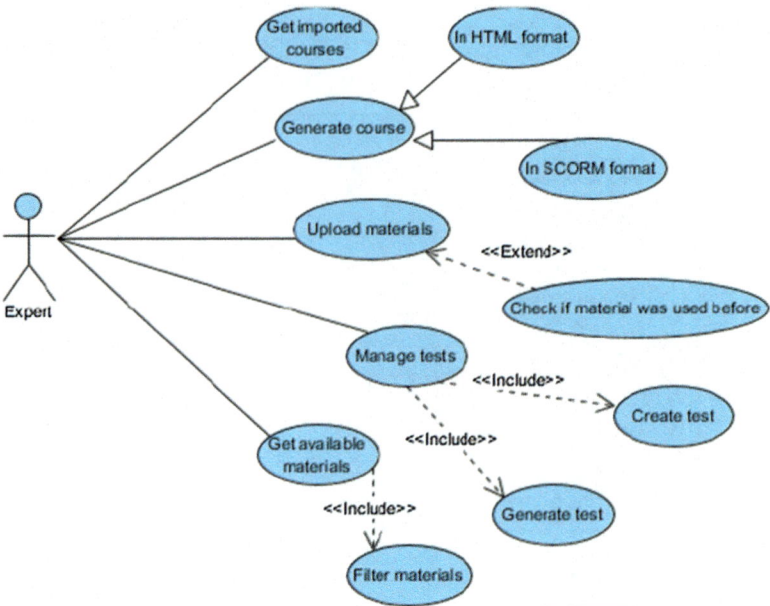

Fig. 3. Use case diagram for the functional requirements.

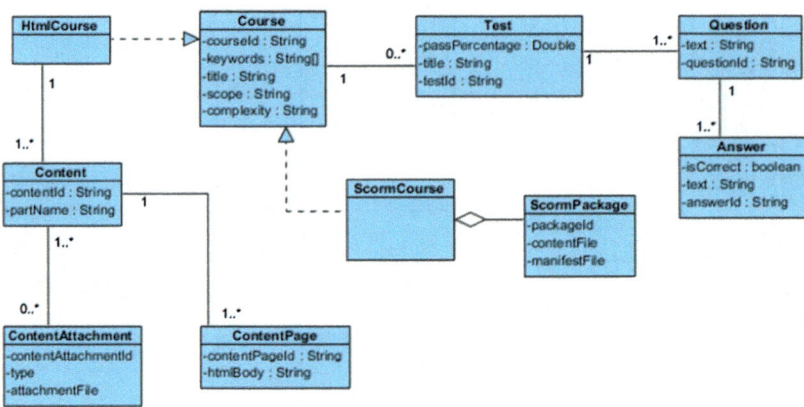

Fig. 4. Class diagram for the elaborated conceptual data model.

Analysis of the subject area allows us to conclude that there is a rather small amount of data. Therefore, queries remain relatively compact to get required information, however it may become case later in the future, in the case of expanding functionality, where system has to process a large number of queries to the database. According to the above features, a satisfactory choice of system architecture is a "thin client". Therefore, taking into account the selected 3-level architecture, you may see deployment diagram of the system on Fig. 5.

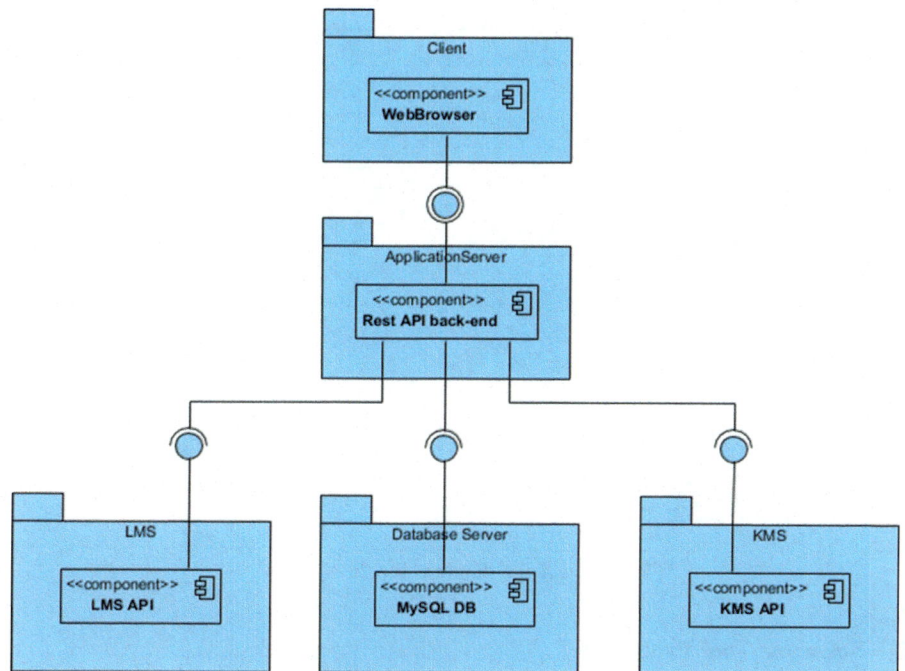

Fig. 5. Deployment diagram for the system architecture.

The deployment diagram (see Fig. 5) shows 5 nodes on which 5 components are located. MySQL is a database component which gets queries from the next component. REST-API back-end is a component that implements database queries and interacts with the LMS API and KMS API, executes business logic and implements REST services. WebBrowser is a component that is a Web-application, uses REST-services. LMS API is a component that is the target LMS. KMS API - a component where knowledge is stored in the enterprise (KMS).

The package diagram (Fig. 6) shows the packages that were designed during software development. All packages belong to certain layers of architecture.

The data layer includes the model and repository packages. The layer of business logic includes service and communicator packages. The representation layer includes the controller package. The model package includes classes that are reflected from the tables in the database. The repository package is a layer of DAO. This package contains classes which interact with the database. The service package is a layer of business logic. It contains classes that process data coming from REST services, communicator API and DAO layer. The controller package is a layer of data presentation. It contains classes that are REST services. The security package includes an application security implementation. It contains the OAuth2 provider configuration and the application access configuration. The communicator package contains classes that implement access to API KMS and LMS systems.

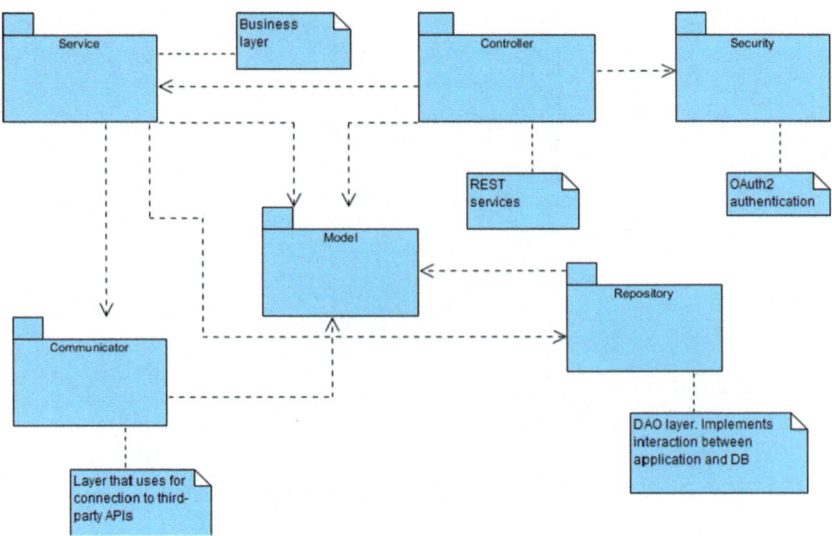

Fig. 6. Package diagram for the system.

4.2 Software Tool Prototype

In accordance with the functional requirements (see Fig. 3), the appropriate database model (Fig. 4) and the chosen architecture (see Fig. 5) software component was developed. Especially, on Fig. 7 the user's interface fragment is shown.

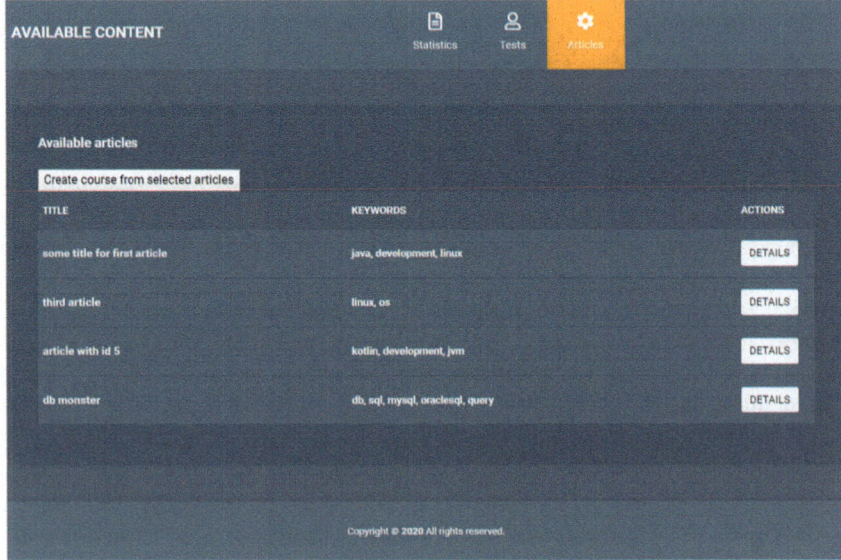

Fig. 7. The list of available content from KMS.

The experiments were performed on the data obtained during the creation of a test course based on data obtained from KMS, which is used in the IT-company "Academy – Smart" LTD, Kharkiv [17]. The system was deployed on a Windows server according to the following parameters:

- CPU Athlon 3000G with frequency 3.5 GHz;
- 4 Gb RAM;
- Windows 10 operating system.

For the experiment we have completed 30 test runs with different data. Evaluation of efficiency has been carried out by means of average time analysis, which is required to complete each stage of creating educational content. The following Table 1 presents experimental results.

Table 1. Results of experiments.

	Manual mode (minutes)	Application (minutes)	Difference (%)
Fetching content list (with certain parameters)	10	7	30
Content copying	7	2	72
Test creation	10	5	50
SCORM package generation	1	1	0
TOTAL	28	15	46

Thus, based on the results of the study and with software component integration in the process of creating educational content, we can conclude that the proposed approach reduces the time of the that process by 46%. Consequently, that leads to increased efficiency of managers or experts responsible for training of new employees. The developed component can be used in the process of creating educational content.

The results of the experiments may differ depending on the details of the course and employees responsible for creating content with help of developed application. Thus, we provide test results only a separate instance of software usage in a specific configuration.

5 Conclusion

This paper includes an overview of Knowledge Management Systems and their features, how they enable accumulation and further usage of employees knowledge in IT companies. Features of educational content creation by means of LMS according to the existing standards in the field of e-learning are overviewed. Proposed approach allows to integrate KMS and LMS by generating learning objects from knowledge base and by importing it into LMS. An appropriate software component has been developed, the test cases shows that the time of creating training courses was reduced by 46%.

References

1. Ouriques, R.A.B., Wnuk, K., Gorschek, T., Svensson, R.B.: Knowledge management strategies and processes in agile software development: a systematic literature review. Int. J. Software Eng. Knowl. Eng. **29**(03), 345–380 (2019). https://doi.org/10.1142/S0218194019500153

2. Kuusinen, K., Gregory, P., Sharp, H., Barroca, L., Taylor, K., Wood, L.: Knowledge sharing in a large agile organisation: a survey study. In: Baumeister, H., Lichter, H., Riebisch, M. (eds.) XP 2017. LNBIP, vol. 283, pp. 135–150. Springer, Cham (2017). https://doi.org/10.1007/978-3-319-57633-6_9

3. Tkachuk, M.V., Sokol, V.E., Bilova, M.O., Kosmachev, O.S.: Classification, typical functionality and application peculiarities of learning management systems and training management systems at IT-companies. Mod. Inf. Syst. **2**(4), 87–95 (2018). (in Ukrainian)

4. Sokol, V.E., Tkachuk, M.V., Vasetka, Y.M.: Adaptive training system for IT-companies personnel: design principles, architectural models and implementation technology. Bull. Nat. Tech. Univ. "KhPI". Ser. Syst. Anal. Manag. Inf. Technol. **51**, 38–43 (2017)

5. Dingsøyr, T.: Knowledge management in medium-sized software consulting companies. An investigation of intranet-based knowledge management tools for knowledge cartography and knowledge repositories for learning software organisations. Dissertation, Norwegian University of Science and Technology (2002)

6. Tessier, D., Dalkir, K.: Implementing Moodle for e-learning for a successful knowledge management strategy. Knowl. Manage. E-Learn. Int. J. **8**(3), 414–429 (2016). https://doi.org/10.34105/j.kmel.2016.08.026

7. Zahari, A.S.M., Salleh, S.M., Baniamin, R.M.R.: Knowledge management and e-Learning in organisations. J. Phys: Conf. Ser. **1529**(2), 022051 (2020). https://doi.org/10.1088/1742-6596/1529/2/022051

8. Emerson, L.C., Berge, Z.L.: Microlearning: knowledge management applications and competency-based training in the workplace. Knowl. Manage. E-Learn. Int. J. **10**(2), 125–132 (2018). https://doi.org/10.34105/j.kmel.2018.10.008

9. Sucahyo, Y.G., Utari, D., Budi, N.F.A., et al.: Knowledge management adoption and its impact on organizational learning and non-financial performance. Knowl. Manage. E-Learn. Int. J. **8**(2), 387–413 (2016). https://doi.org/10.34105/j.kmel.2016.08.025

10. Shariat, Z., Hashemi, S.M., Mohammadi, A.: Researchand compare standards of e-Learning management system: a survey. Int. J. Inf. Technol. Comput. Sci. **6**(2), 52–57 (2014). https://doi.org/10.5815/ijitcs.2014.02.07

11. Behringer, R.: Interoperability standards for microlearning. In: Proceedings of MicroLearning Conference 7.0, Stift Goettweig (Austria), 26–27 September 2013 (2013)

12. Stratakis, M., Christophides, V., Keenoy, K., Magkanaraki, A.: E-Learning standards. SeLeNe (Self E-Learning Networks IST-2001-39045) (2003)

13. Gurunath, R., Kumar, R.A.: SaaS explosion leading to a new phase of a learning management system. Int. J. Curr. Res. Rev. **7**(22), 62–66 (2015)

14. Lim, K.C.: Case studies of xAPI applications to e-Learning. In: The Twelfth International Conference on eLearning for Knowledge-Based Society, pp. 3.1–3.12 (2015)

15. Bakhouyi, A., Dehbi, R., Lti, M.T., Hajoui, O.: Evolution of standardization and interoperability on E-learning systems: an overview. In: 2017 16th International Conference on Information Technology Based Higher Education and Training (ITHET), Ohrid, pp. 1–8. IEEE (2017). https://doi.org/10.1109/ITHET.2017.8067789
16. Victor, S., Werkenthin, A.: Cracking the mobile learning code: xAPI and cmi5. Technical report. Obsidian Learning (2016)
17. "Academy – Smart" IT-company: Official web-site. https://academysmart.com.ua. Accessed 13 July 2020

The Cascading Subsystem of Key Performance Indicators in the Enterprise Performance Management System

Valentina Moskalenko[1]($\boxtimes$) (ID) and Nataliia Fonta[2] (ID)

[1] National Technical University "Kharkiv Polytechnic Institute",
2 Kyrpychova Street, Kharkiv 61002, Ukraine
valentinamoskl7@gmail.com
[2] DataArt, 7/8 Zakhysnykiv Ukrayiny Square, Kharkiv 61000, Ukraine

Abstract. The problems of integration for systems of strategic, tactical, operational planning and budgeting into a single company planning system were analyzed. The company which uses process management and Balanced Scorecard (BSC) to create a Key Performance Indicators (KPIs) system, is considered. The structure of the cascading subsystem for strategic map and BSC is proposed as part of the Enterprise Performance Management System which is being implemented at the company to automate strategic management processes. The cascading method is suggested which is implemented in the process of developing a strategic program for company growth. Strategic KPIs are formed based on the strategic map. Further, these indicators are cascaded in three strategic aspects: by business processes; by company divisions; by planning levels: strategic, tactical and operational (level of budgets). Business processes are considered for the main, investment and financial activities. As a result of the cascading process, the KPIs system is built and the planned values of the strategic KPIs, KPIs over the years for the strategic period and the indicators of the company's strategic budgets are determined. This is the basis for the developing of strategic, tactical, operational plans and strategic budgets of the company and its divisions. This method allows to carry out strategic alignment in the company, coordinate the strategic goals of the company and its divisions, and harmonize the plans of the company and divisions in terms of process management.

Keywords: Key Performance Indicator · Cascading a balanced scorecard · Enterprise Performance Management System · Strategic alignment

1 Analysis of the Problems of Integration for Strategic, Tactical and Operational Planning Systems

One of the main problems of strategic management is the feasibility of strategic plans. This is due to the "gap" in strategic and operational planning, the lack of communication or the presence of a "weak" connection between strategic objectives, strategic indicators, indicators of current activities and budget indicators [1]. Therefore, the main task of the company's top management is to build such a management system that

M. Nechyporuk et al. (Eds.): ICTM 2020, LNNS 188, pp. 704–715, 2021.
https://doi.org/10.1007/978-3-030-66717-7_60

would cover the strategic, tactical and operational levels of management [2]. All management functions at these levels should be implemented within such system. Management function planning and questions of the integration for strategic, tactical and operational planning systems into a single planning system will be considered. We will highlight the main tasks that must be solved for such integration.

As for now, the strategic alignment process is used to solve the problems of strategic plans implementation. Strategic alignment is seen as a tool to achieve the intended goals [1]. The company identifies the processes required for strategies implementation by applying the strategic alignment process [3].

Analyzing the existing research in the field of strategic leveling, it can be concluded that the main focus is made by researchers on creating a relationship between strategies and business processes, projects and operational plans. In recent years, modern companies often use the Balanced Scorecard (BSC) [4]. It allows to analyze systematically the activities of the company, to carry out planning for both short-term and long-term periods, taking into account the interests of all interested stakeholders. The cascading process is carried out for this - the construction of strategic maps for divisions (different levels of the organizational hierarchy) of a company [5]. Strategic maps of individual divisions should fit organically into the general corporate strategic map. When developing a strategic map for a division, first of all, it is necessary to determine (or view) its role and main functions in the overall structure of the company. The following methods of cascading company goals are used depending on the degree of involvement in the process of divisions' strategic planning:

- the method of independent formulation by departments of their strategies and goals, taking into account the strategic framework of their activities and specific tasks on the part of top management;
- the method of direct determination of the divisions' goals by the top management of the company, while the process of coordinating the goals is practically absent;
- the method of forming a standard strategic map (template) with the adaptation of target indicators and strategic measures for the divisions, that is, all the divisions' strategic maps look the same, but have different target values which are determined individually;
- a method of combining standard and individual goals: goals which are supported by specific division, are taken from the level of a company strategic map and individual division's goals which are important only for it, are added;
- the method of coordinated determination of the company's strategic map: the strategic map of the highest level is analyzed from the point of view of the divisions' interests, then the strategic maps of all divisions are determined in the approval process.

All goals of the company and divisions should be measurable, therefore they are described by the systems of Key Performance Indicators (KPI) [6]. The achieved values of these indicators will determine the degree of goals attainability and the effectiveness of the selected strategies. Therefore, within the framework of strategic leveling, the task of adequately presenting goals and strategies through the system of strategic KPIs and the task of cascading these indicators to the divisions' indicators and the performance of the company's personnel are considered [7].

Based on the analysis of the problems for BSC cascading as well as approaches to the integration of strategic, tactical and operational planning systems, the following conclusions were made.

1. The cascading of goals and KPI is considered for structural divisions in the majority of literary sources about the development of company BSC [4]. The most frequently combined system "top-down" and "bottom-up" planning is suggested. The management of the company and the heads of departments agree the strategic maps and divisions KPI. As a result, the company's KPIs of the responsible divisions are assigned to KPI [7]. But company's KPI the most often are aggregated indicators that reflect the work efficiency of many divisions and performers. Therefore, such consolidation leads only to formal strategic planning. Such a distribution can be effective only in small companies, when the activity of a division can be unambiguously described by the same indicators as the efficiency of the entire company. Therefore, to build an adequate KPI system, it is necessary to solve the issue of decomposing aggregated company's KPIs into divisions' KPI and the issue of resolving "conflicts" between divisions' KPI.

2. Cascading company's KPI to KPI of the key employees - heads of departments [8]. Such cascading is considered as the next stage after cascading across functional divisions and as part of developing a company personnel incentive system.

3. Cascading KPIs to KPIs of business processes is conceptually described as part of BSC building, where business processes are considered as one of the perspectives for generating goals and indicators [9, 10]. Here, the main problem is to build a business process system that would cover all the main activities of the company and reflect this activity at the management levels, i.e. business processes of strategic, tactical and operational level.

4. Temporary decomposition – the cascading of strategic KPI values to annual values, to values of indicators characterizing the current company's activity. This is mainly the work of authors who work on the integration of strategic, annual planning and budgeting [11, 12]. Here, the problems of determining the company's annual KPI values and subdivisions KPI and/or business processes KPI remain unsolved. The lack of clear rules for determining the planned values of these KPIs leads to the fact that annual plans are not based on the strategic goals and strategies for achieving them, but on the basis of the achieved state. This ultimately leads to the fact that while the annual plans are fulfilled, the strategic goals are not achieved. At the level of operational management of individual divisions or other responsibility centers, indicators of operational budgets (number of sales, number of clients, costs by budget items, etc.) are taken as indicators of control over their activities. The values of these indicators are not analyzed in terms of their impact on strategic level KPIs. This is one of the main reasons for the "isolation" of operational and strategic management. Therefore, it is necessary to embed in the companies such budgeting systems that are associated with strategic planning systems. And as a result, each budget indicator must be linked to a strategic KPI. The integration of the budgeting system and the system of strategic indicators is most often carried out on the basis of the strategic budgeting subsystem formation. The formation of strategic budgets is carried out on the basis of the results of long-term forecasts, then a certain

aggregation of budget indicators is carried out, on the basis of which the company's strategies are formed. With this approach, "from budgets to strategic goals", strategic KPIs are formed taking into account budgetary indicators. And this will mean that the company will change its current activity on the basis of forecasts adapting itself to the market. However, this will not give reason to build strategic development goals for the future. This approach is effective for small companies that plan their activities only for the short term and consider the survival strategy as the main strategy. It is necessary to combine approaches to the formation of strategic budgets: targeting both sales forecasts and the connection of budget indicators with strategic KPIs for companies that want to develop in the long term.

2 BSC Cascading Process as a Basis for Strategic Leveling

Modern company management is increasingly based on the process approach. The management system is considered as a set of business processes management:

- management business processes that form the upper level, strategic goals are formed here, a strategic map is built, key indicators for a long-term period are determined;
- the main and auxiliary business processes that cover the current activities of the company. The main business processes are carried out within the framework of the operating activities, auxiliary ones ensure the continuity of the main processes and the business processes of the investment and financial activities of the company will be considered as ensuring ones;
- structural divisions which implement business processes;
- business process performers is a staff of structural divisions.
- The system of plans is needed to effectively manage such a company which is built on a KPI system that covers all these levels.

It will be assumed that a strategic map is built at the strategic level and strategic KPIs are formed on the basis of the BSC methodology. Further, it is proposed to carry out KPI cascading in three aspects [13]:

- by business processes, which are formed for the main, investment and financial company's activities;
- by the structural divisions of the company, which are responsible for achieving the planned KPI values of the relevant business processes;
- by terms of planning: the strategic value of KPI, the value of KPI by year of the strategic period, the values of the company's budgets and its divisions.

The cascading method consists of the following steps.

Step 1. Identifying the dependence of the strategic level of KPI from the primary indicators which are reflected in the accounting systems of the company by using numerical methods. Primary indicators include, for example, equity and debt capital, investment, production costs, gross income and others. As the result $FK_i^k = f_i^k(\{PP_j\}_i^k)$, $j \in M_i^k$, $M_i^k \subset M^{PP}$, where $\{PP_j\}_i^k$ are set of primary indicators;

M_i^k is the the set of indexes of primary indicators for the i-th KPI; f_i^k is some function; M^{PP} is set of indexes of all primary indicators $PP = \{PP_j\}, j \in M^{PP}$.

Step 2. Determining the degree of indicators influence $\{PP_j\}, j \in M^{PP}$ on appropriate $\{FK_i^k\}$. All indicators are separated on two groups based on expert analysis. The first group includes primary indicators – parameters, their values do not change during the strategic period, the second group includes variables.

Step 3. Determinating of planned KPI values of strategic level $\{\bar{F}\bar{K}_i^k\}^t$ and values of primary indicators $\{\bar{P}\bar{P}_j\}^t$, $t = \overline{1, T}$. This should be done as follows.

3.1. Streamlining the KPI $\{FK_i^k\}$ by descending priority according to the values of importance coefficients $\{\alpha_i^k\}$, which are determined on the basis of analytic hierarchy process within the k-th perspective.

3.2. Determining of strategic values $\{\bar{F}\bar{K}_i^k\}^T$ KPI and primary indicators $\{\bar{P}\bar{P}_j\}^T$. To do this, a procedure is performed, which is based on the method of sequential concession for ordered criteria of a multicriteria task. The procedure is performed for the KPI within each perspective. It is believed that the prospects are arranged as follows: finance, customers, business processes, staff.

First, the maximum acceptable planned value of the first priority KPI is defined – $\bar{F}\bar{K}_1^{*kT}$, $k = F$ on the set of allowable values. Then the values $\{\bar{P}\bar{P}_j\}_1^T$ are calculated according to $FK_1^{kT} = f_1^k(\{\bar{P}\bar{P}_j\}_1^{kT})$. The values $\{\bar{P}\bar{P}_j\}_1^T$ are used to define the KPI with the lower priority as follows. The value of the second priority KPI $\bar{F}\bar{K}_2^{*FT}$ is determined. If $\{PP_j\}_1^{kT} \cap \{PP_j\}_2^{kT} \neq \varnothing$ and it is impossible to calculate $\bar{F}\bar{K}_2^{*kT}$ based on values $\{\bar{P}\bar{P}_j\}_1^T$, then a concession Δ_1^k is established for indicator FK_1^{*kT} and the best value of $\bar{F}\bar{K}_2^{*kT}$ is determined according to $FK_2^{kT} = f_2^k(\{PP_j\}_2^{kT})$ with the condition that the value of the first KPI will not be less than $\bar{F}\bar{K}_1^{*kT} - \Delta_1^k$ and etc. Therefore, the values $\{\bar{F}\bar{K}_i^k\}^T$ are formed based on solving the sequence of the following tasks:

1) Determine $opt\ FK_1^{kT}(\{PP_j\}_1^{kT}) = \bar{F}\bar{K}_1^{*kT}, j \in M_1^k$;

2) Find $opt\ FK_2^{kT}(\{PP_j\}_2^{kT}) = \bar{F}\bar{K}_2^{*kT}$ in the area specified by the conditions: $FK_1^{kT}(\{PP_j\}_1^{kT}) \geq \bar{F}\bar{K}_1^{*kT} - \Delta_1^k; j \in M_2^k$;

3) Find $opt\ FK_{n^k}^{kT}(\{PP_j\}_{n^k}^{kT}) = \bar{F}\bar{K}_{n^k}^{*kT}$ in the area specified by the conditions: $FK_i^{kT}(\{PP_j\}_i^{kT}) \geq \bar{F}\bar{K}_i^{*kT} - \Delta_i^k$, $i = 1, \ldots n^k - 1$; $j \in M_i^k$, $n^k = |N^k|$. Condition "$\geq$" in the tasks $1..n^k$ corresponds to that value FK_i^{kT} should be maximized, otherwise a sign "$\leq$" is placed. In the case of a "conflict" situation, when it is impossible to define $\bar{F}\bar{K}_s^{*kT}$ for some s-th KPI in area which is given with conditions $FK_i^{kT}(\{PP_j\}_i^{kT}) \geq \bar{F}\bar{K}_i^{*kT} - \Delta_i^k$, $i = 1, \ldots, s - 1$, the decision-maker changes the value of the concession Δ_{s-1}^k first and calculate $\bar{F}\bar{K}_s^{*kT}$. If change Δ_{s-1}^k to an acceptable level does not allow to calculate $\bar{F}\bar{K}_s^{*kT}$, then Δ_{s-2}^k should be changed, then Δ_{s-3}^k until Δ_1^k.

This iterative procedure for adjusting concessions allows to determine the strategic values of $\{\bar{F}\bar{K}_i^k\}^T$ and $\{\bar{P}\bar{P}_j\}^T$.

3.3. Calculated planned KPI values $\{\bar{F}\bar{K}_i^k\}^t$, $i \in N^k$ and primary indicators, $j \in M^{PP}j \in M^{PP}$, $t = \overline{1, T-1}$. To do this, the task of dynamic programming with discrete variables is formed. It is solved by using the follow procedure. Spaces are considered $\{\Pi^t\}$, $t = \overline{0, T}$ with dimension $\bar{M}^P$ of primary indicators $\{PP_j\}^t$, $j \in M^{PP}$, $\bar{M}^P = |M^{PP}|$.

Using the method of concessions similar to the procedure of finding $\{\bar{F}\bar{K}_i^k\}^T$ in 3.2, on the intervals $t = T-1, \ldots 1$ variants of values KPI $\{\bar{F}\bar{K}_{iv}^{kt}\}$ are formed each of which corresponds to a set of primary indicators $\{\{\bar{P}\bar{P}_j\}_v^t\} \in \Pi^t$, $v = \overline{1, V^t}$, V^t is the number of options.

The decision-maker determines the initial concessions for the first priority KPI $\{\Delta_{1v}^{kt}, k = F\}$, $t = T-1, \ldots 1$, $v = \overline{1, V^t}$ depending on the initial state $(\{\bar{F}\bar{K}_i^k\}^0$, $\{\bar{P}\bar{P}_j\}^0) \in \Pi^0$) and strategic goal $(\{\bar{F}\bar{K}_i^{kT}\}, \{\bar{P}\bar{P}_j\}^T \in \Pi^T)$.

Next, experts assess the feasibility of transitions from the values of indicators in the $(t-1)$-th interval to the values in the t-th interval $(t = \overline{1, T-1})$ based on the analysis of company production capacity (see Fig. 1).

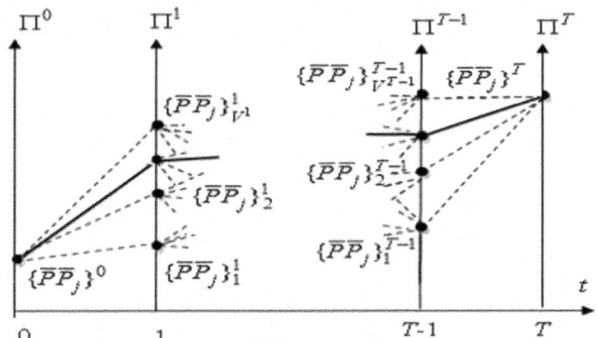

Fig. 1. Variants of primary indicators values.

It is necessary to resolve following task to define $\{\bar{P}\bar{P}_j\}^t$, $t = \overline{1, T-1}$:

$$I_\Sigma^{P*}(PP) = \min_{PP} \{\sum_{t=1}^{T} I^{Pt}(PP^{t-1}, PP^t)\}, \tag{1}$$

$$\sum_{\tau=1}^{t} I^{P\tau}(PP^{\tau-1}, PP^\tau) \le \sum_{\tau=1}^{t} \bar{I}^{P\tau}, \tag{2}$$

$$I^{Pt}(PP^{t-1}, PP^t) \le \bar{\bar{I}}^{Pt}, t = \overline{1, T}, \tag{3}$$

where $I^{Pt}(PP^{t-1}, PP^t)$ is amount of investments which are required to support «transition» from the values of primary indicators in the $(t-1)$-th interval to values in the

t-th interval, $I^{Pt}(PP^{t-1}, PP^t) = \sum_{j \in M^{PP}} I_j^{Pt}(PP_j^{t-1}, PP_j^t)$, where $I_j^{Pt}(PP_j^{t-1}, PP_j^t)$ is amount of investments which are required for changing of primary indicator with value $PP_j^{t-1} \in \{\{\bar{P}\bar{P}_j\}_v^{t-1}, v = \overline{1, V^{t-1}}\}$ to value $PP_j^t \in \{\{\bar{P}\bar{P}_j\}_v^t, v = \overline{1, V^t}\}$; $\bar{I}^{Pt}$ is planned amount of investments in the t-th interval; $\bar{\bar{I}}^{Pt}$ is amount of investments which will be able to spend in the t-th interval.

Task (1)–(3) is resolved by using method of sequential analysis of options solutions. As a result, the optimal trajectory is formed for the company transition from the initial state $\{\bar{P}\bar{P}_j\}^0 \in \Pi^0$ to the target $\{\bar{P}\bar{P}_j\}^T \in \Pi^T$. Thus, sets are formed $\{\bar{P}\bar{P}_j\}^t$ which define $\{\bar{F}\bar{K}_i^k\}^t$, $t = \overline{1, T}$.

Step 4. Distributing indicators $\{PP_j\}, j \in M^{PP}$ by experts into three groups which form sets $\{PP_j, i \in \widehat{N}^s\}$, $s \in \{B^a, I^a, F^a\}$ where indexes s characterize the type of activity: B^a is operating activity, I^a is investment activity and is financial activity $M^{PP} = \widehat{N}^{B^a} \bigcup \widehat{N}^{I^a} \bigcup \widehat{N}^{F^a}$. These activities are recorded separately in the financial statements of management and accounting, as well as in the reporting of divisions which are responsible for individual business processes in the company.

Step 5. Defining of KPI for business processes. It is considered that business processes were developed for each type of activity. They are marked A^s where $s \in \{B^a, I^a, F^a\}$ is set of indexes for business processes. Each primary indicator PP_j , $j \in \widehat{N}^s$, $s \in \{B^a, I^a, F^a\}$ is got in accordance KPI of business process $\{P_b, \ b \in B_j\}$ from experts, where B_j is set of KPI business processes indexes. Each i-th business process is characterized by set of KPI $\{P_b, \ b \in \bar{B}_j, j \in \bar{N}_i^s\}$, $i \in A^s$, where $\bar{N}_i^s$ is set of primary indicators indexes for the s-th type of activity on the basis of which KRI of the i-th business process are formed; $\bar{N}_i^s \subseteq \widehat{N}^s$; $\bar{B}_j$ is set of KPI indexes for the i-th business process which are got based on PP_j, $(\bar{B}_j, \ j \in \bar{N}_i^s) \subseteq B_j$, $i \in A^s$. Experts define planned values of business process KPI $\{\bar{P}_b\}^t$, $t = \overline{1, T}$ according to $\{\bar{P}\bar{P}_j, \ j \in M^{PP}\}^t$.

Step 6. Defining of KPI for structural divisions of company. The procedures of business processes are assigned to responsible persons and performers which belong to the structural divisions of the company. Assume L^s is set of divisions indexes which perform the s-th type of activity, $s \in \{B^a, I^a, F^a\}$. Then, according to the regulations of business processes, employees of divisions are executors of certain procedures and are responsible for execution of planned values of KPI. Thus, business processes KPI $\{P_b, \ b \in B_j\}, j \in \widehat{N}^s$ are assigned to appropriate divisions. As result, divisions KPI are defined which form sets $\{P_b, \ b \in \bar{\bar{B}}_j, j \in \bar{\bar{N}}_l^s\}$, $l \in L^s$, where $\bar{\bar{N}}_l^s$ is set of indicators indexes for the s-th type of activity on the basis of which KPI of the l -th division are formed; $\bar{\bar{N}}_l^s \subseteq \widehat{N}^s$; $\bar{\bar{B}}_j$ is set of KPI indexes for the l-th division which are got based on indicator PP_j. Thus, $\{\bar{\bar{B}}_j, \ j \in \bar{\bar{N}}_l^s, l \in L^s\} = \{\bar{B}_j, \ j \in \bar{N}_i^s, \ i \in A^s\}$, $s \in \{B^a, I^a, F^a\}$. Planned KPI values for divisions $\{\bar{P}_b, \ b \in \bar{\bar{B}}_j, \ j \in \bar{\bar{N}}_l^s\}^t$, $l \in L^s$ are formed based on appropriate planed KPI values for business processes $\{\bar{P}_b , \ b \in \bar{B}_j, \ j \in \bar{N}_i^s\}^t$, $i \in A^s$, $t = \overline{1, T}$. They are the basis for the development of strategic and annual plans of structural divisions.

Steps 5, 6 are implemented in case if the company has a matrix organizational management structure (by function and business processes). If management is carried out only on a functional principle, then step 5 is not implemented, and the divisions KPI are formed based on $\{PP_j, j \in \widehat{N}^s\}$, $s \in \{B^a, I^a, F^a\}$.

Step 7. Determining the values of budget parameters - indicators of the operational level. The budget of income and expenses of the company (financial budget) is considered, where expenditure and income items are divided by types of activity: main, investment and financial. Each business process KPI is set $\{P_b, \ b \in B_j, \ j \in \widehat{N}^s\}$, $s \in \{B^a, I^a, F^a\}$ and calculated based on budget indicators $P_b = f_b(d_\vartheta^s, c_\mu^s)$, where d_ϑ^s, c_μ^s are budget indicators which are reflected in income and expenses items, respectively, $\vartheta \in D^s$, $\mu \in C^s$ where D^s i C^s are sets of indexes of income and expenses items, respective for the s-th type of activity. The values d_ϑ^s, c_μ^s are determined on the basis of solving the problem of choosing the optimal financial budget as a Boolean programming task, provided that the planned KPI values are achieved $\{\bar{P}_b, \ b \in B_j, \ j \in \widehat{N}^s\}$.

Step 8. Coordination and approval of planned values of indicators in accordance with the regulations of business budgeting processes according to which the system of budgets of the company is developed.

Step 9. Procedure for adjusting strategic and annual KPI values. As at each step of decomposition of strategic KPIs the values of other indicators with some errors were calculated, after the approval of budget indicators the planned values of primary indicators and KPI are adjusted.

Thus, as a result of the implementation of the method of forming a system of key indicators of company efficiency is the formation of the following indicators: 1) strategic level KPI; 2) tactical level KPI – KPI for divisions, business processes of the main, investment and financial activity; 3) indicators of the operational level – budget indicators.

Such KPI system is the basis for the plans formation at all levels of management.

3   The Structure of the Cascading Subsystem of KPI

The Enterprise Performance Management System (EPM System) is an enterprise management concept based on a set of information technologies that automate the basic management processes: forecasting, planning, budgeting, monitoring and analysis. It provides an opportunity for a comprehensive analysis of the company's KPI and the trends of their changes, integrates various components for local strategic management tasks. The cascading subsystem of EPM consists of four main packages, as shown Fig. 2. These packages are related. Information from other EPM subsystems and databases of other software systems is required for their functioning (for example: ERP, forecasting systems, accounting, budgeting, etc.).

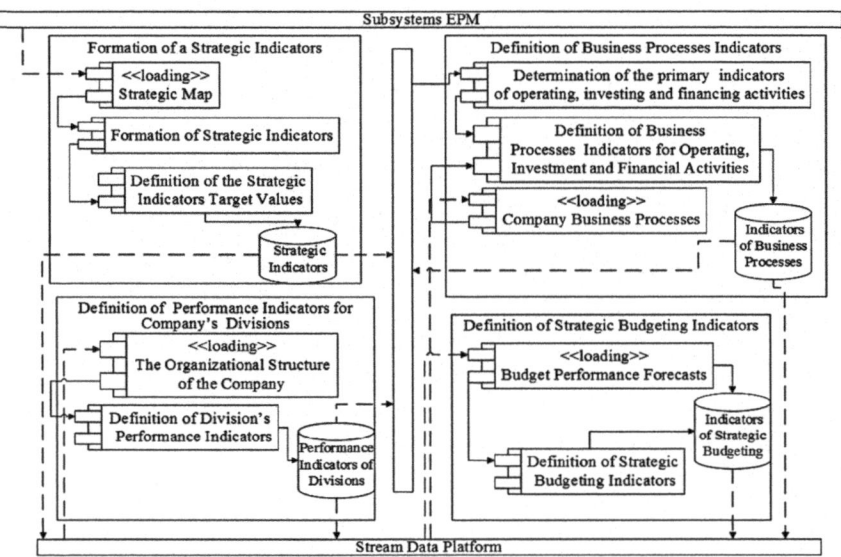

Fig. 2. The structure of the cascading subsystem EPM System.

This communication is carried out through ETL (Extract, Transform, and Load) - means of extracting, transforming, loading data. In [15], it is proposed to build the EPM as a system with a service-oriented architecture (SOA). Then each EPM sub-system will be presented as separate services and Apache Kafka is proposed to be used as the Stream Data Platform and Message Broker. If SOA is selected for EPM then the cascading service will be consisting of microservices: Formation of a Strategic Indicators, Definition of Business Process Indicators and Definition of Strategic Budgeting Indicators. The choice of SOA for EPM is related to the fact that such architecture will increase the productivity of such system, as well as the flexibility and efficiency of strategic planning [16]. However, there are many difficulties in developing systems based on SOA [17].

4   The Balanced Scorecard Development for the Ukrainian Agricultural Holding

The strategic map for the Ukrainian agricultural holding was built based on the con-ducted strategic analysis. The preliminary selected strategic goal is to increase the market value of the business. The strategy was formed to achieve it [18]. A strategic map and a KPI system have been developed using the Balanced Scorecard method-ology [4]. According to the cascading method, the strategic level KPI was decomposed (see Fig. 3) to primary indicators (see Fig. 4). The investments were determined to improve each indicator on the basis of expert assessments. The planned values of the primary indicators were defined based on the implementation of the model (1)–(3).

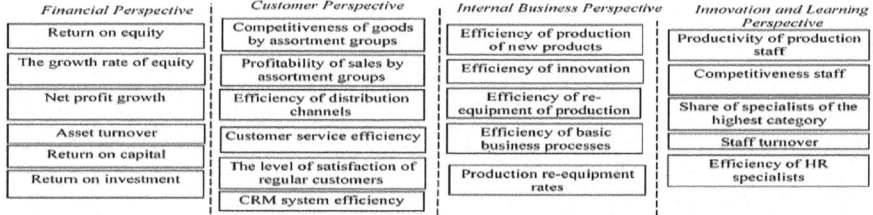

Fig. 3. Fragment of the strategic level KPI system.

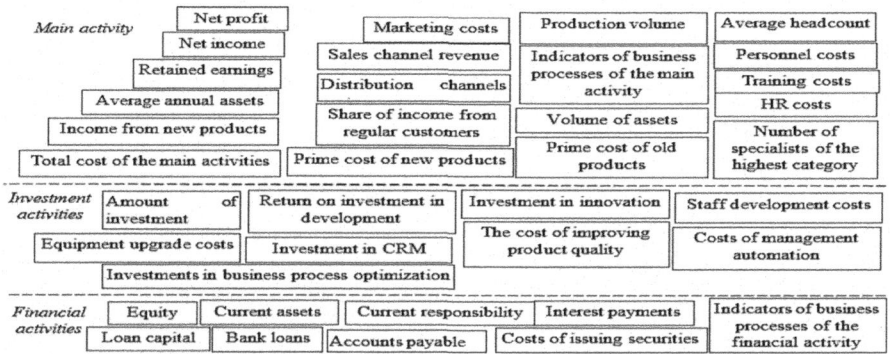

Fig. 4. Fragment of the primary indicators system by type of activity.

The plans were formed for the structural divisions of the agricultural holding on the basis of mentioned indicators. Their implementation will allow reaching the planned KPI values, which will mean the achievement of the strategic goal.

5   Conclusion

Thus, the cascading subsystem is presented as part of the EPM System. This subsystem implements a cascading method which allows for the strategic leveling of development goals by decomposing strategic KPIs into three aspects: by the business processes of the main, investment and financial activities of the company; by structural divisions and by planning levels - strategic, tactical and operational (including budget level). As a result, the KPI system is built, the planned values of all indicators in this system are determined for years of the strategic period. Currently, the authors carry out software implementation of the EPM System based on SOA for the Ukrainian agrarian company.

References

1. Heesen, B.: Effective Strategy Execution. Springer, Berlin, Heidelberg (2012). https://doi.org/10.1007/978-3-662-47923-0
2. Radomska, J.: The concept of sustainable strategy implementation. Sustainability **7**(12), 15847–15856 (2015). https://doi.org/10.3390/su71215790
3. Lederer, M., Schott, P., Huber, S., Kurz, M.: Strategic business process analysis: a procedure model to align business strategy with business process analysis methods. In: Fischer, H., Schneeberger, J. (eds.) S-BPM ONE 2013. CCIS, vol. 360, pp. 247–263. Springer, Heidelberg (2013). https://doi.org/10.1007/978-3-642-36754-0_16
4. Kaplan, R.S., Norton, D.P.: Transforming the balanced scorecard from performance measurement to strategic management: Part I. Account. Horizons **15**(1), 87–104 (2001). https://doi.org/10.2308/acch.2001.15.1.87
5. Hirschel, H.: Cascading performance measures (2009). https://hirschel.wordpress.com/2009/04/05/cascading-performance-measures. Accessed 05 Feb 2019
6. Niven, P.R.: Balanced Scorecard Step-by-Step: Maximizing Performance and Maintaining Results. Wiley, New York (2010)
7. Humphreys, K.A., Trotman, K.T.: The balanced scorecard: the effect of strategy information on performance evaluation judgments. J. Manage. Account. Res. **23**(1), 81–98 (2011). https://doi.org/10.2308/jmar-10085
8. Dean, M.: How to effectively communicate strategy to employees. Peakon (2015). https://peakon.com/resources/guides/how-to-effectively-communicate-strategy-to-employees. Accessed 18 Feb 2019
9. Van Looy, A., Van den Bergh, J.: The effect of organization size and sector on adopting business process management. Bus. Inf. Syst. Eng. **60**(6), 479–491 (2017). https://doi.org/10.1007/s12599-017-0491-3
10. Morrison, E.D., Ghose, A.K., Dam, H.K., Hinge, K.G., Hoesch-Klohe, K.: Strategic alignment of business processes. In: Pallis, G., Jmaiel, M., Charfi, A., Graupner, S., Karabulut, Y., Guinea, S., Rosenberg, F., Sheng, Q.Z., Pautasso, C., Ben Mokhtar, S. (eds.) ICSOC 2011. LNCS, vol. 7221, pp. 9–21. Springer, Heidelberg (2012). https://doi.org/10.1007/978-3-642-31875-7_3
11. Rahschulte, T.: Aligning execution and strategy through program management (2014). https://www.pmi.org/learning/library/aligning-execution-strategy-program-management-9365. Accessed 22 Feb 2019
12. Blumentritt, T.: Integrating strategic management and budgeting. J. Bus. Strategy **27**(6), 73–79 (2006). https://doi.org/10.1108/02756660610710382
13. Moskalenko, V., Zakharova, T., Fonta, N.: Technology of formation of development program as a system of company's annual plans based on key performance indicators. Eur. Coop. **2**(2), 108–124 (2015). (in Russian)
14. Sergienko, I.V.: Methods of Optimization and Systems Analysis for Problems of Transcomputational Complexity. SOIA, vol. 72. Springer, New York (2012). https://doi.org/10.1007/978-1-4614-4211-0
15. Moskalenko, V.V., Berezenko, J.C.: The concept of an architectural solution for the service intended to build an enterprise strategy map. Bull. Nat. Tech. Univ. "KhPI". Ser. Syst. Anal. Control Inf. Technol. **55**, 45–50 (2017)
16. Clark, K.J.: Microservices, SOA, and APIs: Friends or enemies? https://www.ibm.com/developerworks/websphere/library/techarticles/1601_clark-trs/1601_clark.html. Accessed 18 Feb 2019

17. Newman, S.: Building Microservices: Designing Fine-Grained Systems. O'Reilly Media, Inc. (2015)
18. Moskalenko, V.: Module of selecting the company development strategic goals in the Enterprise Performance Management System. In: Proceedings of the International Conference on Computer Algebra and Information Technologies (CAIT 2018), Odessa, pp. 27–30 (2018)

On Using Information and Communication Technologies in Process of Mathematical Specialties Education

Nina Padalko[1]([envelope]) [ID], Halyna Padalko[2] [ID], and Anatoliy Padalko[3] [ID]

[1] Lesya Ukrainka Eastern European National University, 13 Volya Avenue, Lutsk 43025, Ukraine
padalkonina109@gmail.com
[2] National Aerospace University "Kharkiv Aviation Institute", 17 Chkalova Street, Kharkiv 61070, Ukraine
galinapadalko95@gmail.com
[3] Lutsk National Technical University, 75 Lvivska Street, Lutsk 43018, Ukraine
padalkoanatol@gmail.com

Abstract. The study examines the pedagogical conditions for the use of information and communication technologies in the process of professional training of students of mathematical specialties. It is proved that the use of information and communication technologies in the educational process makes it possible to increase the indicators of the formation of students' professional competence. The analysis of the results of experimental studies made it possible to draw a conclusion about the advisability of using information and communication technologies for the formation of professional competence of students of mathematical specialties.

Keywords: Professional education · Mathematics · Information and communication technologies · Model · Process

1 Introduction

The worldwide pandemic of coronavirus infection has made the entire world community feel a new normality. The usual life has changed, and the importance of introducing information technology into all spheres of life has become evident. The forced digitalization has not spared the sphere of education [1]. Over the past half a year, the number of students affected by the closure of schools and universities in 138 countries has nearly quadrupled to reach 1.37 billion [2]. This means that over 3 out of 4 children and young people around the world are not able to attend educational institutions. The closure of educational institutions also affected nearly 60.2 million teachers [3].

Some countries closed schools completely for the first month of the pandemic and then adapted to accepted safety standards [4]. At the same time, in other educational institutions, they worked in the emergency mode, that is, only the children of those parents who worked during the quarantine could attend school or kindergarten: doctors,

© The Author(s), under exclusive license to Springer Nature Switzerland AG 2021
M. Nechyporuk et al. (Eds.): ICTM 2020, LNNS 188, pp. 716–725, 2021.
https://doi.org/10.1007/978-3-030-66717-7_61

sellers, police officers, firefighters, and the like [5]. Norway and Denmark reopened schools in April, about a month after the quarantine was closed, but initially only did so for younger children, leaving the high school distance learning. Israel reopened schools on May 3, first for smaller groups and subsequently for full classes. A month later, about 220 students and teachers were diagnosed with coronavirus infection there. Moreover, more than 100 infected were from one Jerusalem gymnasium.

On March 12, quarantine began in Ukraine, during which educational institutions were forced to switch to distance learning. In this mode, schools and universities worked until the end of the academic year.

However, the pandemic only accelerated the process of digitalization of the educational process, and trends in the transition of educational process to online have been observed for several years [6–8]. And now trends of digitalization not only in organization of educational process, but in Public Health [9–13], medical diagnostics [14–17], security [18–20], management [21–24], financial branch [25] and other spheres of human being are developed with high speed.

Although the coronavirus pandemic has become a catalyst for the digitalization of education in Ukraine, the urgent transfer of education to a distance format in a pandemic has significant differences from a properly planned online education. Educational organizations that are forced to work with students remotely in order to reduce the risks of the spread of coronavirus should be aware of this difference when assessing the effectiveness of the so-called "online learning" using distance learning technologies.

The professional activity of specialist mathematicians is complex, multifaceted and multifunctional. Professional functions require a high level of professionalism, professional skill, intellectual qualities and abilities, especially in modern conditions of the complication of tasks that are posed to future mathematicians. Taking this circumstance into account, the problem of increasing the level and quality of professional training of students of mathematical specialties is urgent [26].

The **aim of our research** is to consider and substantiate the organizational and pedagogical conditions for the use of information and communication technologies (ICT) for the formation of professional competence of students of mathematical specialties.

2 Current Researches Analysis

In the field of distance learning technologies, there is no uniformity of terminology. The conceptual apparatus in this area is in its infancy. The literature actively uses terms such as distance learning, distance education, Internet learning, distance learning technologies, they are used to describe the features of distance learning using modern information technologies.

The study analyzed the results published by the world's leading scientists. The following concepts and theories were used as a theoretical basis:

– methodology for the introduction of information environments in the learning process of mathematics [27];

- general theory of the use of information and communication technologies in education [28];
- methodology for applying factor analysis to the study of the problem of knowledge quality [29];
- design of pedagogical systems [30].

In order to organize the process of teaching mathematics with using distance learning technologies, we highlight the goals of using ICT:

- satisfaction of the individual's need for education (lifelong learning technology);
- improving the quality of education through the introduction of modern technologies, in which targeted mediated or incompletely mediated interaction between the student and the teacher is carried out regardless of their location and distribution in time based on the use of telecommunications;
- strengthening the personal orientation of the learning process, intensifying the student's independent work;
- enhancement learning efficiency through the introduction of innovative educational technologies.
- free use by students of various information resources for the educational process at any convenient time;
- ensuring the advanced nature of the entire education system, its focus on disseminating knowledge among the population, raising its general educational and cultural level;
- creation of conditions for the application of the education quality control system.

In order to determine the requirements for the organization of the educational process using distance educational technologies, the characteristics of learning using ICT are given:

- separation of teaching and learning processes in time and space;
- the use of a modular principle, which involves dividing the subject into logically closed blocks, called modules, within which both the study of new material and control measures to check its mastery take place;
- extensive use of survey training, implemented through survey lectures, helping the student to create a holistic picture of the studied area of knowledge and activities;
- management of the student's independent work by means educational institution through curricula, specially prepared teaching and learning materials and special control procedures;
- mastering by the trainees of educational programs at the place of residence with the dominant of independent work, with periodic meetings of a group of students;
- creation of a special information and educational environment, including various educational products - from a working textbook to computer training programs, slide lectures and audio and video courses;
- the obligatory use of communication technologies for the transfer of knowledge, mediated, dialogue and interactive interaction of learning subjects and solving administrative problems.

According to modern researches, studies of the learning process cannot be proven if they were not subject to experimental verification. In pedagogy, the leading role in the organization of experimental verification of the effectiveness and appropriateness of theoretical provisions is assigned to the pedagogical experiment is a complex method of hypothesis testing, organized by a systemic process.

3   Results

The concept of efficiency is not clearly defined, since efficiency is an evaluation category that is related to the ratio of the value of the result to the value of the cost. By the effectiveness of the formation of professional competence of students of mathematical specialties by means of ICT, we mean qualitative and quantitative improvements in the performance indicators of students in the experimental groups in comparison with the control ones. In our opinion, the purpose of the study allows us to implement the cost-effective concept of determining the didactic effectiveness of organizational and pedagogical conditions: a positive increase in the result obtained relative to the previous result, taking into account the time, technical and psychophysical costs (the effect of the teacher's work, the achievement of pre-predicted learning goals for students).

In the process of research, we tried to prove that an effective methodology for the formation of professional competence of students of mathematical specialties should be based on the implementation of certain organizational and pedagogical conditions for the use of ICT tools.

The results of experimental learning using innovative professionally oriented teaching methods based on ICT in the study of vocational training disciplines confirmed the hypothesis of our research, which is the assumption that the formation of professional competence of students of mathematical specialties by means of ICT is provided by increasing their motivation in vocational training and translation of vocational training. competence in the subjective need and the purpose of the upcoming activity. The effectiveness of this process can be much higher if, when studying disciplines of the cycle of professional (professionally oriented) and special training, the implementation of certain organizational and pedagogical conditions is ensured. Let's consider and justify them in more detail.

The provision of didactic design, construction and implementation of professionally oriented ICT tools (electronic educational and methodological complexes, electronic textbooks, multimedia training courses, virtual laboratories, technical equipment) is based on modeling the professional activities of students of mathematical specialties, taking into account the qualification requirements for graduates.

In terms of informatization, it is necessary to consider the problems associated, firstly, with insufficient staffing of educational institutions with technical support, and, secondly, with imperfection, inadequacy of professionally oriented software for educational purposes. The second problem is characterized by a relatively weak development of instrumental and technological means used in the preparation of students of mathematical specialties, which is typical for the domestic programming industry as a whole.

Our research has confirmed that in order to solve the indicated problems of using ICT tools in the formation of professional competence of students of mathematical specialties, it is necessary to create the appropriate conditions:

- to provide the educational process with appropriate technical equipment, create and ensure the functioning of the internal network;
- to develop and introduce electronic manuals, electronic educational and methodological complexes that provide the formation of professional knowledge of students of mathematical specialties according to the principles of differentiation, individualization of training;
- to build the educational process using innovative technologies based on integrated methods, by introducing modern ICT (modular, distance learning, mobile learning, working on the Internet, etc.);
- to organize a system of advanced training for teachers in the field of ICT.

One of the promising ways to solve the second of these problems, it is advisable, in our opinion, to consider the development and implementation of professionally oriented ICT tools in the educational process.

Professional training of students of mathematical specialties is carried out in the following areas: classroom teaching technologies (multimedia educational complexes of disciplines, multimedia lecture hall), telecommunication and library technologies.

Learning using ICT tools cannot replace a human teacher, but it can not only supplement and improve the teacher's work, but also in some areas where independence, creative thinking develops, in general, it plays a unique role that we cannot now realize fully.

Using ICT in the educational process:

- helps to increase the interest of students, enhance the motivation of learning;
- provides opportunities for using various ways of presenting information;
- allows you to actively involve students in the educational process, focuses their attention on the most important aspects of the material;
- organizes psychologically calm work;
- allows you to use during classes significant amounts of information (information networks, databases, etc.);
- requires continuous professional development of teaching staff, appropriate equipment and improved methodological and software.

The development and implementation of electronic educational and methodological complexes, including electronic training courses, computer testing systems, video demonstrations, etc., allows:

- to submit educational information in various forms;
- to initiate the processes of assimilation of knowledge, acquisition of skills and abilities of educational or professional activity;
- to implement repetition and control of learning outcomes effectively;
- to intensify cognitive activity;
- to form and develop certain types of activities.

To conduct a formative experiment in the disciplines of the mathematical cycle of professional training, we have developed multimedia electronic complexes that contain large volumes of demonstration material based on new pedagogical teaching technologies, adapted to teaching in a telecommunication network.

An electronic textbook is a tool that integrates the main forms of the educational process, such as the presentation of theoretical material, elements of consolidation and control of the assimilation of the material using all the advantages of ICT. The main tasks that are set for such a textbook are the development of the skills of independent experiments, the accumulation of experience in managing modern software and hardware resources, the study and practical development of methods for processing the experimental results obtained. The solution of these problems contributes to the creative development and increase of the educational activity of students, the development of real technology, they will be used by them at the place of service.

The formative experiment was carried out on the basis of the communal institution "Lutsk educational complex No. 26 of the Lutsk city council of the Volyn region". We used ICT elements, which contain large volumes of demonstration material based on new pedagogical teaching technologies, adapted to teaching in telecommunication networks in the process of studying mathematical disciplines.

The results of approbation of the electronic textbooks developed by us allow us to draw the following conclusions:

- the use in such textbooks of a structured functional communication environment that ensures the perception and assimilation of information as a whole (as the perception of an image). The analysis of statistical data shows that the use of an electronic textbook increases the perception of educational material by 45.7%, and assimilation – by 31.6%;
- research of the operation of power plants by means of ICT through the use of a wide range of virtual measuring instruments and mathematical research methods increases the reliability of the results obtained, forms the ability to work with measuring devices (by 29.8%), contributes to the development of students' creative abilities (by 22.4%);
- electronic textbooks can be used in the distance education system, subject to the creation and development of an educational information environment in an educational institution;
- the development of electronic textbooks allows attracting students and teachers who do not have sufficient programming experience to work.

Building an educational process using ICT based on professionally oriented technologies in order to form skills and abilities allows you to make optimal decisions or offer options for solving them in real conditions.

The strategy of innovative learning changes primarily the goals of educational activities. The goal of transferring information in the field of specific disciplines is replaced by the goal of developing students by expanding the existing volume of knowledge and the scope of their cognitive activity. This implies:

- determination of the current level of student development;
- the need for motivation to continue learning;

– the use of developmental teaching methods allows not only transferring new knowledge, but also simultaneously teaching methods, techniques, and self-education skills.

The role positions of the teacher and the student change, the nature of the organization of educational activities.

By activating the educational process through the use of professionally oriented ICT tools, it is possible to solve a complex of didactic and methodological problems, to carry out individual training, motivation, development of thinking and creative abilities of students. The structure of an interactive learning system should develop the cognitive functions of students and at the same time adapt to their needs. Interactive (computer) dialogue provides communication between two partners - an educational tool (computer) and a student. Thus, using the classical provisions of didactics, the introduction of interactive learning technologies based on ICT introduces significant changes not only in practice, but also in theory.

It can be concluded that it is possible and advisable to use ICT tools to support a complex of different types of learning activities, but at the same time it is necessary to adhere to clear boundaries of their use, so that, on the one hand, they facilitate the learning activities of students, and on the other hand, so that they provide the identification of conceptual scientific provisions and guaranteed the possibility of their assimilation.

Practical recommendations for professional training of students for the use of ICT tools:

– conducting training sessions on the basics of ICT application;
– introduction of a topic that reveals the features of the use of computer and other information tools in the educational process;
– mastering by students the method of independent work;
– obligatory preliminary psychological work with students before using new ICT tools.

Certain organizational and pedagogical conditions and principles of interaction in the educational process of the participants in the experimental work made it possible to develop a model of the future specialist's readiness to be included in professional activity. The model includes three structural elements - educational competence, subject-personal competence, motivational orientation, each of which has its own content. These elements were considered in two aspects: from the standpoint of personal readiness and professional readiness.

Educational competence at the level of personal readiness includes professional, general cultural and humanistic components, as well as communication skills, creativity, a system of in-depth knowledge and skills. At the level of professional readiness, these are professional knowledge, skills and abilities to perform functional duties, the ability to creative activity.

Subject-personal competence at the level of personal readiness is considered by us as an adequate self-assessment of physical, mental and creative development, and at the level of professional readiness - as a self-assessment of professionally significant personality traits, the ability to self-regulate work capacity, and self-control in professional activity. Motivational orientation is considered at the level of personal readiness as broad cognitive interests, value-semantic orientations, and at the level of professional readiness – as sustainability of professional choice, professional orientation, development of professional qualities.

4 Conclusions

Full and timely fulfillment of all organizational and pedagogical conditions justified above can contribute to the formation of professional competence by means of ICT among students of mathematical specialties. The analysis of the results of experimental studies made it possible to conclude that it is advisable to use ICT tools for the formation of professional competence of students of mathematical specialties, since, firstly, it has been experimentally proved that the use of ICT in the educational process makes it possible to increase the indicators of the formation of students' professional competence, and secondly, to activate them cognitive activity, to increase the stimulating and motivational component of the educational process.

References

1. Xiao, C., Li, Y.: Analysis on the influence of the epidemic on the education in China. In: 2020 International Conference on Big Data and Informatization Education (ICBDIE), Zhangjiajie, pp. 143–147. IEEE (2020). https://doi.org/10.1109/ICBDIE50010.2020.00040
2. Herasymova, A., Chumachenko, D., Padalko, H.: Development of intelligent information technology of computer processing of pedagogical tests open tasks based on machine learning approach. In: CEUR Workshop Proceedings, vol. 2631, pp. 121–130 (2020)
3. Plancher, K.D., Shanmugam, J.P., Petterson, S.C.: The changing face of orthopedic education: Searching for the new reality after COVID-19. Arthrosc. Sports Med. Rehabil. **2**(4), e295–e298 (2020). https://doi.org/10.1016/j.asmr.2020.04.007
4. Wang, G., Zhang, Y., Zhao, J., et al.: Mitigate the effects of home confinement on children during the COVID-19 outbreak. Lancet **395**(10228), 945–947 (2020). https://doi.org/10.1016/S0140-6736(20)30547-X
5. Sheikh, A., Sheikh, A., Sheikh, Z., Dhami, S.: Reopening schools after the COVID-19 lockdown. J. Glob. Health **10**(1), 010376 (2020). https://doi.org/10.7189/jogh.10.010376
6. Piletskiy, P., Chumachenko, D., Meniailov, I.: Development and analysis of intelligent recommendation system using machine learning approach. In: Nechyporuk, M., Pavlikov, V., Kritskiy, D. (eds.) Integrated Computer Technologies in Mechanical Engineering. AISC, vol. 1113, pp. 186–197. Springer, Cham (2020). https://doi.org/10.1007/978-3-030-37618-5_17
7. Chumachenko, D., Balitskii, V., Chumachenko, T., et. al.: Intelligent expert system of knowledge examination of medical staff regarding infections associated with the provision of medical care. In: CEUR Workshop Proceedings, vol. 2386, pp. 321–330 (2019)
8. Mazorchuck, M., Dobriak, V., Chumachenko, D.: Web-application development for tasks of prediction in medical domain. In: 2018 IEEE 13th International Scientific and Technical Conference on Computer Sciences and Information Technologies (CSIT), Lviv, vol. 1, pp. 5–8. IEEE (2018). https://doi.org/10.1109/STC-CSIT.2018.8526684

9. Chumachenko, D., Dobriak, V., Mazorchuk, M., et al.: On agent-based approach to influenza and acute respiratory virus infection simulation. In: 2018 14th International Conference on Advanced Trends in Radioelecrtronics, Telecommunications and Computer Engineering (TCSET), Slavske, pp. 192–195. IEEE (2018). https://doi.org/10.1109/TCSET. 2018.8336184

10. Chumachenko, D.: On intelligent multiagent approach to viral hepatitis B epidemic processes simulation. In: 2018 IEEE Second International Conference on Data Stream Mining & Processing (DSMP), Lviv, pp. 415–419. IEEE (2018). https://doi.org/10.1109/ DSMP.2018.8478602

11. Polyvianna, Y., Chumachenko, D., Chumachenko, T.: Computer aided system of time series analysis methods for forecasting the epidemics outbreaks. In: 2019 IEEE 15th International Conference on the Experience of Designing and Application of CAD Systems (CADSM), Polyana, pp. 1–4. IEEE (2019). https://doi.org/10.1109/CADSM.2019.8779344

12. Chumachenko, D., Meniailov, I., Bazilevych, K., et al.: Development of an intelligent agent-based model of the epidemic process of syphilis. In: 2019 IEEE 14th International Conference on Computer Sciences and Information Technologies (CSIT), Lviv, vol. 1, pp. 42–45. IEEE (2019). https://doi.org/10.1109/STC-CSIT.2019.8929749

13. Chumachenko, D., Chumachenko, T.: Intelligent agent-based simulation of HIV epidemic process. In: Lytvynenko, V., Babichev, S., Wójcik, W., Vynokurova, O., Vyshemyrskaya, S., Radetskaya, S. (eds.) ISDMCI 2019. AISC, vol. 1020, pp. 175–188. Springer, Cham (2020). https://doi.org/10.1007/978-3-030-26474-1_13

14. Bazilevych, K., Meniailov, I., Fedulov, K., et al.: Determining the probability of heart disease using data mining methods. In: CEUR Workshop Proceedings, vol. 2488, pp. 383–394 (2019)

15. Mashtalir, V.P., Yakovlev, S.V.: Point-set methods of clusterization of standard information. Cybern. Syst. Anal. **37**(3), 295–307 (2001). https://doi.org/10.1023/A:1011985908177

16. Meniailov, I., Bazilevych, K., Fedulov, K., Goranina, S.: Using the K-means method for diagnosing cancer stage using the Pandas library. In: CEUR Workshop Proceedings, vol. 2386, pp. 107–116 (2019)

17. Mashtalir, V.P., Shlyakhov, V.V., Yakovlev, S.V.: Group structures on quotient sets in classification problems. Cybern. Syst. Anal. **50**(4), 507–518 (2014). https://doi.org/10.1007/ s10559-014-9639-z

18. Chumachenko, D., Yakovlev, S.: On intelligent agent-based simulation of network worms propagation. In: 2019 IEEE 15th International Conference on the Experience of Designing and Application of CAD Systems (CADSM), Polyana, pp. 11–14. IEEE (2019). https://doi. org/10.1109/CADSM.2019.8779342

19. Chumachenko, D., Yakovlev, S.: Development of deterministic models of malicious software distribution in heterogeneous networks. In: 2019 3rd International Conference on Advanced Information and Communications Technologies (AICT), Lviv, pp. 439–442. IEEE (2019). https://doi.org/10.1109/AIACT.2019.8847903

20. Chumachenko, D., Chumachenko, K., Yakovlev, S.: Intelligent simulation of network worm propagation using the code red as an example. Telecommun. Radio Eng. **78**(5), 443–464 (2019). https://doi.org/10.1615/TelecomRadEng.v78.i5.60

21. Dotsenko, N., Chumachenko, D., Chumachenko, I.: Modeling of the process of critical competencies management in the multi-project environment. In: 2019 IEEE 14th International Conference on Computer Sciences and Information Technologies (CSIT), Lviv, vol. 3, pp. 89–93. IEEE (2019). https://doi.org/10.1109/STC-CSIT.2019.8929765

22. Dotsenko, N., Chumachenko, D., Chumachenko, I.: Management of critical competencies in a multi-project environment. In: CEUR Workshop Proceedings, vol. 2387, pp. 495–500 (2019)

23. Dotsenko, N., Chumachenko, D., Chumachenko, I.: Project-oriented management of adaptive teams' formation resources in multi-project environment. In: CEUR Workshop Proceedings, vol. 2353, pp. 911–920 (2019)
24. Dotsenko, N., Chumachenko, D., Chumachenko, I.: Modeling of the processes of stakeholder involvement in command management in a multi-project environment. In: 2018 IEEE 13th International Scientific and Technical Conference on Computer Sciences and Information Technologies (CSIT), Lviv, vol. 1, pp. 29–32. IEEE (2018). https://doi.org/10.1109/STC-CSIT.2018.8526613
25. Bazilevych, K., Mazorchuk, M., Parfeniuk, Y., et al.: Stochastic modelling of cash flow for personal insurance fund using the cloud data storage. Int. J. Comput. **17**(3), 153–162 (2018)
26. Meniailov, I., Ugryumov, M., Chumachenko, D., Bazilevych, K., Chernysh, S., Trofymova, I.: Non-linear estimation methods in multi-objective problems of robust optimal design and diagnostics of systems under uncertainties. In: Nechyporuk, M., Pavlikov, V., Kritskiy, D. (eds.) Integrated Computer Technologies in Mechanical Engineering. AISC, vol. 1113, pp. 198–207. Springer, Cham (2020). https://doi.org/10.1007/978-3-030-37618-5_18
27. Chumachenko, D., Meniailov, I., Bazilevych, K., Chumachenko, T.: On intelligent decision making in multiagent systems in conditions of uncertainty. In: 2019 XIth International Scientific and Practical Conference on Electronics and Information Technologies (ELIT), Lviv, pp. 150–153. IEEE (2019). https://doi.org/10.1109/ELIT.2019.8892307
28. Chumachenko, D., Sokolov, O., Yakovlev, S.: Fuzzy recurrent mappings in multiagent simulation of population dynamics. Int. J. Comput. **19**(2), 290–297 (2020)
29. Zabolotnyj, A.V., Koshevoj, N.D.: Development, examination and optimization of the device for quality control of dielectric materials. Pribory i Sistemy Upravleniya **1**, 39–42 (2004)
30. Zabolotnyi, O.V., Zabolotnyi, V.A., Koshevoi, M.D.: Conditionality examination of the new testing algorithms for coal-water slurries moisture measurement. Naukovyi Visnyk Natsionalnoho Hirnychoho Universytetu **1**, 51–59 (2018)

Social and Legal Aspects of the Transition to Industry 4.0

Svitlana Gutsu[(✉)] , Maryna Mkrtchyan ,
and Anastasiia Strielkina

National Aerospace University "Kharkiv Aviation Institute",
17 Chkalova Street, Kharkiv 61070, Ukraine
s.gutsu@khai.edu, marinamkrtcan264@gmail.com,
a.strielkina@csn.khai.edu

Abstract. Discussion questions about the ethical nature and legal regulation have always been an integral part of the rapid development of information technology, robotics and artificial intelligence. Innovation will inevitably lead to social and legal changes, which can cause some difficulties for both business and countries, as well as for individuals. In the paper, we tried to highlight the trends and give an overview of the most common socio-legal issues that the Industry 4.0 concept raises. Working with extra-large amounts of data, machine learning, Internet of things and cloud storage each year increase the number of violations in the field of storage, use and transmission of data. This causes tension in society and creates people's distrust of technology. Also, intellectual property law receives new challenges related to the participation of artificial intelligence in the creation of intellectual property. Robots and artificial intelligence significantly change the labor market and compete with people in the field of employment. Studies and statistical indicators in these areas were examined, on the basis of which was concluded that it is necessary to develop standards and areas for further work.

Keywords: Artificial intelligence · Cybersecurity · Employment · Industry 4.0 · Labor law · Personal data protection · Standard

1 Introduction

1.1 Motivation

Modern technology has a huge impact on human civilization, which is building a global digital society. But the culture in which we live has an impact on the industrial world. And this is the distinguishing characteristic of the digital revolution, which combines various technologies and blurs the boundaries between the physical world, the digital and the biological spheres.

Industry 4.0 covers many technologies and is used in a variety of contexts. Maybe that is why the term "Industry 4.0" does not have a generally accepted definition and a unified vision of the system and its components. But the concepts often associated with it include and not limited to as shown in Fig. 1.

M. Nechyporuk et al. (Eds.): ICTM 2020, LNNS 188, pp. 726–737, 2021.
https://doi.org/10.1007/978-3-030-66717-7_62

Fig. 1. Industry 4.0 concepts.

1.2   State-of-the-Art

Recently, the topic of legal and social aspects of the fourth technical revolution has been increasingly raised by scientists. The author of [1] highlighted that Industry 4.0 affects the development of human resources, creates new production systems and business models, affects the overall value chain, society and the environment.

It was shown that such transformations require a holistic approach that encompasses innovative and sustainable system solutions and not just technological ones in paper [2].

The author of [3] showed legal challenges and methods to overcome them in Industry 4.0. He mentioned that IT security and data protection laws differ from country to country.

The authors of [4] analyzed a set of the European Union Agency for Network and Information Security documents in area of Industry 4.0 security. Emphasized that it is important to combine security features and create a comprehensive security management system, including for industry 4.0.

1.3   Goals and Structure of the Paper

The main task of the Industrial Revolution 4.0 is to manage the convergence of these areas, minimizing various risks, including legal ones. Given the multiplicity of components of Industry 4.0, we distinguish three blocks of social and legal issues that relate to cybersecurity and personal data protection, intellectual property rights (IPR), labor law and employment (see Fig. 2) that are aimed to discuss in this paper.

The section II presents a description of cybersecurity and personal data protection issues and legal solutions. Section III presents an artificial intelligence and intellectual property law legal and social interconnections. Section IV discusses labor law and

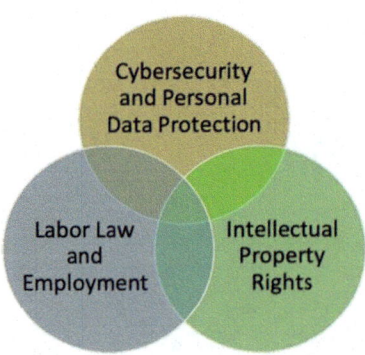

Fig. 2. Social and legal issues over discussion.

employment issues enhanced because of the fourth technical revolution. Section V concludes and describes future research directions.

2 Cybersecurity and Personal Data Protection

Advances in business and technology pose new challenges in ensuring their compliance with regulatory requirements, with the fact that most of them written before the invention of technology. The development and implementation of new legislative standards in the field of cybersecurity is a priority area of lawmaking both in international law and at national levels. As various studies show, the problem of cybersecurity is equally relevant for countries, business and individuals. Thus, the monitoring results of the PwC company "PwC's Global Crisis Survey 2019" show that 38% of the leaders and company executives surveyed consider cybercrime the most serious cause of crisis corporate situations [5].

The well-known American company Check Point Software Technologies in its latest cybersecurity report "The 2020 Cyber Security Report" says the following: "In 2019, we saw an escalation of sophisticated and targeted ransomware exploits. Specific industries were heavily victimized, including state and local government and healthcare organizations" [6]. The criminals began to spend more time collecting information about their victims, increasing the degree of damage and increasing the amount of the ransom. The attacks became so serious that the FBI softened its previous buyback position. Now they recognize the right of companies to evaluate their options in order to protect their shareholders, employees and customers.

2019 was a year of targeted attacks using ransomware. First of all, the list of victims included software, health services and public sectors. In the first half of 2019, the number of attacks by malicious programs for mobile banks increased by 50% compared to 2018.

The bandwidth provided by 5G will lead to a sharp increase in the number of connected devices and sensors. This ever-growing amount of personal data must be protected from hacking and theft.

Many organizations have shifted workloads to the cloud. However, the level of understanding as to securing them remains dangerously low. And our data indicates that 27% of all organizations globally were impacted by cyber-attacks that involved mobile device" [6].

The problems of cybersecurity give rise to social problems of community, and as a consequence of the country. According to Strategy + Business: Currently, there is a high degree of mistrust in society. People have lost confidence in government agencies as sources of reliability and stability. They do not believe that enterprises are able to provide work that will lead a decent life. According to the Edelman Trust Barometer, which measures the level of trust worldwide, in the last decade, less than half of the world's population trusts government, business, or even civil society. Growing inequalities make people doubt that governments or the employer can protect their interests.

The speed of technological change and the unprecedented amount of information available - anytime, anywhere, using smart devices - exacerbates this phenomenon. Since destructive information spreads faster than answers to it. This trend will only grow as 5G mobile networks increase speed and expand connectivity [7].

Research results say that human anxiety has a real reason. The leakage of personal data from the databases of state institutions is as real as from social networks or personal devices. Cybercrime does not discriminate. The victims of cybercrime involve individuals, organizations, and businesses alike - actually everyone from all areas of life [8].

Thus, the problem of cybersecurity requires a comprehensive solution and joint efforts of the state, business and public organizations. We have witnessed the serious legislative work of international institutions and national systems in this area. Table 1 shows the main legislative documents by countries.

The basis of legislative work of European institutions and national governments in the field of cybersecurity is European Parliament resolution of 16 February 2017 with recommendations to the Commission on Civil Law Rules on Robotics (2015/2103 (INL)) [9] and Convention on Cybercrime of 23.11.2001 [10]. The General Data Protection Regulation introduces a tougher enforcement regime and it exposes entities to increased financial liability. Since it was implemented, this regulation has been used to hand million-euro fines to such large companies for example as British Airways and Marriott International.

In the State of California (USA), on January 1, 2020, a new law came into force called the California Consumer Privacy Act (CCPA). The law establishes strict requirements regarding the transparency of the use of personal data [11].

In the State of California (USA), on January 1, 2020, a new law came into force called the California Consumer Privacy Act (CCPA). The law establishes strict requirements regarding the transparency of the use of personal data [11].

In Brazil, in February 2020, came into force on the protection of personal data. Chilean Law 19.628 / 2011 establishes certain data privacy provisions for government organizations. Colombian Law 1581/2012 lays down the principles and obligations that must be observed in order to avoid allegations of non-compliance with Law 1581/2012. Mexico's federal human rights act is one of the most comprehensive

Table 1. Legislative documents in cybersecurity and personal data protection areas.

Region and/or Country	Organization	Implementation date	Title
Europe	European Parliament Resolution	Feb 16, 2017	2015/2103 (INL) Civil Law in robotics
		Nov 23, 2001	Convention on Cybercrime
		May 25, 2018	GDPR
USA	California State Legislature	Jan 1, 2020	California Consumer Privacy Act
Brazil	Brazilian Congress	Feb, 2020	Law 13.709
Chile	State of Privacy	Jan, 2019	Law 19.628/2011
Colombia	Ministry of trade, industry and tourism	June 27, 2013	Law 1581/2012
Mexico	Mexican Congress	July 6, 2010	LFPDPPP
Africa	African Union	June 27, 2014	The African Union Convention on Cyber Security and Personal Data Protection
Singapore	Personal Data Protection Commission	May 22, 2019	Guide on Active Enforcement and Guide to Managing Data Breaches 2.0
Ukraine		June 19, 2019	

personal data protection documents in the region. It provides a specific legal definition for data breach, requires notification of the breach of the data owner and investigation [12].

The African Union Convention on Cyber Security and Personal Data Protection adopted at Malabo, Equatorial Guinea on 27 June 2014. Law 024/2019 approving the ratification of the Malaga Convention covers the areas of: electronic transactions (electronic commerce); protecting personal data and promoting cybersecurity; fight against cybercrime.

The Personal Data Protection Commission (PDPC) issued a Guide on Active Enforcement and Guide to Managing Data Breaches 2.0 on 22 May 2019, which detailed the PDPC's approach to regulating Singapore's data privacy regime [13].

Talking about Ukraine, the Cybersecurity Strategy of Ukraine and the Law of Ukraine "On the Fundamental Principles of Cyber Security of Ukraine" adopted in 2017 were fully operational in 2018. The Resolution of the Cabinet of Ministers of Ukraine "On approval of General requirements for the cyber defense of critical infrastructure facilities" was also adopted on June 19, 2019. However, to date, Ukraine is still far behind in terms of regulatory and technical support for cybersecurity and data protection. The National Cybersecurity Index has dropped by a point (from 26 to 28) for the year [14] (see Fig. 3).

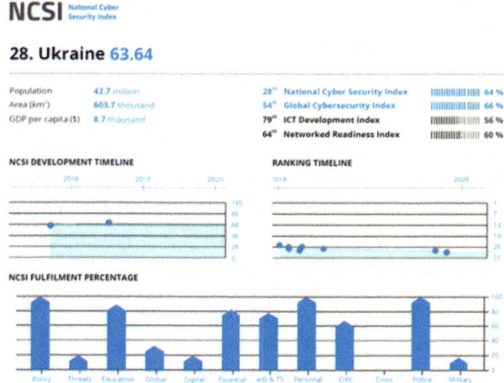

Fig. 3. The National cybersecurity index in Ukraine [14].

It could be concluded that the main legal task in the field of cybersecurity remains the development of common international definitions and standards for their implementation in national legal systems.

In addition, the presented analysis of the legislation on information protection makes it possible to draw a number of conclusions and propose a concept for building information security that would ensure the perfection of legal regulation in this area. First, we see that a number of countries share the protection of personal data and the information protection of the state, and some identify these two concepts. Ukraine, unfortunately, belongs to the second group. We insist on a fundamental distinction between the information security of man, society and the information security of the state. The information security of a person and society is based on the norms of natural law, that is, they are inalienable rights of a person and a citizen, at the same time, the information security parameters of the state are based on the norms of positive law and are established by the state itself. Thus, when considering the problems of information security of man and society and the problems of information security of the state, we must be aware that we are dealing with two completely different in nature phenomena. This means that approaches to legal regulation should differ.

Secondly, the protection of information security is one of the functions of the state, national legislation should determine those minimum necessary conditions and parameters of information processes that can be considered safe for the existence of a person, society and the state. In this regard, Ukraine needs to develop and implement national standards and indicators that would characterize the level of information security in various spheres of life.

3   Artificial Intelligence and Intellectual Property Law

The fourth technical revolution brings together human and machine capabilities and covers a wide range of areas such as artificial intelligence (AI), 3D printing, autonomous vehicles and the Internet of Things. Technology has led to new types of legal conflicts for which, as it turned out, the law was not ready. Extensive discussions about whether artificial intelligence can be considered a subject of law have not yet led to a unified position.

It must be admitted that today there are examples of providing legal personality to new forms. Official recognition of the properties of a subject of law for a robot with developed artificial intelligence took place for the first time in Saudi Arabia, where a humanoid robot designed in China was granted citizenship [15].

Therefore, we can talk about cyberphysical relationships that require legal regulation. In 2017, Mount Taranaki and the Wanganui River received legal entity status in New Zealand, like the Ganges and Yamuna rivers in India. Recently, a Colombian court ruled that the Amazon rainforests have the same rights as humans. However, when it comes to liability, the law is ambiguous. Natural objects are not able to create objects of intellectual property rights.

Is it possible to give rights to an object that obviously cannot be held responsible, and therefore become a subject? We believe that a violation of the principle "there is a right means that there is an obligation" will lead, at least, to the inequality of the subjects of legal relations. Therefore, in this paper, legal problems are considered from the lack of legal personality in artificial intelligence.

To date, two legal approaches to the recognition of authorship of works created by artificial intelligence can be distinguished. According to the first the authorship belongs to the creator (developer) or user of artificial intelligence (computer program or code) and to the second one the author is directly artificial intelligence (the introduction of the concept of "electronic face"). The legislations in such countries as the United Kingdom, India, New Zealand, provide that the author of the work generated by artificial intelligence, are individuals who used the program to create the work.

The current legislation of Ukraine connects authorship and creation of intellectual property rights only with a person, therefore, in the legal field, forms of expression of artificial intelligence are not authors and inventors. This situation has set, in particular, because artificial intelligence is not considered as a separate entity. Neither in Ukraine, nor in other countries works created by artificial intelligence, are not yet objects of intellectual property law. They can be used without restriction. However, there are already precedents when AI is indicated as the author in an application for a title of protection.

Thus, the development of innovative technologies puts us before the need for a legal settlement of this issue. Thus, the World Intellectual Property Organization has developed and tabled the document Impact of Artificial Intelligence on IP Policy. This document just discusses the feasibility of legislatively consolidation the copyright of the authorship of AI and its legal consequences [16].

4 Labor Law and Employment

The rapid development of innovative technologies, which is characterized by a new round of the technological revolution, entails changes in the labor market. The introduction of artificial intelligence, robots, 3D printing, the creation of the Smart Factory largely creates a conflict of interest between the existing workforce and business.

The new world needs new skills. However, the problem is that the qualitative characteristics of the workforce do not keep pace with the enormous speed, scale and influence of technological changes.

One in three jobs is likely to be seriously changed or disappear in the next decade due to technological changes. This trend may affect 44% of workers with low education, one-third of semi-skilled jobs. By the mid-2030s, up to 30% of jobs can be automated. In industry, robots will fully or partially replace many manual jobs, while more men will suffer because their share of employment is higher and women will be at greater risk of automation in clerical and other administrative functions [17]. The European Foundation for the Improvement of Living and Working Conditions in January 2020 published a Game-changing Technologies: Transforming production and employment in Europe report, which examined the impact of eight factors (additive manufacturing, advanced robotics, the Internet of Things, industrial biotechnologies, electric vehicles, autonomous vehicles, blockchain and virtual and augmented reality) across different sectors of the economy. The report forecasts changes in industry, services, labor market and employment under the influence of each of the eight factors [18]. The opportunities offered by Industry 4.0 will only be available to employees if they are able to undergo training, education and training in the areas and skills that will be required. But keep in mind that the level and quality of specialist training and the level of technology and innovation are interrelated. Therefore, the issue of training is relevant for companies and governments.

The World Economic Forum estimates that training of 1.37 million workers in the United States, whose work is currently at risk, will cost $ 34 billion – or $ 24,800 per person. Multiply this by 100 to include the rest of the world, and the amounts will be staggering [7]. Amazon, for example, announced a $700 million investment to upskill 100,000 workers. In September 2019, PwC announced a $3 billion investment to upskill its people. However, not all companies are ready to solve this problem. And this leads to a generation gap. Today, young adults' deep familiarity with digital tools and platforms gives them advantages in the workplace as technological change makes some tasks redundant [7]. Many companies are faced with the problem of a shortage of professional workers, since on the one hand they have workers of pre-retirement age, but with extensive experience, and on the other, young people who are familiar with digital tools, but who do not have practical skills. In recent years, two forms of overcoming the problem have become most widespread, namely the development of corporate training and the cooperation of governmental institutions with employers. Corporate training requires large financial resources, so only large companies implement it on a full-time basis and free of charge for their staff. The disadvantage is that such training usually involves having an appropriate education for the employee, aimed at the short-term goals of the company and highly specialized. An employee is preparing to work for a certain period of time at a certain workplace. At the same time, joint government programs with employers have a broader prospect of improving the skills of employees. In this case, the person receives the knowledge and job opportunities not only in the company where his workplace was changed or lost, but also in other areas of production. The employer, in turn, receives either government grants to retrain staff or the opportunity to get the necessary staff for the new job. There are successful examples of individual countries and regions in implementing their own professional development programs. Yes, Canada has a Job Grant program, a federal government initiative. Provincial management has the right to compensate employers for a significant portion or all of their training costs for new skills in the workplace. The Digital Skills Bridge project in Luxembourg aims to provide technical and financial assistance to the state to improve the skills of workers in companies facing major technological changes. Bridge of Skills helps invest in the future of employees and businesses by developing employee competencies to better anticipate and adapt the organization to new technologies. In Japan, the government adopted the document. "The 5th Science and Technology Basic Plan" in 2016. It indicates that the time has come when special efforts should be directed to unifying into a single set of technologies "physical space" and "cyberspace". To this end, the government, among other things, provides for training measures based on common international standards in cooperation with higher education institutions and research enterprises. There are also positive developments in Ukraine. So in 2020, the Ministry of Education finally allowed teachers to choose their own place and form of professional development. Funding for education is provided by the state, by providing appropriate funds to the school, at the request of the principal. Of course, this practice is a step towards a positive transformation of education in Ukraine, so we consider it appropriate to extend it to teachers of higher education institutions, as well as to employees of other sectors of the economy.

However, the role of government should not be limited to subsidizing digital transformation. Governments must create and enforce public policies, standards and policies; to involve all participants of labor relations in cooperation; to encourage change, learning and innovation.

In addition to the problem of qualifications, labor relations also face other challenges. Besides robots compete with humans. And in this regard, the public faces the issue of protection against discrimination, protection of the right to equal opportunities for all. Implanting technology into the human body is also no longer a fantasy. How equal will the two applicants for the workplace be when the abilities of one of them will be changed/enhanced with the help of technologies implanted into his body? Obviously, lawyers need to discuss ways to regulate such situations today.

As digitalization strengthens and improves the analysis of people and processes, two areas of law conflict: labor law and data protection legislation. Firstly, enterprises are actively introducing technological analysis of labor productivity. In this case, only objective indicators are taken into account, which can significantly affect the competitiveness of individual workers [19].

Secondly, data collection about employees should be carried out only with their consent. This applies not only to the information contained in documentary sources but also to the installation of surveillance cameras and other recording devices with which the workflow is monitored. According to Joanne Bone: "The GDPR protects the employee by setting limits on how employers use this data. Gathering information about the employee, including his productivity without his consent, as well as using such information in the interests of the employer is a violation of the GDPR. In such cases, employers may face enormous fines, and they must be careful in their risk and transparency assessments to avoid this [18].

Thus, there is the need to create new standards for the protection of labor rights, the development of qualification requirements for new professions, the regulation of vocational training and advanced training are the primary legislative goals in cooperation between states and companies in the field of labor law and employment.

5   Conclusions

In the paper, we discussed some of the social and legal issues that society faced during the fourth technical revolution. In our opinion, the basis of most problems lies in the fact that the speed of development of society is significantly behind the speed of development of technology. Innovations do not cover society as a whole, as a result of which social and economic inequality between people and between countries is exacerbated.

Large companies, as well as richer countries, have great tools both for introducing innovations and for protecting their data and information systems. On the one hand, this allows you to develop the concept of Industry 4.0, on the other hand, it widens the gap and the social crisis.

In the modern world, the creation of new legal standards should be carried out in the process of close cooperation between states and non-governmental organizations with significant experience and competence in relevant areas. The law should be more flexible, provide for the possibility of legal engineering, take into account the speed of informatization of society and work for the future. The second important problem that needs to be solved is bringing the national legal systems to uniform standards of legislation. We observe the active legislative work of the governments of different countries (as well as international institutions) in the field of information security, the development of artificial intelligence and other technologies, but they do not always have a single standard. Thus, it is difficult to build a global information world in which uniform rules apply, technologies "see" each other, and participants are protected from risks.

Our suggestions for highlighting issues include:

1. In the field of information security, we insist that the terms "information security of the person" and "information security of the state (society)" be legally separated. The very fact of the identical interpretation of the "security status" of such different subjects of information relations as the state, society and the individual cannot but raise questions. Man as a biological being has properties that are not inherent in such entities as the state and society. Due to biological factors, potential information threats, methods and means of human influence have their own peculiarities. Information security of a person must be determined through the prism of three components: information-technical, social and information-psychological.
2. National legislation should determine those minimum necessary conditions and parameters of information processes that may be considered safe for the existence of a person, society and the state. In this regard, Ukraine needs to develop and implement national standards and indicators that would characterize the level of information security in various life spheres.

3. The copyright laws of Ukraine should be supplemented with a list of rights and obligations of the owner of the artificial intelligence that creates the copyright objects. It is also necessary to determine the grounds and degree of liability of the owner for damage that may be caused by artificial intelligence.
4. In the area of regulation of labor relations and employment, it would be advisable to introduce state support programs for employers that train workers with information skills. For example, by providing tax benefits or offsetting some of the costs.

Professionals in various fields of technology implementation should be invited and involved in the discussion of legislative initiatives to create effective laws and build a fairer and safer world.

References

1. Fonseca, L.M.: Industry 4.0 and the digital society: concepts, dimensions and envisioned benefits. In: Proceedings of the International Conference on Business Excellence, vol. 12, no. 1, pp. 386–397 (2018). https://doi.org/10.2478/picbe-2018-0034
2. Morrar, R., Arman, H., Mousa, S.: The fourth industrial revolution (Industry 4.0): a social innovation perspective. Technol. Innov. Manag. Rev. 7(11), 12–20 (2017). https://doi.org/10.22215/timreview/1117
3. Zimmermann, S.: Industry 4.0 – Legal challenges and methods to overcome them. Atos (2017). https://atos.net/en/blog/industry-4-0-legal-challenges-methods-overcome, Accessed 06 Feb 2020
4. Sklyar, V., Kharchenko, V.: ENISA documents in cybersecurity assurance for Industry 4.0: IIoT threats and attacks scenarios. In: 2019 10th IEEE International Conference on Intelligent Data Acquisition and Advanced Computing Systems: Technology and Applications (IDAACS), vol. 2, pp. 1046–1049. IEEE, Metz (2019). https://doi.org/10.1109/IDAACS.2019.8924452
5. PwC: Global crisis survey 2019 (2019). https://www.pwc.com/globalcrisissurvey, Accessed 05 Feb 2020
6. Check Point Research: The 2020 Cyber security report: Crypto Miners, Targeted ransomware and cloud attacks dominate the threat landscape (2020). https://research.checkpoint.com/2020/the-2020-cyber-security-report/, Accessed 10 Feb 2020
7. Hyde, S., Sheppard, B.: Strengthening the foundations of trust in the digital age. Strategy+Business (2019). https://www.strategy-business.com/article/Strengthening-the-foundations-of-trust-in-the-digital-age, Accessed 11 Feb 2020
8. Crane, C.: 33 Alarming cybercrime statistics you should know in 2019. Hashed Out (2019). https://www.thesslstore.com/blog/33-alarming-cybercrime-statistics-you-should-know/, Accessed 09 Feb 2020
9. European Parliament: European Parliament resolution of 16 February 2017 with recommendations to the Commission on Civil Law Rules on Robotics (2017). https://www.europarl.europa.eu/doceo/document/TA-8-2017-0051_EN.html, Accessed 09 Oct 2020
10. European Parliament: European Parliament convention on cybercrime of 23 November 2001. https://www.europarl.europa.eu/meetdocs/2014_2019/documents/libe/dv/7_conv_budapest_/7_conv_budapest_en.pdf, Accessed 09 Oct 2020
11. State of California Department of Justice: California Consumer Privacy Act (CCPA) (2018). https://oag.ca.gov/privacy/ccpa, Accessed 09 Oct 2020

12. Guillamont, A., Vijil, J.: An overview of cyber legislation in Latin America. Kennedys (2020). https://www.kennedyslaw.com/es/thought-leadership/article/an-overview-of-cyber-legislation-in-latin-america/, Accessed 14 Jan 2020
13. Tham, Y.: Singapore: the privacy, data protection and cybersecurity law review, 6th edn. The Law Reviews (2019). https://thelawreviews.co.uk/edition/the-privacy-data-protection-and-cybersecurity-law-review-edition-6/1210086/singapore, Accessed 18 Feb 2020
14. NCSI: Ukraine (2020). https://ncsi.ega.ee/country/ua/, Accessed 13 Feb 2020
15. Hanson Robotics: Sophia. https://www.hansonrobotics.com/sophia/, Accessed 22 Feb 2020
16. WIPO: Impact of artificial intelligence on intellectual property policy: Call for comments. https://www.wipo.int/about-ip/en/artificial_intelligence/call_for_comments/index.html, Accessed 13 Feb 2020
17. Stubbings, C.: The case for change: New world. New skills. Strategy+Business (2020). https://strategy-business.com/blog/The-case-for-change-New-world-new-skills, Accessed 10 Feb 2020
18. Peruffo, E., Rodriguez Contreras, R., Mandl, I., Bisello, M.: Game-changing technologies: Transforming production and employment in Europe. Eurofound (2020). https://www.eurofound.europa.eu/de/publications/report/2020/game-changing-technologies-transforming-production-and-employment-in-europe, Accessed 28 Feb 2020
19. Going Fourth: data, Industry 4.0 and the future of manufacturing (2019). https://irwinmitchell.turtl.co/story/going-fourth-data-industry-40-and-the-future-of-manufacturing/page/6/1, Accessed 09 Feb 2020

Author Index

Printed by Printforce, the Netherlands